Faults, Fluid Flow, and Petroleum Traps

Edited by

Rasoul Sorkhabi
Yoshihiro Tsuji

AAPG Memoir 85

Co-Published by

The American Association of Petroleum Geologists
Tulsa, Oklahoma
and
Japan Oil, Gas and Metals National Corporation
Japan

ISBN: 0-89181-366-7

Printed in Canada

AAPG Editor: Ernest A. Mancini
Geoscience Director: James B. Blankenship

This publication is available from:

The AAPG Bookstore
P.O. Box 979
Tulsa, OK U.S.A. 74101-0979
Phone: 1-918-584-2555 or 1-800-364-AAPG (U.S.A. and Canada only)
Fax: 1-918-560-2652 or 1-800-898-2274 (U.S.A. and Canada only)
E-mail: bookstore@aapg.org
www.aapg.org

About the Editors

Rasoul Sorkhabi, a native of Iran, obtained his B.Sc. (1983) and M.Sc. (1985) in geology from India, and Ph.D. (1991) in geology from Japan. He was a Post-Doctoral Fellow and Research Associate at Arizona State University, Tempe, (1992–1997), and a Senior Geologist at the Technology Research Center of Japan National Oil Corporation, Chiba (1997–2003), where he mainly worked on the Evaluation of Traps and Seals. He has published on fault seal analysis, thermochronology, and tectonics of Asia. He co-edited *Himalaya and Tibet: Mountain Roots to Mountain Tops* (Geological Society of America Special Paper 328, 1999). He has given short courses on basin tectonics and fault seals to geoscientists from petroleum companies around the world. He is currently a Research Professor at the Energy & Geoscience Institute of the University of Utah, Salt Lake City. He is a member of the American Association of Petroleum Geologists, American Geophysical Union, Geological Society of America, European Association of Geoscientists and Engineers, Geological Society of India, Geological Society of Japan, and the Japanese Association for Petroleum Technology.

Yoshihiro Tsuji, obtained his B.Sc. (1977), M.Sc. (1979), and Ph.D. (1993) in geology from Kanazawa University, Japan. He started his professional career as a geologist with Dia Consultants Co. Ltd., Japan, where he worked for four years on surveys of landslides and active faults, and dating of Quaternary sediments. He joined Japan National Oil Corporation (JNOC) in 1983, and spent nearly ten years researching carbonate sedimentology. This was followed by his work on basin evaluation studies for JNOC's petroleum exploration activities. As the Director of Research Division for Geoscience at JNOC's Technology Research Center, he managed the Project on the Evaluation of Traps and Seals. He has co-edited two JNOC Reports of the Technology Research Center, and a special issue of a Japanese journal on methane hydrates. He is currently the Team Leader for the Methane Hydrate Research Project of the Japan Oil, Gas and Metals National Corporation (JOGMEC; formerly JNOC). He is a member of the Geological Society of Japan, Japan Association for Quaternary Research, Japanese Coral Reef Society, and the Japanese Association for Petroleum Technology.

Acknowledgments

The American Association of Petroleum Geologists thanks the following for their generous contribution to *Faults, Fluid Flow, and Petroleum Traps*

Japan Oil, Gas and Metals National Corporation

Contributions are applied toward the production costs of publication, thus directly reducing the book's price and making the volume available to a larger readership.

REVIEWERS

The editors would like to express their gratitude to the following colleagues who reviewed papers submitted to this volume:

Dr. Salman Bloch
Consultant Geologists, Houston, Texas

Dr. David Bout
New Mexico Institute of Mining and Technology, Socorro, New Mexico

Dr. Jonathan Caine
U.S. Geological Survey, Denver, Colorado

Dr. Nancy Dawers
Tulane University, New Orleans

Dr. David Dettman
University of Arizona, Tucson, Arizona

Dr. William M. Donne
University of Tennessee, Knoxville, Tennessee

Dr. Gloria Eisenstadt
University of Texas at Arlington, Texas

Dr. James Evans
Utah State University, Logan, Utah

Dr. Emma Finch
University of Manchester, Manchester, England

Dr. John Garver
Union College, Schenectady, New York

Dr. Alan Gibbs
Midland Valley Exploration Ltd., Glasgow, Scotland

Dr. Richard Groshong, Jr.
University of Alabama, Tuscaloosa, Alabama

Dr. Yves Gueguen
Ecole normale superieure, Paris, France

Dr. James Handschy
ConocoPhillips, Houston, Texas

Dr. Ezat Heydari
Jackson State University, Jackson, Mississippi

Dr. Hiroaki Komuro
Shimane University, Shimane, Japan

Dr. John Lorenz
Sandia National Laboratories, Albuquerque, New Mexico

Dr. Yuji Kanaori
Yamaguchi University, Yamaguchi, Japan

Dr. Robbert Maddock
Baker Hughes, Aberdeen, Scotland

Dr. Mazlan Madon
PETRONAS, Kuala Lumpur, Malaysia

Dr. Ken McClay
Royal Holloway University of London, London, England

Dr. Yuichiro Miyata
Yamaguchi University, Yamaguchi, Japan

Dr. Joe Moore
University of Utah, Salt Lake City, Utah

Dr. K. Nakayama
JAPEX Geoscience Institute Inc., Tokyo, Japan

Dr. Steve Naruk
Shell Co., Houston, Texas

Dr. Stanley Paxton
Oklahoma State University, Stillwater, Oklahoma

Dr. Mary Roden-Tice
State University of New York, Plattsburgh, New York

Dr. Grant Skerlec
SEALS International Inc., Lakewood, Washington D.C.

Dr. Richard Schultz
University of Nevada, Reno, Nevada

Dr. Toshihiko Shimamoto
Kyoto University, Kyoto, Japan

Dr. Zoe Shipton
Trinity College, Dublin, Ireland

Dr. John Sneider
Sneider Exploration Inc., Clear Lake Shore, Texas

Dr. Charles Stuart
Consultant Geologist, Park City, Utah

Dr. Roberto Suarez-Riverta
TerraTek Co., Salt Lake City, Utah

Dr. Christopher Talbot
Uppsala University, Uppsala, Sweden

Dr. Jack Thomas
AAPG Training Partners, Tulsa, Oklahoma

Dr. James Turner
PTT Exploration & Production Public Co., Bangkok, Thailand

Dr. Mario Wannier
Shell Co., Houston, Texas

Dr. John Wickham
University of Texas at Arlington, Texas

Dr. Jennifer Wilson
New Mexico Institute of Mining and Technology, Socorro, New Mexico

Dr. Teng-fong Wong
State University of New York at Stony Brook, New York

Table of Contents

Foreword

Japan National Oil Corporation (JNOC) (presently Japan Oil, Gas and Metals National Oil Corporation) launched a multi-disciplinary and international project on the Evaluation of Traps and Seals in 1997. As part of the project, in-house research was carried out at JNOC's Technology Research Center; some research work was contracted to outside scientists both in academia and industry; and the project also sponsored several academic-industrial consortia related to faults. The project ended in 2003. This AAPG Memoir has resulted from this project and includes JNOC research articles as well as those contributed from our colleagues in industry and academia.

In recent years, a number of volumes on fault seals and fault-associated fluid flow in sedimentary basins have been published (e.g., Evans and Wong, 1992; Ortoleva, 1992; Hickman et al., 1995; National Research Council, 1996; Møller-Pedersen and Koestler, 1997; Surdam, 1997; Coward et al., 1998; Jones et al., 1998; Haneberg et al., 1999; McCaffrey et al., 1999; Faybishenko et al., 2000; Holdsworth et al., 2001; Koestler and Hunsdale, 2002; Russell and Handschy, 2003; and Alshop, 2004). We hope that the addition of this AAPG Memoir to the list is a small step forward in the know-how of fault analysis in petroleum traps.

This volume includes seventeen papers. The prologue by Sorkhabi and Tsuji (Chapter 1) sets the stage with an overview of the role of faults in petroleum entrapment and the historical development of our knowledge of fault seal assessment.

Sawamura and Nakayama (Chapter 2) describe a methodology to estimate the amount of oil/gas accumulation using the concept of equivalent grain size in seal rock (to quantify capillarity), and oil/gas migration to and spill-point geometry of petroleum traps. These authors apply their method to hydrocarbon fields in Russia.

Two case studies of fault seal assessment applied to normal faults in Tertiary clastic reservoirs in offshore Sarawak (Hasegawa et al.) and offshore Gulf of Thailand (Kachi et al.) constitute Chapters 3 and 4, respectively. Both studies employ clay smear algorithms to evaluate the sealing efficiency of normal faults; these papers make a significant addition to similar studies conducted previously by other researchers in the North Sea, the Niger Delta, and Trinidad.

Physical analog studies of the development of extensional fault systems were carried out by Sims et al. (Chapter 5) with implications for reservoir connectivity across faults, and by Yamada et al. (Chapter 6) with regard to faults associated with processes of salt doming and strike-slip motion in parts of offshore United Arab Emirates. Discrete Element Method is an emerging technique useful for modeling fault development processes in sedimentary basins; Yamada and Matsuoka (Chapter 7) report on some of their modeling results and comparisons to sandbox analog experiments and applications to petroleum basins.

Outcrop characterization of faults in sedimentary sequences contributes significantly to our understanding of these structures in subsurface reservoirs. A number of papers report on new outcrop observations of faults and laboratory analyses of outcrop samples by various methods.

Sedimentary layers with similar dip and adjacent to normal faults have been observed in many areas and on various scales. Ferrill et al. (Chapter 8) describe various fault-related mechanisms for development of these structures which may provide hangingwall closures for hydrocarbons and across-fault barriers or pathways to hydrocarbons depending on stratigraphic and structural relationships.

Sorkhabi and Hasegawa (Chapter 9) describe outcrop observations of faults in the Neogene clastic rocks of Sarawak, Malaysia. Fault zone rocks in the study area are characterized by both shale smear and deformation bands. These authors also report data on permeability changes from undeformed sandstone to deformation bands zones.

In a detailed study of the layered sandstone-shale sequence cut by the Moab Fault in Utah, southwest U.S.A., Davatzes and Aydin (Chapter 10) distinguish two distinct structural assemblages in fault zones that have evolved through two different mechanisms: deformation band-based and joint-based mechanisms. In recent decades, deformation bands have drawn much attention as an important feature in fault fabric that reduces permeability in reservoir rocks. These structures have been observed in porous sandstone rocks in many parts of the world, including Utah, where they were first discovered in the late 1970s. Shipton et al. (Chapter 11) discuss the geometry and thickness of deformation band zones that occur in the Navajo Sandstone along the Big Hole Fault in Utah. Flodin et al. (Chapter 12) report new field and laboratory data from the Aztec Sandstone in Nevada (equivalent of the Navajo Sandstone) that improves our knowledge of the hydraulic properties of sheared-joint type deformation bands.

Kwon et al. (Chapter 13) describe results of triaxial deformation experiments on cores of four sandstones from the Moab area, Utah, which provide insight on permeability changes in porous sandstone reservoir rocks during deformation.

In a careful study of samples from Wyoming, Utah, and Texas, Milliken et al. (Chapter 14) demonstrate that both compaction and cementation take place within deformation bands in porous sandstones; these authors have also attempted to quantify these processes in the sandstone samples they studied.

The application of geochemical techniques enhances our understanding of fluid flow associated with faults. Sorkhabi (Chapter 15) reports on a new study of calcite veins in various thrust sheets of western Wyoming fold and thrust belt, and discusses fluid flow associated with the thrust faults in light of data obtained by stable isotope and fluid inclusion techniques.

Knowledge of fault timing is crucial for understanding basin formation and deformation, in general, and development of fault traps, in particular. Geochronologic techniques can be quite helpful in this regard. Tagami and Murakami (Chapter 16) describe an interesting application of fission-track technique to zircon-mineral-separated-from-fault-zone material and nearby undeformed rocks. Takagi et al. (Chapter 17) discuss how an integration of K-Ar dating and X-Ray diffraction analyses of fault gouge materials have aided in constraining the timing of major active faults in southwest Japan.

The Appendix at the end of the volume is a bibliography of nearly 1000 articles and books published on fault traps, fault seal processes, and fault-related fluid flow in sedimentary basins. The bibliography has been arranged according to specific topics. Interested geoscientists may use it as a reference tool to delve into publications preceding this volume.

Rasoul Sorkhabi and Yoshihiro Tsuji
Japan Oil, Gas and Metals National Oil Corporation,
Technology Research Center, Chiba, Japan

REFERENCES CITED

Alshop, G. I., R. E. Holdsworth, K. J. W. McCaffrey, and M. Hand, eds., 2004, Flow processes in faults and shear zones: Geological Society of London Special Publication 224, 400 p.

Coward, M. P., T. S. Daltaban, and H. Johnson, eds., 1998, Structural geology in reservoir characterization: Geological Society of London Special Publication 127, 266 p.

Evans, B., and T.-F. Wong, eds., 1992, Fault mechanics and transport properties of rocks: New York, Academic Press, 542 p.

Faybishenko, B., P. A. Witherspoon, and S. M. Benson, eds., 2000, Dynamics of fluids in fractured rocks: American Geophysical Union Geophysical Monograph 122, 400 p.

Haneberg, W. C., P. S. Mozley, J. C. Moore, and L. B. Goodwin, eds., 1999, Faults and subsurface fluid flow in the shallow crust: American Geophysical Union Geophysical Monograph 113, 222 p.

Hickman, S., R. Sibson, and R. Bruhn, eds., 1995, Mechanical involvement of fluids in faulting: Journal of Geophysical Research, v. 100B (special section), p. 12831-13132.

Holdsworth, R. E., R. A. Strachan, J. F. Magloughlin, and R. J. Knipe, eds., 2001, The nature and tectonic significance of fault zone weakening: Geological Society of London Special Publication 186, 342 p.

Jones, G., Q. J. Fisher, and R. J. Knipe, eds., 1998, Faulting, fault sealing and fluid flow in hydrocarbon reservoirs: Geological Society of London Special Publication 147, 320 p.

Koestler, A. G., and R. Hunsdale, eds., 2002, Hydrocarbon seal quantification: Norwegian Petroleum Society (NPF) Special Publication 11, Amsterdam, Elsevier, 263 p.

McCaffrey, K., L. Lonergan, and J. Wilkinson, eds., 1999, Fractures, fluid flow and mineralization: Geological Society of London Special Publication No. 155, 328 p.

Møller-Pedersen, P., and A. G. Koestler, eds., 1997, Hydrocarbon seals: Importance for exploration and production: Norwegian Petroleum Society (NPF) Special Publication 7, Amsterdam, Elsevier, 250 p.

National Research Council, 1996, Rock fractures and fluid flow: Contemporary understanding and application: Washington, D.C., National Academy Press, 551 p.

Ortoleva, P. J., ed., 1994, Basin compartments and seals: AAPG Memoir 61, 477 p.

Russell, K. D., and J. W. Handschy, eds., 2003, Fault seals: AAPG Bulletin, v. 87, no. 3 (thematic issue), p. 377-527.

Surdam, R. C., ed., 1997, Seals, traps, and the petroleum system: AAPG Memoir 67, 317 p.

1

Sorkhabi, R., and Y. Tsuji, 2005, The place of faults in petroleum traps, *in* R. Sorkhabi and Y. Tsuji, eds., Faults, fluid flow, and petroleum traps: AAPG Memoir 85, p. 1–31.

The Place of Faults in Petroleum Traps

Rasoul Sorkhabi[1]

Technology Research Center, Japan National Oil Corporation, Chiba, Japan

Yoshihiro Tsuji[2]

Technology Research Center, Japan National Oil Corporation, Chiba, Japan

"The incompleteness of available data in most geological studies traps some geologists."

Orlo E. Childs in *Place of tectonic concepts in geological thinking* (AAPG Memoir 2, 1963, p. 1)

"Although the precise role of faults has never been systematically defined, much has been written that touches on the subject. One thing is certain: we need not try to avoid them."

Frederick G. Clapp in *The role of geologic structure in the accumulation of petroleum* (Structure of typical American oil fields II, 1929, p. 686)

ABSTRACT

Ever since Frederick Clapp included fault structures as significant petroleum traps in his landmark paper in 1910, the myriad function of faults in petroleum migration and accumulation in sedimentary basins has drawn increasing attention. Fault analyses in petroleum traps have grown along two distinct and successive lines of thought: (1) fault closures and (2) fault-rock seals. Through most of the last century, geometric closure of fault traps and reservoir seal juxtaposition by faults were the focus of research and industrial application. These research and applications were made as structural geology developed quantitative methods for geometric and kinematic analyses of sedimentary basins, and plate tectonics offered a unified tool to correlate faults and basins on the basis of the nature of plate boundaries to produce stress. Over the last two decades, compartmentalization of reservoirs by fault seals has been more intensively investigated as three-dimensional seismic images better resolve

[1]*Present address:* Energy & Geoscience Institute, University of Utah, Salt Lake City, Utah, U.S.A.
[2]*Present address:* Technology Research Center, Japan Oil, Gas and Metals National Corporation, Chiba, Japan.

DOI:10.1306/1033713M853128

fault structures. Geometric characterization of fault architecture, identification of various sealing processes in fault zones, and quantitative appraisal of petrophysical properties of fault rocks have significantly advanced in recent decades.

Fault-seal analyses have shifted from two-dimensional fault juxtapositions to three-dimensional models encompassing fault surfaces, fault transmissibility, and juxtaposed reservoir units. Current methodologies for fault-seal assessment mostly address normal faults in clastic reservoirs. Fault sealing processes in thrust faults and in carbonate reservoirs represent important blind spots in our knowledge. Shale smear has been effectively applied for sealing assessment of syndepositional faults in sandstone-claystone successions. However, fault-seal analyses based merely on shale smear ignore other important sealing processes, notably cataclasis and cementation in fault zones. During their active stages, faults are conduits of subsurface fluids, irrespective of any sealing mechanism that operated before fault rupture. Therefore, a comprehensive fault-seal assessment needs to be a four-dimensional model integrating fault motions, fault-zone processes, and fluid flow. This remains a major challenge. However, integration of in-situ fault stress analysis and fault-seal analysis has provided a technological breakthrough. The realization that fault rocks are low-permeability and high-capillarity features in sedimentary basins has given an economic impetus for exploration of fault traps. The shift from modeling of single-phase fluid flow to multiphase or even mixed-phase fluid flow along and across fault zones will be of more value to these exploration efforts. Recent studies have transformed the old polarized view of faults as either leaks or seals into realistic notions of more complex fault-fluid flow behavior. Current shortcomings in fault-seal assessment are largely caused by the scarcity of detailed data and the need for robust calibration of numerical models. This implies that empirical data will form the cornerstone of near-future advances in fault-seal methodologies.

INTRODUCTION

The term "trap" was first used in 19th century courts in the United States on legal grounds that petroleum, like wild animals, has a "fugacious" nature and becomes property of the person on whose land they were trapped (Dott and Reynolds, 1969, chapter 13). Over the past century, the notion of trap has been associated separately or jointly with oil location, reservoir, closure, or seal. In a modern sense (North, 1985; Biddle and Wielchowsky, 1994; Vincelette et al., 1999), a petroleum trap has three components: (1) a three-dimensional (3-D) geometric closure, (2) seal rocks (including a cap rock or top seal, a lateral or side seal, and also possibly a bottom seal) with sufficient capillary pressure to overcome the buoyancy pressure of hydrocarbons, and (3) a reservoir rock with sufficient porosity to store hydrocarbons if charged from a source rock and with sufficient permeability to yield hydrocarbons if drilled (Figure 1). Ideally, the physical properties of subsurface fluids should also be added to this list as a fourth component (Hubbert, 1953; Schowalter, 1979; England et al., 1987; Watts, 1987; Sales, 1993; Vincelette et al., 1999) because "traps are not simple receivers of fluid into otherwise empty space; they are focal points of active fluid exchange" (North, 1985, p. 254). However, in petroleum exploration, detailed knowledge of subsurface fluids is the last data to become available for trap evaluation; hence, the famous phrase attributed to W. C. Finch: "A trap is a trap, whether or not it has a mouse in it" (Rittenhouse, 1972, p. 13).

Petroleum basins are structurally deformed sedimentary basins, in which faults of various sizes and styles occur. Faults also occur in all types of structural traps. Even "a fold trap is seldom completely free from faulting" (Levorsen, 1967, p. 242) and "in very many anticlinal traps the fundamental structure is a fault, not a fold" (North, 1985, p. 253). Much attention has recently been given to these forced folds (Cosgrove and Ameen, 2000) or fault-related folds (Anastasio et al., 1997; Wilkerson et al., 2002). Faults are not passive features; they are significant factors in impeding or enhancing subsurface fluid flow. Faults are associated with all elements and processes of the petroleum system. Figure 2 depicts the myriad functions that faults have in petroleum systems and basins.

The function of faults in petroleum traps may be viewed from two perspectives: (1) fault closure traps and (2) fault-rock seals. These roughly correspond to the juxtaposition faults and membrane-sealing faults of Watts (1987) and to the passive and active fault sealing

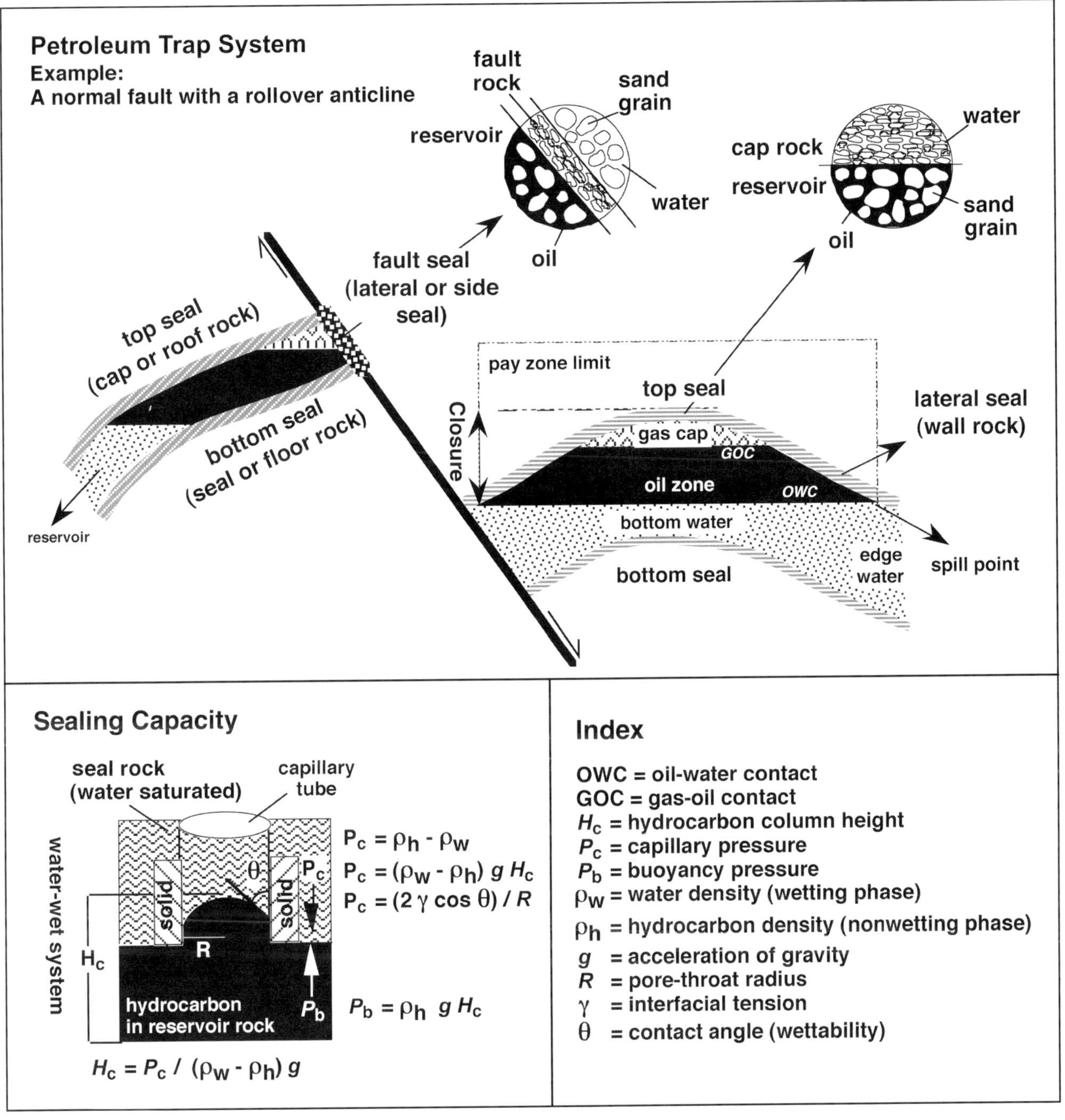

FIGURE 1. Components of a structural trap system and a comparison of a fault trap with an anticlinal trap (a rollover anticline associated with the same normal fault). The sealing capacity of both cap rock (top seal) and fault seal (side seal) is controlled by the capillary pressure of the rock as described quantitatively in the lower diagram. Petroleum is accumulated when the capillary pressure of seal rock is equal to or exceeds the buoyancy pressure of petroleum in the reservoir (Schowalter, 1979; Watts, 1987).

of Weber (1997), respectively. Our usage of "fault closure trap," however, refers not only to lithological juxtaposition across a fault but also to the structural configuration and geometric style of a fault trap. Fault-rock seals result from mechanical and chemical changes that take place along the fault plane and in fault zones as a result of faulting processes; these changes may make the fault plane or zone a barrier to hydrocarbon flow.

The purpose of this chapter is to set the stage for this AAPG Memoir by outlining the historical development of concepts and methods in fault-trap analysis and briefly discussing the functions of faults in petroleum basins at the trap-reservoir scale. In our review, we have focused on fault closures in petroleum traps and petrophysical properties of fault-seal rocks. In preparing this introductory chapter for the volume, we have

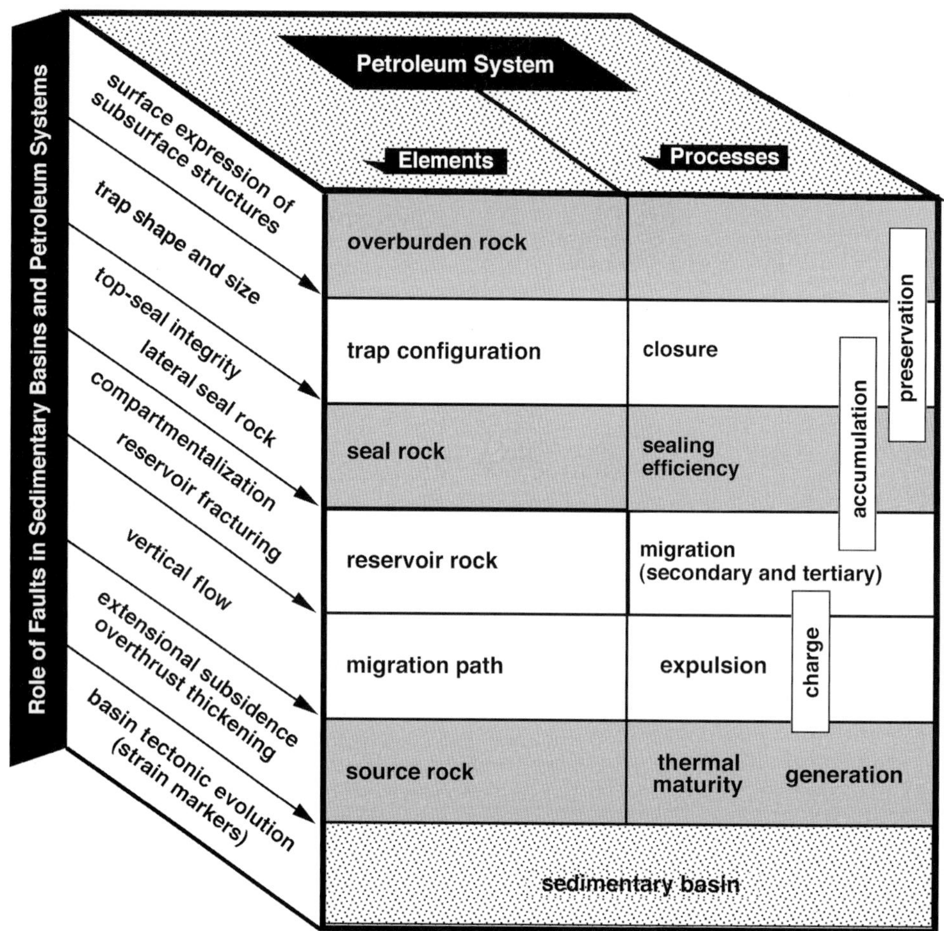

FIGURE 2. A schematic presentation of various elements and processes involved in the petroleum system approach (based on Magoon and Dow, 1994, with some modifications to accommodate the prospect risk evaluation methodologies, e.g., Rose, 2001) and the role of fault analysis in each component.

addressed petroleum geoscientists (instead of fault-seal experts) so that it has a broad readership. This is not an exhaustive review of the subject and "the reader must accept as an apology the increasing difficulty of keeping pace with rapid progress of geology" (Lyell, 1842, p. xiii).

HISTORICAL BACKGROUND

In the second half of the 19th century, anticlines were considered to be major petroleum traps, and faults were commonly ignored. Clapp (1929, p. 686) referred to this period as follows: "The role of faults, although still perhaps imperfectly known, was sadly misunderstood for half a century after the discovery of oil in the United States; and operators, believing faults to be dangerous phenomena, tried to avoid them. The popular superstition against faults was so strong that for years it permeated the geological fraternity, and many government and private sectors recommended the avoidance of faulted structures. The absurdity, as we understand it to be, was commonly expressed in the once-familiar words: 'The country appears too much broken up.'"

Although there is truth in Clapp's remarks, we should not totally blame the oil explorationists of his age given the state of drilling technology and subsurface structural knowledge in the late 19th and early 20th centuries.

The second period began in the first decade of the 20th century, with Clapp's pioneering efforts to classify petroleum traps on the basis of various structural elements, including a distinct category for faults (Clapp, 1910, 1917). During this period, fault closure and juxtaposition traps were identified in the field. A milestone was the publication of *Structure of Typical American Oil Fields* (two volumes, edited by Sidney Powers, 1929), which was the proceedings volume of the "Symposium on the Relation of Oil Accumulation to Structure" held at the AAPG Annual Meeting in Tulsa in March 1927. This was followed by a third volume (edited by J. V. Howell) in 1948, also resulting from a symposium under the same title.

During the second period, several attempts were made to classify petroleum traps, and faults were commonly included in these classifications (Clapp, 1910, 1917; Wilson, 1934, 1942; Heald, 1940; Prison, 1945; Wilhelm, 1945; Levorsen, 1967). The increasing attention to the function of fault structures in petroleum basins during this second period is evident in the papers presented at the Fourth Annual Meeting of the AAPG, New York, March 28–31, 1955, which were later published in a volume of more than 1000 pages, *Habitat of Oil* (Weeks, 1958). It contained 55 papers covering many petroliferous fields around the world, and a cursory review of the volume shows that structural features appear prominent in most of these papers. The editor of the volume, Lewis Weeks, cited 18 questions in his lead article that were important for understanding the "habitat of oil and some factors that control it" and among them was this: "... whether or not faults normally act as conduits or as barriers to migration?"

Despite a few allusions from exploration geologists to the importance of fault seals (McKnight, 1940; Wilhelm, 1945; Willis, 1961), the second stage paid more attention to fault style and juxtaposition (but little attention to fault-rock seals). This latter aspect was highlighted in the third period, which can be traced back to the work of Perkins (1961) and Smith (1966) in the Louisiana fields. Perkins (1961) identified flowage shale barrier (later called shale smear by Weber and Daukoru, 1975) as a fault-sealing mechanism in addition to fault juxtaposition and reverse drag (rollover anticline). Smith (1966) applied the theory of capillary pressure to fault-zone material. Works published by Shell geologists on sealing faults in the Niger Delta (Weber and Daukoru, 1975; Weber et al., 1978) drew attention to the importance of shale smear in growth faults. Pittman's (1981, but first presented at the AAPG Annual Meeting in 1978) work on fault-rock fabrics and petrophysical characteristics was also among the pioneering efforts of the third period. Smith (1980) and Watts (1987, but first presented at the AAPG Annual Meeting in 1985) not only popularized the terms "sealing faults" and "fault seals" with the titles of their articles but also offered elegant theoretical frameworks for fault-seal analysis.

Despite these efforts, fault-sealing analysis in the 1980s was in its infancy (some may argue that it still is). In his 400-page synthesis of knowledge of petroleum traps prior to 1990, Jenyon (1990) devoted only half a page and one figure to fault sealing (with references to the works of Smith and Watts). The past two decades have witnessed a revolutionary development of fault-seal studies both in the academe and in the petroleum industry. As a result of these studies, an appreciable amount of knowledge has been gained on fault-sealing processes and hydraulic properties of faults in sedimentary basins. The Appendix at the end of this chapter chronicles some of these recent developments in fault-seal studies.

FAULTS IN THE CLASSIFICATION OF PETROLEUM TRAPS

Various classification schemes of petroleum traps indicate how the functions of faults have been viewed by different workers, based on field experiences over time. An inherent danger always exists in these classifications because "almost all geological structures, if viewed in enough detail, have unique geometries and histories" (Harding and Lowell, 1979, p. 1016). Nevertheless, as Gould (1989, p. 98) once remarked, "classifications are theories about the basis of natural order, not dull catalogues compiled only to avoid chaos." Classifications and generalizations provide conceptual tools for comparison, analysis, and prediction, which are helpful in the exploration of sedimentary basins where subsurface structures are not directly observed. Therefore, by tracing various schemes of trap classification that have been proposed over the past century, we also get a sense of the progression of theories on the place of faults in petroleum traps and an evaluation of the current theories against their predecessors.

Table 1a presents several trap classifications proposed in the first half of the 20th century (Clapp, 1917; Wilson, 1934, 1942; Heald, 1940; Prison, 1945; Wilhelm, 1945), and Table 1b shows major trap classifications proposed by various workers in the second half of the 20th century (Levorsen, 1967; Harding and Lowell, 1979; North, 1985; Milton and Bertram, 1992; Biddle and Wielchowsky, 1994). The advent of the plate tectonic theory in the 1960s provided a unified framework to correlate sedimentary basins (e.g., Dickinson, 1974) and fault traps (e.g., Harding and Lowell, 1979) on the basis of the nature of plate boundaries as stress engines. The most recent (and a very comprehensive) trap classification has been offered by Vincelette et al. (1999), who have adopted hierarchical levels (system, regime, class, subclass, style, superfamily, and variety) similar to the biological classification of organisms. Figure 3 schematically shows the Vincelette et al.'s (1999) classification of fault-dominated traps, with some modifications in the wording and presentation of details.

A review of petroleum trap classifications shows that researchers have adopted one or a combination of four basic schemes, i.e., basin tectonic, morphological, genetic, and fluid dynamic schemes, and as such, faults have been viewed in these four schools of thought briefly described below:

1) Basin tectonic schemes (Harding and Lowell, 1979; Harding and Tuminas, 1989; Vincelette et al., 1999), in which the tectonic habitat and style of structures on a basin scale is the main component of trap classification. For example, thick-skinned (basement-involved) tectonics vs. thin-skinned (basement-cover-detached) tectonics is the fundamental factor in the trap classification proposed by Harding and Lowell (1979) (Table 1b) and has been followed by Bally (1983) and Lowell (2002).
2) Morphological schemes, in which the shape and spatial relations of trap-forming features are primarily considered. For Wilhelm (1945) and North (1985), the presence or absence of a convex shape in the trap is a fundamental classifying factor. Milton and Bertram (1992) consider the presence of single-seal vs. polyseal traps and the conformable vs. unconformable (tectonic or erosional) relations of the reservoirs to top-, bottom-, and side-sealing surfaces (Table 1b).

Table 1a. Place of faults in the classification schemes of structural traps during the first half of the 20th century.

Clapp (1910, 1917, 1929)	Wilson (1934)	Heald (1940)	Prison (1945)	Wilhelm (1945)
I. Anticlinal structures II. Synclinal structures III. Homoclinal structures IV. Domes V. Unconformities VI. Lenticular sands VII. Rock cavities VIII. Faulting structures a. on the upthrown side b. on the downthrown side c. overthrusts d. horsts	A. Reservoirs closed by deformation of strata A.1. closed by folding A.2. offset by faulting A.3. combined folding and faulting A.4. cutting by salt or igneous intrusions A.5. developed in joint fissures and crush zones B. Reservoirs closed by porosity variations B.1. lenses in sandstones B.2. lenses in limestones and dolomites B.3. lenses in igneous and metamorphic rocks B.4. truncated and sealed strata C. Reservoirs closed by combination of folding and varying porosity D. Reservoirs closed by combination of faulting and varying porosity	I. Structural traps a. synclines b. anticlines c. salt structures d. hydrodynamic e. fault II. Varying permeability traps a. caused by sedimentation b. caused by groundwater c. truncation and sealing	I. Petrologic (stratigraphic) traps II. Combination petrologic and structural traps III. Structural traps III.1. closure by folding III.2. closure by changes in dip III.3. closure by faulting a. normal fault b. reverse fault c. graben d. horst III.4. closure by folding and faulting	A. Convex trap reservoirs B. Permeability trap reservoirs C. Pinch-out trap reservoirs D. Fault trap reservoirs a. single fault segment reservoirs b. parallel fault-block reservoirs c. fault-wedge (intersecting faults) reservoirs F. Piercement trap reservoirs

3) Genetic schemes, in which the origin and formation of closure is the main consideration. Levorsen's (1967) threefold classification of traps as structural, stratigraphic, and combination traps is a textbook example of this scheme. For fault traps, the genetic classifications (Prison, 1945; Biddle and Wielchowsky, 1994; Jenyon, 1990; Vincelette et al., 1999) have commonly followed the slip direction of fault planes, i.e., normal, reverse, wrench, or oblique-slip faults, which, in turn, arise from distinct stress regimes (i.e., tension, compression, couple and shear, and torsion and rotation), acting on faults.
4) Fluid dynamic schemes (Hubbert, 1953; Gussow, 1954; England et al., 1987; Sales, 1993, 1997; Bradley and Powley, 1994; Heum, 1996; Bjørkum et al., 1998; Brown, 2003), in which the migration of hydrocarbons through the basin and the dynamic interactions of hydrocarbons with sealing features are considered.

Sales (1993, 1997) has worked out a classification of petroleum traps as gas-dominated, oil-dominated, and gas-and-oil-dominated classes, considering the multiphase fluid flow in the basin (as initially discussed by Gussow, 1954) in relation to the structural spillpoint and sealing capacity of a trap.

Bradley and Powley (1994) have distinguished between capillary seal and pressure seal. The former is controlled by a capillary tube (interconnected pore-throat size and surface tension between the wetting fluid and hydrocarbon) and prevents the migration of hydrocarbon but allows brine flow. The latter is characterized by effectively closed pore throats and virtually zero permeability, thus inhibiting the movement of both hydrocarbons and brine. Capillary seals fail when the buoyancy pressure of hydrocarbons in the reservoir exceeds the capillary pressure of seal rock; pressure seals fail when the internal fluid pressure of seal rock exceeds its fracture pressure. Pressure seals are thought to enclose abnormally pressured reservoirs, although the processes responsible for the formation of pressure seals are poorly known.

Heum (1996), who first used the term "fluid dynamic classification," has suggested five different components (capillary seal, pressure seal, hydraulic resistance seal, water-derived leakage, and hydraulic-fracturing

Table 1b. Place of faults in the classification schemes of petroleum traps during the second half of the 20th century.

Levorsen (1967)	Biddle and Wielchowsky (1994) Modern Version of Levorsen (1967)	Harding and Lowell (1979) (Based on Structural Styles and Plate Tectonic Habitat)		North (1985)	Milton and Bertram (1992) (Based on Relation of Sealing Surfaces with Top Seals)	
I. Structural traps a. folding b. faulting (both normal and reverse faults) c. fracturing d. salt intrusion e. combination	I. Structural traps IA. Fold dominated Fault related (a) fault bend, (b) fault propagation (c) fault drag, (d) fault drape Fault-free (e) lift off, (f) chevron/kink band (g) diapir, (h) differential compaction IB. Fault dominated (a) basement-involved normal fault (b) detached listric normal fault (c) reverse fault with fault-bend fold (d) reverse fault with ductile deformation IC. Piercement ID. Combination fold and fault	Basement-cover detached styles	1. Decollement fold-and-thrust (stress: compression habitat: convergent transpression) 2. Detached normal faults (stress: extension habitat: passive margins) 3. Salt and shale structures (stress: density contrast habitat: divergent)	I. Convex traps a. buckle- and thrust-fold traps (tangetial movements) b. bending fold traps (vertical movements) c. traps of immobile convexity (buried hills, reefs, drape folding)	One-seal closures	Code C drape anticline, fold anticline, depositional mound [sealing surface conformable with top seal] Code U burried hill, erosional remnant [sealing surface unconformable with top seal] Code CT high-side fault closure [sealing surface conformable + tectonic relation with sealing surface] Code UT high-side fault closure [sealing surface unconformable + tectonic relation with top seal]
II. Stratigraphic traps a. primary (lenses of clastic and igneous rocks) b. secondary (lenses of biochemical rocks) III. Combination traps (including salt domes)	II. Stratigraphic traps IIA. primary (depositional) IIB. unconformities IIIC. secondary III. Combination traps IV. Hydrodynamic traps	Basement-involved styles	1. Wrench fault (stress: couple habitat: transform convergent) 2. Basement thrust blocks (stress: compression habitat: convergent transpression) 3. Extensional faults (stress: extension habitat: divergent transtension) 4. Basement warps and arches (stress: thermal, isostasy habitat: plate interiors)	II. Nonconvex traps a. depositional wedgeout b. erosional wedgeout c. isolated (Lenticular) wedgeout d. permeability pinch-out e. fault cutoff	Poly-seal closures	Code U/C subcrop trap [sealing surface unconformable with top seal and conformable with bottom seal] Code C/U onlap trap, incised-valley fill, lowstand wedge trap [sealing surface conformable with top seal and unconformable with bottom seal] Code CT/C overfull high-side fault closure [sealing surface conformable + tectonic relation with top seal and conformable with bottom seal] Code C/T low-side fault closure [sealing surface conformable relation with top seal and tectonic with bottom seal] Code C/F shale-out, diagenetic seal, fault-gouge seal, tar seal [sealing surface conformable relation with top seal and facies change with bottom seal]

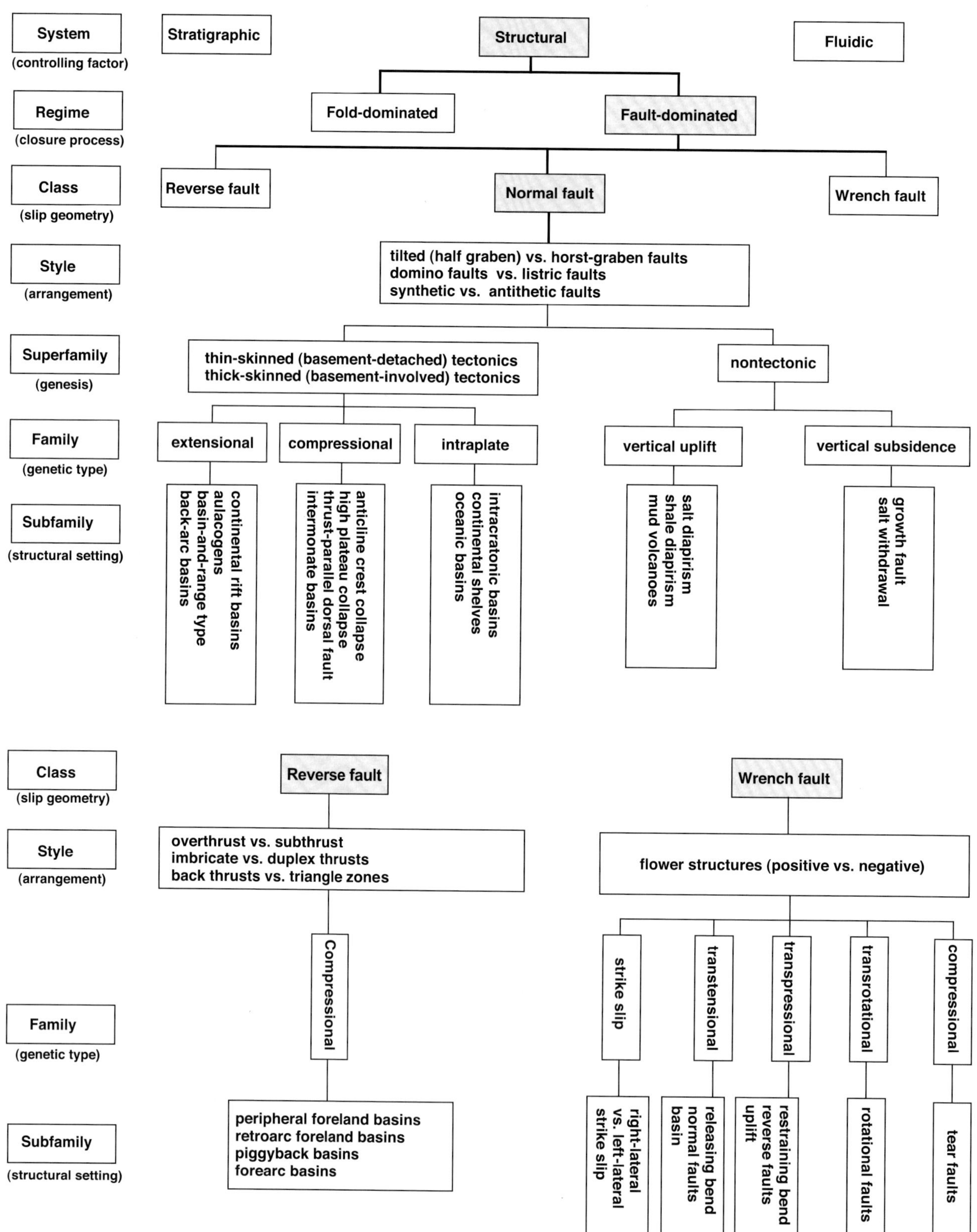

FIGURE 3. A classification of fault traps according to the hierarchical scheme proposed by Vincelette et al. (1999), with more details on the styles of fault structures and the basin types in which the fault structures occur.

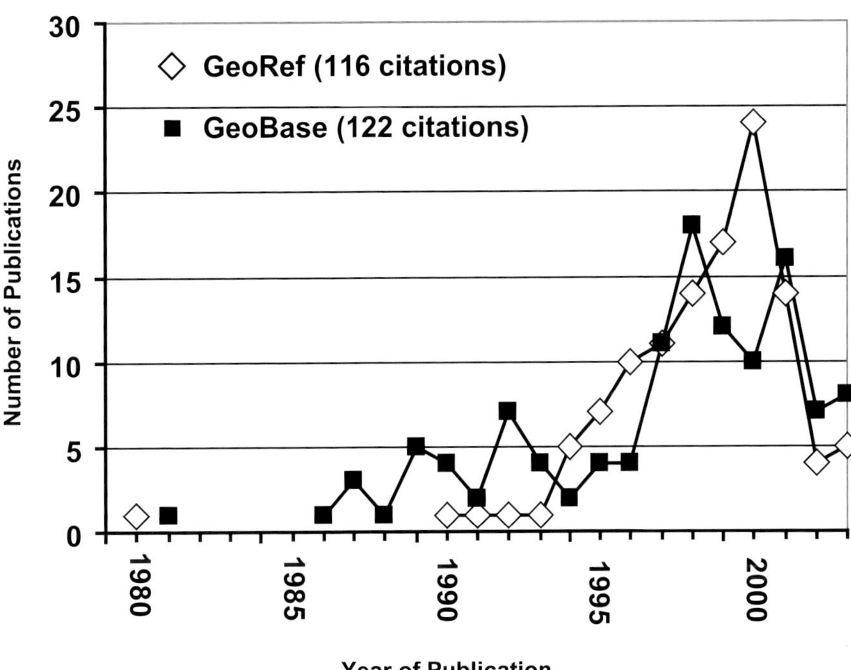

FIGURE 4. A histogram of publications (articles and conference presentations) on fault seal in the reference databases GeoRef and GeoBase (the year 2003 is incomplete).

leakage), with varying importance with respect to water-flow potential or hydrocarbon-flow potential in the trap.

Fluid dynamic classifications are significant because they shift the focus from static to dynamic traps and, thus, better capture the subsurface reality. Nevertheless, fluid dynamic analyses of petroleum traps are less effective as predictive models because of the complexities in integrating petroleum charge (generation and migration), structural movements in the basin, and physical properties of reservoir seal rocks.

RECENT ATTENTION TO FAULT SEALS

Although results of fault-seal analyses are relatively new in exploration and production activities compared to reservoir simulation or seismic interpretation, several petroleum geologists have hinted at the existence and importance of fault seals through the years. Three quotations are noteworthy here. McKnight (1940, p. 133), in his report on the geology of Utah, stated: "The fault has apparently acted as an avenue of escape for the petroliferous material rather than as a seal across the ends of the broken and tilted rocks." Wilhelm (1945, p. 1568) wrote: "Normally a reservoir in a fault trap is sealed at the fault plane by impermeable strata against the reservoir bed by the action of faulting. However, fault planes lined with thinnest veneers of plastic clay or pulverized fault gouge of low permeability may separate oil reservoirs from water-logged porous sands that lie in juxtaposition." Willis (1961, p. 6–24) remarked: "Fault surfaces provide many different possibilities for the development of the proper geometry for a trap. Often there will be a zone of gouge, or broken rock, associated with the fault surface which is capable of acting as an impermeable barrier to oil."

Evaluation of fault sealing and leaking has become a focus of intense research over the past two decades. Figure 4 shows the number of publications (both articles and conference presentations) on fault seal cited in two reference databases (GeoRef® and GeoBase®). Although this is not a complete list (definitely not for the year 2003), both databases show a dramatic increase in fault-seal publications in the second half of the 1990s.

In a 1995 report on technological needs envisioned by oil companies, the National Petroleum Council (1995) identified fault-seal analysis as one of the hot areas in exploration. Based on a survey of 16 oil companies, the Industrial Task Force (2001) noted that the detection, characterization, and prediction of faults and fractures and their impact on production are at the top of the agenda for developments in petroleum geology. This recent attention to fault-seal analysis begs the question "why?"

Characterization of faults in petroleum basins has traditionally posed a major challenge because of the inherent complexity of fault zones, scarcity of quantitative knowledge on petrophysical properties of faults, less accessibility to sampling and examining fault zones in petroleum wells, and difficulties in correlating the outcrop observations of faults to subsurface conditions. Although none of these factors has waned, three other factors, in turn, have motivated fault-seal studies as follows:

1) It has been increasingly recognized that faults are too important to be ignored in petroleum exploration, reservoir management, and production strategy. Fault geometry (size and shape), for instance, partly controls the volumetric estimation of petroleum reserves in a given pool. Moreover, fault-seal evaluation helps drilling plans because if a fault is known to provide a lateral barrier to hydrocarbon flow at given pay zones, wells drilled parallel to the fault (along fault dip) and down to the deepest fault-bounded accumulation may tap multiple

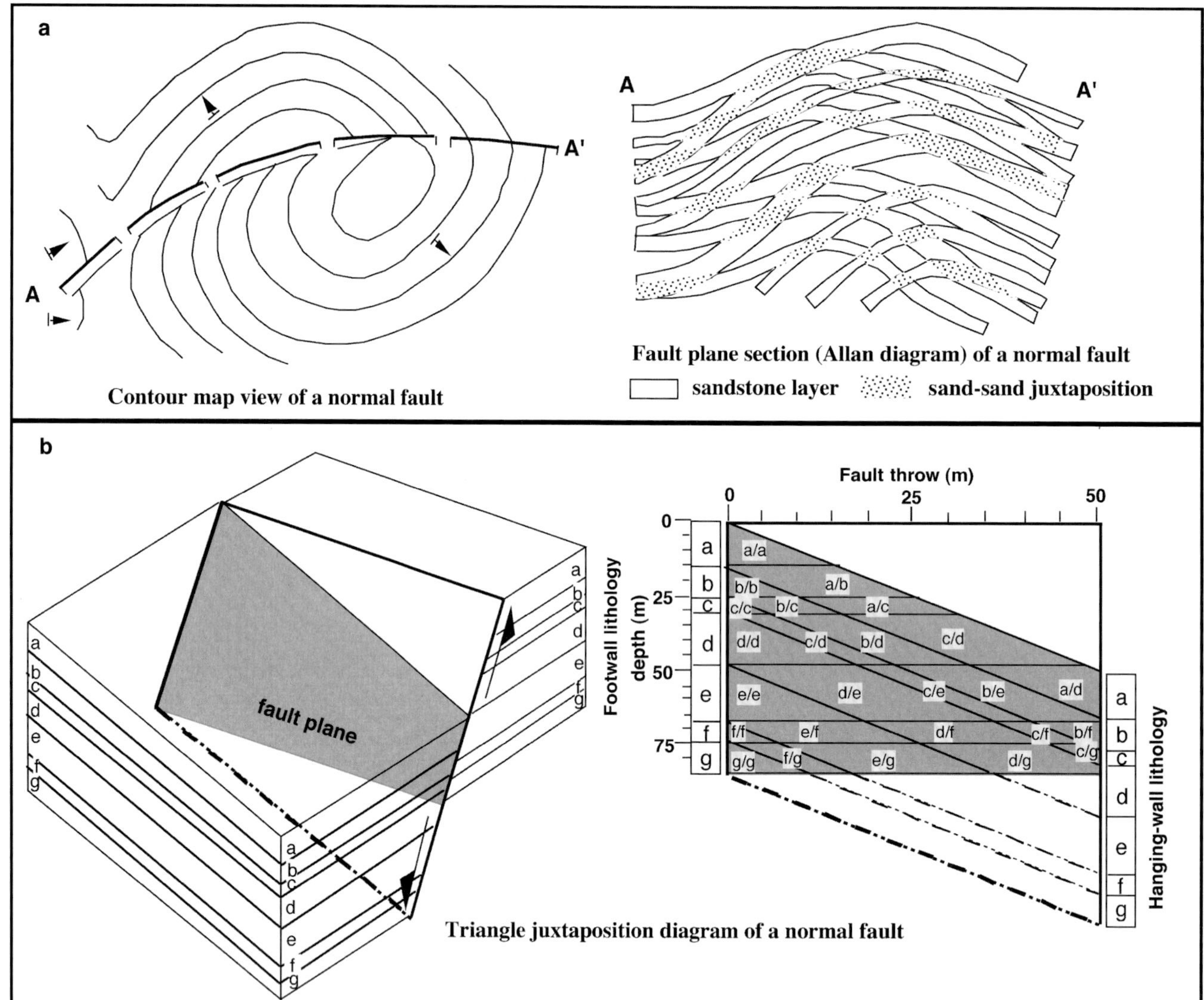

FIGURE 5. Two schemes of juxtaposition diagrams on a fault surface used currently in petroleum industry: (1) fault plane section or Allan diagram (Allan, 1989); (b) triangle juxtaposition diagram (Knipe, 1997).

hydrocarbon pay zones (if present) on both hanging-wall and footwall sides.

2) With a decrease in discovery of giant oil fields in the recent decades, reservoirs compartmentalized by faults have become economically attractive targets.
3) The development of high-resolution, 3-D seismic technology and borehole image logs has refined the identification and visualization of subsurface fault structures.

FAULT JUXTAPOSITION TRAPS

Fault juxtaposition of a reservoir rock against a low-permeable rock was recognized as early as 1910 (Clapp, 1910), and diagrams of fault traps appeared in the first textbook on petroleum geology (Hager, 1915). Evaluation of fault juxtaposition traps as depicted on structural cross sections is still a useful method, especially because it considers the entrapment on the scale of pay zones (Bailey and Stoneley, 1981). However, a comprehensive fault-trap evaluation should not rely simply on two-dimensional fault juxtaposition because petroleum traps have a 3-D closure geometry.

A powerful technique for fault juxtaposition analysis was developed by the late Urban Allan. It constructs the juxtaposition of footwall and hanging-wall sedimentary layers on a fault plane (along-strike fault surface). The resulting graphs are called fault plane sections (Allan, 1989) or, simply, Allan diagrams (Figure 5a). This technique was first presented as "composite cross sections" by Read and Watson (1962) in their classic geology textbook and was extensively improved and applied to petroleum exploration by Allan as early as 1967 in Texas. Allan first presented his graphical technique in 1980 at the AAPG Research Conference on Seals for Hydrocarbons held in Colorado and finally published it in 1989. The computerization of Allan diagrams in the last decade (e.g., Hoffman et al., 1996) has provided a quick-view tool for evaluating the potential areas of fault trapping (caused by sandstone-shale juxtaposition) or potential reservoir communication (caused by contact between sandstone layers). Allan (1989) contended that a fault plane itself has no sealing properties, and that sandstone-sandstone contacts are simply pathways through which hydrocarbons migrate vertically as they move back and forth across the fault. His notion may have been true for the growth faults he studied; however, more recent studies on fault-rock properties and shale smear phenomena refute such a simplified notion. Allan diagrams provide a useful tool for (1) lithological mapping of the fault plane and (2) displaying various fault-rock properties (such as shale gouge ratio and permeability) on the juxtaposed layer contacts.

Triangle-shaped juxtaposition diagrams have also been employed for fault-sealing assessment by Bentley and Barry (1991) (fault-type panel), Childs et al. (1997) (sequence-throw juxtaposition diagram), and Knipe (1997) (juxtaposition diagram). In these diagrams (Figure 5b), a slice of fault plane is depicted in the form of a triangle, in which the horizontal axis shows fault throw (from 0 to the maximum fault throw in the study area), and the vertical axis on the left of the diagram shows the thickness (depth section) of stratigraphic layers on the upthrown side of the fault. The stratigraphic layers of the downthrown section are overlain (juxtaposed) on the triangle for different fault throw intervals. Similar to Allan diagrams, fault-rock properties, such as permeability or shale gouge ratio, can also be displayed (color-coded) on the triangle diagrams. However, unlike Allan diagrams, the triangle juxtaposition diagram is a hypothetical slice of a fault plane, which simulates fault-rock properties for juxtaposed layers for questions like "what if fault throw is *n*?"

Precise fault juxtaposition analyses should consider (1) the type of the juxtaposed seal rock and (2) the nature of fault rock between the juxtaposed units, because different types of juxtaposed seal rocks or fault rocks will have different capillary pressures.

The juxtaposed seal rock can be ranked according to its position on the scale of ductility-brittleness, i.e., halite, anhydrite, organic-rich shale, clay-rich shale, silty shale, certain limestones, sandy shale, clay-rich sandstone, tight dolomite, cemented sandstone, and quartzite, in an order of more ductile toward more brittle (Downey, 1984; Skerlec, 1999). Robert Sneider's scale of these seal rocks (Sneider et al., 1991, 1997) is a useful tool to rank the juxtaposed seal rocks on a fault surface because his quantitative scheme includes various types of seal rocks in terms of their capillarity based on numerous measurements.

Furthermore, the existence of fault rock between the juxtaposed rocks should also be considered. Figure 6 shows an example from Iran where an anhydrite formation (the Miocene Mishan formation, which is a regional seal rock in the Zagros foreland basin) is juxtaposed against a fractured limestone (the Oligocene–Miocene Asmari Formation, which is a major reservoir rock in the region), but fault gouge has also developed between the juxtaposed formations. In such a case (which is a norm instead of an exception as field observations of fault zones testify), the sealing capacity is controlled by the rock (either juxtaposed seal rock or the intervening fault rock), which has a higher capillary pressure to counteract the maximum buoyancy pressure of hydrocarbons exerted against fault rock or juxtaposed seal rock (if fault rock fails by capillarity). Therefore, a quantitative appraisal of fault rocks is necessary for fault-trap analysis, and this theme is reviewed in the following section.

FIGURE 6. The Khurgu fault in the Zagros foreland basin, southern Iran. Fault juxtaposition of Asmari limestone (a major reservoir rock in the subsurface) against Mishan anhydrite (a regional seal rock) with intervening fault gouge (about 1 m [3 ft] thick). In such juxtaposition traps, the higher capillary pressure of either juxtaposed seal rock or fault rock determines the sealing capacity of the fault. This high-angle reverse fault was not reactivated in the 1977 Khurgu earthquake (magnitude 7). Drawn from a photograph in Berberian et al. (1977).

FAULT-ZONE ARCHITECTURE, FAULT-SEALING PROCESSES, AND PETROPHYSICAL PROPERTIES

Several published articles discussing fault-sealing mechanisms and assessment are available (e.g., Smith, 1966, 1980; Weber, 1986, 1997; Watts, 1987; Mitra, 1988; Nybakken, 1991; Knipe, 1992, 1993; Knott, 1993; Berg and Avery, 1995; Harper and Lundin, 1997; Knipe et al., 1997, 1998; Yielding et al., 1997; Skerlec, 1999; Aydin, 2000; Wehr et al., 2000; Sorkhabi et al., 2002, 2003). In this section, we will briefly discuss how the current knowledge quantifies the petrophysical properties and sealing capacity of fault zones.

Fault Architecture

A working concept of fault architecture is necessary as a first step in fault-sealing assessment. Numerous outcrop mapping (e.g., Bruhn et al., 1990; Knipe, 1992; Antonellini and Aydin, 1994; Caine et al., 1996; Burhannudinnur and Morley, 1997; Walsh et al., 1998) as well as subsurface investigations (e.g., Wallace and Morris, 1986; Gibson, 1994; Shipton et al., 2002) have demonstrated that a fault is rarely a simple slip plane but a relatively narrow zone of deformation dissecting the host rock. The fault zone may be divided into fault core and fault damage zone (Caine et al., 1996). The fault core contains the main slip plane and is characterized by one or more types of fault rocks (Sibson, 1977; Knipe et al., 1997). The fault damage zone surrounds the fault core and is characterized by abundant fractures (joints, shear fractures, veins, deformation bands, etc.) that form before and during faulting and act as a transition zone from undeformed host rock to fault rock.

This two-component (fault core and fault damage zone) model is a simplified view of fault architecture at the outcrop scale. Fault architecture is further complicated by factors and processes such as fault-associated folding, fault stepping at heterogeneous layer boundaries, occurrence of multiple slip planes in a fault zone, sharp changes in fault dip angle, incorporation of rock blocks (fault lenses) from wall rock into the fault zone, and so on.

Fault-sealing Processes and Fault Rocks

Several distinct processes that generate fault rocks have been identified in sedimentary basins (Sibson, 1977; Watts, 1987; Mitra, 1988; Knipe, 1989, 1992; Knipe et al., 1997; Weber, 1997; Fisher and Knipe, 1998; Gibson, 1998). These processes include cataclasis, development of deformation bands in porous sandstones and in unconsolidated sandstones (disaggregation zones), clay smearing, development of clay matrix gouge (framework phyllosilicate) in impure, immature sandstones, cementation (chemical diagenesis), and pressure solution. These fault-rock types resulting from these processes are summarized as follows:

1) Cataclasite: Cataclasis is a friction-dependent mechanism of brittle deformation involving both fracturing and rigid-body rotation (Engelder, 1974). Cataclastic fault rocks are characterized by grain-size reduction and porosity collapse caused by fracturing and crushing of grains and fragments in the host rock. Sibson (1977) distinguishes between fault gouge (with visible lithic fragments of <30% in rock matrix) and fault breccia (with visible fragments of >30%) on a textural basis. Cataclasites may become lithified or remain incohesive rock material after faulting, depending on confining pressure and other subsurface environmental conditions.
2) Deformation bands: Deformation bands (Aydin, 1978) have been observed as millimeter-thick planar structures in porous sandstones and have been variously called "granulation seams" (Pittman, 1981) or "cataclastic slip bands" (Fowles and Burley, 1994). Bruhn et al. (1990) and Antonellini and Aydin (1994) demonstrated in their studies of Jurassic sandstones in Utah, United States, that deformation bands occur as solitary features far away from fault zones but increase in abundance toward the fault and form an anastomosing pattern in the fault zone surrounding the main slip plane. This seems to suggest that faulting in porous sandstones is an evolutionary process beginning with solitary deformation bands and culminating in the main slip plane (Antonellini and Aydin, 1994).
3) Disaggregation zones: As defined by Knipe et al. (1997), these are deformation bands that show little fracturing and occur in little consolidated, clean sandstones.
4) Clay smear or shale smear: Faulting in a sand-mud succession may cause smearing of soft clay material from source layers into the fault zone (Weber et al., 1978; Lindsay et al., 1993; Lehner and Pilaar, 1997). Lindsay et al. (1993) identified three mechanisms for clay smear, including clay abrasion on a sandstone surface during fault movement, shale shearing along fault planes, and ductile injection of clay material into fault zones. In the last decade, several algorithms, such as shale smear factor (Lindsay et al., 1993), clay smear potential (Bouvier et al., 1989; Fulljames et al., 1997), shale gouge ratio (Yielding et al., 1997; Freeman et al., 1998), and smear gouge ratio (Skerlec, 1999), have been proposed to use this phenomenon for evaluating the sealing capacity of normal faults in clastic reservoirs. These algorithms are not equivalent but are based on different

assumptions of shale smear; therefore, their results should be interpreted with particular attention to what the algorithms mean in terms of smear processes (Naruk et al., 2002). The current calibrations for these algorithms are based on empirical evidence from petroleum fields in deltaic environments and syndepositional faults and can be applied to this type of field (Sorkhabi et al., 2002).

5) Framework-phyllosilicate fault rock (Knipe et al., 1997) or clay matrix gouge (Gibson, 1998): This type of fault rock occurs in impure, immature sandstones (with a detrital clay content of 15–40% according to Knipe et al., 1997, or containing >30% clay + mica + lithics and <60% quartz according to Gibson, 1998) by smearing and mixing of clay minerals during faulting. It is a clay and shale smear process operating on a small scale in a sandstone unit.
6) Cemented fault rock: The porosity and pore connectivity in rocks are effectively reduced by cementation, which occurs in sedimentary reservoir rocks during burial and chemical diagenesis (Bjørlykke, 1983), as well as in fault rocks (Knipe, 1993; Mozley and Goodwin, 1995; Hippler, 1997; Knipe et al., 1997; Sverdrup and Bjørlykke, 1997; Hadizadeh and Foit, 2000) because of flow of hydrothermal fluids through fault-zone fractures or growth of new mineral phases in fault rocks as a result of fault-induced changes in temperature and pressure. A variety of cements, including quartz, calcite, dolomite, kaolin, anhydrite, pyrite, and barite, may fill fractures and pore spaces in fault-zone rocks. Fault-associated cementation ("precipitation sealing") has been studied both experimentally (Tenthorey et al., 1998) and theoretically (Aharonov et al., 1998).
7) Stylolite: Pressure solution is a process of diffusive mass transfer along grain contacts. It has been widely observed in fine-grained quartz and calcite-bearing rocks at elevated temperatures, as well as in shear zones (Rutter, 1983; Hadizadeh, 1994). Pressure solution is a combined mechanical and chemical process of porosity reduction and crack sealing that may be associated with fault zones (e.g., Gratier et al., 1994; Peacock et al., 1998).

Permeability Distribution in Faulted Reservoirs

Understanding the pattern of permeability heterogeneity that faults induce in reservoir rocks is of great importance for reservoir simulation. Here, we describe the results of two studies that shed light on this issue.

Evans et al. (1997) reported gas permeability measurements on plugs from the Precambrian granitic rocks in the Washakie Range, Wyoming, United States, where the East Fork thrust faults occur. Measured at an effective stress of 3.4 MPa (493 psi), the protolith samples had permeabilities in the range of 10^{-18}–10^{-17} m^2 (10^{-6}–10^{-5} d); samples from fault damage zone showed higher permeabilities in the range of 10^{-16}–10^{-14} m^2 (10^{-4}–10^{-2} d) (i.e., 10–1000 times greater than protolith permeability); samples from fault core (indurated gouge or clay-rich foliated cataclasite) had the lowest permeabilities (<10^{-20}–10^{-17} m^2; <10^{-8}–10^{-5} d). Furthermore, these authors demonstrated that permeability in fault core samples measured perpendicular to the fault plane was lower than that measured parallel to the fault plane.

Fowles and Burley (1994) reported permeability and porosity data (measured on plugs) from the Lower Permian sandstones (Penrith Formation) in the Vale of Eden Basin in northwest England and their equivalent rocks (the Locharbriggs Sandstone) in southwest Scotland. These sandstones are dissected by normal faults of mainly late Paleozoic age, and the fault rocks are characterized by anastomosing cataclastic slip bands (similar to deformation bands first observed by Aydin, 1978, in eolian sandstones in Utah). These authors found that both porosity and permeability increased from the undeformed sandstone toward the faults but decreased drastically in deformation band fault rock, and that the permeability of the fault rock perpendicular to the fault plane was the lowest.

Overall, the results of studies by Fowles and Burley (1994) and Evans et al. (1997) are consistent in terms of permeability distribution in a faulted area. These results are significant, considering the differences in the study areas, rock types and ages, and fault types. The slight permeability increase from undeformed rock to fault damage zone observed in both studies may be attributed to the abundance of fractures in this zone. Indeed, outcrop studies on faulted sandstones in the southwestern United States (Anders and Wiltschko, 1994) and faulted limestones in Israel (Becker and Cross, 1996), as well as a subsurface study on a faulted carbonate reservoir in the Arabian Gulf (Ericsson et al., 1998), have shown that the frequency of fractures increases from undeformed rock to fault damage zone. This may explain the slightly higher permeability of the fault damage zone, as also documented by Caine and Tomusiak (2003) in brittle faults in the Precambrian crystalline rocks of the Colorado Rocky Mountain Front Range. However, as Knipe (1993) has noted, cementation processes in the fault damage zone may heal fractures and hence reduce permeability, and this possibility should also be considered, especially in deeper areas of basins having significant hydrothermal fluid flow. The drastic decrease of permeability in fault rocks is attributed to textural and mineralogical changes (e.g., Knipe, 1992, 1993).

In view of these results and interpretations, a simple model for permeability distribution in a faulted

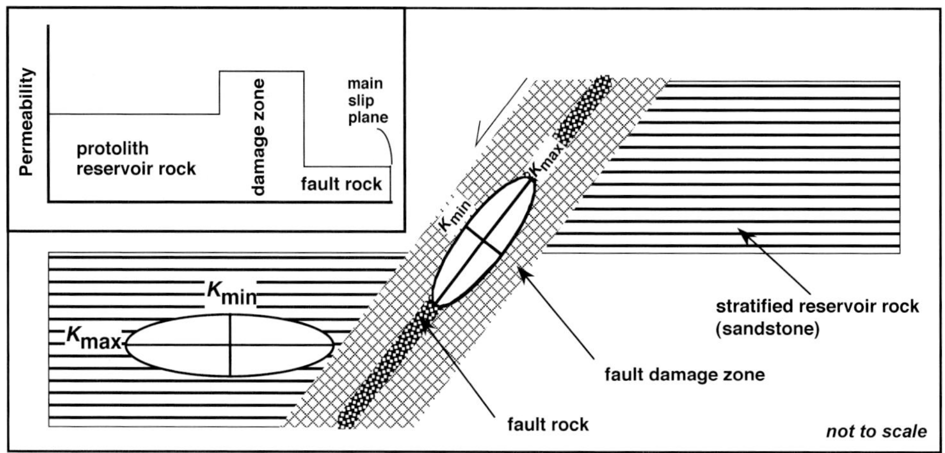

FIGURE 7. A simplified model for distribution of single-phase permeability in a faulted reservoir (clastic) rock. The insert shows that permeability increases from protolith (undeformed reservoir rock) to fault damage zone (caused by fault-associated fractures) but decreases rapidly in the fault rock (based on studies by Fowles and Burley, 1994; Evans et al., 1997). The diagram shows permeability ellipsoids for both reservoir rock and fault zone, in which the fault induces a permeability inversion in the rock. Note that maximum permeability in a stratified (laminated) rock is horizontal but becomes parallel to slip plane in a fault zone caused by fracture network, whereas minimum permeability is vertical in sedimentary layers but becomes horizontal in a fault zone caused by fault rock fabric. The figure is a schematic presentation of concepts; it is not to scale.

clastic reservoir is depicted in Figure 7. In this model, the fault induces a permeability inversion in the reservoir rock characterized by two distinct permeability ellipsoids. In bedded reservoir rocks, horizontal permeability is commonly several times larger than vertical permeability (North, 1985; Al-Qahtani and Ershaqi, 1999), but the permeability ellipsoid is inverted in fault zones where horizontal permeability (perpendicular to fault plane) is the minimum, whereas maximum permeability is parallel to the fault plane.

Fault-rock Permeability

In the last decade, several researchers have attempted to quantify the permeability of fault rocks. One definite result from their studies is that the permeability depends largely on the type of fault rock. Figure 8 is a plot of 322 fault-rock permeability measurements (on normal faults in clastic reservoirs) compiled from Antonellini and Aydin (1994), Fowles and Burley (1994), Knipe et al. (1997), Leveille et al. (1997), Fisher and Knipe (1998), Gibson (1998), and Sperrevik et al. (2002) and categorized according to fault-rock types. Overall, the data show that deformation bands have the highest permeability, whereas cemented fault rocks and clay smear samples show the lowest permeability.

The capability to predict fault-rock permeability is very helpful for reservoir simulation and field production (e.g., Lia et al., 1997; Wehr et al., 2000). Two empirical approaches to estimate fault-rock permeability in normal faults and clastic reservoirs are discussed here.

The first approach considers the reduction of permeability from undeformed (host) reservoir rock to fault rock (Leveille et al., 1997; Fisher and Knipe, 1998). Figure 9 is a compilation of permeability measurements on 66 pairs of host rock and fault rock reported by Antonellini and Aydin (1994), Fowles and Burley (1994), Leveille et al. (1997), Fisher and Knipe (1998), and Gibson (1998). All data pertain to normal faults in sandstone rocks. The fault rocks show reduced permeability by one to three orders of magnitude compared to the host rocks. Cemented fault rocks have the lowest permeability, deformation bands show the highest permeability, and cataclastic fault rocks have intermediate permeability. To employ this permeability reduction approach in the oil fields, we need data on reservoir rock permeability, empirical calibrations for various fault-rock types and fault types, and a realistic understanding of what fault-rock type is expected to be present in a given case. This approach, although still in its infancy, has promising potential as a predictive tool.

The second approach uses relationships between clay content (determined mainly by the x-ray diffraction analysis of samples) and one-phase permeability measured in fault rocks, and assumes that a higher content of clay minerals or clay-size minerals in a rock results in lower permeability. Two empirical calibrations for this approach have been reported as follows:

1) Manzocchi et al. (1999)

$$\log K_f = -4\mathrm{CCR} - 0.25\log(D)(1 - \mathrm{CCR})^5$$

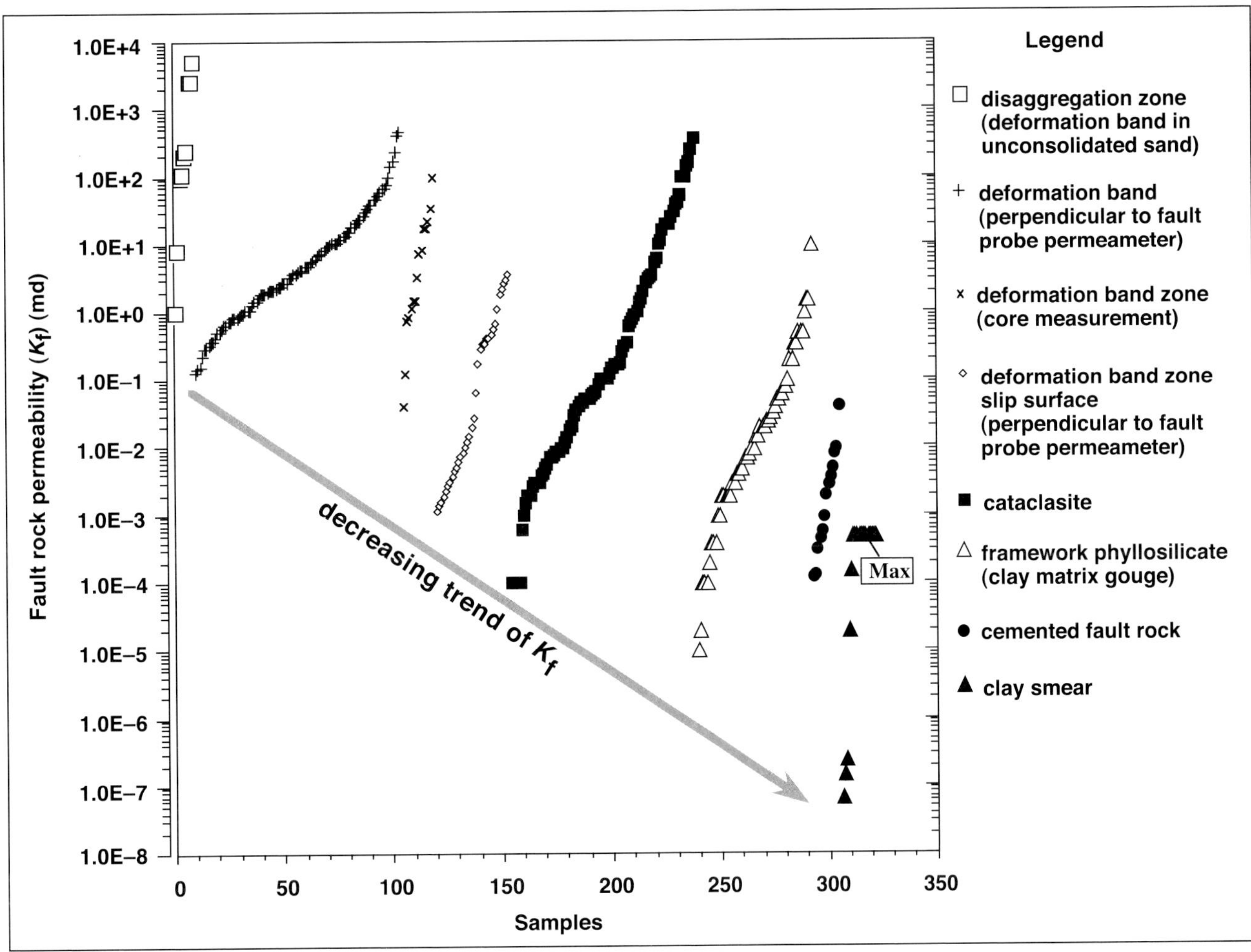

FIGURE 8. Data on 322 fault-rock permeability measurements categorized according to fault rock types: disaggregation zone (deformation bands in unconsolidated sands, as defined by Knipe et al., 1997); deformation bands (individual bands) (as defined by Aydin, 1978); deformation band anastomosing zone; fault-slip planes in deformation band zones; cataclasites; framework phyllosilicate (clay matrix gouge) in immature sandstones (Knipe et al., 1997; Gibson, 1998); cemented fault rock; and clay smear (Lindsay et al., 1993). Permeability data on deformation bands were mostly obtained from measurements on outcrop samples using a probe minipermeameter (gas permeability); all other permeabilities were determined on plug samples (perpendicular to fault) using either gas or water flow. The small box Max for clay smear data indicates that permeability measurements hit the detection limit; therefore, permeability values may be even smaller. Overall, the data show a trendline (marked by the gray arrow) according to fault-rock type, with the clay smear samples having the lowest permeability. Data sources include Antonellini and Aydin (1994), Fowles and Burley (1994), Knipe et al. (1997), Leveille et al. (1997), Fisher and Knipe (1998), Gibson (1998), and Sperrevik et al. (2002).

where K_f is the fault-rock permeability (millidarcys), CCR is the clay content ratio in the fault rock (fraction 0.0–1.0), and D is the fault displacement (meters). This calibration is based on a compilation of reported data (references given in Manzocchi et al., 1999) and indicates that higher CCR and D values yield lower permeability values in fault rocks.

2) Sperrevik et al. (2002)

$$K_f = a\exp\{-[b\mathrm{CCR} + cZ_{\max} + (dZ_f - e)(1 - \mathrm{CCR})^7]\}$$

where K_f is the fault-rock permeability (millidarcys), $Z_{\max}$ is the maximum subsurface depth (meters) of the fault rock under consideration, Z_f is the depth (meters) at the time of faulting, CCR is the clay content ratio (fraction), and other symbols are empirically derived constants: $a = 80{,}000$; $b = 19.4$; $c = 0.00403$; $d = 0.0055$; and $e = 12.5$. This calibration is based on data from nearly 100 normal faults in the North Sea and shows that higher clay contents and greater depths result in lower permeabilities in fault rocks. The incorporation of a depth factor in this calibration is important,

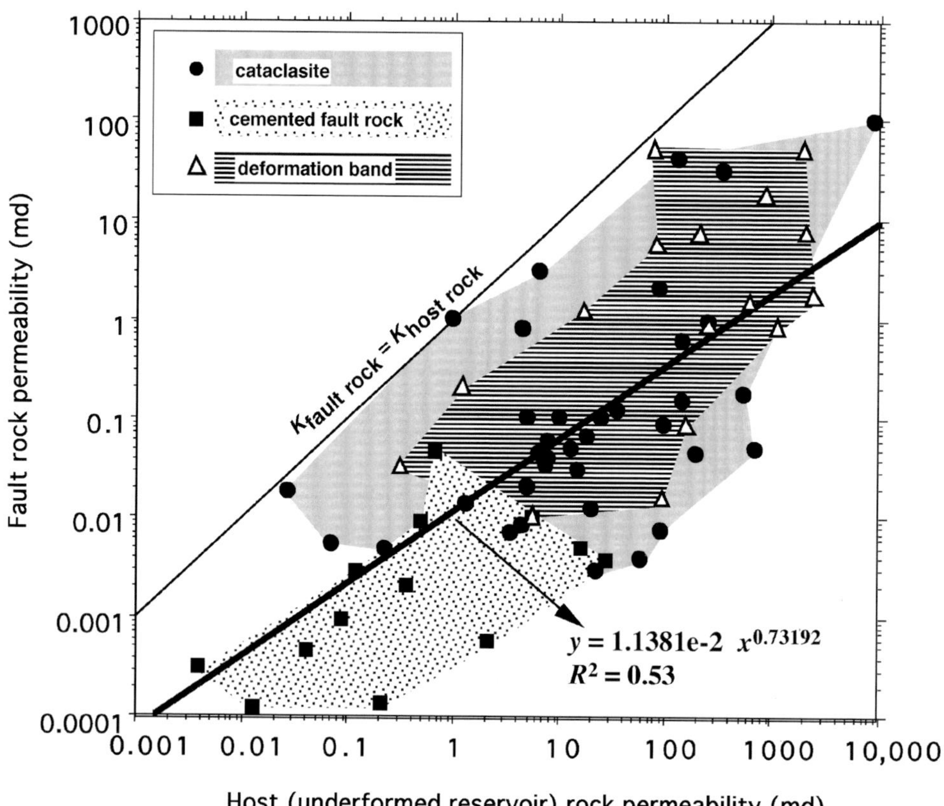

FIGURE 9. Plot of a data set containing 66 pairs of host rock and fault rock permeability (compiled from Antonellini and Aydin, 1994; Fowles and Burley, 1994; Leveille et al., 1997; Fisher and Knipe, 1998; Gibson, 1998). All data pertain to normal faults in sandstone rocks. The fault rocks show reduced permeability by one to three orders of magnitude compared to the associated undeformed rocks. Cemented fault rocks have the lowest permeability, and deformation bands show the highest permeability.

because several studies have demonstrated that fault-rock permeability is partly controlled by confining pressure (e.g., Morrow et al., 1984; Evans et al., 1997; Faulkner and Rutter, 1998) similar to test results of intact rock samples that show an inverse relationship between permeability and confining pressure (e.g., Brace, 1978).

The clay content permeability approach is simplistic because it ignores other processes such as cementation in fault zones, but it is a quick-evaluation tool that can be applied to normal faults in mudstone-sandstone sequences from which the empirical calibrations have been derived. It should be noted that the uncertainty range in both the above-mentioned calibrations is currently one to two orders of magnitude because of uncertainty in the empirical data.

Data on fault-rock permeability also provide insight into the sealing probability of fault rocks because very good cap rocks commonly have intrinsic permeabilities of less than 1 md (about $<10^{-16}$ m^2) (Cossé, 1993), and an unpublished database of top-seal rocks at the Japan National Oil Corporation (JNOC [presently Japan Oil, Gas and Metals National Corporation]) shows that more than 90% of them have permeabilities of less than 5 md ($<5 \times 10^{-15}$ m^2).

Fault-rock permeability data can also be used in reservoir simulation. Knowing reservoir rock permeabilities and the lengths of two reservoir blocks separated by a fault, fault-rock permeability, and fault-zone thickness derived from an empirical relationship between fault displacement, D, and fault-zone thickness, T_f (where $T_f = D/66$, according to data compilation by Manzocchi et al., 1999), transmissibility multipliers (the permeability ratio of $K_{with\ fault}/K_{without\ fault}$) can be derived for reservoir grids and incorporated in reservoir-simulation models (Knai and Knipe, 1998; Manzocchi et al., 1999, 2002).

Hydrocarbon Column Heights in Fault Traps and Capillary Pressure of Fault Rocks

The accumulation and volume of hydrocarbons in fault traps depend on the following factors:

1) Capillary pressure of the cap rock: A higher capillary pressure will hold more hydrocarbons in the reservoir (e.g., Berg, 1975; Schowalter, 1979).
2) Integrity of the cap rock: Hydraulic fracturing (Hubbert and Rubey, 1959; Capuano, 1993) in overpressured cap rocks as well as structural (fold-related or fault-related) fracturing (Ingram and Urai, 1999) in overstressed cap rocks leads to a breaching of the top seal. For this reason, given the same capillary pressure, thicker cap rocks provide better sealing probability than thinner layers.
3) Thickness of the juxtaposed seal rock in relation to the amount of fault offset: If fault entrapment is caused by the juxtaposition of low-permeable (sealing) rock against the reservoir, a thicker seal

rock with a fault displacement equal to its thickness will hold more hydrocarbons than one with a fault having less displacement ("self-juxtaposed reservoirs" of Gibson, 1994).

4) Capillary pressure of the fault rock: A higher capillary pressure in the fault rock (Watts, 1987) will hold more hydrocarbons in the reservoir. However, comparing the capillary pressures of cap rock and fault rock, the lower capillary pressure will determine hydrocarbon column height.
5) Water pressure in the fault fill: Recently, Brown (2003) has highlighted the importance of water-overpressured or water-underpressured fault rock (compared to reservoir water pressure) as a factor in addition to capillarity in controlling hydrocarbon column height. Overpressured fault rock increases the height of the sealed petroleum column.
6) Structural spillpoint: Traps with deeper spillpoints than their crests will provide larger closure for holding hydrocarbons in the reservoir.
7) Thickness of reservoir (pay zone): Given similar capillary conditions of seal rock and similar porosity of reservoir rock, a thicker reservoir holds a larger volume of hydrocarbons charged to the reservoir.
8) Oil/gas ratio: Hydrocarbon gases have about three times higher buoyancy pressure than liquid petroleum. Therefore, oil column height also partly depends on the ratio of later-migrated gas that has displaced oil in the reservoir.

Capillary pressure is a key factor in petroleum accumulation on a geological timescale (e.g., Berg, 1975; Schowalter, 1979; Watts, 1987; Vavra et al., 1992). Ideally, the capillary (displacement or entry) pressure of a fault rock should be directly determined (by the mercury-injection method). However, with no access to fault-zone materials in drill cores, the ultimate goal of fault-seal assessment should be the ability to predict the capillary pressure of the fault rock. Petroleum traps are commonly assumed to be water-wet systems. This is true for the majority of cases (Cossé, 1993), although it is not always the case. Therefore, the water-wet assumption is a limitation of the prediction methods of capillary pressure in seal rocks.

One method currently used for predicting the capillary pressure of fault rocks is based on the relationship between clay content and capillary pressure (Gibson, 1998; Childs et al., 2002; Yielding, 2002). Recently, Bretan et al. (2003) have proposed the following calibration between clay content (computed shale gouge ratio in petroleum fields) and capillary pressure (estimated from across-fault pressure differences in the same studied fields):

$$\mathrm{FRP_c} = 10^{(\mathrm{SGR}/27-C)}$$

where $\mathrm{FRP_c}$ is the fault-rock capillary pressure (in bar); SGR is the shale gouge ratio (Yielding et al., 1997); and C is a calibration factor related to depth: C is 0.5 for burial depths of less than 3000 m (9850 ft); C is 0.25 for depths of 3000–3500 m (9850–11,500 ft); and C is 0 for depths of greater than 3500 m (11,500 ft).

A second empirical approach is based on the relationship between single-phase permeability and capillary (displacement) pressure. This assumption is valid because both permeability and capillary pressure are petrophysical properties of a porous medium, in which permeability is directly related to connected porosity, and capillary pressure is inversely proportional to connected porosity. Beginning with the work of Purcell (1949), reservoir engineers have tried to quantify the relationship between permeability and capillarity.

Harper and Lundin (1997) have proposed the following theoretical equation relating capillary pressure (P_c) to permeability (K):

$$P_c \propto K^{-0.5} \quad \text{and} \quad \log P_c = \log 65.453K^{-0.5056}$$

Sperrevik et al. (2002) have determined single-phase permeability and mercury-injection capillary pressure for numerous samples from North Sea oil fields, including host rocks (sandstones) and fault rocks (disaggregation zone, phyllosilicate framework, and cataclastic fault rocks). Figure 10 plots the data of Sperrevik et al. (2002), as well as those reported previously by Schowalter (1979), Swanson (1981), Berg and Avery (1995), Harper and Lundin (1997), Fisher and Knipe (1998), and Gibson (1998). Two significant results emerge from these data.

First, a power-law regression line is obtained for the relationship between fault-rock permeability (K_f in millidarcys) and fault-rock capillary (displacement) pressure ($\mathrm{FRP_c}$ in pounds per square inch) as follows:

$$\mathrm{FRP_c} = 55.402 \times K_f^{-0.4273}$$

This is very similar to the theoretical equation proposed by Harper and Lundin (1997).

Second, as noted by Sperrevik et al. (2002), the power-law regression lines for both sandstone reservoir rocks and fault rocks are very similar (Figure 10), indicating that both faulting and sedimentary processes in sandstones produce rock materials in which permeability and capillarity follow the general percolation properties of porous media (Sperrevik et al., 2002). The validity of this relationship and interpretation is unknown for cemented fault rocks (which are absent in the data set). Nevertheless, it appears that the theoretical and empirical relationships between petrophysical properties (porosity, permeability, and capillarity), which have been developed for clastic reservoirs,

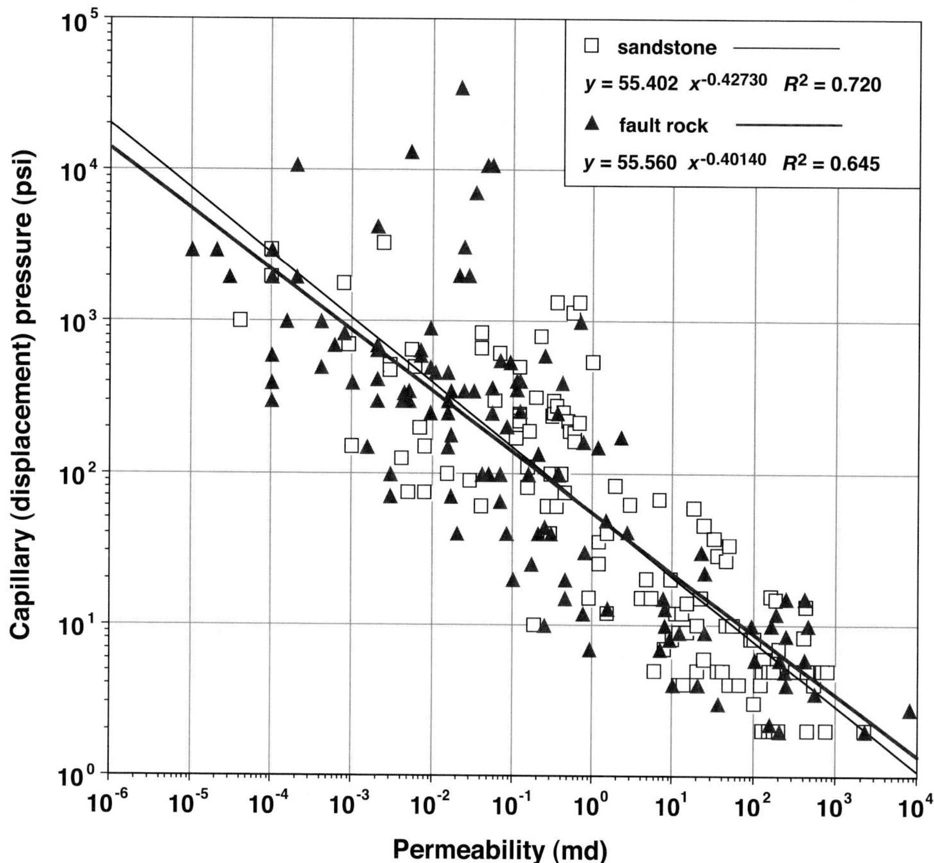

FIGURE 10. Plot of a data set containing 244 pairs of gas permeability (md) and capillary pressure (psi) (displacement pressure in mercury injection tests) for fault rocks and sandstone rocks compiled from Schowalter (1979), Swanson (1981), Berg and Avery (1995), Harper and Lundin (1997), Fisher and Knipe (1998), Gibson (1998), and Sperrevik et al. (2002). Note that both fault rocks and reservoir rocks exhibit very similar regression lines, implying that both sedimentary and faulting processes in sandstones produce rock materials whose hydraulic properties follow the same percolation laws of porous media.

may also be applicable to cataclastic fault rocks produced by normal faults in clastic reservoirs.

Hippler (1997) used the following equation relating permeability to capillary pressure, first proposed by Thompson et al. (1987):

$$K = C(L_c)^2(\sigma/\sigma_0)$$

where K is permeability (darcys); C is a constant that relates the conductivity of the capillary structure of equivalent size L_c to the permeability of the same path (the work of Thompson et al., 1987, indicates that C is 1/226); L_c is the capillary pore-throat diameter (in micrometers) associated with the first connected path in a mercury injection test; σ is the electrical conductivity of the rock saturated in brine solution; and σ_0 is brine conductivity. According to Archie's (1942) law for homogenous saturated media, σ/σ_0 can be approximated by φ^2, where φ is porosity.

Manzocchi et al. (2002) suggested the following equation for permeability-capillarity relationship in fault rock, derived from a previous analysis by Ringrose et al. (1993) for sediments:

$$P_c = C(1 - S^5)S^{-2/3}(\varphi/K)^{0.5}$$

where C = 3 gives capillary pressure in bar; S is the effective wetting phase saturation; φ is porosity; and K is single-phase permeability (millidarcys).

DIRECT FAULT-SEAL EVALUATION

Where data are available, a comparison of the depths of the oil-water contacts and/or oil-gas contacts on both sides of a fault gives direct evidence for the sealing or leaking behavior of the fault (Smith, 1980). Equal oil-water contact levels on both sides of a fault, for instance, indicate cross-fault leaking. Smith (1980) has worked out hypothetical scenarios for fluid-fluid contact levels across faults and the implications for fault sealing or leakage.

Similarly, a comparison of pore pressure trends in reservoir rocks on both sides of a fault provides direct insight of the sealing (if a pressure difference is present at the same depth) or leaking (if pressure is the same at a given depth) behavior of the fault (Skerlec, 1999). These direct, data-intensive approaches to fault-sealing assessment can be used to calibrate other methods of fault-seal prediction (e.g., Yielding et al., 1997).

BARREN FAULT TRAPS AND AN INTEGRATED FAULT-TRAP EVALUATION

In the beginning of this chapter, we quoted the well-known phrase, "A trap is a trap, whether or not it has a mouse in it." Nevertheless, given the high costs of drilling wells, petroleum explorationists try to avoid barren traps as much as possible. There are many reasons for dry wells. A database of Japan National Oil Corporation, including 92 exploration targets around the world, indicates the following causes for exploration failures and dry wells (Figure 11):

1) Source rock factors (15%), such as insufficient volume or poor quality of source rock (in terms of organic content, thermal maturity, or expulsion efficiency) to generate petroleum in the basin
2) Reservoir rock factors (17%), which mainly include the poor quality of reservoir rock to hold or transmit petroleum
3) Closure factors (10%), i.e., lack or misinterpretation of 3-D closure geometry necessary for entrapment
4) Seal rock and preservation factors (17%), including the lack or inefficiency of seal rocks and, hence, the leakage of hydrocarbons
5) Migration factors (41%), such as the wrong timing of migration and structural development or lack of effective migration pathways (i.e., faults, fractures, and permeable carrier beds), although there were good source rocks, reservoirs, and traps in the explored basin

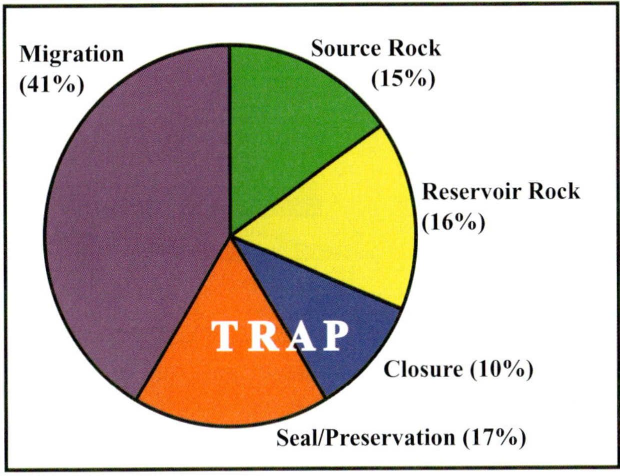

FIGURE 11. Causes of exploration failures and dry wells according to a database of 92 exploration targets at the Japan National Oil Corporation. See text for explanation of the factors. Although closure and seal and preservation have been grouped as trap factors, part of migration related to trap timing may also be considered as a trap factor.

Fault-trap analysis should particularly consider the following three factors:

1) Fault timing: Productive traps with fault seals are those in which the fault formed prior to the last charge of hydrocarbons to the reservoir. This implies that the chronology of structural development with respect to source rock maturation and hydrocarbon migration is crucial for fault entrapment. Faults in sedimentary basins can be reasonably dated from stratigraphic comparison of sedimentary layers on both sides of a fault. In addition, the application of low-temperature radiometric techniques to fault-zone materials (either recovered from drill cores or collected from exposed sections of a given fault), such as the $^{40}Ar/^{39}Ar$ dating of potassium feldspar and of cleavage-filling mica, fission-track dating of apatite, zircon, or pseudotachylyte, (U-Th)/He dating of apatite, thermoluminescence, and electron-spin resonance methods, can be helpful. Note that the last two methods have a dating range of $10^3 - 10^5$ yr and are thus useful for dating active faults, whereas the first three methods are applicable to greater than 10^5-yr events. Noller et al. (2000) give up-to-date information on various dating methods applicable to faults in sedimentary basins.
2) Fault-seal failure: Although a fault may have been a sealing feature in the geological past (as inferred from shale gouge ratio data), it is equally crucial to know if the fault seal has remained intact or has been breached under the contemporary stress conditions of the basin in which the fault occurs. Faults are highly conductive of fluids during their active times. For example, after the 1995 Kobe earthquake in Japan, many of the nearby springs showed a sevenfold increase in flow rate (Sato et al., 2000). In the last two decades, a powerful technology has emerged for in-situ stress analyses using borehole images (borehole breakouts and drilling-induced tensile fractures) obtained by acoustic and electrical televiewers, in conjunction with drilling pressure data (drilling mud injection and leak-off tests) (e.g., Zoback and Healy, 1992). In-situ stress data can be used to evaluate the failure potential of faults (e.g., Bell, 1990; Finkbeiner et al., 1997, 2001; Ferrill et al., 1999; Wiprut and Zoback, 2000; Jones and Hillis, 2003).
3) Top-seal failure: Corcoran and Doré (2002) have discussed four top-seal leakage mechanisms, namely, tectonic breaching (fold-related tensile fractures and minor faults in top seal), capillary leakage, overpressure leakage (hydraulic fracturing in cap rock), and molecular transport (diffusion) of natural gas through the cap rock. Ingram and Urai (1999) have demonstrated that top-seal leakage through subseismic faults and fractures depends partly on mudrock properties such as brittleness and partly on fracture behavior (shear vs. dilational fractures).

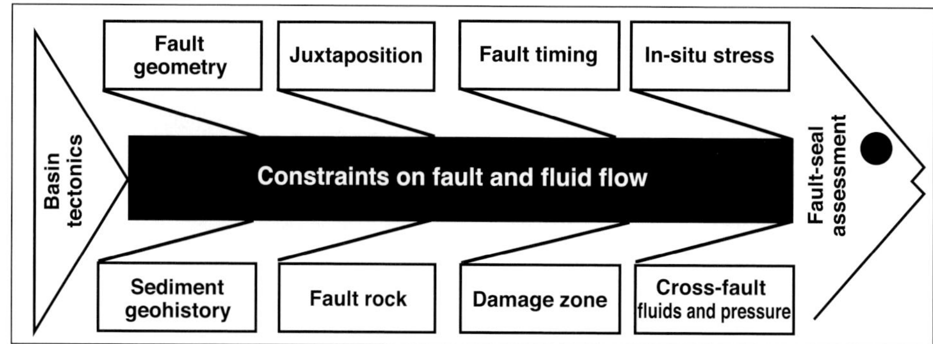

FIGURE 12. A fish-bone schematic presentation of a strategy for fault-seal assessment. Nine components are identified, which contribute directly or indirectly to the fluid-flow properties of fault structures. Components are arranged from left to right roughly according to the scale and the stagewise availability of data to evaluate each component. For example, basin tectonics is placed at the tail of the fish-bone diagram because of its largest scale and early-stage available data in exploration; in-situ stress and cross-fault pressure analyses are placed in the rightmost boxes because of their more refined scale and last-stage available data.

A science-based petroleum exploration strategy requires that fault-seal evaluation be integrated with basin tectonics and basin modeling (geohistory analysis and source rock maturation modeling).

Information gained from basin tectonics is helpful for understanding the evolution, kinematic interactions, geometric styles, and robust structural/seismic interpretation of faults. Faults formed in intraplate, continental collision, subductional, passive-margin, continental rift, back-arc, and wrench basins represent particular tectonic styles that place first-order constraints on petroleum traps (e.g., Harding and Lowell, 1979; Harding and Tuminas, 1989).

Geohistory analysis (Van Hinte, 1978; Allen and Allen, 1990) provides information on sedimentary evolution of a basin and thus helps us understand the initial conditions of sediments subjected to faulting processes. Knipe et al. (2000) have argued that geohistory should be incorporated as a fundamental parameter in fault-seal assessments.

Various schemes have been proposed for an integrative, systematic approach to fault-seal assessment (e.g., Mitra, 1988; Allard, 1997; Knipe et al., 1997; Skerlec, 1999; Grauls et al., 2002; Jones and Hillis, 2003). Figure 12 is our attempt of summing up this chapter and of supporting the previous authors' concepts.

PROGRESS, PROBLEMS, AND PROSPECTS

Faults are known to have dual functions in subsurface fluid flow in sedimentary basins; they can both enhance and impede fluid flow. Some lines of evidence, such as the occurrence of oil and gas seeps and geothermal springs in areas of active faults and the presence of hydrothermal deposits and mineral veins in fault zones, demonstrate the function of faults in fluid migration. However, numerous measurements of low-permeability fault rocks and field evidence for fault compartmentalization of petroleum reservoirs indicate that faults can form effective seals.

Although much more remains to be learned about fault-sealing processes, the recent studies have transformed the old polarized views of faults as either barriers or conduits into a more realistic framework. The realization that a fault can seal at one point but leak at another point on its along-strike surface, that it may seal laterally but allow migration vertically, that it can leak while active but seal later, and that it may seal one fluid phase but transmit another is no small achievement of the last decade (Aydin et al., 1998; Skerlec, 1999).

Methodologies for fault-seal assessment are associated with uncertainties, as discussed by Hesthammer and Fossen (2000), Wehr et al. (2000), and Yielding (2002). These uncertainties arise from several sources, including the mapping of fault geometry from seismic images, characterization of subsurface stratigraphy and subsurface pressure regimes from well logs, and application of outcrop geology to subsurface structures and rocks. Therefore, the validity of fault-sealing assessment is dependent on how to best reduce these uncertainties.

Currently, fault-sealing assessment is reasonably applicable to normal faults in clastic reservoirs for which data are available. Even in these cases, field calibrations of clay-smear fault rocks surpass our quantitative appraisal of other sealing processes. Furthermore, our knowledge of fault-sealing processes in thrust faults and for faults in carbonate reservoirs is very poor. It is therefore expected that fault-sealing processes in thrust faults and carbonate reservoirs, in addition to more sophisticated analyses and calibrations for normal faults in clastic reservoirs, will be a focus of research and development in the coming years.

Studies of fault-sealing processes and fault-trap structures have and will continue to render a great service to petroleum exploration and production, as well as contribute to our scientific understanding of faults and faulting processes. These structures and processes have shaped our planetary environment through devastating earthquakes operating on the scale of seconds to the formation of high mountain ranges and deep basins persisting for millions of years. During the last century, the petroleum industry has developed such fields as micropaleontology, sequence stratigraphy, seismic imaging, and basin modeling as powerful tools in geoscience. Fault-seal research continues this noble tradition and affirms the motto of AAPG's 2002 Annual Convention in Houston: "Our heritage is the key to global discovery."

ACKNOWLEDGMENTS

We express our gratitude to numerous researchers in the academe and industry who have contributed to Japan National Oil Corporation's Project on Evaluation of Traps and Seals (1997–2003); some of them are represented by their articles in this volume. Our thanks also are extended to our colleagues who have supported the project in managing positions, including Satoshi Sasaki, Hisashi Ishida, Masanori Okamoto, Uko Suzuki, H. Hasegawa, Daichi Sato, and Masamichi Fujimoto. We also greatly appreciate John Lorenz, Jack Thomas, Jim Handschy, Steve Naruk, Raymond Levey, and David Curtiss for reading this article and for their comments to improve it; we alone, however, are responsible for any error.

APPENDIX: MILESTONES (SYMPOSIA, CONSORTIA, AND PUBLICATIONS) ON FAULT SEALS, FAULT ROCKS, FAULT TRAPS, AND FAULT-ASSOCIATED FLUID FLOW (1980–2003)

Abbreviations

AAPG	American Association of Petroleum Geologists
AGU	American Geophysical Union
EAGE	European Association of Geoscientists and Engineers
EUG	European Union of Geosciences
GSA	Geological Society of America
GSL	Geological Society (London)
ISRM	International Society for Soil and Rock Mechanics
NPF	Norsk Petroleums-forening
SPE	Society for Petroleum Engineers
USGS	United States Geological Survey

1980

Downey, M. W., and T. T. Schowalter, convenors, 1980, AAPG Research Conference on "Seals for Hydrocarbons," Keystone, Colorado.

Deformation Mechanisms, Rheology, and Tectonics Conference on "The Effect of Deformation on Rocks," Göttingen, Germany: Proceedings, *in* Lister, G. S., H.-J. Behr, K. Weber, and H. J. Zwart, eds., 1981, Tectonophysics, v. 78, p. 1–698.

Carreras, J., and P. R. Cobbold, eds., 1980, Sheared zones in rocks: Journal of Structural Geology, v. 2, p. 1–287.

1981

Smith, D., 1981, Sealing and non-sealing faults in Louisiana Gulf Coast salt basin: AAPG Bulletin, v. 64, p. 145–172.

Carter, N. L., M. Friedman, J. M. Logan, and D. W. Stearns, 1981, Mechanical behavior of crustal rocks, "Handin Volume," AGU Geophysical Memoir 49, 236 p.

1982

Deformation Mechanisms, Rheology, and Tectonics Conference on "Planar and Linear Fabrics of Deformed Rocks," Zürich, Switzerland: Proceedings, *in* P. L. Hancock, E. M. Kalper, N. S. Macktelow, and J. G. Ramsay, eds., 1984, Journal of Structural Geology, v. 6, p. 1–221.

International Conference on "Strain Patterns in Rocks," Rennes, France: Proceedings, *in* P. R. Cobbold and W. M. Schwerdtner, eds., 1983, Journal of Structural Geology, v. 5, p. 255–470.

Workshop on "Chemical Effects of Water on the Deformation and Strengths of Rocks," Carmel, California: Proceedings, *in* S. H. Kirby and C. H. Scholz, eds., 1984, Journal of Geophysical Research, v. 89, p. 3991–4358.

AGU Chapman Conference on "Fault Behavior and the Earthquake Generation Process," Snowbird, Utah: Proceedings, *in* K. J. Coppersmith and D. P. Schwarz, eds., 1984, Journal of Geophysical Research, v. 89, p. 5669–5927.

1983

Bally, A. W., ed., 1983, Seismic expressions of structural styles, three volumes, AAPG Studies in Geology 15, The Layered Earth, v. 1, 195 p.; Tectonics of Extensional Provinces, v. 2, 336 p.; Tectonics of Compressional Provinces/Strike-slip Tectonics, v. 3, 411 p.

1984

International Conference on "Multiple Deformations in Ductile and Brittle Rocks," Bermagui, Australia, organized by the Geological Society of Australia: Proceedings, *in* P. C. Hancock and C. M. Powell, eds., 1985, Journal of Structural Geology, v. 7, p. 269–501.

International Conference on "Thrusting and Deformation," Toulouse, France, to mark the 100th anniversary of Archibald Geikie's definition of 'thrust fault:' Proceedings, *in* J. P. Platt, M. P. Coward, J. Deramond, and J. Hossack, eds., 1986, Journal of Structural Geology, v. 8, p. 215–492.

Stanford Rock Physics and Borehole Geophysics Consortium, started at Stanford University, California (leaders: Amos Nur for rock physics and Mark Zoback for borehole geophysics).

1985

Deformation Mechanisms, Rheology, and Tectonics Conference on "Tectonic and Structural Processes on a Macro-, Meso-,and Micro-scale," Utrecht, The Netherlands: Proceedings, *in* H. J. Zwart, M. Martens, I. Van Der Molen, C. W. Passchier, C. J. Spiers, and R. L. M. Vissers, eds., 1987, Tectonophysics, v. 135, p. 1–251.

International Conference on "Deformation of Sediments and Sedimentary Rocks," University College, London: Proceedings, *in* M. E. Jones and R. M. F. Preston, eds., 1987, Deformation of sediments and sedimentary rocks: GSL Special Publication 29, 350 p.

International Conference on "Fluid Flow in Sedimentary Basins and Aquifers," Geological Society (London): Proceedings, *in* J. C. Goff and B. P. Williams, eds., 1987, Fluid flow in sedimentary basins and aquifers: GSL Special Publication 34, 230 p.

Fault Analysis Group founded at University of Liverpool, England (moved to University College, Dublin, Ireland, in 2000) (leaders: Jun Watterson at Liverpool and John Walsh at Dublin).

1986

International Conference on "Shear Strain in Rocks," Imperial College, London: Proceedings, *in* P. R. Cobbold, D. Gapais, W. D. Means, and S. H. Treagus, eds., 1987, Journal of Structural Geology, v. 9, p. 521–778.

1987

Watt, N. L., 1987, Theoretical aspects of cap-rock and fault seals for single- and two-phase hydrocarbon columns (originally presented at the AAPG Annual Meeting in 1985): Marine and Petroleum Geology, v. 4, p. 274–307.

Deformation Mechanisms, Rheology, and Tectonics Conference on "Geological Kinematics and Dynamics: In Honor of the 70th Birthday of Hans Ramberg," Uppsala, Sweden: Proceedings, *in* C. Talbot, ed., 1988, Acta Universitatis Upsaliensis, Bulletin of the Geological Institution of the University of Uppsala, New Series 14, 286 p.

Conference on "Deformation of Crustal Rocks," Mount Buffalo, Victoria, organized by the Specialist Group in Tectonics and Structural Geology of the Geological Society of Australia: Proceedings, *in* A. Ord, ed., 1989, Tectonophysics, v. 158, p. 1–354.

1988

Foster, N. H., and E. A. Beaumont, compilers, 1988, Traps and seals: AAPG Treatise of Petroleum Geology Reprint Series 6–7 (reprint of 42 articles); and Foster, N. H., and E. A. Beaumont, compilers, 1988, Structural concepts and techniques (reprints of 93 articles): AAPG Treatise of Petroleum Geology Reprint Series 9–11, v. 1, 723 p.; v. 2, 479 p.; v. 3, 651 p.

International Conference on "Friction Phenomena in Rocks," Fredericton, United Kingdom: Proceedings, *in* J. G. Spray and P. J. Hudleston, eds., 1989, Journal of Structural Geology, v. 11, p. 783–831.

Applied Geodynamics Laboratory Research (focusing on salt tectonics) founded at the Bureau of Economic Geology, University of Texas at Austin (leader: Martin Jackson).

1989

Allan, U., 1989, Construction fault plane-juxtaposition diagrams (nicknamed "Allan Diagram"): AAPG Bulletin, v. 73, p. 803–811.

NPF Conference on "Structural and Tectonic Modelling and Its Application to Petroleum Geology," Stavanger, Norway: Proceedings, *in* R. M. Larsen, H. Brekke, B. T. Larsen, and E. Talleraas, eds., 1992, Structural and tectonic modelling and its application to petroleum geology: Amsterdam, Elsevier, 549 p.

International Conference on "The Geometry of Normal Faults," London: Proceedings, *in* A. M. Roberts, G. Yielding, and B. Freeman, eds., 1991, The geometry of normal faults: GSL Special Publication 56, 264 p.

Deformation Mechanisms, Rheology, and Tectonics Conference, Leeds University, United Kingdom: Proceedings, *in* R. J. Knipe and E. H. Rutter, eds., 1990, Deformation mechanics, rheology, and tectonics: GSL Special Publication 54, 535 p.

ISRM/SPE International Symposium on "Rock at Great Depth," Pau, France: Proceedings, *in* V. Maury and D. Fourmaintraux, eds., 1990, Rock at great depth, three volumes: Rotterdam, A. A. Balkema, 1620 p.

1990

Beaumont, E. A., and N. H. Foster, eds., 1990–1993, Structural traps, 8 volumes: AAPG Atlas of Oil and Gas Fields, v. 1, 232 p.; v. 2, 267 p.; v. 3, 355 p.; v. 4, 382 p.; v. 5, 305 p.; v. 6, 304 p.; v. 7, 347 p.; v. 8, 328 p.

The Brace symposium on "Fault Mechanics and Transport Properties of Rocks," Massachusetts Institute of Technology, Cambridge: Proceedings, *in* B. Evans and T.-F. Wong, eds., 1992, Fault mechanics and transport properties of rocks: New York, Academic Press, 542 p.

Duba, A. G., W. B. Durham, J. W. Handin, and H. F. Wu, eds., 1990, The brittle-ductile transition in rocks (the "Heard Volume"): AGU Geophysical Memoir 56, 243 p.

International Conference on "Mechanics of Jointed and Faulted Rock," Technical University of Vienna, Austria: Proceedings, *in* H. P. Rossmanith, ed., 1990, Mechanics of jointed and faulted rock: Rotterdam, A. A. Balkema.

GSL Meeting on "Absolute Dating of Fault Movements," London: Proceedings, *in* R. H. Maddock, ed., 1992, Journal of Geological Society (London), v. 149, p. 249–301.

Rock Fracture Project Consortium began at Stanford University, California (leaders: David Pollard and Atilla Aydin).

1991

Sessions on "Fluid Seals" and "Quantification of Faults and Their Effect on Fluid Flow," AAPG Annual Convention in Dallas.

Deformation Mechanisms, Rheology, and Tectonics Conference on "Mechanical Instabilities in Rocks and Tectonics," Montpellier, France: Proceedings, *in* J.-P. Burg, D. Mainprice, and J. P. Petit, eds., 1992, Journal of Structural Geology, v. 14, p. 893–1109.

International Conference on "The Geometry of Naturally Deformed Rocks," ETH, Zurich, Switzerland, in celebration of John Ramsay's 60th birthday: Proceedings, *in* M. Casey, D. Dietrich, M. Ford, and A. J. Watkinson, eds., 1993, Journal of Structural Geology, v. 15, p. 243–671.

Petroleum Group of the Geological Society of London's Meeting on "Minipermeametry in Reservoir Rocks," Edinburgh, United Kingdom: Proceedings in Marine and Petroleum Geology, v. 10, p. 298–351.

Fault Dynamics Group started at Royal Holloway University of London, focusing on analog modeling of faults (leader: Ken McClay).

1992

Session on "Pressure Seals and Abnormally Pressured Reservoirs," AAPG Annual Convention, Calgary, Canada.

International Conference on "Fractured and Jointed Rock Masses," Lake Tahoe, California: Proceedings, *in* L. R. Myer, C.-F. Tsang, N. G. W. Cook, and R. E. Goodman, eds., 1995, Fractured and jointed rock masses: Rotterdam, A. A. Balkema, 772 p.

GSA Penrose Conference on "Applications of Strain: From Microstructures to Orogenic Belts," Halifax, Canada: Proceedings, *in* M. T. Brandon, J. R. Henderson, W. D. Means, and S. Patreson, eds., 1994, Journal of Structural Geology, v. 16, p. 437–612.

Symposium on "Influence of Fluids on Deformation Processes in Rocks," International Geological Congress, Kyoto, Japan: Proceedings, *in* C. J. Spiers and T. Takeshita, eds., 1995, Tectonophysics, v. 254, nos. 3–4, p. 121–297.

Rock Deformation Research group started at University of Leeds, United Kingdom (leader: Robert Knipe).

1993

Session on "Reservoirs, Traps, Seal Integrity," AAPG Annual Convention, New Orleans.

AAPG Hedberg Research Conference on "Seals and Traps: A Multidisciplinary Approach," Crested Butte, Colorado.

USGS Conference on "Mechanical Effects of Fluids in Faulting," Fish Camp, California: Proceedings, *in* S. Hickman, R. Sibson, and R. Bruhn, eds., 1993, USGS Open File Report, 94-228; and in S. Hickman, R. Sibson, and R. Bruhn, eds., 1995, Journal of Geophysical Research, v. 100, p. 12,831–13,132.

Deformation Mechanisms, Rheology, and Tectonics Conference on "Structures and Tectonics at Different Lithospheric Levels" held in Graz, Austria: Proceedings, *in* E. Wallbercher, W. Unzong, and M. Brandmayr, eds., 1994, Journal of Structural Geology, v. 16, p. 1495–1575.

GeoFluids Conference, Torquay: Proceedings, *in* J. Parnell, ed., 1994, GeoFluids: Origin, migration, and evolution of fluids in sedimentary basins: GSL Special Publication 78, 372 p.

1994

Ortoleva, P. J., ed., 1994, Basin compartments and seals (project results of Gas Research Institute in 1992): AAPG Memoir 61, 477 p.

AAPG Hedberg Research Conference on "Abnormal Pressures in Hydrocarbon Environments," Golden, Colorado: Proceedings, *in* B. E. Law, G. F. Ulmishek, and V. I. Slavin, eds., 1998, Abnormal pressures in hydrocarbon environments: AAPG Memoir 70, 270 p.

Sessions on "Evaluation of Traps and Seal" and "Seal Types and Characteristics," AAPG Annual Convention, Denver: Proceedings, *in* R. C. Surdam, ed., 1997, Seals, traps, and the petroleum system: AAPG Memoir 67, 317 p.

ISRM/SPE International Conference EUROCK94, Delft, The Netherlands: Proceedings, *in* Rock mechanics in petroleum engineering, Rotterdam, A. A. Balkema, 992 p.

Seal Evaluation Consortium started at Pennsylvania State University, Philadelphia (leader: Terry Engelder).

Research consortium entitled GeoPop (Geoscience Project into Overpressure) launched by three British universities (Durham, Newcastel, and Herriot-Watt) (leader: Richard Swarbrick).

1995

Meeting on "Structural Geology in Reservoir Characterization," Imperial College, London, organized by the Petroleum Group of GSL: Proceedings, *in* M. P. Coward, T. S. Daltaban, and H. Johnson, eds., 1998, Structural geology in reservoir characterization: GSL Special Publication 127, 266 p.

Meeting on "Scaling Laws for Fault and Fracture Populations— Analysis and Applications," Royal Society of Edinburgh, United Kingdom: Proceedings, *in* P. A. Cowie, R. J. Knipe, and I. G. Main, eds., 1996, Journal of Structural Geology, v. 18, p. 1–383.

Deformation Mechanisms, Rheology, and Tectonics Conference on "Thermal and Mechanical Interactions in Deep-seated Rocks," Prague, Czechoslovakia: Proceedings, *in* K. Schulmann, ed., 1997, Tectonophysics, v. 280, p. 1–197.

Second International Conference on "Mechanics of Jointed and Faulted Rock," Vienna, Austria: Proceedings, *in* H. P. Rossmanith, ed., 1995, Mechanics of jointed and faulted rock: Rotterdam, A. A. Balkema, 1068 p.

GSA Penrose Conference on "Fault-Related Folding," Banff, Alberta, Canada: Proceedings, *in* D. J. Anastasio, E. A. Erslev, and D. M. Fisher, eds., 1997, Journal of Structural Geology, v. 19, p. 243–602.

1996

NPF Conference on "Hydrocarbon Seals: Importance for Exploration and Production," Trondheim, Norway: Proceedings, *in* P. Møller-Pedersen and A. G. Koestler, 1997, Hydrocarbon seals: Importance for exploration and production: Amsterdam, Elsevier, 263 p.

Conference on "Faulting, Fault Sealing and Fluid Flow in Hydrocarbon Reservoirs," University of Leeds, Edinburgh: Proceedings, *in* G. Jones, Q. J. Fisher, and R. J. Knipe, eds., 1998, Faulting, fault sealing, and fluid flow in hydrocarbon reservoirs: GSL Special Publication 147, 320 p.

AAPG/EAGE Research Symposium on "Compartmentalized Reservoirs: Their Detection, Characterization, and Management," Woodlands, Texas.

Session on "Risking Fault Seal: Examples from Exploration and Production," AAPG Annual Convention, San Diego.

International Conference on "Structures and Properties of High Strain Zones in Rocks," Verbania-Pallanza, Italy: Proceedings, *in* E. H. Rutter, A. Boriani, K. H. Brodie, and L. Burlini, eds., 1998, Journal of Structural Geology, v. 20, p. 111–320.

Rock fractures and fluid flow: Washington, D.C, National Academy of Sciences, 551 p.

1997

Session on "Diagenesis Associated with Faults, Folds, and Fractures," AAPG Annual Convention, Dallas.

GSA Penrose conference on "Faults and Subsurface Fluid Flow," Taos, New Mexico: Proceedings, *in* W. C. Haneberg, P. S. Mozely, J. C. Moore, and L. B. Goodwin, eds., 1999, Faults and subsurface fluid flow in the shallow crust: AGU Geophysical Monograph 113, 222 p.

Meeting on "Structural Controls and Genesis of Economic Resources: Mineral and Hydrocarbon Deposits," Dublin, Ireland: Proceedings, *in* K. McCaffrey, L. Lonergan, and J. Wilkinson, eds., 1999, Fractures, fluid flow, and mineralization: GSL Special Publication 155, 328 p.

Deformation Mechanisms, Rheology, and Tectonics Conference on "Deformation Mechanisms in Nature and Experiment," Basel, Switzerland: Proceedings, *in* S. M. Schmid, R. Heilbronner, and H. Stünitz, eds., 1999, Tectonophysics, v. 303, p. 1–319.

GeoFluids II: Second International Conference on "Fluid Evolution, Migration and Interaction in Sedimentary Basins and Orogenic Belts," Belfast, Ireland: Proceedings, *in* J. Parnell, ed., 1998, Dating and duration of fluid flow and fluid-rock interaction: GSL Special Publication 144, 284 p.

Consortium on Utah Faults, Fluids, and Fractures (UF^3) started at Utah State University and University of Utah (leaders: James Evans and Craig Forster).

Reactivation Research Group began at University of Durham, United Kingdom (leader: Robert Holdsworth).

Project on Evaluation of Traps and Seals at Japan National Oil Corporation, 1997–2003.

1998

AAPG Hedberg Research Conference on "Reservoir Scale Deformation: Characterization and Prediction," Bryce, Utah.

Special Session on "Faults: Seals or Migration Pathways?," AAPG Annual Convention, Salt Lake City, Utah.

International Forum on "Pressure Regimes in Sedimentary Basins and Their Prediction," Houston, and sponsored by the Houston chapter of the American Association of Drilling Engineers: Proceedings, *in* A. R. Huffman and G. L. Bowers, eds., 2002, Pressure regimes in sedimentary basins and their prediction: AAPG Memoir 76, 238 p.

International Conference on "Evolution of Structures in Deforming Rocks," Canmore, Alberta, in honor of Paul F. Williams: Proceedings, *in* W. Blecker, C. Elliott, S. Lin, and C. R. van Staal, eds., 2001, Journal of Structural Geology, v. 23, p. 843–1178.

ISRM/SPE Symposium EUROCK98, Trondheim, Norway: Proceedings, *in* Rock Mechanics in Petroleum Engineering, two volumes, Richardson, Texas: Society of Petroleum Engineers, 952 p.

Third International Conference on "Mechanics of Jointed and Faulted Rock," Vienna, Austria: Proceedings, *in* H. P. Rossmanith, ed., 1998, Mechanics of jointed and faulted rock: Rotterdam, A. A. Balkema, 658 p.

Chester, F. M., T. Engelder, and T. Shimamoto, eds., 1998, Rock deformation: The Logan volume: Tectonophysics, v. 295, p. 1–257.

Fracture Research and Application Consortium (FRAC) began at the Bureau of Economic Geology, University of Texas at Austin.

Consortium Project on "GeoFluids" began at Pennsylvania State University, Philadelphia (leader: Peter Flemings).

Consortium Project on "Stress and Diagenesis as Controls on Fault Flow" (STADIA) began by the Edinburgh Rock Mechanics Consortium at University of Edinburgh and Heriot-Watt University (leaders: Bryne Ngwenya and Brian Crawford).

1999

Beaumont, E. A., and N. Foster, eds., 1999, AAPG handbook on exploring for oil and gas traps, 1150 p.

Symposium on "Dynamics of Fluids in Fractured Rocks," Lawrence Berkeley National Laboratory, California: Proceedings, *in* B. Faybishenko, P. A. Witherspoon, and S. M. Benson, eds., 2000, Dynamics of fluids in fractured rocks: American Geophysical Union Geophysical Monograph 122, 400 p.

"Deformation Mechanisms, Rheology, and Tectonics Conference," Neustadt an der Weinstrasse, Germany: Proceedings, *in* G. Dresen and M. Handy, eds., 2001, International Journal of Earth Sciences (Geologische Rundschau), v. 90, p. 1–210.

International Conference on "Textural and Physical Properties of Rocks," University of Göttingen, Germany: Proceedings, *in* B. Leiss, K. Ullemeyer, and K. Weber, eds., 2000, Journal of Structural Geology, v. 22, p. 1527–1873.

Session on "Fault-related Folds: The Transition from 2-D to 3-D" at GSA Annual Meeting, Denver: Proceedings, *in* M. S. Wilkerson, M. P. Fischer, and T. Apotria, eds., 2002, Journal of Structural Geology, v. 24, p. 591–904.

Session on "Fault Rocks," EUG 10, Strassburg, France.

2000

NPF Conference on "Hydrocarbon Seal Quantification," Stavanger, Norway: Proceedings, *in* A. G. Koestler and R. Hunsdale, eds., 2002, Hydrocarbon seal quantification: Amsterdam, Elsevier, 263 p.

Joint GSA/GSL Conference on "The Nature and Tectonic Significance of Fault Zone Weakening," London: Proceedings, *in* R. E. Holdsworth, R. A. Strachan, J. F. Magloughlin, and R. J. Knipe, eds., 2001, The nature and tectonic significance of fault zone weakening: GSL Special Publication 186, 342 p.

Session on "Successful Techniques to Evaluate Traps and Pressure Compartments," AAPG Annual Convention, New Orleans.

GeoFluids III: Third International Conference on Fluid Evolution, Migration, and Interaction in Sedimentary Basins and Orogenic Belts (July 12–14), Barcelona, Spain: Proceedings, *in* Journal of Geochemical Exploration, v. 69–70, p. 1–714.

Consortium on Hydrocarbon Sealing Potential of Faults and Cap Rocks began at Australian Petroleum Cooperative Research Center (APCRC) (leader: John Kaldi).

Short Course and Field Trip on "Characterization and Modeling Fluid Flow in Fault and Fracture Zone: The Reality and the Idealized" (by J. Caine, J. Evans, and C. Forster), GSA Annual Meeting (November 13–16), Reno, Nevada.

2001

Session on "Fault Seal Analysis Best Practices," AAPG Annual Convention, Denver.

GeoFluids, a new journal launched by Blackwell Science Publishers.

"Deformation Mechanisms, Rheology, and Tectonics Conference," Noordwijkerhout, The Netherlands: Proceedings, *in* S. De Meer, M. R. Drury, J. H. P. DePresser, and G. M. Pennock, eds., 2002, Deformation mechanisms, rheology, and tectonics: Current status and future perspectives: GSL Special Publication 200, 424 p.

Boland, J., and A. Ord, 2001, Deformation processes in the Earth's crust, in honor of B. E. Hobbs: Tectonophysics, v. 335, nos. 1–2, p. 1–228.

2002

Session on "Pathways of Hydrocarbon Migrations, Faults as Conduits or Seals," AAPG Annual Meeting, Houston.

AAPG Hedberg Conference on "Deformation History, Fluid Flow Reconstruction, and Reservoir Appraisal in Foreland Fold-Thrust Belts (FFTB): Exploration in High Risk Areas," Sicily, Italy.

AAPG Hedberg Conference on "Evaluating the Hydrocarbon Sealing Potential of Faults and Caprocks," Adelaide, Australia.

Gordon Research Conference on "Rock Deformation: Deformation Mechanism and Mode of Failure Transitions in Rocks," Barga, Italy.

GSL/GSA Joint International Research Meeting on "Transport and Flow Processes within Shear Zones," London, United Kingdom.

2003

Davies, R., and J. Handschy, eds., 2003, Fault seals: AAPG Bulletin, v. 87, no. 3, p. 377–527.

GeoFluids IV: Fourth International Conference on "Fluid Evolution, Migration, and Interaction in Sedimentary Basins and Orogenic Belts," Utrecht, The Netherlands.

EAGE International Conference and Fieldtrip on "Fault and Top Seals: What Do We Know and Where Do We Go?," Montpellier, France, and sponsored by Shell and TotalFinaElf.

Session on "Why Do Traps Fail?," at AAPG International Meeting, Barcelona, Spain.

GSL Meeting on "Micro-to-Macro Finale: Understanding the Micro-to-Macro Behaviour of Rock-Fluid Systems," London, United Kingdom.

REFERENCES CITED

Aharonov, E., E. Tenthorey, and C. H. Scholz, 1998, Precipitation sealing and diagenesis: 2. Theoretical analysis: Journal of Geophysical Research, v. 103B, p. 23,969–23,981.

Allan, U. S., 1989, Model for hydrocarbon migration and entrapment within faulted structures: AAPG Bulletin, v. 73, p. 803–811.

Allard, D. M., 1997, Fault seal controlled trap fill: Rift basin examples, *in* R. C. Surdam, ed., Seals, traps, and the petroleum system: AAPG Memoir 67, p. 135–142.

Allen, P. A., and J. R. Allen, 1990, Basin analysis: Principles and applications: Oxford, Blackwell Science, 451 p.

Al-Qahtani, M. Y., and E. Ershaqi, 1999, Characterization and estimation of permeability correlation structure from performance data, *in* R. Schatzinger and J. Jordan, eds., Reservoir characterization— Recent advances: AAPG Memoir 71, p. 343–358.

Anastasio, D. J., E. A. Erslev, and D. M. Fisher, eds., 1997, Fault-related folding: Journal of Structural Geology, v. 19, nos. 3–4 (special issue), p. 243–602.

Anders, M. H., and D. V. Wiltschko, 1994, Microfracturing, paleostress and the growth of faults: Journal of Structural Geology, v. 16, p. 795–815.

Antonellini, M., and A. Aydin, 1994, Effect of faulting on fluid flow in porous sandstones: Petrophysical properties: AAPG Bulletin, v. 78, p. 355–377.

Archie, G. E., 1942, The electrical resistivity log as an aid in determining some reservoir characteristics: American Institute of Mechanical Engineering Transactions, v. 146, p. 54–61.

Aydin, A., 1978, Small faults formed as deformation bands in sandstone: Pure and Applied Geophysics, v. 116, p. 913–930.

Aydin, A., 2000, Fractures, faults, and hydrocarbon migration and flow: Marine and Petroleum Geology, v. 17, p. 797–814.

Aydin, A., R. Myers, and A. Younes, 1998, Faults: Seal or migration pathways? Yes, no, some are but some aren't, and some faults are but only sometimes (abs.): AAPG Annual Meeting Program, v. 8, p. A37.

Bailey, R. J., and R. Stoneley, 1981, Petroleum: Entrapment and conclusions, *in* D. H. Tarling, ed., 1981, Economic geology and geotectonics: Oxford, Blackwell, p. 73–97.

Bally, A. W., ed., 1983, Seismic expressions of structural styles— A picture and work atlas: AAPG Studies in Geology 15, three volumes, 942 p.

Becker, A., and M. R. Cross, 1996, Mechanism for joint saturation in mechanically layered rocks: An example from southern Israel: Tectonophysics, v. 257, p. 224–237.

Bell, J. S., 1990, Investigating stress regimes in sedimentary basins using information form oil industry wire-line logs and drilling records, *in* A. Hurst, M. A. Lovell, and A. C. Morton, eds., Geological applications of wireline logs: Geological Society (London) Special Publication 48, p. 305–325.

Bentley, M. R., and J. J. Barry, 1991, Representation of fault sealing in a reservoir simulation: Cormorant block IV UK North Sea, *in* 66th Annual Technical Conference and Exhibition of the Society of Petroleum Engineers, Dallas, Texas, p. 119–126.

Berberian, M., D. Papastamatiou, and M. Qoraishi, 1977, Khurgu (north Bandar Abbas-Iran) earthquake of March 21, 1977: A preliminary field report and a seismogenic discussion, *in* M. Berberian, ed., Contribution to the seismotectonics of Iran (part III): Geological and Mining Survey of Iran Report No. 40, p. 7–49.

Berg, R. R., 1975, Capillary pressure in stratigraphic traps: AAPG Bulletin, v. 59, p. 939–956.

Berg, R. R., and A. H. Avery, 1995, Sealing properties of Tertiary growth faults, Texas Gulf Coast: AAPG Bulletin, v. 79, p. 375–393.

Biddle, K. T., and Wielchowsky, C. C., 1994, Hydrocarbon traps, *in* L. B. Magoon and W. G. Dow, eds., The petroleum system—From source to trap: AAPG Memoir 60, p. 219–235.

Bjørkum, P. A., O. Walderhaug, and P. H. Nadeau, 1998, Physical constraints on hydrocarbon leakage and trapping revisited: Petroleum Geoscience, v. 43, p. 273–239.

Bjørlykke, K., 1983, Diagenetic reactions in sandstones, *in* A. Parker and B. W. Shellwood, eds., Sediment diagenesis: Dordrecht, Reidel, p. 169–213.

Bouvier, J. D., C. H. Kaars-Sijpersteijn, D. F. Kluesner, C. C. Onyejekwe, and R. C. van der Pal, 1989, Three-dimensional seismic interpretation and fault sealing investigations, Nun River field, Nigeria: AAPG Bulletin, v. 73, p. 1397–1414.

Brace, W. F., 1978, A note on permeability changes in geologic material due to stress: Pure and Applied Geophysics, v. 116, p. 627–633.

Bradley, J. S., and D. E. Powley, 1994, Pressure compartments in sedimentary basins: A review, *in* P. J. Ortoleva, ed., Basin compartments and seals: AAPG Memoir 61, p. 3–26.

Bretan, P., G. Yielding, and H. Jones, 2003, Using calibrated shale gouge ratio to estimate hydrocarbon column heights: AAPG Bulletin, v. 87, p. 396–414.

Brown, A., 2003, Capillary effects on fault-fill sealing: AAPG Bulletin, v. 87, p. 381–396.

Bruhn, R. L., W. A. Yonkee, and W. T. Parry, 1990, Structural and fluid-chemical properties of seismogenic normal faults: Tectonophysics, v. 175, p. 139–157.

Burhannudinnur, M., and C. K. Morley, 1997, Anatomy of growth fault zones in poorly lithified sandstones and shales: Implications for reservoir studies and seismic interpretation: Part 1. Outcrop study: Petroleum Geoscience, v. 3, p. 211–224.

Caine, J. S., and R. A. Tomusiak, 2003, Brittle structures and their role in controlling porosity and permeability in a complex Precambrian crystalline-rock aquifer system in the Colorado Rocky Mountain Front Range: Geological Society of America Bulletin, v. 115, p. 1410–1424.

Caine, J. S., J. P. Evans, and C. B. Forster, 1996, Fault zone architecture and permeability structure: Geology, v. 24, p. 1025–1028.

Capuano, R. M., 1993, Evidence for fluid flow in microfractures in geopressured shales: AAPG Bulletin, v. 77, p. 1303–1314.

Childs, O. E., 1963, Place of tectonic concepts in geological thinking, *in* O. E. Childs and B. W. Beebee, eds., Backbone of the Americas: Tectonic history from pole to pole: AAPG Memoir 2, p. 1–3.

Childs, C. J., J. J. Walsh, and J. Watterson, 1997, Complexity in fault zone structure and implications for fault seal prediction, *in* P. Møller-Pedersen and A. G. Koestler, eds., Hydrocarbon seals: Importance for exploration and production: Norwegian Petroleum Society Special Publication 7, p. 61–72.

Childs, C., T. Manzocchi, P. A. R. Nell, J. J. Walsh, J. A. Strand, A. E. Heath, and T. H. Lygren, 2002, Geological implications of a large pressure difference across a small fault in the Viking graben, *in* A. G. Koestler and R. Hunsdale, eds., Hydrocarbon seal quantification: Norwegian Petroleum Society Special Publication 11, p. 187–201.

Clapp, F. G., 1910, A proposed classification of petroleum and natural gas fields based on structure: Economic Geology, v. 5, p. 503–521.

Clapp, F. G., 1917, Revision of the structural classification of petroleum and natural gas fields: Geological Society of America Bulletin, v. 28, p. 553–602.

Clapp, F. G., 1929, The role of geologic structure in the accumulation of petroleum, *in* S. Powers, ed., Structure of typical American oil fields II: Tulsa, AAPG, p. 667–716.

Corcoran, D. V., and A. G. Doré, 2002, Top seal assessment in exhumed basin settings—Some insights from Atlantic margin and borderland basins, *in* A. G. Koeslter and R. Hunsdale, eds., Hydrocarbon seal quantification: Norwegian Petroleum Society Special Publication 11, p. 89–107.

Cosgrove, J. W., and M. S. Ameen, eds., 2000, Forced folds and fractures: Geological Society (London) Special Publication 169, 225 p.

Cossé, R., 1993, Basics of reservoir engineering: Paris, Editions Technip, 372 p.

Dickinson, W. R., 1974, Plate tectonics and sedimentation: SEPM Special Publication 22, p. 1–27.

Dott, R. H., and M. J. Reynolds, 1969, Sourcebook for petroleum geology: AAPG Memoir 5, chapter 13, p. 342–440.

Downey, M. W., 1984, Evaluating seals for hydrocarbon accumulations: AAPG Bulletin, v. 68, p. 1752–1763.

Engelder, J. T., 1974, Cataclasis and the generation of fault gouge: Geological Society of America Bulletin, v. 85, p. 1515–1522.

England, W. A., A. S. Mackenzie, D. M. Mann, and T. M. Quigley, 1987, The movement and entrapment of petroleum fluids in the subsurface: Journal of the Geological Society (London), v. 144, p. 327–347.

Ericsson, J. B., H. C. McKean, and R. J. Hooper, 1998, Facies and curvature controlled 3-D fracture models in a Cretaceous carbonate reservoir, Arabian Gulf, *in* G. Jones, Q. J. Fisher, and R. J. Knipe, eds., Faulting, fault sealing and fluid flow in hydrocarbon reservoirs: Geological Society (London) Special Publication 147, p. 299–312.

Evans, J. P., C. B. Forster, and J. V. Goddard, 1997, Permeability of fault-related rocks, implications for hydraulic structure of fault zones: Journal of Structural Geology, v. 19, p. 1393–1404.

Faulkner, D. R., and E. H. Rutter, 1998, The gas permeability of clay-bearing fault gouge at 20°C, *in* G.

Jones, Q. J. Fisher, and R. J. Knipe, eds., Faulting, fault sealing and fluid flow in hydrocarbon reservoirs: Geological Society (London) Special Publication 147, p. 147–156.

Ferrill, D. A., J. Winterle, G. Witmeyer, D. Sims, S. Colton, A. Armstrong, and A. P. Morris, 1999, Stressed rock strains groundwater at Yucca mountain, Nevada: Geological Society of America Today, v. 9, no. 5, p. 1–8.

Finkbeiner, T., C. A. Barton, and M. D. Zoback, 1997, Relationship between in-situ stress, fractures and faults, and fluid flow in the Monterey formation, Santa Maria Basin, California: AAPG Bulletin, v. 81, p. 1975–1999.

Finkbeiner, T., M. D. Zoback, B. Stump, and P. Flemings, 2001, Stress, pore pressure, and dynamically-constrained hydrocarbon column heights in the South Eugene Island 330 field, Gulf of Mexico: AAPG Bulletin, v. 85, p. 1007–1031.

Fisher, Q. J., and R. J. Knipe, 1998, Fault sealing processes in siliciclastic sediments, *in* R. J. Knipe, G. Jones, and Q. J. Fisher., eds., Faulting, fault sealing, and fluid flow in hydrocarbon reservoirs: Geological Society (London) Special Publication 147, p. 117–134.

Fowles, J., and S. D. Burley, 1994, Textural and permeability characteristics of faulted, high porosity sandstones: Marine Petroleum and Geology, v. 11, p. 608–623.

Freeman, B., G. Yielding, D. T. Needham, and M. E. Badley, 1998, Fault seal prediction: The gouge ratio method, *in* M. P. Coward, T. S. Daltaban, and H. Johnson, eds., Structural geology in reservoir characterization: Geological Society (London) Special Publication 127, p. 19–25.

Fulljames, J. R., L. J. J. Zijerveld, R. C. M. W. Franssen, G. M. Ingram, and P. D. Richard, 1997, Fault seal processes, *in* P. Møller-Pedersen and A. G. Koestler, eds., Hydrocarbon seals: Importance for exploration and production: Norwegian Petroleum Society Special Publication 7, p. 51–59.

Gibson, R. G., 1994, Fault-zone seals in siliclastic strata of the Columbus basin, offshore Trinidad: AAPG Bulletin, v. 78, p. 1372–1385.

Gibson, R. G., 1998, Physical character and fluid-flow properties of sandstone-derived fault zones, *in* M. P. Coward, T. S. Daltaban, and H. Johnson, eds., Structural geology in reservoir characterization: Geological Society (London) Special Publication 127, p. 83–97.

Gould, S. J., 1989, Wonderful life: The Burgess Shale and the nature of history: New York, W. W. Norton, 256 p.

Gratier, J. P., T. Chen, and R. Hellmann, 1994, Pressure solution as a mechanism for crack sealing around faults, *in* S. Hickman, R. Sibson, and R. Bruhn, eds., Proceedings of the U.S. Geological Survey Red Book Conference on the Mechanical Involvement of Fluids in Faulting: U.S. Geological Survey Open-file Report 94-228, p. 279–300.

Grauls, D., F. Pascaud, and T. Rives, 2002, Quantitative fault seal assessment in hydrocarbon-compartmentalised structures using fluid pressure data, *in* A. G. Koeslter and R. Hunsdale, eds., Hydrocarbon seal quantification: Norwegian Petroleum Society Special Publication 11, p. 141–156.

Gussow, W. C., 1954, Differential entrapment of oil and gas: A fundamental principle: AAPG Bulletin, v. 38, p. 816–853.

Hadizadeh, J., 1994, Interaction of cataclasis and pressure solution in a low-temperature carbonate shear zone: Pure and Applied Geophysics, v. 143, p. 255–280.

Hadizadeh, J., and F. F. Foit, 2000, Feasibility of estimating cementation rate in a brittle fault zone using some precepts of sedimentary diagenesis: Journal of Structural Geology, v. 22, p. 401–409.

Hager, D., 1915, Practical oil geology: New York, McGraw-Hill, 150 p.

Harding, T. P., and J. D. Lowell, 1979, Structural styles, their plate-tectonic habitats, and hydrocarbon traps in petroleum provinces: AAPG Bulletin, v. 63, p. 1016–1058.

Harding, T. P., and A. C. Tuminas, 1989, Structural interpretation of hydrocarbon traps sealed by basement normal block faults at stable flanks of foredeep basins and at rift basins: AAPG Bulletin, v. 73, p. 812–840.

Harper, T. R., and E. R. Lundin, 1997, Fault seal analysis: Reducing our dependence on empiricism, *in* P. Møller-Pederson and A. G. Koestler, eds., Hydrocarbon seals: Importance for exploration and production: Norwegian Petroleum Society Special Publication 7, p. 149–165.

Heald, K. C., 1940, Essentials for oil pools, *in* E. DeGolyer, ed., Elements of the petroleum industry: American Institute of Mining and Metallurgical Engineers, p. 26–62.

Hesthammer, J., and H. Fossen, 2000, Uncertainties associated with fault sealing analysis: Petroleum Geoscience, v. 6, p. 37–45.

Heum, O. R., 1996, A fluid dynamic classification of hydrocarbon entrapment: Petroleum Geoscience, v. 2, p. 145–158.

Hippler, S. J., 1997, Microstructures and diagenesis in North Sea fault zones: Implications for fault-seal potential and fault-migration rate, *in* R. C. Surdam, ed., Seals, traps, and the petroleum system: AAPG Memoir 67, p. 103–113.

Hoffman, K. S., D. R. Taylor, and R. T. Schnell, 1996, 3-D improves/speeds up fault plane analysis: Leading Edge, v. 15, p. 117–122.

Hubbert, M. K., 1953, Entrapment of petroleum under hydrodynamic conditions: AAPG Bulletin, v. 37, p. 1954–2026.

Hubbert, M. K., and W. W. Rubey, 1959, Role of pore fluid pressures in the mechanics of overthrust faulting: Geological Society of America Bulletin, v. 70, p. 115–205.

Ingram, G. M., and J. L. Urai, 1999, Top-seal leakage through faults and fractures: The role of mudrock properties, *in* A. C. Aplin, A. J. Fleet, and J. H. S. Macquaker, eds., Muds and mudstones: Physical and fluid flow properties: Geological Society (London) Special Publication 158, p. 125–135.

International Task Force, 2001, Bringing research and technology to optimize reservoir performance: First Break, v. 19, p. 351–353.

Jenyon, M. K., 1990, Oil and gas traps: New York, John Wiley, 398 p.

Jones, R. M., and R. R. Hillis, 2003, An integrated, quantitative approach to assessing fault-seal risk: AAPG Bulletin, v. 87, p. 507–524.

Knai, T. A., and R. J. Knipe, 1998, The impact of faults on fluid flow in the Heidrun field, *in* R. J. Knipe, G. Jones, and Q. J. Fisher, eds., Faulting, fault sealing and fluid flow in hydrocarbon reservoirs: Geological Society (London) Special Publication 147, p. 269–282.

Knipe, R. J., 1989, Deformation mechanisms— Recognition from natural tectonites: Journal of Structural Geology, v. 11, p. 127–146.

Knipe, R. J., 1992, Faulting processes and fault seal, *in* R. M. Larsen, H. Brekke, B. T. Larsen, and E. Talleraas, eds., Structural and tectonic modelling and its application to petroleum geology: Norwegian Petroleum Society Special Publication 1, p. 325–342.

Knipe, R. J., 1993, The influence of fault zone processes and diagenesis on fluid flow, *in* A. D. Horbury and A. G. Robinson, eds., Diagenesis and basin development: AAPG Studies in Geology 36, p. 135–154.

Knipe, R. J., 1997, Juxtaposition and seal diagrams to help analyze fault seals in hydrocarbon reservoirs: AAPG Bulletin, v. 81, p. 187–195.

Knipe, R. J., Q. J. Fisher, G. Jones, B. Clennell, A. B. Farmer, B. Kidd, E. McAllister, and E. White, 1997, Fault seal analysis: Successful methodologies, application and future directions, *in* P. Møller-Pederson and A. G. Koestler, eds., Hydrocarbon seals: Importance for exploration and production: Norwegian Petroleum Society Special Publication 7, p. 15–40.

Knipe, R. J., G. Jones, and Q. J. Fisher, 1998, Faulting, fault sealing and fluid flow in hydrocarbon reservoirs: An introduction, *in* G. Jones, Q. J. Fisher, and R. J. Knipe, eds., Faulting, fault sealing and fluid flow in hydrocarbon reservoirs: Geological Society (London) Special Publication 147, p. vii–xxi.

Knipe, R. J., et al., 2000, Quantification and prediction of fault seal parameters: The importance of geohistory (abs.): Hydrocarbon Seal Quantification, Norwegian Petroleum Society Conference (Stavanger) Extended Abstracts, p. 39–42.

Knott, S. D., 1993, Fault seal analysis in the North Sea: AAPG Bulletin, v. 77, p. 778–792.

Lehner, F. K., and W. F. Pilaar, 1997, The emplacement of clay smears in synsedimentary normal faults: Inferences from field observations near Frechen, Germany, *in* P. Møller-Pedersen and A. G. Koestler, eds., Hydrocarbon seals: Importance for exploration and production: Norwegian Petroleum Society Special Publication 7, p. 39–50.

Leveille, G. P., R. J. Knipe, C. More, D. Ellis, G. Dudley, G. Jones, and Q. J. Fisher, 1997, Compartmentalization of Rotliegendes gas reservoirs by sealing faults, Jupiter area, southern North Sea, *in* K. Ziegler, P. Turner, and S. R. Daines, eds., Petroleum geology of the southern North Sea: Future potential: Geological Society (London) Special Publication 123, p. 87–104.

Levorsen, A. I., 1967, Geology of petroleum, 2d ed.: San Francisco, W. H. Freeman, chapters 6–7, p. 232–384.

Lia, O., H. More, H. Tjelmeland, L. Holden, and T. England, 1997, Uncertainties in reservoir production forecasts: AAPG Bulletin, v. 81, p. 775–801.

Lindsay, N. G., F. C. Murphy, J. J. Walsh, and J. Watterson, 1993, Outcrop studies of shale smears on fault surfaces: International Association of Sedimentologists Special Publication 15, p. 113–123.

Lowell, J. D., 2002, Structural styles in petroleum geology, 4th ed.: Tulsa, Oil and Gas Consultants Inc., 311 p.

Lyell, C., 1842, Principles of geology, v. 1: Boston, Hilliard, Grey & Co., 511 p.

Magoon, L. B., and W. G. Dow, 1994, The petroleum system, *in* L. B. Magoon and W. G. Dow, eds., The petroleum system— From source to trap: AAPG Memoir 60, p. 3–24.

Manzocchi, T., J. J. Walsh, P. Nell, and G. Yielding, 1999, Fault transmissibility multipliers for flow simulation models: Petroleum Geoscience, v. 5, p. 53–63.

Manzocchi, T., A. E. Heath, J. J. Walsh, and C. Childs, 2002, The representation of two phase fault-rock properties in flow simulation models: Petroleum Geoscience, v. 8, p. 119–132.

McKnight, E. T., 1940, Geology of area between Green and Colorado rivers, Grand and San Juan Counties, Utah: U.S. Geological Survey Bulletin, v. 908, 147 p.

Milton, N. J., and G. T. Bertram, 1992, Trap styles— A new classification based on sealing surfaces: AAPG Bulletin, v. 76, p. 983–999.

Mitra, S., 1988, Effects of deformation mechanisms on reservoir potential in central Appalachian overthrust belt: AAPG Bulletin, v. 72, p. 536–554.

Morrow, C. A., L. Q. Shi, and J. D. Byerlee, 1984, Permeability of fault gouge under confining pressure and shear stress: Journal of Geophysical Research, v. 89B5, p. 3193–3200.

Mozley, P. S., and L. B. Goodwin, 1995, Patterns of cementation along a Cenozoic normal fault: A record of paleoflow orientations: Geology, v. 23, p. 539–542.

Naruk, S. J., et al., 2002, Common characteristics of proven and leaking faults: AAPG Hedberg Research Conference on "Evaluating the Hydrocarbon Sealing Potential of Faults and Caprocks," December 1–5, 2002, Barossa Valley, Australia, p. 71–74.

National Petroleum Council, 1995, Research, development, and demonstration needs of the oil and gas industry, three volumes, 1100 p.

Noller, J. S., J. N. Sowers, and W. R. Lettis, eds., 2000, Quaternary geochronology: Methods and applications: Washington, D.C., American Geophysical Union, 582 p.

North, F. K., 1985, Petroleum geology: Boston, Allen & Unwin, chapter 16, p. 253–341.

Nybakken, S., 1991, Sealing fault traps— An exploration concept in a mature petroleum province: Tampen Spur, northern North Sea: First Break, v. 9, p. 209–222.

Peacock, D. C. P., Q. J. Fisher, E. J. M. Willemse, and A. Ayden, 1998, The relationship between faults and pressure solution seams in carbonate rocks and the implications for fluid flow, *in* G. Jones, Q. J. Fisher, and R. J. Knipe, eds., Faulting, fault sealing, and fluid flow in hydrocarbon reservoirs: Geological Society (London) Special Publication 147, p. 105–115.

Perkins, H., 1961, Fault-closure type fields, southeast Louisiana: Gulf Coast Association of Geological Societies Transactions, v. 11, p. 177–196.

Pittman, E. D., 1981, Effect of fault-related granulation on porosity and permeability of quartz sandstones, Simpson Group (Ordovician), Oklahoma: AAPG Bulletin, v. 65, p. 2381–2387.

Prison, S. J., 1945, Genetic and morphologic classification of reservoirs: Oil Weekly, CXVIII (June 18, 1945), p. 54–59.

Purcell, W. R., 1949, Capillary pressure— Their measurement using mercury and the calculation of permeability therefrom: American Institute of Mining and Metallurgical Engineers Transactions, v. 186, p. 39–48.

Read, H. H., and J. Watson, 1962, Introduction to geology, v. 1: London, Macmillan, p. 487–490.

Ringrose, P. S., K. S. Srobie, P. W. M. Corbett, and J. L. Jensen, 1993, Immiscible flow behavior in laminated and cross-bedded sandstones: Journal of Petroleum Science and Engineering, v. 9, p. 103–124.

Rittenhouse, G., 1972, Stratigraphic trap classification: AAPG Memoir 16, p. 14–28.

Rose, P. R., 2001, Risk analysis and management of petroleum exploration ventures: AAPG Methods in Exploration Series 12, 164 p.

Rutter, E. H., 1983, Pressure solution in nature, theory and experiment: Journal of Geological Society (London), v. 144, p. 725–740.

Sales, J. K., 1993, Closure vs. sealing strength— A fundamental control on the distribution of oil and gas, *in* A. G. Dore, J. H. Auguston, C. Hermanrud, O. Sylta, and D. J. Stewart, eds., Basin modelling: Advances and applications: Norwegian Petroleum Society Special Publication 3, p. 399–414.

Sales, J. K., 1997, Seal strength vs. trap closure— A fundamental control on the distribution of oil and gas, *in* R. C. Surdam, ed., Seals, traps, and the petroleum system: AAPG Memoir 67, p. 57–83.

Sato, T., R. Sakai, K. Furuya, and T. Kodama, 2000, Coseismic spring flow changes associated with the 1995 Kobe earthquake: Geophysical Research Letters, v. 27, p. 1219–1222.

Schowalter, T. T., 1979, Mechanisms of secondary hydrocarbon migration and entrapment: AAPG Bulletin, v. 63, p. 723–760.

Shipton, Z. K., J. P. Evans, K. R. Robeson, C. B. Forster, and S. Snelgrove, 2002, Structural heterogeneity and permeability in faulted eolian sandstone: Implications for subsurface modeling of faults: AAPG Bulletin, v. 86, p. 863–883.

Sibson, R. H., 1977, Fault rocks and fault mechanisms: Journal of the Geological Society, v. 133, p. 199–213.

Skerlec, G. M., 1999, Evaluating top and fault seal, *in* E. A. Beaumont, and N. H. Foster, eds., Exploring for oil and gas traps: AAPG Treatise of Petroleum Geology, chapter 10, p. 1–94.

Smith, D. A., 1966, Theoretical considerations of sealing and non-sealing faults: AAPG Bulletin, v. 50, p. 363–374.

Smith, D. A., 1980, Sealing and non-sealing faults in Louisiana Gulf Coast salt basin: AAPG Bulletin, v. 64, p. 145–172.

Sneider, R. M., K. K. Stopler, and J. S. Sneider, 1991, Petrophysical properties of seals (abs.): AAPG Bulletin, v. 75, p. 673–674.

Sneider, R. M., J. S. Sneider, G. W. Bolger, and J. W. Neasham, 1997, Comparison of seal capacity determinations: Conventional cores vs. cuttings, *in* R. C. Surdam, ed., Seals, traps, and the petroleum system: AAPG Memoir 67, p. 1–12.

Sorkhabi, R. B., S. Hasegawa, S. Iwanaga, and M. Fujimoto, 2002, Sealing assessment of normal faults in clastic reservoirs: The role of fault geometry and shale smear parameters: Journal of the Japanese Association of Petroleum Technology, v. 67, p. 576–589.

Sorkhabi, R. B., S. Iwanaga, M. Fujimoto, and S. Hasegawa, 2003, Sealing assessment of normal faults in clastic reservoirs: Modeling the petrophysical and stress attributes of faults: Journal of the Japanese Association of Petroleum Technology, v. 68, p. 291–304.

Sperrevik, S., P. A. Gillespie, Q. J. Fisher, T. Halvorsen, and R. J. Knipe, 2002, Empirical estimation of fault rock properties, *in* A. G. Koeslter and R. Hunsdale, eds., Hydrocarbon seal quantification: Norwegian Petroleum Society Special Publication 11, p. 109–125.

Sverdrup, E., and K. Bjørlykke, 1997, Fault properties and the development of cemented fault zones in sedimentary basins: Field examples and predictive models, *in* P. Møller-Pedersen and A. G. Koestler, eds., Hydrocarbon seals: Importance for exploration and production: Norwegian Petroleum Society Special Publication 7, p. 91–106.

Swanson, B. F., 1981, A simple correlation between permeabilities and mercury capillary pressures: Journal of Petroleum Technology, v. 33, p. 2498–2504.

Tenthorey, E., E. Aharonov, and C. H. Scholz, 1998, Precipitation sealing and diagenesis: 1. Experimental results: Journal of Geophysical Research, v. 103B, p. 23,951–23,967.

Thompson, A. H., A. J. Katz, and C. E. Krohn, 1987, The microgeometry and transport properties of sedimentary rocks: Advances in Physics, v. 36, p. 625–694.

Van Hinte, J. E., 1978, Geohistory analysis— Application of micropaleontology in exploration geology: AAPG Bulletin, v. 62, p. 201–222.

Vavra, C. L., J. G. Kaldi, and R. M. Sneider, 1992, Geological application of capillary pressure: A review: AAPG Bulletin, v. 76, p. 840–850.

Vincelette, R. R., E. A. Beaumont, and N. H. Foster, 1999, Classification of exploration traps, *in* E. A. Beaumont and N. H. Foster, eds., Exploring for oil and gas traps: AAPG Treatise of Petroleum Geology, chapter 2, p. 1–42.

Wallace, R. E., and H. T. Morris, 1986, Characteristics of faults and shear zones in deep mines: Pure and Applied Geophysics, v. 124, p. 107–125.

Walsh, J. J., J. Watterson, A. E. Heath, and C. Childs, 1998, Representation and scaling of faults in fluid flow models: Petroleum Geoscience, v. 4, p. 241–251.

Watts, N. L., 1987, Theoretical aspects of cap-rock and fault seals for single- and two-phase hydrocarbon columns: Marine and Petroleum Geology, v. 4, p. 274–307.

Weber, K. J., 1986, How heterogeneity affects oil recovery, *in* L. W. Lake, and H. B. Carroll, eds., Reservoir characterisation: Orlando, Academic Press, p. 487–544.

Weber, K. J., 1997, A historical overview of the efforts to predict and quantify hydrocarbon trapping features in the exploration phase and in field development planning, *in* P. Møller-Pedersen and A. G. Koestler, eds., Hydrocarbon seals: Importance for exploration and production: Norwegian Petroleum Society Special Publication 7, p. 1–13.

Weber, K. J., and E. Daukoru, 1975, Petroleum geology of the Niger delta, *in* Proceedings of 9th World Petroleum Congress, v. 2: London, Applied Science Press, p. 209–221.

Weber, K. J., G. Mandl, W. F. Pilaar, F. Lehner, and R. G. Precious, 1978, The role of faults in hydrocarbon migration and trapping in Nigerian growth fault structures: Society of Petroleum Engineers, 10th Annual Offshore Technology Conference Proceedings, v. 4, p. 2643–2653.

Weeks, L. G., ed., 1958, Habitat of oil: Tulsa, AAPG, 1384 p.

Wehr, F. L., L. H. Fairchild, M. R. Hudec, R. K. Shafto, W. T. Shea, and J. P. White, 2000, Fault seal: Contrasts between the exploration and production problem, *in* M. R. Mello and B. J. Katz, eds., Petroleum systems of south Atlantic margins: AAPG Memoir 73, p. 121–132.

Wilhelm, O., 1945, Classification of petroleum reservoirs: AAPG Bulletin, v. 29, p. 1537–1579.

Wilkerson, M. S., M. P. Fischer, and T. Apotria, eds., 2002, Fault-related folds: The transition from 2-D to 3-D: Journal of Structural Geology, v. 24, p. 591–904.

Willis, D. G., 1961, Entrapment of petroleum, *in* G. B. Moody, ed., Petroleum exploration handbook: New York, McGraw-Hill, p. 6-1, 6-68.

Wilson, W. B., 1934, Proposed classification of oil and gas reservoirs, *in* Problems of petroleum geology: A symposium: Tulsa, AAPG, p. 433–445.

Wilson, W. B., 1942, Classification of oil reservoirs: AAPG Bulletin, v. 26, p. 1291–1292.

Wiprut, D., and M. D. Zoback, 2000, Fault reactivation and fluid flow along a previously dormant normal fault in the Norwegian North Sea: Geology, v. 28, p. 595–598.

Yielding, G., 2002, Shale gouge ratio— Calibration by geohistory, *in* A. G. Koeslter and R. Hunsdale, eds., Hydrocarbon seal quantification: Norwegian Petroleum Society Special Publication 11, p. 1–15.

Yielding, G., B. Freeman, and D. T. Needham, 1997, Quantitative fault seal prediction: AAPG Bulletin, v. 81, p. 897–917.

Zoback, M. D., and J. H. Healy, 1992, In situ stress measurements to 3.5 km depth in the Cajon Pass scientific research boreholes— Implications for the mechanics of crustal faulting: Journal of Geophysical Research, v. 97B, p. 5039–5057.

2

Sawamura, F., and K. Nakayama, 2005, Estimating the amount of oil and gas accumulation from top seal and trap geometry, *in* R. Sorkhabi and Y. Tsuji, eds., Faults, fluid flow, and petroleum traps: AAPG Memoir 85, p. 33–42.

Estimating the Amount of Oil and Gas Accumulation from Top Seal and Trap Geometry

Fuminori Sawamura[1]
Geology Department, JGI, Inc., Tokyo, Japan

Kazuo Nakayama
Geology Department, JGI, Inc., Tokyo, Japan

ABSTRACT

Oil and gas volumes are controlled by top-seal capillary properties, spillpoints, and trap geometry. The top-seal capillary properties and seal capacity can be estimated from the equivalent grain size (EGS) method. The EGS method uses an experimentally derived relationship between pore-throat size, porosity, and grain size to evaluate seal capacity. A "pure spillpoint-limited trap" is one in which the hydrocarbon column height is determined solely by the spillpoints. The observed hydrocarbon column in this trap is less than that which can be held by top-seal capacity. This trap type will be dominated by gas. In a "capillary and spillpoint mixed trap," where both oil and gas can be filled down to the spillpoint, both top-seal capacity and spillpoint control relative oil and gas column heights. A "pure capillary-limited trap" is that where the oil and gas are not filled down to the spillpoint.

Top seal and spillpoint have been the focus of seal analyses; however, a case study for fields referred to as AN and YA in this chapter demonstrates an important relationship between trap geometry and top-seal capacity. These two fields have the same top-seal capacity, but the total column heights, as well as the relative oil and gas columns, are very different. This is explained by the different ratios of the base area to its relief in the two fields. The ratio of the area to its relief of the AN field is smaller, whereas that of the YA field is much larger. Given the same top-seal capacity, a trap with a higher area-to-relief ratio can hold a larger gas column because the oil pushed down by the migrated gas reduces its column height remarkably.

Thus, the EGS method can provide new insights into understanding hydrocarbon fill patterns in fields and prospects, including fault traps.

[1]*Present address:* Chiba University Graduate School of Science and Technology, Chiba, Japan.

DOI:10.1306/1033714M853130

INTRODUCTION

A seal analysis is important in order to understand the oil and gas accumulation in fields and prospects, including fault traps (e.g., Gussow, 1954; Downey, 1994; Sales, 1997). The seal capacity of a cap rock (the maximum hydrocarbon column height that a cap rock can hold) does not control the hydrocarbon column heights in all traps, even if sufficient amounts of hydrocarbons were supplied to saturate the trap. Hydrocarbons only leak through the top seal in traps where the height of the trap is larger than the maximum hydrocarbon column height that the top seal can hold (Figure 1). However, in traps where the maximum hydrocarbon column height is larger than the height of the trap, the overcharged hydrocarbons will migrate out from the trap through the spillpoint because these traps are fully filled with hydrocarbons before the seal failure occurs (Figure 1).

Sales (1997) suggested that traps are characterized as oil prone or gas prone by the relationship between the seal capacity and the height of the trap described as above. A practical method is required to evaluate the seal capacity for applying Sales' concept to petroleum exploration effectively. The authors modified and applied the equivalent grain size (EGS) method, which gives a macro view of the effective sealing of a trap (Nakayama and Sato, 2002). The aims of this chapter are (1) to introduce the concept and methodology of the EGS method and (2) to discuss the factors controlling oil and gas accumulation through a case study of a discovered field using the EGS method. It is also hoped that this newly developed method to evaluate the top-seal capacity should help the prediction of the oil and gas accumulation in fault traps.

OVERVIEW: TRAP CLASSIFICATION

Figure 2 shows three types of traps classified by the relationship between the height of a trap and the maximum hydrocarbon column height to be held by the top seal (Sales, 1997). According to Sales, traps are also characterized as oil prone or gas prone by this relationship. Gas with a lower density and higher mobility than oil preferentially fills a trap from the top. Even if a trap has been fully filled with oil, gas that migrated in a later stage would accumulate in the trap by replacing the preexisting oil.

In traps where the spillpoint is located above the level that is determined by the maximum gas column that the cap rock can hold (called "pure spillpoint-limited trap" in this chapter), the gas that migrated in the upper part of the trap pushes down the oil reserves, so that the oil tends to migrate out into the adjacent reservoirs in the updip direction through the spillpoint. The whole trap will eventually be occupied by gas. In traps where the spillpoint is located below the level that is determined by the maximum oil column that the cap rock can hold (called "pure capillary-limited trap" in this chapter), oil that is pushed down by migrated gas does not leak out through the spillpoint. The gas that is present above the oil may migrate upward by top-seal failure during the stage of gas migration. In traps where the spillpoint is located between the levels of the maximum possible oil and gas columns that the cap rock can hold (called "capillary and spillpoint mixed trap" in this chapter), overcharged gas migrates out from the trap through the top seal before the gas occupies the whole trap. Therefore, the upper part of the trap is filled with gas, but the oil still remains in the lower part of the trap, even if some oil may have migrated out through the spillpoint.

THE EQUIVALENT GRAIN SIZE METHOD

Concept of the EGS Method: Hydrostatic Trap-equilibrium Equation

Over the geological timescale of hydrocarbon migration, only capillary-pressure properties are needed to understand the static fluid-flow model. Recent geochemical research suggests that in most of the economical accumulations, hydrocarbons have migrated in a separate oil or gas phase (Barker, 1978; Hunt, 1979). Water, being the wetting phase, can migrate through any medium if there is a pore, whereas hydrocarbons can migrate through media only if the driving force exceeds the capillary entry pressure of that media.

Migrating hydrocarbons from the source rocks are trapped when a downward force resulting from the capillary pressure caused by the pore throat of the cap rock and interfacial tension of fluids restrains the upward buoyancy force of the hydrocarbon at the sealing surface (Figure 3). Hydrocarbons will be trapped until the buoyancy force reaches equilibrium with the capillary force at the top seal; the buoyancy force increases as the hydrocarbon column height increases. The situation where hydrocarbons have accumulated to the maximum possible column height is described by the hydrostatic trap-equilibrium equation (Nakayama and Van Siclen, 1981):

$$2\gamma \cos\theta / R = gH_c(\rho_w - \rho_h) \quad (1)$$

where γ is the interfacial tension; θ is the wettability (interfaced angle); R is the pore-throat radius; g is the

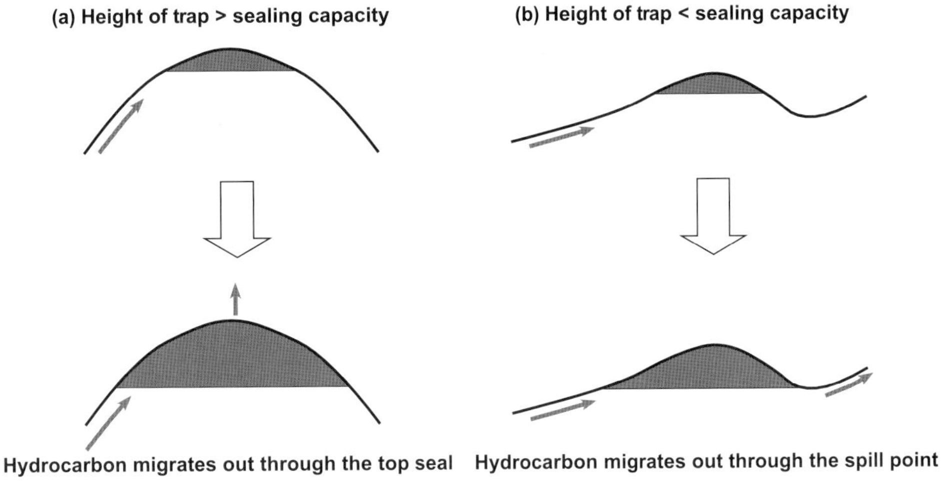

Figure 1. Schematic model showing different processes of hydrocarbon accumulation controlled by the relationship between the height of a trap and seal capacity. Overcharged hydrocarbons migrate out from a trap through either a top seal (a) or a spillpoint (b).

(a) Height of trap > sealing capacity

(b) Height of trap < sealing capacity

Hydrocarbon migrates out through the top seal

Hydrocarbon migrates out through the spill point

acceleration of gravity; ρ_w is the density of formation water; ρ_h is the density of hydrocarbons; and H_c is the maximum hydrocarbon column height.

In the traps where the buoyancy of hydrocarbon reaches a balance with the capillary forces, using the observed original hydrocarbon columns, it is possible to back-calculate an effective pore-throat radius assuming a proper consideration of interfacial tension, wettability, and fluid densities. This means that hydrocarbons cannot leak until a certain amount is trapped at the top of reservoirs, assuming a water-wet condition in cap rocks (Berg, 1975; Schowalter, 1979). Because water is the wetting phase, it can always go through permeable media.

Concept of the EGS Method: Conversion of Pore-throat Size to Grain Size

The ideal grain size, EGS, of the cap rock can be used as a parameter in the evaluation of seal capacity. The sealing capacity for a top seal can be quantified by the hydrostatic trap-equilibrium equation (equation 1); therefore, it can be related to the pore-throat size in the cap rock. If a cap rock consists of equal-size spherical grains, then the pore-throat size is a function of grain size and porosity. Figure 4 shows that the pore-throat sizes vary with the packing geometry even if the grain sizes are equal (Nakayama and Van Siclen, 1981). According to Nakayama and Van Siclen (1981), the ratio of pore throat to grain diameter (COEF) is estimated from the theoretical packing geometry as follows (Figure 5):

$$\text{COEF} = 2R/D_{\text{m}} = 1.92\phi^2 + 0.0882\phi \qquad (2)$$

where R is the pore-throat radius; D_{m} is the grain diameter; and ϕ is the porosity.

To evaluate this theoretical relationship, physical experiments measured the maximum oil column in a sample of artificial glass beads with a known grain-size distribution (Figure 6) (Nakayama and Sato, 2002). The apparatus consisted of pipelines and a glass cylinder with an artificial seal disk that is made of artificial glass beads in the center. Two inlet tubes (one for oil and the other for water) were attached at the bottom of the cylinder (Figure 7).

At first, the whole system was filled with water; the oil line was then opened forming an oil accumulation under the seal disk. For the oil, dodecane was used as a measurable oil sample. Because the theoretical column height was too high for this apparatus, the water line was pressurized with a certain column height of water (water manometer), so that the pressure at the bottom of this artificial trap could be controlled (Figure 7). The water column above the seal disk (water capillary) was used to record the leakage of oil through the

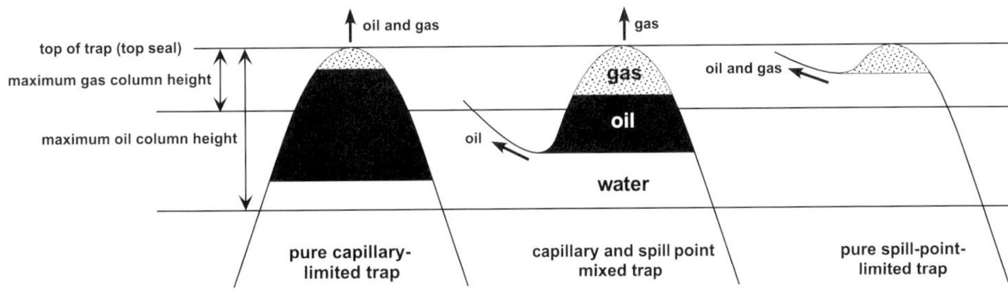

Figure 2. Schematic cross section showing three types of traps classified by the relation between the height of a trap and the maximum hydrocarbon column height to be held by the top seal (modified after Sales, 1997).

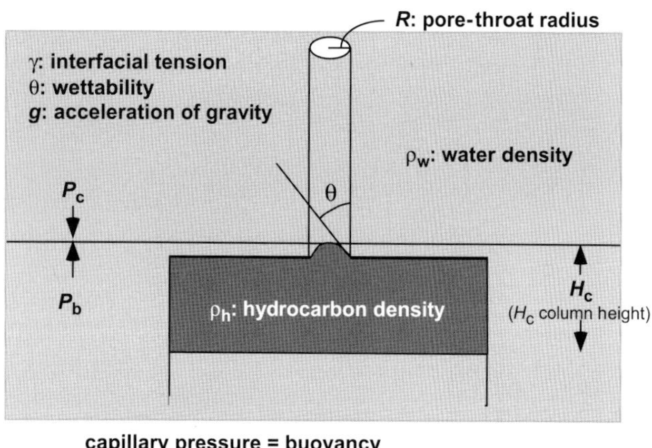

FIGURE 3. Schematic diagram showing hydrostatic equilibrium condition. Assuming that hydrocarbons have migrated in a separate hydrocarbon phase, the upward buoyancy must be smaller than the downward capillary pressure for the hydrocarbon accumulation in a trap. At the maximum hydrocarbon height, the buoyancy should be equal to the capillary pressure (hydrostatic trap-equilibrium equation).

top of the seal. Figure 8 is a plot of the column height measurements. The initial increase of pressure is caused by the oil invading the pore space in the artificial seal. The next phase of the experiment (pressure plateau) corresponds to hydrocarbon sealing. The phase with overpressure corresponds to seal leakage. The measured pressure of leakage is then converted into an equivalent oil column height.

The experiment used four artificial seal disks representing different grain sizes. From the hydrostatic trap-equilibrium equation (equation 1), the observed capillary-pressure values that were measured by the physical experiments were converted to the pore-throat size (Table 1). Using these results and the measured porosities of the artificial seals, the corrected relation between porosity and COEF (the ratio of pore-throat diameter to grain diameter) shown in Figure 5 was modified as

$$\mathrm{COEF} = 2R/D_\mathrm{m} = 3.085\phi^2 + 0.2087\phi \qquad (2a)$$

where R is the pore-throat radius; D_m is the grain diameter; and ϕ is the porosity.

The change in the values of constants in equation 2a from equation 2 can be explained by the drag effect at the grain surface for the case of the experiment. It is suggested to be closer to nature than to the theoretical case.

Assuming the relation between porosity and COEF that is derived from the experiment described above, the authors developed a spreadsheet to easily calculate the maximum oil and gas column height (Figure 9). In this spreadsheet, physical conditions, such as temperature, density, and interfacial tension, are automatically determined as a function of depth. After the observed oil and gas column height is obtained, the EGS is calculated on this sheet using an iterative technique of trial and error. The EGS is expressed in the phi scale (phi = $-\log_2 d$, where d is the grain size in millimeters); a higher grain size value in the phi scale corresponds to finer grain size and indicates higher sealing capacity.

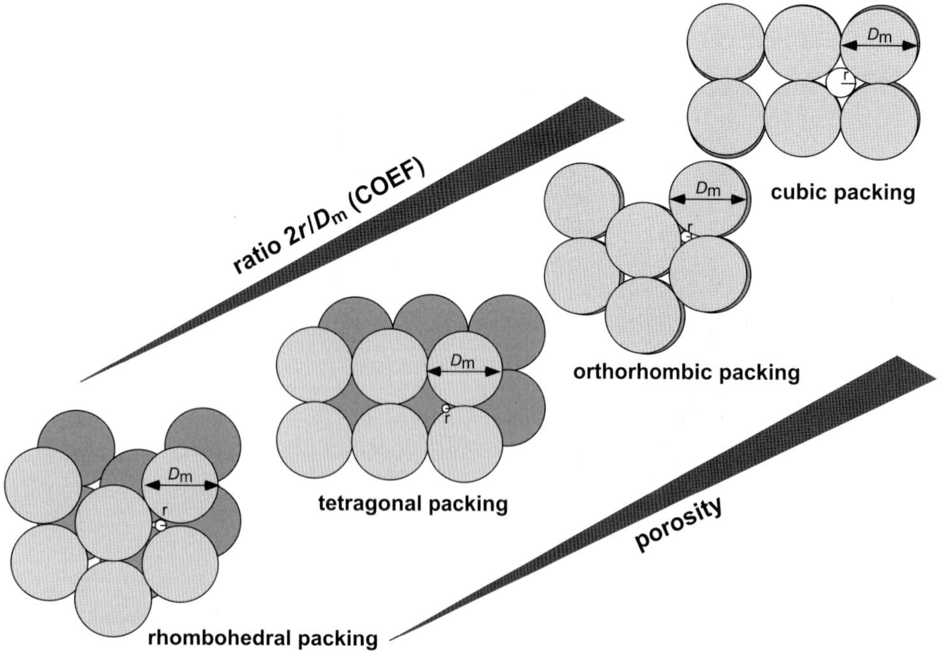

Methodology

The advantages of the EGS method are that (1) it requires no rock samples; (2) it gives a macro view of the effective sealing of a trap; and

FIGURE 4. The theoretical relation of pore throat and porosity according to the packing types. COEF indicates the ratio of pore-throat diameter ($2r$) to grain size (D_m). COEF is a function of the porosity according to the geometrical calculation if the grains consist of equally-sized spheres.

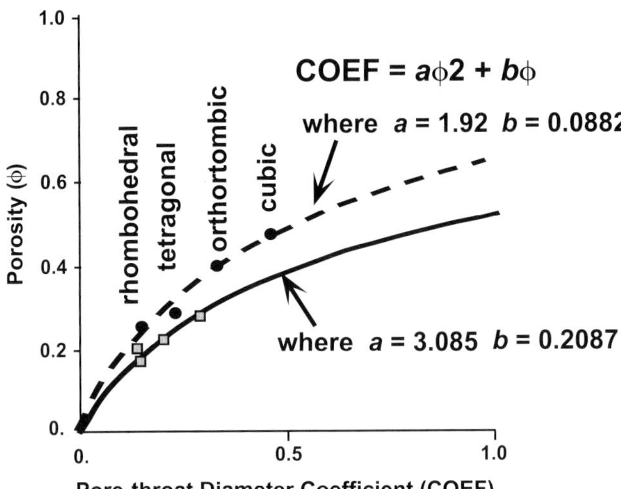

FIGURE 5. Approximations of the relationship between COEF and porosity obtained from the theoretical case using the different packing types (a = 1.92, b = 0.0882; after Nakayama and Van Siclen, 1981) and from the case that resulted from the physical experiment (a = 3.085, b = 0.2087).

(3) it indicates the primary seal property of the cap rock that is related to the lithological facies.

Conventionally, the seal property of the cap rock is estimated by measuring the pore-throat size of shale samples. Two conditions are required for this method. First, a seal sample must be available, and second, the sample must be representative of the entire trap (Downey, 1994). If the cap rock is not uniform over the entire trap, then the hydrocarbon accumulation should be affected by this situation. In other words, the observed hydrocarbon column height results from the effective seal capacity of the entire trap. The EGS determined from the observed hydrocarbon column height does not correspond to the actual grain size, but it indicates this effective seal capacity.

The EGS is more useful for the discussion of the primary seal property of the cap rock than for the discussion of the pore-throat size varying with the burial depth, because the grain size is independent of the porosity. The seal capacity is related to the grain size (not the pore size), which relates to lithological facies controlled by depositional environments in the EGS method. Advancing this concept, the EGSs that are calculated from the observed hydrocarbon column heights, where overlying cap rocks are the same in horizon and lithology, should be identical to other accumulations in the area. Consequently, when the suitable grain size holding the observed oil and gas column height in a field is determined, the maximum possible column height for the other traps within or near this field can be estimated using that grain size.

The EGS method also enables a quantitative evaluation or prediction of the relationship between the maximum oil and gas column heights and the heights of traps. Figure 10 shows the relations between the heights and top-seal depths of the assumed traps A and B, and the maximum oil and gas column heights derived from modeled EGS (8.0 phi), temperature, density, and porosity. Trap A plots between the lines of the maximum gas column height and the maximum oil column height, indicating that it is a capillary and spillpoint mixed trap, whereas trap B plots below the line showing the maximum gas column height, indicating that trap B is a gas-prone, pure spillpoint-limited trap. This technique indicates that trap A has more oil potential than trap B.

Two factors must be clarified for the prediction of the seal capacity using the EGS method. The first is the type of the trap that is used for the determination of the grain size, and the second is the spillpoint.

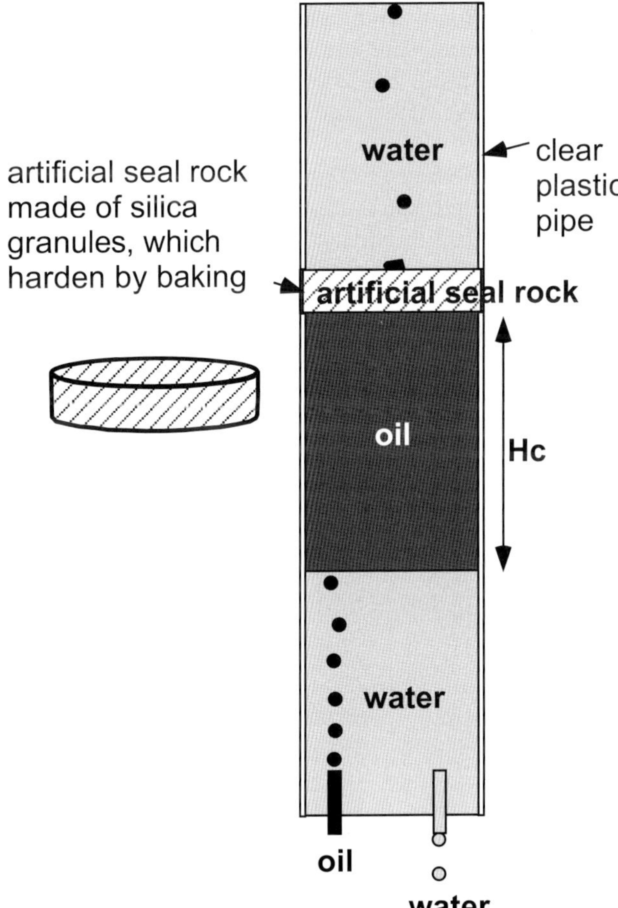

FIGURE 6. Schematic diagram showing the main cylinder of the physical experimental apparatus used to measure maximum oil column (H_c) with the artificially created seal. The artificial seals are made of silica beads with a known grain size.

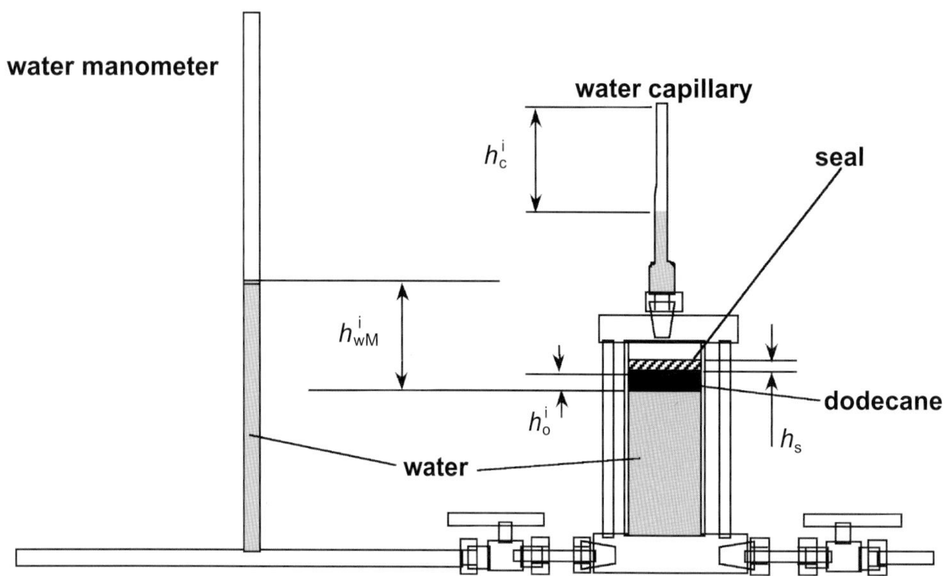

FIGURE 7. Schematic diagram showing the concept of the experiment for artificial seal measurement. For the actual measurement, the hydrodynamic force separately pushes the oil column as necessary. Setting up the seal-oil (dodecane)-water sequence (from the upper position) in the cylinder, the water column (h_{WM}) increases gradually in the water manometer. To detect the leaking, the water column in the water capillary (h_c) above seal is measured.

An EGS must be determined from the observed column heights of hydrocarbons. As described in the previous section, the total buoyancy of oil and gas reaches a balance with the capillary forces in a pure capillary-limited trap and a capillary and spillpoint mixed trap. In such cases, the estimated EGS can be used as an effective parameter to indicate the seal capacity. However, the hydrocarbon column height in a pure spillpoint-limited trap is commonly much shorter than the maximum column height that the top seal can hold. Therefore, the back-calculated EGS from the observed hydrocarbon column shows only the minimum value for the actual seal capacity.

The spillpoint may not correspond to the bottom of the maximum closure in all traps, because hydrocarbons can migrate out across a fault through sand-to-sand juxtaposition (Allan, 1989) or the point of smear-gauge failure. The crest and bottom of the maximum closure is available from seismic data; however, an across-fault sealing assessment (e.g., Skerlec, 1999; Sorkhabi et al., 2002) is also required for the estimation of the across-fault spillpoint in a trap bounded by faults.

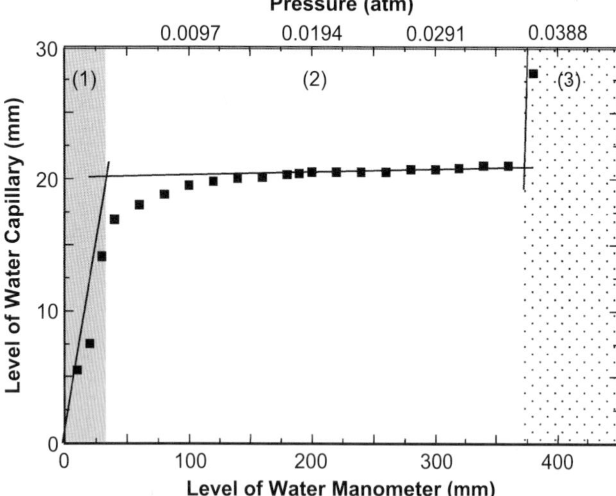

FIGURE 8. Graph showing a typical response of the experiment. As h_{WM} increases gradually, h_c increases correspondingly: Area 1 = the oil migrating into the pore space of the seal; area 2 = oil accumulating; and area 3 = oil leaking.

CASE STUDY: FACTORS CONTROLLING OIL AND GAS ACCUMULATION

This chapter concerns a study of seals in actual discovered oil and gas fields in Russia, based on the concept of the EGS method. This case study indicates that the geometry of a trap is also an important factor in controlling the oil and gas column height in a pure capillary-limited trap. This is in addition to the functions of the seal capacity and the height of the trap emphasized by Sales (1997).

In the fields referred to as YA, AN, and DU in this chapter, oil and gas have accumulated in the reservoir, which dips southward in a homoclinal feature. Reservoir facies in these fields are very variable. If the reservoir is near low-permeable facies, then the lateral change in reservoir facies acts as a seal, forming a stratigraphic trap in those fields (Figures 11, 12). All three traps are not filled with the hydrocarbons to the synclinal spillpoints. Consequently, the traps in those fields

Table 1. Result of pore-throat estimation with four kinds of artificial seal material consisting of different-sized glass beads. Two kinds of pore-throat radiuses are shown: estimated from the capillary failure in the experiment and from the theoretical relation of pore throat and porosity according to the packing types (after Nakayama and Van Siclen, 1981).

Sample	*Porosity (%)*	*Average Grain Size (μm)*	*Standard Deviation (μm)*	*Experimental Capillary Pressure (atm)*	*Calculated Pore-throat Radius from Sealing Pressure (μm)*	*Theoretical Pore-throat Radius (μm)*
A (100)	28.0	79.33	10.78	0.0674	12.204	8.008
B (200)	22.4	185.90	16.28	0.0431	18.985	11.818
C (300)	17.2	280.88	28.71	0.0379	21.601	11.587
D (400)	20.3	357.12	22.94	0.0327	24.991	19.056

are pure capillary-limited traps, assuming sufficient amounts of hydrocarbon have migrated into the traps.

The EGS of the cap rock for the YA field is estimated as 8.06 phi. The oil and gas distribution map shows that the top seal is 2085 m (6841 ft) below sea level, and gas and oil column heights are 51 and 21 m (167 and 69 ft), respectively, in the YA field (Figure 11). Fluid densities at the depth of the top seal that are determined from production tests are 1.23 g/cm^3 for the formation water, 0.76 g/cm^3 for the oil, and 0.30 g/cm^3 for the gas. Gas interfacial tension contact with the top seal is also calculated as 46.84 mN/m. Because total buoyancy (P_b) balances the capillary pressure (P_c) in pure capillary-limited traps, the pore throat is estimated as 0.000333 mm (1×10^{-5} in.) in diameter (2*R*) from the hydrostatic trap-equilibrium equation (equation 1).

$$P_b = \{21\ \text{m} \times 9.81\ \text{m/s}^2 \times (1.23\ \text{g/cm}^3 - 0.76\ \text{g/cm}^3)\} + \{51\ \text{m} \times 9.81\ \text{m/s}^2 \times (1.23\ \text{g/cm}^3 - 0.30\ \text{g/cm}^3)\}$$

$$P_c = \frac{2 \times 46.84\ \text{mN/m}}{R\ (\text{in mm})/1000}$$

The pore-throat size is converted into the EGS using the relationship between porosity and COEF in equation 2a. The porosity of the top seal is calculated from

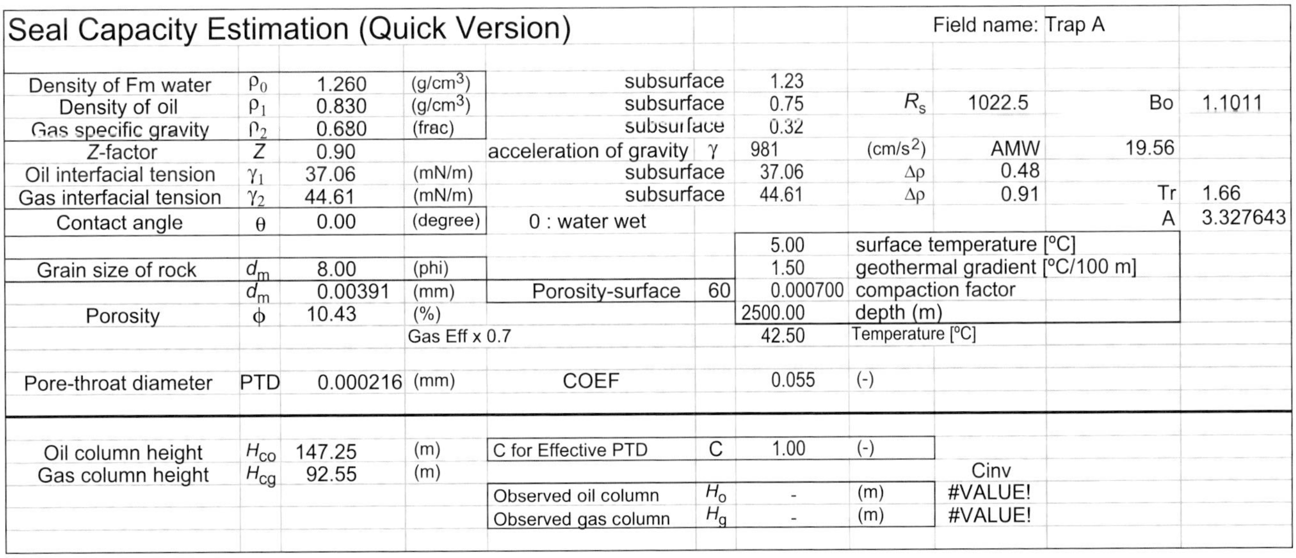

Seal Capacity Estimation (Quick Version)								Field name:	Trap A	
Density of Fm water	ρ_0	1.260	(g/cm^3)	subsurface		1.23				
Density of oil	ρ_1	0.830	(g/cm^3)	subsurface		0.75	R_s	1022.5	Bo	1.1011
Gas specific gravity	ρ_2	0.680	(frac)	subsurface		0.32				
Z-factor	Z	0.90		acceleration of gravity	γ	981	(cm/s^2)	AMW	19.56	
Oil interfacial tension	γ_1	37.06	(mN/m)	subsurface		37.06	$\Delta\rho$	0.48		
Gas interfacial tension	γ_2	44.61	(mN/m)	subsurface		44.61	$\Delta\rho$	0.91	Tr	1.66
Contact angle	θ	0.00	(degree)	0 : water wet					A	3.327643
						5.00	surface temperature [°C]			
Grain size of rock	d_m	8.00	(phi)			1.50	geothermal gradient [°C/100 m]			
	d_m	0.00391	(mm)	Porosity-surface	60	0.000700	compaction factor			
Porosity	ϕ	10.43	(%)			2500.00	depth (m)			
			Gas Eff x 0.7			42.50	Temperature [°C]			
Pore-throat diameter	PTD	0.000216	(mm)	COEF		0.055	(-)			
Oil column height	H_{co}	147.25	(m)	C for Effective PTD	C	1.00	(-)			
Gas column height	H_{cg}	92.55	(m)					Cinv		
				Observed oil column	H_o	-	(m)	#VALUE!		
				Observed gas column	H_g	-	(m)	#VALUE!		

FIGURE 9. Spreadsheet (Excel®) for calculating oil and gas column height from equivalent grain size. The given fluid densities at the surface are converted into the subsurface condition, which is calculated from the depth, geothermal gradient, surface temperature, etc. The porosity is calculated from the exponential function of depth provided the compaction factor is given. The other parameters (interfacial tensions, contact angle, and *Z*-factor) are also estimated in the subsurface condition. The sheet calculates the oil and gas column height corresponding to the given equivalent grain size. If the observed maximum column height is known, we can find a suitable equivalent grain size using this sheet by trial and error.

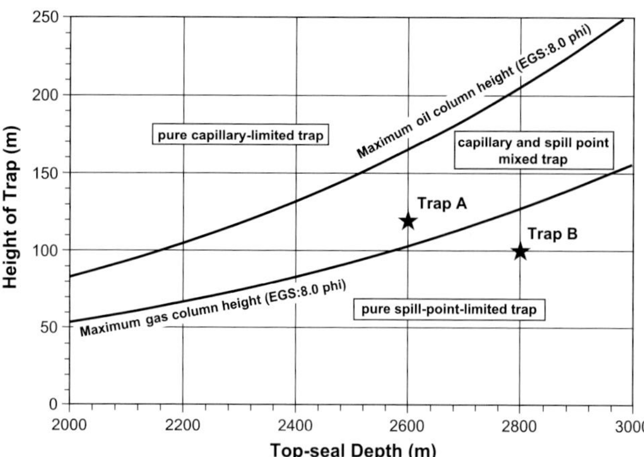

FIGURE 10. Crossplot of heights and top-seal depths of the virtual traps (A and B) overlying the maximum oil and gas column heights that the top seal can hold assuming the equivalent grain size of 8.0 phi. Trap B, plotted below the line showing the maximum gas column height, is classified as a gas-prone, spillpoint-limited trap. Trap A, plotted between the lines showing the maximum oil and gas column height, is classified as a mixed trap.

the compaction factor that is derived from the core analyses. The EGSs of cap rocks, which is estimated by same sequence as above, are 8.00 phi for the AN field and 8.61 phi for the DU field, respectively (Figure 11).

Although EGSs and top-seal depths of the YA and AN fields are similar, oil and gas column heights are different from each other. The oil column height is 21 m (69 ft), and the gas column height is 51 m (167 ft) in the YA field. In the AN field, the oil column height is 45 m (148 ft), and the gas column height is 35 m (115 ft). Figure 13 shows the estimated hydrocarbon-accumulating processes in the YA and AN fields. It is generally suggested that oil accumulated first; then gas migrated in from the deeper part of basin center. If this scenario is correct, then the traps were filled with oil until the maximum column height that the top seal could hold (phase I in Figure 13). Overcharged oil leaked upward through the top seal during this first phase. Because seal capacities of two fields are nearly equal, oil column heights should also be nearly equal; however, the oil volume was controlled by the trap geometry.

In the second phase (phase II in Figure 13), oil generation ceased, and gas was generated and migrated into the traps, replacing oil from the top of the trap, pushing down the oil column. During this phase, gas made contact with the top seal, which held the total buoyancy of oil and gas. Although the volume of oil did not change during the accumulation of gas, the oil column height decreased because of the geometry of the trap. In the AN field, the area of the reservoir does not

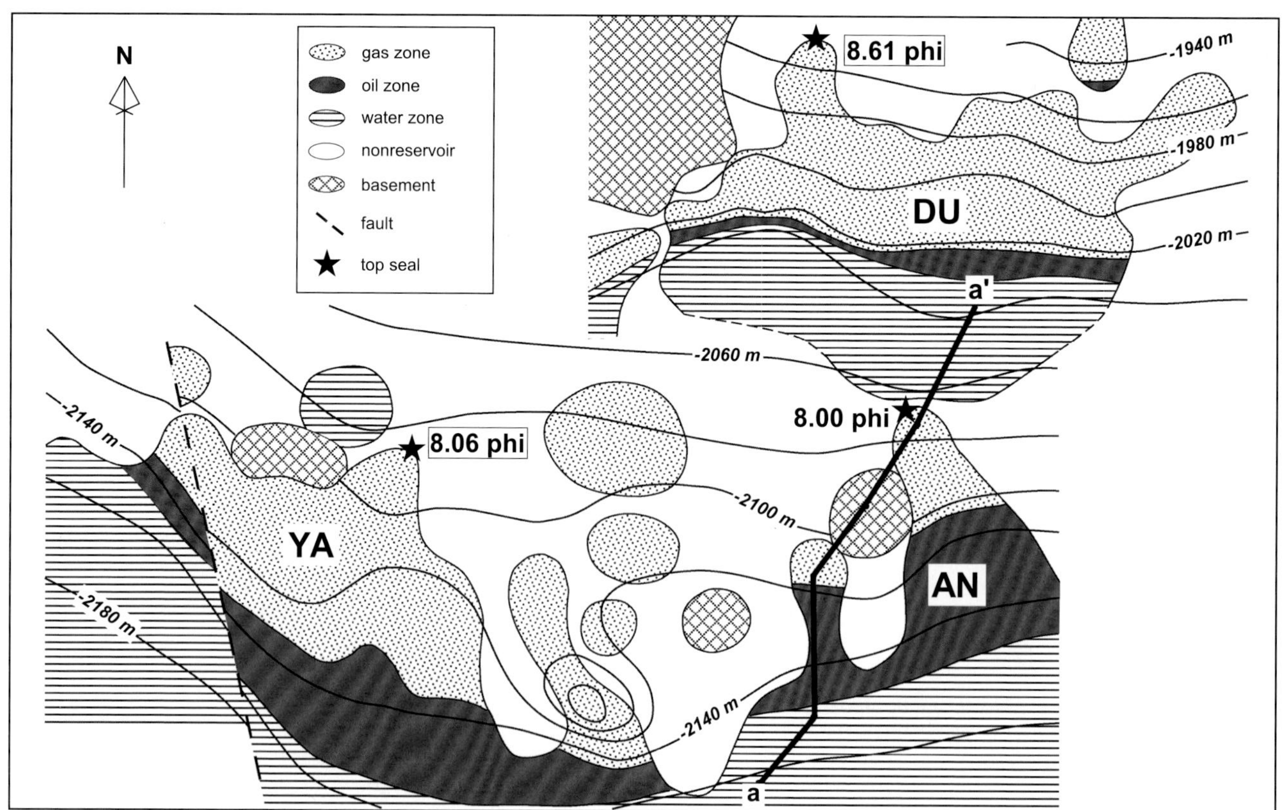

FIGURE 11. Depth contour map of the top reservoir horizon, showing the distribution of the oil and gas pools in the AN, YA, and DU fields.

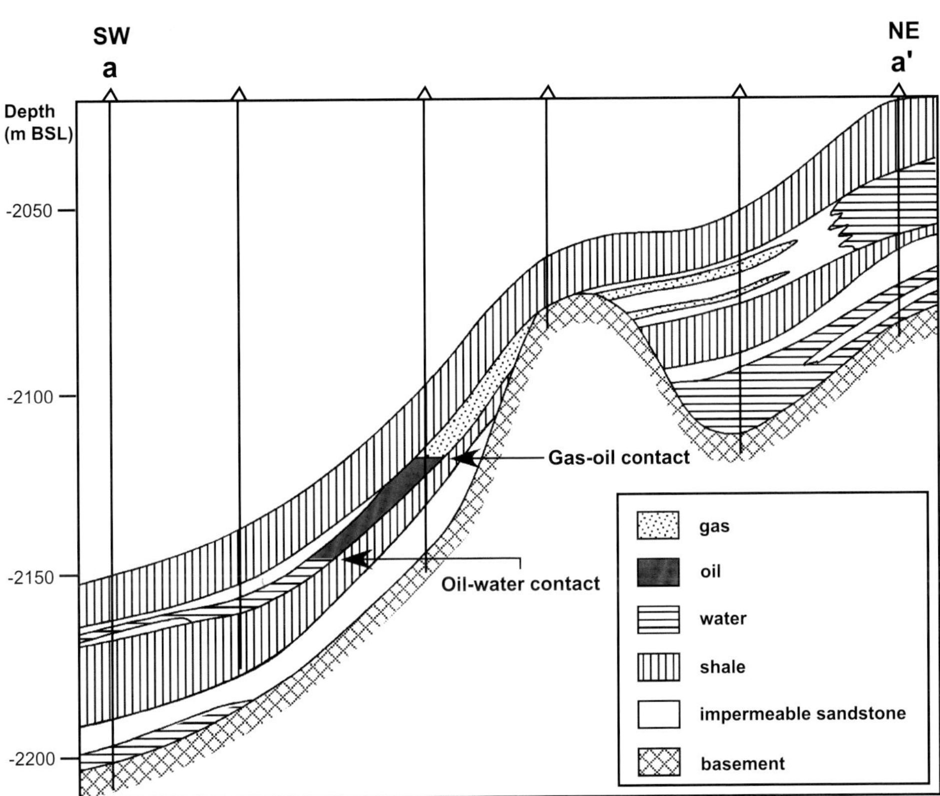

FIGURE 12. Geologic cross section across the AN–DU fields (vertical scale is exaggerated). Cross section aa′ corresponds to that in Figure 11.

change much with depth; therefore, the oil column height did not change much before and after the accumulation of gas. In the YA field, where the area of the reservoir remarkably increased with depth, the oil column height decreased dramatically despite the fact that no change occurred in oil volume.

As mentioned above, the total buoyancy of the oil and gas that the top seal can hold in the AN field is almost equal to that in the YA field, but oil column heights are

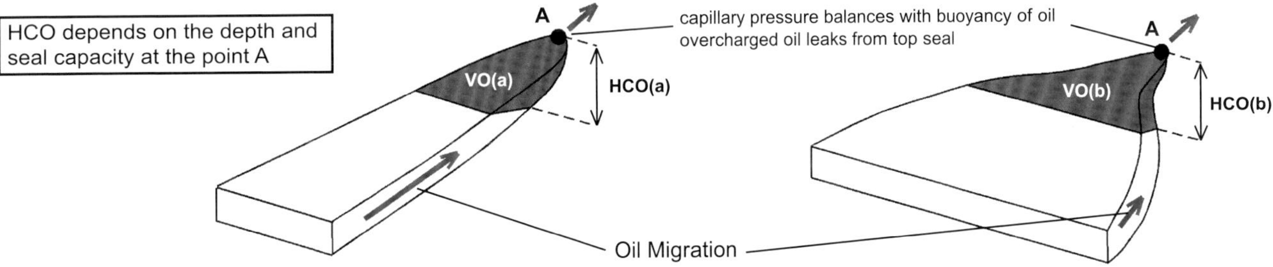

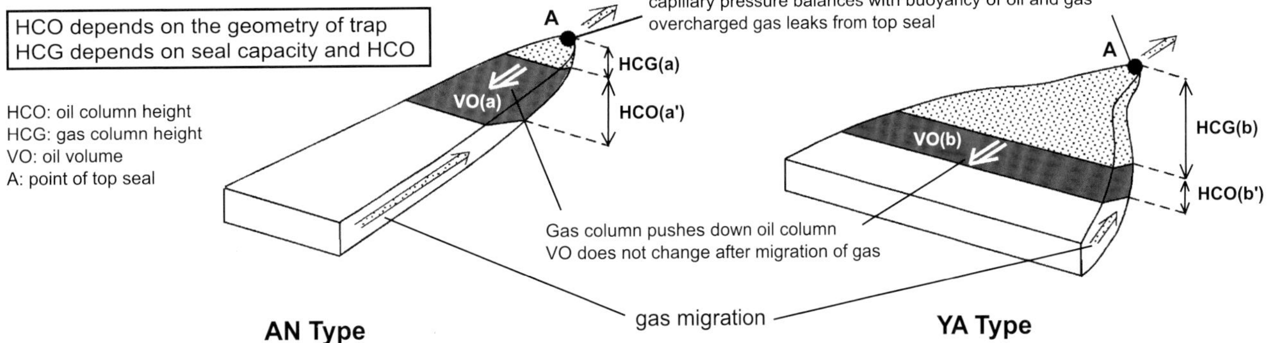

FIGURE 13. Schematic model explaining the differences of the oil and gas distributions in the AN and YA fields. Note that the geometries of the traps can affect the oil and gas distributions, even if the sealing capacities are equal, assuming that oil has accumulated in the traps before the inflow of gas.

different after the accumulation of gas. As a result, the accumulated volume and column height of the gas in the AN field are different from that in the YA field. The final gas column height in the AN field is relatively small because of the small decrease in oil column after the inflow of gas. In contrast, the final gas column height is much larger in the YA field, because the oil column height decreases remarkably after the inflow of gas caused by the trap geometry. Once the hydrostatic equilibrium is established, the overcharged gas leaks upward through the top seal.

CONCLUSIONS

The height of trap, the maximum oil and gas column height to be held by a cap rock, and the trap geometry are essential factors in predicting or interpreting oil and gas accumulations. A pure spillpoint-limited trap is one in which the hydrocarbon column height is determined solely by the spillpoints. The hydrocarbon column is less than that determined solely by the top-seal capacity. This trap will be dominated by gas. Pure capillary-limited trap is that not filled down to the spillpoint. A case study of the AN and YA fields demonstrates that both the trap geometry and the top-seal capacity control hydrocarbon column heights in these traps. In capillary and spillpoint mixed trap, both the top-seal capacity and the height of the trap control hydrocarbon column heights.

The EGS method quantifies seal capacity in terms of the ideal grain size of a cap rock. If the EGS can be inferred to the depositional environment, the calculated grain size holding the observed oil and gas column height in a given field can be used in the estimation of the possible maximum column height in the traps nearby. In combination with the determination of the spillpoint from seismic data and across-fault sealing assessment, this method helps to understand how the trap characteristics would control the oil and gas accumulation in fields and prospects, including fault traps. Furthermore, the concept of the EGS method may be applied not only to the top seal but also to the clay-rich fault gauge in the light of further studies in the future.

ACKNOWLEDGMENTS

The authors thank the Technology Research Center of the Japan National Oil Corporation (presently Japan Oil, Gas and Metals National Corporation) for being part of the development of equivalent grain size method. The authors thank Rasoul Sorkhabi of University of Utah for encouraging the publication of this article. The AN and YA fields are located in Russia; however, the exact location cannot be disclosed because of confidentiality. We are highly thankful to Mark M. Mandelbaum (Ministry of Natural Resources, Russia) for the permission to publish Figures 11 and 12. We also express our gratitude to Grant Skerlec and John Sneider whose comments greatly improved this chapter.

REFERENCES CITED

Allan, U. S., 1989, Model for hydrocarbon migration and entrapment within faulted structures: AAPG Bulletin, v. 73, p. 803–811.

Barker, C., 1978, Primary migration— The importance of water-mineral-organic matter interactions in the source rock, *in* Physical and chemical constraints on petroleum migration: AAPG Continuing Education Course Note Series 8, p. D1–D19.

Berg, R. R., 1975, Capillary pressure in stratigraphic traps: AAPG Bulletin, v. 59, p. 939–956.

Downey, M. W., 1994, Hydrocarbon seal rocks, *in* L. B. Magoon and W. G. Dow, eds., The petroleum system— From source to trap: AAPG Memoir 60, p. 159–164.

Gussow, W. G., 1954, Differential entrapment of oil and gas— A fundamental principle: AAPG Bulletin, v. 38, p. 816–853.

Hunt, J. M., 1979, Petroleum geochemistry and geology: San Francisco, W. H. Freeman and Co., 617 p.

Nakayama, K., and D. Sato, 2002, Prediction of sealing capacity by the equivalent grain size method, *in* A. G. Koestler and X. Hunsdale, eds., Hydrocarbon seal quantification: Norwegian Petroleum Society Special Publication 11, p. 51–60.

Nakayama, K., and D. C. Van Siclen, 1981, Simulation model for petroleum exploration: AAPG Bulletin, v. 65, p. 1230–1255.

Sales, J. K., 1997, Seal strength vs. trap closure— A fundamental control on the distribution of oil and gas, *in* R. C. Surdam, ed., Seals, traps, and the petroleum system: AAPG Memoir 67, p. 57–83.

Schowalter, T. T., 1979, Mechanics of secondary hydrocarbon migration and entrapment: AAPG Bulletin, v. 63, p. 723–760.

Skerlec, G. M., 1999, Evaluating top and fault seal, *in* E. A. Beaumont and N. H. Foster, eds., Exploring for oil and gas traps: Tulsa, AAPG, chapter 10, p. 1–94.

Sorkhabi, R., S. Hasegawa, S. Iwanaga, and M. Fujimoto, 2002, Sealing assessment of normal faults in clastic reservoirs— The role of fault geometry and shale smear parameters: Journal of the Japanese Association for Petroleum Technology, v. 67, p. 576–589.

3

Hasegawa, S., R. Sorkhabi, S. Iwanaga, N. Sakuyama, and O. A. Mahmud, 2005, Fault-seal analysis in the Temana field, offshore Sarawak, and Malaysia, *in* R. Sorkhabi and Y. Tsuji, eds., Faults, fluid flow, and petroleum traps: AAPG Memoir 85, p. 43–58.

Fault-seal Analysis in the Temana Field, Offshore Sarawak, Malaysia

Shutaro Hasegawa,[1] Rasoul Sorkhabi,[2] Shoji Iwanaga,[3] and Naofumi Sakuyama[4]

Technology Research Center, Japan National Oil Corporation, Chiba, Japan

Othman Ali Mahmud[5]

Petroleum Management Unit, Petroliam Nasional Berhad, Kuala Lumpur, Malaysia

ABSTRACT

The Temana field is located on a structural high in the Balingian province, offshore Sarawak, Malaysia. Fault-sealing assessment of a normal fault in the Tertiary clastic rocks of the Temana field was carried out using shale smear parameters. Shale smear factor values of less than 6 and clay content ratio of greater than 30% on the fault surface indicate across-fault sealing of the reservoir rocks on sand-sand interfaces. Hydrocarbon column height estimated for sandstone pay zones from across-fault pressure difference is comparable to that calculated from the structural spillpoint (76 m; 249 ft). The fault seal thus appears to be efficient enough to support hydrocarbon columns filled down to the structural spillpoints of the reservoirs. Fault-rock permeability calculated from the available calibrations of the clay content-permeability relationship shows lower permeabilities of less than 0.3 md. Taking the Temana fault as a case in point, a new approach to evaluate fault-rock permeability probability (based on integration of clay content, fault displacement, and depth factors) is presented.

INTRODUCTION

Many hydrocarbon reservoirs in Southeast Asia have been dismembered by faults, and a critical issue for field development, reservoir management, and drilling strategy is whether these faults are barriers or conduits for hydrocarbons. This chapter reports on the sealing evaluation of a normal fault in the Tertiary clastic reservoir of the Temana field, offshore Sarawak, Malaysia. This study was jointly carried out by the Japan National Oil Corporation (JNOC [presently Japan Oil, Gas and Metals National Corporation]) and the Petroliam Nasional Berhad (PETRONAS), and the fault-seal analysis software used in this study (FAULTAP) has been developed at

[1]*Present address:* Idemitsu Oil & Gas Co., Tokyo, Japan.
[2]*Present address:* Energy & Geoscience Institute, University of Utah, Salt Lake City, Utah, U.S.A.
[3]*Present address:* Geoscience Research Laboratory, Yamato, Kanagawa, Japan.
[4]*Present address:* Idemitsu Oil & Gas Co., Tokyo, Japan.
[5]*Present address:* Petronas Carigali, Kuala Lumpur, Malaysia.

DOI:10.1306/1033715M853128

JNOC. As background information for this chapter, we also describe the geological framework of the Temana field and the methodology for fault-sealing analysis adopted in this study.

GEOLOGIC OUTLINE OF THE TEMANA FIELD

General Information

Detailed geologic information on Sarawak is given in, e.g., Ho (1978), Hussin and Bait (1992), and PETRONAS (1999). The Temana field is located 22 km (14 mi) west-northwest of Bintulu, offshore Sarawak, Malaysia (Figure 1). The discovery well, Temana-1, was first reported in 1962, but the existence of commercial volumes of oil in the field was not proven until Temana-9 and Temana-10 were drilled in 1972. In 1995, the oil in place and the gas in place were estimated to be 460 million bbl and 230 bcf, respectively. Production from the Temana field started in 1979, and the current field production is more than 9000 bbl/day after 21 yr of production. A total of 69 wells have been drilled in the field to date. The Temana field is subdivided into three structural domains: Temana west, Temana central, and Temana east.

Regional Geology

The Balingian province, offshore Sarawak, is a tectonostratigraphic province bounded by the West Baram Line on the northeast side and by the Lupar Line to the southwest (Figure 1). The region is characterized by many anticlinal folds and faults (both normal and reverse faults). A northwest–southeast-oriented structural high subdivides the Balingian province into northeast and southwest parts. The Temana field is located on this structural high and along a major east-northeast-trending thrust-bounded zone (Figure 2).

During the Miocene, two distinct deformation phases affected the Balingian province. (1) The first phase occurred in the early Miocene and is characterized by highly faulted low-relief ridges and depressions. Numerous deep

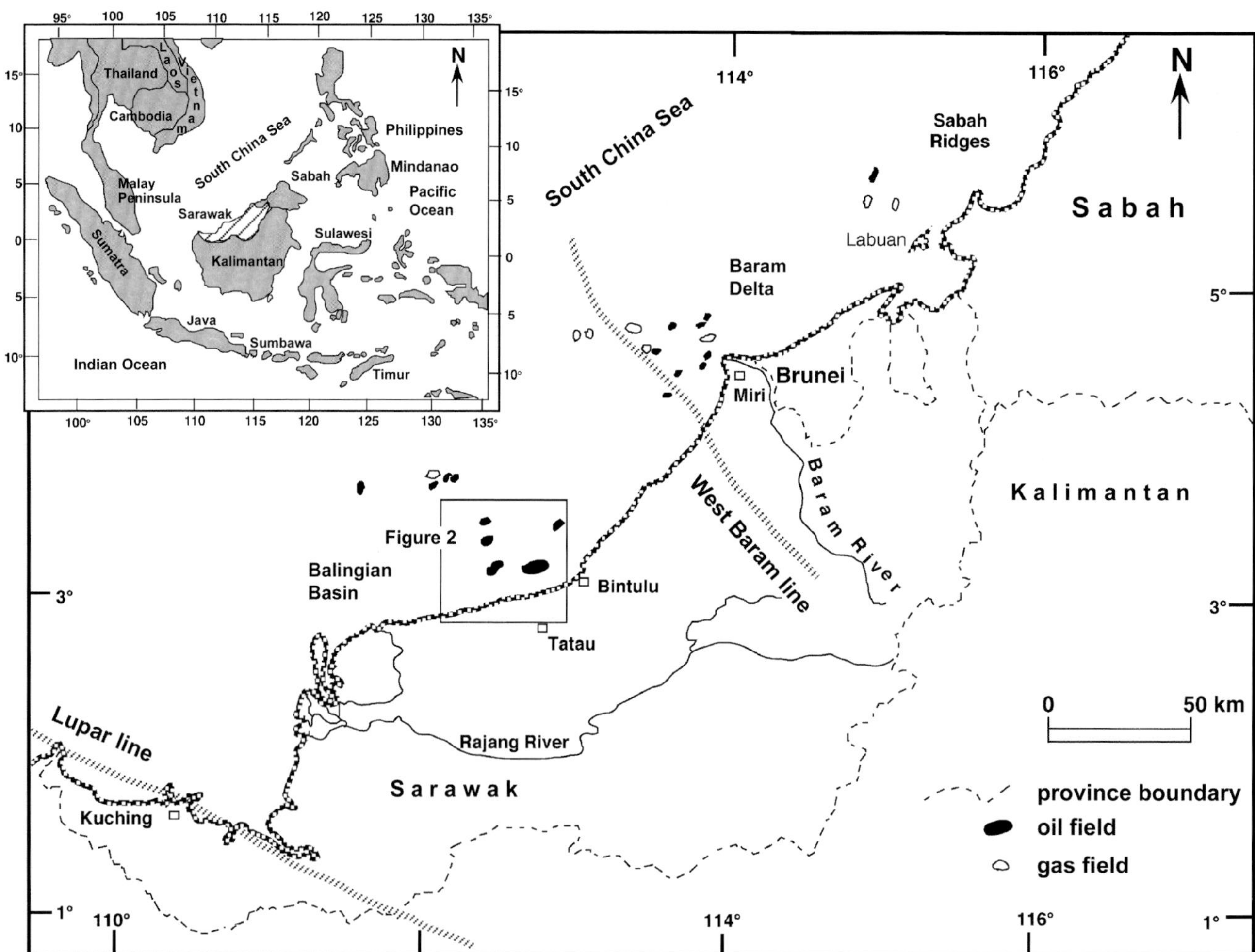

FIGURE 1. Geographic location of the Temana field, offshore Sarawak, Malaysia.

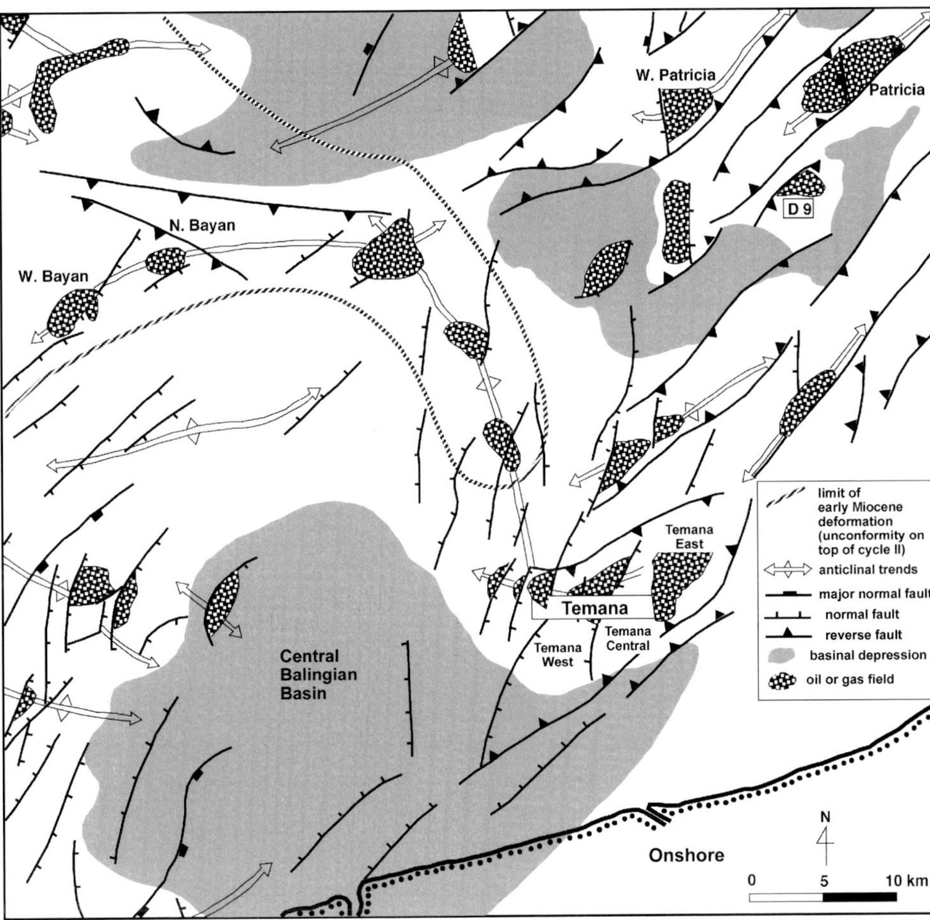

FIGURE 2. Structural setting of the Temana field on a structural high in the Balingian delta basin (compiled from PETRONAS, 1999).

and narrow half grabens were generated during this phase, which are well observed in the western part of the Balingian province. (2) The second tectonic deformation occurred in the late Miocene and is characterized by compressional features typically observed in the eastern part of Balingian province.

Stratigraphy

The onshore Tertiary strata in Sarawak have been divided into several specific lithostratigraphic units, and their equivalent offshore subsurface strata have been divided into eight sedimentary cycles based on seismic stratigraphy. For this study, cycles I, II, and III (upper Oligocene–lower Miocene) were the target zones (Figure 3). Based on stratigraphic data from the Temana-37 well, several key surfaces were recognized, and the strata were accordingly divided into the following series.

The K-series consists of alternating shale and mostly channel sandstone. The lower K-series consists of fluvial-dominated deposits overlain by thick marine shale in the middle part of the K-series. The upper K-series is characterized by thick, fining-upward sandstones intercalated with shallow-marine, prograding sequences.

The J-series consists of alternating shale and sandstone with thin coal beds. Correlation based on well data suggests a rapid lateral change in facies caused by the limited areal extent of the channel sandstone.

The I-series is predominantly marine shale intercalated with sandstones of variable thickness and thin coal beds. Although individual sandstone layers are relatively thin, the sandstone has good continuity, and its porosity ranges from 20 to 35%. The I60 sandstone is a very good reservoir rock occurring near the base of the I-series.

The H-series consists of alternating sandstone and shale (of variable thicknesses) and thin coal layers (reaching a maximum thickness of 10 ft [3 m]). Many of the sandstone strata show blocky to bell-shaped patterns and contain coal layers. This suggests that the deposition of the H-series was largely controlled by fluviomarine to fluvial processes. Porosity in the reservoir ranges from 25 to 35%. The top of the H-series is defined by a major angular unconformity, along which cycle VI cuts through the underlying cycles I and II and part of cycle III sections.

Figure 4 shows the gamma-ray log of the interval selected for this study, together with the location of the seismic horizons.

Structural Style

The Temana field is located on a structural ridge at the northeast part of the Balingian subbasin (Figure 2). The Temana field is a westerly plunging fold structure bounded by two major reverse faults both in the north and southeast. The two reverse faults merge to form a single fault to the northeast of the Temana field, and they diverge to the southwest. The field consists of smoothly curved structures with northeast–southwest-oriented normal faults. The primary trapping configurations are associated with the normal faults.

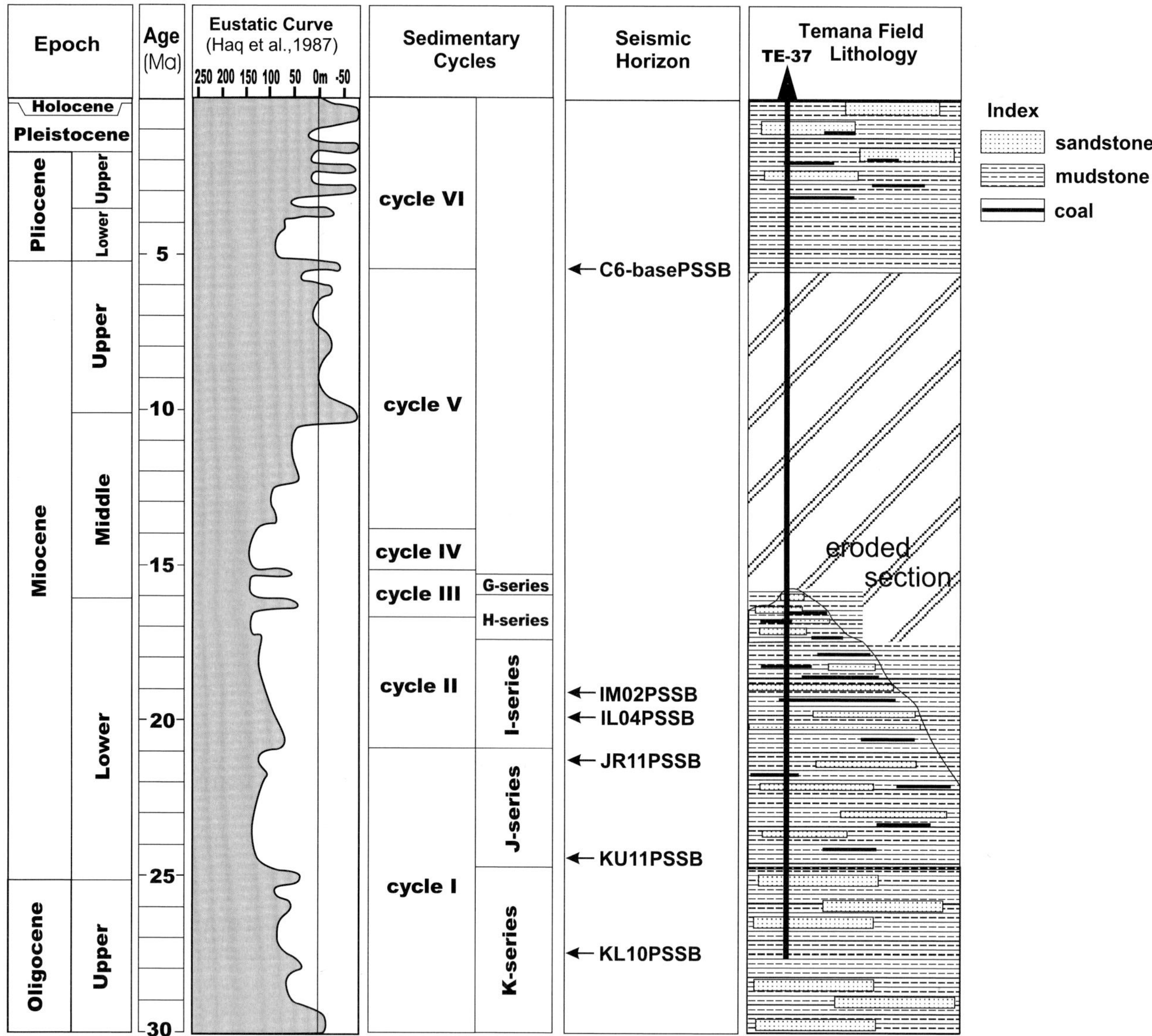

FIGURE 3. Stratigraphic column of the Temana field. Seismic horizons are depicted in Figure 8. TE-37 (Temana-37) is a well drilled in 1985 and has been used in this study for stratigraphic characterization. PSSB = parasequence sets sequence stratigraphy.

Two culminations in the Temana field exist. The eastern culmination is the southwestern extension of the compressional features typically observed in the northeast area of the Balingian province, whereas the western culmination is characterized by normal faults typically developed in the southwest of the province. The western culmination is further subdivided into the Temana central and the Temana west structures by a northeast–southwest-trending normal fault. Many small normal fault systems are observed in the Temana field, but they are all dissected by a major reverse fault. A large unconformity exists at the base of cycle VI, where the entire middle–late Miocene section is missing. Clayey sediments at the base of the cycle VI transgression provide a regional top seal for the hydrocarbons in the Temana field.

FAULT-SEAL ANALYSIS: METHODOLOGY

The salient features of the FAULTAP software have been reported by Sorkhabi et al. (2002, 2003) Here, a brief introduction is given for the purposes of this chapter. FAULTAP is a personal-computer-based software (written in C++ language) for assessing the across-fault-sealing capacity of normal faults in sandstone-shale sequences (Figure 5).

The input data include the following:

1) fault traces and seismic pick horizons, which are interpreted by seismic interpretation tools and converted from time domain to depth domain and are imported to FAULTAP in the form of an ASCII file;

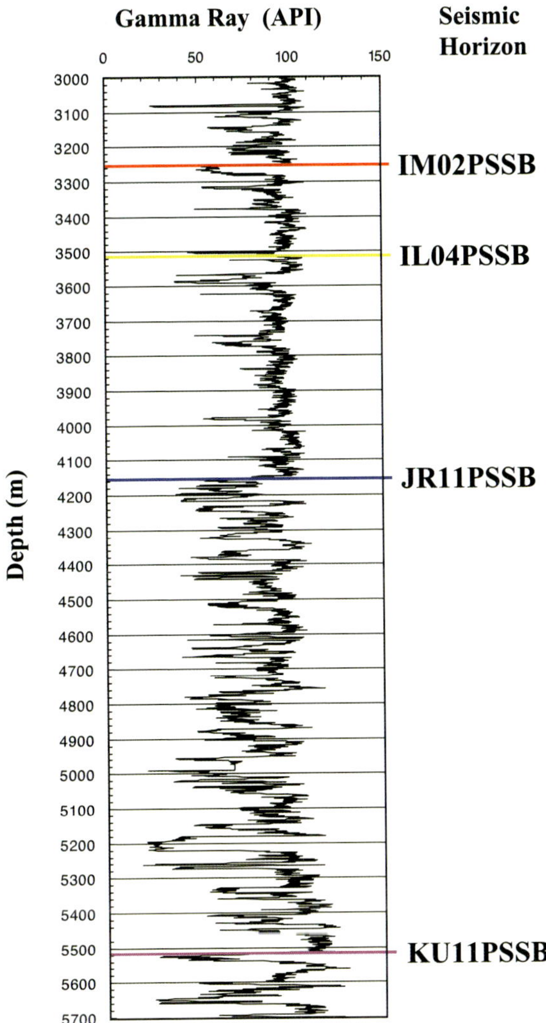

FIGURE 4. Gamma-ray log from well Temana-37 used to construct the lithology on fault surface. PSSB = parasequence sets sequence stratigraphy.

2) well data for characterizing the lithology, stratigraphic markers, bed thickness, rock density, subsurface fluid saturation, and information of well deviation and location; and
3) subsurface pressure data (repeat formation tester [RFT] data) to reconstruct the subsurface pressure gradients for water, oil, and gas.

The software contains modules for the calculation of (1) fault geometrical parameters (displacement, throw, heave, hade, and dip angle), (2) shale smear parameters, (3) fault-rock petrophysical parameters, and (4) rock stress parameters. The input data are processed through the modules, and the output data are visualized in the following forms: (1) data tables and crossplots, (2) fault surface profiles (along-strike fault section), and (3) fault juxtaposition of the sedimentary layers on the upthrown and downthrown sides (Allan diagrams; Allan, 1989).

In the past decade, shale smear in normal faults in clastic reservoirs has been recognized as an efficient mechanism of across-fault sealing and reservoir compartmentalization (e.g., Bouvier et al., 1989; Jev et al., 1993; Gibson, 1994; Fristad et al., 1997; Fulljames et al., 1997; Yielding et al., 1997, 1999; Alexander and Handschy, 1998). This methodology has been primarily used for this study. Outcrop observations of Tertiary strata in Borneo (Burhannudinnur and Morley, 1997) and Sarawak (Sorkhabi, 2000) have already demonstrated that shale smear occurs in many normal faults in this region; therefore, the adoption of shale smear parameters for fault-sealing analysis in this study is a realistic approach. Figure 6 shows five algorithms for shale smear parameters that have been used throughout the past two decades for assessment of fault sealing and which the FAULTAP software employs.

In this chapter, we describe our study of fault O in the Temana field (Figure 7). A total of six seismic horizons were interpreted for the reservoir sections covering the Temana central area. The footwall block of fault O was penetrated by the Temana-37 well. The stratigraphic units of the Temana-37 well were determined and used as input data for the footwall and hanging-wall blocks of fault O. The east–west seismic profile across Temana central, including Temana-37 well and fault O, is shown in Figure 8.

For the data preparation of the fault-seal analysis, detailed lithologic characterization was made using electric logs from the Temana central area. Seismic-well ties and sequence-stratigraphic interpretations were also conducted. The three-dimensional (3-D) seismic

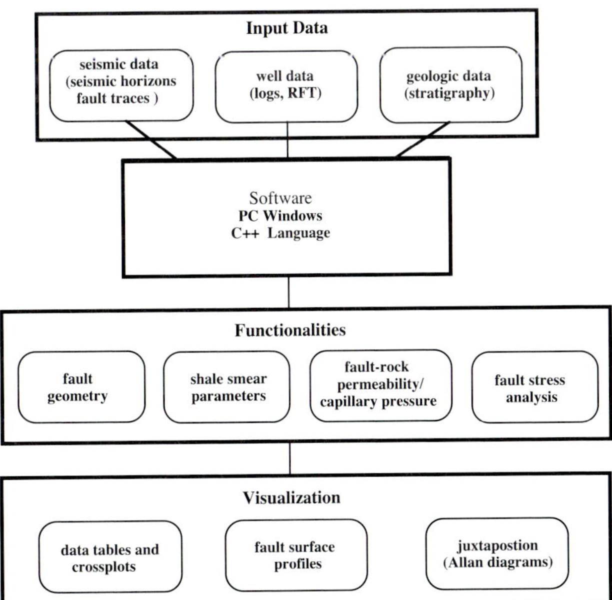

FIGURE 5. The architecture of the software used for fault-sealing assessment in this study. RFT = repeated formation test.

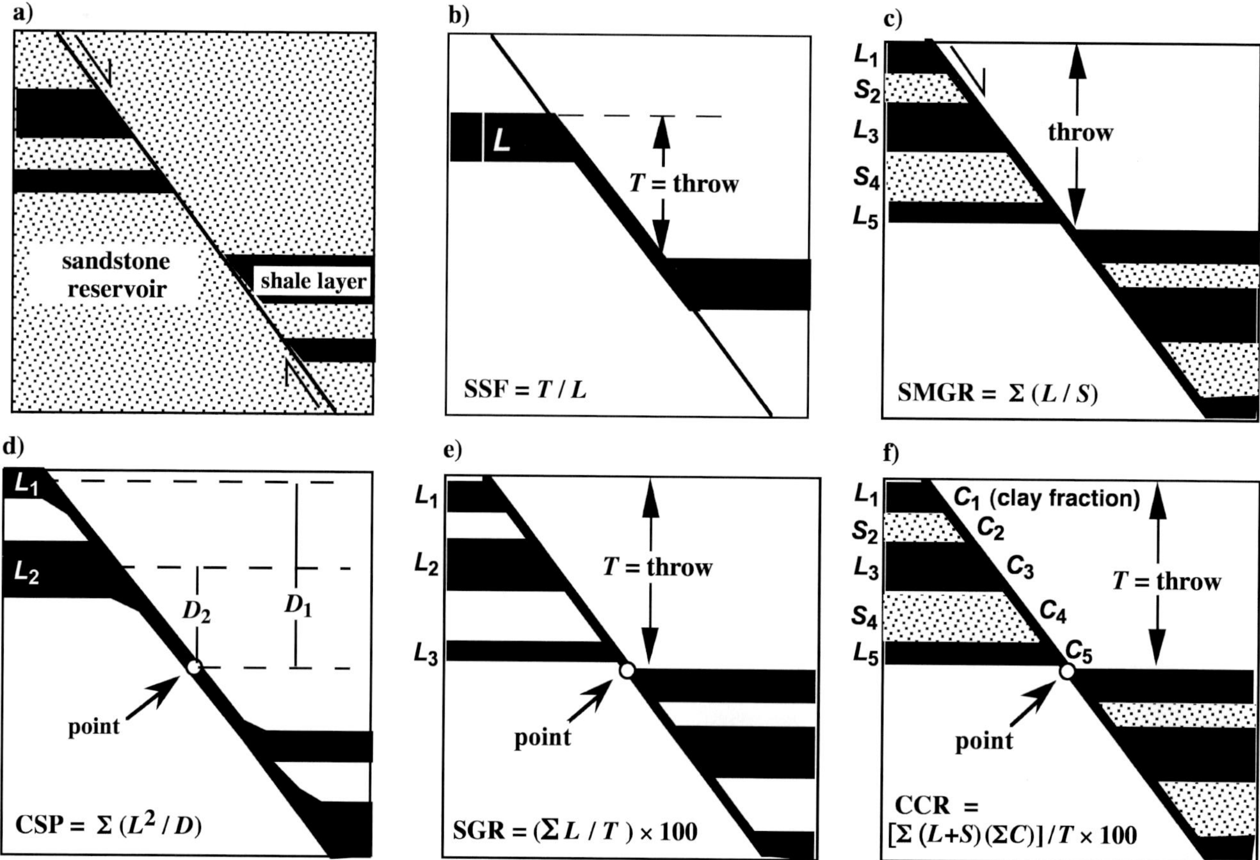

Figure 6. Various algorithms for shale smear parameters currently used in the petroleum industry for (a) normal faults in clastic reservoirs; (b) shale smear factor (Lindsay et al., 1993); (c) smear gouge ratio (Skerlec, 1999); (d) clay smear potential (Fulljames et al., 1997; Yielding et al., 1997); (e) shale gouge ratio (Yielding et al., 1997); and (f) clay content ratio (same as the gouge ratio of Fristad et al., 1997; Yielding et al., 1997, 1999). L = shale layer; S = sandstone reservoir; C = clay fraction.

data were available to interpret seismic horizons and faults in the Temana central area.

Fault interpretation of the seismic data was conducted on the workstation with seismic interpretation software to accurately construct the fault surface. The two-way-travel time data of seismic horizons and fault traces were converted to subsea depths using the corrected time-depth relations from check-shot data of the field. Then, the data on depth-converted horizon and fault traces were imported to FAULTAP to construct the fault surface and to conduct a fault-sealing assessment. Petrophysical interpretation was performed using electric logs to identify the fluid type of the sandstone reservoir zones.

RESULTS AND DISCUSSION

Fault O is a northeast–southwest-trending normal fault downthrown to the southeast. Temana-37 well is a deviated well drilled through the footwall block of fault O. No well has been drilled through the hanging-wall block of the fault. Temana-37 penetrated several hydrocarbon-bearing sandstones from the H-series at the top, down through the K-series at the total depth of the well.

The depth-converted fault files of fault O and horizon files of four seismic horizons, IM02PSSB pick (I-28 marker), IL04PSSB pick (I-60 marker), JR11PSSB pick (J-15 marker), and KU11PSSB pick (J-80 marker), were imported to the software (PSSB = parasequence sets sequence stratigraphy). The reservoir interval for the fault-seal analysis was between the I-28 and J-80 markers. The lithological data of the Temana-37 well (constructed from gamma-ray logs, Figure 4) and the top depth for each stratigraphical marker were entered into the software. Pressure, density, and depth of water and hydrocarbon horizons were also used in the software.

Structural Attributes

Figure 9 shows fault surface profiles of the lithology and the calculated geometrical parameters (fault throw, dip angle, and fault heave) of fault O, projection-viewed from the southeast to the fault surface. The depth range of the fault-seal study was from about

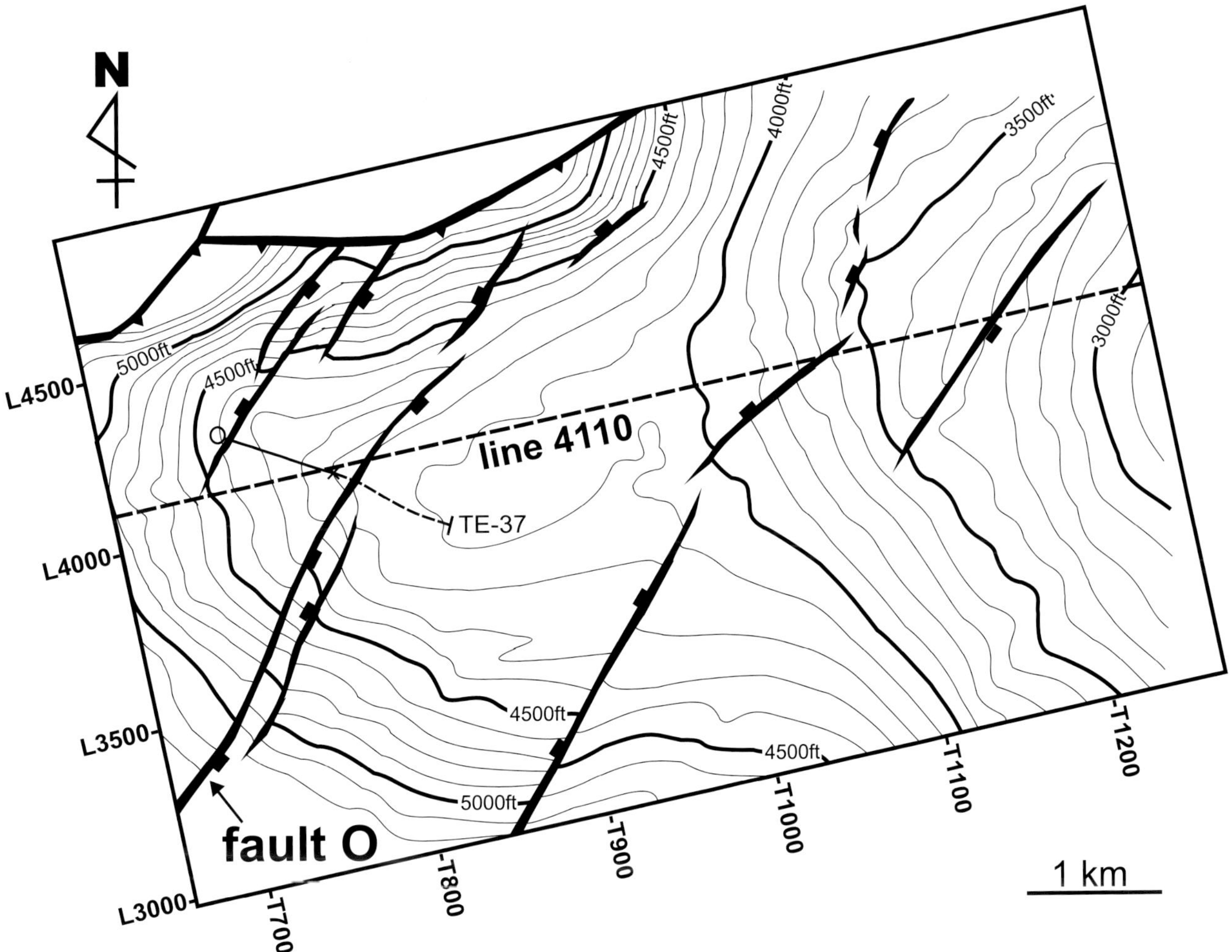

FIGURE 7. Planar map view of fault O in the Temana field and the seismic line 4110 used for this study with depth contours (in feet) of JR11PSSB seismic pick horizon. TE-37 = Temana-37.

1000 to 1800 m (3300 to 5900 ft) at the crest of the fault block. The maximum throw of fault O is about 200 m (660 ft) located in the southwest end of the 3-D seismic data set (Figure 9b). The fault dip profile shows that the fault surface has relatively high angles in the range of 60–80°, although the shallow fault surface at the northeast part of fault O has lower angles of 40–60° (Figure 9c). Fault heave quantifies the horizontal separation of the fault. The fault heave profile of fault O shows that there are two high heave areas separated by a low heave area at the crest of the fault structure (Figure 9d). The shape of the fault surface and the pattern of fault throw, fault heave, and dip angle suggest that fault O consists of two originally separate faults that were linked together during fault growth.

Fault Smear Attributes

Here, the results of shale smear factor (SSF) and clay content ratio (CCR) are discussed (Figure 10). The SSF and CCR values are depicted on the fault surface in Figures 9a and c, and the crossplots of the SSF and CCR values against across-fault pressure differences are given in Figures 9b and d.

Shale smear factor is obtained by dividing the fault throw by the thickness of source shale layer (Figure 6b). Outcrop studies show that SSF values of as much as 7 are valid for continuous shale smear on the fault (Lindsay et al., 1993), and subsurface data from oil fields also demonstrate that sealing faults have SSF values below 7 (Gibson, 1994; Yielding et al., 1997; Koledoye, 1998) Figure 10a shows that the reservoirs of I-series have very low SSF values (less than 2), which indicate efficient fault sealing. Most of these seem to be juxtaposition traps because an SSF value of 1 implies fault juxtaposition of shale and sandstone. However, the reservoir sands of JR01 and JR02 just above the KU11PSSB have SSF values of 3–6, which are lower than the SSF threshold value of 7. The reservoir sands of JR01 and JR02 at the southwestern end of fault O surfaces have high SSF values (colored in red, Figure 10a) because of large values of fault throw. It indicates nonsealing area on

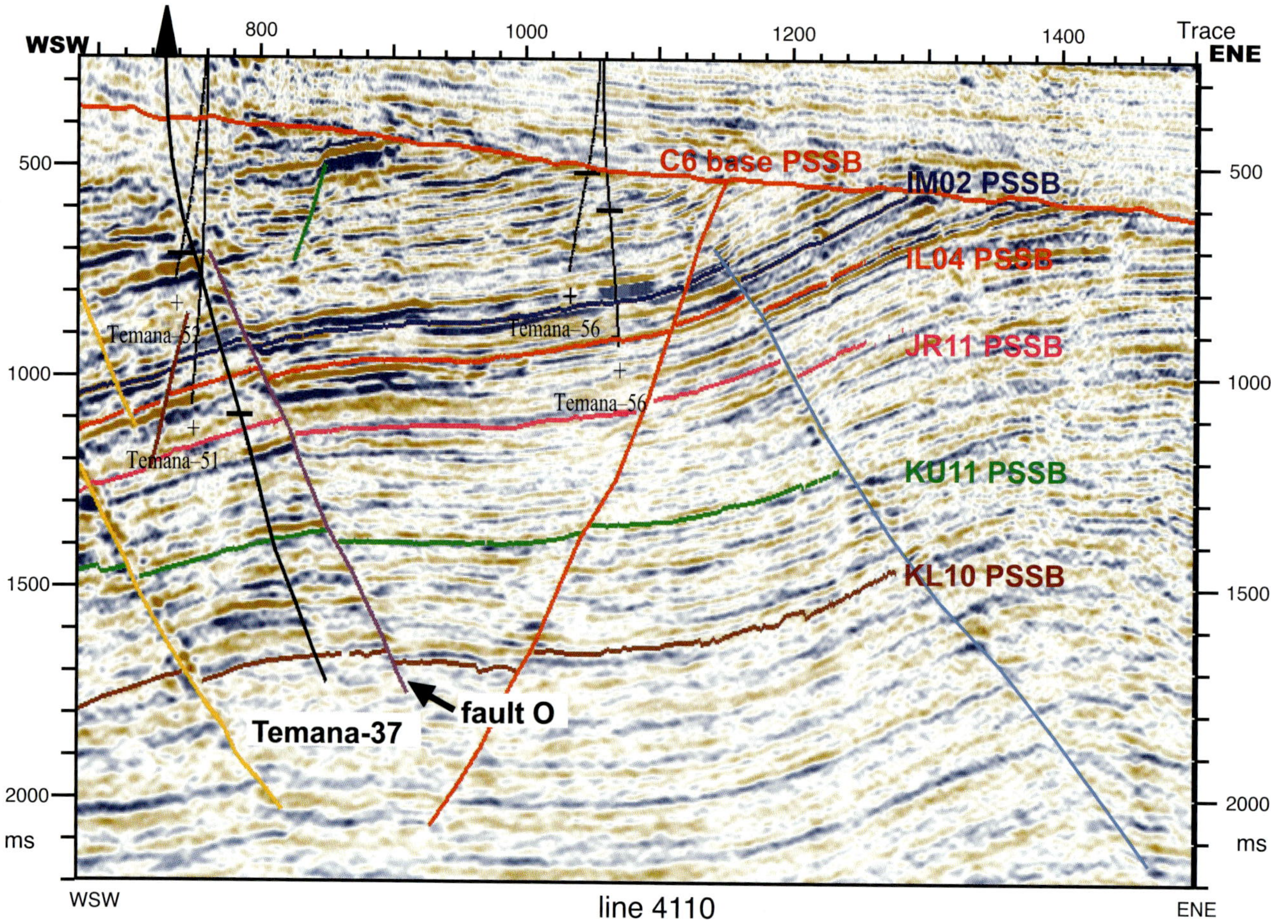

FIGURE 8. Seismic section of the Temana field along the seismic line 4110 showing fault O and the drilling direction of Temana-37.

the fault surface, although the reservoir facing the leak area is outside of fault closure. The black area on the fault O profile of SSF (Figure 10a) means an absence of SSF data, because the area on the fault plane is out of the range of shale smearing.

Clay content ratio is the percentage of clay that may be contributed from the various faulted layers into the fault rock (Figure 6f). This is the same as the gouge ratio of Fristad et al. (1997) and Yielding et al. (1997, 1999); however, we prefer the term "clay content ratio" to avoid confusion with the structural term "fault gouge." Higher values of CCR indicate a more efficient fault seal. Reported data from oil fields in the North Sea and the Niger Delta (Fristad et al., 1997; Yielding et al., 1997, 1999) indicate a CCR value of about 20% as a minimum threshold value for sealing faults.

Figure 10c shows the CCR values on the surface of fault O. The reservoir interval of the I-sands has CCR values more than 60%, suggesting very efficient fault sealing. In contrast, the reservoir interval of the lower J-sands has CCR values in the range of 30–60%, suggesting moderate fault-sealing potential. No sandstone horizons exist with CCR values less than 20% in the calculated interval between IM02PSSB and KU11PSSB, which indicates that fault O has sealing potential for almost all sandstone reservoir zones, even for those sandstone horizons in the footwall block that are in contact with sandstones on the hanging wall.

The results of this fault-sealing analysis are consistent with petrophysical interpretations using well logs, which suggest that almost all the sandstone horizons have some hydrocarbon saturation. Therefore, the fault-sealing analysis reported in this chapter is a good calibration of SSF (<6) and CCR (>30%) parameters, which can be applied to similar faults and basins.

Estimation of Hydrocarbon Column Height

Figure 10b and d show the crossplots of across-fault pressure difference against the SSF and CCR values for hydrocarbon-bearing sands with RFT pressure data. The across-fault pressure differences were obtained from RFT pressure data for the footwall block (penetrated by Temana-37) and assuming (in the absence of RFT data)

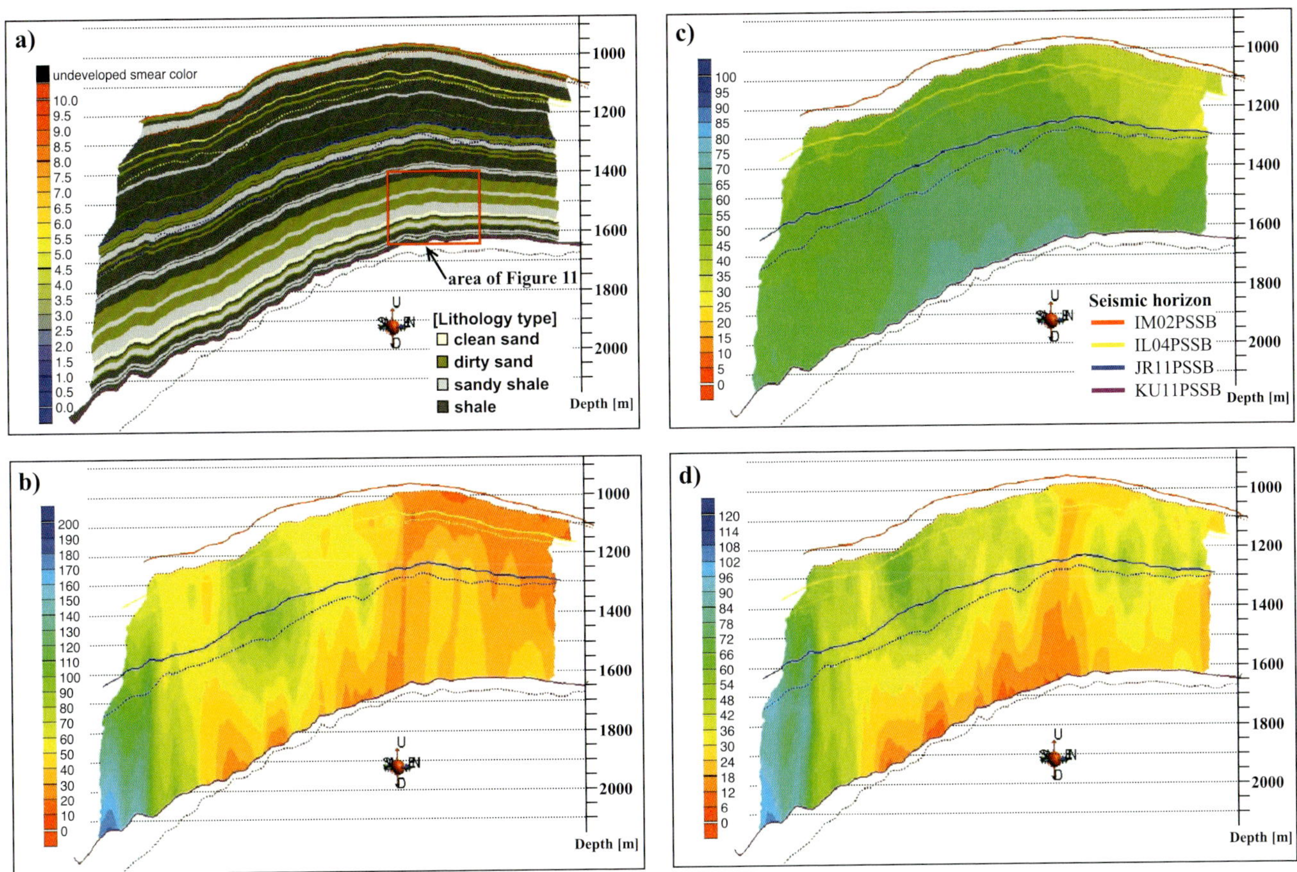

FIGURE 9. Display of lithology and structural geometric parameters for fault O on the fault surface (along the strike): (a) lithology; (b) fault throw (in meters); (c) fault dip angle; and (d) fault heave (in meters).

water saturation for all reservoirs in the hanging-wall block through which no well has been drilled.

As a methodological contribution of this study, we also employed the following equation to calculate hydrocarbon column heights from across-fault pressure differences:

$$HC = \Delta P/(\rho_w - \rho_h)g$$

where HC is the potential column of fault-trapped hydrocarbon; ΔP is the across-fault pressure difference; ρ_w and ρ_h are, respectively, densities of water and hydrocarbon (oil or gas); and g is a constant value for gravitational acceleration.

The above equation was used to estimate the gas column heights for reservoir JR11PRSS using the crossplots of SSF (Figure 10b) and CCR (Figure 10d) vs. across-fault pressure differences, and the results are plotted along the vertical axes on the right side of Figure 10b and d. It is assumed that sandstone horizons in the hanging-wall block are all water saturated, with the same water gradient and pressure profiles as those of the footwall block. The across-fault pressure differences (the vertical axis on the left side in Figure 9b, d) were calculated by subtracting the hanging-wall pressures from the footwall pressures of the same depths and were then plotted on the fault surface. The hydrocarbon gas column heights (the vertical axis on the right side of Figure 10b, d) were calculated using the above equation and the values of pressure gradients for gas and water in selected reservoirs.

The crossplots in Figure 10b and d show that the SSF values for the hydrocarbon-bearing sands are less than 6, and that the CCR values are greater than 30%. These values are within the domain of sealing faults as reported from other oil fields (e.g., Gibson, 1994; Fristad et al., 1997; Yielding et al., 1997, 1999) and correspond to fault sealing of hydrocarbon columns with an across-fault pressure difference of 3 bar.

The depth contour map of JR11PSSB (Figure 7) shows that the structural height of the fault trap on the footwall side of fault O is about 250 ft (76 m). Assuming the pressure gradient of gas and water to be 0.009 and 0.097 bar/m, respectively, the pressure difference across the fault would be 6.69 bar, if gas filled the JR11PSSB horizon to its structural spillpoint. The hydrocarbon column height estimated from the across-fault pressure differences and depicted in Figure 10d for JR11PSSB also shows a gas column height of 80 m (260 ft). These similar results of gas column heights

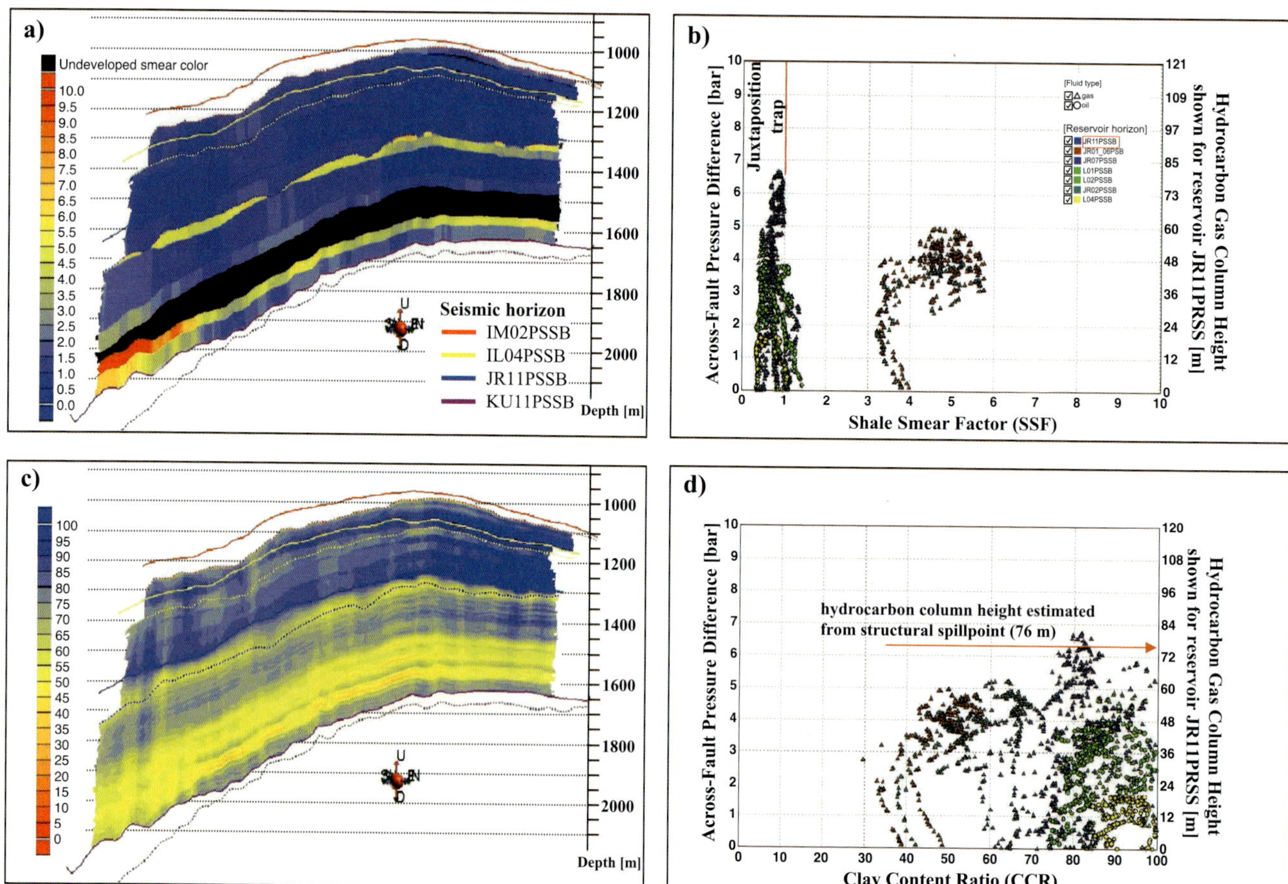

FIGURE 10. Fault-sealing assessment of fault O in the Temana field: (a) display of SSF on the fault surface (color code 0-10, black color indicates the absence of SSF data); (b) crossplot of SSF and across-fault pressure difference (in bar) for the same fault; (c) display of CCR on the fault surface (color code from 0 to 100%); (d) crossplot of CCR and across-fault pressure (in bar) for the same fault. Also shown in the crossplots (the vertical axes on the right) is the hydrocarbon gas column height (in meters) for reservoir JR11PRSS, estimated from the across-fault pressure difference. The methodology for this estimation is described in the text.

obtained by two different methods (structural spillpoints and across-fault pressure differences) indicate that fault O has sufficient sealing capacity to support the maximum hydrocarbon column heights controlled by the structural spillpoint.

Sand-sand Juxtaposition Analysis

Allan (1989) suggested a way of showing the layer lithologies of the hanging wall and footwall of a fault as juxtaposed on the fault plane. The Allan juxtaposition diagrams are helpful to identify the sand-sand interfaces or shale-sand traps along the strike of a fault. We used this method to investigate a sand-dominated area on fault O (specified box in Figure 9a). Figure 11 shows an Allan diagram for sand-sand interface (colored in red, Figure 11a) for an area of fault O (box in Figure 9a). The SSF and CCR profiles for the same area are shown in Figure 11b and c, respectively. Because the SSF algorithm is applicable for the fault section between the hanging-wall and footwall cutoff of a displaced shale layer (Figure 6b) and because the sand-sand interfaces depicted in Figure 11a largely lie out of the range of the displaced shale layers, SSF values could not be calculated for the majority of sand-sand interfaces on the fault plane (colored black in Figure 11b). This shows a

FIGURE 11. Allan-type juxtaposition diagram for a selected area (see box in Figure 9a) of fault O surface showing sand-sand interfaces (a) together with SSF profile (b) and CCR profile (c) for the same area. The black color in (b) indicates the absence of SSF data because of the limitation of the SSF algorithm. Clay content ratio values of 35–65% indicate across-fault sealing on the sand-sand interfaces.

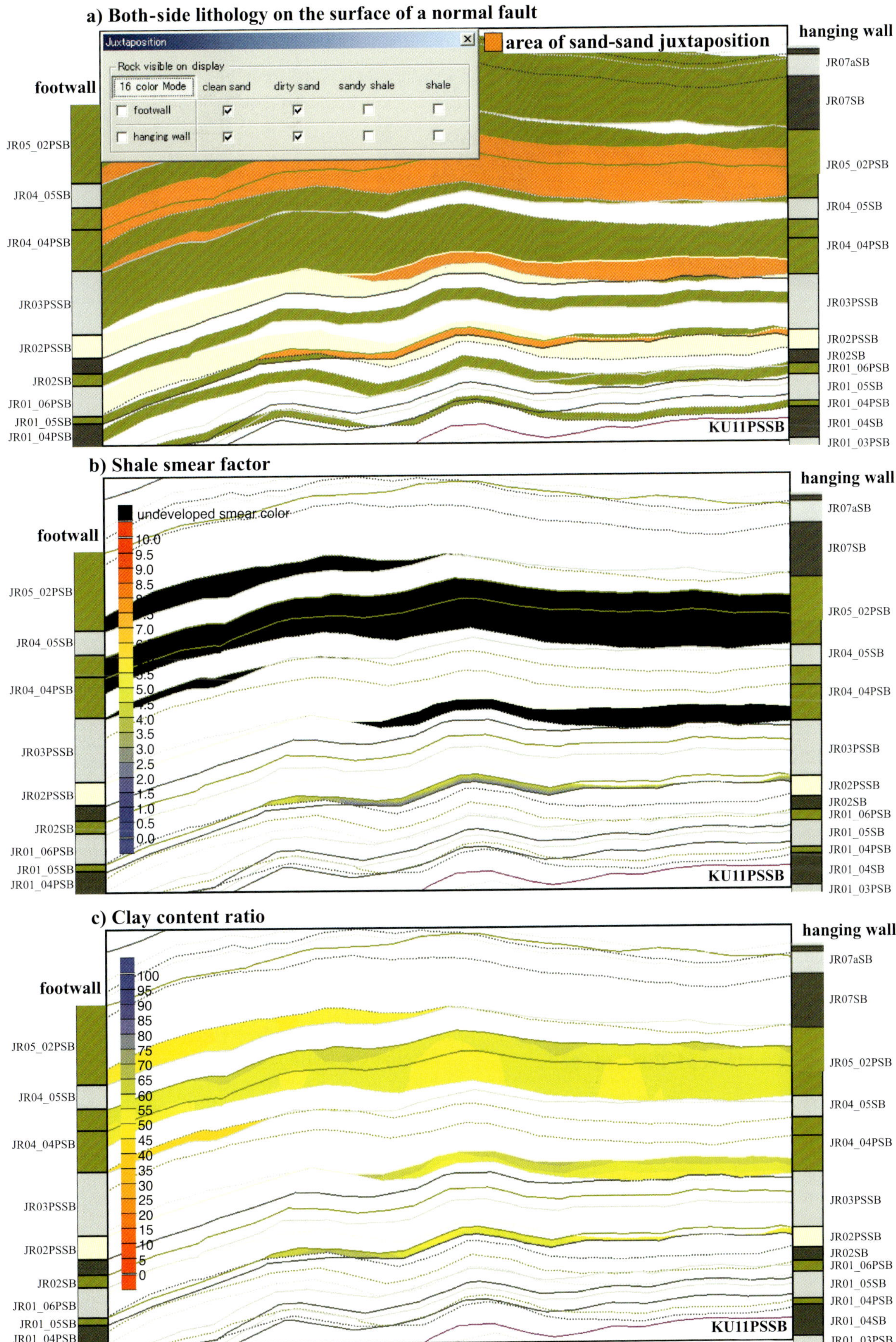
a) Both-side lithology on the surface of a normal fault
Juxtaposition
Rock visible on display
16 color Mode
clean sand
dirty sand
sandy shale
shale
footwall
hanging wall
area of sand-sand juxtaposition
footwall
hanging wall
JR07aSB
JR07SB
JR05_02PSB
JR04_05SB
JR04_04PSB
JR03PSSB
JR02PSSB
JR02SB
JR01_06PSB
JR01_05SB
JR01_04PSB
JR01_04SB
JR01_03PSB
KU11PSSB
b) Shale smear factor
undeveloped smear color
10.0
9.5
9.0
8.5
7.0
5.0
4.0
3.5
3.0
2.5
2.0
1.5
1.0
0.5
0.0
c) Clay content ratio
100
95
90
85
80
75
70
65
60
55
50
45
40
35
30
25
20
15
10
5
0

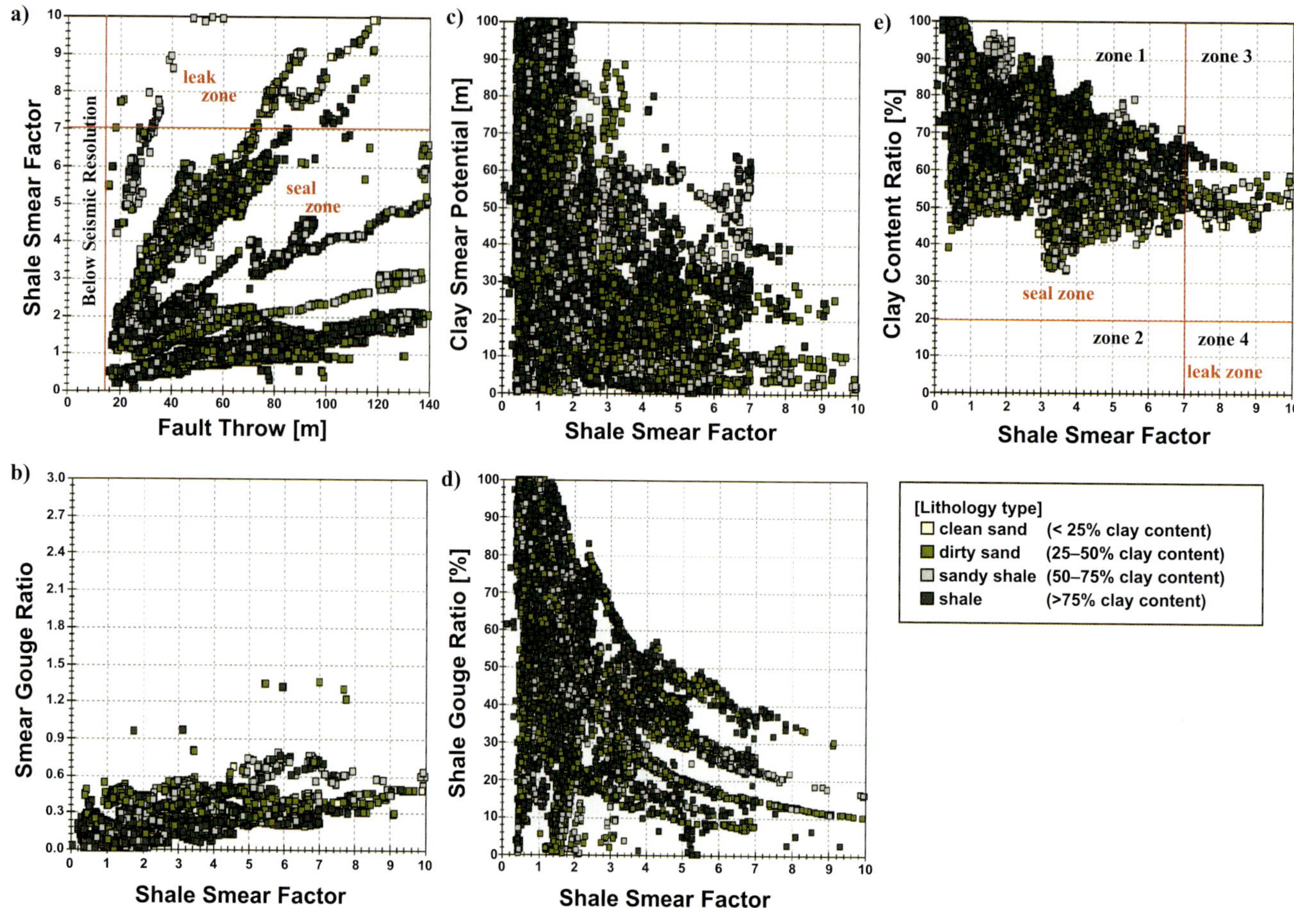

FIGURE 12. Crossplots of SSF against fault throw of fault O in the Temana field (a); smear gouge ratio (SMGR) (b); clay smear potential (CSP) (c); shale gouge ratio (SGR) (d); and clay content ratio (CCR) (e) for the same fault. These crossplots can be used for fault-seal ranking as shown in (e), where the diagram is divided into four zones. Zone 1 is definitely a sealing zone inferred from SSF and CCR values, whereas zone 4 is clearly a leak zone according to both parameters; zones 2 and 4 are mixed zones, which need to be further constrained by other data.

limitation of the SSF algorithm. However, because the CCR algorithm is a point-by-point calculation, CCR values could be calculated for the sand-sand interfaces on the fault plane (Figure 11c). The CCR values are between 35 and 65%, which suggests fault-sealing potential on the juxtaposed sand-sand parts of fault O.

Crossplots of Fault-seal Parameters

An important approach in fault-sealing assessment is to compare the results of various structural or shale smear parameters with one another. This can be done by crossplots, as shown in Figure 12. Figure 12a shows the relationship between fault throw and SSF values for fault O. Fault throws less than 20 m (65 ft) could not be plotted because they were below seismic resolution (i.e., subseismic faults). Leaking faults with SSF values greater than 7 have various fault throws, and a proportional relationship does not seem to exist between fault throw and fault-sealing capacity.

Figure 12b–e show the relationships between SSF and the other shale smear parameters. These crossplots can be used for drawing probability zones of fault sealing or leaking. This is illustrated in Figure 12e for the crossplot of SSF and CCR values for fault O. The diagram can be divided into four zones: zone 1 is definitely a seal zone according to both SSF and CCR parameters; zone 2 is a seal zone indicated by SSF but a leak zone according to CCR; zone 3 is a seal zone indicated by CCR but a leak zone according to SSF; and zone 4 is definitely a leak zone according to both parameters. By locating the points of these zones on the fault surface in the software, we can better rank the sealing potential of the fault.

Fault-rock Permeability Attributes

Fault-rock permeability is a key attribute for reservoir management, although it is extremely difficult to predict it for subsurface faults. The available methodologies for estimation of fault-rock permeability in normal faults in clastic reservoirs use the empirical relationship between the clay content and the permeability of the fault rock.

Based on a compilation of fault-rock data, Manzocchi et al. (1999) proposed the following empirical relationship between fault-rock permeability, clay content, and fault displacement:

$$\log K_f = -4\text{CCR} - 1/4\log(D)(1-\text{CCR})^5$$

where K_f is the fault-rock permeability; CCR is the clay content ratio in the fault rock (expressed in decimal numbers of 0.0–1.0); and D is fault displacement (in meters).

In a recent study of the North Sea fields, Sperrevik et al. (2002) suggested an alternative calibration for the relationship between fault-rock clay content and permeability. In their equation, fault displacement is replaced by depth as follows:

$$K_f = a\exp\{-[b\text{CCR} + cZ_{max} + (dZ_f - e)\,(1-\text{CCR})^7]\}$$

where K_f is the fault-rock permeability (in millidarcys); Z_{max} is the maximum subsurface depth of the target fault rock (in meters); Z_f is the depth at the time of faulting (in meters); CCR is the clay content (fraction) in fault rock; and the others are constants: $a = 80{,}000$; $b = 19.4$; $c = 0.00403$; $d = 0.0055$; and $e = 12.5$. (The meaning of these constant values was not described in Sperrevik et al., 2002.)

Both these equations give fault-rock permeability in absolute values (in millidarcys), although these values should be considered median values with an uncertainty range of two orders of magnitude. A different approach, called "fault-rock permeability probability" (FRP), proposed by Sorkhabi et al. (2003), integrates clay content ratio, fault-rock depth, and displacement of target points on the fault surface and expresses the fault-rock permeability in a probabilistic (normalized) manner as follows:

$$\text{FRP} = 1 - [(\text{CCR}Z_{max}D)/(\text{Max CCR}Z_{max}D)]$$

where FRP is the fault-rock permeability probability (1 being the maximum permeability); CCR is the clay content ratio (fraction); Z_{max} is the maximum subsurface depth of the target point on the fault surface (in meters); and D is the fault displacement (in meters).

Figure 13 shows the FRP values depicted on the fault O surface (Figure 13a), as well as the fault-rock permeability values calculated by the methods of Manzocchi et al. (1999) (Figure 13b) and Sperrevik et al. (2002) (Figure 13c). Interestingly, the FRP values are highest (red color code in Figure 13a) for the parts of the fault surface where fault throw (see Figure 9b), fault heave (see Figure 9d), and CCR values (Figure 10c) are relatively lower. The fault-rock permeability calculated by the equations of Manzocchi et al. (1999) and Sperrevik et al. (2002) indicates quite low permeability values (<0.3 md) for the fault surface.

Fault Slip and Dilation Tendency

To understand the behavior of faults in the contemporary (in-situ) stress regime, Ferrill et al. (1999) suggested two algorithms— slip tendency and dilation tendency— and used them to quantify the probability of fault failure (by dilation or by shear) in the Yucca Mountain area of Nevada, where a high-radioactive waste disposal site has been planned. The concepts and equations of slip tendency (shear failure) and dilation tendency (failure by extension fracturing) for a normal fault are depicted in Figure 14a. Both algorithms are normalized to 1 as the maximum probability of fault slip or dilation.

The fault surface profiles of slip tendency and dilation tendency for fault O are shown in Figure 14b and c, respectively. The lower central part of the fault surface exhibits a high tendency for slip and dilation, indicating that this is the weakest part of fault O. This does not necessarily mean that fault O has been reactivated in recent times. Indeed, the seismic image data of the Temana central (Figure 7) show that seismic-scale normal faults do not cut the major unconformity ("C6 base pssb" in Figure 7) at the base cycle VI. This means that normal faults, including fault O, have not been active on a seismic scale since the early Pliocene. Nonetheless, the slip and dilation tendency profiles show the relative weakness and strength of various parts of the fault surface under the contemporary stress regime, and these data may be used to forecast the response of the fault surface to subsurface fluid-pressure changes during the production stage.

SUMMARY AND CONCLUDING REMARKS

The fault-sealing assessment of a normal fault (fault O) in Tertiary clastic reservoirs was conducted in the Temana field, offshore Sarawak, Malaysia. The conclusions of the study are as follows:

1) The fault surface profiles of the calculated geometrical parameters suggest that fault O consists of two separate faults that were linked together during the structural development of the fault.
2) The calculated shale smear parameters show that the sandstone reservoirs of I-series have very good

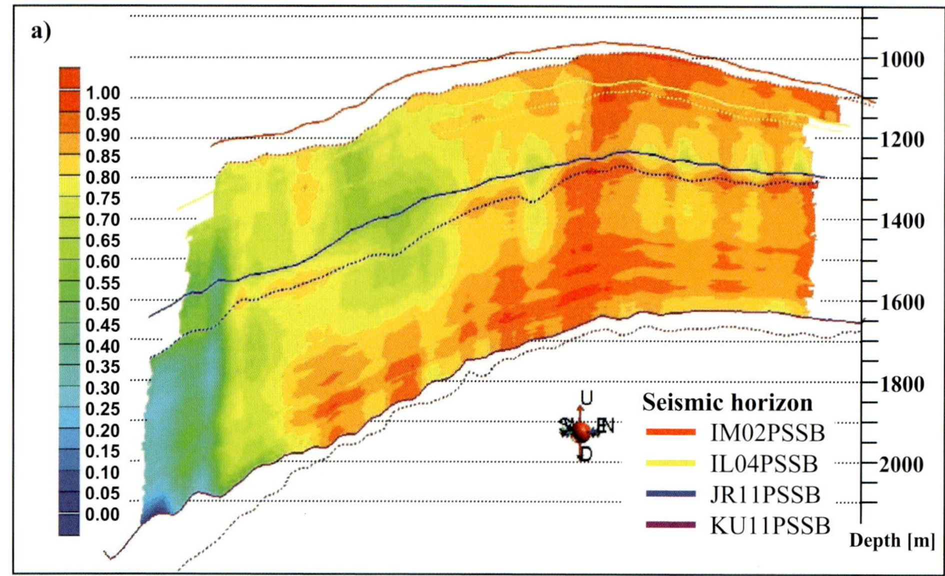

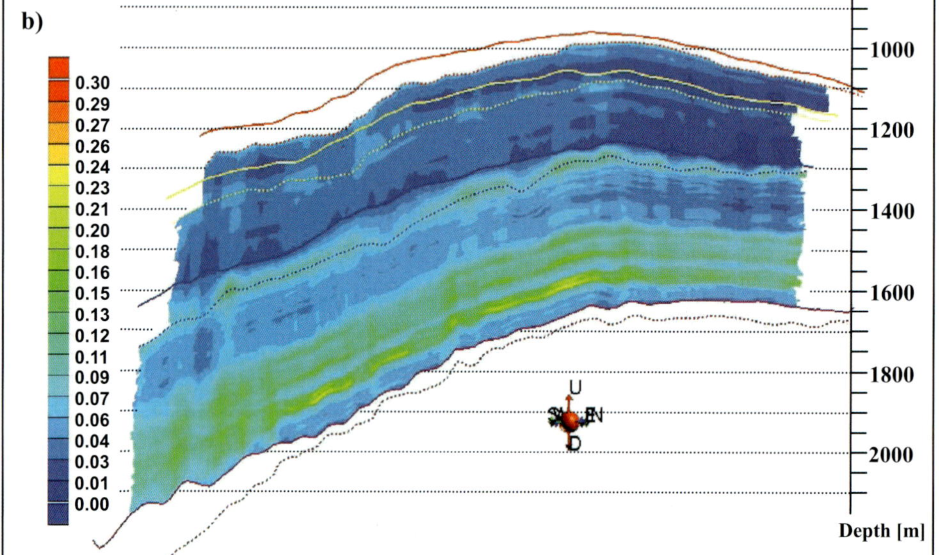

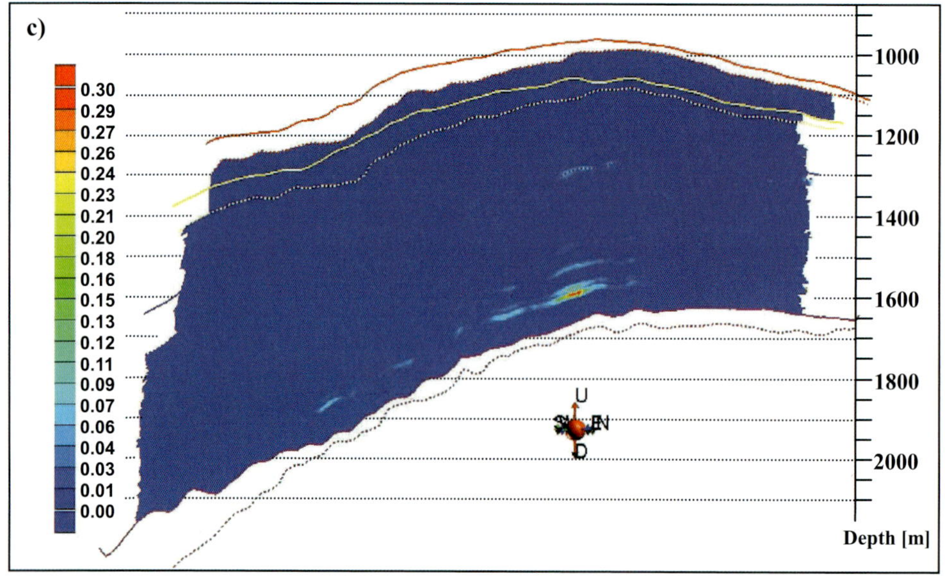

FIGURE 13. Fault rock permeability of fault O surface: Display of fault rock permeability probability (1 being the highest permeability probability) (a), and fault rock permeability estimation after the methods of Manzocchi et al. (1999) (b) and Sperrevik et al. (2002) (c). Explanations of the methods are given in the text. Both (b) and (c) yield fault rock permeability values of <0.3 md.

sealing efficiency with SSF values of less than 2 and CCR values of greater than 60%, and that those of the J-series have relatively moderate sealing efficiency with SSF values of 3–6 and CCR values of 30–60%. The results of these fault smear attributes are consistent with the hydrocarbon habitat inferred from well logs, which suggests that almost all of the sandstone horizons show some degree of hydrocarbon saturation, including the sandstone reservoirs in the footwall block, which are juxtaposed against the sandstone layers in the hanging wall.

3) The hydrocarbon column height was estimated from the crossplots of the calculated pressure difference across fault against the CCR and was compared with the maximum hydrocarbon column height inferred from the subsurface structural contour map. These two different methods yield similar results, indicating that fault O has sufficient sealing capacity to support hydrocarbon

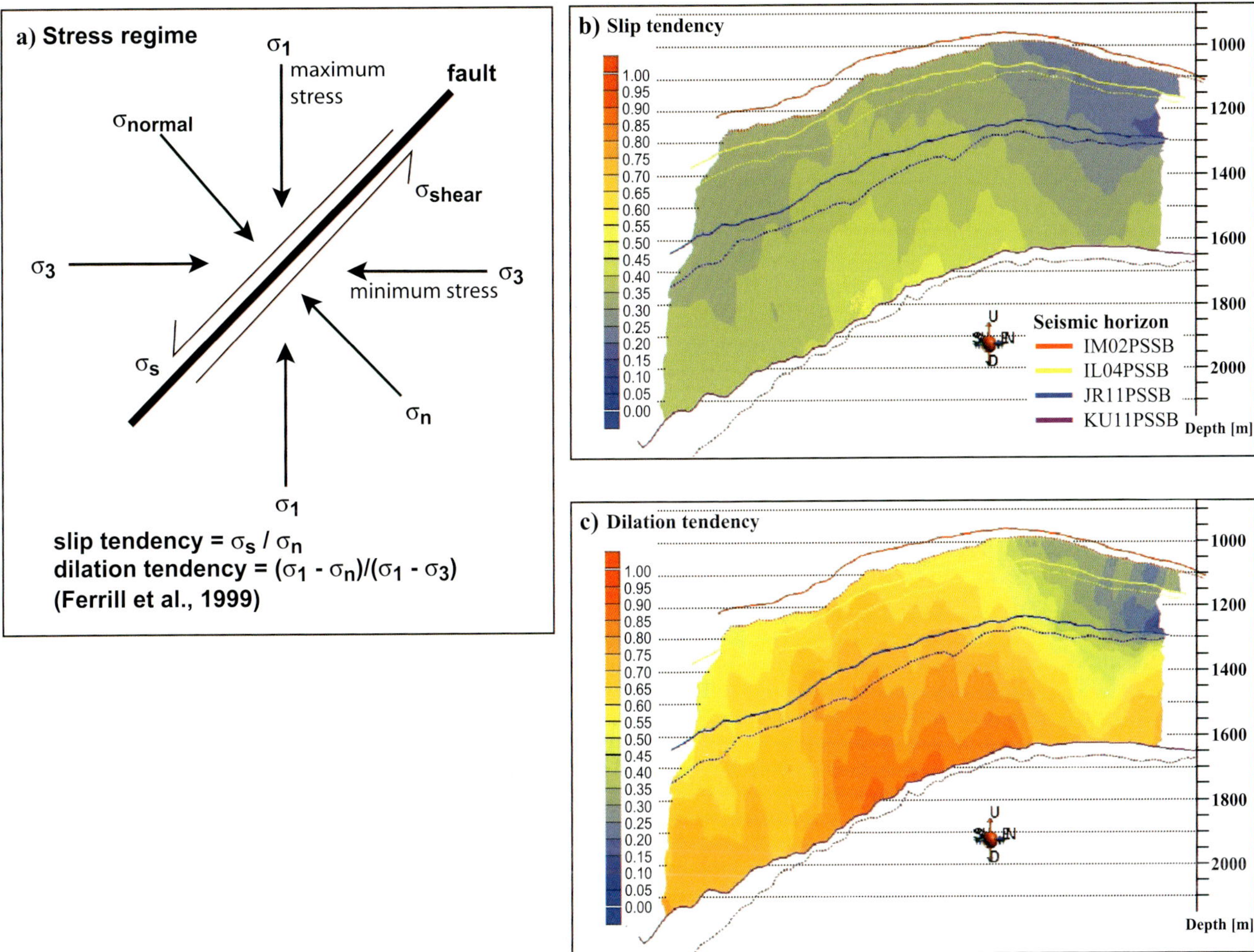

FIGURE 14. Stress field for a two-dimensional normal fault showing the concepts of slip tendency and dilation tendency (proposed by Ferrill et al., 1999) (a); display of slip tendency (normalized to the maximum of 1) on the fault surface of fault O in the Temana field (b); display of dilation tendency (normalized to the maximum of 1) on the same fault surface.

columns filled down to the structural spillpoints of the reservoir units.

4) Fault-rock permeability was calculated for the fault surface using the available calibrations of the clay content-permeability relationship (Manzocchi et al., 1999; Sperrevik et al., 2002), and both results yield fault-rock permeability values less than 0.3 md.
5) To further assist the fault-sealing assessment, new approaches were applied to the study of fault O, including fault-rock permeability probability (based on the integration of fault-rock clay content, fault displacement, and depth factors) and fault-seal ranking zones on the crossplots of various shale smear parameters.
6) Profiles of slip tendency and dilation tendency on the fault surface demonstrate the relative strength or weakness of the fault for failure by shear or extension under the contemporary stress regime. This information may be important to forecast the behavior of the fault surface to subsurface fluid-pressure changes during the production stage.

ACKNOWLEDGMENTS

This study was jointly supported by the Japan National Oil Corporation (JNOC) and the Petroliam Nasional Berhad (PETRONAS). The authors express their gratitude to the management of both organizations for permission to publish this chapter, to K. Nakayama of Japex Geophysical Institute (JGI)-Tokyo and Mazlan Madon of PETRONAS for reviewing the chapter, and Y. Tsuji of JNOC and David Curtis of the University of Utah for reading the manuscript and for their informative comments.

REFERENCES CITED

Alexander, L. L., and J. W. Handschy, 1998, Fluid flow in a faulted reservoir system: Fault trap analysis for the Block 330 field in Eugene Island, south Addition, offshore Louisiana: AAPG Bulletin, v. 82, p. 387–411.

Allan, U. S., 1989, Model for hydrocarbon migration and entrapment within faulted structures: AAPG Bulletin, v. 73, p. 803–811.

Bouvier, J. D., C. H. Kaars-Sijpersteijin, D. F. Kluesner, C. C. Onyejekwe, and R. C. van der Pal, 1989, Three-dimensional seismic interpretation and fault sealing investigations, Nun River field, Nigeria: AAPG Bulletin, v. 73, p. 1397–1414.

Burhannudinnur, M., and C. K. Morley, 1997, Anatomy of growth fault zones in poorly lithified sandstones and shales: Implications for reservoir studies and seismic interpretation: Part 1. Outcrop study: Petroleum Geoscience, v. 3, p. 211–224.

Ferrill, D. A., J. Winterle, G. Witmeyer, D. Sims, S. Colton, A. Armstrong, and A. P. Morris, 1999, Stressed rock strains groundwater at Yucca Mountain, Nevada: Geological Society of America Today, v. 9, p. 1–8.

Fristad, T., A. Groth, G. Yielding, and B. Freeman, 1997, Quantitative fault seal prediction: A case study from Oseberg Syd, *in* P. Moller-Pedersen and A. G. Koestler, eds., Hydrocarbon seals: Norwegian Petroleum Society Special Publication 7, p. 107–124.

Fulljames, J. R., L. J. J. Zijerveld, R. C. M. W. Franssen, G. M. Ingram, and P. D. Richard, 1997, Fault seal processes, *in* P. Moller-Pedersen and A. G. Koestler, eds., Hydrocarbon seals: Norwegian Petroleum Society Special Publication 7, p. 51–59.

Gibson, R. G., 1994, Fault-zone seals in siliclastic strata of the Columbus Basin, offshore Trinidad: AAPG Bulletin, v. 78, p. 1372–1385.

Ho, K. F., 1978, Stratigraphic framework for oil exploration in Sarawak: Geological Society of Malaysia Bulletin, v. 10, p. 1–13.

Hussin, A., and B. Bait, 1992, Geology of Sarawak: Guide for Field Trip 1, Symposium on Tectonic Framework and Energy Resources of the Western Margin of the Pacific Basin: Geological Survey of Malaysia, 21 p.

Jev, B. I. C., C. H. Kaars-Sijpesteijn, M. P. A. M. Peters, N. L. Watts, and J. T. Wilkie, 1993, Akaso field, Nigeria: Use of integrated 3-D seismic, fault slicing, clay smearing, and RFT pressure data on fault trapping and dynamic leakage: AAPG Bulletin, v. 77, p. 1389–1404.

Koledoye, B., 1998, Quantitative shale smear analysis, using well logs: Stanford University, Stanford Rock Fracture Project Annual Report, v. 9, p. D1–D16.

Lindsay, N. G., F. C. Murphy, J. J. Walsh, and J. Watterson, 1993, Outcrop studies of shale smears on fault surfaces: International Association of Sedimentologists Special Publication 15, p. 113–123.

Manzocchi, T., J. J. Walsh, P. Nell, and G. Yielding, 1999, Fault transmissibility multipliers for flow simulation models: Petroleum Geoscience, v. 5, p. 53–63.

PETRONAS, 1999, The petroleum geology and resources of Malaysia: Kuala Lumpur, PETRONAS, 665 p.

Skerlec, G. M., 1999, Evaluating top and fault seal, *in* E. A. Beaumont and N. H. Foster, eds., Exploring for oil and gas traps: AAPG Treatise of Petroleum Geology, chapter 10, p. 1–94.

Sorkhabi, R. B., 2000, Outcrop observations on normal faults in a Tertiary sandstone reservoir, Miri, Malaysia: Annual Reports of Technology Research Center, Japan National Oil Corporation, v. 12, p. 129–133.

Sorkhabi, R. B., S. Hasegawa, S. Iwanaga, and M. Fujimoto, 2002, Sealing assessment of normal faults in clastic reservoirs: The role of fault geometry and shale smear parameters: Journal of the Japanese Association of Petroleum Technology, v. 67, p. 576–589.

Sorkhabi, R. B., S. Iwanaga, M. Fujimoto, and S. Hasegawa, 2003, Sealing assessment of normal faults in clastic reservoirs: Modeling the petrophysical and stress attributes of faults: Journal of the Japanese Association of Petroleum Technology, v. 68, p. 291–304.

Sperrevik, S., P. A. Gillespie, Q. J., Fisher, T. Halvorsen, and R. J. Knipe, 2002, Empirical estimation of fault rock properties, *in* A. G. Koeslter and R. Hunsdale, eds., Hydrocarbon seal quantification: Norwegian Petroleum Society Special Publication 11, p. 109–125.

Yielding, G., B. Freeman, and D. T. Needham, 1997, Quantitative fault seal prediction: AAPG Bulletin, v. 81, p. 897–917.

Yielding, G., J. A. Overland, and G. Byberg, 1999, Characterization of fault zones for reservoir modeling: An example from the Gullfaks fields, northern North Sea: AAPG Bulletin, v. 83, p. 925–951.

4

Kachi, T., H. Yamada, K. Yasuhara, M. Fujimoto, S. Hasegawa, S. Iwanaga, and R. Sorkhabi, 2005, Fault-seal analysis applied to the Erawan gas-condensate field in the Gulf of Thailand, *in* R. Sorkhabi and Y. Tsuji, eds., Faults, fluid flow, and petroleum traps: AAPG Memoir 85, p. 59–78.

Fault-seal Analysis Applied to the Erawan Gas-condensate Field in the Gulf of Thailand

Tokio Kachi
Mitsui Oil Exploration Co., Ltd. (MOECO), Tokyo, Japan

Hideki Yamada
Mitsui Oil Exploration Co., Ltd. (MOECO), Tokyo, Japan

Kiyoshi Yasuhara
Mitsui Oil Exploration Co., Ltd. (MOECO), Tokyo, Japan

Masamichi Fujimoto[1]
Technology Research Center, Japan National Oil Corporation, Chiba, Japan

Shutaro Hasegawa[2]
Technology Research Center, Japan National Oil Corporation, Chiba, Japan

Shoji Iwanaga[3]
Technology Research Center, Japan National Oil Corporation, Chiba, Japan

Rasoul Sorkhabi[4]
Technology Research Center, Japan National Oil Corporation, Chiba, Japan

ABSTRACT

The Erawan field in the Gulf of Thailand is characterized by a series of east- and west-dipping normal faults displacing the Miocene clastic reservoirs. The fault-seal capacity of these faults was assessed using sand-shale juxtaposition diagrams, shale smear parameters, and fault-seal failure probability (FSFP) (based on in-situ stress conditions). For this study, five east-dipping faults in the Erawan N Platform area were selected (faults E-16, E-17, E-18, E-20, and E-27). Several deviated wells have been drilled through the footwall blocks of these faults. Low values of the shale smear factor (SSF < 6) and high values of the clay content ratio (CCR > 30%) of four of the

[1]*Present address:* INPEX Co., Tokyo, Japan.
[2]*Present address:* Idemitsu Oil & Gas Co., Tokyo, Japan.
[3]*Present address:* Geoscience Research Laboratory, Yamato, Kanagawa, Japan.
[4]*Present address:* Energy & Geoscience Institute, University of Utah, Salt Lake City, Utah, U.S.A.

DOI:10.1306/1033716M853131

faults suggest that faults seal along their planes. In contrast to these four faults, fault E-27 appears to act as an across-fault conduit for some intervals and seal in others. Intervals without trapped hydrocarbons have higher SSF values, suggesting that the fault leaks locally. These five faults trap 15 gas pay zones. Eight of the pay zones have sand-shale juxtaposition across the faults, which may explain 8 of the 15 accumulations. Shale smear parameters can account for all 15 accumulations. Fault-seal failure probability was derived for one of these faults (E-16) by integrating the CCR values and the probability of fault slip tendency and fault dilation tendency under the current stress regime in the Erawan field. Low FSFP values indicate that the fault seals do not appear to have been breached given the in-situ stress field acting on the normal fault.

INTRODUCTION

The Erawan gas-condensate field lies in 60 m (197 ft) of water offshore of Thailand about 400 km south of Rayong (Figure 1). Field production began in 1981. Most of the reservoirs in the Erawan field are discontinuous fluvial and fluvial-deltaic sandstones that are segmented by numerous faults. Recent information on this giant field in the Gulf of Thailand is summarized in Turner (2002), Yawwapapong et al. (2002), and Turner and Lambert (2003). Because of the structural and stratigraphic complexities of the field, fault-sealing assessment is crucial for identifying potential fault traps and fault migration pathways and for exploration, drilling, development, and production in the field. This chapter discusses the results of fault-sealing analysis of normal faults in the Erawan N Platform area, located in the southwestern part of the Erawan field (Figure 1).

GEOLOGICAL OUTLINE OF THE STUDY AREA

The Gulf of Thailand contains a series of Tertiary-age north–south-trending elongate extensional basins. Most geologists consider these basins to be rift related, although some have argued for an interior sag basin as well (Turner, 2002). Located near the geographic center of the Gulf of Thailand is the Pattani Basin, which is the largest of these rift basins; it contains a more than 25,000-ft (7600-m) thickness of predominantly nonmarine fluvial-deltaic sediments along its main axial depocenter (Figure 2). A brief description of the stratigraphy, petroleum systems, and structural regime is given as background information for this chapter. A more complete review of the geology of the Pattani Basin and the tectonic setting of the Erawan gas-condensate field has been discussed by various authors, e.g., Lian and Bladley (1986), Western and Davis (1993), Jardin (1997), Leo (1997), Kongwung and Ronghe (2000), Mial (2002), and Morley (2002).

Stratigraphy

Five main formations comprise the Tertiary sedimentary section of the Pattani Basin; they are numbered 1–5 (Figure 3) from the oldest to the youngest. Within the sedimentary section, two periods of nondeposition and/or erosion occur, one at 25 Ma, named the middle Tertiary unconformity (MTU), and the second at 10 Ma, called the middle Miocene unconformity (MMU). No Tertiary sediments older than Oligocene have been identified. Sequence 1 strata were deposited in the Oligocene during the initial rifting phase that created the Pattani Basin. However, the age of sequence 1 is not precisely known, because the fossils and spores contained in it are not restricted to the Oligocene epoch; therefore, sequence 1 may be as old as late Eocene. Sequences 2 (lower red beds), 3 (lower gray beds), and 4 (upper red beds) largely consist of interbedded sandstones and shales, with numerous coal beds predominantly in sequences 2 and 3. These three sequences contain the main hydrocarbon-bearing reservoirs located in the Unocal/MOECO's producing fields such as the Erawan gas-condensate field.

Three seismic horizons (markers) were defined for this study. The K-83-2 marker, corresponding to top sequence 2, marks the boundary between fluvial-dominated red bed sequence and a more marine-influenced sequence with continuous shale and coal beds. The F68-3 marker forms a very continuous high-amplitude seismic event in sequence 3. The C61-3 marker, corresponding to top sequence 3, is also represented by a high-amplitude seismic event. The seismic horizons MMU and O were interpreted in this study as continuous seismic events in the study area (Figure 4).

Pay sands in upper sequence 2 and sequences 3 and 4 are mainly thin, fluvial layers and are located at depths from 5000 to 9000 ft (1500 to 2700 m) below sea level in the study area (Figure 5). Gas is trapped in the sandstone beds, terminating updip against the faults (Figure 6). Shale beds overlying the sandstone layers form effective top seals.

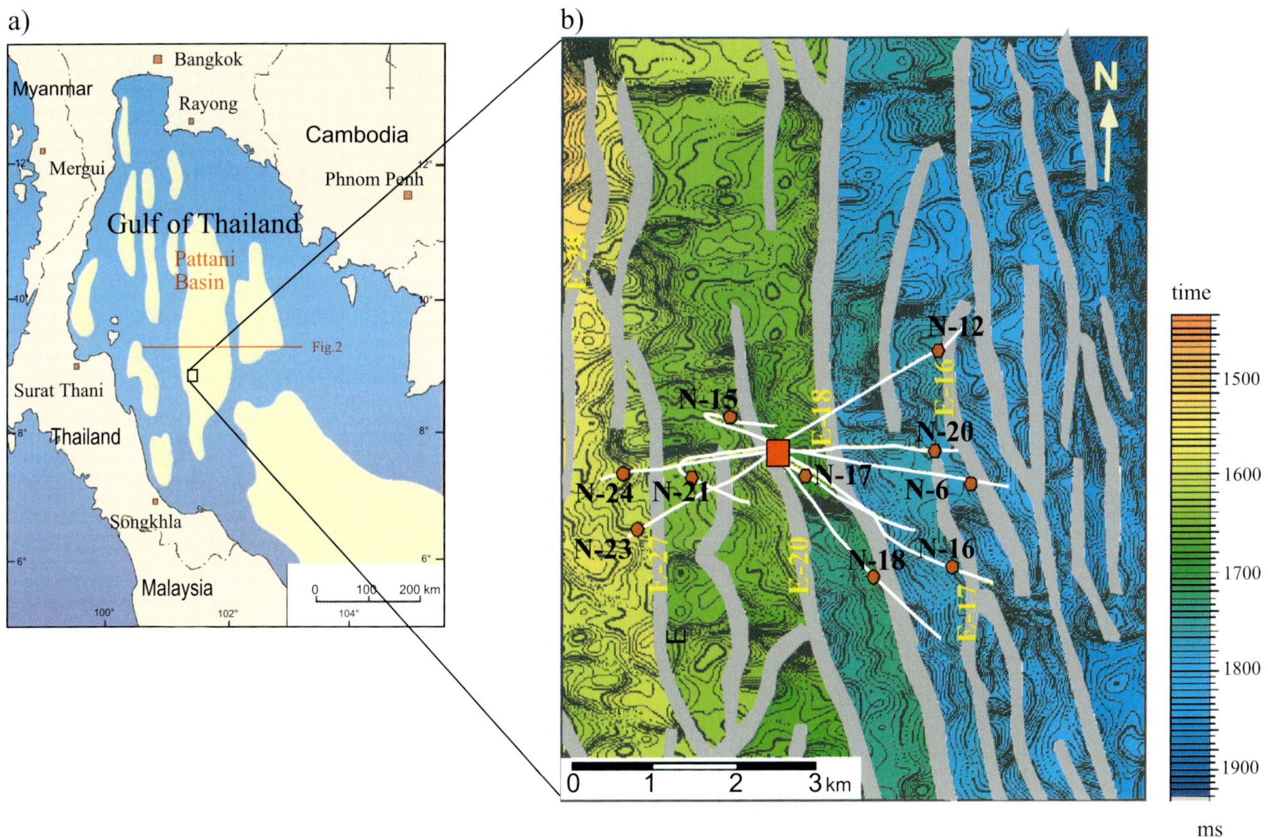

FIGURE 1. Location map of (a) the Pattani Basin in the Gulf of Thailand and (b) the Erawan N Platform area viewed on the structural contour map (plan view at the level of F68-3 seismic horizon). Faults E-16 to E-29 are east-dipping normal faults; N-15 to N-24 are deviated wells.

Amplitude extraction maps were generated to help define the aerial extent of seismic anomalies corresponding to channel sands and to help correlate sandstone layers between wells. Well-to-well sand correlation was difficult along an east–west transect through the study area, because the channel sandstones trend roughly north–south in the Erawan field. The continuity of sandstone layers is not clear at

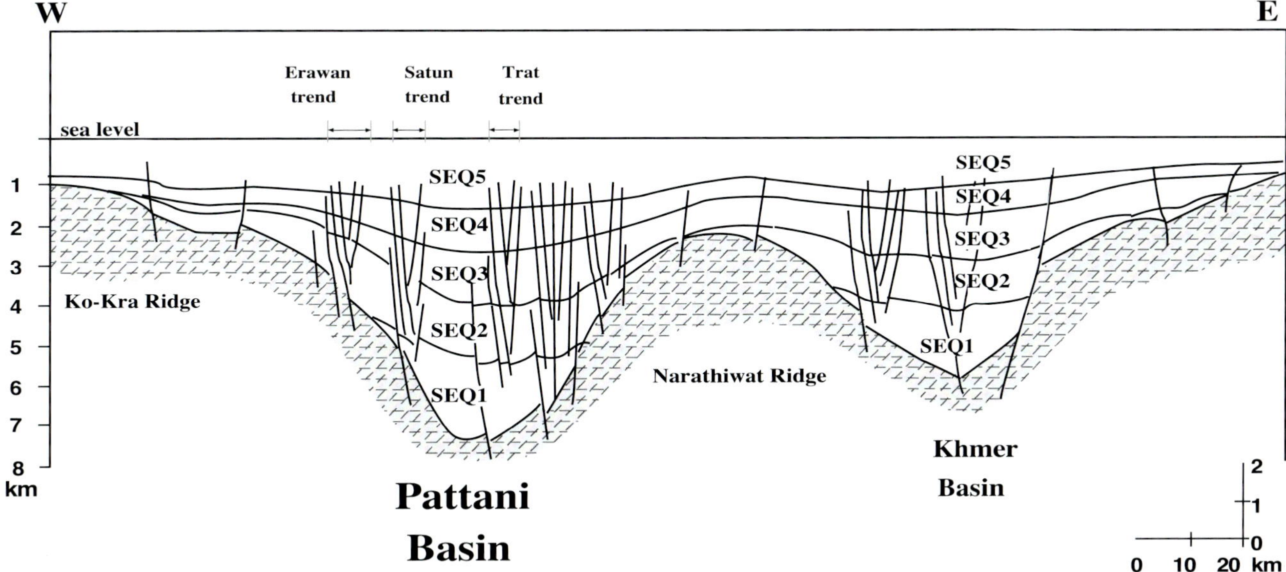

FIGURE 2. Structural cross section across the Pattani Basin (modified after Minezaki and Moriyama, 2002). SEQ = sequence.

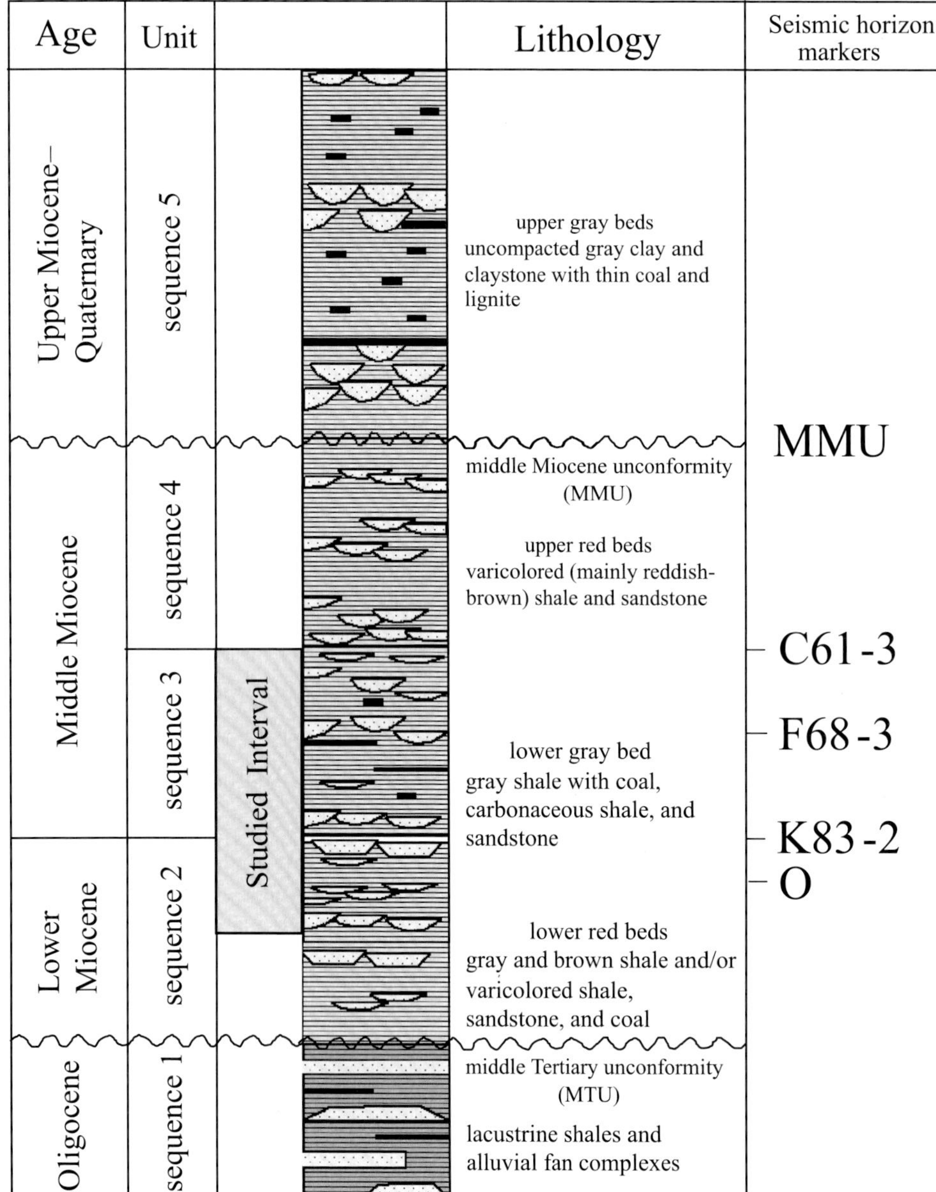

FIGURE 3. Stratigraphic summary of the Erawan field, showing the studied interval and location of seismic horizons used for fault-sealing assessment.

levels deeper than the C61-3 seismic marker, probably because only discontinuous sandstone layers are present.

Petroleum Systems

Two principal petroleum systems are active in the Pattani Basin (Jardin, 1997). The first is fed by the Miocene gas-prone terrestrial coals and shales currently at a mature stage in the central part of the basin, and the second is sourced from the Oligocene oil-prone lacustrine shale, which is currently mature in the basin flank areas. The Oligocene oil-prone shale is also present in the basin center, but it is overmature.

These two petroleum systems include a gas-condensate-source family of fluvial-deltaic coaly origin in sequence 3 and upper sequence 2, and a lacustrine shale algal oil-source family identified in sequence 1 and lower sequence 2. The gas system is volumetrically more significant in the basin center. Oil tends to be preserved on basin flanks, where it has migrated into the overlying Miocene reservoirs via basement onlap and fault conduits. In the majority of the fields, structure closely controls hydrocarbon accumulations.

Minezaki and Moriyama (2002) carried out two-dimensional basin modeling along an east–west transect through the central Pattani Basin, including the Erawan gas-condensate field, using SIGMA2D basin-modeling software developed at the Japan National Oil Corporation (JNOC [presently Japan Oil, Gas and Metals National Corporation]) (Okui et al., 1996). The basin-modeling results show that in the early Miocene, oil-prone sequence 1 source rocks began to generate oil in the central part of the basin, and then an oil expulsion window propagated toward the basin flanks. Gas generation from the gas-prone sequence 2 source rocks began in the early Miocene in the deepest parts, and then in the middle Miocene, the gas generation zone propagated through the basin. Replacement of previously accumulated oil occurred in the early middle Miocene by gas that either migrated from sequence 1 or generated from the cracking of oil in the trap. At present, oil generation and accumulation areas are confined to the shallower parts along the basin flanks. The gas-prone sequence 3 source rocks began to generate gas in the late Miocene–Pliocene. At present, sequence 3 continues to generate gas across the axial part of the basin, and both gas and oil accumulate in sequence 3.

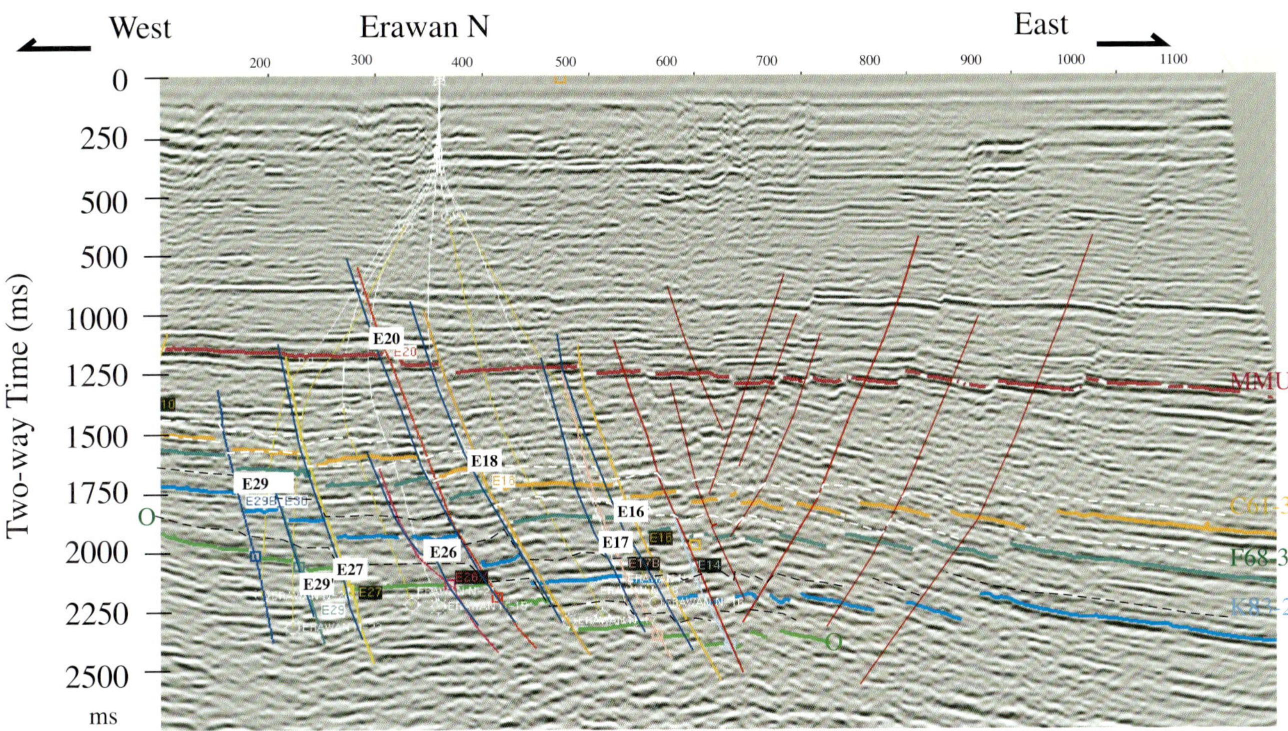

FIGURE 4. Seismic section along east–west transect near the Erawan N Platform area contains the seismic horizons used for this study.

Rifting commenced in the Oligocene. This was followed by a regional sag after the formation of the MTU. The main phase of rifting was in the Oligocene, and the sag phase of basin subsidence occurred in the early middle Miocene. Therefore, the major trapping faults formed after the rift phase ended and trapped gas expelled from sequences 2 and 3 throughout a relatively long period of time from the middle Miocene to the present.

The interval discussed in this chapter ranges from top sequence 3 down to uppermost sequence 2, and this section has generated hydrocarbons in the Erawan N Platform area since the late Miocene (Minezaki and Moriyama, 2002). This implies that in-situ gas from sequence 3 source rocks very likely filled the interbedded sandstone reservoirs, taking relatively short migration pathways (Figure 5).

The distribution of net pay sand thickness in the Erawan N Platform area shows that the thicker accumulations concentrate in the structural closures in the footwall blocks associated with a series of east-dipping normal faults in the study area (Figures 1 and 6).

Structural Style

The sedimentary succession in the Pattani Basin is cut by a multitude of closely spaced north–south-trending normal faults, which form a pattern of elongate horsts and grabens and tilted fault blocks. Many of the graben systems in the Pattani Basin contain sets of en echelon normal synthetic and antithetic faults flanked by larger graben-bounding faults.

The larger hydrocarbon accumulations on the west side of the Pattani Basin generally occur on the western flanks of the producing fields that are controlled by faults dipping toward the basin depocenter (Figure 2). These faults have penetrated deeper than their associated antithetic faults and have played prominent roles in the upward migration of hydrocarbons during their active stage and in the trapping of hydrocarbons during their closed stage.

METHODOLOGY OF INVESTIGATION

The software used for this study (FAULTAP^{TM}) is PC software (C++ code) developed at JNOC and has been described in detail by Sorkhabi et al. (2002, 2003). The software provides model results on fault geometry, shale smear parameters, permeability, and stress attributes depicted on fault surface sections and on Allan-type juxtaposition diagrams (Allan, 1989). Here, we describe the input data prepared for fault-sealing assessment.

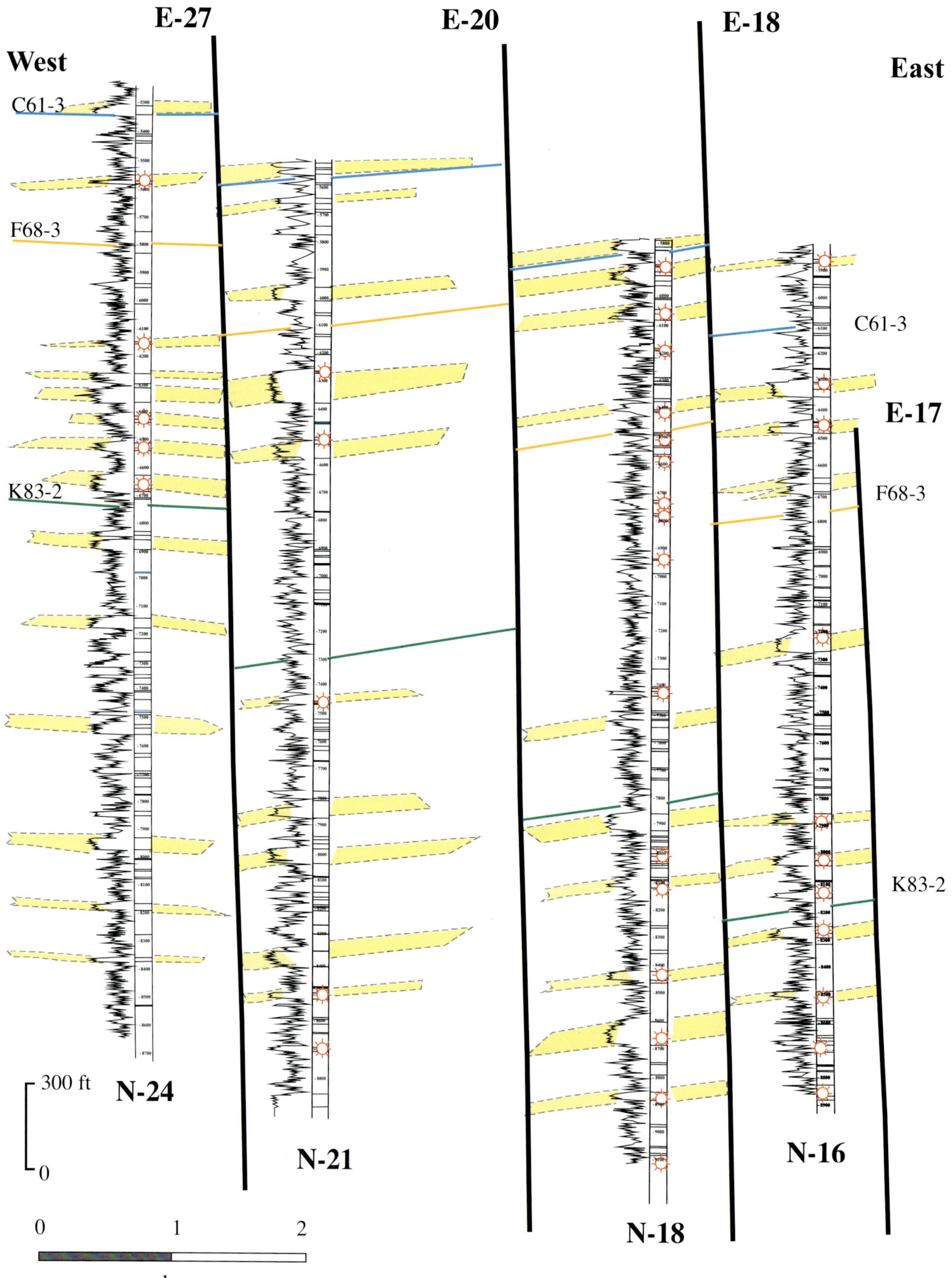

West
East
E-27
E-20
E-18
E-17
C61-3
F68-3
K83-2
C61-3
F68-3
K83-2
N-24
N-21
N-18
N-16
300 ft
0
0
1
2
km

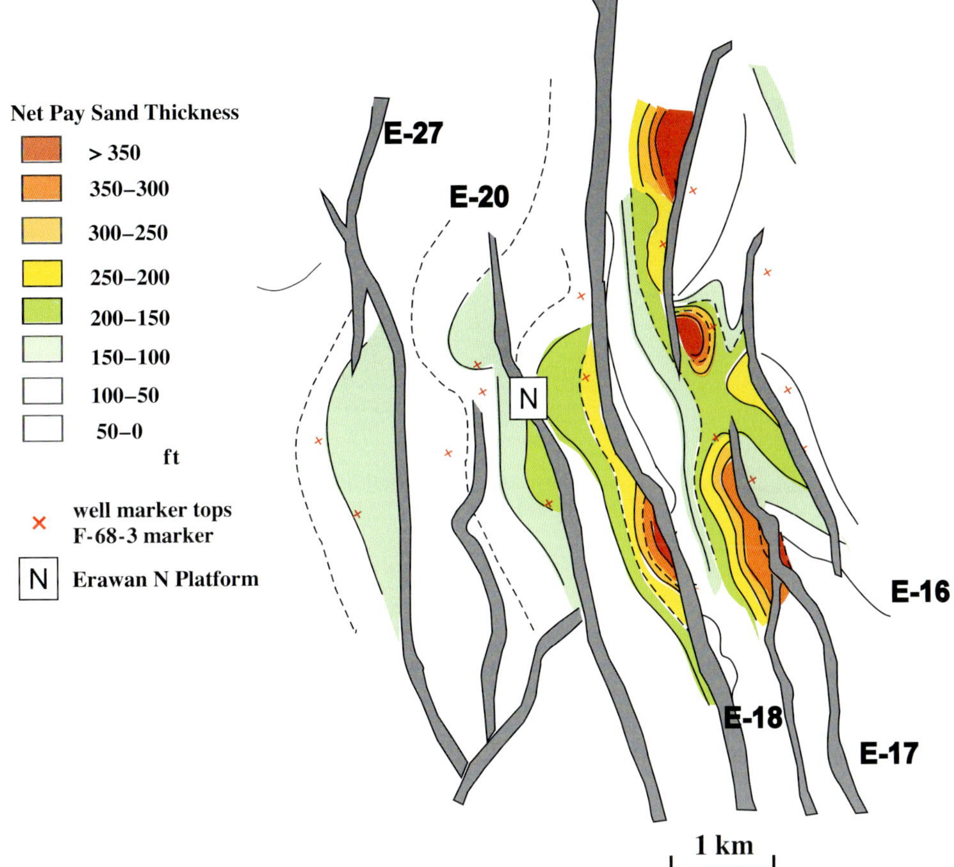

FIGURE 6. Thickness of net pay reservoir sands in the Erawan N Platform area.

Database

The following data sets were prepared:

- three-dimensional (3-D) seismic data: Erawan 1979–1980 3-D seismic cube in the Erawan N Platform area
- seismic horizon data: interpreted horizons C61-3, F68-3, K83-2, and O
- well data: gamma-ray logs for wells N-15, N-16, N-17, N-18, N-20, N-21, N-23, and N-24
- fault data: interpreted fault traces E-16, E-17, E-18, E-20, and E27

Seismic Interpretation

Three seismic horizons (C61-3, F68-3, and K83-2) previously interpreted in 3-D seismic volume were refined to the scale of this study, and an additional horizon (O) was interpreted so as to extend the study interval to a deeper level and to the graben center of Erawan. Trace line data for five faults (E16, E17, E18, E20, and E27) were also created. The interpreted horizons and fault traces picked from the 3-D seismic cube of the Erawan N Platform area were generated in ASCII format, converted to depth scale using a time-depth table, and then imported to the FAULTAP software. Fault trace lines were then edited, so that the selected points of seismic horizons joined the fault trace line, and finally, fault planes were constructed.

Lithology Construction

The lithology of sedimentary layers that was interpreted from gamma-ray logs was listed in the form of a table in the software. The lithologies of the layers were projected onto the fault surfaces and were expressed in terms of clay percentage: clean sand (clay content, 0–25%), dirty sand (25–50%), sandy shale (50–75%), and shale (75–100%). The lithological data were depicted on the fault surfaces using seismic horizons as footwall and hanging wall cutoffs. Figure 7 schematically shows this procedure of the lithological construction for fault E-27 using log data from two wells (N-21 and N-24).

Pressure Data

Repeat formation tester (RFT) data were used to reconstruct the subsurface pressure gradients for water, gas, and oil, if any.

RESULTS AND DISCUSSION

Drilling in the Erawan N Platform area began in 1988, and a total of 24 wells were drilled by 2000. The wells are commonly drilled directionally, parallel to the faults, and through their footwalls (Figures 1 and 8).

FIGURE 5. Well correlation (using wells N-16, N-18, N-21, and N-24) across faults E-17, E-18, E-20, and E-27 in the Erawan N Platform field. Seismic horizons include C61-3, F68-3, and K83-2. Gamma-ray logs are displayed in true vertical depth (numbers in feet); yellow layers are sandstone; gas zones are also shown (red circles). Note that the wells are shown vertically for simplification in the figure; the wells are not vertical.

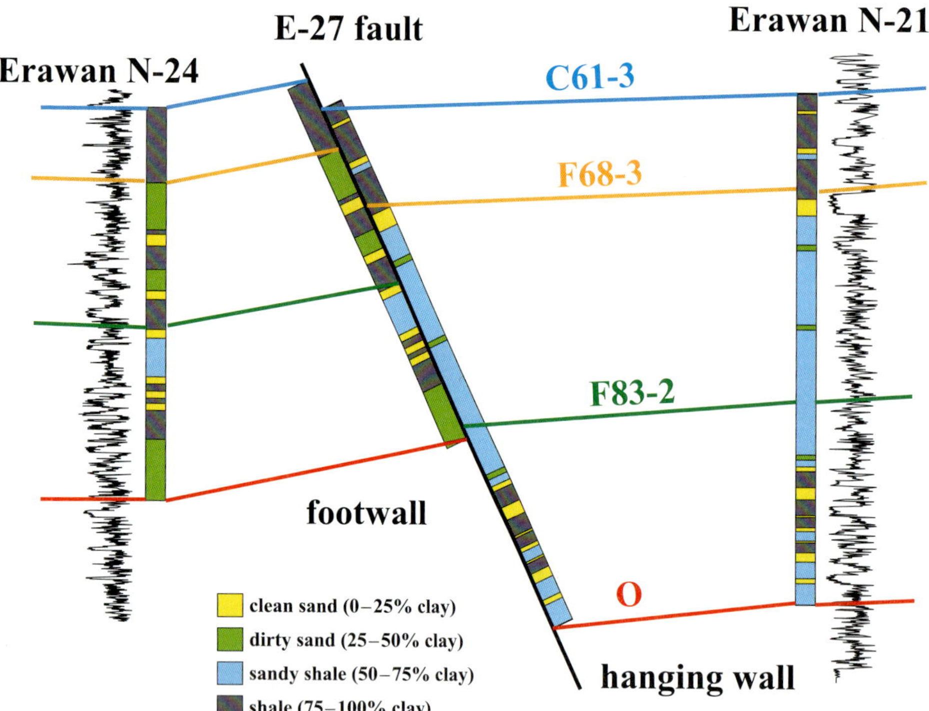

FIGURE 7. An example of the construction of lithology on fault E-27 using gamma-ray log data (clay percentage) and seismic horizons as markers of fault displacement.

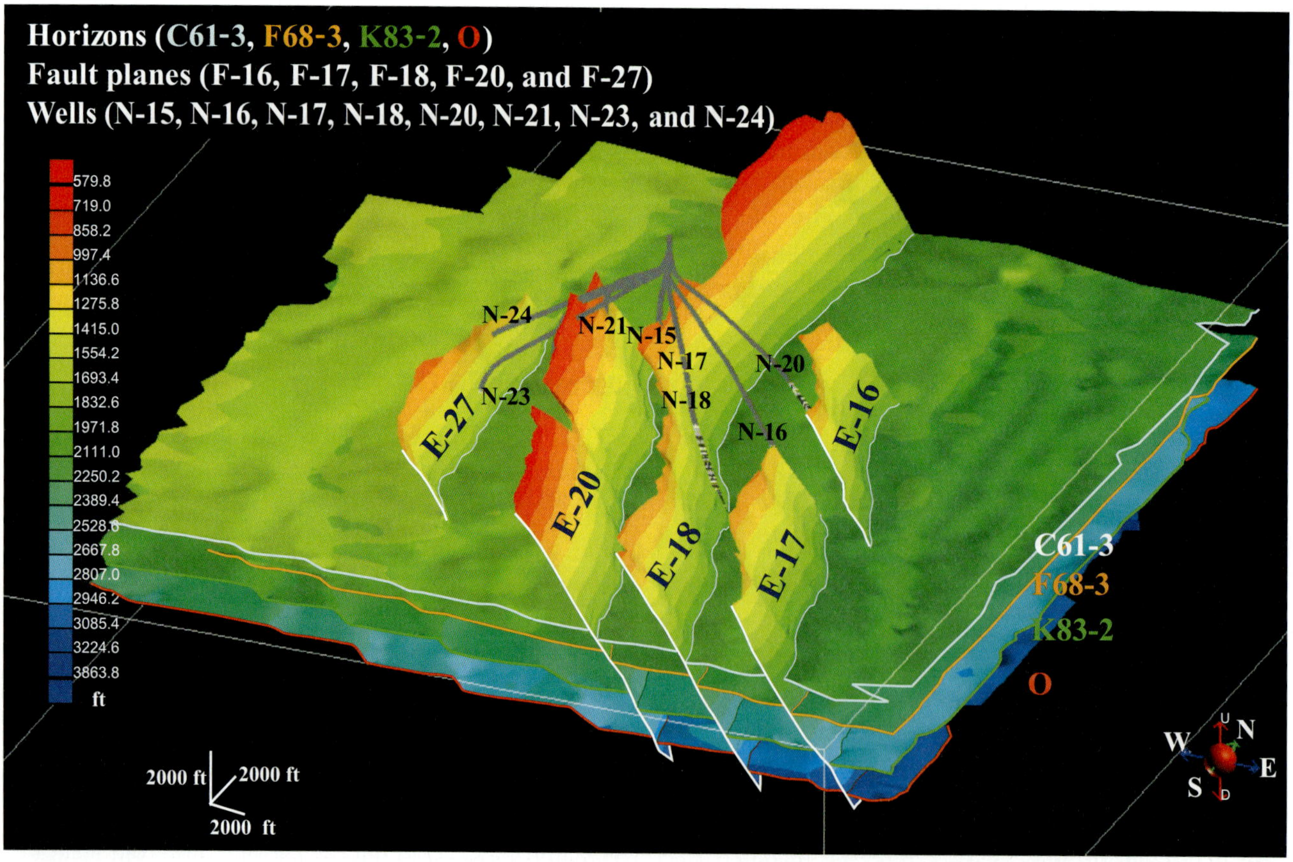

Horizons (C61-3, F68-3, K83-2, O)
Fault planes (F-16, F-17, F-18, F-20, and F-27)
Wells (N-15, N-16, N-17, N-18, N-20, N-21, N-23, and N-24)

FIGURE 8. View of fault planes, wells, and seismic horizons constructed in this study of the Erawan N Platform area. The color scale indicates depth in feet.

Fault Displacement

Fault displacement (the amount of fault movement along the dip direction) was among the geometric parameters calculated for each fault and displayed on the fault surface. Figure 9 shows fault displacement profiles for faults E-27 and E-16, in which fault displacement variation is color coded from red (small) to blue (large). The hanging-wall and footwall cutoffs for the interpreted seismic horizons (C61-3, F68-3, K83-2, and O) are also depicted on the fault surface in Figure 9. As can be seen in the figure, the deeper parts of the fault surface corresponding to the O seismic horizon show the largest displacement. The faults in the Erawan field are extensional graben complexes of synthetic and antithetic faults, and some of them have listric shapes. The shale smear parameters, which have been calibrated for normal faulting in soft shale layers, are applicable to the normal faults in the Erawan field.

Shale Smear

Shale smear parameters (Yielding et al., 1997) programmed in FAULTAP (Sorkhabi et al., 2002, 2003) were applied for fault-sealing assessment in the Erawan field (Figure 10). In this chapter, we show the model results of clay content ratio (CCR) and shale smear factor (SSF) on each fault surface in the study field.

Clay content ratio has been termed as shale gouge ratio (SGR) and gouge ratio by Fristad et al. (1997) and Yielding et al. (1997, 1999), respectively. Clay content ratio theoretically considers the percentage of clay that has slipped along fault displacement. In this algorithm, individual layers are multiplied by clay content and then divided by their throw. This is done for each layer, and then it is summed and multiplied by 100 (Figure 10). Various studies in oil fields show that a CCR of at least 20% is required for fault sealing (Yielding et al., 1997; Sorkhabi et al., 2002).

As can be seen in Figure 11, of the five faults analyzed (E-16, E-17, E-18, E-20, and E-27), CCR values of the E-27 fault surface for the interval between seismic horizons F68-3 and K83-2 are relatively lower (yellowish in color code), but those for the E-20 fault surface are relatively higher (bluish in color code), indicating the relative difference in sealing capacity of these faults for the same interval. Nevertheless, CCR values for all faults are greater than 30%, indicating effective fault sealing for the most part. The lowest CCR values (reddish in color code) are found in the northern portions of faults E-18 and E-27, indicating less clay content on fault surface and, hence, more likelihood of across-fault fluid communication.

Shale smear factor is defined as fault throw divided by the thickness of the displaced shale layer (Figure 10) (Lindsay et al., 1993; Gibson, 1994). Outcrop studies show that an SSF value of less than 8 is the threshold for continuous shale smear (Lindsay et al., 1993), and oil fields studies indicate that an SSF value of less than 7 is required for fault sealing (Gibson, 1994; Yielding et al., 1997; Sorkhabi et al., 2002). As can be seen in Figure 12, the fault surfaces for the five faults analyzed (E-16, E-17, E-18, E-20, and E-27) have SSF values of less than 6, indicating sealing potential, except for fault E-27 in which the interval between seismic horizons K83-2 and O demonstrates higher SSF values (reddish in color code) and, hence, the high probability of across-fault communication. The SSF results corroborate the CCR results and, hence, cross-check the model results for fault-sealing assessment.

Clay content ratio and shale smear factor values were plotted against across-fault pressure difference from the RFT data for various sandstone layers. This is illustrated in Figure 13 for CCR values and in Figure 14 for SSF values. Repeat formation tester pressure data for each fault were obtained from wells that were drilled through the footwall block (for example, well N-24 in the case of fault E-27). It is assumed that sandstone horizons in the hanging-wall block were all water saturated and have the same water gradient as the footwall block. Across-fault pressure differences were calculated by subtracting the hanging-wall pressures from the footwall pressures at the same depth. Figure 13 shows that (1) CCR values largely exceed 30–40%, indicating high sealing probability for the faults; and (2) across-fault pressure difference for fault E-27 is the lowest, indicating relatively lower sealing capacity of this fault. The lower pressure difference across fault E-27 is consistent with its lower CCR. Similarly, Figure 14 shows that SSF values for the fault planes are mostly less than 6, indicating the continuity of shale smear and good sealing potential except for fault E-27 (which is similar to CCR results).

Shale-sand Juxtaposition

Wells N-23 and N-24 were drilled through the footwall block of fault E-27. Based on the lithological data from these wells and fault displacement data from seismic horizons, the Allan-type juxtaposition diagram for fault E-27 is shown in Figure 15. Sandstone intervals on the footwall are shown in yellow, and shaly layers are shown in blue. It should be noted that Figure 15 depicts continuous sandstone layers extending over a distance of 1100 m (3600 ft). This is not necessarily true, because the sandstone layers may be discontinuous laterally. Figure 15 has been constructed from the projection of lithologies on the fault plane and the lithology has been derived from two wells' data. Nevertheless, the constructed diagram shows the statistical

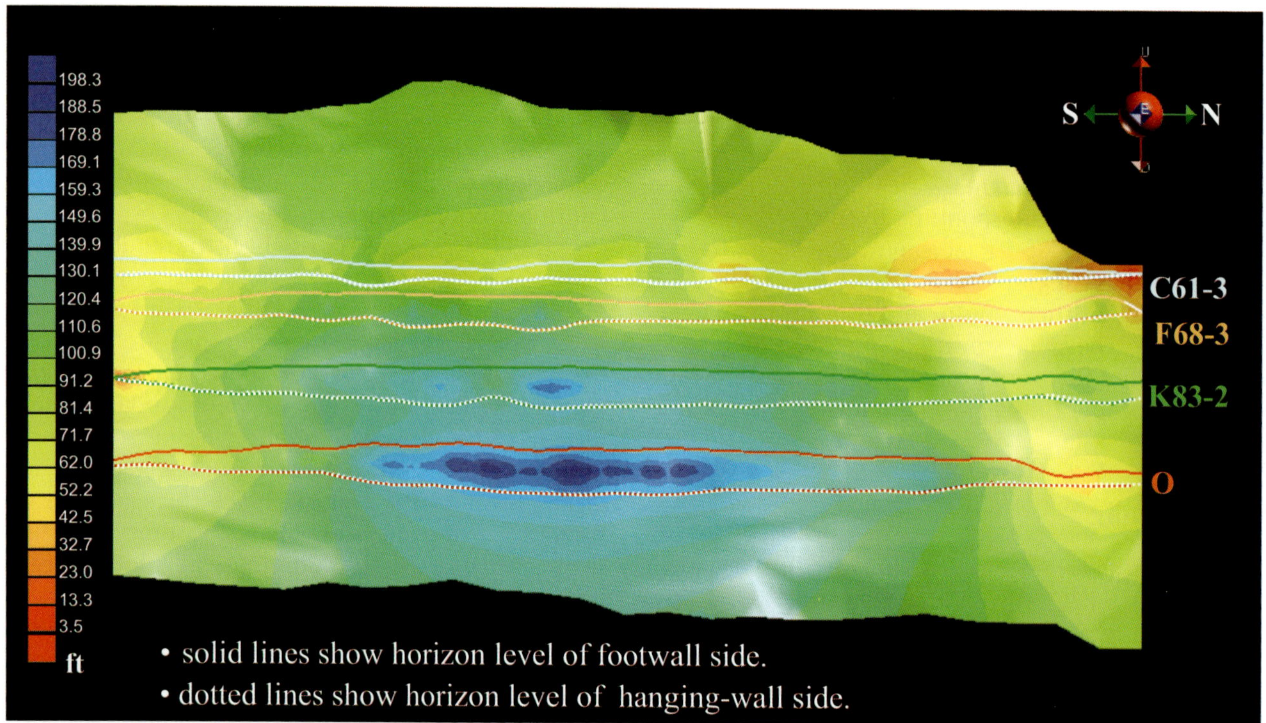

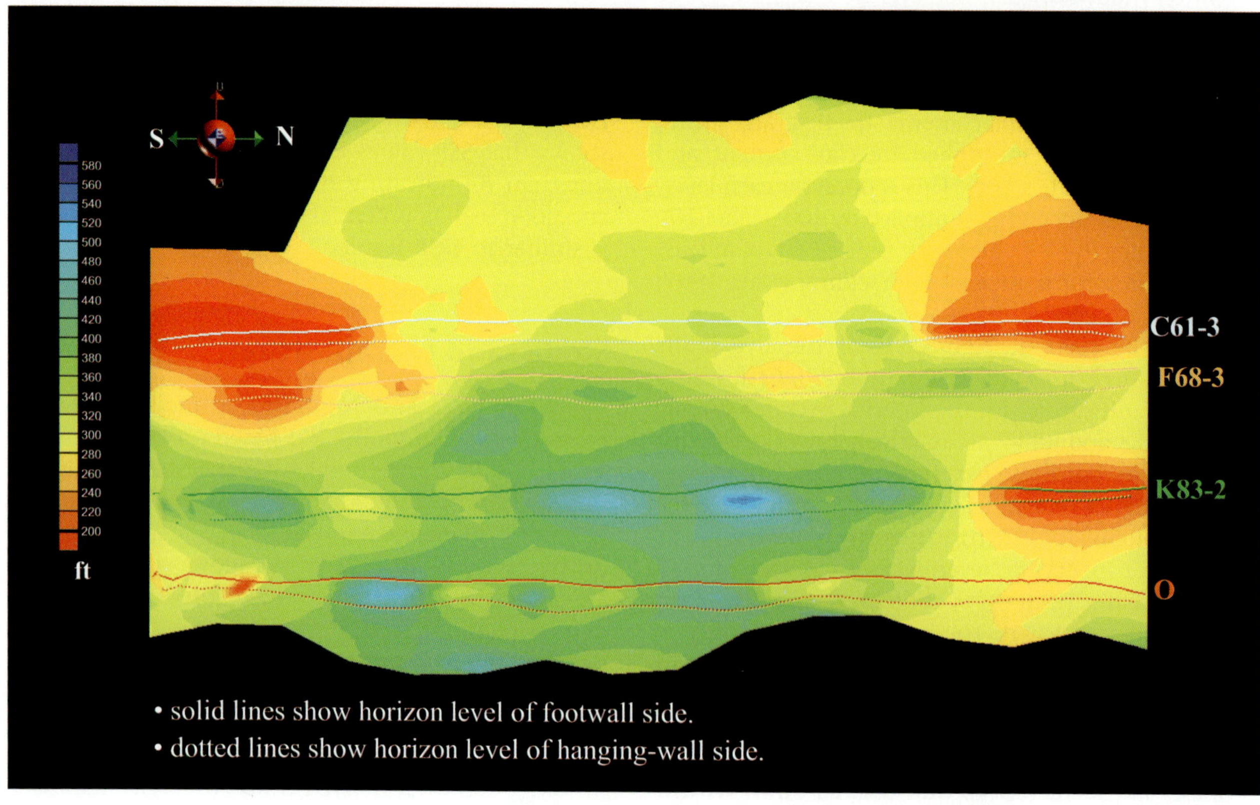

FIGURE 9. Display of fault displacement (in feet) on (a) the E-27 fault and (b) E-16 fault planes.

FIGURE 10. Schematic diagrams showing the concepts and algorithms of shale smear parameters used to assess fault sealing.

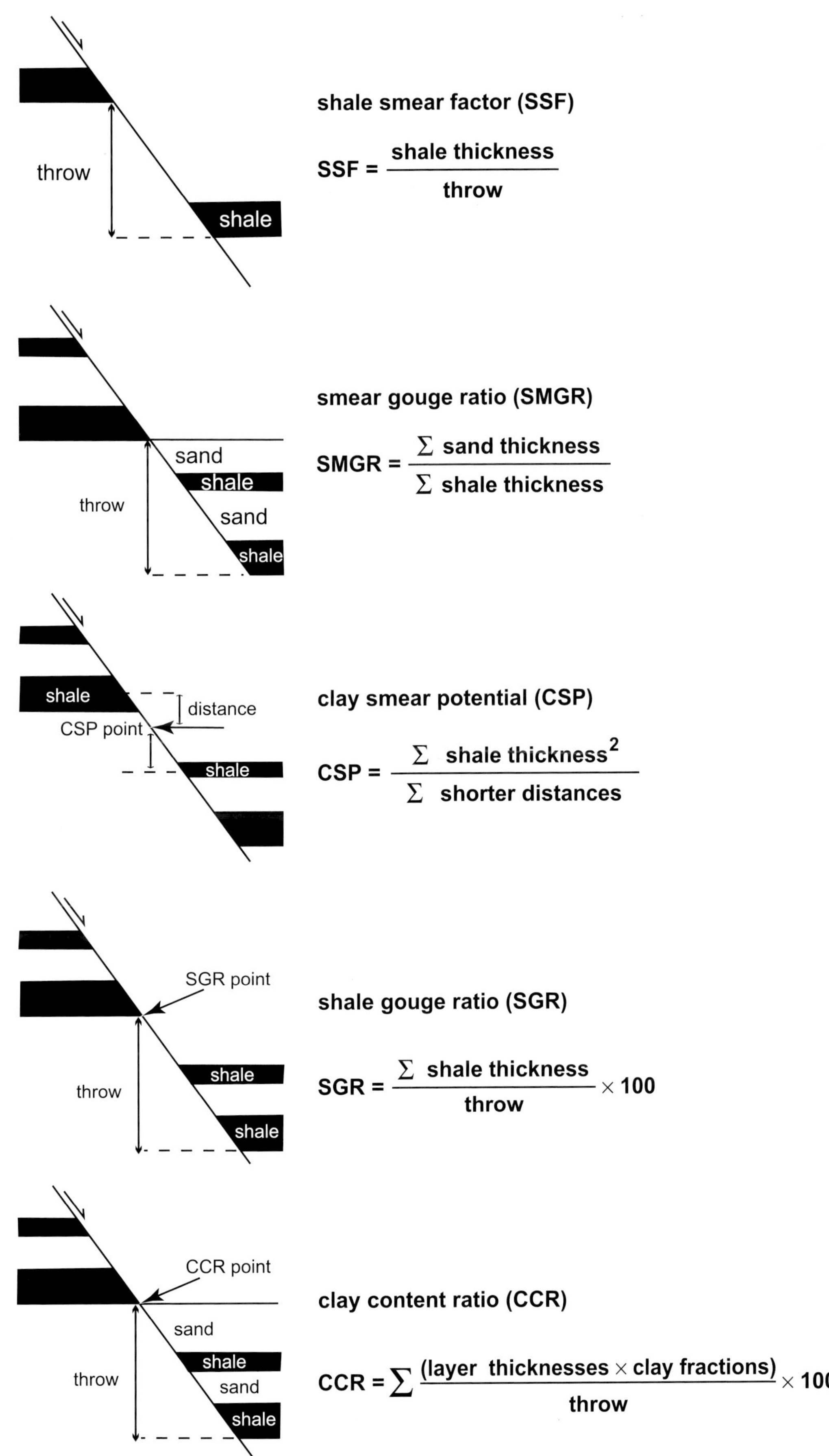

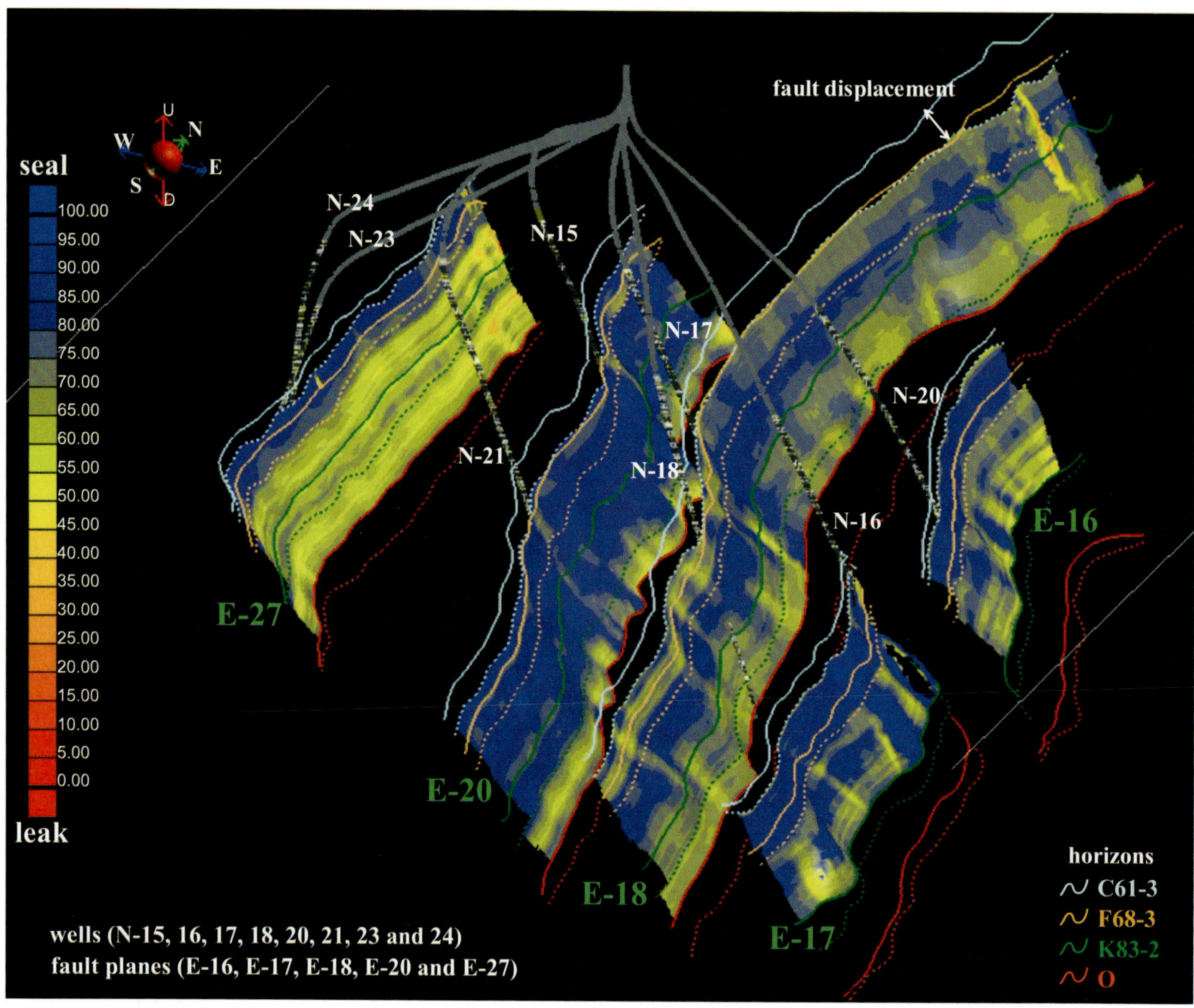

FIGURE 11. Clay content ratio (1–100%) on the footwall blocks of faults E-16, E-17, E-18, E-20, and E-27 in the Erawan N Platform area. Well pathways (N-15 to N-24) and seismic horizons (C61-3, F68-3, K83-2, and O) are also depicted. Clay content ratio values of at least 20% are necessary for across-fault sealing.

tendency of sand distribution on the E-27 fault surface and can be evaluated by drilling additional wells. If our visualization of shale-sandstone juxtaposition on fault E-27 is valid, two potential zones of gas accumulation are expected (shown as ellipsoids in Figure 15). These prospects have not yet been drilled but should be tested. Furthermore, it can be seen in Figure 15 that in the interval between K83-2 and O seismic horizons, water-bearing zones are present, but gas-bearing zones are not. This is probably caused by across-fault leaking via sand-sand contact; this interpretation is consistent with higher values of SSF for this interval (see Figure 12, fault E-27).

Because fault entrapment may be caused by sand-shale juxtaposition as well as shale smear, it is important to sort out the control of these two factors on the trapping or leaking potential of the faults. Table 1 shows the results of juxtaposition and shale smear parameters for several gas-bearing and water-bearing sandstone zones in fault E-27 (16 zones), fault E-20 (12 zones), fault E-18 (12 zones), and fault E-16 (7 zones). The shale smear parameters include SSF, smear gouge ratio (SMGR), clay smear potential (CSP), SGR, and CCR. Algorithms for calculating these parameters are shown in Figure 10. Table 1 shows how shale smear parameters and sand-shale juxtaposition can or cannot explain the gas accumulations for each fault trap. For example, in the case of fault E-27, out of 16 sandstone units, 3 are gas-bearing zones (evaluated from well N-24). The juxtaposition can explain two of these, but the third gas accumulation can only be explained by shale smear sealing expressed by SSF, SMGR, CSP, SGR,

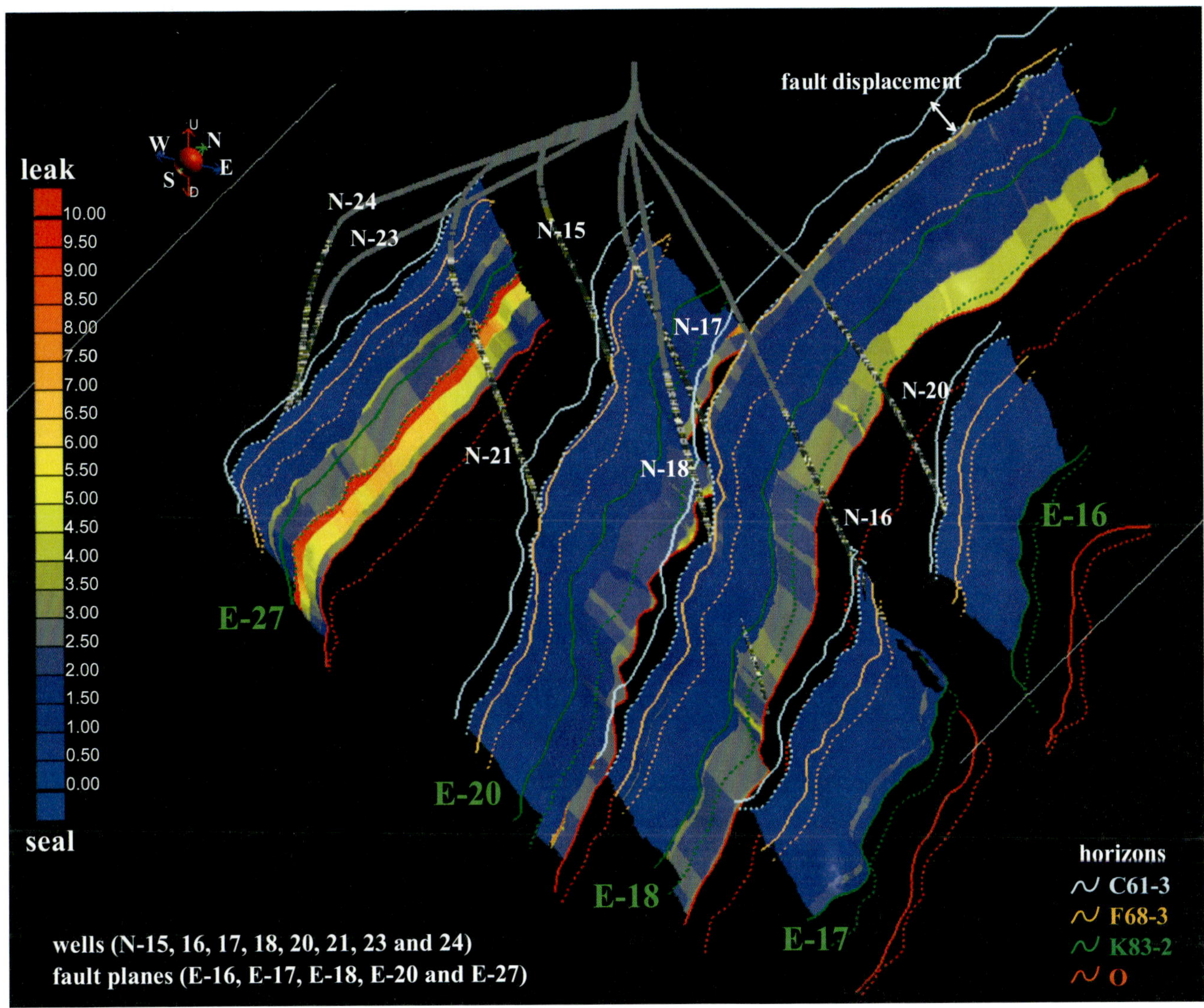

FIGURE 12. Shale smear factor values displayed on the footwall blocks of faults E-16, E-17, E-18, E-20, and E-27 in the Erawan N Platform. Well pathways (N-15 to N-24) and seismic horizons (C61-3, F68-3, K83-2, and O) are also depicted. Shale smear factor values less than 7 are necessary for across-fault sealing.

and CCR values. The minimum threshold value for each shale smear parameter was used to interpret the sealing or leaking likelihood as follows: SSF greater than 7 leak; SMGR greater than 2 leak; CSP greater than 10 leak; and SGR and CCR less than 20% leak (Sorkhabi et al., 2002). Out of 15 gas pay zones, 8 can be explained on account of sand-shale juxtaposition, but the shale smear parameters can explain all of the accumulations.

Fault-seal Failure Probability

Shale smear parameters such as CCR are useful to assess fault sealing associated with the formation of a fault. However, by reference only to shale smear parameters, it is not known if a given fault or various portions of a fault are still sealing or leaking features under the current stress regime. To address this problem, Sorkhabi et al. (2003) suggested an algorithm called the fault-seal failure probability (FSFP) that integrates CCR with the tendency of the fault to slip or dilate in the contemporary stress field.

The following equations are used to calculate FSFP:

$$\mathrm{FSFP_{slip}} = \left(1 - \frac{\mathrm{CCR}}{\max(\mathrm{CCR})}\right)\frac{T_s}{\max(T_s)}$$

$$\mathrm{FSFP_{dilation}} = \left(1 - \frac{\mathrm{CCR}}{\max(\mathrm{CCR})}\right)\frac{D_s}{\max(D_s)}$$

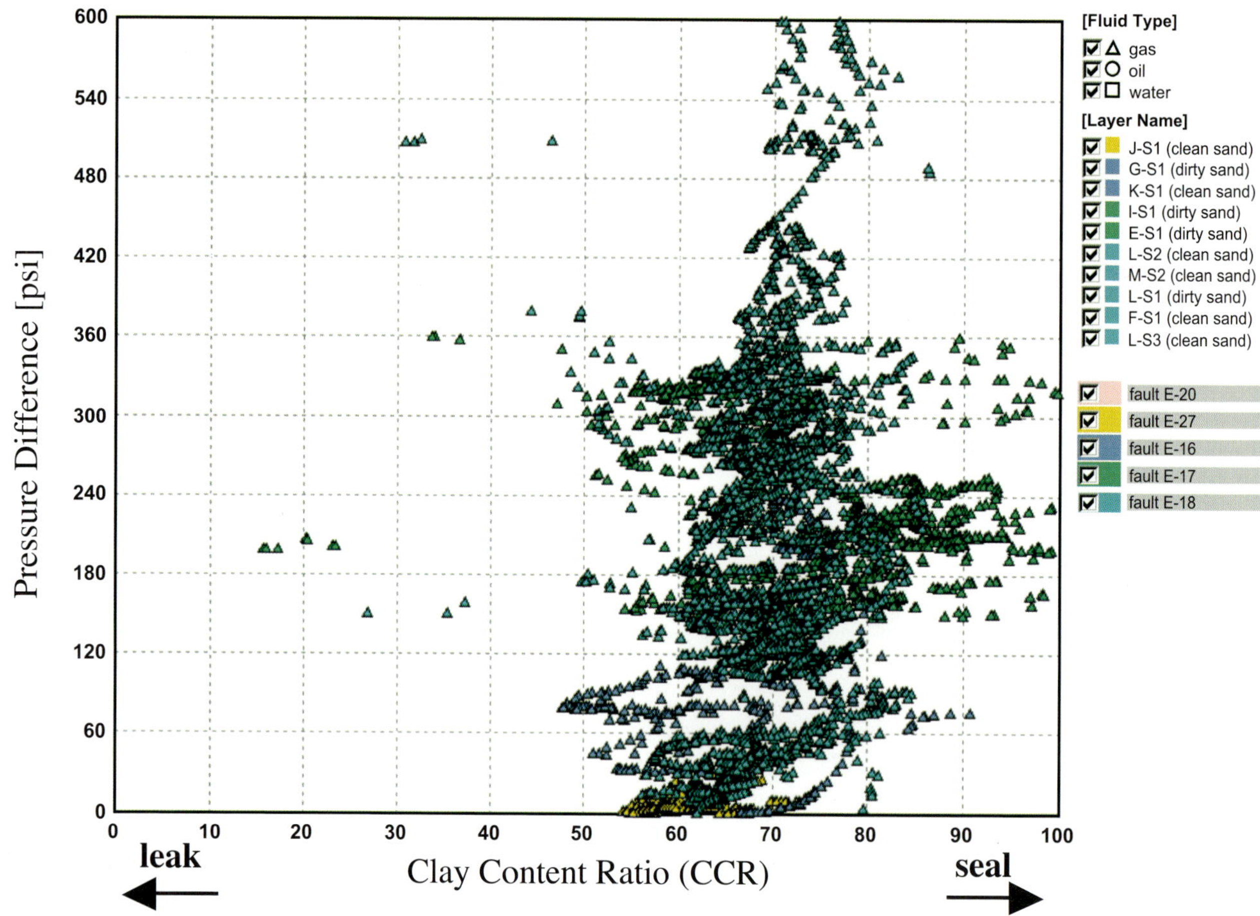

FIGURE 13. Plot of CCR values against across-fault pressure difference for the studied faults E-16, E-17, E-18, E-20, and E-27 in the Erawan gas-condensate field. Clay content ratio values of at least 20% are necessary for across-fault sealing.

where CCR is clay content ratio (in fraction); T_s is slip tendency; and D_s is dilation tendency. Max values for CCR, T_s, and D_s are maximum values obtained in each data set (i.e., dividing each CCR, T_s, and D_s by its maximum value yields a normalized value for each factor). Fault-seal failure probability values range from 0 to 1, with 1 being the maximum failure probability of fault seal either caused by fault slip ($FSFP_{slip}$) or dilation ($FSFP_{dilation}$).

Slip tendency (T_s) and dilation tendency (D_s) are algorithms suggested by Ferrill et al. (1999) to evaluate the relative strength or weakness of fault seal under in-situ stress conditions.

Slip tendency (shear failure) is the ratio of shear stress and the normal stress (Ferrill et al., 1999):

$$T_s = \sigma_s/\sigma_n$$

Dilation tendency (failure by extension fracturing) is derived as follows (Ferrill et al., 1999):

$$T_d = (\sigma_1 - \sigma_n)/(\sigma_1 - \sigma_3)$$

where σ_s is shear stress; σ_n is normal stress; and σ_1, σ_2, and σ_3 are three principal stress axes acting on the fault surface.

In this study, σ_s and σ_n were calculated as follows:

$$\sigma_n = \sigma_v \sin^2 \alpha + \sigma_h \cos^2 \alpha$$

$$\sigma_s = (\sigma_v - \sigma_h) \sin 2\alpha/2$$

where σ_v is the vertical stress; σ_h is the horizontal stress; and α is the angle between σ_n and the fault plane.

Because we deal with normal faults, σ_1 or the maximum principal stress was taken as the vertical stress (σ_v) exerted by the overburden thickness, and the horizontal stress (σ_h) was considered as the minimum principal axis (σ_3) and was calculated as follows:

$$\sigma_h = K_0\ \sigma_v$$

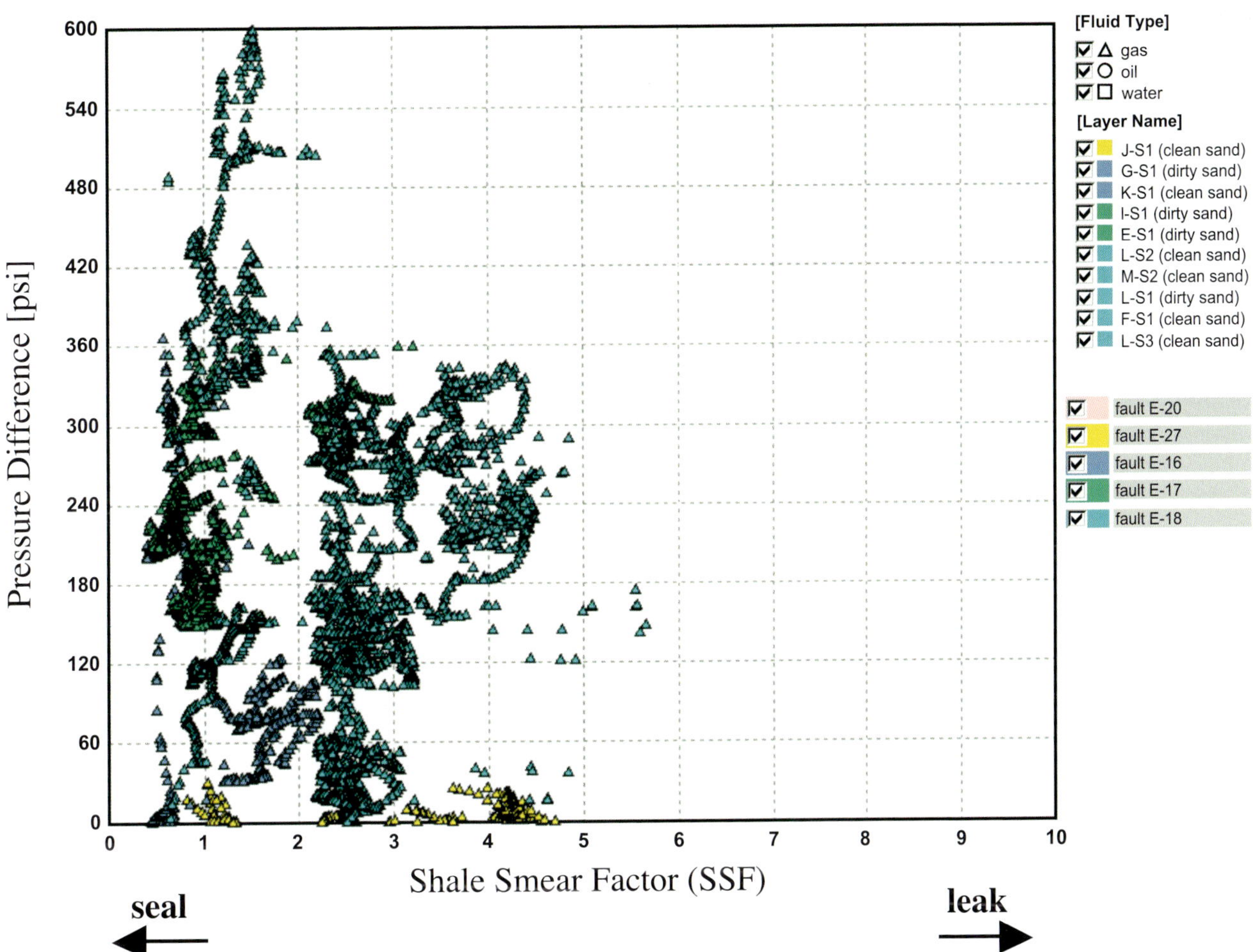

FIGURE 14. Plot of SSF values against across-fault pressure difference for the studied faults E-16, E-17, E-18, E-20, and E-27 in the Erawan gas-condensate field. Shale smear factor values less than 7 are necessary for across-fault sealing.

where K_0 is a calibration factor depending on depth and rock type; its value was determined experimentally at JNOC and was incorporated in the software.

Figure 16 shows the model results of slip tendency and dilation tendency (normalized to 1) for fault E-16. Figure 17 shows the CCR values (Figure 17a) and the model results of $FSFP_{slip}$ (Figure 17b) and $FSFP_{dilation}$ (Figure 17c) for the same fault (E-16).

Well N-20 penetrated the gas reservoirs in the footwall block of the east-dipping E-16 fault. This long-reach well targeted a small dip closure between faults E-16 and E-17 to the east of the N Platform. Clay content ratio values for fault E-16 range from 50 to 70 (Figure 17a), which are within the fault-sealing envelope. Nevertheless, apart from the gas-bearing sandstones, water-bearing zones also occur along the fault (Figure 17a), although CCR values are high enough to indicate sealing. Therefore, it is important to know whether the water-bearing zones are breached seals. The model results of $FSFP_{slip}$ and $FSFP_{dilation}$ can be helpful in this regard. These results indicate lower values for both $FSFP_{slip}$ and $FSFP_{dilation}$ on fault E-16, implying that water-bearing sandstone is less likely to have been breached across the fault. In other words, water accumulations in the fault-seal zone reflect the migration of water in the fault compartment in the basin. Because the gas-bearing sandstones lie below the water-bearing sandstones (Figure 17a), and because gas has higher buoyancy pressure than water, the gas does not seem to have migrated along its maximum vertical direction (to shallowest levels in the basin). This may reflect a relatively recent generation of the gas in the basin and/or the existence of effective cap rocks in gas-bearing sandstones to prevent the accumulated gas from upward migration.

In Figure 17a and b, we also suggest pods of gas accumulation at the same depth levels of producing zones but on the southern and northern portions of fault E-16. This suggestion is based on the fact that these portions have very low $FSFP_{dilation}$ values (Figure 17b) and, hence, higher probability of fault-seal integrity.

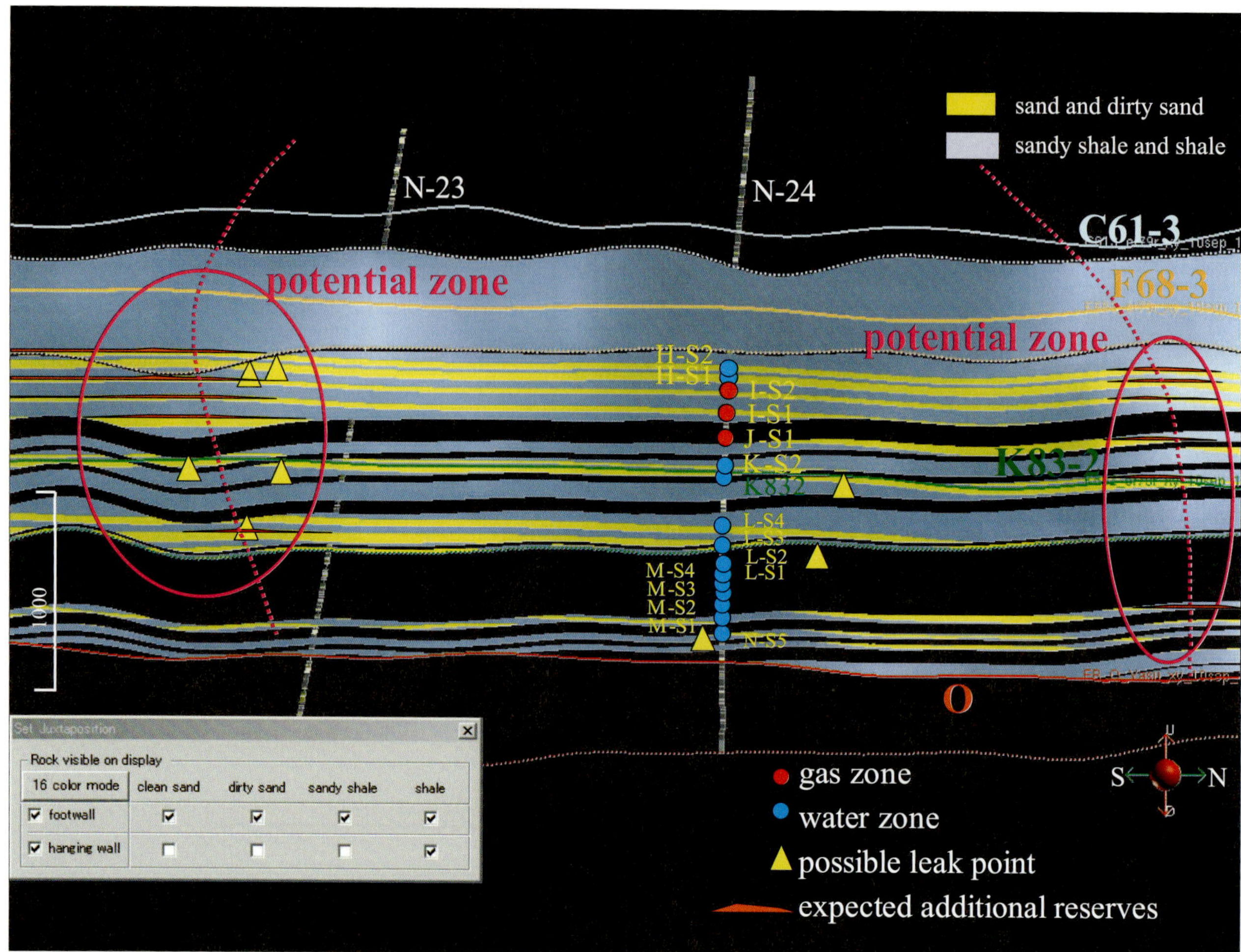

FIGURE 15. Allan-type diagram for fault E-27 showing the shale-sand juxtaposition of footwall vs. hanging-wall sides. Based on juxtaposition, possible (untested) accumulation zones are shown, and dashed red lines show the direction of suggested well drilling to produce these gas accumulations. Possible leak points on the fault surface are also shown.

Table 1. Explanation of gas-bearing fault traps based on juxtaposition and shale smear parameters in the Erawan field, Gulf of Thailand.

Fault Block	*Number of Zones*		*Can be Explained by*					
	Water Zone	*Gas Zone*	*JUXT*	*SSF < 7*	*SMGR < 2*	*CSP > 10*	*SGR > 20*	*CCR > 20*
E-27	13	3	2	3	3	3	3	3
E-20	9	3	3	3	3	3	3	3
E-18	6	6	1	6	6	6	6	6
E-16	4	3	2	3	3	3	3	3

JUXT = shale-sand juxtaposition; SSF = shale smear factor; SMGR = smear gouge ratio; CSP = clay smear potential; SGR = shale gouge ratio; CCR = clay content ratio. See Figure 10 for the algorithms of the shale smear parameters. The number of gas zones is valid for the interval investigated and controlled by well data. The actual number may be higher.

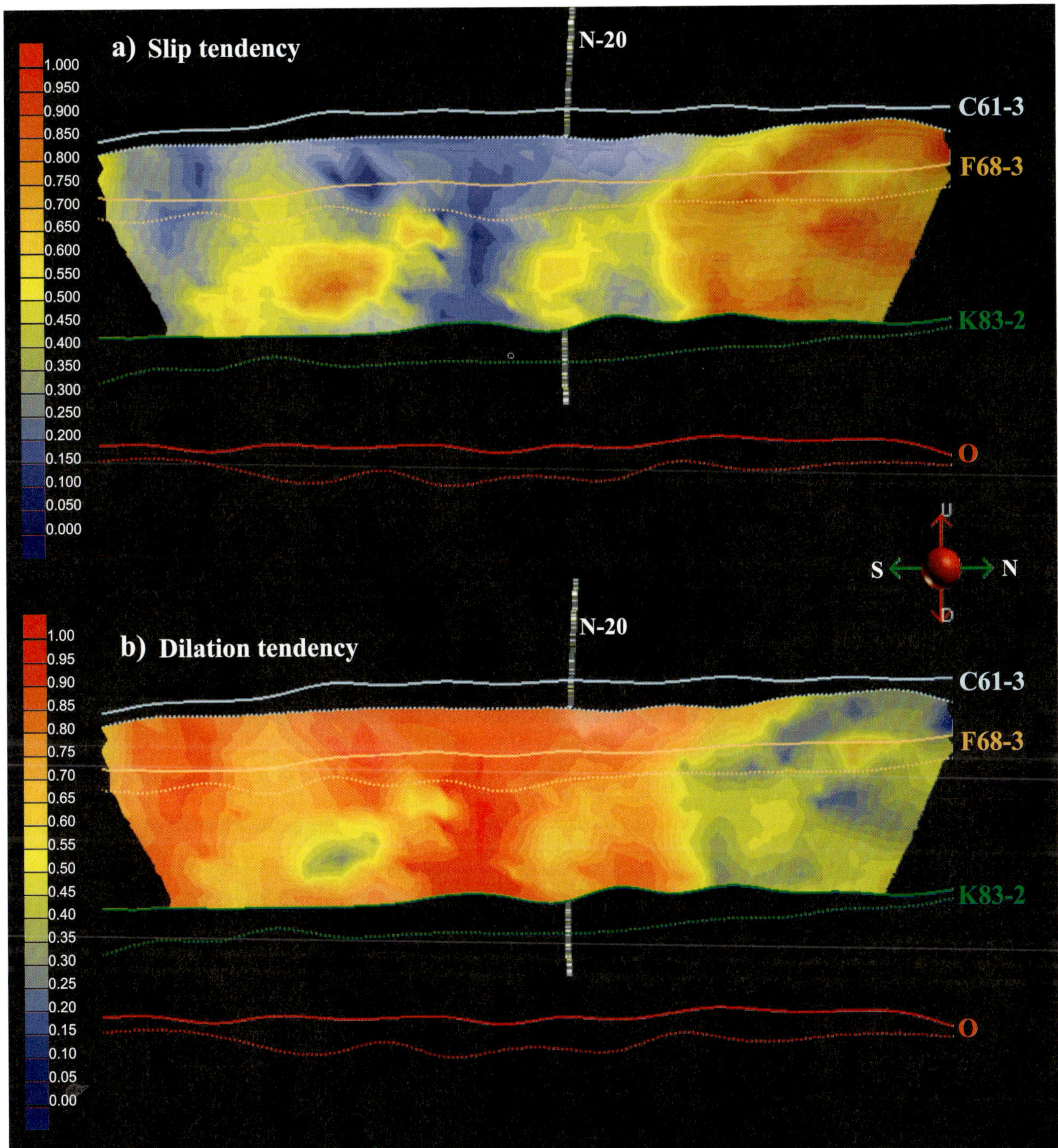

FIGURE 16. Slip tendency (a) and dilation tendency (b) for fault E-16. N-20 is the well drilled on the footwall of the fault. Both slip tendency and dilation tendency are normalized values (1 being maximum tendency). Algorithms for calculating slip tendency and dilation tendency (after Ferrill et al., 1999) are discussed in the text.

SUMMARY AND CONCLUSIONS

Fault-sealing analysis was carried out for five faults in the Erawan gas-condensate field in the Pattani Basin offshore the Gulf of Thailand. This field features a series of east- and west-dipping normal faults cutting through the Miocene clastic sediments. The trapping faults formed after the rift phase ended, in response to sag tectonics. Older, rift-related faults are commonly truncated by the middle Tertiary unconformity (MTU). In the Erawan N Platform area, several deviated wells have been drilled through the footwall blocks of the studied fault (E-16, E-17, E-18, E-20, and E-27). Fault-sealing assessment was made through the sand-shale juxtaposition

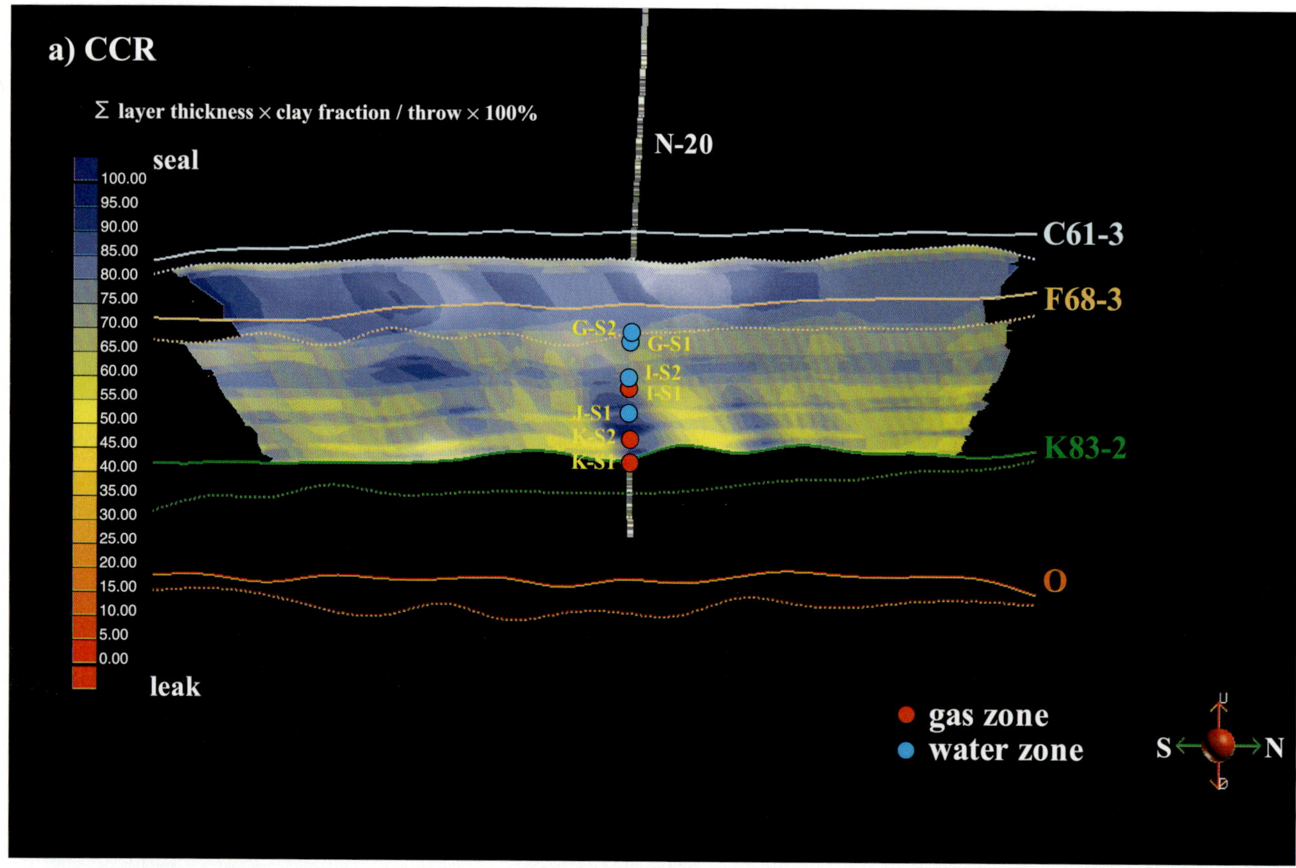

FIGURE 17. (a) Display of CCR values (in percentage) on fault E-16. Fault-seal failure probability values normalized to 1 (1 being the maximum failure probability) are also displayed for the same fault based on $FSFP_{slip}$ (integration of CCR and fault slip probability) (b) and $FSFP_{dilation}$ (integration of CCR and fault dilation probability) (c). Note that both gas-bearing (red circles) and water-bearing zones (green circles) show low FSFP, indicating the integrity of across-fault seal under the contemporary stress regime in the Erawan field.

diagram, shale smear parameters, and FSFP. The shale smear parameters, including SSF values of less than 6 and CCR values of greater than 30%, indicate that all the faults have provided across-fault barriers and along-strike reservoir compartments, except for some intervals on fault E-27, where SSF values are higher (>7), and only water-bearing zones occur. The results of shale smear parameters, suggesting across-fault sealing in many of the sandstone reservoirs, are important for the prediction of potential prospects and for the drilling strategy of infill wells for further development of the field. The trapping efficiency of sand-shale juxtaposition and shale smear parameters were compared for 15 gas pay zones trapped by faults E-16, E-18, E-20, and E-27. The juxtaposition could explain 8 of these gas accumulations, whereas the shale smear could account for all of them. The CCR parameter was combined with fault slip tendency and fault dilation tendency to derive FSFP in the contemporary stress regime of the field. This approach was applied to fault E-16, in which both gas pay zones and water-bearing zones occur in reservoirs with higher CCR values (i.e., higher sealing efficiency). Low values of $FSFP_{slip}$ and $FSFP_{dilation}$ suggest less likelihood of fault-seal breaching on this fault.

ACKNOWLEDGMENTS

This case study could not have been possible without the support rendered by Unocal Thailand, Inc. (the operator for the Erawan gas-condensate field); we thank the staff of Unocal Thailand who contributed to the preparation of seismic and log data necessary for this study. We also appreciate the permission given by Mitsui Oil Exploration Co. Ltd (MOECO), Unocal Thailand, Inc., and Japan National Oil Corporation (JNOC) (presently Japan Oil, Gas and Metals National Corporation) to publish this chapter. This study was carried out using the FAULTAP© software designed at the JNOC. We are grateful to Kazuo Nakayama, John Sneider, James Turner, and Yoshihiro Tsuji for reviewing this

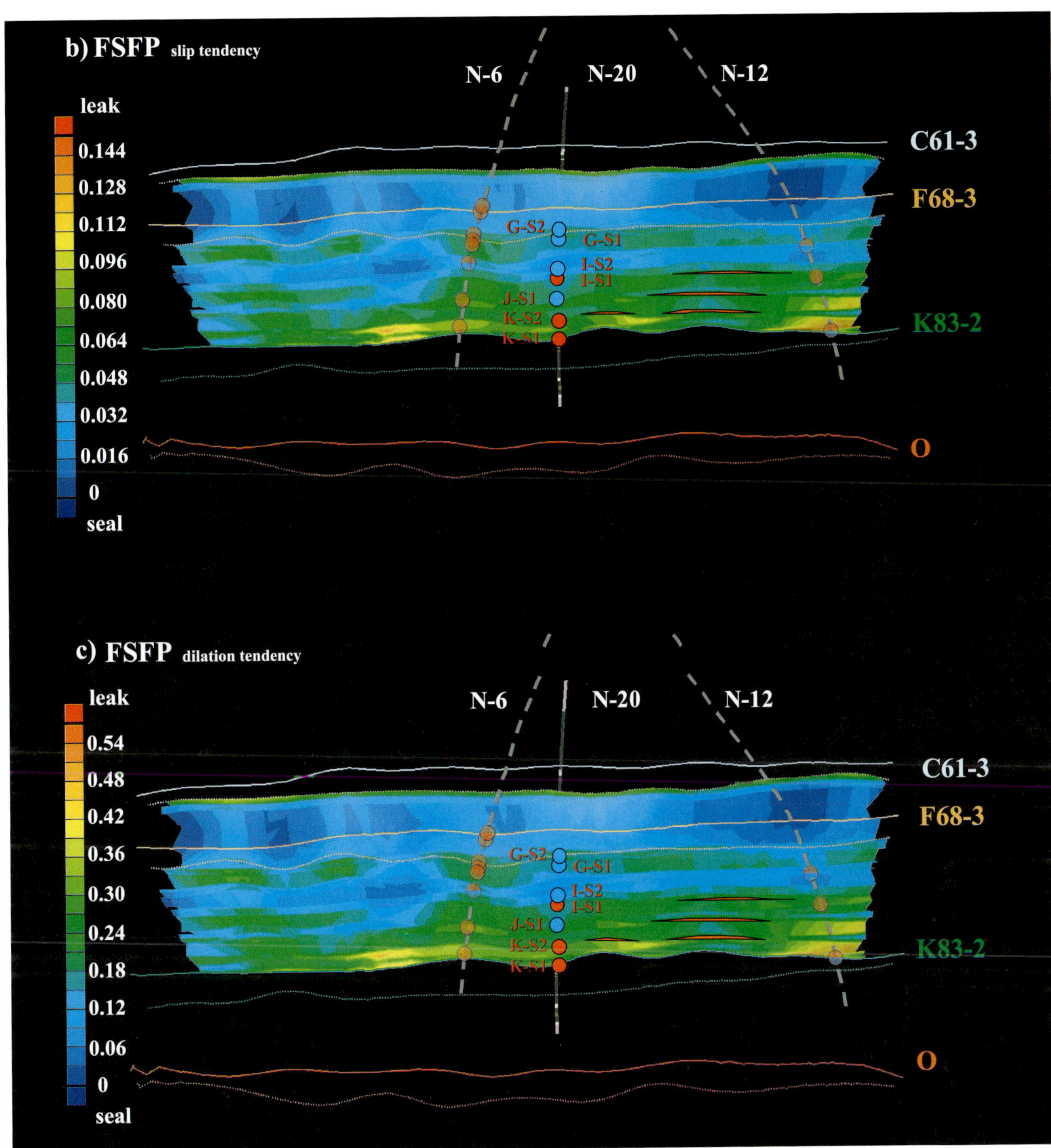

Figure 17. (cont.).

chapter and for their comments that improved the accuracy and clarity of the chapter; nevertheless, we alone are responsible for the content of the chapter.

REFERENCES CITED

Allan, U. S., 1989, Model for hydrocarbon migration and entrapment within faulted structures: AAPG Bulletin, v. 73, p. 803–811.

Ferrill, D. A., J. Winterle, G. Witmeyer, D. Sims, S. Colton, A. Armstrong, and A. P. Morris, 1999, Stressed rock strains groundwater at Yucca Mountain, Nevada: Geological Society of America Today, v. 9, p. 1–8.

Fristad, T., A. Groth, G. Yielding, and B. Freeman, 1997, Quantitative fault seal prediction: A case study from Oseberg Syd, *in* P. Moller-Pedersen and A. G. Koestler, eds., Hydrocarbon seals: Norwegian Petroleum Society Special Publication 7, p. 107–124.

Gibson, R. G., 1994, Fault-zone seals in siliciclastic strata of the Columbus Basin, offshore Trinidad: AAPG Bulletin, v. 78, p. 1372–1385.

Jardin, E., 1997, Dual petroleum systems governing the prolific Pattani Basin, offshore Thailand: Proceedings of the Petroleum Systems of Southeast Asia and Australasia Conference, May 21–23, 1997: The Indonesian Petroleum Association, p. 351–363.

Kongwung, B., and S. Ronghe, 2000, Reservoir identification and characterization through sequential horizon mapping and geostatistical analysis: A case study from the Gulf of Thailand: Petroleum Geoscience, v. 6, p. 46–57.

Leo, C. T. A. N., 1997, Exploration in the Gulf of Thailand in deltaic reservoirs, related to the Bongkot field, *in* A. J. Fraser, S. J. Matthews, and R. W. Murphy, eds., Petroleum geology of Southeast Asia: Geological Society (London) Special Publication 126, p. 77–87.

Lian, H. M., and K. Bladley, 1986, Exploration and development of natural gas, Pattani Basin, Gulf of Thailand: Transactions of the Fourth Circum-Pacific Energy and Mineral Resources Conference, August 17–22, 1986, Singapore: AAPG, p. 171–181.

Lindsay, N. G., F. C. Murphy, J. J. Walsh, and J. Watterson, 1993, Outcrop studies of shale smears on fault surfaces: International Association of Sedimentologists Special Publication 15, p. 113–123.

Mial, A. D., 2002, Architecture and sequence stratigraphy of Pleistocene fluvial systems in the Malay Basin, based on seismic time-slices analysis: AAPG Bulletin, v. 86, p. 1201–1216.

Minezaki, T., and K. Moriyama, 2002, The origin of hydrocarbon and carbon dioxide in the gas fields for the Pattani trough, the Gulf of Thailand (in Japanese with English abstract): Journal of the Japanese Association for Petroleum Technology, v. 67, p. 16–29.

Morley, C. K., 2002, Evolution of normal faults: Evidence from seismic reflection data: AAPG Bulletin, v. 86, p. 961–978.

Okui A., M. Hara, H. Fu, and K. Takayama, 1996, SIGMA-2D: A simulator for the integration of generation, migration, and accumulation of oil and gas: Proceeding of the VIIIth International Symposium on the Observation of the Continental Crust through Drilling, Tsukuba, Japan: National Research Institute for Earth Science and Disaster Prevention/Science and Technology Agency, Geological Survey of Japan/Agency of Industrial Science and Technology, p. 365–368.

Sorkhabi, R., S. Hasegawa, S. Iwanaga, and M. Fujimoto, 2002, Sealing assessment of normal faults in clastic reservoirs: The role of fault geometry and shale smear parameters: Journal of the Japanese Association for Petroleum Technology, v. 67, p. 576–589.

Sorkhabi, R., S. Iwanaga, M. Fujimoto, and S. Hasegawa, 2003, Sealing assessment of normal faults clastic reservoirs: Modeling the petrophysical and stress attributes of faults: Journal of the Japanese Association for Petroleum Technology, v. 68, p. 291–304.

Turner, J. W., 2002, Gas in antithetic fault blocks of the Pattani trough, Thailand: An underappreciated play? (abs.): AAPG Annual Meeting Program, v. 12, p. A179.

Turner, J. W., and S. P. Lambert, 2003, Erawan field: Looking back through thirty years of geologic thinking (abs.): Southeast Asia Petroleum Exploration Society (SEAPEX), Conference Proceedings, September 15–17, 2003, Singapore, (CD-ROM).

Western, P. G., and B. Davis, 1993, Erawan field: A young elephant, the first twenty years: The 5th Asian Council on Petroleum Conference Proceedings 2–6 November 1993, Bangkok, p. 95–123.

Yawwapapong, C., S. P. Lambert, J. Turner, and S. Jeeradete, 2002, Erawan gas field, challenge and victory (abs.): Thailand Petroleum Conference 2002 Unocal Technical Papers, September 26–27, 2002, Department of Mineral Fuels, Bangkok.

Yielding, G., B. Freeman, and D. T. Needham, 1997, Quantitative fault seal prediction: AAPG Bulletin, v. 81, p. 897–917.

Yielding, G., J. A. Overland, and G. Byberg, 1999, Characterization of fault zones for reservoir modeling: An example from the Gullfaks fields, northern North sea: AAPG Bulletin, v. 83, p. 925–951.

5

Sims, D. W., A. P. Morris, D. A. Ferrill, and R. Sorkhabi, 2005, Extensional fault system evolution and reservoir connectivity, *in* R. Sorkhabi and Y. Tsuji, eds., Faults, fluid flow, and petroleum traps: AAPG Memoir 85, p. 79–93.

Extensional Fault System Evolution and Reservoir Connectivity

Darrell W. Sims

Center for Nuclear Waste Regulatory Analyses (CNWRA), Southwest Research Institute, San Antonio, Texas, U.S.A.

Alan P. Morris

Department of Earth and Environmental Science, University of Texas at San Antonio, San Antonio, Texas, U.S.A.

David A. Ferrill

Center for Nuclear Waste Regulatory Analyses (CNWRA), Southwest Research Institute, San Antonio, Texas, U.S.A.

Rasoul Sorkhabi[1]

Technology Research Center, Japan National Oil Corporation, Chiba, Japan

ABSTRACT

Sandbox analog modeling experiments provide new insights into the effects of fault geometry on reservoir connectivity. During progressive distributed extension, three phases of fault system evolution are apparent. In Phase I, geometrically simple faults nucleate rapidly at a large number of sites throughout the deforming region. This is followed by Phase II, in which faults link and increase in trace length. Phase III is characterized by a quasi-steady-state nucleation and linkage of faults. Reservoir connectivity has many components; here, we focus on fault-controlled connectivity, which can be viewed from two complementary perspectives: rock mass connectivity (continuity of rock between and around faults) and fault network connectivity. Which of these perspectives is adopted depends on whether faults cutting the reservoir act as barriers to flow (e.g., in highly porous sandstone reservoirs) or conduits for flow (e.g., in fractured carbonate reservoirs). We use two measures of fault-controlled connectivity: (1) a fault density measure derived from the number of intersections

[1]*Present address:* Energy & Geoscience Institute, University of Utah, Salt Lake City, Utah, U.S.A.

DOI:10.1306/1033717M853132

between faults and potential flow paths and (2) the ratio of the number of fault tips to the number of faults. Taken together, these characteristics convey both the transmissivity characteristics and the ultimate leakiness of the reservoir.

INTRODUCTION

Connectivity analyses are performed on spatial systems to determine which features are connected (Jackson, 1997). Reservoir connectivity has many components; for example, lithological heterogeneities, represented by facies variation, and structural heterogeneities, such as faults and fractures. Here, we focus on fault-controlled reservoir connectivity. Reservoir connectivity is of critical importance to hydrocarbon exploration (modeling migration pathways and charge characteristics) and production (calculating drained volumes, production rates, and enhanced recovery strategies). The deterministic approach to understanding reservoir connectivity is to build a virtual model of the reservoir, using all available data, then perform model realizations under different simulated conditions. Using percolation theory, flow characteristics of reservoirs can be simulated using a probabilistic approach (King et al., 2001). Both of these methods have limitations: deterministic models are complex, are difficult to build, and require long computation times; probabilistic models require that a model be built from simplifications and averages of the field data and provide little geometric information about the reservoir. In our method, we attempt to fill the gap between these modeling strategies using a geometric approach to extract connectivity information in the form of size, shape, and orientation of fault-bounded blocks.

The rock mass and fault network contributing to reservoir connectivity form a complex continuum. We define rock mass connectivity as the degree to which the rock mass is connected between and around faults and the fault network connectivity as the degree to which the fault system is connected. An unfaulted reservoir has maximum rock mass connectivity and zero fault network connectivity. Rock mass and fault network connectivities are complementary; as fault systems develop, rock mass connectivity decreases as fault network connectivity increases. Fault-controlled reservoir connectivity (rock and faults as an integrated whole) is the result of these opposing tendencies, as expressed through the reservoir rock type, and is an important determinant of the ease with which fluid can flow through a reservoir in any direction.

Regardless of whether faults act as barriers to or conduits for fluid flow, they impart a directional quality to the fluid-flow characteristics of the reservoir (Ferrill et al., 2000). In the case of sealing faults, dominant flow will occur in the rock mass, whereas in the case of transmissive faults, dominant flow will occur in the fault zones. To evaluate reservoir flow characteristics, it is necessary to analyze both rock mass and fault network connectivities.

In the short-term case of hydrocarbon production, directional variation and complexity of flow paths is important for quantifying production rates and drained volumes. In the long-term case of natural hydrocarbon migration, the mere existence of pathways may be more important than the complexity or tortuosity of those paths.

In this chapter, we present a quantitative analysis of the directional character of fault-controlled reservoir connectivity (Figure 1). The analysis can be viewed from the perspective of either rock mass connectivity or fault network connectivity. We demonstrate the approach by applying it to a physical analog model experiment simulating extensional faulting and to natural examples of extensional fault systems from the Volcanic Tableland in Owens Valley (California, United States) and the Canyonlands National Park (Utah, United States). Our analysis is applicable to any situation in which the connectivity of either the rock mass or the fault network is of interest.

INTERSECTION ANALYSIS AND ROCK MASS CONNECTIVITY

Where faults are present in a reservoir, they modify the flow paths. Where faults are barriers to flow, the more faults that need to be crossed or circumvented, the less efficiently fluid flows. The resulting transmissivities are likely to be anisotropic, especially where faulting has imparted a directional fabric. To evaluate this effect, we analyze the number of faults encountered by different hypothetical flow paths. We refer to this purely geometric analysis as intersection analysis. Flow paths and faults are considered in plan view to be two populations of lines with consistent orientations and spacing. This is directly analogous to the analysis used by Waiting et al. (2003) to evaluate seismic rupture risk to underground drift tunnels. The number of intersections between the two populations of lines in a sample circle depends on the spacing of the two populations and the angle between them and is approximated by the relationship (Waiting et al., 2003)

$$N_i \approx \text{round}\left\{\frac{\sin\alpha}{d_B}\left(\sum L_A\right)\right\} \tag{1}$$

where N_i is the number of intersections in the sample circle; α is the angle between two line populations; ΣL_A

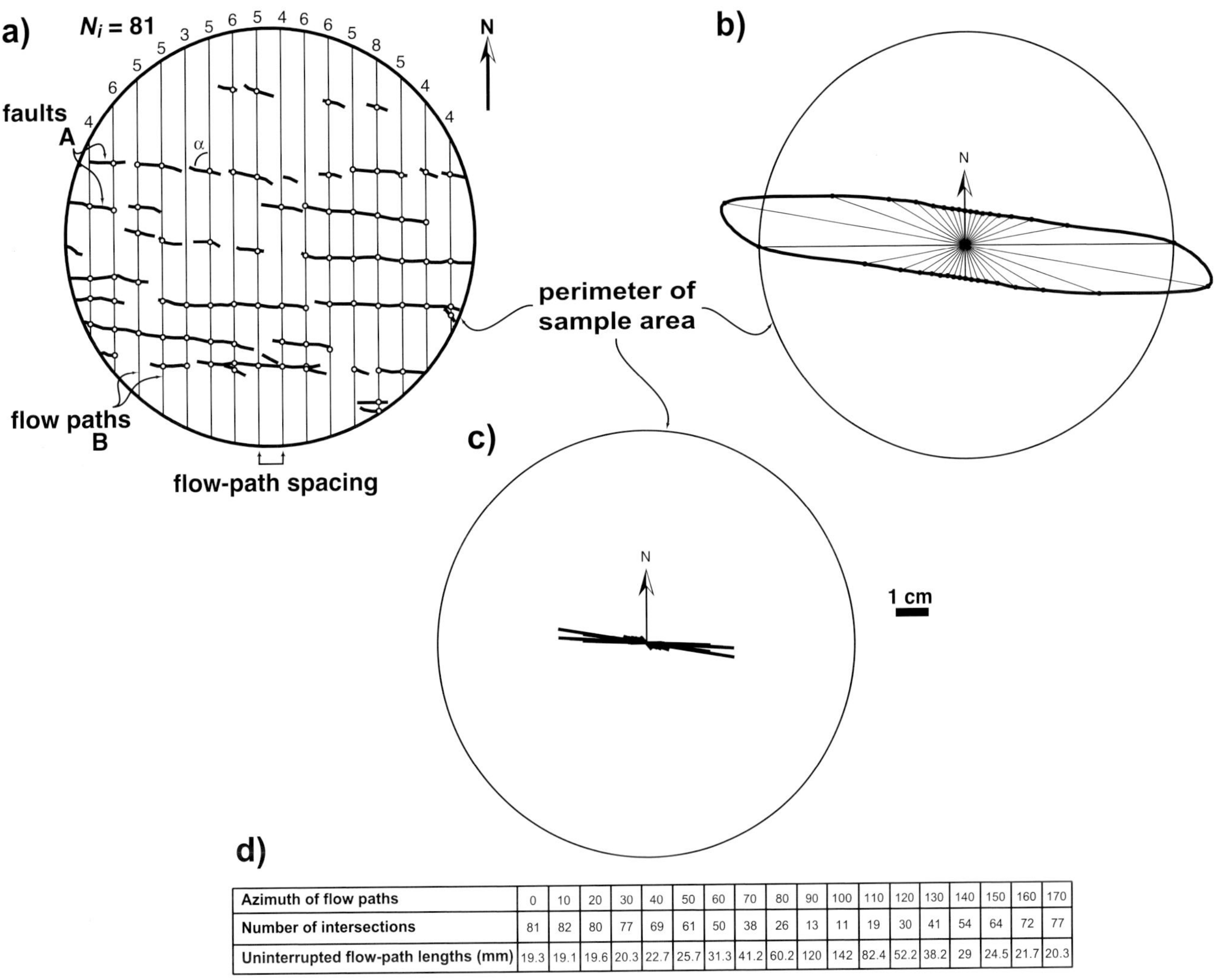

Azimuth of flow paths	0	10	20	30	40	50	60	70	80	90	100	110	120	130	140	150	160	170
Number of intersections	81	82	80	77	69	61	50	38	26	13	11	19	30	41	54	64	72	77
Uninterrupted flow-path lengths (mm)	19.3	19.1	19.6	20.3	22.7	25.7	31.3	41.2	60.2	120	142	82.4	52.2	38.2	29	24.5	21.7	20.3

FIGURE 1. Flowpath-fault trace intersection analysis. (a) Simple geometric basis for the analysis. The 16 sampling lines are evenly spaced, straight, hypothetical flow paths with an azimuth of 0°. Small, open circles indicate where the flow paths intersect with simulated fault traces (dark, discontinuous lines), and the numbers at the north end of each flow path are the number of intersections along that flowpath; the total (N_i) is given at the top left. (b) Azimuthal plot of flowpath lengths in (d) for a fixed fracture pattern and variable flowpath azimuths. The lengths of the rays are proportional to the uninterrupted flowpath length in that direction. (c) Azimuthal plot of fault trace length by segment-length-weighted mean azimuth for simulated fault traces in (a). Each ray has a length proportional to a fault length; its orientation is obtained by dividing the fault into segments, measuring each segment length and orientation, and then taking the weighted average of orientation. (d) Table of results for the analysis using 16 flow paths and the technique described in the text.

is the total length of line population A in the sample circle; and d_B is the spacing of line population B.

Shifting the line populations such that they are not symmetric in the circle has little effect on the resulting number of intersections (±1) except in the limiting cases where the lines of both populations coincide. In addition, maintaining the total length and orientation of population A but permitting variable spacing and distribution (i.e., lines terminating in the circle and not at the perimeter) has little effect on the number of intersections (±1).

Equation 1 can be applied to natural fault traces and hypothetical flow paths (line populations A and B, respectively; Figure 1a). Grouping the fault trace population into orientation bins and summing the trace length in each bin provide the value ΣL_A in equation 1. The value for d_B is arbitrary and only affects results if it is too large with respect to the sample circle to effectively characterize the fault population. In general, smaller magnitude values of d_B produce more detailed results. For comparative purposes, the ratio of d_B to the diameter of the sample circle is the same for all fault systems being compared here (d_B/diameter = 0.059). The angle α is the difference between the median fault trace orientation in the orientation bin being analyzed and the azimuth of the flow-path population. Each orientation bin is analyzed for flow paths that cover the entire range of flow-path azimuths (in 10° increments from 0 to 170°), and the number of intersections for

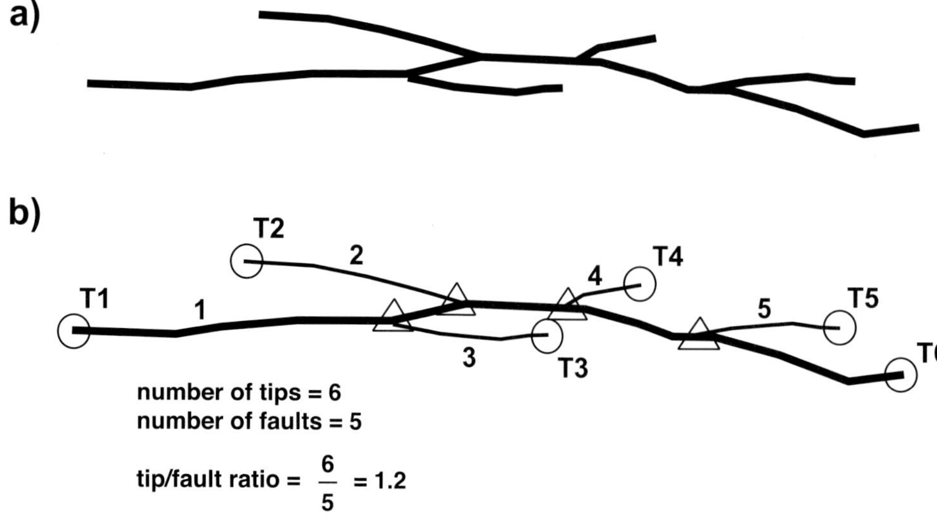

Figure 2. Convention for counting multisegment, linked faults. (a) Fault traces used in (b). (b) The bold line, shown as 1, represents the trace of the longest, continuous fault; at its zero-displacement terminations are tips [circles T1 and T6]. Splays [fine lines (2, 3, 4, 5)] are counted separately, and if they terminate in the sample area, these points are considered tips [circles T1–T6]. Branch points (triangles) are not counted. Thus, the tip-to-fault ratio here is 1.2. No a priori knowledge of fault evolution is required to apply this technique, and it can be used on any fault system.

each flow-path orientation is summed. This number represents the number of potential interruptions a given unidirectional flow path would experience as a result of faulting. Dividing the total length of flow paths by this number yields the average, uninterrupted length of flow path and is a measure of rock mass connectivity in that direction (Figure 1d). Plotting these lengths vs. their azimuth yields both an intuitively useful and quantitatively transferable plot of the size, shape, and orientation of interfault blocks (Figure 1b). The data in Figure 1d can be used to calculate the eigenvectors of the distribution to yield an ellipse. However, a simple azimuthal plot (Figure 1b) clearly shows the connectivity trends and details that may be informative in the interpretation of results.

FAULT NETWORK CONNECTIVITY

Although fault density and orientation affect reservoir connectivity, whether the rock mass or the fault network is ultimately connected across the entire sample area depends on the continuity of faults. Faults that are consistent in orientation but with short trace lengths may have a high fault density, but if they are not continuously linked, the rock mass remains connected. This is an important consideration when analyzing migration pathways, because efficiency of flow is less important than whether a flow path exists. Fault network connectivity can be represented by an azimuthal plot of fault length (Figure 1c). In this plot, each ray represents a fault; the length of the ray is proportional to the fault's length. The orientation of the ray is obtained by dividing the fault into segments, measuring each segment length and orientation, and then taking a weighted average of orientation. In this study, we also use the ratio of fault tips to the number of individual faults as a measure of the extent to which faults are connected (fault network connectivity). Linked faults are counted as illustrated in Figure 2. Fault tips are points of no displacement, and branch points are not counted (Figure 2).

DATA ANALYSIS

So that data from fault network and intersection analyses are comparable, they should be collected from the same area and at the same scale. A square oriented so that one side is at a low angle to the dominant fault strike is drawn on the fault trace interpretation. All faults and fault segments falling inside this square are used to generate the tip-to-fault ratio, fault length, and fault length vs. frequency data sets. A circle inscribed in this square is used to extract the fault length vs. orientation data sets, which are required for intersection analysis and fault orientation plots. Sample area size is chosen so that many features of interest are included. For example, in the case of the analog models, the sample area was chosen to include as many faults developed in the area of distributed extension as possible but to exclude the larger magnitude, graben-bounding faults (Figure 3a, b).

SANDBOX MODEL EXPERIMENTS

In these experiments, an aluminum base plate was overlain by two plastic sheets, one of which was mobile and the other fixed (Figure 3a). The fixed sheet was cut with an asymmetrical curvature, and the plastic sheets were underlain by a rubber sheet that was attached to both plastic sheets. A 5-cm-thick sand pack composed of alternating colored (acrylic dyed) and plain white (not dyed) layers of quartz sand (commercially named Oklahoma #1) was sifted over this to simulate a mechanically brittle portion of Earth's crust. The length scale for the model is 10^{-5}, so that 1 cm (0.4 in.) in the model represents 1 km (0.62 mi) in nature. A rectilinear grid was added to the uppermost layer to provide a reference frame to track displacements during model evolution. The model was photographed from above at regular time and displacement intervals during deformation to document the map-view structural evolution. After completion of deformation, the model was dampened with water, sliced, and photographed to document the internal structure at the time of maximum displacement.

MODEL RESULTS

Model deformation produced an interconnected network of closely spaced imbricate normal faults that developed between the bounding graben systems (Figure 3b). Faults in this system formed with dip predominantly in the direction of movement of the mobile sheet (Figure 3c) and with geometries similar to imbricate normal fault systems in the North Sea (e.g., Rouby et al., 1996) and the Basin and Range Province (e.g., Ferrill et al., 1998). Propagation of fault tips past each other in opposite directions and the subsequent formation and breaching of relay ramps were the primary mechanism by which fault sinuosity evolved (e.g., Ferrill et al., 1999a) and led to the development of a complex, anastomosing system of normal faults.

Our experimental design results in distributed extensional faulting (Figure 3d) and the development of a regional structural gradient (Figure 3c). Fault system evolution falls into three phases: (1) rapid nucleation of faults (Phase I); (2) growth and linking of faults (Phase II); and (3) displacement accumulation on large faults with steady-state initiation and linkage of small faults (Phase III). These phases are similar to the stages of Cowie (1998), although our Phase II incorporates elements of her stages 2 and 3, and our Phase III represents a more evolved system than her stage 3. Concurrent with and contingent on these fault development phases are changes in fault-controlled reservoir connectivity and local structural relief. Our model configuration causes a deformation front to sweep through the model from south to north (in the reference frame of the model; see Figure 3c). This is manifest in the first three steps of Figure 4a (model elongations 2.0–3.0 cm [0.8–1.2 in.]) and is preserved throughout the model run as a regional slope to the south (Figure 3c). Both the number and the lengths of the longest faults increase through the deformation sequence. However, the longest measurable fault length in the sample window (truncation length) is approximately 120 mm (4.7 in.), and the maximum length saturates at this value while the number of long faults continues to increase (Figure 4b).

Phase I

Rapid nucleation of faults is characterized by a similar increase in the number of fault tips (Figure 5a, b) and total fault length (Figure 5c). This phase of model evolution is also characterized by a rapid decrease in the maximum rock mass connectivity of the system (Figures 4a, 5d). These characteristics are the result of rapid and widespread nucleation of new faults throughout the deforming medium. Distinctions between phases of fault system evolution are arbitrary and not perfectly delineated. However, we choose the peak of the tip-to-fault ratio curve as the boundary between rapid nucleation and incipient linking of faults.

Phase II

Once a distribution of faults has developed, continued deformation is accomplished primarily by growth and linkage of existing faults. The longest faults in the distribution grow at a faster rate than the smaller faults (Figure 4b). The number of new fault tips generated decreases from its Phase I maximum, as does the ratio of fault tips to the number of faults (Figure 5a, b). Total fault length continues to increase, but more slowly than during the initial burst of activity (Figure 5c). Maximum rock mass connectivity decreases more slowly in Phase II than in Phase I (Figure 5d). The number of faults present in the population stabilizes in this phase, although new faults are forming (Figure 5a). This is a result of fault linkage and is a pattern that persists into Phase III.

Phase III

The distinction between phases II and III is gradational, but in our analysis, it is marked by long, multisegment faults traversing the area of analysis and being truncated against the margins of the sample area (throughgoing faults; Figures 4b, 5a). Concurrently, interactions between these complex faults renew fault nucleation to accommodate displacement anomalies. This is reflected in the low-amplitude fluctuations in the number of new fault tips and in the tip-to-fault ratio, which decreases very slowly in the tip-to-fault ratio range near 1 (Figure 5a, b). The rate of increase in total fault length also shows a slight increase in Phase III,

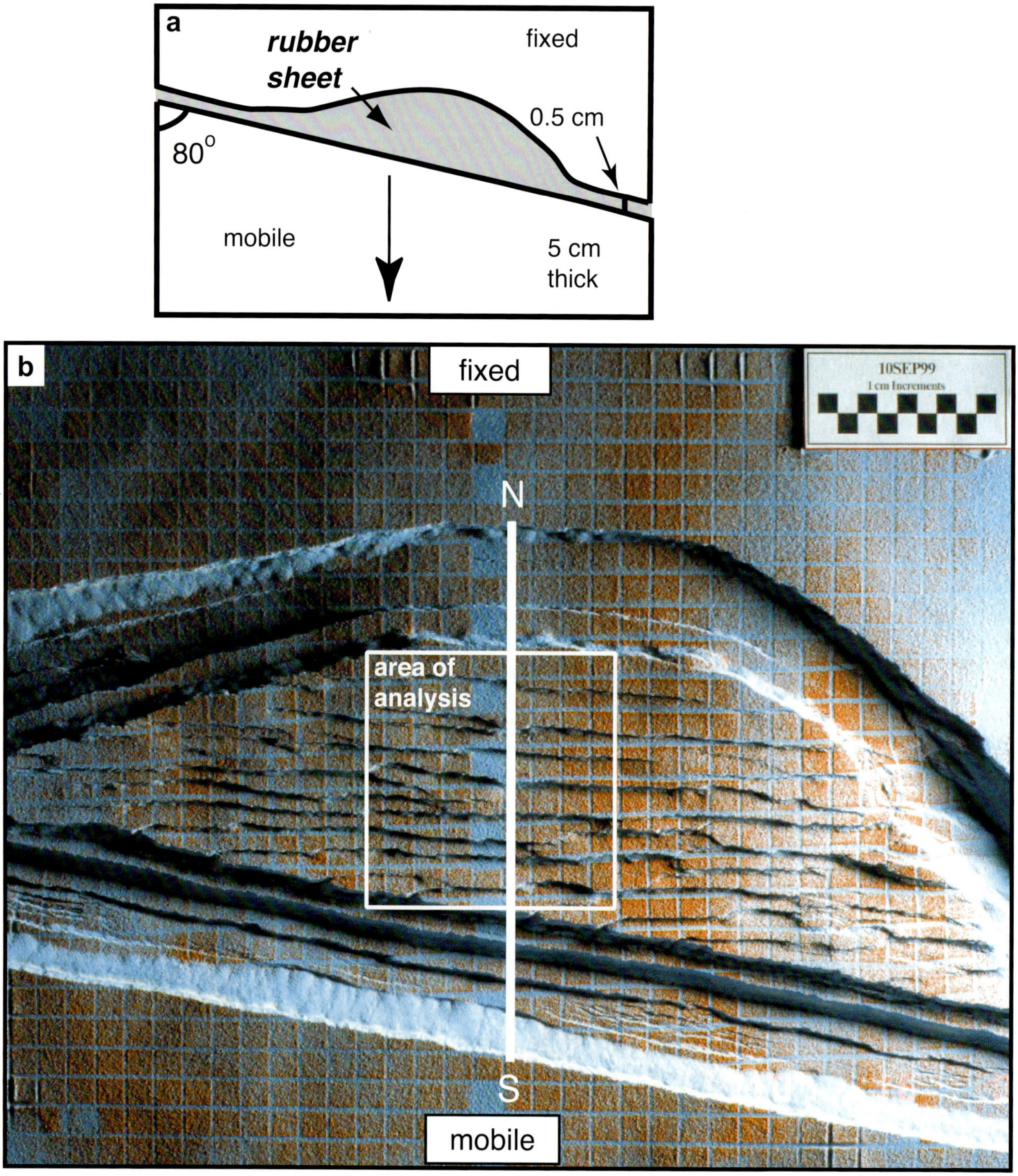

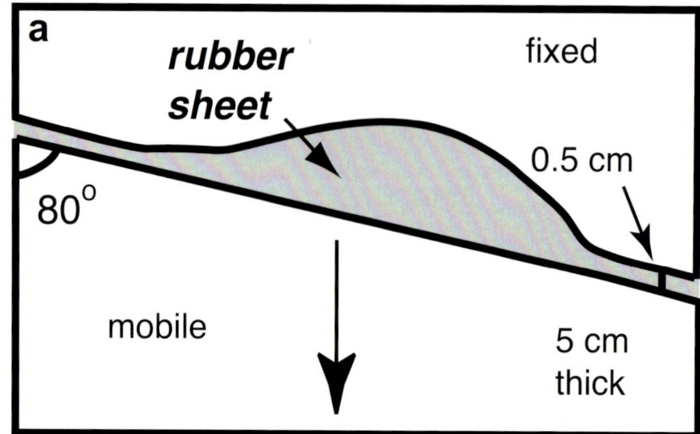

Figure 3. Sandbox model 10SEP99. (a) Plan view of predeformation model configuration. (b) Plan view of model after 6-cm (2.4-in.) total model displacement. (c) Cross section along line NS. Lowermost white and blue layers are prekinematic layers, yellow and red are synkinematic layers simulating basin fill, and the uppermost white layer is postkinematic. Scale is 5-cm (2-in.) increments. (d) Low-angle, oblique view of the model surface after 6.5-cm (2.6-in.) displacement. The view is from right to left of (b).

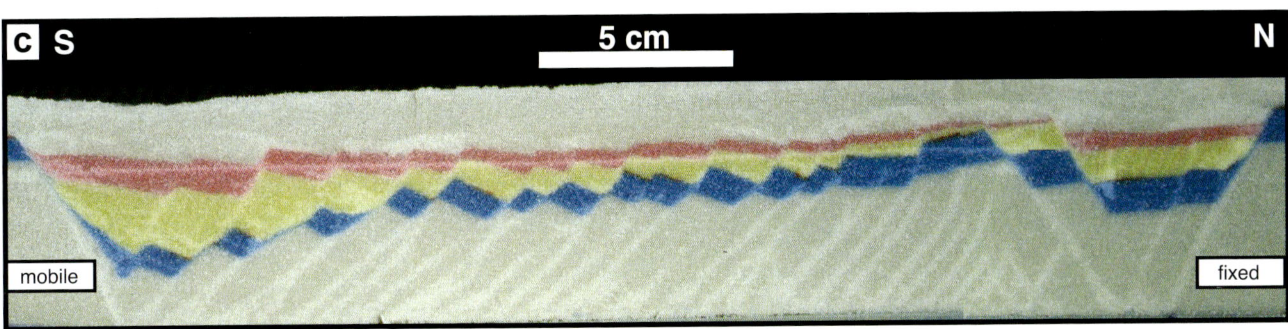

FIGURE 3. (cont.).

and the decline in maximum rock mass connectivity steepens somewhat (Figure 5c, d), both reflecting the formation of small faults accommodating local stress perturbations.

DISCUSSION OF MODEL RESULTS

After the initial phase of rapid fault nucleation, the fault system evolves by lateral propagation of existing faults, linkage of faults, and a continuous, slow nucleation of faults in response to local stress perturbations. Large faults grow at the expense of small faults and come to dominate the system (Figure 4b). The rock mass connectivity plot (Figure 4a) provides a good representation of the size and shape of a closely connected rock mass and the fault network connectivity plot (Figure 4a). Throughout the evolution of the fault system there is a clear decrease in the rock mass connectivity, and a concomitant increase in the fault network connectivity exists (Figure 4a).

In the case where faults are barriers to flow and primary fluid flow is within the rock mass, information regarding directional rock mass connectivity is required to evaluate drained volumes and the well distribution required to produce hydrocarbons economically from a field. A randomly placed well in the modeled fault system after 2.5 cm (1 in.) of model length change would potentially drain an area 7 cm (2.7 in.) long parallel to the dominant fault strike and 1 cm (0.4 in.) perpendicular to the dominant fault strike (rock mass connectivity plot for model elongation in Figure 4a).

Where faults are transmissive, information about the connectivity of the fault network is required. Model results indicate that throughgoing faults do not develop until Phase III of fault system development, which is marked by a tip-to-fault ratio near 1 (Figure 5a, b) and an increasing dominance of long faults (Figure 4b). A well that intersects a fault after 2.5 cm (1 in.) of model elongation will potentially drain a fault network that extends 5 cm (2 in.) parallel to the dominant fault strike and less than 0.5 cm (0.2 in.) perpendicular to the dominant fault strike (fault network connectivity plot for 2.5-cm [1-in.] model elongation in Figure 4a).

These are two end members of a spectrum of behavior, and real reservoirs will have both rock mass and faults of variable permeability. In these cases, the connectivity analysis should be weighted by relative permeability.

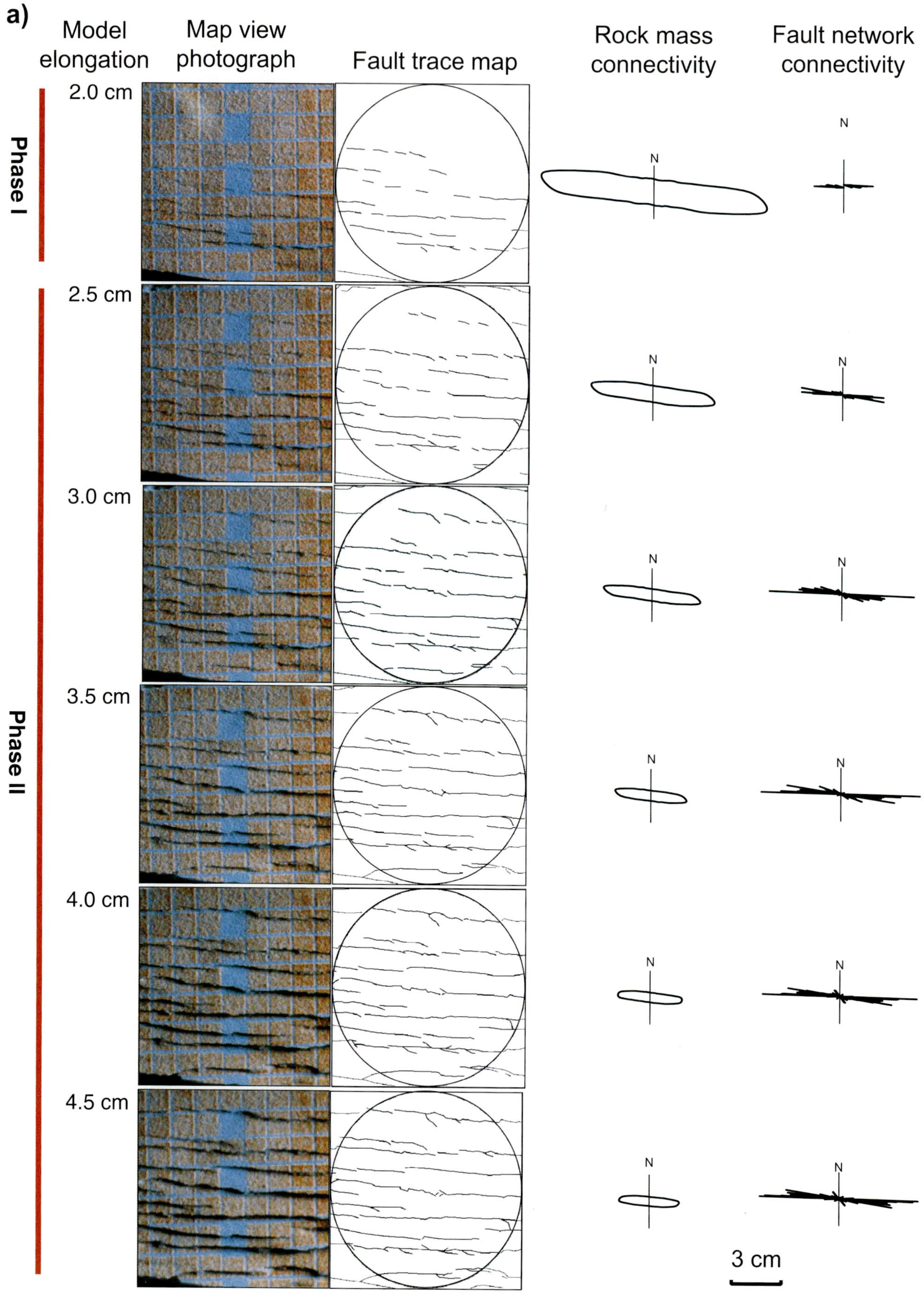
a)
Model elongation
Map view photograph
Fault trace map
Rock mass connectivity
Fault network connectivity
2.0 cm
2.5 cm
3.0 cm
3.5 cm
4.0 cm
4.5 cm
Phase I
Phase II
N
3 cm

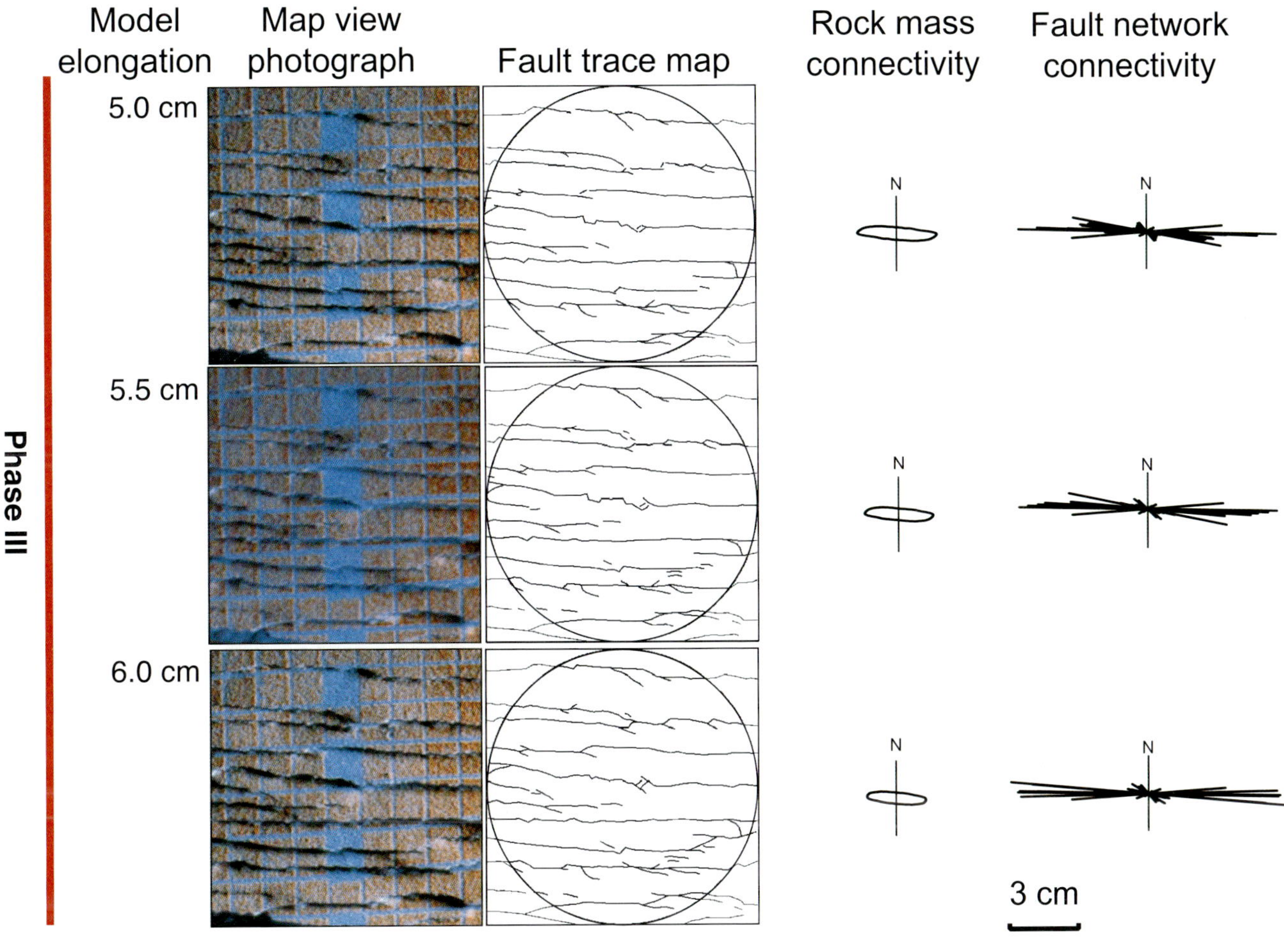

FIGURE 4. Summary of model results. (a) Vertical views of the sand pack at 0.5-cm (0.2-in.) intervals of total model elongation. Line drawings are fault traces interpreted from the photographs. Square outlines (119 mm [4.7 in.] on a side) are the sample areas used in fault length analysis (b) and tip vs. fault analysis (Figure 5a–c). Circle outlines (119-mm [4.7-in.] diameter) are the sample areas used in the intersection analysis. Each interpretation-photograph pair is labeled with its model elongation. Rock mass connectivity is characterized for each model step by an azimuthal plot of uninterrupted flow-path lengths. Fault network connectivity is characterized for each model step by an azimuthal plot of fault length vs. segment-length-weighted mean azimuth. (b) Fault length vs. percentile. Individual fault lengths are plotted in accordance with their contribution to total fault length, ordered from small faults to the left and larger faults to the right. The approximate maximum length of a single fault measurable in the sample box (truncation length) is shown. Lines are labeled according to model elongation. Because intervals between model elongations are equal, each line can be thought of as representing a time increment. The upper portions of the lines have progressively shallower slopes (until the truncation length is reached), indicating that the longer faults gain more length with each time step than the smaller faults. Dashed lines show the approximate boundaries of the three fault system development phases defined in the text.

NATURAL EXAMPLES

Volcanic Tableland (California, United States)

The Volcanic Tableland in northern Owens Valley, California, is a plateau capped by the 760-ka Bishop Tuff (Sarna-Wojcicki et al., 2000), deposited on air fall ash and fluvially reworked Quaternary volcaniclastic deposits and other sedimentary basin fill (Bateman, 1965). After deposition, the region was extended, and more than 500 predominantly north–south-trending normal faults have formed to accommodate this extension. Growth across some faults indicates that faulting was active prior to tuff deposition (Bateman, 1965). The southern part of the Volcanic Tableland is a gently east-sloping, faulted homocline with rollover into the Fish Slough fault (Ferrill et al., 1999a). Because of the resistant nature of the welded component of the Bishop Tuff, the predominantly dip-slip faults are clearly

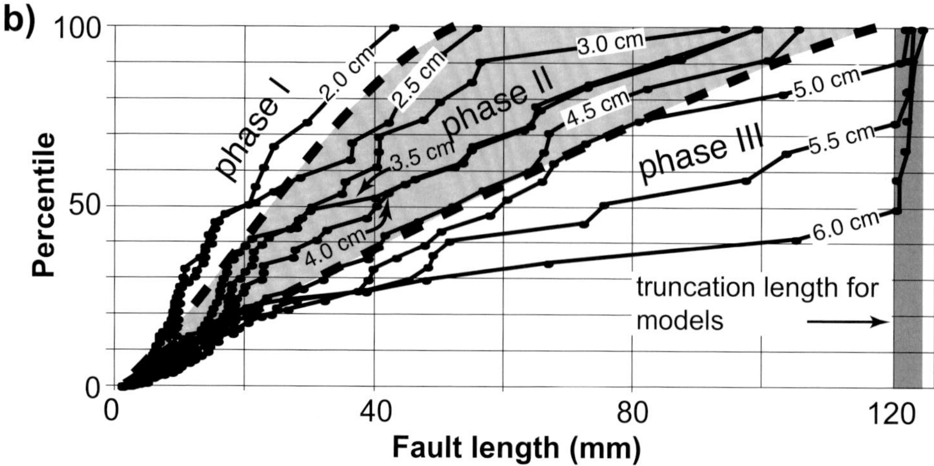

Figure 4. (cont.).

exposed (Bateman, 1965; Dawers et al., 1993; Dawers and Anders, 1995; Pinter, 1995; Ferrill et al., 1999a; Ferrill and Morris, 2001).

Fault traces in the Volcanic Tableland were interpreted on a side-looking airborne radar (SLAR) image (Figure 6a). Radar illumination used to create the image was from the east, so that east-dipping fault scarps appear brightly illuminated, and west-dipping fault scarps appear in shadow. Because of this, the downthrown traces of west-dipping faults cannot be resolved directly from the image, and we have based our fault trace interpretation on the well-resolved upthrown or footwall traces of both west- and east-dipping faults. Prior to interpretation, the image was georeferenced using 64 reference points (Environmental Systems Research Institute, 1991) to remove the distortion in the original image.

Analysis of the data yields a tip-to-fault ratio of 1.85 and a length distribution as illustrated in Figure 6c. Comparison with the analog modeling results (Figures 4b, 5b) indicates that the sampled area from the Volcanic Tableland is within Phase I of fault system evolution. This is because of the lack of throughgoing faults in the sampled area. The rock mass connectivity plot for the Volcanic Tableland (Figure 6d) has an almost elliptical shape with a long-to-short-axis ratio of 4:1. In the case where faults are barriers to flow, a randomly placed well in the sampled area would potentially drain an area 4 km (2.5 mi) along the predominant fault strike (north–south) by 1 km (0.62 mi) perpendicular to the predominant fault strike. If the faults were transmissive, a well intersecting a fault would be connected to a fault network 2 km (1.2 mi) long parallel to the dominant fault strike by 0.3 km (0.2 mi) perpendicular to this (Figure 6d).

Canyonlands (Utah, United States)

The grabens area lies within and to the south of Canyonlands National Park (Utah) and consists of an extensive network of graben and half-graben valleys in Pennsylvanian and Permian sedimentary strata (McGill and Stromquist, 1975; Stromquist, 1976; McGill and Stromquist, 1979; Trudgill and Cartwright, 1994; Cartwright et al., 1995; Schultz and Moore, 1996; Cartwright and Mansfield, 1998; Moore and Schultz, 1999; Cartwright et al., 2000; McGill et al., 2000). These structural basins formed in response to lateral unloading, caused by entrenchment and erosion by the Colorado River in Cataract Canyon, and displacement down the gentle (~4°) decollement to the west-northwest over the Pennsylvanian Paradox evaporites (McGill and Stromquist, 1975; Stromquist, 1976; McGill and Stromquist, 1979; Huntoon, 1982; Doelling et al., 1988; McGill et al., 2000). Normal faults and grabens are in various stages of interaction and linkage in the area. Faults exposed at the surface are nearly vertical (e.g., McGill and Stromquist, 1979) and probably dip at 75° at depths of 460 m (1510 ft) (Cartwright et al., 2000; McGill et al., 2000).

The geological map of the Canyonlands National Park (Huntoon et al., 1982) provides the basis for fault trace data in the Canyonlands area (Figure 6b). A tip-to-fault ratio of 1.1 and a fault length distribution as illustrated in Figure 6c indicate that the sampled area is just within Phase III of fault system evolution (compare with Figures 4b, 5b). This conclusion is also supported by the occurrence of one throughgoing fault in the sampled area (Figure 6c). In the case of sealing faults, a randomly placed well would potentially drain an area 1.5 km (0.9 mi) in a direction parallel to the dominant fault strike by 0.4 km (0.25 mi) perpendicular to this (Figure 6d). Conversely, if the faults are transmissive, a well intersecting a fault would be connected to a fault network 4.7 km (2.9 mi) long parallel to the dominant fault strike and, possibly, as much as 1 km (0.6 mi) perpendicular to the dominant fault strike.

DISCUSSION OF NATURAL EXAMPLES

Normal fault systems in the Volcanic Tableland (California) and the Canyonlands (Utah) have contrasting characteristics. Comparison with our analog modeling results indicates that the Volcanic Tableland is currently at an early stage in fault system development

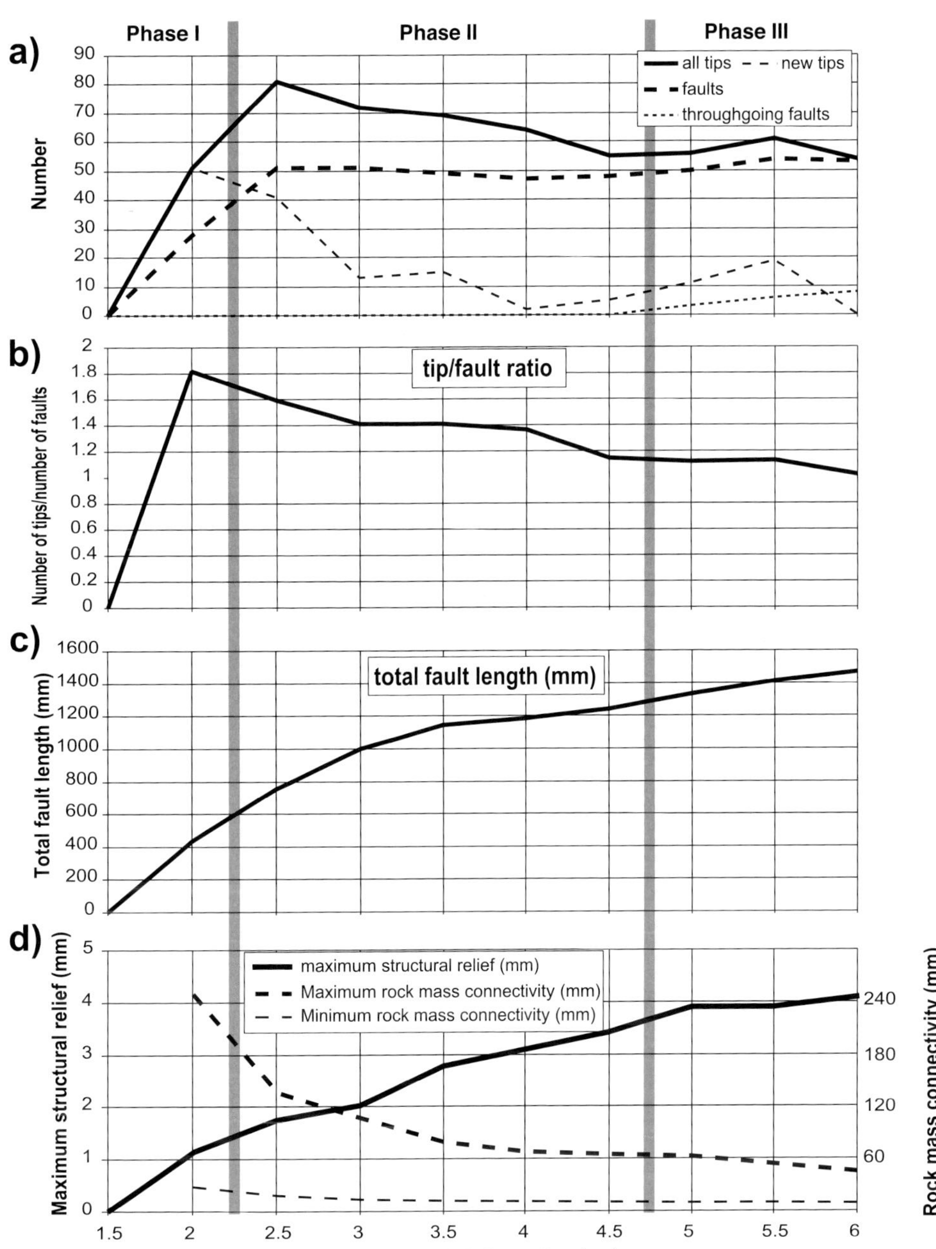

FIGURE 5. Summary of model results; all data except (d) were collected from the square sample areas. The horizontal axis, which is the total model elongation, is common to all graphs. (a) Fault, tip, and throughgoing fault data for the evolution of the model fault system. (b) Tip-to-fault ratio for the evolving model fault system. (c) Total fault length vs. total model elongation. (d) Maximum structural relief, measured as the largest fault throw measurable in the sample area. Maximum and minimum rock mass connectivity measured as described in the text.

(our Phase I) and is characterized by a relatively well-connected rock mass and a poorly connected fault network. The fault system in the Canyonlands is considerably more developed than the fault system in the Volcanic Tableland and is entering our Phase III. Rock mass connectivity is relatively low, and fault system connectivity is relatively high.

Several recent studies have investigated the importance of faults as either barriers to or conduits for fluid flow. The permeability behavior of faults depends on the deformation mechanisms active in the fault zones. These mechanisms, in turn, depend on the rock properties and deformation conditions. Antonellini and Aydin (1994) show that in porous sandstones, fault zone deformation at shallow crustal depths is commonly accomplished by grain comminution, which reduces permeability. Other processes, such as clay smearing and cementation, are also common in sedimentary rocks and reduce permeability. In contrast, deformation in already lithified or low-porosity rocks, such as many massive limestones, quartzite, and crystalline basement rocks, commonly experiences permeability enhancement caused by brecciation and development of large and connected pores in fault zones (e.g., Caine et al., 1996, and references therein; Ferrill et al., 1999b). In the case of permeability-reducing faults, faults act as barriers or baffles to fluid movement that may partition the reservoir, and fluid movement occurs predominantly in the rock mass. In the case of permeability-enhancing faults, fluid movement will occur preferentially in fault zones. As noted by Ferrill et al. (2000), the shape of the bulk permeability tensor may be similar for each case, but the magnitudes and details of fluid movement will differ.

SUMMARY

Our analog modeling results indicate that extensional fault systems evolve in at least three phases. In Phase I, faults rapidly nucleate at a variety of sites, and

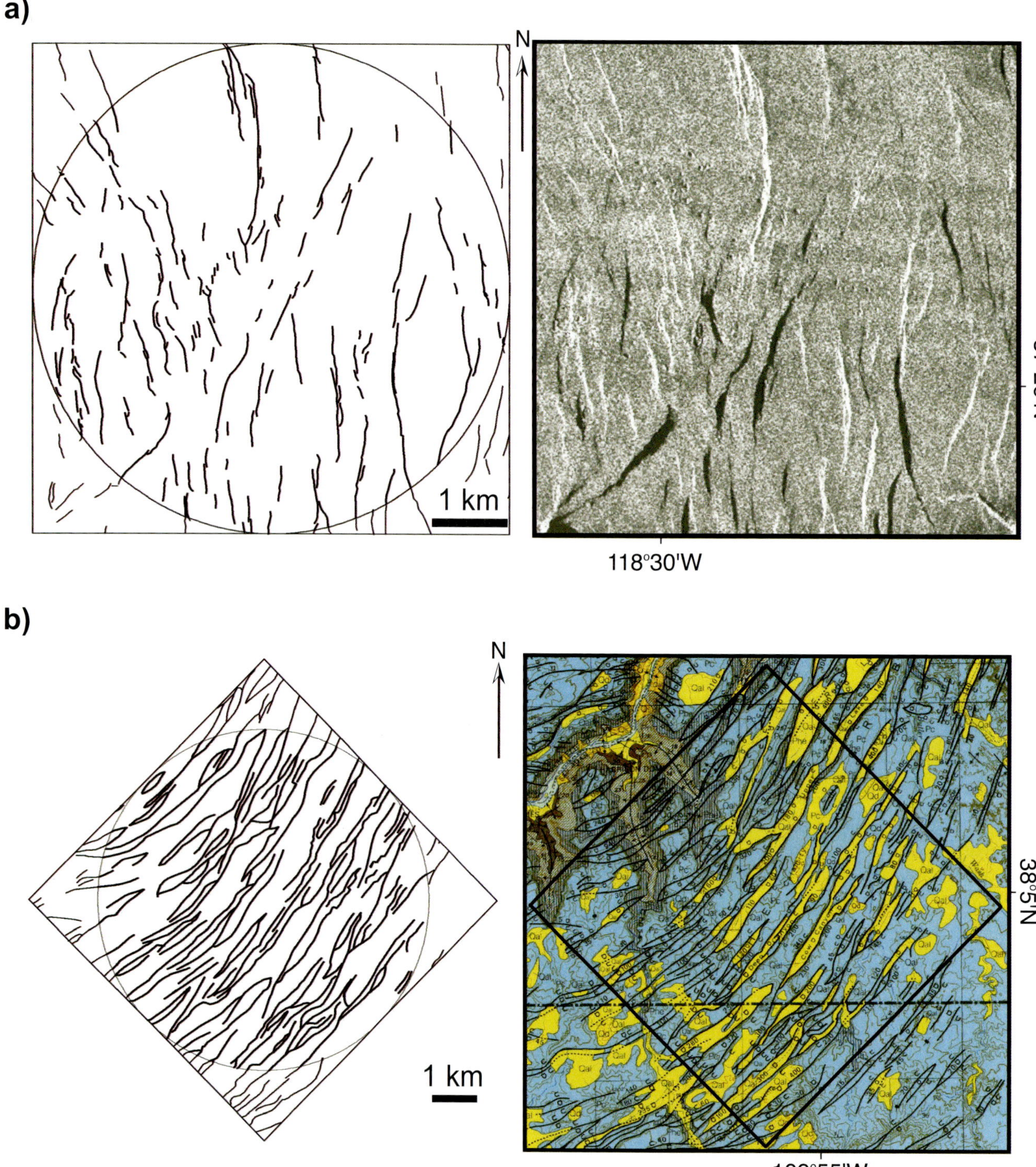

FIGURE 6. Fault length and rock mass connectivity analysis. (a)Volcanic Tableland sample areas. The square (6.25 km [3.88 mi] on a side) was sampled for data used in (c), and the circle (6.25-km [3.88-mi] diameter) for data in (d). Fault traces were interpreted on the side-looking airborne radar image using footwall (upthrown) cutoff traces only. (b) Canyonlands sample areas. The square (7.28 km [4.52 mi] on a side) was sampled for data used in (c), and the circle (7.28-km [4.52-mi] diameter) for data in (d). Fault traces were taken from the geologic map by Huntoon et al. (1982), a portion of which is shown. (c) Fault length percentile plot. Individual fault lengths are plotted in accordance with their contribution to total fault length. The approximate maximum length of a single fault measurable in the sample box (truncation length) is shown for each population sampled. (d) Rock mass connectivity and fault network connectivity azimuthal plots for Volcanic Tableland and Canyonlands. Note that the scale is uniform, and all plots in (d) may be directly compared.

c)

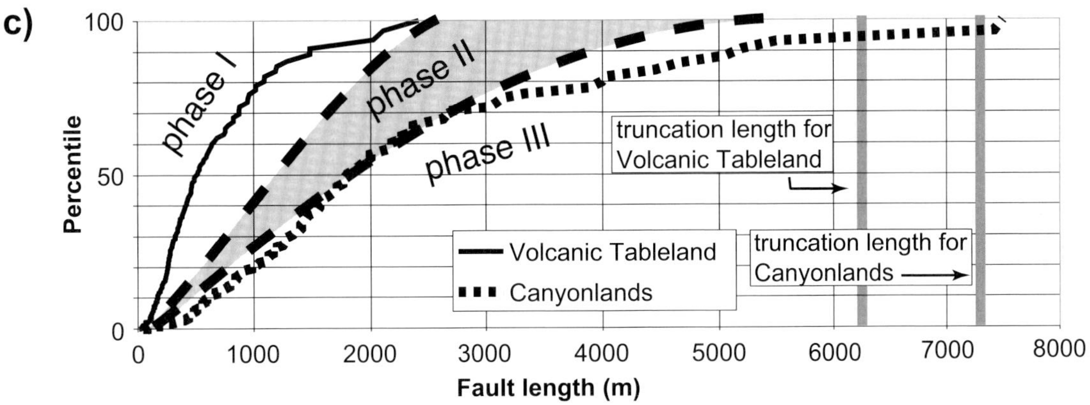

d)

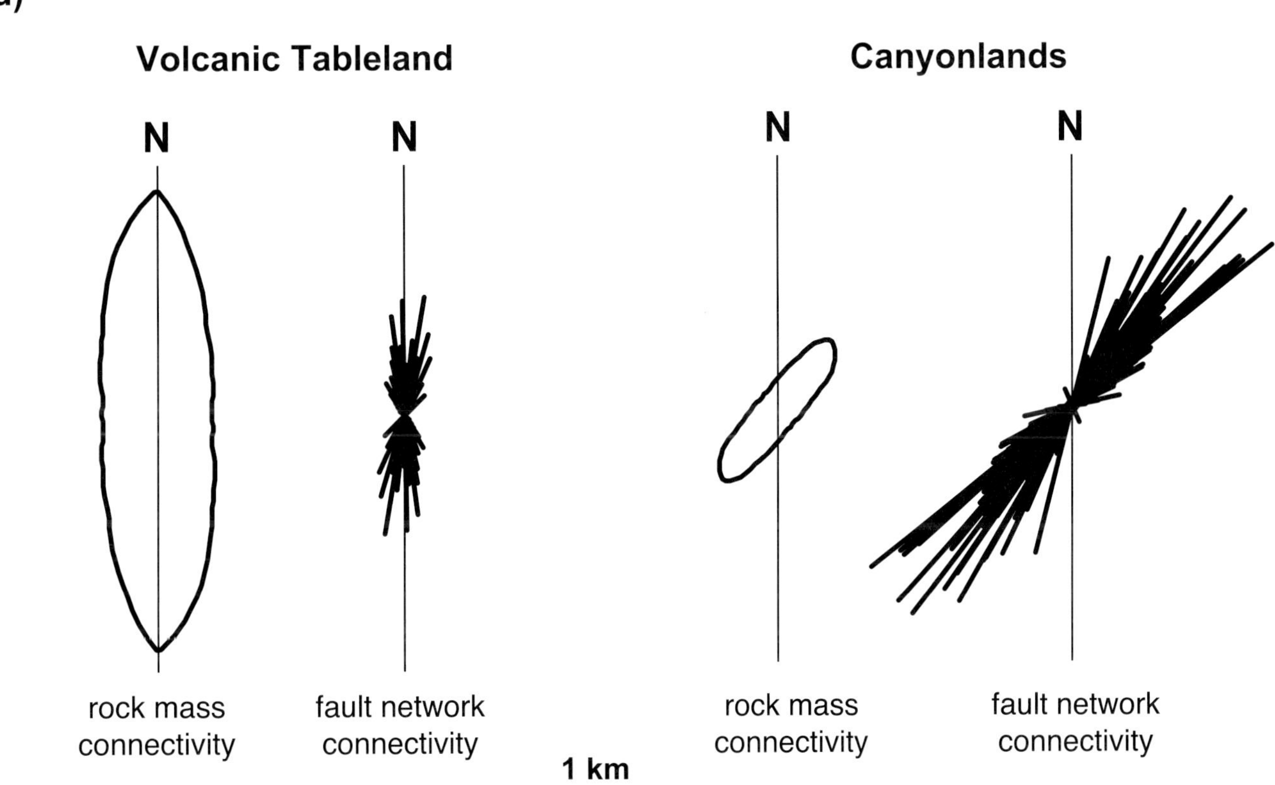

FIGURE 6. (cont.).

new fault formation dominates the system. Phase II is characterized by fault propagation, linkage, and the increasing dominance of larger faults. Phase III sees a slight resurgence in new fault formation and the increasing dominance of large, throughgoing faults. These evolutionary phases can be described in terms of measurable parameters such as tip-to-fault ratio, fault length distribution, average lengths of uninterrupted flow paths, and fault azimuth vs. length analysis. In addition, fault system development has consequences for rock mass connectivity (the degree to which rock is connected between and around faults) and fault network connectivity. Rock mass connectivity decreases as the fault system evolves, and fault network connectivity increases concomitantly. Analysis of uninterrupted flow paths provides a measure of rock mass connectivity, and tip-to-fault ratios combined with azimuthal plots of fault lengths provide information about fault network connectivity.

Reservoir connectivity is complex, and here, we only concern ourselves with its component that is controlled by faulting. Depending on whether a reservoir has sealing faults (e.g., porous sandstones) or transmissive faults (e.g., carbonates), the ease of fluid flow through the reservoir will be largely determined by rock mass (sealing faults) or fault network (transmissive faults) connectivity. Our measures of connectivity permit rapid evaluation of the extent to which a randomly placed well is uninterruptedly connected to transmissive reservoir in any horizontal direction from simple fault trace maps, important information for estimating drained volumes and well spacing in a new field. The method can also be extended into three dimensions if sufficient data are available.

Application of our analysis to natural examples of extensional faulting in the Volcanic Tableland (California) and the Canyonlands (Utah) indicates that these fault systems are in different stages of development. The Volcanic Tableland is at an early stage in fault system development and exhibits high rock mass connectivity and low fault network connectivity, whereas the Canyonlands have a more evolved fault system characterized by low rock mass connectivity and high fault network connectivity.

ACKNOWLEDGMENTS

We thank the Japan National Oil Corporation (presently Japan Oil, Gas and Metals National Corporation) for financial support of this work. Michael Ferguson assisted with preparing, running, and disassembling of model 10SEP99. Larry McKague, Wes Patrick, Nancye Dawers, Gloria Eisenstadt, and John Wickham greatly improved the chapter by their constructive reviews of the manuscript. Nick Davatzes georeferenced the Volcanic Tableland SLAR image, and Katherine Murphy, Rebecca Emmot, and Sharon Odam assisted with figure and manuscript preparation.

REFERENCES CITED

Antonellini, M., and A. Aydin, 1994, Effect of faulting on fluid flow in porous sandstones: Petrophysical properties: AAPG Bulletin, v. 78, no. 3, p. 355–377.

Bateman, P. C., 1965, Geology and tungsten mineralization of the Bishop District, California, U.S.A.: U.S. Geological Survey Professional Paper 470, 280 p.

Caine, J. S., J. P. Evans, and C. B. Forster, 1996, Fault zone architecture and permeability structure: Geology, v. 24, p. 1025–1028.

Cartwright, J. A., and C. S. Mansfield, 1998, Lateral displacement variation and lateral tip geometry of normal faults in the Canyonlands National Park, Utah: Journal of Structural Geology, v. 20, p. 3–19.

Cartwright, J. A., B. D. Trudgill, and C. S. Mansfield, 1995, Fault growth by segment linkage: An explanation for scatter in maximum displacement and trace length data from the Canyonlands grabens of SE Utah: Journal of Structural Geology, v. 17, p. 1319–1326.

Cartwright, J., B. Trudgill, and C. Mansfield, 2000, Fault growth by segment linkage: An explanation for scatter in maximum displacement and trace length data from the Canyonlands grabens of SE Utah: Reply: Journal of Structural Geology, v. 22, p. 141–143.

Cowie, P. A., 1998, A healing-reloading feedback control on the growth rate of seismogenic faults: Journal of Structural Geology, v. 20, p. 1075–1087.

Dawers, N. H., and M. H. Anders, 1995, Displacement-length scaling and fault linkage: Journal of Structural Geology, v. 17, p. 607–614.

Dawers, N. H., M. H. Anders, and C. H. Scholz, 1993, Growth of normal faults: Displacement-length scaling: Geology, v. 21, p. 1107–1110.

Doelling, H. H., C. G. Oviatt, and P. W. Huntoon, 1988, Salt deformation in the Paradox Basin: Utah Geological and Mineral Survey Bulletin, v. 122, 93 p.

Environmental Systems Research Institute, 1991, ARC/INFO® Version 6.0 for UNIX: Environmental Systems Research Institute, Inc., Redlands, California (CD-ROM).

Ferrill, D. A., and A. P. Morris, 2001, Displacement gradient and deformation in normal fault systems: Journal of Structural Geology, v. 23, p. 619–638.

Ferrill, D. A., A. P. Morris, S. M. Jones, and J. A. Stamatakos, 1998, Extensional layer parallel shear and normal faulting: Journal of Structural Geology, v. 20, p. 355–362.

Ferrill, D. A., J. A. Stamatakos, and D. Sims, 1999a, Normal fault corrugation: Implications for the growth and seismicity of active normal faults: Journal of Structural Geology, v. 21, p. 1027–1038.

Ferrill, D. A., J. Winterle, G. Wittmeyer, D. Sims, S. Colton, A. Armstrong, and A. P. Morris, 1999b, Stressed rock strains groundwater at Yucca Mountain, Nevada: Geological Society of America Today, v. 9, no. 5, p. 1–8.

Ferrill, D. A., A. P. Morris, J. A. Stamatakos, and D. W. Sims, 2000, Crossing conjugate normal faults: AAPG Bulletin, v. 84, p. 1543–1559.

Huntoon, P. W., 1982, The meander anticline, Canyonlands, Utah: An unloading structure resulting from horizontal gliding on salt: Geological Society of America Bulletin, v. 93, p. 941–950.

Huntoon, P. W., G. H. Billingsby, Jr., and W. J. Breed, 1982, Geologic map of Canyonlands National Park and vicinity, Utah: The Canyonlands Natural History Association, Moab, Utah, scale 1:62,500, 2 sheets.

Jackson, J. A., ed., 1997, Glossary of Geology: Alexandria, Virginia, American Geological Institute, 769 p.

King, P. R., S. V. Buldyrev, N. V. Dokholyan, S. Havlin, Y. Lee, G. Paul, H. E. Stanley, and N. Vandesteeg, 2001, Predicting oil recovery using percolation theory: Petroleum Geoscience, v. 7, p. S105–S107.

McGill, G. W., and A. W. Stromquist, 1975, Origin of graben in the Needles District, Canyonlands National Park, Utah, *in* J. E. Fassett, ed., Canyonlands country: Four Corners Geological Society Guidebook, Durango, Colorado, p. 235–243.

McGill, G. E., and A. W. Stromquist, 1979, The grabens of Canyonlands National Park, Utah: Geometry, mechanics, and kinematics: Journal of Geophysical Research, v. 84, p. 4547–4563.

McGill, G. E., R. A. Schultz, and J. M. Moore, 2000, Fault

growth by segment linkage: An explanation for scatter in maximum displacement and trace length data from the Canyonlands grabens of SE Utah: Discussion: Journal of Structural Geology, v. 22, p. 135–140.

Moore, J. M., and R. A. Schultz, 1999, Processes of faulting in jointed rocks of Canyonlands National Park, Utah: Geological Society of America Bulletin, v. 111, p. 808–822.

Pinter, N., 1995, Faulting on the Volcanic Tableland, Owens Valley, California: Journal of Geology, v. 103, p. 73–83.

Rouby, D., H. Fossen, and P. R. Cobbold, 1996, Extension, displacement, and block rotation in the larger Gullfaks area, northern North Sea: Determined from map view restoration: AAPG Bulletin, v. 80, p. 875–890.

Sarna-Wojcicki, A. M., M. S. Pringle, and J. Wijbrans, 2000, New Ar-40/Ar-39 age of the Bishop Tuff from multiple sites and sediment rate calibration for the Matuyama–Brunches boundary: Journal of Geophysical Research, v. 105, no. B9, p. 21,431–21,443.

Schultz, R. A., and J. M. Moore, 1996, New observations of grabens from the Needles District, Canyonlands National Park, Utah, *in* A. C. Huffman, Jr., W. R. Lund, and L. H. Godwin, eds., Geology and resources of the Paradox Basin: Utah Geological Association Guidebook, v. 25, p. 295–302.

Stromquist, A. W., 1976, Geometry and growth of grabens, lower Red Lake Canyon area, Canyonlands National Park, Utah: Department of Geology and Geography, University of Massachusetts Contribution, v. 28, p. 118.

Trudgill, B., and J. Cartwright, 1994, Relay-ramp forms and normal-fault linkages, Canyonlands National Park, Utah: Geological Society of America Bulletin, v. 106, p. 1143–1157.

Waiting, D. J., J. A. Stamatakos, D. A. Ferrill, D. W. Sims, A. P. Morris, P. S. Justus, and K. I. Abou-Bakr, 2003, Methodologies for the evaluation of faulting at Yucca Mountain, Nevada: Proceedings of the 10th International High-Level Radioactive Waste Management Conference (IHLRWM), March 30–April 2, 2003. CD-ROM version. La Grange Park, Illinois: American Nuclear Society, Inc.

Yamada, Y., H. Okamura, Y. Tamura, and F. Tsuneyama, 2005, Analog Models of Faults Associated with Salt Doming and Wrenching: Application to offshore United Arab Emirates, *in* R. Sorkhabi and Y. Tsuji, eds., Faults, fluid flow, and petroleum traps: AAPG Memoir 85, p. 95–106.

Analog Models of Faults Associated with Salt Doming and Wrenching: Application to offshore United Arab Emirates

Yasuhiro Yamada[1]
Japan Petroleum Exploration Co., Ltd., Research Center, Chiba, Japan

Hitoshi Okamura[2]
Technology Research Center, Japan National Oil Corporation, Chiba, Japan

Yoshihiko Tamura
Japan Oil Development Co., Ltd., Tokyo, Japan

Futoshi Tsuneyama[3]
Technology Research Center, Japan National Oil Corporation, Chiba, Japan

ABSTRACT

Regional stress has a significant impact on fault development during the formation of salt dome structures. To examine such effects of wrenching, we conducted a series of analog experiments of updoming using dry, granular materials and observed the deformation on the top free surface. The experiments included three deformation styles: (1) updoming followed by wrenching, (2) simultaneous updoming and wrenching, and (3) wrenching followed by updoming. In the first series of the experiments, the faults produced by simple updoming were overprinted by two strike-slip fault systems that were generated by the subsequent wrenching. The second series of experiments with the configuration of simultaneous updoming and wrenching generated normal faults in a direction perpendicular to relative extension by the wrench. In the third series of experiments, the Riedel and anti-Riedel shear faults formed by

[1]*Present address:* Department of Earth Resources Engineering, Kyoto University, Kyoto, Japan.
[2]*Present address:* INPEX Co., Tokyo, Japan.
[3]*Present address:* Department of Geophysics, Stanford University, Stanford, California, U.S.A.

DOI:10.1306/1033718M852969

wrenching were deformed by the subsequent updoming and were overprinted by the faults related to the updoming.

These experimental results are applied to the fault systems observed above dome structures in the United Arab Emirates region, where extensive faults in the northwest–southeast direction have developed. By analogy, these faults were probably formed during an updoming and simultaneous wrenching. The direction of simple shear inferred from a comparison of real faults and experimental results suggests that dextral wrenching caused by the Oman stress regime during the Late Cretaceous affected the region at the time of the updoming.

INTRODUCTION

Fault systems above salt domes have been investigated by analog experiments for decades (e.g., Parker and McDowell, 1951, 1955; Cloos, 1955; Withjack and Scheiner, 1982; Davison et al., 1993; Alsop, 1996). These experiments provide an ideal situation in which both the experimental configuration and the resulting deformation are known. Several geometric and kinematic models for salt domes have been proposed based on analog experiments and have been subsequently applied to real salt-related structures (e.g., Vendeville and Jackson, 1992a, b). These experiments have expanded our knowledge of fault system development around salt domes and are also useful for predicting subsurface fluid flow associated with faults acting as migration pathways.

Experiments on simple updoming have been performed by many researchers following the pioneer works of Parker and McDowell (1951, 1955) and Cloos (1955). These researchers used wet clay and dry sand as sedimentary overburden and a pushing rigid plug (Cloos, 1955) or asphalt (Parker and McDowell, 1951, 1955) to simulate salt. Despite this difference in the material for salt, both experiments demonstrated that radial normal faults are a common structural feature developed above a dome. Parker and McDowell (1955) have also noticed that fault patterns are controlled by the size and shape of the salt dome, the overburden thickness, and the amount of uplift. The elliptical dome experiments of these authors showed that major normal faults generated along the long axis of the antiform.

The effects of regional stress on fault patterns above a dome were investigated by Withjack and Scheiner (1982). They performed experiments having two dome shapes, circular and elliptical, with simultaneously applied extensional or compressional stress. Their wet clay models showed that regional stress also affects the fault patterns produced by updoming. Updoming with extension generated normal faults perpendicular to the extension direction, as well as minor strike-slip faults at the rim. With compression, doming produced normal faults that are parallel to the compression direction at the center of the dome and reverse faults that are perpendicular to the compression at the periphery. The normal faults curved at their tips and showed a strike-slip element of the displacement.

The current research on dynamically scaled experiments of salt-related structures generally uses substrates of viscous fluids such as polydimethylsiloxane to model the ductile behavior of salt, and the results of these experiments have been successfully applied to a large number of real structures (e.g., Koyi, 1988, 1996; Jackson et al., 1990; Vendeville and Jackson, 1992a; Jackson, 1995, and references therein). These works recognized that most salt diapers initiate and grow by differential loading of the overburden, and the buoyancy of salt is inadequate as the driving force of salt movement. Waltham (1997) demonstrated with numerical simulations that the mechanism can be well explained by differential loading, folding of the overburden, and drag by a moving overburden.

The modeling technique using viscous fluids is, however, difficult to control particular geometries of salt structures to form. This is why we used a geometrically constrained balloon instead of ductile substrates for the updoming of salt.

Fault patterns generated in a wrench tectonic setting have also been investigated by analog experiments (e.g., Wilcox et al., 1973; Naylor et al., 1986). The shear-box experiments by Wilcox et al. (1973) showed characteristic fault patterns, including the Riedel and anti-Riedel faults, normal faults, and reverse faults. The direction of these faults agrees with that predicted by the presumable stress field: the normal faults perpendicular to the relative extension and the reverse faults perpendicular to the relative compression. The Riedel and anti-Riedel faults are a conjugate set of strike-slip faults having an axis in the relative compression direction.

To examine the effects of regional wrench on faults above a dome structure, we have conducted a series of analog experiments with dry granular material. The effects of deformation sequence of updoming and wrenching are also examined. Six representative results are described and compared with the fault system observed above a dome structure in the United Arab Emirates (U.A.E.) region.

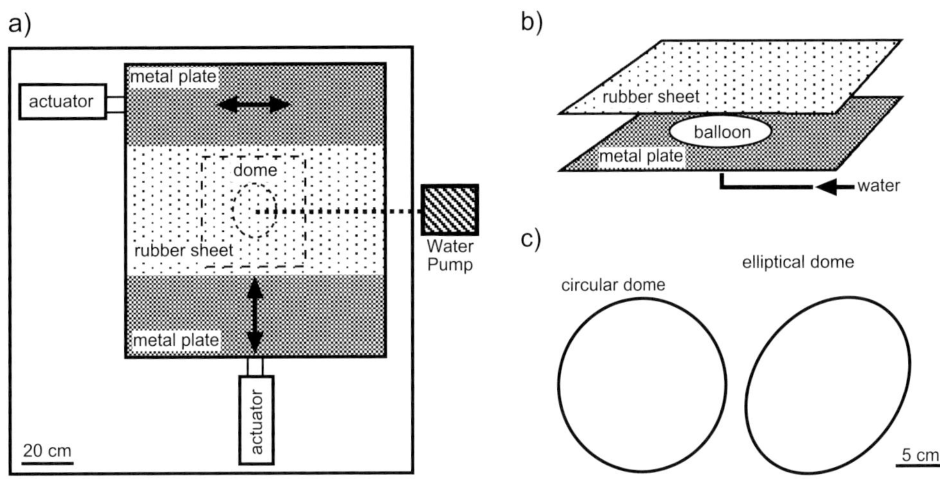

FIGURE 1. Experimental rig: (a) apparatus; (b) updoming apparatus; (c) dome shapes.

EXPERIMENTAL PROCEDURE

Apparatus

The experimental apparatus used for this study had a rectangular rubber sheet, creating regional stress, attached to two metal base plates that could be moved independently by two oil-pressured actuators (Figure 1a). One actuator generated extension and contraction by pulling and pushing a plate, and the other actuator moved the second base plate perpendicular to the extension and contraction direction, thus producing simple shear. For this study, the displacement rates of the actuator were between 1.33×10^{-3} and 5.0×10^{-3} cm s^{-1} (5.2×10^{-4} and 1.97×10^{-3} in. s^{-1}). It was ensured that no unnecessary folding of the rubber sheet occurred during experiments. This was achieved by extending the rubber sheet 10%, and then, the actuator for extension was fixed at the position throughout the experimental program. Each side of the base plate had fixed vertical end-walls, forming a deformation box, and its initial dimension was $100 \times 100 \times 20$ cm ($39 \times 39 \times 8$ in.) (length × width × height).

Updoming was produced by an apparatus with a rubber balloon inflated by a water pump (Figure 1b). The balloon was sandwiched with two metal plates, one of which has a hole simulating circular or elliptical dome geometry. The apparatus was placed under the rubber sheet described above, and its top surface was entirely coated with fine powder to reduce friction to the overlain sheet on which experimental material was deposited. The size of the circular dome was 17.8 cm (7 in.) in diameter, and that of the elliptical dome was 20 cm (8 in.) in the long axis and 15.6 cm (6.14 in.) in the short axis (Figure 1c). The direction of the elliptical dome in the experiments is shown in Figure 1c. The rate of updoming was between 2.78×10^{-4} and 1.56×10^{-3} cm s^{-1} (1.1×10^{-4} and 6.1×10^{-4} in. s^{-1}).

Combinations of updoming and shear produced by the rubber sheet can create a variety of experimental configurations. To analyze the effects of the deformation sequence of updoming (2-cm [0.8-in.] height) and wrenching (20% shear strain), three types of kinematics were applied to the experiments. The first kinematic type (type I) was simple updoming, followed by dextral wrenching. In the second kinematic type (type II), updoming was simultaneous with dextral wrenching. In the third kinematic type of experiment (type III), dextral wrenching was followed by updoming.

Modeling Material

The modeling material employed was homogeneous dry glass beads, which is a cohesionless spherical material and fails in accordance with the Navier–Coulomb criteria. Such granular materials (e.g., dry sand) are appropriate to model the brittle deformation of the upper crust and have been successfully used for various tectonic environments (see Koyi, 1996; Cobbold and Castro, 1999, and references therein). Artificial microspheres are also commonly used where the shear strength of dry sand is too strong (e.g., McClay et al., 1998). Physical properties of such experimental materials have been measured by shear tests, and the microspheres are now regarded as an appropriate material for analog experiments (Schellart, 2000; Rossi and Storti, 2003).

The glass beads we used were of 45–63 μm in diameter, with a grain density of 2.50 g/cm^3. The internal friction angle and the friction coefficient were 25.5° and 0.48, respectively. The glass beads were evenly sprinkled on the rubber sheet, and the top free surface was flattened with a scraper every few millimeters. This procedure was repeated until the material reached the required thickness. The initial thickness of the material was 4 cm (1.6 in.) in our experiments. The scaling ratio of the models to natural structures is approximately 10^{-5}; thus, 1 cm (0.4 in.) in the model scales to 1 km (0.6 mi) in nature.

Limitations

Physical modeling requires simplification of structural deformation processes, which are products of various processes over geologic timescales, so that experimental models can be constructed in the laboratory (Vendeville and Cobbold, 1988). This inherent limitation must be kept in mind, especially when applying the model results to natural examples. The experiments in this chapter include the following constraints and assumptions: use of granular material, no variation in the overburden thickness, assumption of homogeneous brittle nature of the overburden, constant rate of deformation, no mechanical compaction or chemical diagenesis during deformation, and exclusion of the effects of fluid flow and heat.

The updoming apparatus we used in this study may be regarded as classic because it does not agree with the recent knowledge of salt movement mechanism (e.g., Waltham, 1997). However, the purpose of our experiments is to examine the fault geometry above a salt dome where local extension is generally produced by the updoming. Such local extension can be similarly produced with our simple updoming apparatus, and we can use the experimental results to assume the background tectonics of a region as a first-order approximation.

EXPERIMENTAL RESULTS

Circular Dome Experiments

Kinematic Type I

Structures produced on the top free surface of the experiments during updoming included radial normal faults above the dome and circular reverse faults on the periphery (Figure 2a). The radial faults were generally short (about 3 cm [1.2 in.] or less) and developed particularly at the dome's flank, which approximately corresponds to the point of minimum curvature in the topographic relief of the experimental surface. The reverse faults were generally short and segmented and were observed in the gently dipping region on the periphery of the dome.

During the subsequent dextral wrench stage, two strike-slip fault systems corresponding to the Riedel (dextral) and the anti-Riedel (sinistral) shear were observed on the free surface of the experiments (Figure 2b). These were relatively long (as much as 30 cm [12 in.]) and almost straight, but not clearly developed above the dome. The faults slightly changed their direction at the flank of the dome, where the preexisting normal faults were cut by the latter strike-slip faults.

Kinematic Type II

When updoming was produced simultaneously with dextral wrenching, a distinct normal fault system developed on the dome (Figure 2c). These faults were generally long (as much as 10 cm [4 in.]) and dipped toward the center of the dome. Their strike was perpendicular to the direction of relative extension that was generated by the wrench. At the flank of the dome, these faults branched into strike-slip fault systems, which were clearly developed in the region free from the influence of updoming. No reverse faults were observed around the dome.

Kinematic Type III

During the dextral wrench stage, two strike-slip fault systems were generated on the free surface (Figure 2d). They had almost the same geometries as those of the strike-slip faults observed in the region apart from the dome in the types I and II experiments. Subsequent updoming produced radial normal faults of short length and occurring on the dome, similar to those of the type I experiments (Figure 2e). The reverse faults observed in the type I experiment were not developed here. The preexisting strike-slip faults were slightly deformed by the updoming and were overprinted by the radial faults.

Elliptical Dome Experiments

Kinematic Type I

Updoming produced normal faults above the dome and circular reverse faults on the periphery (Figure 3a). The normal faults formed with a trend roughly parallel to the long axis, and they splayed outward at the flanks. The reverse faults were short and segmented, similar to those of the circular dome experiments.

Subsequent dextral wrench produced strike-slip fault systems except at the crest of the dome (Figure 3b). On the flanks, the radial normal faults were overprinted by the Riedel and anti-Riedel faults. These are similar to those of the circular dome experiments.

Kinematic Type II

Simultaneous updoming and dextral wrenching generated a distinct normal fault system above the dome (Figure 3c). These faults were generally long (as much as about 10 cm [4 in.]) and dipped toward the center of the dome. Their strike was perpendicular to the relative extension generated by the wrench. At the flank of the dome, the fault branched into two strike-slip fault systems, which were also developed in the region free from the influence of updoming.

Kinematic Type III

The results of the dextral wrench stage were identical to the circular dome experiments (Figure 3d; also see Figure 2d) because of similarities in the experimental configuration. This indicates good reproducibility of the experiments. Subsequent updoming produced short normal faults, similar to those of the type I experiment (Figure 3e). The reverse faults observed in

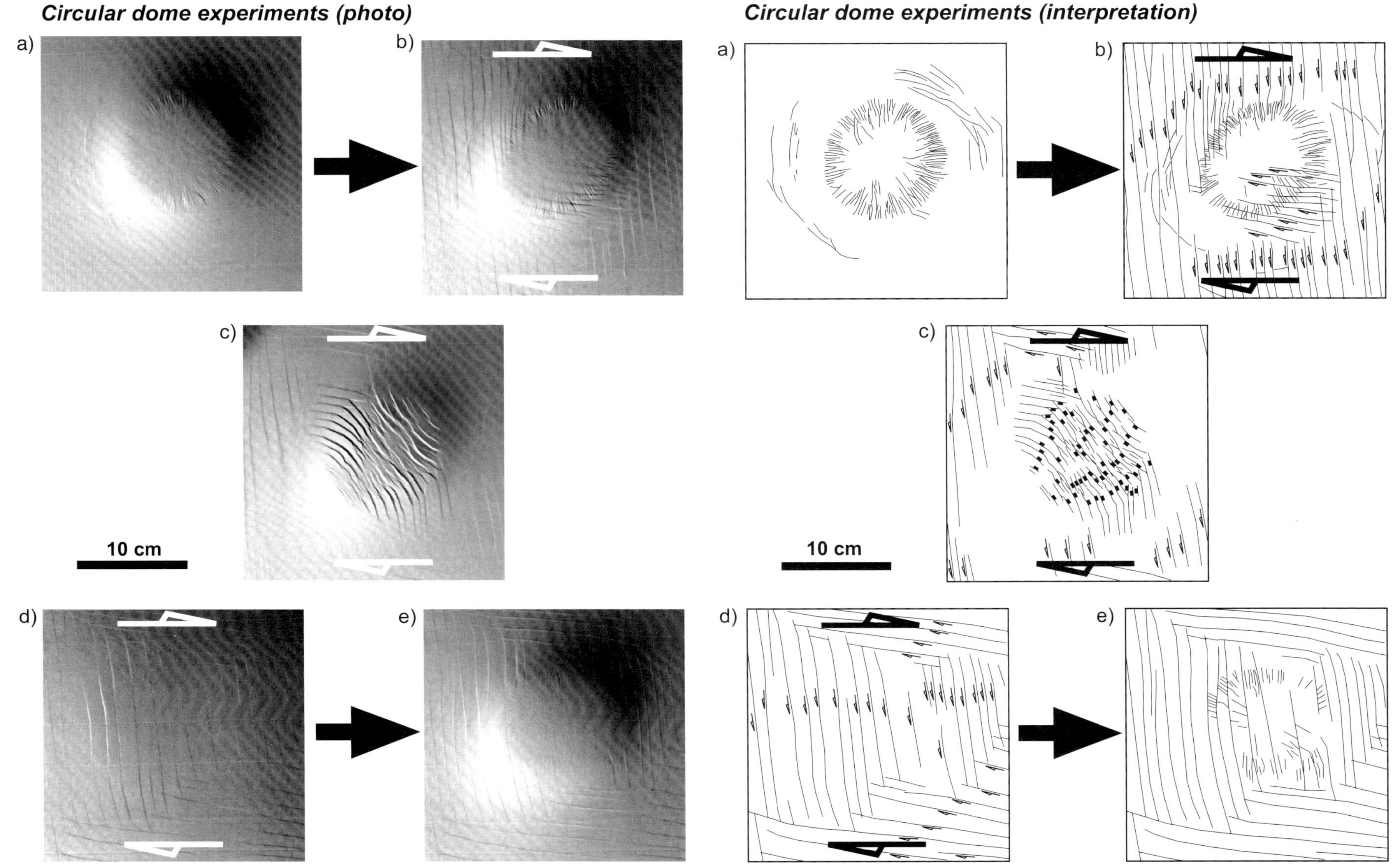

FIGURE 2. Results of the circular dome experiments: (a) after simple updoming stage (the kinematic type I); (b) after subsequent dextral wrench (the kinematic type I); (c) after updoming with simultaneous dextral wrench (the kinematic type II); (d) after dextral wrench (the kinematic type III); (e) after subsequent simple updoming (the kinematic type III).

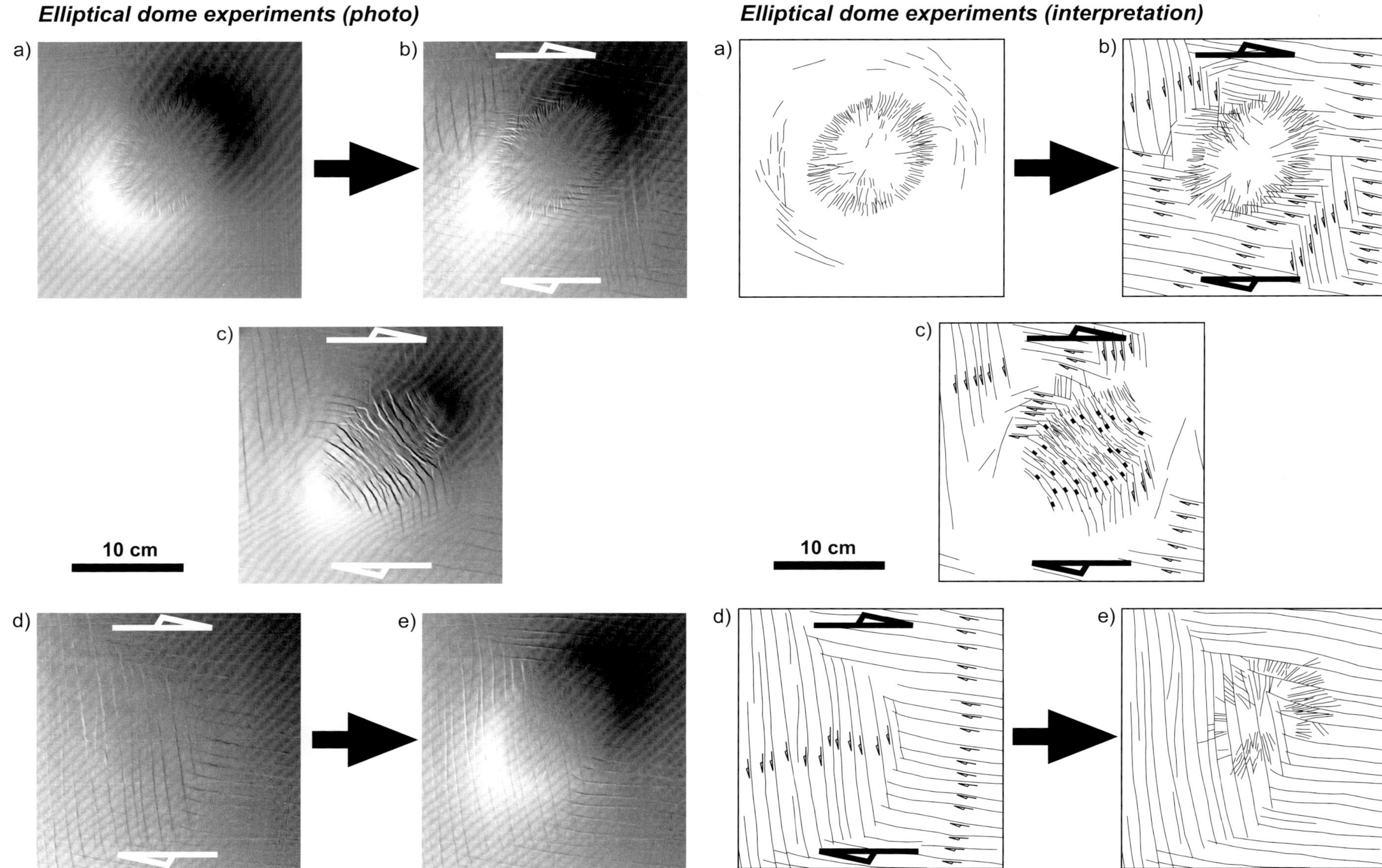

FIGURE 3. Results of the elliptical dome experiments: (a) after simple updoming stage (the kinematic type I); (b) after subsequent dextral wrench (the kinematic type I); (c) after updoming with simultaneous dextral wrench (the kinematic type II); (d) after dextral wrench (the kinematic type III); (e) after subsequent simple updoming (the kinematic type III).

the type I experiments were not developed. The preexisting strike-slip faults were slightly deformed by the updoming and were overprinted by the short normal faults.

Summary and Discussions of Experimental Results

Our experiments clearly showed that faulting was controlled by the regional stress applied to the experiment. During the updoming in the types I and III experiments, faults observed on the free surface were similar to those reported from simple updoming experiments (e.g., Parker and McDowell, 1951, 1955; Cloos, 1955; Withjack and Scheiner, 1982). With wrenching, two strike-slip fault systems were developed as also observed in the previous shear-box experiments (e.g., Cloos, 1955; Naylor et al., 1986; Richard et al., 1995).

In the experiments of the kinematic type II, the stress field above the dome was a combination of updoming and wrenching. The updoming obviously generated layer-parallel extension around the dome, whereas the wrenching produced a stress field in which the intermediate principal stress axis was in the vertical direction. Their combined stress should have the least principal compressive stress axis horizontally, parallel to the relative extension caused by the wrench, and the maximum principal compressive stress axis vertically (the gravity). This stress orientation explains the geometry of the characteristic normal faults above the dome in the kinematic type II experiments.

Types I and III model results also showed that a change in the stress field could cause interactions of faults that were generated earlier and those that were formed later. In the experiments of kinematic type III, the reverse faults observed in the type I experiments were less developed. Because the difference in the wrenching stage of type I and that of type III is only the existence of the strike-slip faults, the contractional stress around the dome in the type III experiment might be partly consumed by minor reactivation of the strike-slip faults during the updoming. Such effects of preexisting faults have been particularly observed in the inverted basins (e.g., Cooper and Williams, 1989; Buchanan and Buchanan, 1995; Yamada, 1999), where extensional faults have significant impacts on the subsequent inversion geometry.

APPLICATION TO FAULTS ABOVE SALT DOMES OFFSHORE UNITED ARAB EMIRATES

Several domal structures exist in the Middle East, and many of them are faulted at their reservoir horizons (Al-Husseini and Chimblo, 1994; Cosgrove and Jubralla, 1994; Mendeck and Al-Madani, 1994; Carman, 1996; Edgell, 1996; Honda et al., 1996; Alsharhan and Salah, 1997; Grutsch et al., 1998; Johnson et al., 2002). Figure 4 shows schematic diagrams of fault patterns above such dome structures around the United Arab Emirates (U.A.E.) region. It is apparent that the structures are traversed by many northwest–southeast-trending normal faults, sometimes in an oblique direction to the longitudinal axis of the structure (Figure 4a, d). This suggests that the faults and their geometries were generated under a regional tectonic regime. To discuss the origin of these faults, the regional tectonics of the U.A.E. is briefly reviewed, and then the experimental results of this study are compared.

Tectonic Outline of the U.A.E. Region

The U.A.E. is situated in the southeastern part of the Arabian Plate and located in a Hormuz salt basin (Figure 5). The distribution of the Cambrian salt is strongly controlled by the tectonic environment of the basin (Edgell, 1996; Alsharhan and Salah, 1997) and bounded by three main orientations (Talbot and Alavi, 1996). Two boundaries in the north–south and northeast–southwest directions are of the 1000–600-Ma terrane-accreted fabric, and the northwest–southeast-trending boundaries are parallel to the Najd transcurrent faults that were formed about 600 Ma in the Pan-African basement (Husseini and Husseini, 1990; Talbot and Alavi, 1996). These tectonic lines place major controls on the petroleum provinces of the region and also may have affected the local stress fields during the subsequent tectonic history.

The regional tectonics have been investigated especially from the Iranian side of the salt basin (e.g., Talbot and Alavi, 1996), whereas information from the U.A.E. side is mainly obtained by petroleum companies and thus is not widely published. Alsharhan and Salah (1997) summarized the regional tectonics that affected the structural configuration of the U.A.E. and suggested that the basin has experienced six main tectonic stages from the late Paleozoic to the present. The first two stages older than the Late Jurassic were extensional tectonics related to the opening of the Neotethys. Then the tectonics changed to a compressional regime (including four stages) because of subduction of the Neotethys and collision of the Afro-Arabia Plate with the Sanandaj–Sirjan block of the Persian Plate. Because of the subduction, the Neotethys closed, and the subsequent collision formed the Oman Mountains and the Zagros fold belt.

Marzouk and El Sattar (1994) examined the geometry and location of the dome structures in the U.A.E. region and proposed that the region was affected by two major stress regimes: the Oman stress and the Zagros stress. The Oman stress was named after the tectonic

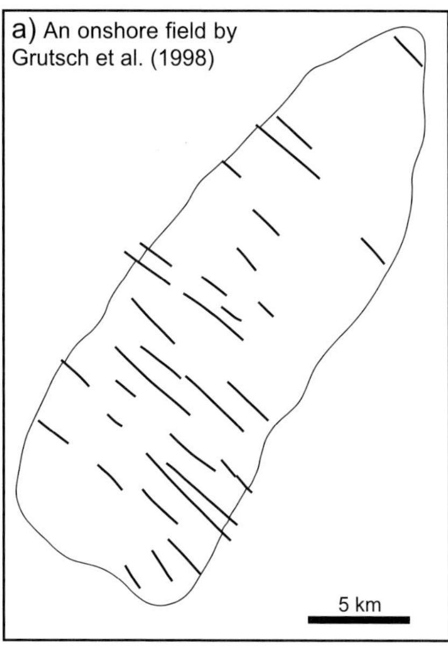

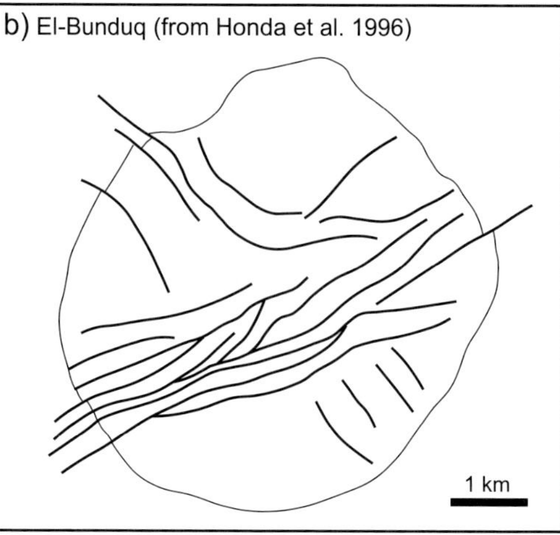

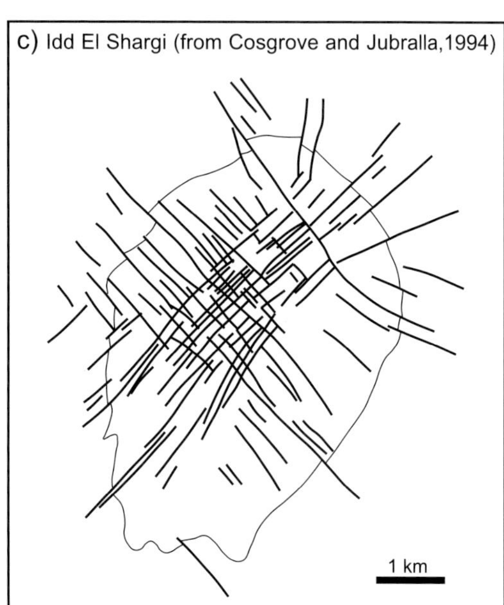

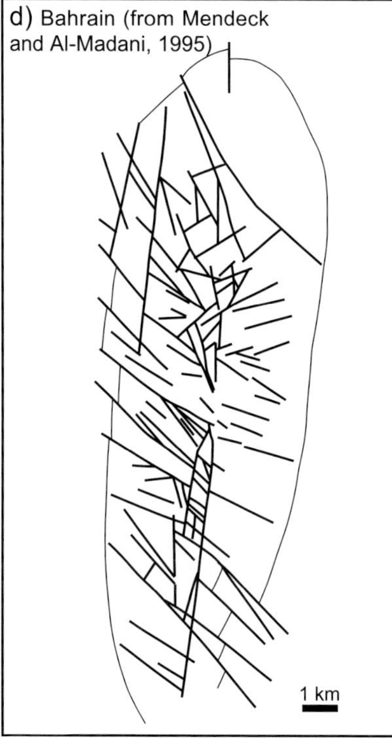

FIGURE 4. Fault patterns above dome structures around the U.A.E. region.

Detailed observations of the structural geometry and the subsidence and uplifting history of the U.A.E. region suggest that most of the preserved strains are not products of compressional stress but are related to shear deformation (Marzouk and El Sattar, 1994). This is supported by a recent three-dimensional (3-D) seismic survey revealing that the faults are actually fault zones composed of arrays of minor en echelon faults, and that the fault zones can be interpreted as conjugate shears (Johnson et al., 2002). This shear stress should have strong impact on the fault development in the U.A.E. region.

activity during the Late Cretaceous, which formed the Oman Mountains. The principal horizontal stress was east–west, and the primary shear directions of the Oman stress are illustrated in Figure 6a. Because the U.A.E. region is relatively close to the convergent plate boundary east of the Oman Mountains, the indications of strain by the Oman stress are well preserved. During the Cenozoic, the direction of the principal horizontal stress in the region has been northeast–southwest, which concords with the current Arabian Plate motion vector relative to the Eurasian Plate. This stress is referred to as the Zagros stress because it was active during the formation of the Zagros Mountains. Figure 6b shows the primary shear directions of the Zagros stress. The indications of this shear strain are, however, not obvious in the U.A.E. region, probably because of its distant location from the deformation front of the Zagros Mountains.

Updoming

The sedimentary rocks in the U.A.E. region are generally flat with few gentle structural undulations (Alsharhan and Salah, 1997), and even at salt domes, the reservoir horizon dips mostly dip less than 2° (e.g., Alsharhan, 1990).

To understand the timing and magnitude of uplifting of dome structures, the lateral variation in sedimentary thicknesses was examined on seismic sections. The target dome structures are located in the offshore U.A.E. These results (Figure 7) suggest that the rate of the updoming of each structure in the region generally reached its maximum in the Late Cretaceous. Some structures ceased updoming activity in the early Tertiary, whereas others have continued until the Quaternary. In conjunction with arguments by Talbot and Alavi (1996) and Alsharhan and Salah (1997) that the Hormuz Salt did not start to rise until the Jurassic or the Early Cretaceous, we suggest that the dome structures rapidly developed during the Late Cretaceous. This updoming history suggests that the Oman stress significantly influenced the U.A.E. region and triggered the salt movement.

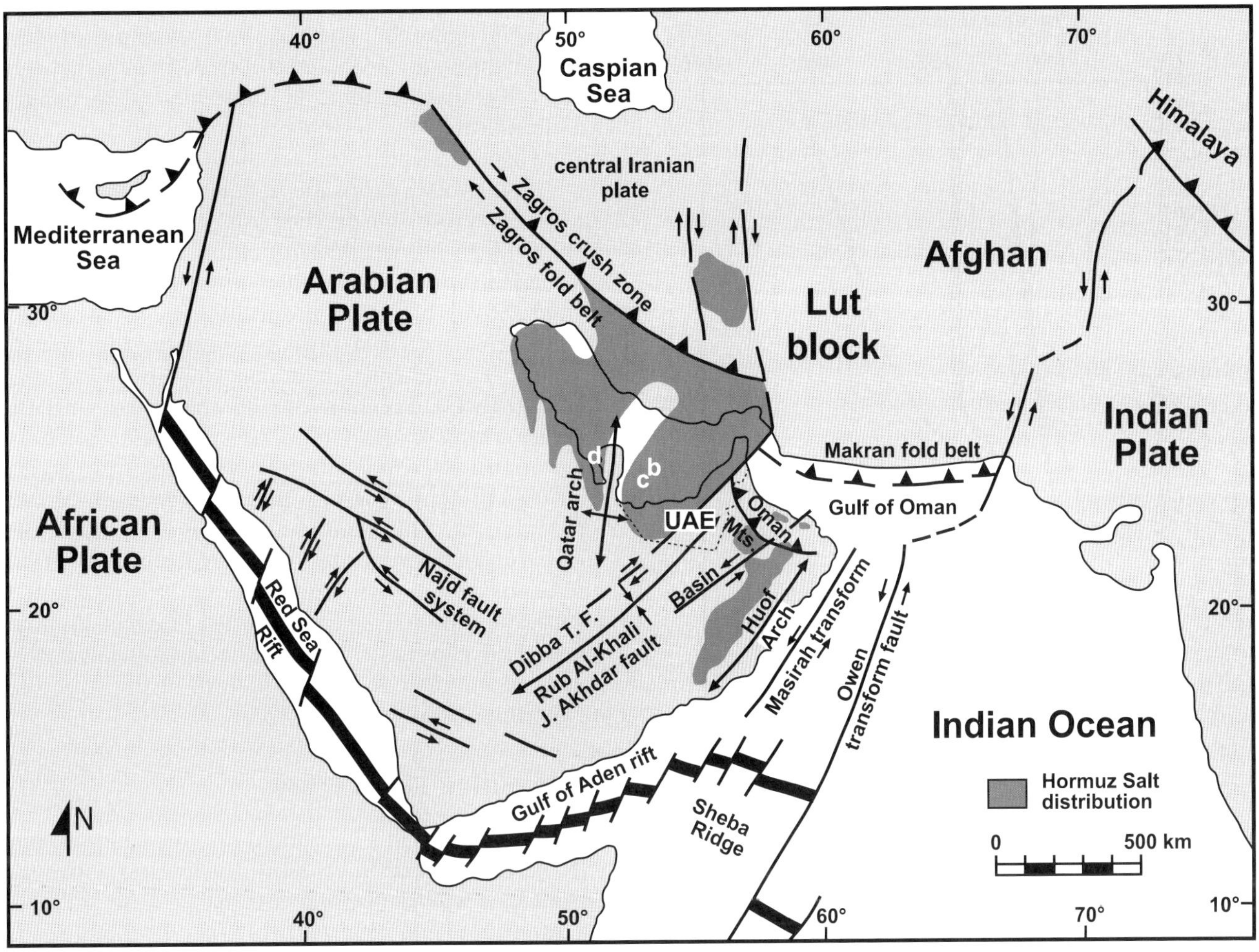

FIGURE 5. The current plate tectonic setting around the Arabian Plate (modified from Marzouk and El Sattar, 1994).

Comparison with Experimental Results

Faults that developed above the salt domes in the U.A.E. region are highly directional and show little influence of updoming geometry (Figure 4). In addition, no clear evidence exists for radial faults. These facts suggest that the domes in the U.A.E. region generally have not experienced an "updoming-only" stage corresponding to the kinematic types I and III of the experiments. This implies that the regional tectonics were significantly active and strongly affected the stress field when the dome structures formed. This tectonic activity, as well as the possible reactivation

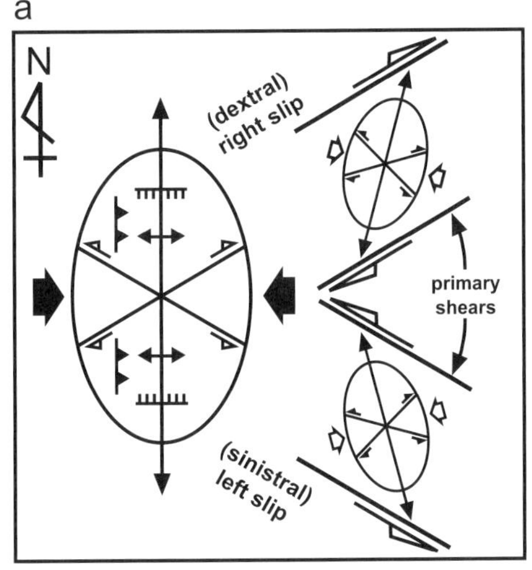

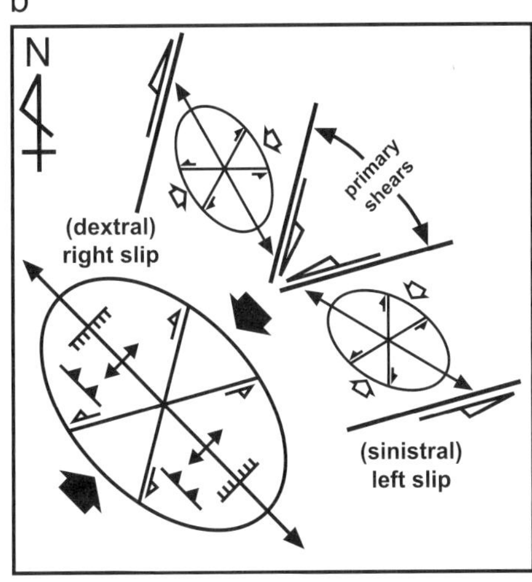

FIGURE 6. Two stress regimes that affected the U.A.E. region (modified from Marzouk and El Sattar, 1994): (a) the Oman stress with the maximum horizontal compression in the east–west direction; (b) the Zagros stress with the maximum horizontal compression in the northeast–southwest direction.

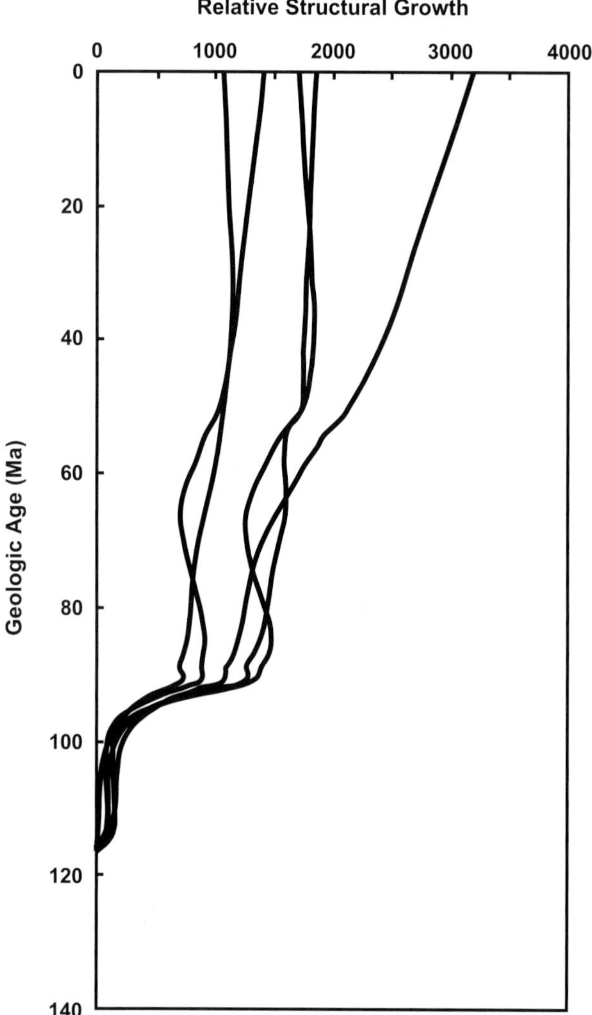

FIGURE 7. Relative structural growth of dome structures offshore the U.A.E. The curvature was estimated from sediment thickness changes on seismic sections. Note that the doming was most profound between 100 and 90 Ma.

of preexisting fault systems and deformation of the overburden, could have triggered the active diapirism of the salt in the U.A.E. region.

The northwest–southeast-trending faults of the U.A.E. domes (Figure 4) can be compared with the fault pattern in the kinematic type II experiment, which also produced a dominant fault system in the direction of the relative extension caused by the wrenching. If the faults in the U.A.E. region were generated in a similar tectonic setting as that of the type II experiment, the direction of the wrench can be inferred from the comparison of the fault patterns observed in the field and those produced by the experiments. Because the type II experiment included a dextral wrenching, a direct correlation of the faults in the real domes and those in the experiments indicates the direction of the possible dextral wrench to be east-northeast–west-southwest (Figure 8b). The experimental results of sinistral wrenching can be obtained by making mirror images of the original experiments. By using these mirror images, the direction of the possible sinistral wrench is determined as north-northwest–south-southeast (Figure 8c).

These two scenarios are geometrically equivalent, and both are possible on the basis of the experimental results. More refined determination can be made by comparison with the regional tectonic history. As mentioned above, the region was influenced by two major stresses, resulting in a primary shear having four directions. The dextral wrench in the east-northeast–west-southwest direction can be correlated with the primary shear of the Oman stress (Figure 6a); however, no sinistral wrench in the north-northwest–south-southeast direction can be found in the tectonic history (cf. Figure 6). This strongly suggests that the faults above the dome structures in Figure 4 formed during an updoming event with simultaneous dextral wrench, which is a primary shear caused by the Cretaceous Oman stress. This timing of the updoming agrees with the results of the relative growth pattern of the structural development shown in Figure 7.

Subsurface Fluid Flow

The experiments presented in this chapter provide basic information on fault distribution above salt domes in a wrench tectonic setting. Because faults are an important migration pathway of subsurface fluids, the knowledge of fault development processes and geometries derived from the experiments helps to predict possible fault systems and migration pathways in the basin even in the absence of 3-D seismic data. The deformation styles observed in the experiments can also be used to approximate the internal strain in each block segmented by the faults. This knowledge is, in turn, useful to predict possible fracture systems on subseismic scales and to construct fractured reservoir models, including reservoir heterogeneity induced by faults.

CONCLUSIONS

The geometry of faults that developed above salt domes is controlled by regional stress. Our experiments with combinations of updoming and wrenching demonstrated that the faults are generated in a direction perpendicular to the relative extension caused by the wrench. The faults obviously change their geometry if the regional stress is converted from simple updoming to wrench or vice versa. The faults generated after the stress conversion are affected by the preexisting faults; thus, the resultant geometry of the faults is determined by the sequence of the stresses.

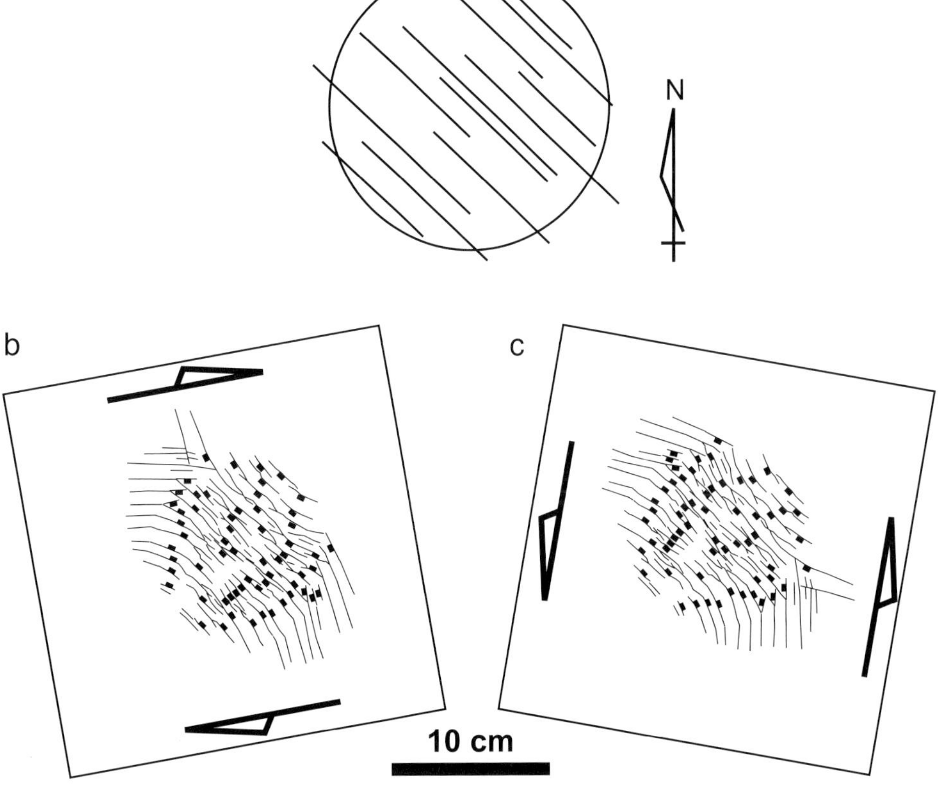

FIGURE 8. Two possible scenarios of shear directions to explain the faults above the dome structures around the U.A.E.: (a) schematic illustration of a dome and faults of the region; (b) shear direction in case of dextral shear; (c) shear direction in case of sinistral shear. Note that all fault directions are all northwest–southeast.

The experiments also suggest that the dome structures characterized by a series of northwest–southeast-directed normal faults in the offshore U.A.E. were developed under a regional stress field that triggered the active diapirism of salt at deeper structural levels.

A comparison with the experimental results (updoming with dextral wrench) and their mirror images (updoming with sinistral wrench) suggests that the faults observed in the U.A.E. dome structures were probably generated by updoming with simultaneous dextral wrench in an east-northeast–west-southwest direction. This wrench component may correspond to the primary shear of the Oman stress regime that was active during the Late Cretaceous. The timing of updoming inferred from sedimentary thickness changes on seismic sections supports this interpretation.

These experimental results are useful to analyze faults above dome structures in the regions even with poor seismic quality, to predict possible fault systems and migration pathways, and to construct models of fractured reservoir in petroleum fields.

ACKNOWLEDGMENTS

This research was conducted as a part of the Carbonate Reservoir Research Project at Japan National Oil Corporation–Technology Research Center (presently Japan Oil, Gas and Metals National Corporation). We acknowledge permissions to publish this work from the Abu Dhabi National Oil Company, the Japan Oil Development Co., Ltd., the Japan National Oil Corporation, and the Japan Petroleum Exploration Co., Ltd. This chapter was reviewed by Hiroaki Komuro, Christopher Talbot, Ken McClay, and Rasoul Sorkhabi, whose comments greatly improved the manuscript.

REFERENCES CITED

Al-Husseini, M. I., and R. D. Chimblo, 1994, 3-D seismology in the Arabian Gulf region, *in* M. I. Al-Husseini, ed., Middle East Petroleum Geosciences, GEO'94: Gulf Petrolink, v. 1, p. 93–102.

Alsharhan, A. S., 1990, Geology and reservoir characteristics of Lower Cretaceous Kharaib Formation in Zakum field, Abu Dhabi, United Arab Emirates, *in* J. Brooks, ed., Classic petroleum provinces: Geological Society (London) Special Publication 50, p. 299–316.

Alsharhan, A. S., and M. G. Salah, 1997, Tectonic implications of diapirism in hydrocarbon accumulation in United Arab Emirates: Bulletin of Canadian Petroleum Geology, v. 45, p. 279–296.

Alsop, G. I., 1996, Physical modelling of fold and fracture geometries associated with salt diapirism, *in* G. I. Alsop, D. J. Brundell, and I. Davison, eds., Salt tectonics: Geological Society (London) Special Publication 100, p. 227–241.

Buchanan, J. G., and P. G. Buchanan, 1995, Basin inversion: Geological Society (London) Special Publication 88, 596 p.

Carman, G. J., 1996, Structural elements of onshore Kuwait: GeoArabia, v. 1, p. 239–266.

Cloos, E., 1955, Experimental analysis of fracture patterns: Geological Society of America Bulletin, v. 66, p. 241–256.

Cobbold, P. R., and L. Castro, 1999, Fluid pressure and

effective stress in sandbox models: Tectonophysics, v. 301, p. 1–19.
Cooper, M. A., and G. D. Williams, 1989, Inversion tectonics: Geological Society (London) Special Publication 44, 376 p.
Cosgrove, P., and A. F. Jubralla, 1994, The development of a tight chalky limestone guided by 3-D seismic and the FMS in the Idd El Shargi field, offshore Qatar, *in* M. I. Al-Husseini, ed., Middle East Petroleum Geosciences, GEO'94: Gulf Petrolink, v. 1, p. 335–343.
Davison, I., M. Insley, M. Harper, P. Weston, D. Blundell, K. McClay, and A. Quallington, 1993, Physical modelling of overburden deformation around salt diapirs: Tectonophysics, v. 228, p. 255–274.
Edgell, H. S., 1996, Salt tectonism in the Persian Gulf, *in* G. I. Alsop, D. J. Blundell, and I. Davison, eds., Salt tectonics: Geological Society (London) Special Publication 100, p. 129–151.
Grutsch, J., O. Al-Jelani, and Y. Al-Mehairi, 1998, Integrated reservoir characterization of a giant Lower Cretaceous oil field, Abu Dhabi, UAE: 8th Abu Dhabi International Petroleum Exhibition and Conference (ADIPEC), Society of Petroleum Engineers Paper 49454.
Honda, N., M. K. M. Abouelenein, C. Loader, and M. Akbar, 1996, Fault interpretation and fracture morphology: A case study of Jurassic carbonate reservoir in El-Bunduq oil field, offshore Abu Dhabi and Qatar: 7th Abu Dhabi International Petroleum Exhibition and Conference (ADIPEC), Society of Petroleum Engineers Paper 36202.
Husseini, M. I., and S. I. Husseini, 1990, Origin of the infracambrian salt basins of the Middle East, *in* J. Brooks, ed., Classic petroleum provinces: Geological Society (London) Special Publication 50, p. 279–292.
Jackson, M. P. A., 1995, Retrospective salt tectonics, *in* M. P. A. Jackson, D. G. Roberts, and S. Snelson, eds., Salt tectonics: A global perspective: AAPG Memoir 65, p. 1–28.
Jackson, M. P. A., R. R. Cornelius, C. H. Craig, A. Gansser, J. Stöcklin, and C. J. Talbot, 1990, Salt diapirs of the Great Kavir, central Iran: Geological Society of America Memoir 177, 139 p.
Johnson, C. A., M. A. Satter, R. Rosell, F. Al-Shekaili, N. Al-Zaabi, and A. Gombos, 2002, Structure and regional context of onshore fields in Abu Dhabi, U.A.E.: Proceedings of the Society of Petroleum Engineers 10th Abu Dhabi International Petroleum Exhibition and Conference, SPE Paper 78488.
Koyi, H., 1988. Experimental modelling of the role of gravity and lateral shortening in Zagros Mountain belt: AAPG Bulletin, v. 72, p. 1381–1394.
Koyi, H., 1996, Salt flow by aggrading and prograding overburdens, *in* G. I. Alsop, D. J. Brundell, and I. Davison, eds., Salt tectonics: Geological Society (London) Special Publication 100, p. 243–258.
Marzouk, I., and M. A. El Sattar, 1994, Wrench tectonics in Abu Dhabi, United Arab Emirates, *in* M. I. Al-Husseini, ed., Middle East Petroleum Geosciences, GEO'94: Gulf Petrolink, v. 1, p. 655–668.
McClay, K. R., T. Dooley, and G. Lewis, 1998, Analogue modelling of progradational delta systems: Geology, v. 26, p. 771–774.
Mendeck, M. F., A. Al-Madani, 1994, Bahrain field: Challenges for reservoir characterization, *in* M. I. Al-Husseini, ed., Middle East Petroleum Geosciences, GEO'94: Gulf Petrolink, v. 1, p. 669–677.
Naylor, M. A., G. Mandl, and C. H. K. Sijpesteijn, 1986, Fault geometries in basement-induced wrench faulting under different initial stress states: Journal of Structural Geology, v. 7, p. 737–752.
Parker, T. J., and A. N. McDowell, 1951, Scale models as guide to interpretation of salt-dome faulting: AAPG Bulletin, v. 35, p. 2076–2086.
Parker, T. J., and A. N. McDowell, 1955, Model studies of salt-dome tectonics: AAPG Bulletin, v. 39, p. 2384–2470.
Richard, P. D., M. A. Naylor, and A. Koopman, 1995, Experimental models of strike-slip tectonics: Petroleum Geoscience, v. 1, p. 71–80.
Rossi, D., and F. Storti, 2003, New analogue materials for analogue laboratory experiments: Siliceous and aluminum microspheres: Journal of Structural Geology, v. 25, p. 1893–1899.
Schellart, W. P., 2000, Shear test results for cohesion and friction coefficients for different granular materials: Scaling implications for their usage in analogue modelling: Tectonophysics, v. 324, p. 1–16.
Talbot, C. J., and M. Alavi, 1996, The past of a future syntaxis across the Zagros, *in* G. I. Alsop, D. J. Blundell, and I. Davison, eds., Salt tectonics: Geological Society (London) Special Publication 100, p. 89–109.
Vendeville, B., and P. R. Cobbold, 1988, How normal faulting and sedimentation interact to produce listric fault profiles and stratigraphic wedges: Journal of Structural Geology, v. 10, p. 649–659.
Vendeville, B. C., and M. P. A. Jackson, 1992a, The rise of diapirs during thin-skinned extension: Marine and Petroleum Geology, v. 9, p. 331–353.
Vendeville, B. C., and M. P. A. Jackson, 1992b, The fall of diapirs during thin-skinned extension: Marine and Petroleum Geology, v. 9, p. 354–371.
Waltham, D., 1997, Why does salt start to move?: Tectonophysics, v. 282, p. 117–128.
Wilcox, R. E., T. P. Harding, and D. R. Seely, 1973, Basic wrench tectonics: AAPG Bulletin, v. 57, p. 74–96.
Withjack, M., and C. Scheiner, 1982, Fault patterns associated with domes— An experimental and analytical study: AAPG Bulletin, v. 66, p. 302–316.
Yamada, Y., 1999, 3-D analogue modelling of inversion structures: Ph.D. thesis, Royal Holloway, University of London, London, 743 p.

Yamada, Y., and T. Matsuoka, 2005, Digital sandbox modeling using distinct element method: Applications to fault tectonics, *in* R. Sorkhabi and Y. Tsuji, eds., Faults, fluid flow, and petroleum traps: AAPG Memoir 85, p. 107–123.

Digital Sandbox Modeling Using Distinct Element Method: Applications to Fault Tectonics

Yasuhiro Yamada[1]
Japan Petroleum Exploration Co. Research Center, Chiba, Japan

Toshifumi Matsuoka
Department of Earth Resources Engineering, Kyoto University, Kyoto, Japan

ABSTRACT

Structural deformation by faulting and folding has been analyzed by sandbox experiments that appropriately model the brittle behavior of the upper crust. This type of physical experiment using granular materials can also be done by numerical simulation (digital modeling) using the distinct element method (DEM). This chapter presents a set of two-dimensional simulation results of structural deformations using the DEM in basic tectonic settings of extension and contraction and also in the indentation tectonics as a consequence of continental collision of India. By comparing with analog experiments, the simulations reproduced fault systems similar to those of the experiments in terms of the overall deformation geometry and their development sequence. In particular, characteristic features of the Indian collision tectonics, such as fragmentation and rotation of continental blocks, were clearly identified in our DEM simulation. Because the DEM digital modeling is cheaper and faster than conventional sandbox experiments and it can incorporate discontinuity surfaces properly, the method can prove to be a powerful tool to simulate fault-related phenomena, such as structural traps in sedimentary basins. The DEM is a forward-modeling technique and can provide useful information on possible deformation pathways.

INTRODUCTION

Sandbox experiments have been a key technique in petroleum exploration to understand the development processes of faults, petroleum traps, and sedimentary basins. The knowledge of experimental results forms the core of modern structural geology and is vital to interpret geophysical data, particularly seismic sections. The significance of these experiments is increasing, because the frontier targets of exploration and

[1]*Present address:* Department of Earth Resources Engineering, Kyoto University, Kyoto, Japan.

DOI:10.1306/1033719M852969

development include reservoirs that do not easily yield seismic images of good quality, such as those occurring in 5–6-km (3.1–3.7-mi)-deep, severely deformed structures or those under a high topographic relief.

The sandbox experimental technique is relatively simple but requires some skills to reduce the effects of boundary conditions of the apparatus (e.g., Yamada, 1999), which severely affect the results. The technique also needs an experimental apparatus specially designed for each experimental program, and the apparatus sometimes needs to be altered to overcome minor problems arising during the preliminary modeling. In addition, the experiments cannot consider some fundamental parameters, such as effects of heat flow and pore fluid pressure. High-resolution images of internal deformation of three-dimensional sandbox experiments are also difficult to obtain during an experiment.

To solve these problems, scientists have employed numerical simulation. Because sandbox experiments use quartz grains as the material, deformation can be simulated on the computer using the distinct element method (DEM), which also approximates a geologic body comprising a large number of particles. As the scaled sandbox experiments are reproducing miniatures of natural geologic deformation, the DEM has the potential ability to simulate geological deformation. This can be done in two ways. One is to simulate real deformation with representative values; this approach is called "true-scale simulation." The other employs scaled-down values of deformation, similar to scaled physical experiments; this approach is called "scaled-down simulation." The most significant aspect of the DEM is that we can follow the structural history of the area under consideration and discuss the tectonic environment of rock deformation, including fault development, which is important in petroleum exploration. In addition, because the DEM is a forward modeling technique, it can also provide information on possible deformation features. This information can be used to assess environmental stabilities for the purpose of underground storage and waste disposal.

The DEM simulation has been applied to various research fields (see Oda and Iwashita, 1999, and references therein) including fault systems (Saltzer and Pollard, 1992; Saltzer, 1993; Donze et al., 1994; Antonelli and Pollard, 1995; Place and Mora, 1999). These models are generated to simulate structural deformation using simplified equations of physically realistic processes, and they improve our understanding of the complex evolution of such structures. However, because geological phenomena cannot be fully described by simple equations, properly scaled laboratory experiments place important constraints on the knowledge of processes and provide a framework of comparison for simulation results. In this chapter, we have employed the results of sandbox experiments as simplified geologic models to determine the boundary conditions of our DEM simulations and have investigated a variety of fault-related deformation phenomena, which are of vital importance in basin tectonics.

SANDBOX MODELING OF FAULTS AND PETROLEUM TRAPS

According to current geophysical knowledge, the mechanics of the upper crust can be approximated by Navier–Coulomb brittle behavior. This states that rock strength increases with depth as overburden pressure increases. By using the scaling theory of Hubbert (1937), rock mechanics can be properly scaled down to design laboratory experiments. McClay (1990a) argued that cohesionless dry sand is the most appropriate material to model such brittle deformation of the upper crust. The geometric and kinematic features found in analog experiments are therefore scaled-down representations of tectonic and rock deformation processes. Such sandbox experiments have been successfully performed for modeling various tectonic styles (see Koyi, 1997; Cobbold and Castro, 1999, and references therein).

In sandbox experiments, the study of fault-related structures is most informative, because they include discontinuity phenomena, which are difficult to treat by continuum mechanics of the crust. Fault structures are important petroleum traps, and analog experiments on faults are thus of interest to petroleum exploration. In particular, normal faulting and its hanging-wall deformation have been extensively investigated, and the results have been applied to petroleum traps in rift basins and passive continental margins (e.g., Horsfield, 1977; McClay and Ellis, 1987; McClay, 1990b; Buchanan and McClay, 1991). Fold and thrust belts are also important petroleum provinces, and their geometry and kinematics have been examined in detail by sandbox experiments (e.g., Colletta et al., 1991; Huiqi, et al., 1992; Lallemand et al., 1992; Yamada and McClay, 2003a, b).

A typical result of two-dimensional (2-D) experiments for pure shear-type extension is presented in McClay (1990b). Extension was achieved using a rubber sheet that is attached to two pieces of plastic sheet at each end; thus, deformation was restricted to the central part of the model. After stretching the rubber sheet, the models produced a graben with the greatest subsidence and extension at the edge where stretching initiated (Figure 1a). The graben is clearly bounded by a planar fault dipping toward the stretching basal detachment.

Several sandbox experiments have targeted master detachment faults to examine the deformation of its hanging wall (e.g., McClay and Ellis, 1987; Buchanan and McClay, 1991; Yamada and McClay, 2003c).

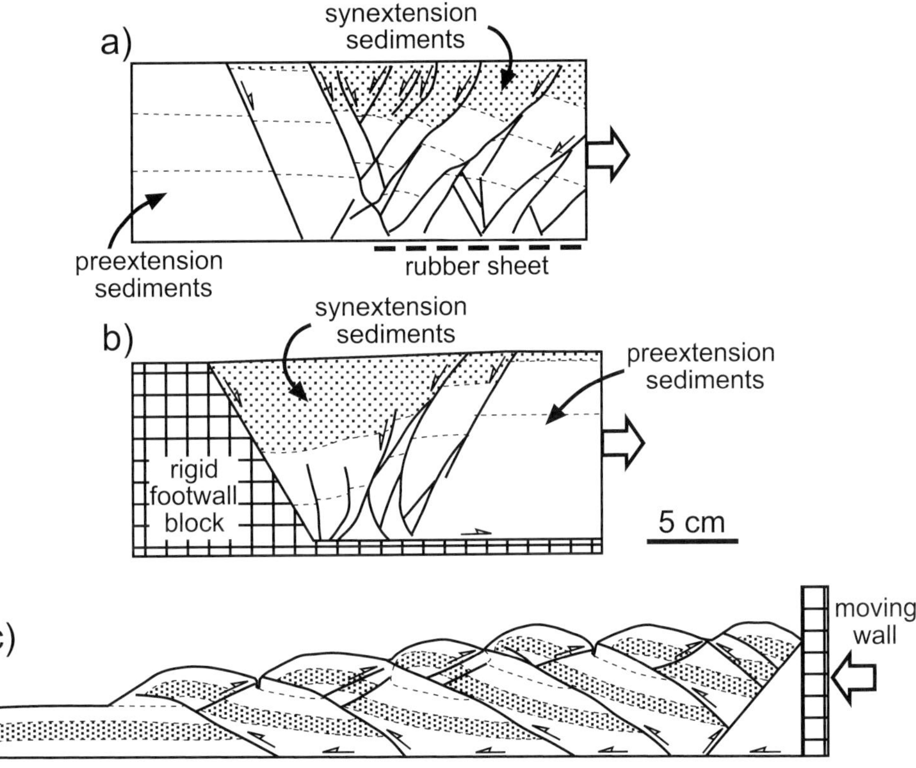

FIGURE 1. Typical results of sandbox experiments of (a) pure-shear-type extension (modified from McClay, 1990a), (b) extension over a kinked planar master fault (modified from Buchanan and McClay, 1991), and (c) a thrust wedge (modified from Huiqi et al., 1992).

Buchanan and McClay (1991) used a kinked planar master fault with the cutoff angle to the horizontal being 60°. In their experiments, extension produced major antithetic faults that occurred above the soling-out point of the planar detachment and a graben bounded by antithetic faults and the master fault (Figure 1b). The antithetic faults formed with a basinward-propagating sequence. All faults that developed in the hanging wall were characteristically planar.

Sandbox experiments of thrust wedges have also significantly expanded our knowledge. A typical example is shown in Huiqi et al. (1992), who used a vertical rigid backstop and a plastic sheet to simulate a detachment underneath sand (Figure 1c). As a result of the buttressing force of the backstop, a foreland-propagating sequence (piggyback sequence) of thrusts formed a Coulomb wedge with a taper that is mainly dependent on the friction of detachment fault.

DIGITAL SANDBOX TECHNIQUE USING THE DISTINCT ELEMENT METHOD

Numerical solution techniques, such as the finite element method, finite difference, and boundary element method, are based on a continuum assumption and are extremely useful for geological sciences and geotechnical engineering. For modeling fractures and faults, however, these numerical methodologies are less powerful, because the continuum assumption prevents the separation of elements from each other, and this makes large amounts of strain localization very difficult (Burbidge and Braun, 2002). The DEM (Cundall and Strack, 1979) can overcome these difficulties, because the DEM approximates that a geologic body consists of numerous particles, and each particle is displaced independently. Therefore, the DEM allows discontinuous surfaces to form in the particle assembly.

The DEM simulation is based on the interactions of numerous particles. There are three stages in each calculation cycle: the first stage is to examine the contact relationship for each particle; the second stage is to evaluate the interaction forces for each particle; and the third stage is to move all particles by numerical integration of Newton's equation of motion for the evaluated external forces.

First of all, the DEM needs information on the contact of particles to calculate the interparticle forces. A simple method to check whether two particles in the assembly are in contact with each other requires $n(n-1)/2$ calculations, where n is the number of particles (e.g., Pande et al., 1990). Because this is impractical for a large number of particles, most DEM studies reduce the number of calculations by checking only the nearby particles (e.g., Pande et al., 1990; Carrillo et al., 1999; Burbidge and Braun, 2002). Our method uses a lattice defined by the minimum radii of the particles to avoid two particles being in the same square. When particle i is in a square, only particles in the surrounding squares have the possibility of contact with particle i. The size of the search area can be determined by the maximum radii of the particle assembly; a wider area should be searched when the assembly includes a large particle size. Then, the particles in the search area are examined for their contact situation (Carrillo et al., 1999).

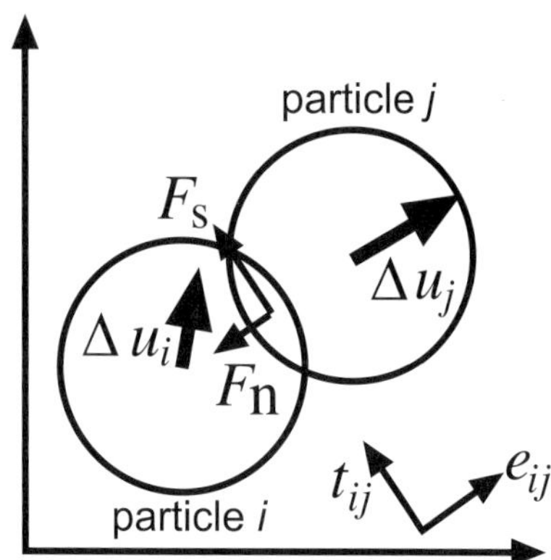

FIGURE 2. The force displacement law for two particles (modified after Cundall and Strack, 1979).

As the second stage of the computation cycle, the interaction forces are calculated according to the force-displacement law (Cundall and Strack, 1979), which is shown for two particles in contact in Figure 2.

The normal and tangential components of the relative displacement are expressed as $\Delta u^{\mathrm{n}}_{ij}$ and $\Delta u^{\mathrm{s}}_{ij}$ with

$$\Delta u^{\mathrm{n}}_{ij} = (\Delta u_i - \Delta u_j) \cdot e_{ij}$$

$$\Delta u^{\mathrm{s}}_{ij} = (\Delta u_i - \Delta u_j) \cdot t_{ij}$$

where Δu_i and Δu_j are displacements at a discrete time step for particles i and j. e_{ij} and t_{ij} are unit vectors in the normal and shear directions, respectively. From these relative displacements, the increments of normal and shear forces, $\Delta F^{\mathrm{n}}_{ij}$ and $\Delta F^{\mathrm{s}}_{ij}$, respectively, become

$$\Delta F^{\mathrm{n}}_{ij} = -k_{\mathrm{n}} \Delta u^{\mathrm{n}}_{ij}$$

$$\Delta F^{\mathrm{s}}_{ij} = -k_{\mathrm{s}} \Delta u^{\mathrm{s}}_{ij}$$

where k_{n} and k_{s} represent the normal and shear stiffness coefficients, respectively. These increment forces are added to the forces determined at the previous time step.

$$[F^{\mathrm{n}}_{ij}]_t = [F^{\mathrm{n}}_{ij}]_{t-\Delta t} + \Delta F^{n}_{ij}$$

$$[F^{\mathrm{s}}_{ij}]_t = [F^{\mathrm{s}}_{ij}]_{t-\Delta t} + \Delta F^{\mathrm{s}}_{ij}$$

Note that the shear force $[F^{\mathrm{s}}_{ij}]_t$ should be lower than the frictional limit defined by the normal force $[F^{\mathrm{s}}_{ij}]_t$ and the frictional coefficient μ.

$$[F^{\mathrm{s}}_{ij}]_t < \mu [F^{\mathrm{n}}_{ij}]_t$$

When the calculated shear force exceeds the frictional limit, the shear force is reduced to the limit value. The above procedure is carried out for all particles in the computational region.

In the third stage of computations, particles are moved because of the total acting forces, which were separately evaluated at the second stage. The Newtonian equation of motion

$$m_i \ddot{u}_i - C \dot{u}_i + F_i = 0$$

can be solved by explicit numerical integration, where m_i is the mass of particle i and C is the normal viscous damper coefficient equivalent to a restitution coefficient. The meaning of this viscous damper is difficult to explain geologically, but it might include the effects of energy consumption because of heating caused by interparticle friction. In our simulations of basic tectonics described later, gravity force was implemented to the integration in the vertical direction. From the velocity solutions of the equation, displacements at a discrete time step are obtained for all particles.

As an initial condition, the particles are bonded to each other to produce cohesion. When two neighboring particles have a distance of less than a maximum limit called a "bonding radius," the particles experience a tensional bonding force, and hence, the distance decreases. Once the particle distance exceeds the bonding radius, the bonding breaks, and no such pulling force will be further generated.

Faults are defined as a series of breakages of the bond between particles and can be detected in the calculation (e.g., Wang et al., 2000). They are expressed as discontinuity surfaces in the particle assembly with some displacement in marker horizons, which were appropriately constructed at the beginning of the deformation.

SIMULATIONS OF BASIC TECTONICS

Extension Simulations

As a basic simulation study, structural deformation under extension was examined. Figure 3a shows the initial closest packing condition of the homogeneous particle assembly, so that each particle is in contact with six neighboring particles, and the assembly contains preferred orientations for failure in three directions, along horizontal layers (bedding planes) and the 60° dipping planes. The assembly contains 3146 elements in an array of 61 (or 60) × 52. Two kinematic types were simulated: removing a fixed wall promptly (E-1, E-2) and pulling the wall gradually (E-3). To examine the shear stiffness coefficient k_{s}, two values for the parameter were employed; high k_{s} (10^4 kg/s^2) in E-1 and low k_{s} (10^3 kg/s^2) in E-2. In E-3, the shear stiffness coefficient k_{s} was the same as that in E-1 (10^4 kg/s^2). The velocity of the moving wall in E-3 was $1.0 \times 10^{-4}R$ for each time step (R is the particle radius; 0.1 m [0.33 ft]); thus, the deformation can be approximated as quasi-static. The frictional coefficient μ was 0.5, and the normal

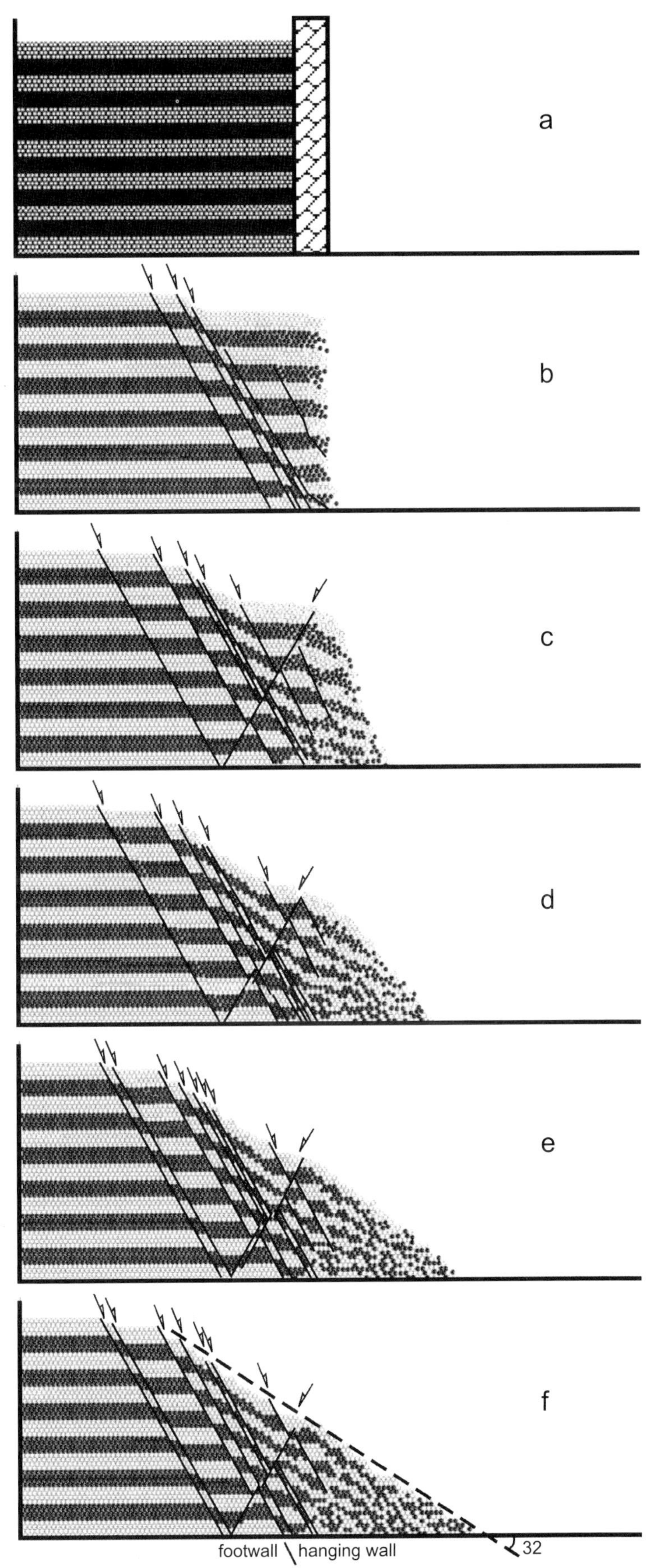

FIGURE 3. E-1 extension simulation results: (a) initial state; (b) after 100,000 time steps; (c) after 200,000 time steps; (d) after 300,000 time steps; (e) after 400,000 time steps; and (f) after 600,000 time steps (repose state).

stiffness coefficient k_n was 2.0×10^7 kg/s^2 in all simulations. Gravity force (9.8 m/s^2; 32 ft/s^2) was also applied to all simulations. Particle density was 2000 kg/m^3. To make the physical properties of the particles equivalent to those of the materials of sandbox experiments, no initial bonding was applied to the particles.

The E-1 simulation produced a series of synthetic faults dipping to the right (Figure 3). Most of the faults were concentrated in the hanging wall of a fault that divides the particle assembly into deformed and undeformed segments. This bounding fault was formed at a relatively early stage (200,000 time steps; Figure 3c), and the footwall of the fault remained undeformed afterward. After 600,000 time steps, the surface topography of the deformed segment showed a repose angle of 32° (Figure 3f).

In the E-2 simulation, normal faults were produced in two directions (Figure 4). The synthetic faults showed a simple planar geometry dipping 60° to the right, whereas the antithetic faults tended to produce broad shear zones dipping 45–50° to the left. After 1,400,000 time steps (Figure 4f), the topography attained a repose angle of 20°.

The E-3 simulation also produced normal faults in two directions (Figure 5), similar to those of the E-2 simulation. In the E-3 simulation, however, the deformation style adjacent to the moving wall was different because of different kinematics. A triangular region that is defined by the moving wall and the normal fault propagating from the base of the moving wall was characteristically generated (Figure 5f). A series of normal faults were generated from the base of the moving wall, and they consumed the triangular region from the left; thus, the area of the triangle became smaller as the moving wall displaced (Figure 5f).

Geologic Interpretation of Extension Simulations and Comparison with Sandbox Model

The extension simulations resulted in a series of planar normal faults, because the initial packing condition has a hexagonal geometry.

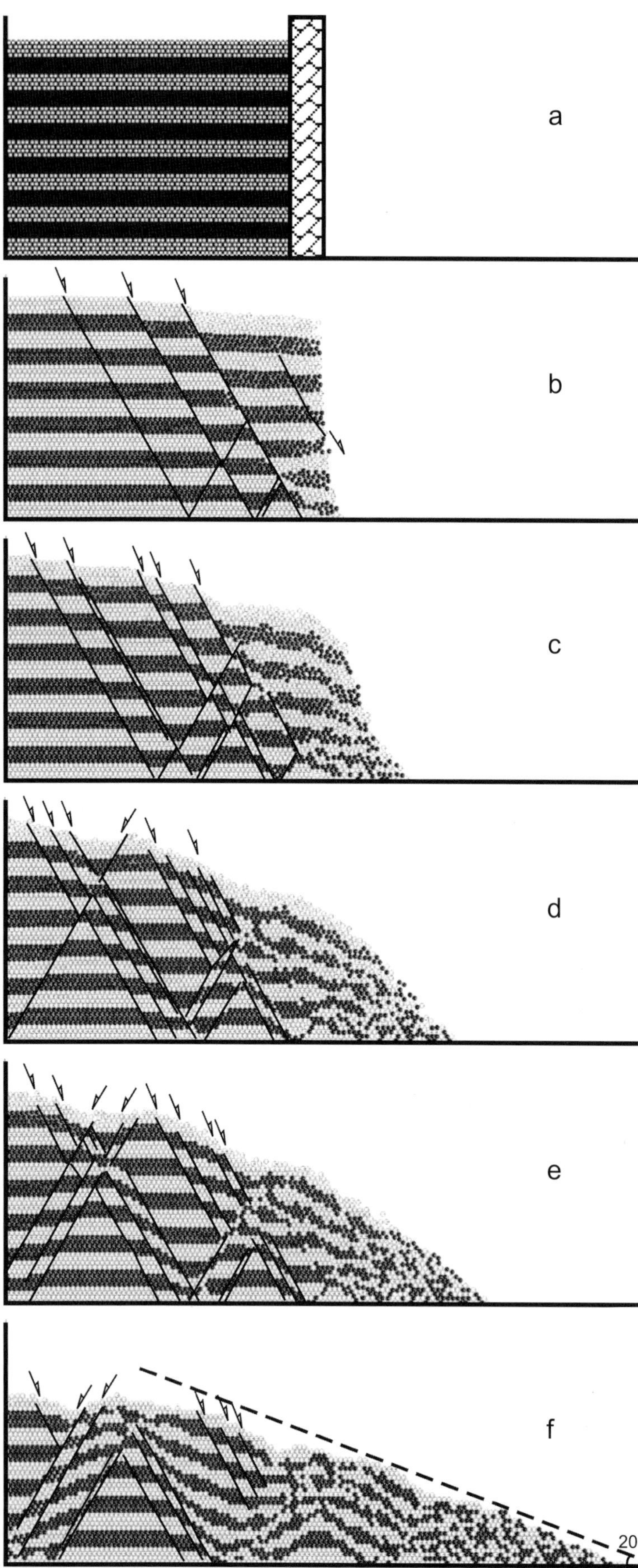

FIGURE 4. E-2 extension simulation results: (a) initial state; (b) after 100,000 time steps; (c) after 200,000 time steps; (d) after 300,000 time steps; (e) after 400,000 time steps; (f) after 1,400,000 time steps (repose state).

Faulting was triggered by the removal of a fixed backstop that rendered the strata unstable in the E-1 and E-2 simulations. As a result, E-1 produced a repose angle of 32°, whereas E-2 showed a repose angle of 20°. Because some of the actual values of input parameters can hardly be interpreted geologically, the bulk deformation styles represented by the repose angle, for example, should be employed to control the mechanics of the particle assembly. In E-1, the hanging wall of the deformation-bounding fault was characterized by ductile deformation behavior during sliding, whereas the footwall remained in a solid phase. This feature may be compared with natural landslides. Sandbox experiments for structural geology are commonly designed to simulate static deformation, and to the best of the author's knowledge, no experiment has so far been published to model dynamic deformation processes, like landslides, in accordance with the scaling theory of Hubbert (1937).

The E-3 simulation generated a deformation-bounding fault at an early stage, and conjugate sets of normal faults (synthetic and antithetic faults) are concentrated in the hanging wall. This feature is well correlated with a sandbox experiment shown in Figure 6, which is performed under similar boundary conditions. In the experiment, the bounding fault generated with a steep inclination (about 70°) immediately after the end wall started to be displaced (Figure 6b). The triangular assembly of sand grains, which formed by the same mechanism as that of the E-3 simulation, was gradually reduced in size in a similar manner as the simulation. Because the repose angle is much smaller than the dip of the bounding fault, the uppermost segment of the fault collapsed to form the slope (Figure 6e). This characteristic can also be found in sandbox experiments by Cloos (1968, their figures 20–23). Deformation in the E-3 simulation was static, and therefore, no ductile deformation was observed, again, similar to the experiment shown in Figure 6.

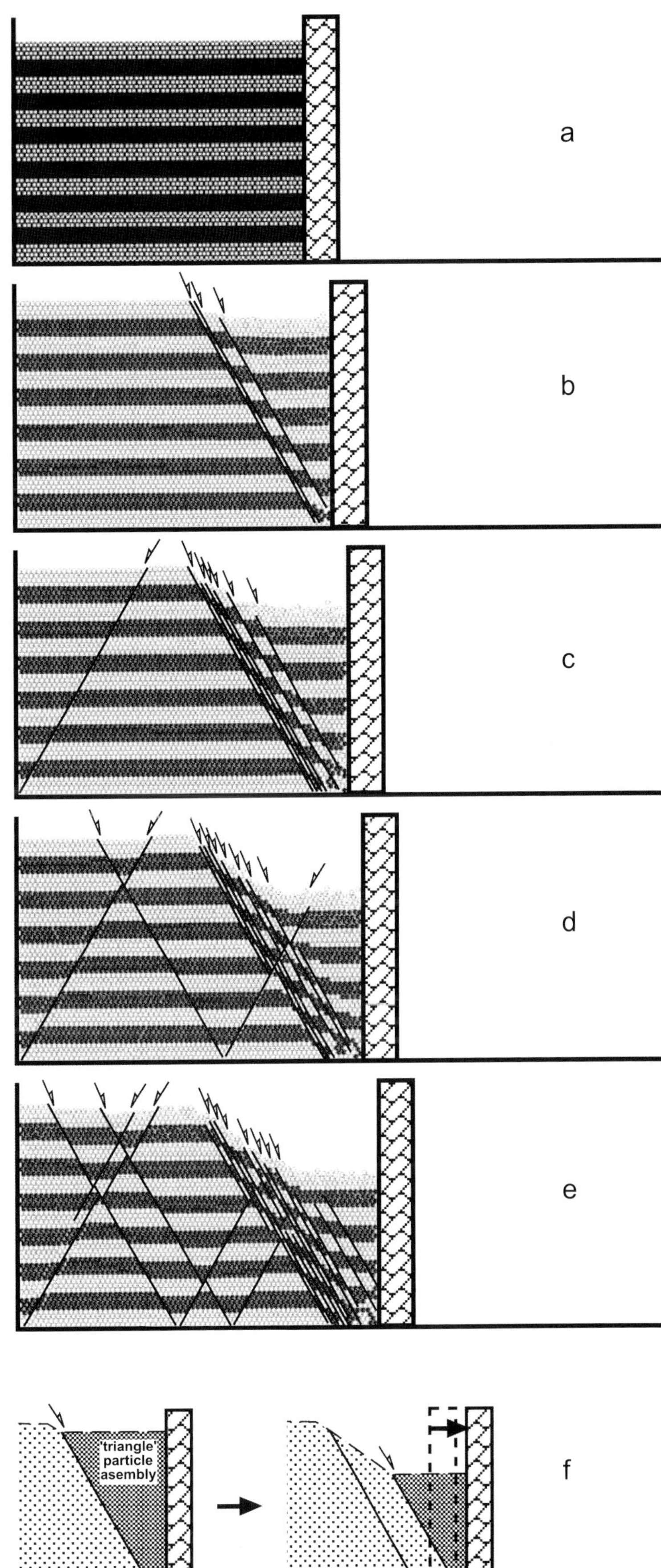

FIGURE 5. E-3 extension simulation results: (a) initial state; (b) after 120,000 time steps; (c) after 240,000 time steps; (d) after 360,000 time steps; (e) after 480,000 time steps; (f) simplified model of deformation adjacent to the moving wall.

Thrust Fault Simulations

A fold and thrust belt-type deformation can be simulated by pushing a moving wall into the particle assembly. In our thrust simulations, the initial arrangement of the particles was 200 (or 201) × 40 (i.e., 8020 particles in total) in the closest packing condition, and the particle size was homogeneous (Figure 7a). Two simulations were conducted (T-1 and T-2) with different input parameters of shear stiffness coefficient k_S: high stiffness (10^4 kg/s^2) in T-1 and low stiffness (10^3 kg/s^2) in T-2. The velocity of the moving wall was $5.0 \times 10^{-5}R$ for each time step; thus, the deformation can be approximated as quasistatic. Other parameters were the same as those of the extension simulations.

Figure 7 shows the progressive development of the deformation seen in the T-1 simulation. The first thrusts were of a conjugate set of a foreland-vergent thrusts (dipping to the right) and a hinterland-vergent thrust (dipping to the left), both initially dipping about 60° (Figure 7b). The foreland-vergent thrust continued to be active throughout, whereas the hinterland-vergent thrust ceased its activity by 200,000 time steps and was deformed by later folding. Then, new foreland-vergent thrusts formed on the footwall of the previous thrust, forming a piggyback sequence (Figure 7c). In the later stages of the T-1 simulation (Figure 7f), a thrust fault with a relatively gentle dip (about 30–35°) was generated in the highly deformed segment of the particle assembly. As the deformation proceeded, the top surface inclined toward the foreland, and the fold belt formed a wedge shape.

The T-2 simulation (Figure 8) produced deformation geometry similar to that of T-1; however, the deformation tends to propagate further away when compared to that at the same time step in T-1. For instance, the deformation reached the fixed end wall after 2,000,000 time steps in T-2 (Figure 8f). The deformation style of T-1 (Figure 7f) is, by contrast, similar to that after 1,250,000 time steps of T-2 (Figure 8e).

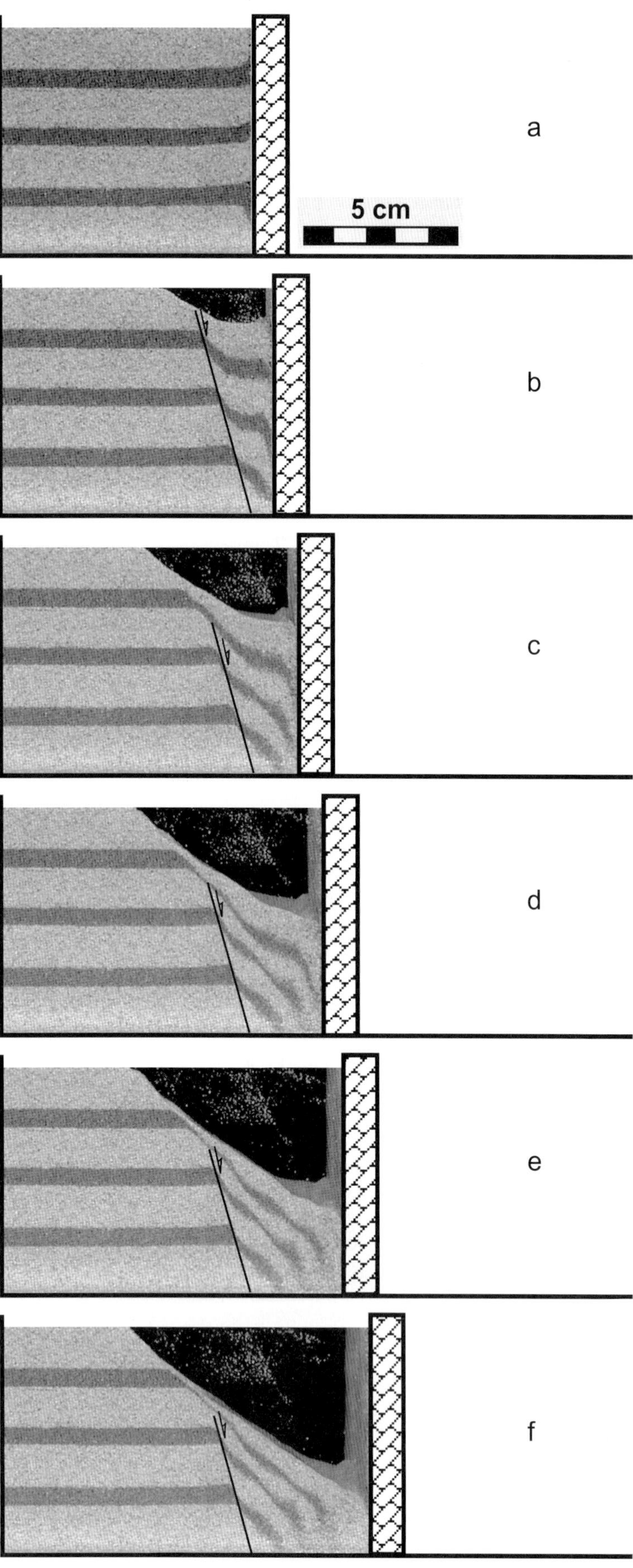

Figure 6. Extension experiment: (a) initial state; (b) after 0.7 cm (0.27 in.) of extension; (c) after 1.4 cm (0.55 in.) of extension; (d) after 2.2 cm (0.87 in.) of extension; (e) after 2.9 cm (1.14 in.) of extension; (f) after 3.8 cm (1.5 in.) of extension.

Geologic Interpretation of Thrust Simulations and Comparison with Analog Models

The T-1 and T-2 simulations generated thrust faults in two directions, which can also be seen in sandbox experiments (e.g., Figure 1c). In sandbox experiments, however, each fault initiates with a lower inclination (about 20–30°) and then rotates to an increased dip as the deformation progresses and the next fault develops in its footwall. The numerical simulation does not capture this feature, presumably because of the existence of preferred orientations for failure by the packing condition and the isotropic particle size in the simulation. These orientations constrain the faults to have a planar geometry with a dip of 60° and the hanging wall to undergo minor rotation. This rotation is supported by the fact that the low-angle thrust was formed in the disturbed segment of the particle assembly where the initial preferred orientations were destroyed. The overall geometry of the deformed strata shows a wedge shape, similar to the deformation geometry of the thrust wedge sandbox experiments (Figure 1c).

Effects of Input Parameters

The effects of the shear stiffness coefficient k_s can be examined with the simulation results. By comparing E-1 and E-2 results, our simulations demonstrated that the shear stiffness coefficient k_s has a strong control on the final deformation geometry. That is, a small value of the shear stiffness coefficient k_s produces gentle topography of the upper surface of the particle assembly. This suggests that the shear stiffness coefficient k_s in the DEM was a controlling factor in deformation styles similar to the internal friction angle of granular materials. This is further supported by the results of the T-1 and T-2 simulations, which indicate that deformation with a smaller value of k_s propagates further away.

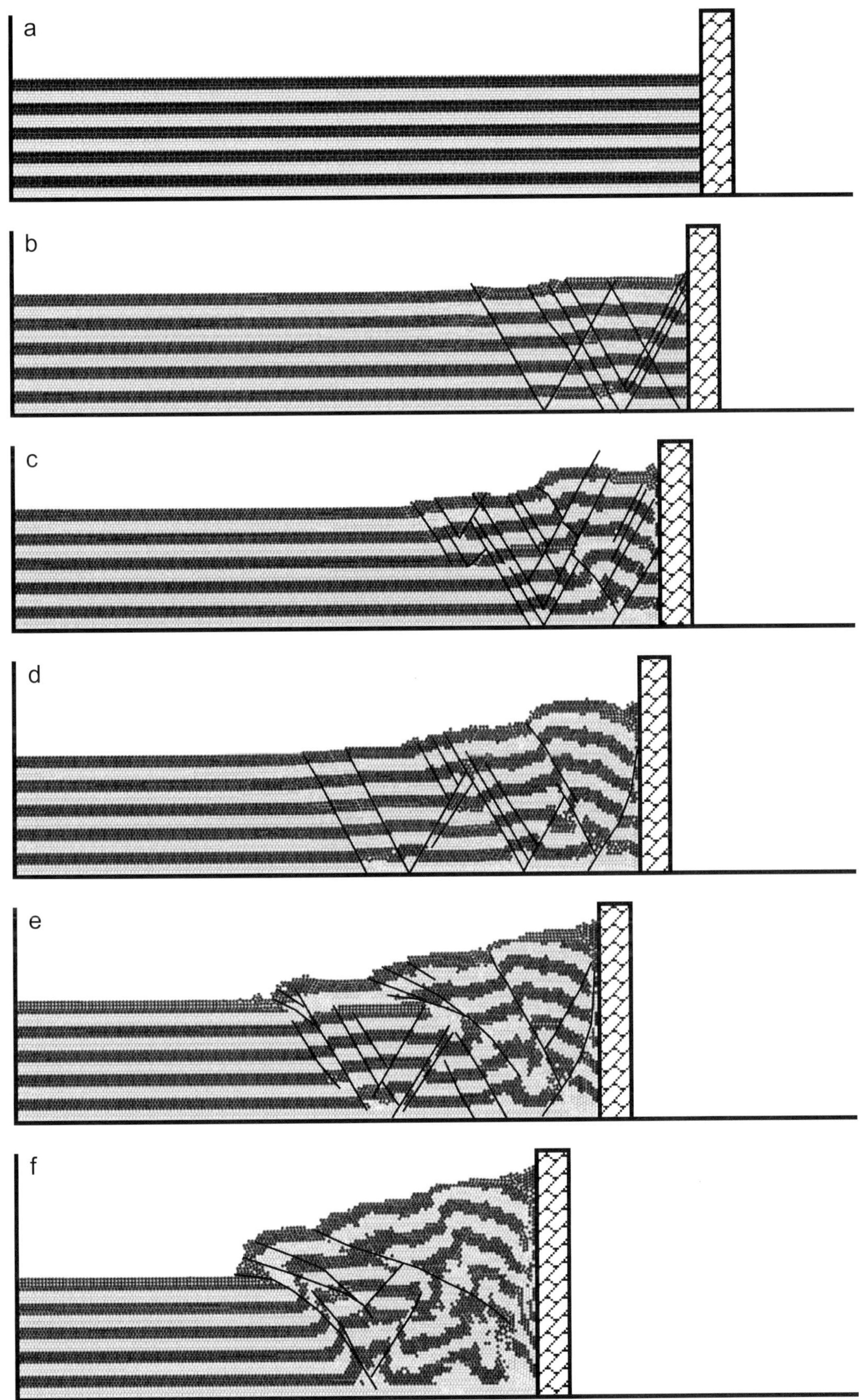

Figure 7. T-1 thrust simulation results: (a) initial state; (b) after 150,000 time steps; (c) after 500,000 time steps; (d) after 750,000 time steps; (e) after 1,250,000 time steps; (f) after 2,000,000 time steps.

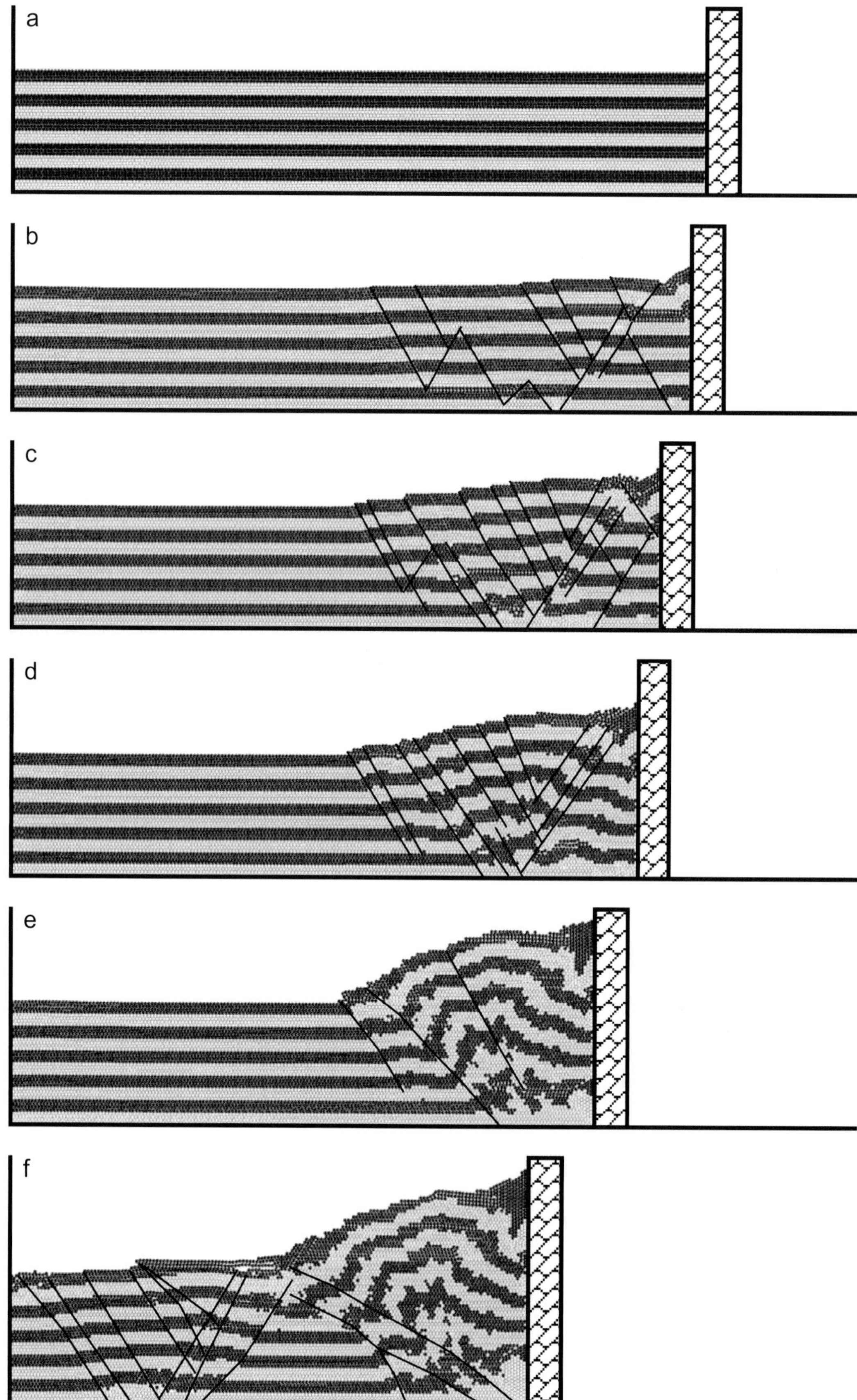

FIGURE 8. T-2 thrust simulation results: (a) initial state; (b) after 150,000 time steps; (c) after 500,000 time steps; (d) after 750,000 time steps; (e) after 1,250,000 time steps; (f) after 2,000,000 time steps.

FIGURE 9. Major tectonic features in the eastern Asia (modified from Tapponnier et al., 1986).

SIMULATION OF INDENTATION TECTONICS

Continental Collision of India

Here, we present an application of the DEM to the collision-escape tectonics of the Indian Plate to the Eurasian Plate, which is one of the best examples of active tectonics on the Earth (e.g., Sorkhabi and Macfarlane, 1999). This collision-escape tectonics has also been referred to as "indentation tectonics" (Figure 9) (Tapponnier et al., 1982, 1986), and a series of physical experiments using analog materials, such as combinations of dry quartz sand and silicone polymer (Figure 10a) (Davy and Cobbold, 1988) and layered plasticine (Figure 10b) (Tapponnier et al., 1982, 1986), have been conducted to investigate the lithospheric deformation of indentation tectonics. These experimental results have constrained the analysis of the India Asia tectonics and have aided our understanding of basin formation and deformation in Asia as a consequence of the indentation tectonics.

DEM Simulations

In our simulations, parameters were defined to reproduce the deformation configuration of the analog experiments on indentation tectonics (Figure 11). The two plates were modeled as rectangular particle assemblies at the initial stage. The small (203 particles) assembly corresponding to India was essentially rigid (having relatively little internal deformation), and its initial dimension was 4000 × 2000 km (2485 × 1243 mi). The rigidity of the Indian block was achieved by ignoring the calculated forces and using the initial velocity throughout the deformation process. The large (2014 particles) assembly corresponding to Eurasia was allowed to deform, and its initial dimension was 10,000 × 8000 km (6214 × 4971 mi). The assembly was placed at the corner of a folded fixed wall but not bonded to the wall. The

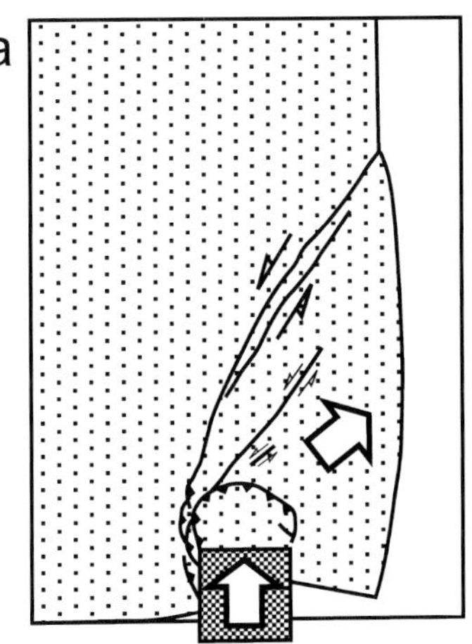

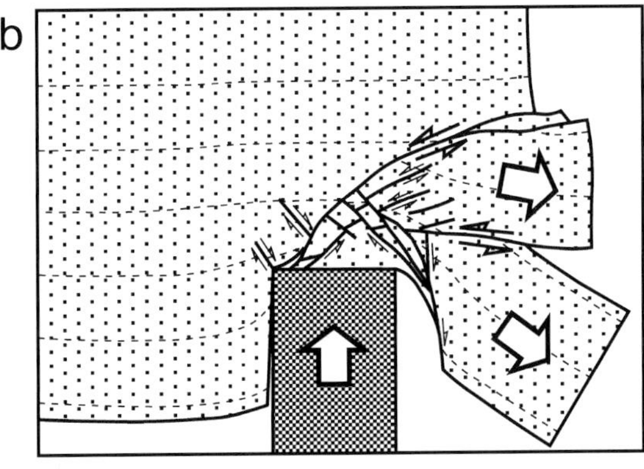

FIGURE 10. Examples of analog experiments: (a) sand and polymer model by Davy and Cobbold (1988); (b) layered clay model by Tapponnier et al. (1986).

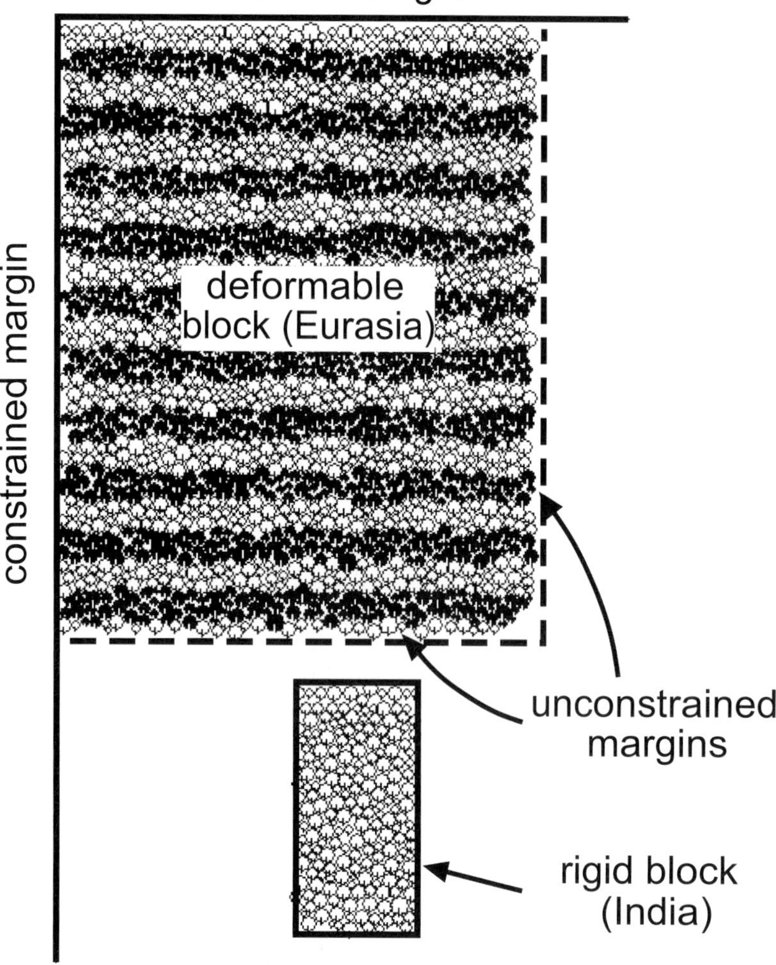

FIGURE 11. Initial boundary conditions of the indentation simulations.

eastern and southern margins of the assembly were unconstrained (Figure 11). The diameter of the circular particles was randomized ±40% from 200 km (124 mi) (the average) to avoid the preferred orientations for failure in three directions observed in the basic tectonic simulations. The idea of this variation in the particle size has been also employed by other authors (e.g., Scott, 1996; Strayer and Suppe, 2002; Finch et al., 2003). The normal and shear stiffness coefficients (k_n and k_s) are 2.0×10^7 and 1.0×10^4 kg/s^2, respectively. Each time step for calculation was 600 yr, which corresponded to the movement of the small assembly 30 m (100 ft) to the north (5 cm/yr; 2 in./yr). The bonding radius was $0.1R$ for each particle. Because this is 2-D simulation in plan view, no gravity force was applied. Other parameters were the same as those of basic tectonics presented earlier.

The results of four DEM simulations on indentation tectonics (I-1 to I-4) are described here. The differences among them are caused by the relative position of the indenter and the bonding condition between particles. The three particle assemblies shown in Figure 12a–c are homogeneously bonded at the beginning, and the black-and-white layers are for visualization purpose only. Particles in the fourth simulation shown in Figure 12d are also bonded to each other but are unbonded at the black-and-white boundaries, allowing flexural slip during deformation.

The I-1 indentation simulation with a homogeneous bonding condition shows that the indentation of the small block, whose center was positioned 3500 km (2175 mi) from the eastern margin of the large assembly, produced a progressive development of fault systems (Figure 12a). The faults, represented as the lines of gaps between the particles, are seen in the eastern region of the layers, and the activity was generally intermittent. The assembly to the east of the indenter is segmented into two blocks and displaced to the east. The block close to the indenter has clearly rotated in a clockwise direction. The outline of the eastern margin of the particle assembly is kinked at the fault and broadened to the east to compensate for the space problem because of the indentation. The layers in front of the indenter decrease in thickness, suggesting that the region has been highly compressed by the indentation.

In the I-2 simulation (Figure 12b), the location of the indenter is moved 500 km (310 mi) to the east, and the center of the indenter is 3000 km (1860 mi) from the eastern margin of the large assembly. This produces two major sinistral (left-lateral) strike-slip faults initiating from the western corner of the indenter and propagating to the eastern margin of the particle assembly. As the displacement of the faults increases, the particle assemblies segmented by the faults are displaced toward the east and rotated in a clockwise direction. The outline of the assembly is gently curved but very similar to that of the I-1 simulation.

In the I-3 simulation (Figure 12c), the indenter position is moved to the west, and the deformation shows a major dextral (right-lateral) strike-slip fault initiated from the eastern corner of the indenter, propagating to the western margin of the particle assembly. As the displacement of the fault increases, the particle assembly on the western side of the fault moves forward and pushes the fixed wall. This causes the particle assembly to the east of the fault to move to the east. A circular-shaped gap along the fault is also a notable feature (Figure 12c). Other faults are not clearly observed.

The outline of the particle assembly is deformed on the western side to accommodate displacement of the indenter.

The I-4 result (Figure 12d) was obtained under conditions that the layer boundaries had no bonding, and that the indenter position was the same as that of I-3. The faults are not clearly developed, but a flexural slip surface can be seen at a layer boundary in front of the compressed region. The region fragmented by the slip surface has been pushed to the east.

Geometric Comparison with Analog Models

Tapponnier et al. (1982, 1986) conducted experiments under a boundary condition similar to our DEM simulations, i.e., the fixed wall on the left and back sides of the particle assembly and the unconstrained right and front sides. The models are characterized by a sequential development of major fault systems and the rotation of fault-bounded continental blocks (Figure 10b). This feature was reproduced by the I-1 simulation (Figure 12a). The overall geometries at the final stage of the deformation are also similar in both these models (sandbox and DEM simulation).

The analog experiments by Davy and Cobbold (1988) employed sand and polymer materials, which were properly scaled for lithospheric deformation. In their models, a major strike-slip fault was commonly developed that was associated with minor faults and basins (Figure 10a). The major fault initiated on the left corner of the indenter and propagated to the unconstrained right margin of the sand layer. These features, as well as the overall deformation geometry, are similar to those of the I-2 results (Figure 12b), and they are a mirror image of the I-3 results (Figure 12c). The relative position of the basin formed along the fault is also similar to that seen in I-3. In the frontal region of the indenter, thrust faults were produced perpendicular to the indentation in the experiment (Figure 10a). This is similar to that observed in the I-2 and I-3 results, where the layers became thinner in the frontal side.

The I-4 simulation produced a layer-parallel slip, and the deformation hardly reached the northern margin of the assembly (Figure 12d). This might explain why the major faults in the models by Tapponnier et al. (1982, 1986) are commonly subperpendicular to the right margin of the experimental material. In other words, the boundary surfaces of the layered plasticine are probably weak and easily cause slippage on them. Therefore, indentation produced a sequence of faults subparallel to the clay layers in the experiments. In I-4, similar to the experiments of Tapponnier et al. (1982), faults developed with a curved geometry caused by layer-parallel slip.

Several different features exist between the analog experiments and our simulation results. Most of them are brought by the particle size problem that significantly affects the resolution of deformation structures. For example, the secondary fault systems that developed in the analog experiments were absent in the simulation probably because of the larger size of the particles.

APPLICABILITY OF DEM SIMULATIONS TO PETROLEUM TRAPS AND BASIN FORMATION AND DEFORMATION

Our DEM simulation results reproduced the geometric features of structural deformation consistent with both analog experiments and natural examples. Here, we focus our discussion on the applicability of these results to petroleum basins that have experienced similar tectonics.

The basic simulations of extensional and compressional faults are still in a preliminary stage, and the parameters need to be further refined, but the results can be applied to natural structures. Similar to the analog sandbox experiments by Cloos (1968) and Keep and McClay (1997), the E-3 simulation produced a series of planar normal faults that dip to the basin center. This feature can be compared with the basin-bounding faults in the Gulf of Suez (e.g., Lyberis, 1988; Khalil and McClay, 2001), where characteristic planar normal faults and their associated hanging-wall deformation structures strongly control the petroleum traps in the basin. The thrust simulations also show characteristic features found in fold and thrust belts in petroleum provinces, such as the piggyback sequence of thrusting in Papua New Guinea (e.g., Buchanan, 1996).

The indentation simulations produced the extrusion of continental blocks along major strike-slip fault systems that can be correlated with the natural deformation architecture. For example, a left-lateral strike-slip fault in the I-1 simulation that separates a fragment of the particle assembly (Figure 12a) corresponds to the Red River fault in northern Vietnam (Figure 9). The Red River fault is a major tectonic line and significantly controls the structural styles and, thus, the petroleum potential in the Red River delta and the Tonkin Gulf, Vietnam (e.g., Rangin et al., 1995). The intermittent activity of the fault observed in the simulation may also represent a stick-slip motion along the fault system. The relationship between the activity along the Red River fault and the opening of the South China Sea has been hotly debated (e.g., Tapponnier et al, 1986; Roques et al., 1997), and the simulation results might support direct links between these two tectonic events, because the oceanic lithosphere offshore Vietnam constrained

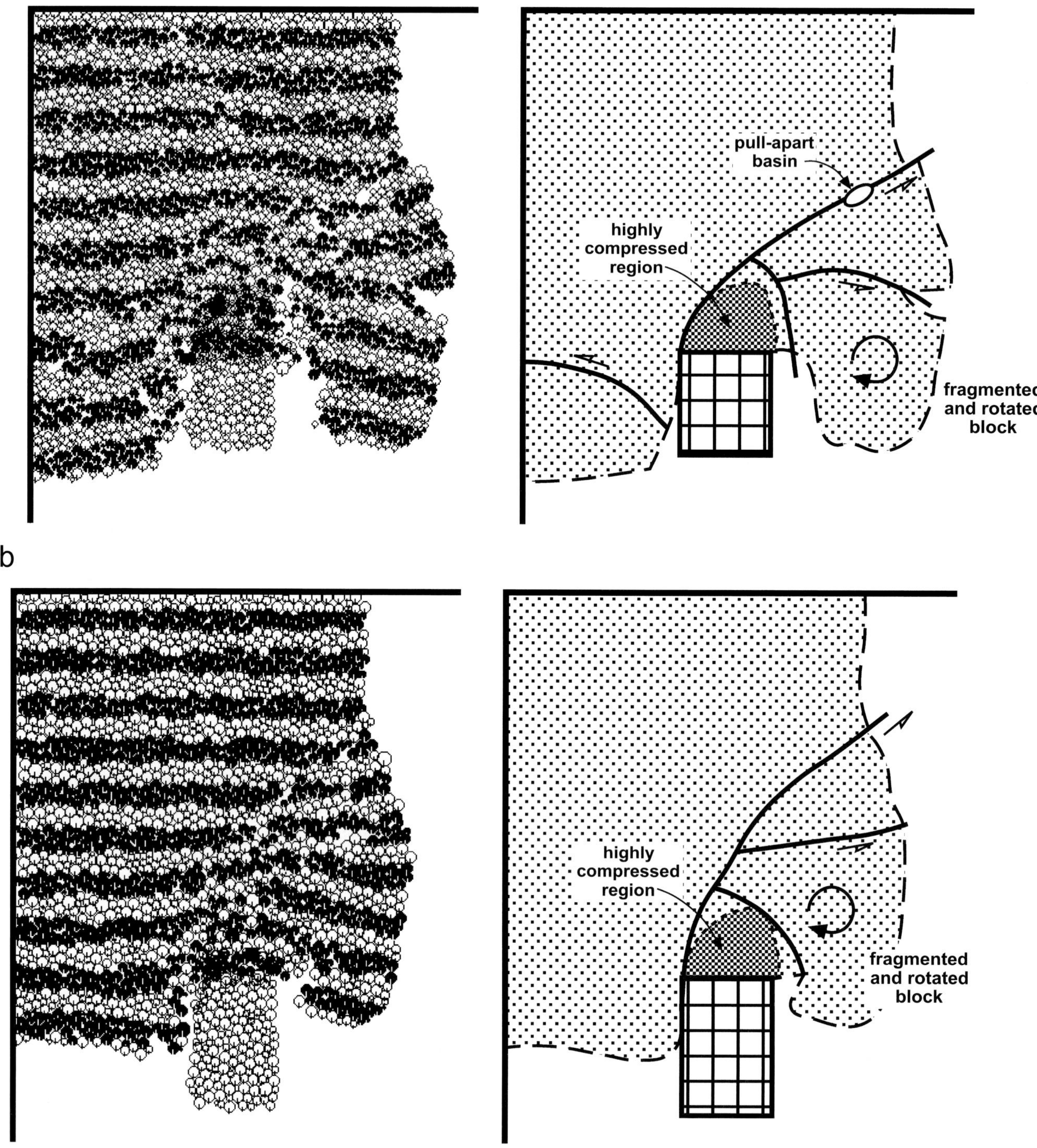

FIGURE 12. Indentation simulation results: (a) I-1; the indenter position is at the center; (b) I-2; the indenter position is moved to the east; (c) I-3; the indenter position is moved to the west; (d) I-4; layered particle configuration with the indenter being at the west.

the deformation at the time of the South China Sea opening. The shearing along the east coast of Vietnam, which might be responsible for the opening of the South China Sea, was not clearly observed in our simulations, but this needs further study with a change in the boundary condition at the unconstrained margin of the particle assembly.

A significant advantage of the DEM technique, in comparison to other simulation methods based on continuum assumptions, is that faulting can be appropriately

c

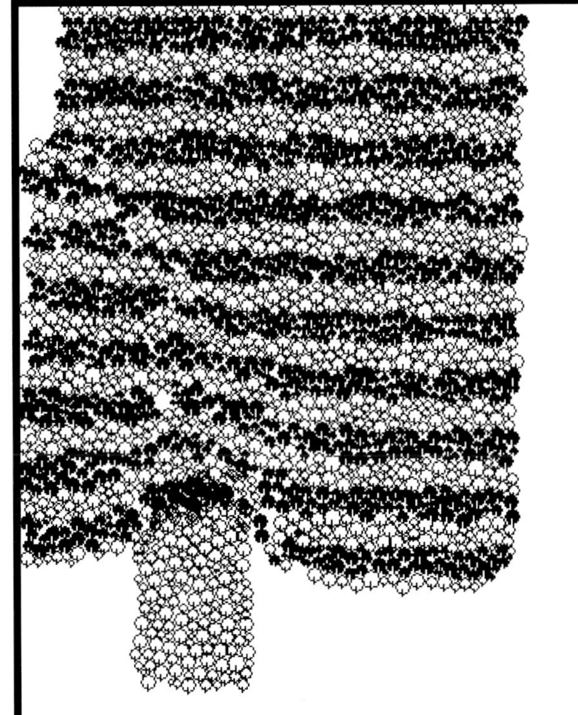

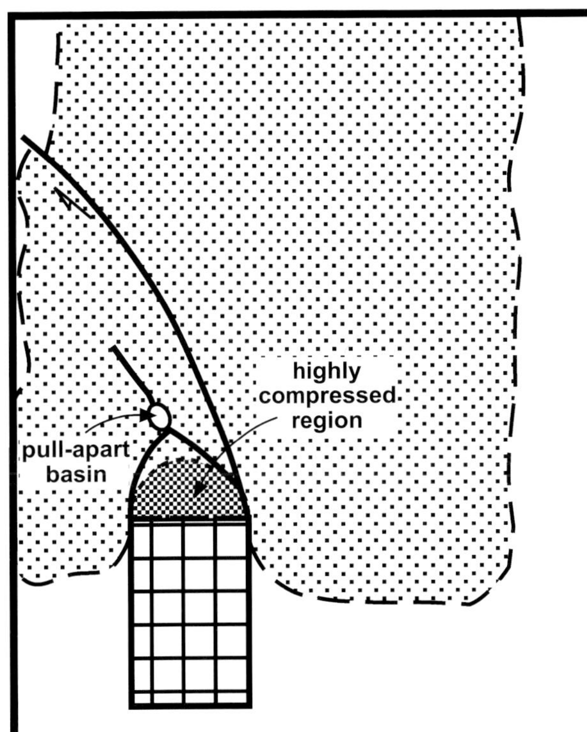

d

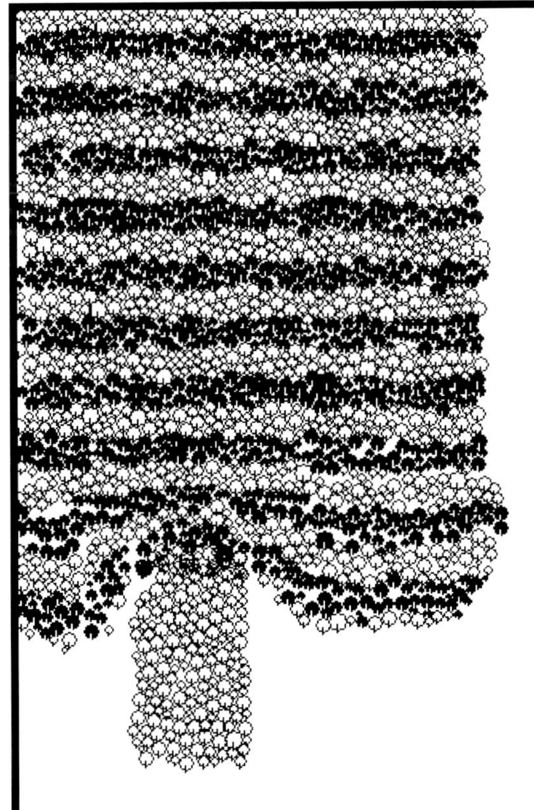

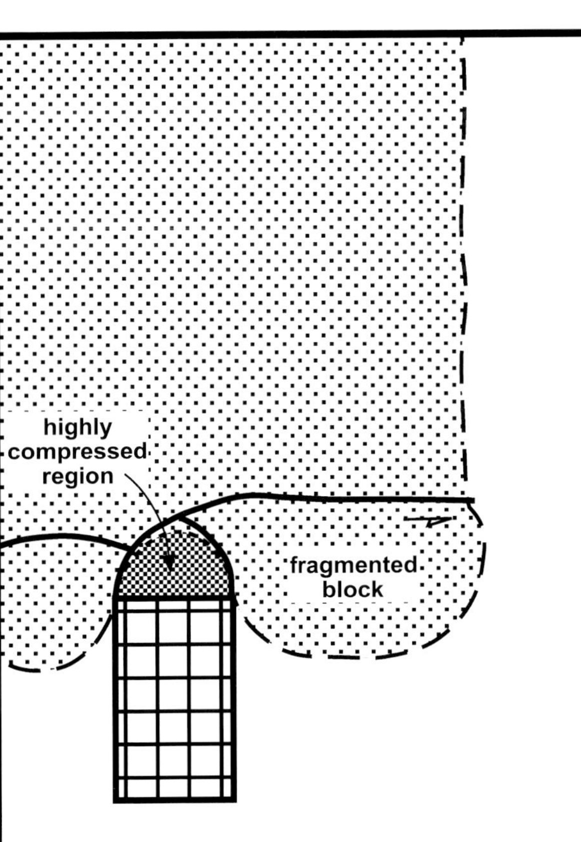

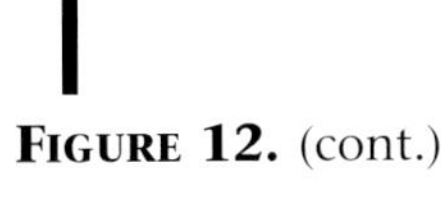

Figure 12. (cont.).

modeled. This is achieved by the granular nature of the particle assembly, which follows the brittle deformation behavior of the upper crust.

CONCLUSIONS

The overall deformation styles observed in the digital simulation results of this study correlate well with those of the analog experiments carried out under similar boundary conditions. In particular, geometry and development process of the faults in analog experiments were properly reproduced in the DEM simulations. This strongly suggests that DEM can be a useful tool to simulate fault-related tectonics under extensional, contractional, and strike-slip tectonic regimes.

The characteristic features of the indentation tectonics, such as fragmentation and rotation of continental blocks, were clearly reproduced in our DEM simulation. Our results demonstrated that the geometric variations in the experiments of Tapponnier et al. (1982, 1986) and Davy and Cobbold (1988) could be generated by the position of the colliding India, under conditions of our input parameters. The results also demonstrated that the well-known experiments by Tapponnier et al. (1982, 1986) might be significantly affected by the layered nature of the material, which caused layer-parallel slip and produced curved faults that were subperpendicular to the right unconstrained margin of the experimental material.

The significance of the DEM digital modeling technique is that the method can incorporate discontinuous surfaces properly and is thus a powerful tool to simulate fault-related phenomena, such as structural traps in sedimentary basins. The method is a forward modeling technique and can be used to predict deformation features. Another advantage of the method is the fact that the technique can overcome the problems of the conventional sandbox experiments that need specially designed apparatus and skills.

ACKNOWLEDGMENTS

We acknowledge grants funded by the Ministry of Education, Culture, Sports, Science, and Technology of Japan and the Japan National Oil Corporation (presently Japan Oil, Gas and Metals National Corporation) for this study. Permission from the Japan Petroleum Exploration Co., Ltd. to publish this work is also acknowledged. Emma Finch, David Boutt, and Rasoul Sorkhabi reviewed the chapter, and their constructive comments greatly improved the manuscript. The authors alone are, however, responsible for the content of this chapter.

REFERENCES CITED

Antonelli, M. A., and D. D. Pollard, 1995, Distinct element modeling of deformation bands in sandstone: Journal of Structural Geology, v. 17, p. 1165–1182.

Buchanan, P. G., 1996, Petroleum exploration, development and production in Papua, New Guinea: Proceedings of the Third Papua New Guinea (PNG) Petroleum Convention, Port Moresby, Chamber of Mines and Petroleum, September 9–11, 1996, 822 p.

Buchanan, P. G., and K. R. McClay, 1991, Sand box experiments of inverted listric and planar fault systems: Tectonophysics, v. 188, p. 97–115.

Burbidge, D. R., and J. Braun, 2002, Numerical models of the evolution of accretionary wedges and fold-and-thrust belts using the distinct element method: Geophysics Journal International, v. 148, p. 542–561.

Carrillo, A. R., J. E. West, D. A. Horner, and J. F. Peters, 1999, Interactive large-scale soil modeling using distributed high performance computing environments: International Journal of High Performance Computing Applications, v. 13, p. 33–48.

Cloos, E. 1968, Experimental analysis of Gulf Coast fracture patterns: AAPG Bulletin, v. 52, p. 420–444.

Cobbold, P. R., and L. Castro, 1999, Fluid pressure and effective stress in sandbox models: Tectonophysics, v. 301, p. 1–19.

Colletta, B., J. Letouzey, R. Pinedo, J. F. Ballard, and P. Bale, 1991, Computerized x-ray tomography analysis of sandbox models: Examples of thin-skinned thrust systems: Geology, v. 19, p. 1063–1067.

Cundall, P. A., and O. D. L. Strack, 1979, A discrete numerical model for granular assemblies: Geotechnique, v. 29, p. 47–65.

Davy, P., and P. R. Cobbold, 1988, Indentation tectonics in nature and experiment: 1. Experiments scaled for gravity: Bulletin of the Geological Institutions of Uppsala, N. S., v. 14, p. 129–141.

Donze, F., P. Mora, and S. Magnier, 1994, Numerical simulation of faults and shear zones: Geophysical Journal International, v. 116, p. 46–52.

Finch, E., S. Hardy, and R. Gawthorpe, 2003, Discrete element modelling of contractional fault-propagation folding above rigid basement fault blocks: Journal of Structural Geology, v. 25, p. 515–528.

Horsfield, W. T., 1977, An experimental approach to basement controlled faulting: Geologie en Mijnbouw, v. 56, p. 363–370.

Hubbert, M. K., 1937, Theory of scaled models as applied to the study of geological structures: Bulletin of Geological Society of America, v. 48, p. 1459–1520.

Huiqi, L., K. R. McClay, and D. Powell, 1992, Physical models of thrust wedges, *in* K. R. McClay, ed., Thrust tectonics: London, Chapman and Hall, p. 71–81.

Keep, M., and K. R. McClay, 1997, Analog modeling of multiphase rift systems: Tectonophysics, v. 273, no. 3–4, p. 239–270.

Khalil, S., and K. R. McClay, 2001, Tectonic evolution of the northwestern Red Sea–Gulf of Suez rift, *in* R. C. L. Wilson, R. B. Whitmarsh, B. Taylor, and N. Froitzheim, eds., Non-volcanic rifting of continental margins: A comparison of evidence from land and sea: Geological Society (London) Special Publication 187, p. 453–473.

Koyi, H., 1997, Analogue modelling; from a qualitative to a quantitative technique, a historical outline: Journal of Petroleum Geology, v. 20, p. 223–238.

Lallemand, S. E., J. Malavieille, and S. Calassou, 1992, Effects of oceanic ridge subduction on accretionary wedges: Experimental modelling and marine observations: Tectonics, v. 11, no. 6, p. 1301–1313.

Lyberis, N., 1988, Tectonic evolution of the Gulf of Suez and the Gulf of Aqaba: Tectonophysics, v. 153, p. 209–220.

McClay, K. R., 1990a, Deformation mechanics in analogue models of extensional fault systems, *in* E. H. Rutter and R. J. Knipe, eds., Deformation mechanisms, rheology and tectonics: Geological Society (London) Special Publication 54, p. 445–454.

McClay, K. R., 1990b, Extensional fault systems in sedimentary basins: A review of analogue model studies: Marine and Petroleum Geology, v. 7, p. 206–233.

McClay, K. R., and P. G. Ellis, 1987, Geometries of extensional fault systems developed in model experiments: Geology, v. 15, p. 341–344.

Oda, M., and K. Iwashita, 1999, Mechanics of granular materials: Rotterdam, A. A. Balkema, p. 383.

Pande, G. N., G. Beer, and J. R. Williams, 1990, Numerical methods in rock mechanics: Chichester, John Wiley & Sons, 327 p.

Place, D., and P. Mora, 1999, The lattice solid model to simulate the physics of rocks and earthquakes: Incorporation of friction: Journal of Computer and Physics, v. 150, p. 332–372.

Rangin, C., M. Klein, D. Roques, X. Le Pichon, and T. Le Van, 1995, The Red River fault system in the Tonkin Gulf, Vietnam: Tectonophysics, v. 243, p. 209–222.

Roques, D., C. Rangin, and P. Huchon, 1997, Geometry and sense of motion along the Vietnam continental margin: Onshore/offshore Da Nang area: Bulletin de la Societe Géologique de France, v. 168, p. 413–422.

Saltzer, S. D., 1993, Boundary conditions in sandbox models of crustal extension: An analysis using distinct elements: Tectonophysics, v. 215, p. 349–362.

Saltzer, S. D., and D. D. Pollard, 1992, Distinct element modeling of structures formed in sedimentary overburden by extensional reactivation of basement faults: Tectonics, v. 11, p. 165–174.

Scott, D. R., 1996, Seismicity and stress rotation in a granular model of the brittle crust: Nature, v. 381, p. 592–595.

Sorkhabi, R. B., and A. Macfarlane, 1999, Himalaya and Tibet: Mountain roots to mountain tops, *in* A. Macfarlane, R. B. Sorkhabi, and J. Quade, eds., Himalaya and Tibet: Mountain roots to mountain tops: Geological Society of America Special Paper 328, p. 1–8.

Strayer, L. M., and J. Suppe, 2002, 3-D mechanical modeling of thrust propagation: Distinct-element method: Journal of Structural Geology, v. 24, p. 637–650.

Tapponnier, P., G. Peltzer, R. Armijo, A.-Y. Le Dain, and P. Cobbold, 1982, Propagating extrusion tectonics in Asia: New insights from simple experiments with plasticine: Geology, v. 10, p. 611–616.

Tapponnier, P., G. Peltzer, and R. Armijo, 1986, On the mechanics of the collision between India and Asia, *in* M. P. Coward and A. C. Ries, eds., Collision tectonics: Geological Society (London) Special Publication 19, p. 115–157.

Wang, Y. C., X. C. Yin, F. J. Ke, and M. F. Xia, 2000, Numerical simulation of rock failure and earthquake process on mesoscopic scale: Pure and Applied Geophysics, v. 157, p. 1905–1926.

Yamada, Y., 1999, 3-D analogue modelling of inversion structures: Ph.D. thesis, Royal Holloway, University of London, London, 743 p.

Yamada, Y., and K. R. McClay, 2003a, Application of geometric models to inverted listric fault systems in sandbox experiments: 1. 2-D hanging-wall deformation and section restoration: Journal of Structural Geology, v. 25, p. 1551–1560.

Yamada, Y., and K. R. McClay, 2003b, Application of geometric models to inverted listric fault systems in sandbox experiments: 2. Insights for possible along strike migration of material during 3-D hanging-wall deformation: Journal of Structural Geology, v. 25, p. 1331–1336.

Yamada, Y., and K. R. McClay, 2004, 3-D analogue modelling of inversion thrust structures, *in* K. R. McClay, ed., Thrust tectonics and hydrocarbon systems: AAPG Memoir 82, p. 276–301.

Ferrill, D. A., A. P. Morris, D. W. Sims, D. J. Waiting, and S. Hasegawa, 2005, Development of synthetic layer dip adjacent to normal faults, *in* R. Sorkhabi and Y. Tsuji, eds., Faults, fluid flow, and petroleum traps: AAPG Memoir 85, p. 125–138.

Development of Synthetic Layer Dip Adjacent to Normal Faults

David A. Ferrill

Center for Nuclear Waste Regulatory Analyses (CNWRA), Southwest Research Institute, San Antonio, Texas, U.S.A.

Alan P. Morris

Division of Earth and Environmental Science, University of Texas at San Antonio, San Antonio, Texas, U.S.A.

Darrell W. Sims

Center for Nuclear Waste Regulatory Analyses (CNWRA), Southwest Research Institute, San Antonio, Texas, U.S.A.

Deborah J. Waiting

Center for Nuclear Waste Regulatory Analyses (CNWRA), Southwest Research Institute, San Antonio, Texas, U.S.A.

Shutaro Hasegawa[1]

Technology Research Center, Japan National Oil Corporation, Chiba, Japan

ABSTRACT

Field analyses of normal faulting illustrate that synthetic layer dip associated with normal faults is a common feature of extensional fault systems. These synthetic dip panels are developed where layers on upthrown, downthrown, or both sides of a normal fault dip toward the downthrown side of the fault. Synthetic dip panels adjacent to normal faults should be expected at some scale in all normal fault systems. In addition to faults that developed in the strata with a regional dip, five fault-related mechanisms for the development of synthetic dip are faulted monocline (fault tip-line folding), antilistric fault bend, distributed shear, shear in relay zone of vertically and/or laterally segmented faults, and fault block impingement and lateral contraction. Development of synthetic dip accommodates a component of throw by tilting or folding, thereby reducing the offset or true displacement on the related normal faults. Fault

[1]*Present address:* Idemitsu Oil & Gas Co., Tokyo, Japan.

DOI:10.1306/1033720M853133

block deformation is strongly dependent on the mechanisms that produce synthetic dip panels and may influence fault zone and fault block permeability. Depending on stratigraphic and structural relationships, synthetic dip panels can produce a downthrown closure for hydrocarbon trapping, provide fluid migration and/or production communication pathways across faults, or produce barriers to fluid communication across faults.

INTRODUCTION

In many cases, the strata in the hanging walls or footwalls of normal faults dip in the direction opposite (antithetic) to the faults. This widely recognized antithetic dip commonly occurs as large-scale dip domains and is typically produced by slip on listric fault surfaces or slip on an array of parallel normal faults (also known as domino or bookshelf faulting) (cf. Hamblin, 1965; Wernicke and Burchfiel, 1982; Mandl, 1988; Rouby et al., 1996; Ferrill and Morris, 1997; Ferrill et al., 1998). Local domains of bed dip in the same direction as the associated normal faults also occur (e.g., Cloos, 1968; Bengtson, 1981; Xiao and Suppe, 1992; Reches and Eidelman, 1995; Schlische, 1995; Janecke et al., 1998; Khalil and McClay, 2002; Stewart and Reeds, 2003). This synthetic dip is commonly interpreted to be related to drag produced by frictional slip along faults and has also been ascribed to prefaulting or fault-propagation deformation. We show here that synthetic dip of the strata displaced by normal faults can be produced by five different mechanisms.

Synthetic dip panels associated with normal faults can affect hydrocarbon exploration and production. Because synthetic dip panels locally decrease the effective throw of a normal fault, they may preserve reservoir connectivity across a fault and thus prevent compartmentalization of a field or compromise trap integrity. Where displacement is sufficient to break this connection between footwall and hanging wall, synthetic dip panels can form downthrown fault traps.

In this chapter, we describe the mechanisms for development of synthetic dip panels associated with normal faults. We illustrate these mechanisms using field examples and consider related secondary deformation. Finally, we summarize the importance of synthetic dip panels for hydrocarbon migration, accumulation, and production.

MODELS FOR DEVELOPMENT OF SYNTHETIC DIP ASSOCIATED WITH NORMAL FAULTS

Five normal-fault-related mechanisms lead to the development of synthetic dip panels adjacent to normal faults (Figure 1): (1) faulted monocline (fault tip-line folding); (2) antilistric (convex upward) fault bend; (3) distributed shear; (4) shear in relay zone of vertically and/or horizontally segmented fault; and (5) fault block impingement and lateral contraction. Here, we briefly describe these five mechanisms, discuss conditions that lead to their development, and summarize their diagnostic characteristics (e.g., structural and stratigraphic associations and deformation mechanisms).

Extensional Monocline (Fault Tip-line Folding)

Monoclines commonly develop structurally above blind faults, where the rate of fault propagation is relatively slow with respect to the rate of fault displacement. Consequently, folding occurs above and laterally beyond fault tip lines (e.g., Reches and Eidelman, 1995; Schlische, 1995; Withjack et al., 1995; Childs et al., 1996). A monocline developed above a normal fault is characterized by a fold limb that dips toward the downthrown side of the normal fault (Figure 1a). Continued fault propagation may cut the monocline limb. Depending on the location of breakthrough, this process will leave synthetic dip panels in either the hanging wall or footwall or both blocks.

Antilistric Fault Bend

The vertical component of displacement (throw) for a given increment of normal fault slip depends on the fault dip. A given amount of slip on a steep fault produces more throw than an equivalent amount of slip on a more gently dipping fault. This dependency of fault throw and fault dip is essentially what drives the development of rollovers in the hanging walls of listric normal faults (Hamblin, 1965; Groshong, 1989; Xiao and Suppe, 1992). In the case of a downward steepening (antilistric) fault, differential throw that is controlled by the downward-steepening normal fault produces increasing throw in the hanging wall above the steeper part of the controlling fault. This throw pattern produces a synthetic dip panel adjacent to the fault that is controlled by the shape of the normal fault (Figure 1b) (Ferrill and Morris, 1997; Morris and Ferrill, 1999). Localized internal deformation (e.g., layer-parallel shear, layer-oblique shear, layer thinning, and tilting) is required in the dipping panel to maintain continuity

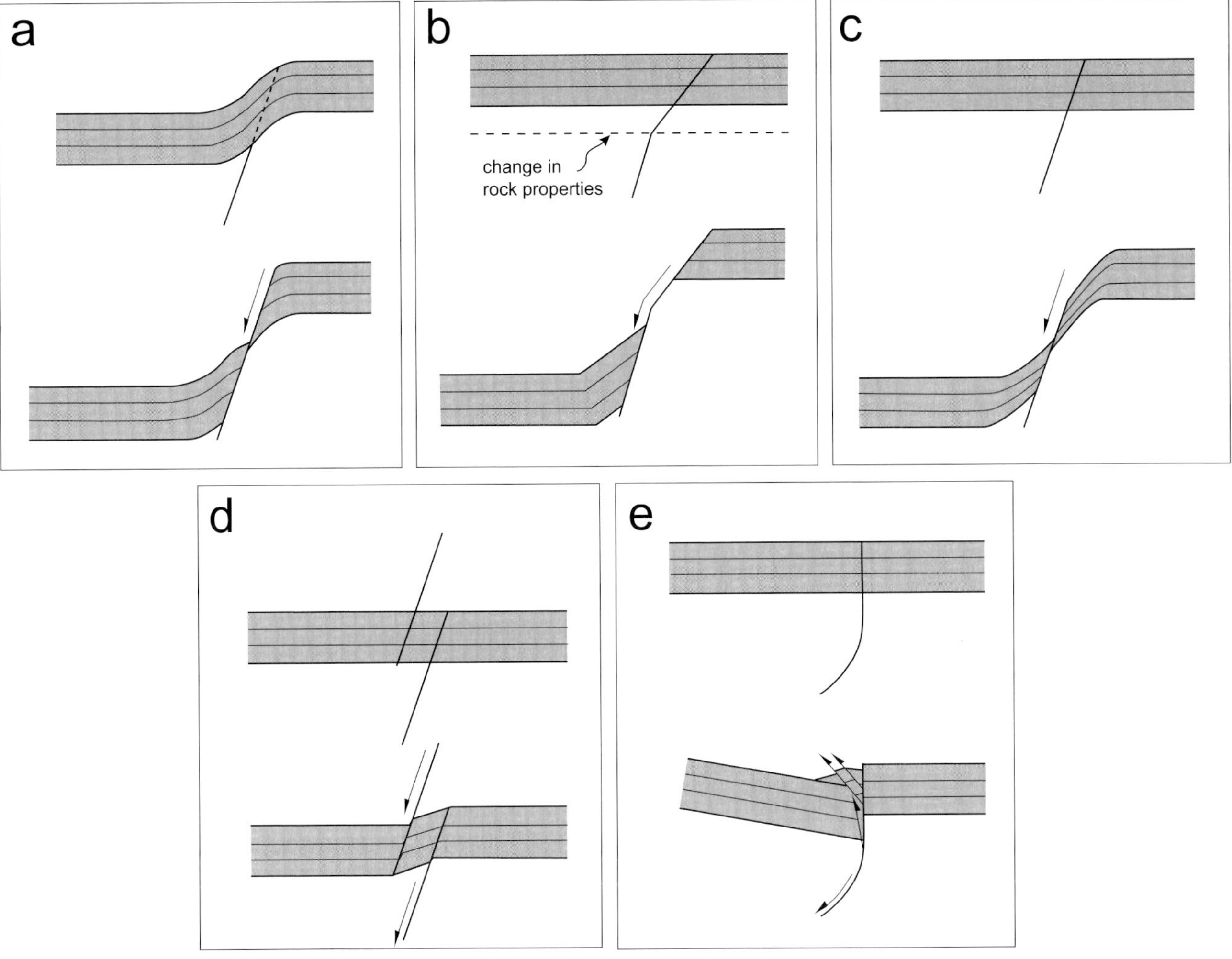

FIGURE 1. Schematic illustration of five mechanisms for the development of synthetic dip panels associated with normal faults: (a) faulted monocline resulting from fault tip-line folding; (b) antilistric fault bend; (c) distributed shear; (d) shear in overlap of vertically segmented fault; and (e) fault block impingement and contraction.

and conserve volume (Suppe, 1983; Groshong, 1989; Dula, 1991; Ferrill and Morris, 1997; Morris and Ferrill, 1999).

Distributed Shear

Distributed shear is used here as a general term for fault-related shear deformation distributed in the hanging wall or footwall of a fault (Figure 1c). This zone of deformation is part of the fault damage zone or process zone (e.g., Caine et al., 1996). In upper crustal rocks, zones of distributed shear tend to be narrowly localized along faults. Fault slip is more efficient than distributed shear for accommodating displacement. Consequently, distributed shear tends to be localized immediately adjacent to faults and is commonly associated with restraining bends along faults. Deformation may develop in zones of linkage between faults that are segmented in the slip direction (vertical segmentation).

Shear in Relay Zone of Vertically or Laterally Segmented Fault

Large faults typically grow by the lateral propagation and intersection of individual fault segments (Peacock and Sanderson, 1991; Childs et al., 1995; Ferrill et al., 1999a; Ferrill and Morris, 2001). Vertical segmentation of normal faults is also relatively common, particularly in anisotropic rock sequences composed of mechanically contrasting layers (Peacock and Sanderson, 1992, 1994; Peacock and Zhang, 1994; Childs et al., 1996; Mansfield and Cartwright, 1996; Aydin and Eyal, 2002). Displacement on vertically overlapping fault segments (as shown in Figure 1d) produces shear in the relay zones between overlapping faults and is expected to produce layer tilting and layer deformation (e.g., layer extension and thinning and/or layer-parallel shear).

Fault Block Impingement and Lateral Contraction

Slip of a rigid fault block along a broadly and continuously curved (listric) fault surface will produce antithetic fault block tilting but may not require any additional deformation in proximity to the controlling fault. In contrast, slip of a rigid fault block above a listric normal fault that has a long, straight, steep segment encounters a space problem (geometric overlap) in the upper part of the hanging wall and footwall adjacent to the fault (e.g., Gibbs, 1983, and figure I.2-79 in Mandl, 1988). This space problem does not arise in geometric and kinematic models of nonrigid fault block deformation, where fault blocks are assumed to deform by layer-parallel, inclined, or vertical shear (cf. Suppe, 1983; Groshong, 1989; Dula, 1991; Ferrill and Morris, 1997; Morris and Ferrill, 1999). One mechanism for accommodating this space problem in a rigid block is by shortening and thickening of layers adjacent to the fault, for example, by contractional faulting and/or folding in the hanging-wall fault block where it impinges on the footwall (Figure 1e).

NATURAL EXAMPLES

Brushy Canyon, Sierra Del Carmen, Texas

Tertiary crustal extension caused normal faulting throughout the western United States (Zoback et al., 1981). In the Big Bend region, along the Texas–Mexico border, normal faulting and tilting of Cretaceous rocks formed numerous elongate ridge systems underlain by the massive Santa Elena Limestone. The Sierra del Carmen Range is an example of this geomorphic style, consisting of hundreds of normal faults bounding linear ridges and valleys with as much as 400 m (1310 ft) of relief (Maxwell et al., 1967; Maxwell, 1968; Moustafa, 1988).

In Big Brushy Canyon, in the northeastern part of the Sierra del Carmen, a typical north-northwest–south-southeast-trending fault displaces the Santa Elena Limestone down to the east by at least 30 m (100 ft). The clay-rich Del Rio Formation, which overlies the Santa Elena Limestone, is not completely cut by the fault, but is dramatically thinned over the fault tip (Figure 2). Overlying the Del Rio is the Buda Formation, which contains limestone beds 1–2 m (3.3–6.6 ft) thick. The Buda Limestone beds are not faulted but form a monoclinal fold over the fault and exhibit layer-parallel extension and interbed uplimb shear and bedding plane slip to accommodate the bending (Figure 2b). In this case, the controlling fault abruptly loses all of the more than 30-m (100-ft) displacement where the fault tips in the Del Rio Formation. With continued displacement, the fault would eventually propagate upward through the Buda Formation. Fault localization is most likely where the Buda Limestone is impinging on the top of the Santa Elena Limestone across the fault. Continued fault propagation would eventually break across the Buda Limestone in the monocline and leave a synthetic dip panel in the hanging wall.

Galera Point, Trinidad

The Northern Range of Trinidad is an east–west-trending mountain range along the Caribbean–South American Plate boundary zone. The range is composed of Mesozoic–aged sedimentary and metasedimentary strata and is deeply dissected by erosion along much of its length. Exposures of the unmetamorphosed Upper Cretaceous Galera Formation (Galera grit) are accessible at Galera Point (Figure 3), the northeasternmost corner of Trinidad (Kugler, 1961). At this location, an antilistric (downward-steepening) normal fault is visible in profile. The fault cuts a massive coarse-grained sandstone layer overlain by thinly bedded interlayered sandstone and shale layers. Although the hanging-wall dip is generally toward (antithetic to) the normal fault, the interbedded sandstone and shale layers in the hanging wall locally dip in the same direction as (synthetic to) the underlying fault. Based on the association of the synthetic panel with pronounced antithetic fault bend, we interpret the synthetic hanging-wall dip panel related to the Galera Point normal fault shown in Figure 3 to be the product of slip on the antilistric fault surface. The fault shape itself is likely the product of the mechanical properties and stress conditions producing different failure angles in the heterolithic section (cf. Ferrill and Morris, 2003).

Cedar Pocket, Arizona

The western Colorado Plateau of northern Arizona is cut by several large north–south-trending normal faults (e.g., the Hurricane, Sullivan Canyon, and Grand Wash faults) and numerous smaller extensional faults (Billingsley and Workman, 2000). These faults define a broad transition between the Colorado Plateau and the Basin and Range Province in this region. A road cut along Interstate 15 exposes Lower Permian red sandstone in the footwall (approximately 400 m [1310 ft] east) of the Sullivan Canyon fault near Cedar Pocket, Arizona. This sandstone is mapped as Esplanade Sandstone by Bohannon et al. (1991) and Hermit Formation by Billingsley and Workman (2000). The exposure includes an array of west-dipping normal faults that progressively step beds of the Hermit Formation down to the west. This fault array represents a zone of distributed shear. Individual faults have displacements of centimeters to meters. Although individual faults have

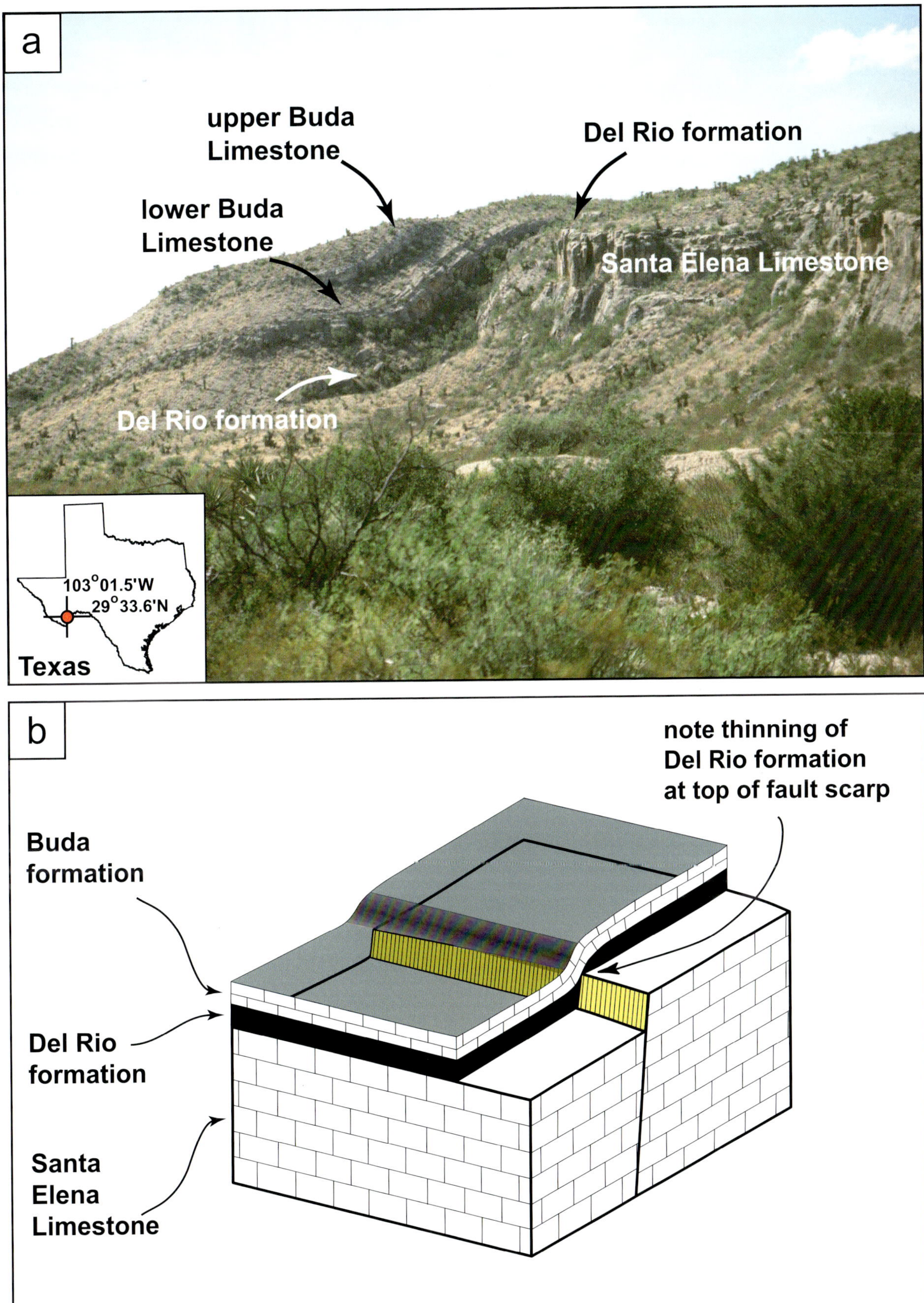

Figure 2. (a) Monocline in limestones of Buda Formation over Del Rio Formation and Santa Elena Limestone (fault scarp in foreground), Brushy Canyon, Sierra Del Carmen, Texas. (b) Schematic diagram of fault system shown in (a).

Figure 3. Synthetic dip panel above an antilistric bend in a normal fault in sedimentary strata exposed at Galera Point in the northeast corner of Trinidad (West Indies). Height of exposure is approximately 10 m (33 ft).

produced little tilting of fault blocks, an enveloping surface traced through the offset layers across the fault array shows an overall synthetic (westward) dip associated with the array of west-dipping faults (Figure 4).

Moab Fault, Utah

The Colorado Plateau of southeastern Utah includes an assemblage of classic extensional fault systems related to deformation above evaporites of the Pennsylvanian Paradox Formation. One such structure is the northeast-dipping Moab fault. The Moab fault is a major salt-related normal fault that includes numerous segments that link to form a fault system with a surface trace approximately 45 km (28 mi) long and a maximum displacement of approximately 960 m (3150 ft) (Foxford et al., 1998). In exposures along Bartlett Wash northwest of Moab, Utah, the Moab fault juxtaposes the Brushy Basin Member (Jmb) of the Morrison Formation in the hanging wall with the Moab Tongue Member (Jem) of the Entrada Sandstone in the footwall (Figure 5). Throw across the fault at Bartlett Wash is 210 m (690 ft) (Foxford et al., 1998).

The hanging wall and footwall of the Moab fault at Bartlett Wash have synthetic dip panels adjacent to the fault (Figure 5). The synthetic dip panel in the hanging wall extends through the entire width of the hanging-wall outcrop adjacent to the fault, a lateral distance of more than 100 m (330 ft) (Figure 5). The macroscopic structure and presence within and above mudstone-rich facies of the Morrison Formation are consistent with the synthetic dip panel having formed as a faulted monocline (faulted tip-line fold formed above the fault tip).

Synthetic dip in the footwall is more localized and is developed between vertically overlapping en echelon fault segments (see dashed lines in Figure 5). These en echelon segments are now connected and offset by the throughgoing Moab fault. Within the footwall exposure on the southwest side of Bartlett Wash, layer dip steepens toward the fault (Figure 5). Exposures of the footwall of the Moab fault on the northwest side of Bartlett Wash have well-developed swarms of small faults (manifested as cataclastic deformation bands; Figure 5). These deformation bands generally have strikes parallel with the Moab fault. Although the dominant dip direction of these deformation bands is synthetic to the Moab fault, oppositely dipping (conjugate) deformation bands also occur and mutually crosscut deformation bands that are synthetic to the Moab fault (Figure 5). Development of synthetic dip in the footwall of the Moab fault at Bartlett Wash is apparently a result of vertical fault segmentation and related distributed shear.

Bobby's Hole Fault, Utah

The grabens area is a region within and south of Canyonlands National Park that consists of an extensive network of valleys formed by grabens and half grabens in Pennsylvanian and Permian sedimentary strata (McGill and Stromquist, 1975, 1979; Huntoon et al., 1982; Trudgill and Cartwright, 1994; Cartwright et al., 1995, 2000; Cartwright and Mansfield, 1998; Moore and Schultz, 1999; McGill et al., 2000). These structural basins formed in response to lateral unloading, caused by entrenchment and lateral erosion by the Colorado River in Cataract Canyon, and gliding down the gentle (~4°) regional dip to the west-northwest above the Pennsylvanian Paradox evaporites (McGill and Stromquist, 1975, 1979; Huntoon, 1982, 1988; Doelling, 1988; McGill et al., 2000). Faults exposed at the surface are nearly vertical (e.g., McGill and Stromquist, 1979). Although

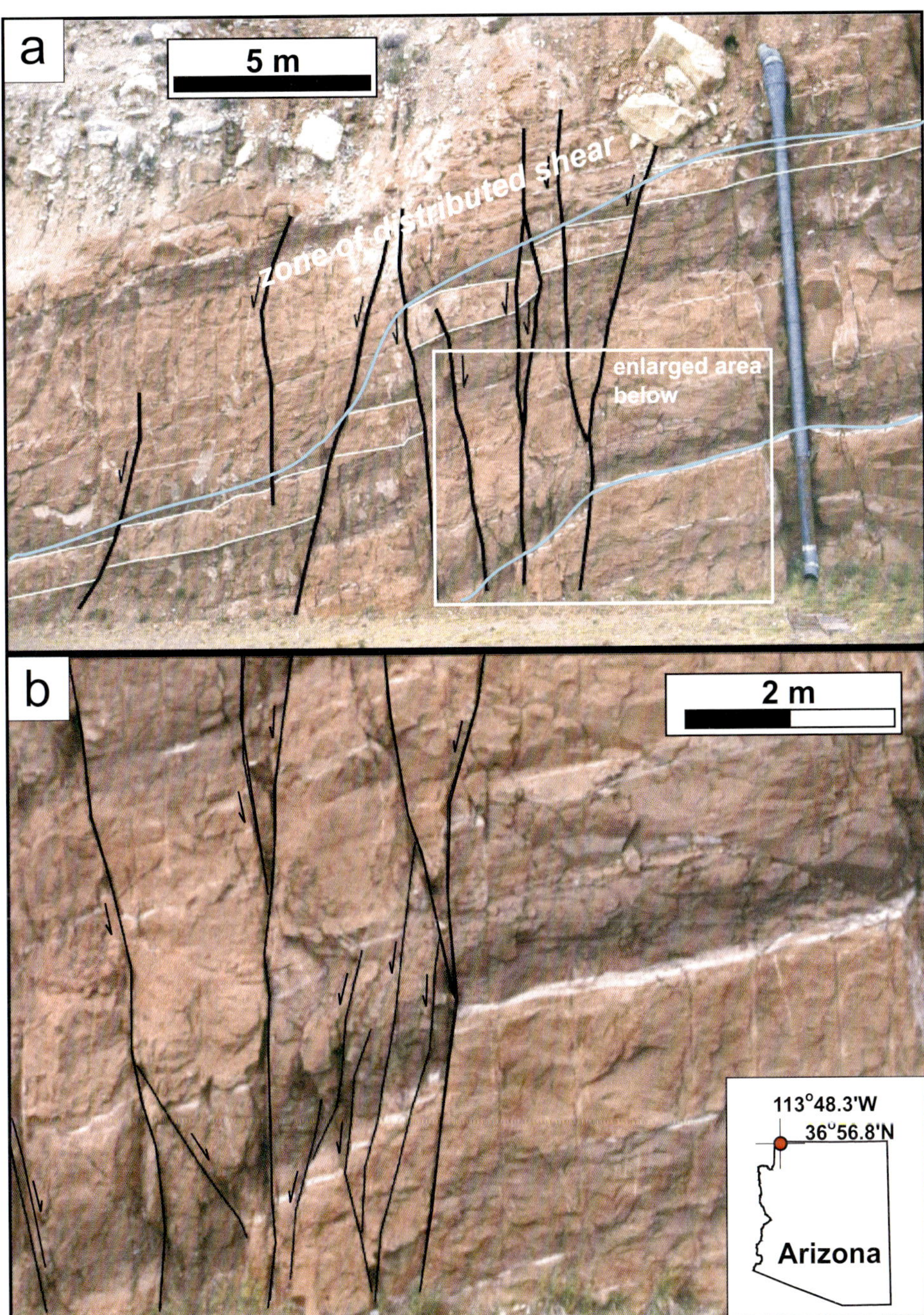

FIGURE 4. A series of small down-to-the-west normal faults produce overall synthetic dip of layering at Cedar Pocket, Arizona. (a) Envelope surfaces (gray lines) illustrate the synthetic dip trend developed by this mechanism. (b) Detail of exposure reveals closely spaced slip surfaces that impart the overall synthetic dip to the lower bedding envelope.

Cartwright et al. (1995) and Cartwright and Mansfield (1998) interpreted the faults as nearly vertical through the entire brittle plate, other workers have shown that the dip on at least some of the faults decreases from vertical at the surface to about 75° at a depth of 100 m (330 ft), which persists to about a 460-m (1510-ft) depth (see comment and reply by McGill et al., 2000; Cartwright et al., 2000).

The Bobby's Hole fault is a near-vertical, northwest-dipping normal fault in the southeastern part of the grabens area, south of the Canyonlands National Park (Figure 6). Hanging-wall strata generally dip gently (4–15°) southeastward toward the fault (Figure 6). In a sandstone knob in the hanging wall of the Bobby's Hole fault, layering adjacent to the fault dips to the northwest (5–37°NW), synthetic to the fault (Figure 6). A red sandstone layer in the hanging wall at this location is thickened and laterally shortened by two reverse faults. These contractional faults have moderate dips (40–50°SE), are antithetic to the Bobby's Hole fault, and are the only deformation features evident in the dip panel.

The synthetic dip panel associated with the Bobby's Hole fault has the following characteristics: (1) restricted to the hanging wall; (2) only locally developed immediately adjacent to the fault; (3) adjacent to nearly vertical faults in antithetically dipping hanging-wall block; (4) contains reverse faults that are antithetic to the Bobby's Hole fault; and (5) neither the synthetic dip panel nor the adjacent footwall has developed significant deformation bands, although the footwall contains conjugate sets of open fractures with a strike perpendicular to that of the main fault. The footwall fault surface at the sandstone knob displays subvertical striations, indicating dip slip on the fault (Figure 6). The antithetic dip through most of the hanging wall is indicative of an upward-steepening (listric) fault geometry. Based on these observations, we interpret the synthetic dip panel in the hanging wall of the Bobby's Hole fault to be caused by fault block impingement and contraction produced by slip on a listric Bobby's Hole fault, with a relatively long, straight, and steep segment near the surface (Figure 7).

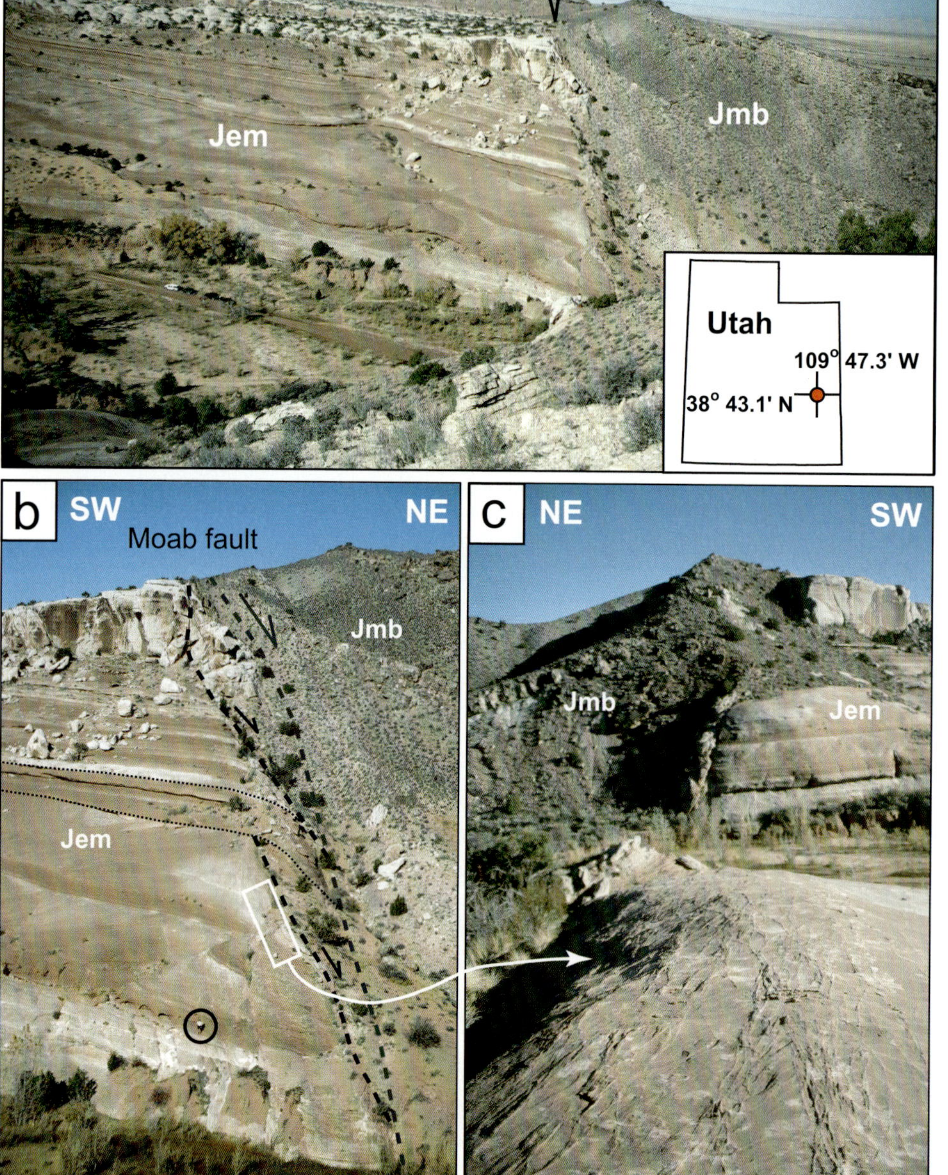

FIGURE 5. Synthetic dip in the footwall of the Moab fault, Bartlett Wash, Utah. (a) The view looking west (note the white vehicle on the dirt road in the canyon at the lower left for scale). (b) Moab fault interpreted on detailed view of (a). Note the synthetic dip of footwall strata. Circle shows a standing geologist for scale. White box shows viewpoint for photo in (c). (c) Swarm of deformation bands in the footwall of the Moab fault, on the west side of Bartlett Wash. The view is to the east. Labeled stratigraph units are the (1) Moab Tongue Member (Jem) of the Entrada Sandstone Formation and (2) Brushy Basin Member (Jmb) of the Morrison Formation (Foxford et al., 1998).

DISCUSSION

Conditions that favor the formation of synthetic dip panels associated with normal faults are as follows:

1) Structures that accumulate throw prior to fault breaking through (Figure 1a). This mechanism is favored in situations where fault tip propagation is arrested, for example, where a fault tip line intersects a ductile layer.
2) Antilistric fault shape (Figure 1b). This develops because of vertical variations in failure angle in mechanically heterogenous rock sequences or in situations where compaction occurs in a heterogenous stratigraphic sequence (e.g., sand and shale) after formation of a planar fault (dip of fault segment-cutting shale will be preferentially reduced because of greater compaction of the shale layer).
3) Faults with broad damage zones (footwall and/or hanging wall) (Figure 1c). This may be produced, for example, by zones of cataclastic deformation bands, which strain harden the deformed region by grain crushing and porosity loss (Antonellini et al., 1994) and lead to the formation of a throughgoing discrete slip surface. Alternatively, this distributed shear may develop during progressive slip on the fault, commonly referred to as "drag." This is consistent with the commonly recognized positive correlation between fault zone thickness and fault displacement (Robertson, 1983; Evans, 1990).

Figure 6. Field photographs of Bobby's Hole half graben south of Canyonlands National Park, Utah. (a) The view looking northeast along the half graben. Note the nearly vertical fault scarp along southeast side of the basin and the southeast dip of strata in the downdropped hanging-wall strata. (b) Field photograph of localized synthetic dip panel along the Bobby's Hole fault. The view is looking eastward, oblique to the fault. Red lines show fault traces, and heavy green lines show bedding traces.

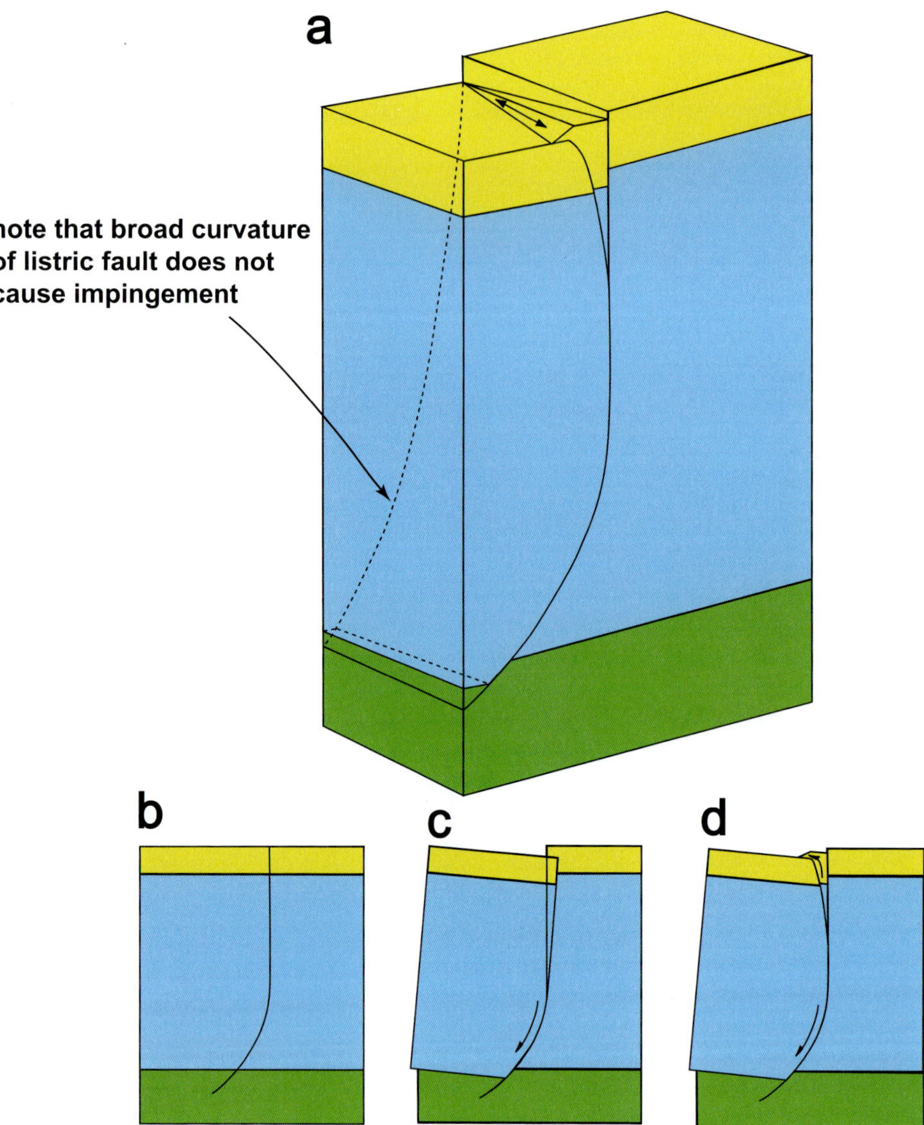

Figure 7. (a) Schematic block diagram showing relatively elevated part of the hanging-wall block adjacent to fault where hanging-wall impingement has produced lateral shortening and thickening of the hanging wall. Note that the fault changes shape laterally to a more broadly curved fault geometry for which there is no hanging-wall space problem. Therefore, the corresponding hanging-wall high plunges laterally and diminishes. The figures in (b–d) show geometry of the developmental sequence.

4) Faulting in a heterogenous lithologic sequence. Faults in stratigraphic sections composed of layers that have contrasting mechanical properties are likely to initiate as vertically segmented fault zones, where individual segments may nucleate in brittle layers (Figure 1d). As slip accumulates, and prior to complete linkage, horizons in relay zones between fault segments will acquire synthetic dip.
5) Presence of a long, steep section on a listric normal fault (Figure 1e). Formation of a synthetic dip panel by hanging-wall impingement is most likely to occur near the Earth's surface in well-lithified rocks where listric faults have very steep sections. These very steep sections commonly form at low confining pressure by tensile or hybrid-mode failure (Ferrill and Morris, 2003). Changes in fault dip can be driven by very high competence contrast between layers that have differing mechanical properties.

Implications for Hydrocarbon Exploration and Production

In general, normal fault systems that involve synthetic dip panels have greater vertical structural displacement than net fault throw (Figure 8). Synthetic dip associated with normal faults can (1) provide migration or production pathways for hydrocarbons; (2) produce structural traps for hydrocarbon accumulation; and (3) be sites of highly heterogenous permeability in reservoirs. These specific details vary as a function of stratigraphy and mechanism of synthetic dip generation and influence both geometry and permeability (e.g., Caine et al., 1996; Ferrill et al., 1999b, and references therein; Ferrill et al., 2000). As examples to illustrate the potential importance of geometry and permeability on hydrocarbon migration, we consider the geometry and deformational style of the Bobby's Hole and Moab faults.

Geometry

The vertical thickening of the red sandstone layer in the hanging wall of the Bobby's Hole fault reduces the effective throw across the fault. The total vertical displacement across the structure (Figures 6, 8) is more than twice the throw across the fault. In a stratigraphic section consisting of interlayered reservoir and seal

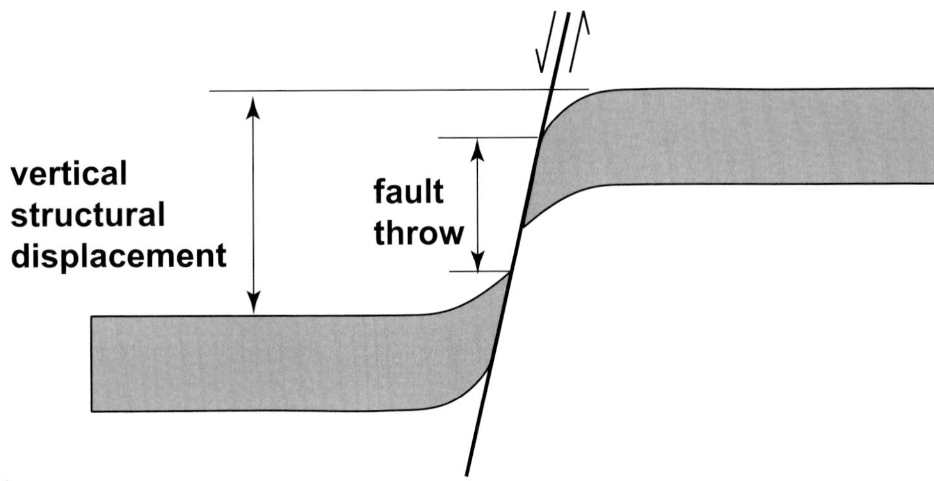

FIGURE 8. Vertical structural displacement is produced by a combination of fault throw and synthetic dip of strata adjacent to the fault.

strata or where throw is small compared with reservoir thickness, a local throw minimum may produce a fluid communication pathway for migration or production of hydrocarbons across an otherwise sealing fault if the throw is less than the reservoir thickness (Figure 9). In another case, a local hanging-wall culmination may develop adjacent to a normal fault that is produced by a local anomaly (e.g., a long, straight, and steep segment shallow on a listric fault) along an otherwise broadly curved fault. Where fluid communication across the fault culmination is inhibited by juxtaposition of the reservoir layer (e.g., sandstone) against a sealing layer (e.g., shale; see Figure 9) or presence of a sealing fault zone (caused by clay smear, porosity loss, or grain size reduction) (Aydin and Eyal, 2002; Dewhurst and Jones, 2002), the hanging-wall culmination could be a trapping structure.

Permeability

The footwall of the Moab fault at and near Bartlett Wash has intensively developed deformation bands associated with synthetic dip panels between en echelon (vertically segmented) fault segments. As has been shown by previous studies of deformation bands in porous sandstone in the Moab area (Antonellini and Aydin, 1994, 1995; Antonellini et al., 1994; Dewhurst and Jones, 2002), porosity and permeability of cataclastic deformation bands are low with respect to the undeformed host rock. The localization of deformation band development in the footwall synthetic dip panels should lead to considerably reduced permeability and the development of permeability anisotropy in the footwall adjacent to the main Moab fault, with the maximum principal permeability parallel to the intersection of the crossing conjugate deformation bands (and approximately parallel to the strike of the Moab fault). Thus, intensive development of deformation bands associated with synthetic dip development by distributed shear may inhibit fluid communication across the fault, even in situations of sand-on-sand juxtaposition.

Deformation in the footwall of the Bobby's Hole fault contrasts sharply with the intense development of cataclastic deformation bands in the footwall of the Moab fault. The two sets of steeply dipping unfilled fractures in the footwall of the Bobby's Hole fault define a conjugate fracture pattern with an intersection perpendicular to the Bobby's Hole fault. These open (unfilled) fractures increase both porosity and permeability with respect to the host sandstone in a zone along the main fault. With maximum bulk permeability perpendicular to the main fault, these unfilled fractures would not impede fluid communication across the fault. As shown by these examples, the presence of synthetic dip panels and the details of deformation within and associated with the development of synthetic dip adjacent to normal faults may strongly influence hydrocarbon movement and trapping.

CONCLUSIONS

Synthetic dip panels can form by five different mechanisms and should be expected at some scale in all normal fault systems, where they can influence trapping and/or migration of hydrocarbons. Synthetic dip panels reduce the effective throw on their related normal faults and may provide fluid communication pathways across faults. Depending on the geologic setting and structural details, synthetic dip panels may produce hydrocarbon traps.

ACKNOWLEDGMENTS

This chapter is part of research work carried out for the Traps and Seals Project at the Japan National Oil Corporation, Technology Research Center (presently Japan Oil, Gas and Metals National Corporation). Uko Suzuki was then manager of the Traps and Seals Project, and Rasoul Sorkhabi acted as coordinator on behalf of the Traps and Seals Project. Daichi Sato and Kiyofumi

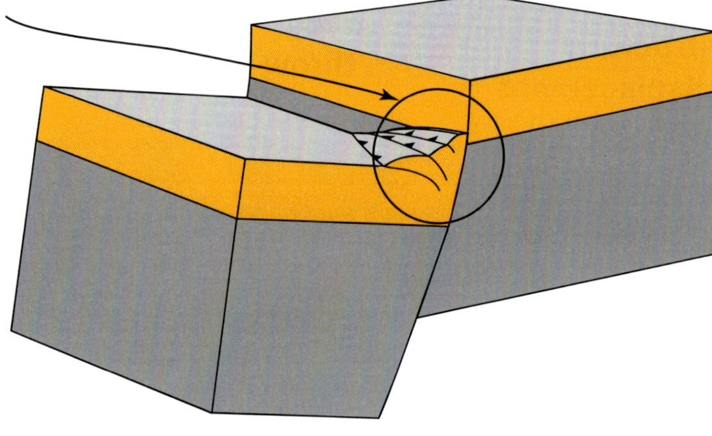

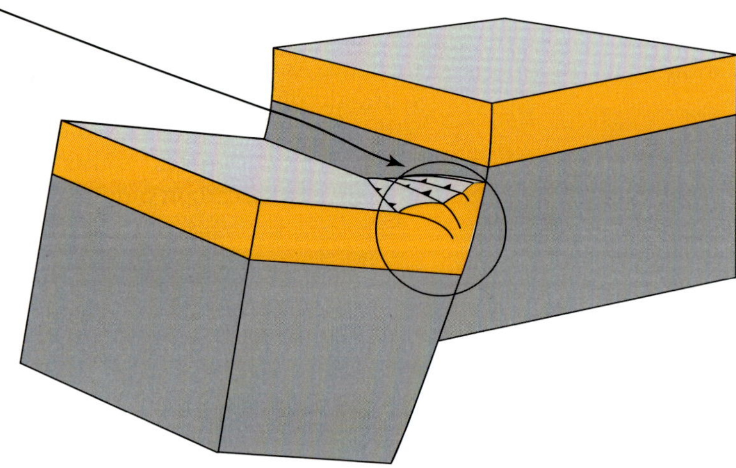

Figure 9. Schematic block diagram showing relatively elevated part of the hanging wall adjacent to fault, where impingement has produced lateral shortening and thickening of the hanging wall. (a) At relatively small displacements, the shortening and thickening of the hanging wall may result in locally diminished throw on the normal fault, hence causing a local sand-on-sand (orange layer) juxtaposition across the fault, where the sand (orange layer) in the hanging wall has (at the same total displacement) been juxtaposed elsewhere against the underlying shale across the fault. (b) With increasing displacement, the relatively elevated part of the hanging-wall block becomes a potential trapping location.

Suzuki participated in the fieldwork in November 1999, and we thank them for technical discussions and their camaraderie. This chapter benefited from technical reviews by Larry McKague, Wesley Patrick, James Evans, Richard Groshong, Rasoul Sorkhabi, and Alan Gibbs. We thank Rebecca Emmot and Janie Gonzalez for preparing the manuscript.

REFERENCES CITED

Antonellini, M., and A. Aydin, 1994, Effect of faulting on fluid flow in porous sandstones. Petrophysical properties: AAPG Bulletin, v. 78, p. 355–377.

Antonellini, M., and A. Aydin, 1995, Effect of faulting on fluid flow in porous sandstones: Geometry and spatial distribution: AAPG Bulletin, v. 79, p. 642–671.

Antonellini, M. A., A. Aydin, and D. D. Pollard, 1994, Microstructure of deformation bands in porous sandstones at Arches National Park, Utah: Journal of Structural Geology, v. 16, p. 941–959.

Aydin, A., and Y. Eyal, 2002, Anatomy of a normal fault with shale smear: Implications for fault seal: AAPG Bulletin, v. 86, p. 1367–1381.

Bengtson, C. A., 1981, Statistical curvature analysis techniques for structural interpretation of diameter data: AAPG Bulletin, v. 65, p. 312–332.

Billingsley, G. H., and J. B. Workman, 2000, Geologic Map of the Littlefield 30′ × 60′ Quadrangle, Mohave County, northwestern Arizona, Miscellaneous Investigations Series Map I-2628, version 1.0: Denver, Colorado, U. S. Geological Survey, scale 1:100,000, 1 sheet.

Bohannon, R. G., I. Lucchitta, and R. E. Anderson, 1991, Geologic map of the Mountain Sheep Springs Quadrangle, Mohave County, Arizona, Miscellaneous Investigations Map I-2165: Denver, Colorado, U. S. Geological Survey, scale 1:24,000, 1 sheet.

Caine, J. S., J. P. Evans, and C. B. Forster, 1996, Fault zone architecture and permeability structure: Geology, v. 24, p. 1025–1028.

Cartwright, J. A., and C. S. Mansfield, 1998, Lateral displacement variation and lateral tip geometry of normal faults in the Canyonlands National Park, Utah: Journal of Structural Geology, v. 20, p. 3–19.

Cartwright, J. A., B. D. Trudgill, and C. S. Mansfield, 1995, Fault growth by segment linkage: An explanation for scatter in maximum displacement and trace length data from the Canyonlands grabens of SE Utah: Journal of Structural Geology, v. 17, p. 1319–1326.

Cartwright, J., B. Trudgill, and C. Mansfield, 2000, Fault growth by segment linkage: An explanation for scatter in maximum displacement and trace length data from the Canyonlands grabens of SE Utah: Reply: Journal of Structural Geology, v. 22, p. 141–143.

Childs, C., J. Watterson, and J. J. Walsh, 1995, Fault overlap zones within developing normal fault systems: Journal of the Geological Society (London), v. 152, p. 535–549.

Childs, C., A. Nicol, J. J. Walsh, and J. Watterson, 1996, Growth of vertically segmented normal faults: Journal of Structural Geology, v. 18, p. 1389–1397.

Cloos, E., 1968, Experimental analysis of Gulf Coast fracture patterns: AAPG Bulletin, v. 52, no. 3, p. 420–444.

Dewhurst, D. N., and R. H. Jones, 2002, Geomechanical, microstructural, and petrophysical evolution in experimentally reactivated cataclasites: Applications to fault seal prediction: AAPG Bulletin, v. 86, p. 1383–1405.

Doelling, H. H., 1988, Geology of Salt Valley anticline and Arches National Park, Grand County, Utah, *in* H. H. Doelling, C. G. Oviatt, and P. W. Huntoon, eds., Salt deformation in the Paradox basin: Utah Geological and Mineral Survey Bulletin, v. 122, p. 1–58.

Dula, W. F. Jr., 1991, Geometric models of listric normal faults and rollover folds: AAPG Bulletin, v. 75, p. 1609–1625.

Evans, J. P., 1990, Thickness-displacement relationships for fault zones: Journal of Structural Geology, v. 12, p. 1061–1065.

Ferrill, D. A., and A. P. Morris, 1997, Geometric considerations of deformation above curved normal faults and salt evacuation surfaces: The Leading Edge, v. 16, p. 1129–1133.

Ferrill, D. A., and A. P. Morris, 2001, Displacement gradient and deformation in normal fault systems: Journal of Structural Geology, v. 23, p. 619–638.

Ferrill, D. A., and A. P. Morris, 2003, Dilational normal faults: Journal of Structural Geology, v. 25, p. 183–196.

Ferrill, D. A., A. P. Morris, S. M. Jones, and J. A. Stamatakos, 1998, Extensional layer parallel shear and normal faulting: Journal of Structural Geology, v. 20, p. 355–362.

Ferrill, D. A., J. A. Stamatakos, and D. Sims, 1999a, Normal fault corrugation: Implications for the growth and seismicity of active normal faults: Journal of Structural Geology, v. 21, p. 1027–1038.

Ferrill, D. A., J. Winterle, G. Wittmeyer, D. Sims, S. Colton, A. Armstrong, and A. P. Morris, 1999b, Stressed rock strains groundwater at Yucca Mountain, Nevada: Geological Society of America Today, v. 9, no. 5, p. 1–8.

Ferrill, D. A., A. P. Morris, J. A. Stamatakos, and D. Sims, 2000, Crossing conjugate normal faults: AAPG Bulletin, v. 84, no. 10, p. 1543–1559.

Foxford, K. A., J. J. Walsh, J. Watterson, I. R. Garden, S. C. Guscott, and S. D. Burley, 1998, Structure and content of the Moab fault zone, Utah, U.S.A., and its implications for fault seal prediction, *in* G. Jones, Q. J. Fisher, and R. J. Knipe, eds., Faulting, fault sealing and fluid flow in hydrocarbon reservoirs: Geological Society (London) Special Publication 147, p. 87–103.

Gibbs, A. D., 1983, Balanced cross-section construction from seismic sections in areas of extensional tectonics: Journal of Structural Geology, v. 5, p. 153–160.

Groshong, R. H. Jr., 1989, Half-graben structure: Balanced models of extensional fault-bend folds: Geological Society of America Bulletin, v. 101, p. 96–105.

Hamblin, W. K., 1965, Origin of "reverse drag" on the downthrown side of normal faults: Geological Society of America Bulletin, v. 76, p. 1145–1164.

Huntoon, P. W., 1982, The Meander anticline, Canyonlands, Utah: An unloading structure resulting from horizontal gliding on salt: Geological Society of America Bulletin, v. 93, p. 941–950.

Huntoon, P. W., 1988, Late Cenozoic gravity tectonic deformation related to the Paradox salts in the Canyonlands area of Utah, *in* H. H. Doelling, C. G. Oviatt, and P. W. Huntoon, eds., Salt deformation in the Paradox basin: Utah Geological and Mineral Survey Bulletin, v. 122, p. 80–93.

Huntoon, P. W., G. H. Billingsby, Jr., and W. J. Breed, 1982, Geologic map of Canyonlands National Park and vicinity, Utah: Moab, Utah, The Canyonlands National History Association, scale 1:62,500, 1 sheet.

Janecke, S. U., C. J. Vandenburg, and J. J. Blankenau, 1998, Geometry, mechanisms and significance of extensional folds from examples in the Rocky Mountain Basin and Range Province, U.S.A.: Journal of Structural Geology, v. 20, p. 841–856.

Khalil, S. M., and K. R. McClay, 2002, Extensional fault-related folding, northwestern Red Sea Egypt: Journal of Structural Geology, v. 24, p. 743–762.

Kugler, H. G., 1961, Geological map and sections of Trinidad: Zurich, Orell Fussli, scale 1:100,000, 1 sheet (also in Treatise on the Geology of Trinidad, H. G. Kugler, compiler, part 2 or part of part 3. Natural History Museum, Basel, Switzerland).

Mandl, G., 1988, Mechanics of tectonic faulting: New York, Elsevier Science Publishing Company, Inc., 407 p.

Mansfield, C. S., and J. A. Cartwright, 1996, High resolution fault displacement mapping from three-dimensional seismic data: Evidence for dip linkage during fault growth: Journal of Structural Geology, v. 18, p. 249–263.

Maxwell, R. A., 1968, The Big Bend of the Rio Grande: A guide to the rocks, landscape, geologic history, and settlers of the area of Big Bend National Park: Bureau of Economic Geology, The University of Texas at Austin, Guidebook 7, 139 p.

Maxwell, R. A., J. T. Lonsdale, R. T. Hazzard, and J. A. Wilson, 1967, Geology of the Big Bend National Park, Brewster County, Texas: University of Texas at Austin, Bureau of Economic Geology, Publication 6711, 320 p.

McGill, G. W., and A. W. Stromquist, 1975, Origin of graben in the Needles District, Canyonlands National Park, Utah, *in* J. E. Fassett, ed., Canyonlands country: Four Corners Geological Society Guidebook, Durango, Colorado, p. 235–243.

McGill, G. E., and A. W. Stromquist, 1979, The grabens of Canyonlands National Park, Utah: Geometry, mechanics, and kinematics: Journal of Geophysical Research, v. 84, p. 4547–4563.

McGill, G. E., R. A. Schultz, and J. M. Moore, 2000, Fault growth by segment linkage: An explanation for scatter in maximum displacement and trace length data from the Canyonlands grabens of SE Utah: Discussion: Journal of Structural Geology, v. 22, p. 135–140.

Moore, J. M., and R. A. Schultz, 1999, Processes of faulting in jointed rocks of Canyonlands National Park, Utah: Geological Society of America Bulletin, v. 111, p. 808–822.

Morris, A. P., and D. A. Ferrill, 1999, Constant thickness deformation above curved normal faults: Journal of Structural Geology, v. 21, p. 67–83.

Moustafa, A. R., 1988, Structural geology of Sierra del Carmen, Trans-Pecos Texas, Geologic Quadrangle Map No. 54: Austin, Texas, Bureau of Economic Geology, The University of Texas at Austin, scale 1:48,000, 3 sheets.

Peacock, D. C. P., and D. J. Sanderson, 1991, Displacements, segment linkage, and relay ramps in normal fault zones: Journal of Structural Geology, v. 13, p. 721–733.

Peacock, D. C. P., and D. J. Sanderson, 1992, Effects of layering and anisotropy on fault geometry: Journal of the Geological Society (London), v. 149, p. 793–802.

Peacock, D. C. P., and D. J. Sanderson, 1994, Geometry and development of relay ramps in normal fault systems: AAPG Bulletin, v. 78, p. 147–165.

Peacock, D. C. P., and X. Zhang, 1994, Field examples and numerical modeling of oversteps and bends along normal faults in cross-section: Tectonophysics, v. 234, p. 147–167.

Reches, Z., and A. Eidelman, 1995, Drag along faults: Tectonophysics, v. 247, p. 145–156.

Robertson, E. C., 1983, Relationship of fault displacement to gouge and breccia thickness: Mining Engineering, v. 35, p. 1426–1432.

Rouby, D., H. Fossen, and P. R. Cobbold, 1996, Extension, displacement, and block rotation in the larger Gullfaks area, northern North Sea: Determined from map view restoration: AAPG Bulletin, v. 80, p. 875–890.

Schlische, R. W., 1995, Geometry and evolution of fault-related folds in extensional settings: AAPG Bulletin, v. 79, p. 1661–1678.

Stewart, S. A., and A. Reeds, 2003, Geomorphology at kilometer-scale extensional fault scarps: Factors that impact seismic interpretation: AAPG Bulletin, v. 87, p. 251–272.

Suppe, J., 1983, Geometry and kinematics of fault-bend folding: American Journal of Science, v. 283, p. 684–721.

Trudgill, B., and J. Cartwright, 1994, Relay-ramp forms and normal-fault linkages, Canyonlands National Park, Utah: Geological Society of America Bulletin, v. 106, p. 1143–1157.

Wernicke, B., and B. C. Burchfiel, 1982, Modes of extensional tectonics: Journal of Structural Geology, v. 4, p. 105–115.

Withjack, M. O., Q. T. Islam, and P. R. La Pointe, 1995, Normal faults and their hanging-wall deformation: An experimental study: AAPG Bulletin, v. 79, p. 1–18.

Xiao, H., and J. Suppe, 1992, Origin of rollover: AAPG Bulletin, v. 76, p. 509–529.

Zoback, M. L., R. E. Anderson, and G. A. Thompson, 1981, Cainozoic evolution of the state of stress and style of tectonism of the Basin and Range Province of the western United States: Philosophical Transactions of the Royal Society of London, v. A300, p. 407–434.

9

Sorkhabi, R., and S. Hasegawa, 2005, Fault zone architecture and permeability distribution in the Neogene clastics of northern Sarawak (Miri Airport Road outcrop), Malaysia, *in* R. Sorkhabi and Y. Tsuji, eds., Faults, fluid flow, and petroleum traps: AAPG Memoir 85, p. 139–151.

Fault Zone Architecture and Permeability Distribution in the Neogene Clastics of Northern Sarawak (Miri Airport Road Outcrop), Malaysia

Rasoul Sorkhabi[1]
Technology Research Center, Japan National Oil Corporation, Chiba, Japan

Shutaro Hasegawa[2]
Technology Research Center, Japan National Oil Corporation, Chiba, Japan

ABSTRACT

The Miri Airport Road outcrop in Miri, Sarawak, exposes a weakly consolidated sandstone-mudstone sequence of Miocene age in the form of a gentle anticline cut by a series of normal faults. An outcrop study of the normal faults shows that fault zones in porous sandstones are characterized by a combination of shale smear and anastomosing deformation bands. The continuity of shale smear on fault offset was observed as having a shale-smear factor (fault throw divided by shale layer thickness) of at least 5. Deformation bands occur as solitary planar structures in the host sandstone away from fault zones but increase markedly in density and linkage toward the fault slip plane, possibly indicating that faulting evolves from individual bands to a high-strain zone characterized by anastomosing deformation bands and culminating in the fault slip plane. Gas-permeability measurements show that individual deformation bands have an order of magnitude lower permeability than in the nearby sandstone matrix, and that the lowest permeability of fault zones defined by anastomosing deformation bands is in traverses nearly perpendicular to fault planes. It was found that the sandstone matrix in the fault zone has a lower permeability than individual deformation bands outside the fault zone, indicating that as a whole, the fault zone undergoes tectonic compaction and porosity collapse. The development of major normal faults was also accompanied by reactivation of bedding-perpendicular joints as small-displacement, simple-shear faults between the major faults.

[1]*Present address:* Energy & Geoscience Institute, University of Utah, Salt Lake City, Utah, U.S.A.
[2]*Present address:* Idemitsu Oil & Gas Co., Tokyo, Japan.

DOI:10.1306/1033721M853128

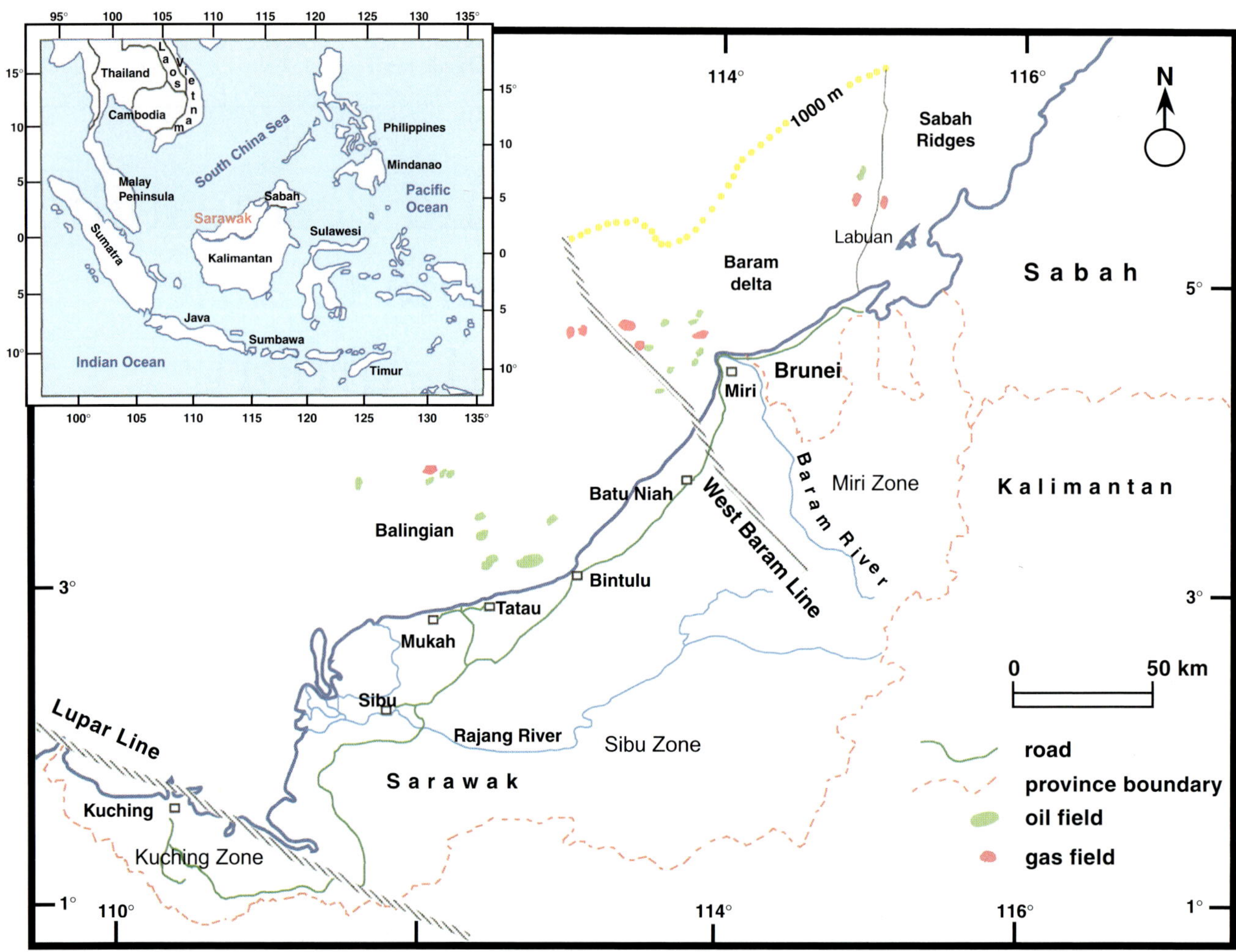

Figure 1. Sketch map of Sarawak showing the location of the study area in Miri, north of the West Baram line. The Miri area is a part of the Baram delta that has been exposed. The map has been compiled from Haile and Ho (1991) and Hutton (1998).

INTRODUCTION

In the past two decades, the assessment of sealing or conductive behavior of faults has drawn much attention from the petroleum industry. Studies of outcrop geology (e.g., Lindsay et al., 1993; Antonellini and Aydin, 1994, 1995; Childs et al., 1997; Garden et al., 2001; Hammond and Evans, 2003) as well as subsurface fields (e.g., Childs et al., 2002; Gibson, 1994; Berg and Avery, 1995; Yielding et al., 1997; Alexander and Handschy, 1998; Naruk et al., 2002; Shipton et al., 2002) have generated useful concepts and significant data that characterize and quantify fault zone processes for normal faults in clastic reservoirs. Characterization of fault zone architecture and quantification of the impact of faulting on permeability heterogeneity in reservoir rocks still remain major challenges in evaluating fault traps. Exhumed faults offer direct clues to understanding the geometry and genesis of fault zones and the fabric and petrophysical properties of fault rocks. This chapter presents new data on a set of normal faults that have cut a Tertiary sandstone-mudstone sequence in Sarawak, Malaysia (Figure 1). Fieldwork for this study took place in December 2000, and the results of this study add to the previous works of Burhannudinnur and Morely (1997) and Van der Zee (2001), who described outcrop observation of fault zones and faulting styles in Brunei and Miri, respectively. In this chapter, we report data on permeability distribution from sandstone reservoir rock to fault rocks in Miri, in addition to data on structural characteristics of fault zones in the study area.

TECTONOSTRATIGRAPHY OF THE STUDY AREA

Miri is located in Sarawak, on the eastern shore of the South China Sea (Figure 1). Pioneer mapping of this area was made by geologists from Shell Oil Company, the University of Malaysia, the Geological Survey of

Malaysia, and Petroliam Nasional Bhd. (e.g., Liechti et al., 1960; Haile, 1969, 1974; Ho, 1978; Bait et al., 1991; Haile and Ho, 1991; Hussin et al., 1992).

Haile (1974) and Haile and Ho (1991) divided the surface geology of Sarawak into three zones (Figure 1):

1) The Kuching Zone lies to the southwest of the Lupar line and includes Paleozoic metasedimentary rocks (schist and phyllite) overlain by volcanic rocks and marine and continental rocks of Mesozoic age and the Tertiary plateau group.
2) The Sibu Zone of central Sarawak is bounded by the Lupar line and the West Baram line and consists of the Late Cretaceous to Eocene deepwater Rajang Group overlain by Oligocene–Miocene clastic sediments with coal deposits at the base (the Nyalau Formation).
3) The Miri Zone lies north of the West Baram line, where post-Eocene, mainly shallow-marine and deltaic sediments crop out.

Overall sedimentation becomes younger from the Kuching Zone in the southwest to the Miri Zone in the northeast, indicating that basins migrated in time, although the tectonic setting of the basins has varied (e.g., Hutchinson, 1989; Hall, 1996). The Miri Zone is the onshore part of the Baram deltaic basin, which has been filled with a 10–15-km (6–9-mi)-thick succession of Neogene clastic sediments. Onshore Tertiary strata in the Baram Basin are divided into several lithostratigraphic formations; their equivalent offshore strata are divided into eight sedimentary cycles. Figure 2 shows the stratigraphy of the onshore and offshore strata, including petroleum source rocks and reservoirs (after Haile and Ho, 1991).

The Tertiary tectonic history of the Miri Zone can be divided into two major phases (Hutchinson, 1989; Haile and Ho, 1991; Hussin et al., 1992): (1) late Miocene extension gave rise to a series of southeast-dipping normal faults. These normal faults provide important structural traps in the Baram deltaic basin. Extension was followed by (2) Pliocene compression, which folded the strata in the Baram Basin into a series of subparallel anticlines and thrust faults such as the Canada Hill thrust (Figure 3). Compression was accompanied by a thrust-parallel normal fault (the Shell Hill fault) and collapse of the anticlinal crest along normal faults as exposed in the Miri Airport Road outcrop.

The Miri Airport Road outcrop, which is studied in this chapter, is located southwest of the town of Miri, Sarawak (Figure 3). Although the first oil-producing well in Sarawak was not drilled in Miri until 1910 (Shell-Miri well 1), oil in the nearby Miri field was known to native peoples of the region for centuries. The Miri field was abandoned in 1972, having produced more than 80 million bbl of oil (Hussin et al., 1992). All of this production came from a thick interval of sandstone that was interbedded with thin mudstone and silty shale comprising the upper Miri Formation (Figure 3). The Sand 456 reservoir in the Miri field is exposed in the Miri Airport Road outcrop. The grain size of the sandstone is 0.2–0.4 mm (0.008–0.02 in.) (fine to medium sand on the Wentworth size classification).

NORMAL FAULTS IN THE MIRI AIRPORT ROAD OUTCROP

The Miri Airport Road outcrop is located along the collapsed crest of an anticline bounded by the Shell Hill fault on the northwest and the Canada Hill thrust on the southeast (Figure 3). The dip of beds on both limbs of the anticline does not exceed 10°.

The Miri Airport Road outcrop offers excellent opportunities to study normal faults in clastic reservoirs because it contains fresh exposures of a Neogene sandstone-mudstone sequence laterally equivalent to nearby subsurface reservoirs and because field observations can be made on a three-dimensional (3-D) exposure, i.e., in both cross-sectional and bedding-plane views.

Eight major faults (numbered 1–8) studied in the Miri Airport outcrop are shown in Figure 4. Some of these normal faults intersect (e.g., faults 2 and 3; faults 4 and 5; and faults 5 and 6 in Figure 4).

The faults trend northeast and dip at moderate to high angles. They can be divided into two groups: those that strike N40–45°E and dip to the northwest; and those that strike N50–55°E and dip to the southeast (Figure 4).

Fault zones are characterized by shale smear and deformation bands as described below.

SHALE SMEAR

Shale smear or clay smear along normal faults has been recognized by several researchers as an efficient across-fault seal in clastic rocks (Yielding et al., 1997; Aydin and Eyal, 2002; Doughthy, 2003; Kolddoye et al., 2003). Shale smear occurs by several mechanisms (Lindsay et al., 1993); a variety of these mechanisms were interpreted for the Miri outcrop (Figure 5) as follows:

1) injection of soft clay into dilating fractures
2) simple shear involving brittle faulting of mud layers with en echelon fault steps in the fault zone
3) dragging of a parallel set of thin mud layers into the fault plane by a combination of shear and injection
4) abrasion whereby the passage of a displaced mud layer leaves a thin veneer of clay on a sandstone surface along the fault plane.

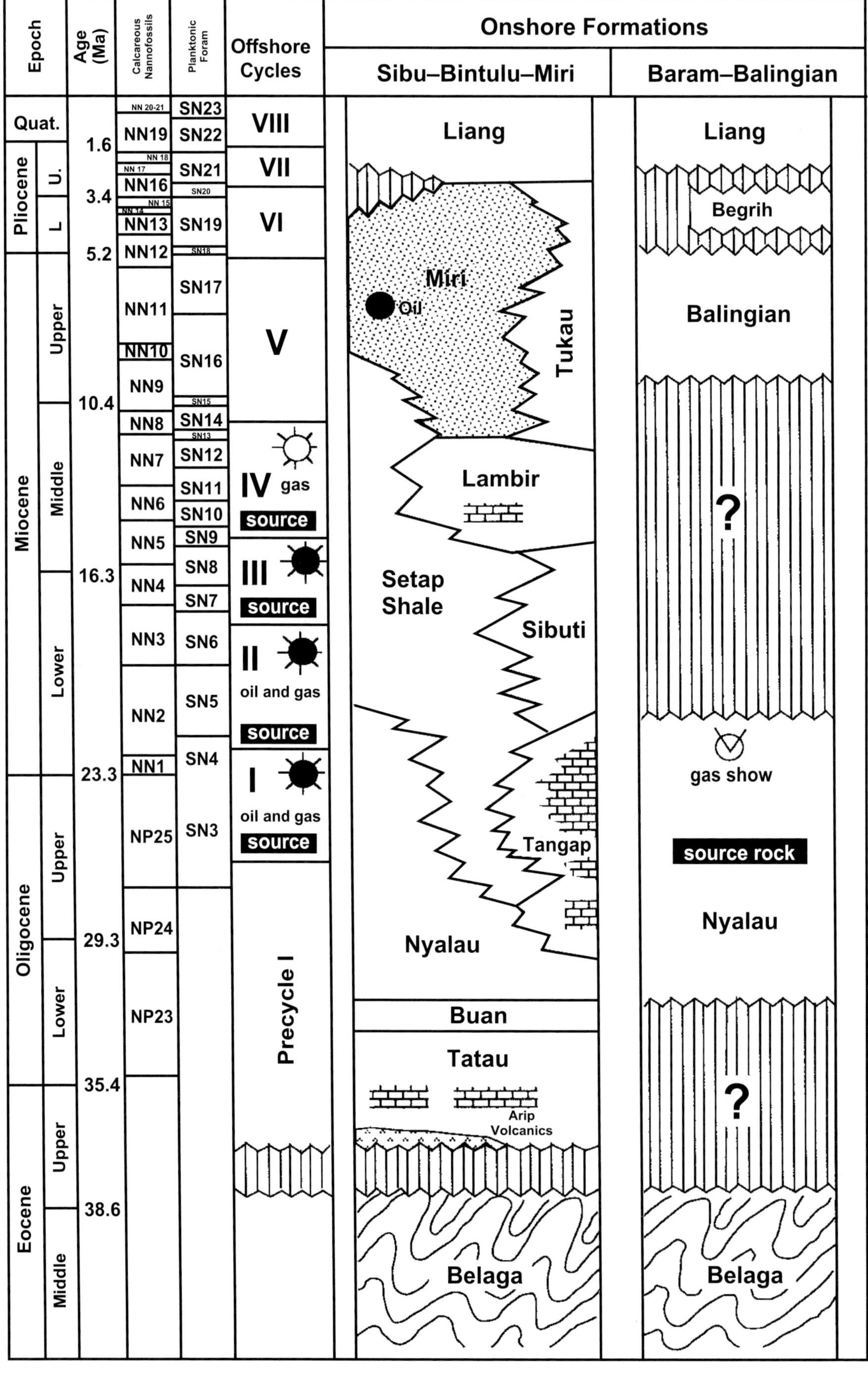
Epoch
Age (Ma)
Calcareous Nannofossils
Planktonic Foram
Offshore Cycles
Onshore Formations
Sibu–Bintulu–Miri
Baram–Balingian
Quat.
Pliocene
U.
L
Miocene
Upper
Middle
Lower
Oligocene
Upper
Lower
Eocene
Upper
Middle
1.6
3.4
5.2
10.4
16.3
23.3
29.3
35.4
38.6
NN 20-21
NN19
NN 18
NN 17
NN16
NN 15
NN 14
NN13
NN12
NN11
NN10
NN9
NN8
NN7
NN6
NN5
NN4
NN3
NN2
NN1
NP25
NP24
NP23
SN23
SN22
SN21
SN20
SN19
SN18
SN17
SN16
SN15
SN14
SN13
SN12
SN11
SN10
SN9
SN8
SN7
SN6
SN5
SN4
SN3
VIII
VII
VI
V
IV gas
source
III
source
II
oil and gas
source
I
oil and gas
source
Precycle I
Liang
Miri
Oil
Tukau
Lambir
Setap Shale
Sibuti
Tangap
Nyalau
Buan
Tatau
Arip Volcanics
Belaga
Liang
Begrih
Balingian
?
gas show
source rock
Nyalau
?
Belaga

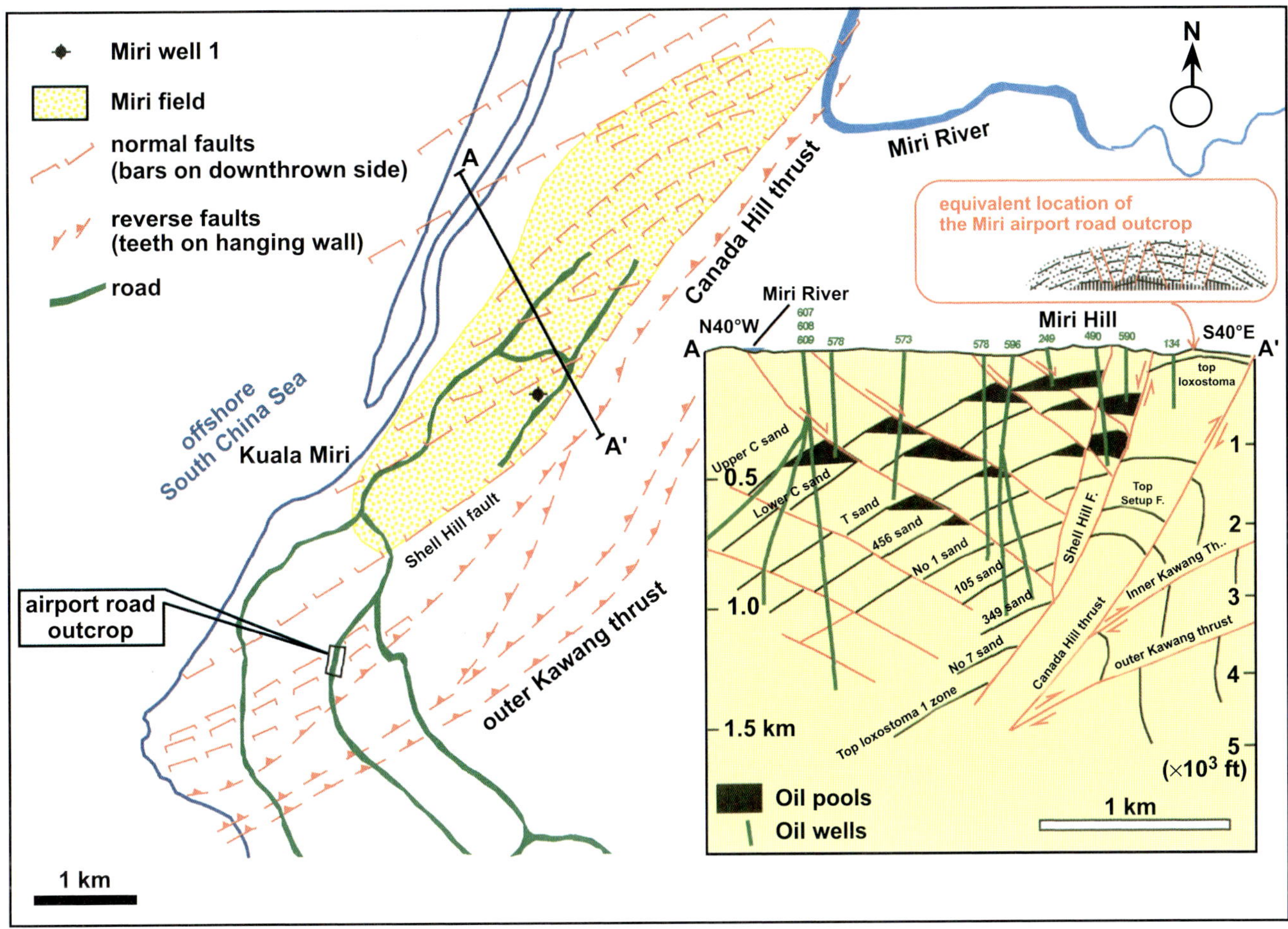

FIGURE 3. Structural position of the Miri (originally drawn in 1946 by von Schumbacher of Shell Company and modified from Haile and Ho, 1991). The Miri Airport Road outcrop lies on the crest of a gentle anticline formed during Pliocene compression and is bounded by the Canada Hill thrust and the Shell Hill fault. The normal faults formed during the collapse of the anticline crest.

It appears that a combination of these processes took place in a given fault zone.

The efficiency of shale smear as an across-fault seal depends largely on its continuity and thickness on the fault plane. This is quantified as shale-smear factor (SSF), which is a ratio of fault throw to the thickness of the source shale layer (Lindsay et al., 1993). Shale-smear factor values of less than or equal to 1 imply juxtaposition of sandstone against the displaced shale layer. Based on measurements of shale smear on two-dimensional (2-D) outcrops of faults in the United Kingdom, Lindsay et al. (1993) found that discontinuous shale smear does not occur below SSF values of 7. In addition, several studies (Yielding et al., 1997) of oil fields have calibrated the SSF threshold value to be about 7. Recently, Gibson and Bentham (2003) examined sealing faults in offshore Trinidad and concluded that an SSF value of 4 characterizes 100% sealing faults, whereas an SSF value of 7 is associated with 50% sealing faults.

In all faults observed in the Miri outcrop (Figure 4), a veneer of black clay was present on fault planes, indicating that shale smear was continuous at least in 2-D cross sections. Most faults were of juxtaposition type, but continuous shale smear was observed for an SSF value of 5.3 (Figure 4), which is consistent with the previous estimations (Gibson, 1994; Yielding et al., 1997). Furthermore, a fault-seal analysis of normal faults in

FIGURE 2. Stratigraphic column of north Sarawak showing the position of the Miri Formation (modified after Haile and Ho, 1991).

Fault	Strike	Dip Angle	Displacement (cm)	Shale Smear Factor (SSF) (Throw / Shale Layer) (cm)	
1	N40–45°E	65°NW	280	250/70 = 3.6	**continuous smear**
2	N40–45°E	65°NW	50	40/70 = 0.6 40/17 = 2.4	**juxtaposition** **continuous smear**
3	N55°E	65°SE	85	77/80 = 1.0	**juxtaposition**
4	N55°E	80°NW	30	29/80 = 1.0	**juxtaposition**
5	N50°E	60°NW (steep plane) 45°NW (listric plane)	500	430/80 = 5.4	**continuous smear**
6	N55°E	60°SE	30	26/55 = 0.5	**juxtaposition**
7	N50–55°E	65°SE	35	31/55 = 0.5	**juxtaposition**
8	N55°E	65–70°SE	35	31/55 = 0.6	**juxtaposition**

FIGURE 4. Photograph of the Miri Airport Road outcrop and the major faults studied. The table provides the structural information and SSF values measured for eight normal faults (F 1–8).

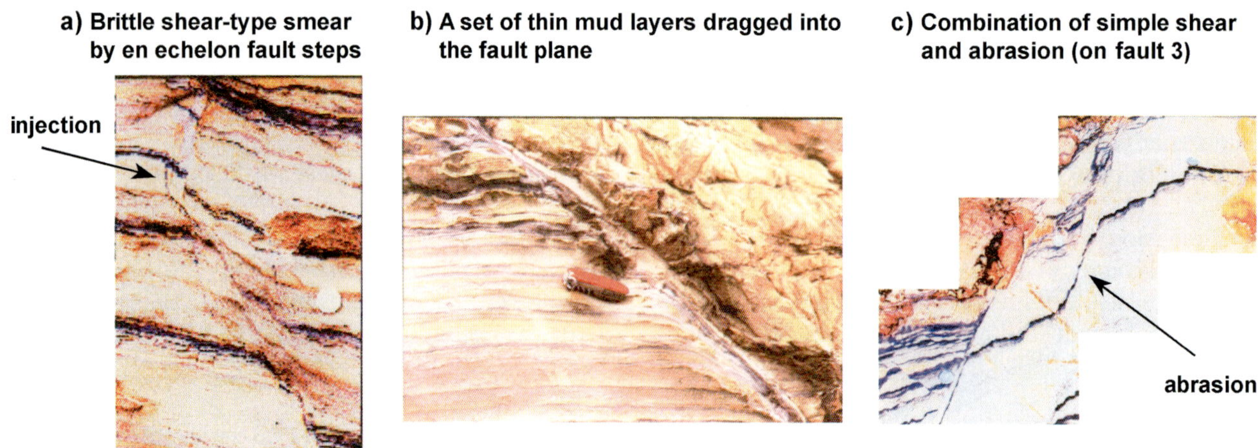

FIGURE 5. Styles of clay smear observed in the fault zones of the Miri Airport Road outcrop.

offshore Sarawak (Hasegawa et al., 2005) indicates that sealing faults have SSF values of less than 6, which is consistent with our outcrop observation.

DEFORMATION BANDS

Deformation bands are thin (millimeter-wide) planar structures in faulted sandstones first documented by Aydin (1978) in Utah. Since then, other geologists have also mapped deformation bands in porous sandstones in Utah (Bruhn et al., 1990; Antonellin and Aydin, 1994, 1995; Shipton et al., 2002), Nevada (Flodin et al., 2003), onshore United Kingdom (Fowles and Burley, 1994), and the North Sea (Gabrielsen and Koestler, 1987). They have also been termed "granulation seams" (Pittman, 1981) and "cataclastic slip bands" (Fowles and Burley, 1994), although in this chapter, we keep the original term for convenience. Deformation bands associated with normal faults in porous sandstone are the smallest visible shear fractures, where the host sandstone has undergone fracturing, crushing, and porosity collapse (Antonellini et al., 1994). Deformation bands have lower permeability and therefore are important for creating across-fault barriers in sandstone reservoirs (Antonellini et al., 1994; Fowles and Burley, 1994; Fossen and Hesthammer, 1998).

The Miri outcrop sandstone has a porosity of about 25–30%. We measured the frequency (occurrence number) of deformation bands along several transects perpendicular to fault slip planes, and the results are depicted in Figure 6. For all the faults studied, it appears that the number of deformation bands increases from the host sandstone toward the fault-slip plane. Deformation bands in the host sandstone away from the fault-slip plane are solitary features but form an anastomosing pattern (linked and narrowly spaced deformation bands) in fault zones (see the photograph in Figure 6). This observation seems to support the interpretation of Antonellini and Aydin (1994) that faulting in porous sandstone is an evolutionary process beginning with solitary deformation bands and culminating with the main slip plane (highest strain). Anastomosing deformation bands form the bulk of fault rock in the Miri outcrop with a thickness of about 20 cm (8 in.) for these faults.

PERMEABILITY MEASUREMENTS

Permeability data for both undeformed sandstone rock and fault-zone rocks were collected at the outcrop using a probe minipermeameter (Temco TM-401), which measures N_2 gas permeability at selected points on a rock. Detailed description of this type of device is given in Goggin et al. (1988) and Goggin (1993), and probe minipermeametry has been used in a variety of structures, including laminated sandstones (Corbett and Jensen, 1993), deltaic strata (Prosser and Carter, 1997), and faulted sandstones (e.g., Antonellini and Aydin, 1994). Data collection by probe minipermeameter ranges from 1 to 10^4 md, although we found the practical minimum measurement limit to be about 5 md. Gas from a high-pressure cylinder was injected onto a flat surface of the rock through a probe with a tip diameter of 1/8 in. (3.18 mm).

Two faults were selected for this study: faults 6 and 3 (Figure 4). Measurements were made along a transect from undeformed sandstone rock through the fault zone. We characterize fault zone as either a fault slip plane surrounded by anastomosing deformation bands (in the case of fault 6) or a zone of anastomosing deformation bands surrounded by two discrete slip planes (in the case of fault 3). For the purpose of these measurements, we also define a rock matrix as sandstone rock sample points devoid of visible deformation bands. These points occur in both the host (undeformed) sandstone rock outside the fault zone and in the fault zone. Solitary deformation bands also occur outside the fault zone and in undeformed sandstone.

Three transects (A, B, and C) across fault 6 and two transects (D and E) across fault 3 were measured to collect permeability data at numerous points. The frequency (number of occurrences) of deformation bands was also counted along each transect. The data obtained are presented in Figure 7 for fault 6 and in Figure 8 for fault 3. The transects are almost perpendicular to the fault plane, and permeability measurements (gas flow) are parallel to the fault plane.

For both faults, it was found that permeability decreases from host sandstone to the fault zone. This permeability reduction is consistent with an increase in the number of deformation bands in the same direction (Figures 7, 8).

Upon further analysis, the permeability data indicate that permeability reduction occurs in two patterns (Figure 9):

1) Permeability reduction from rock matrix to solitary deformation bands, as shown in Figure 9a for fault 6 and in Figure 9c for fault 3. Both figures show that permeability decreases by approximately one order of magnitude from rock matrix to solitary deformation bands in the immediate vicinity of the rock matrix sample point.
2) Permeability reduction from the host (undeformed) sandstone to the fault zone, as shown in Figure 9b for fault 6 and in Figure 9d for fault 3. It is remarkable that permeability of deformation bands

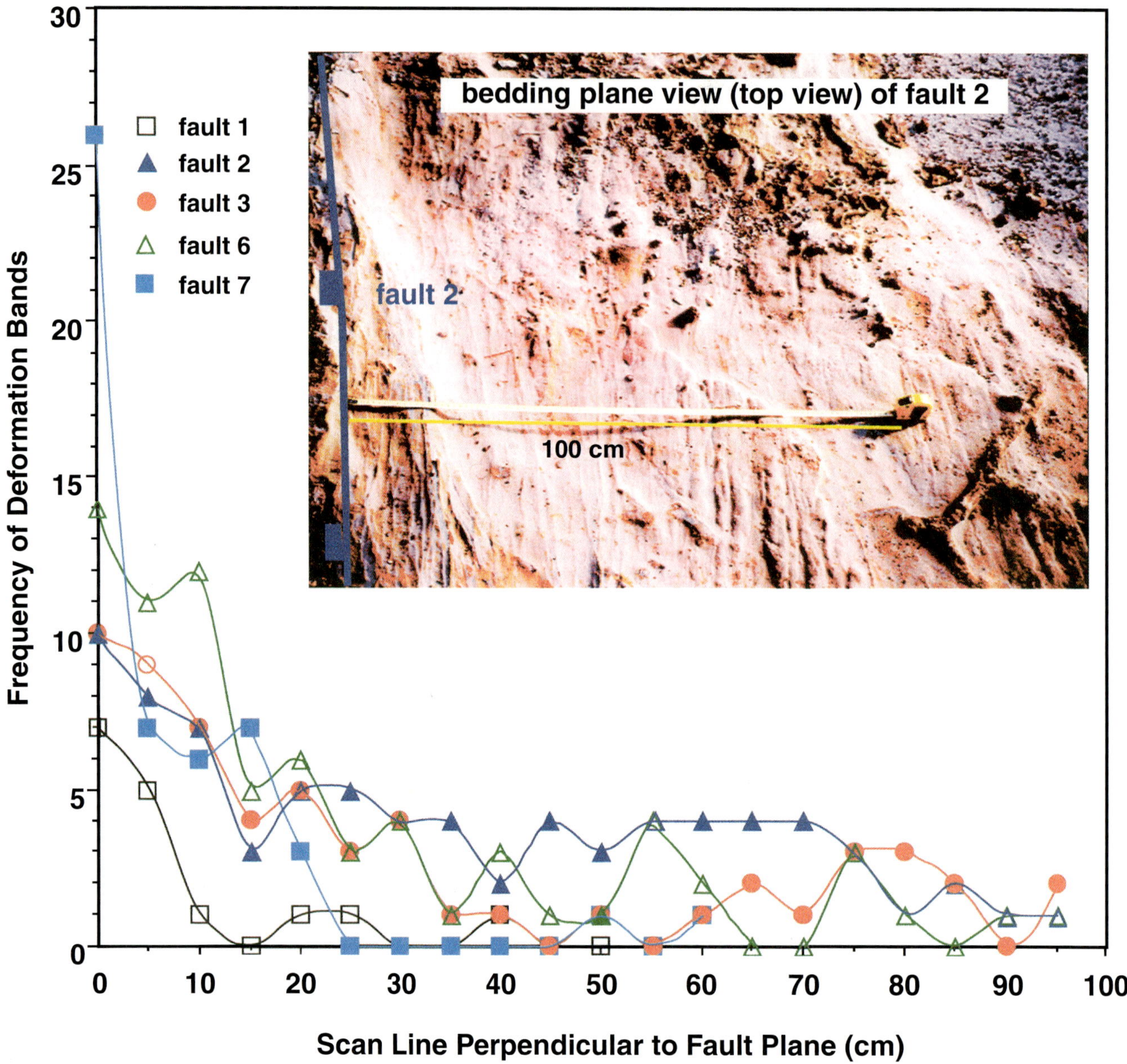

FIGURE 6. Increase in the number of occurrences (frequency) of deformation bands measured along scan lines nearly perpendicular to faults 1, 2, 3, 6, and 7 in the Miri Airport Road outcrop. Deformation bands make an anastomosing pattern in fault zones as shown in the photo insert for fault 2. Fault zones include the main slip planes and a zone of high strain characterized by anastomosing deformation bands.

in the host sandstone zone was found to be higher than or similar to the rock matrix in the fault zone. This indicates that the rock matrix in the fault zone also experiences permeability reduction (probably caused by tectonic compaction) even at points where no visible deformation band occurs. This is why the rock matrix in the fault zone has lower permeability values when compared to those in host sandstone outside the fault zone by a factor of 2–5 as observed for both faults (Figure 9b, d).

The reduction in permeability coincides with the increase in deformation bands toward the main slip plane.

POPULATION ANALYSIS OF SMALL FAULTS

In the Miri Airport Road outcrop, a scan line of 80.9 m (265.4 ft) was selected to study the relationship of small fault population to major faults. All the faults

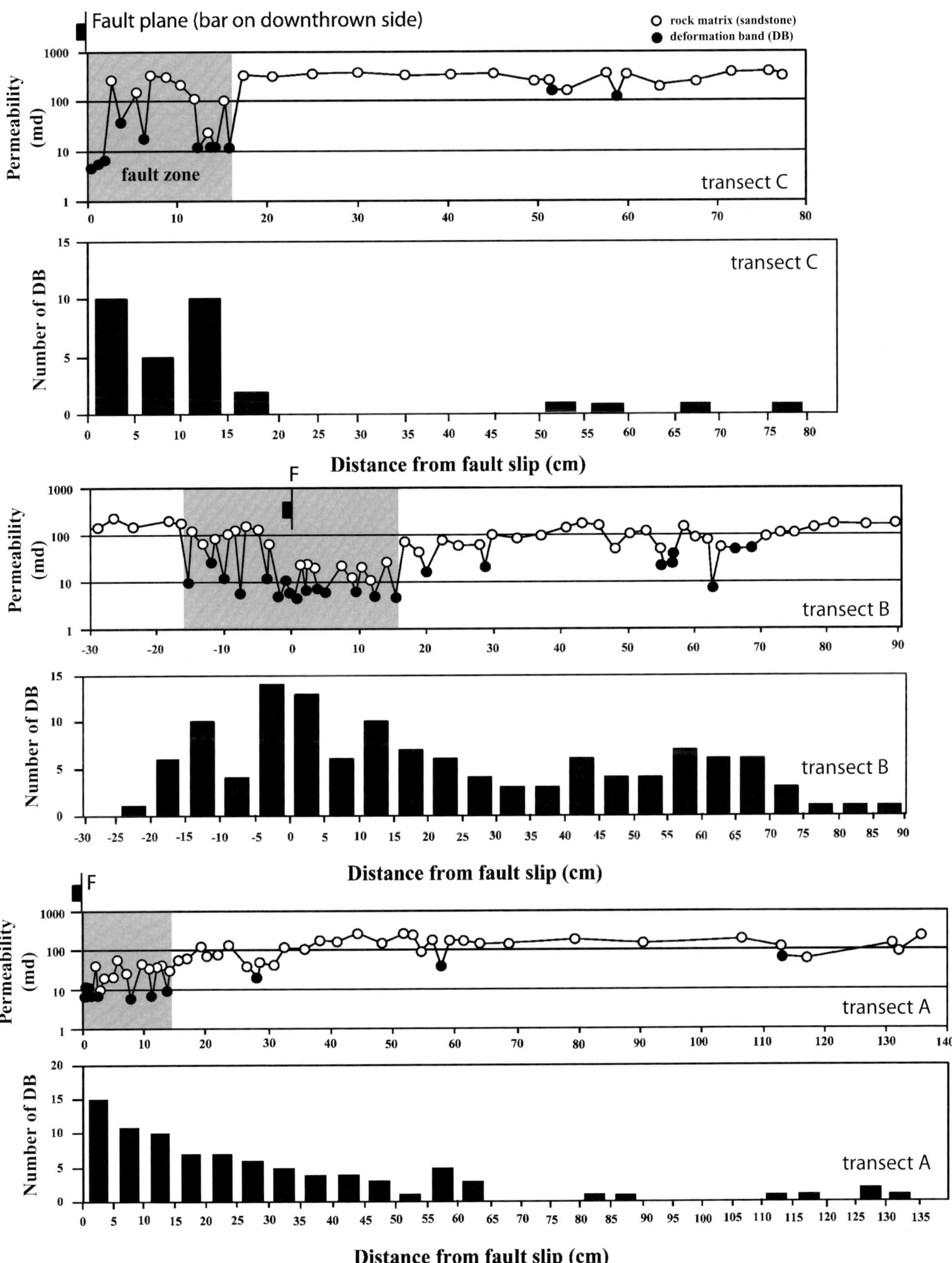

FIGURE 7. Distribution of gas-permeability data obtained by a probe minipermeameter and frequency of deformation bands along three transects (A, B, and C) almost perpendicular to the slip plane of fault 6. Note that the increase in the number of deformation bands (DBs) in the fault zone coincides with a significant decrease in permeability.

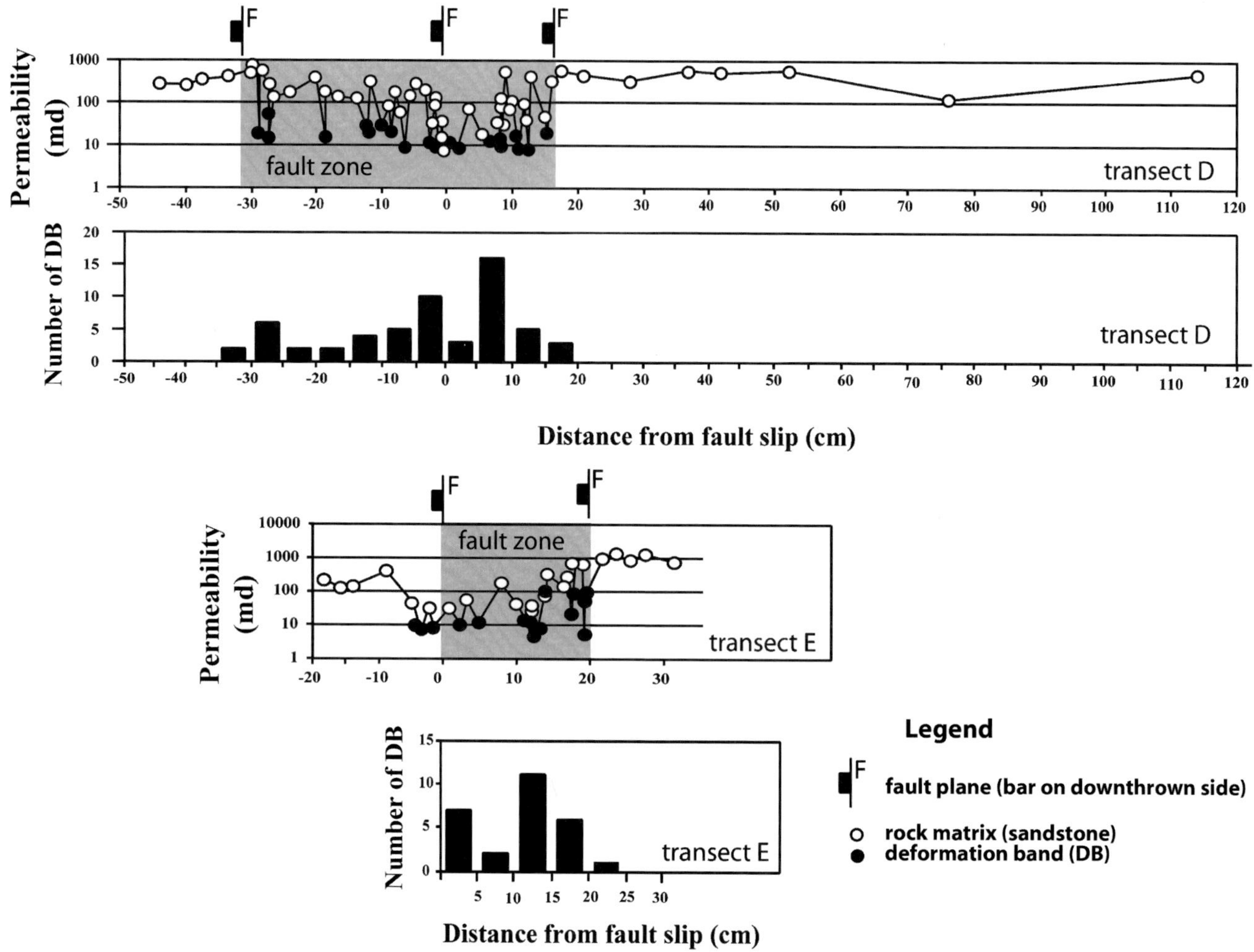

FIGURE 8. Distribution of gas-permeability data obtained by a probe minipermeameter and frequency of deformation bands along two transects (D, E) almost perpendicular to the slip plane of fault 3. Fault zone is defined as a zone of anastomosing deformation bands bounded by two discrete normal faults, in addition to the main slip plane in the center. Note that the increase in the number of deformation bands in the fault zone coincides with a significant decrease in permeability.

measured were normal faults, and mud layers with thicknesses of 25–35 cm (10–14 in.) were used as offset markers. Small faults between major normal faults had offsets of less than 5 cm (2 in.). The number of occurrences (frequency) of small faults is plotted against the displacement of the major faults in Figure 10a and against the distances between the major faults in Figure 10b.

We intended to see if the number of small faults between two major thrust faults was related to the displacement of the major faults. The data (Figure 10a) showed no such relationship. However, these data provide a very good correlation between the population of small faults and the distance between major faults. This can be explained by considering that the small faults were originally joints and were reactivated as faults during the development of the major normal faults. The ratio of bed thickness to median joint spacing is called fracture spacing ratio (FSR), and various studies have shown that FSR values for unfaulted rocks are typically 0.8–1.3 (Gross, 1993; Becker and Gross, 1996; Gross et al., 1997). Fracture spacing ratio values calculated for the small faults along the scan line in Miri are 0.7–1.0, given the bed thickness of 25–35 cm (10–14 in.) and spacing of 25–30 cm (10–12 in.) for throughgoing fractures visible as faults. These values are similar to typical FSR values for unfaulted beds saturated by joints. Reactivation of joints as normal faults has also been observed elsewhere, for example, at Splint Mountain, Utah, which was described by Wilkins et al. (2000). These observations are consistent with a fault mechanism in which fault reactivation in a mechanically heterogeneous rock mass occurs at locations having less cohesive strength, such as along preexisting fractures (Sibson, 1985).

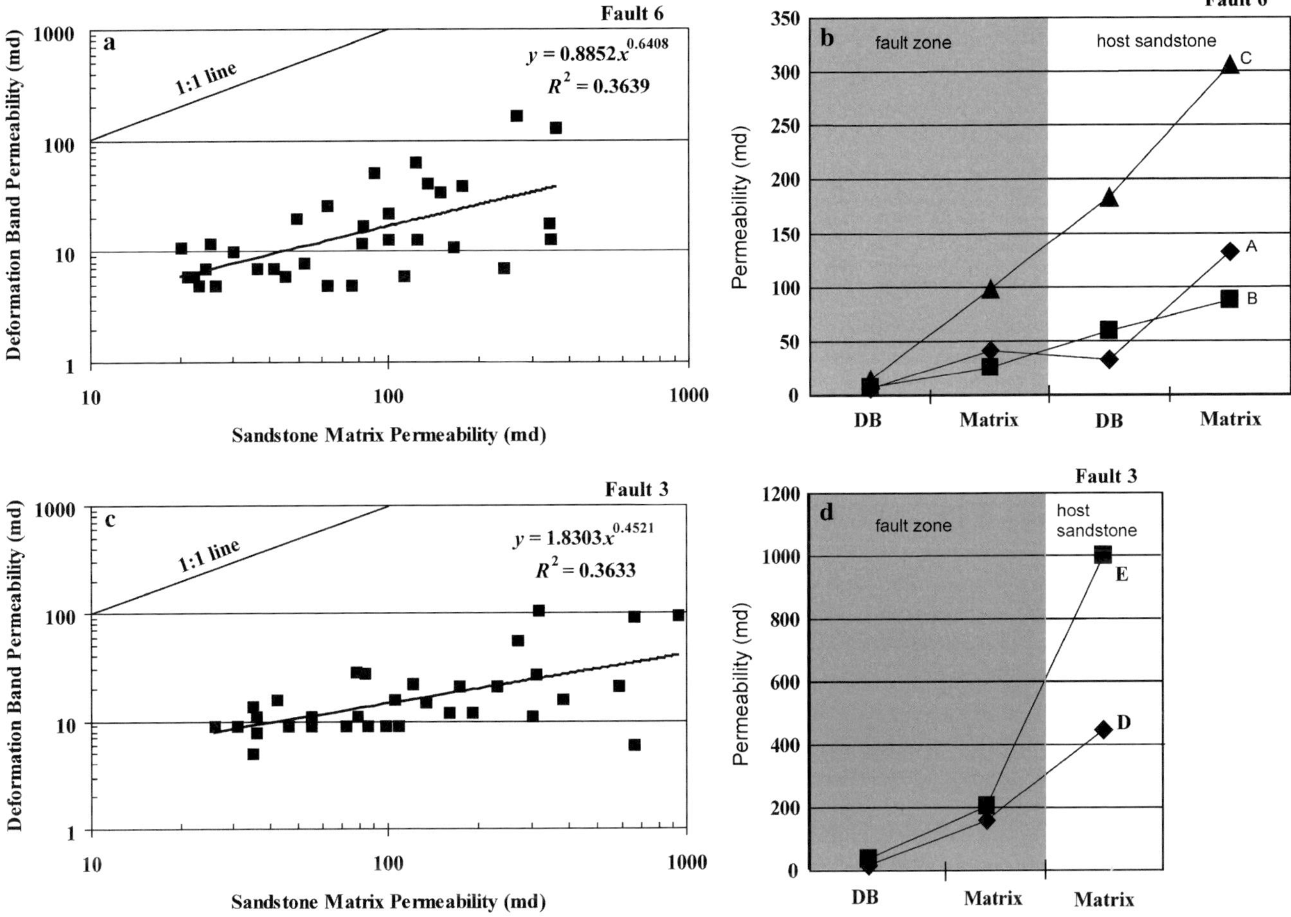

FIGURE 9. (a) Permeability of sandstone matrix plotted vs. that of nearby deformation bands (DB) in the three transects (A, B, and C) nearly perpendicular to fault 6. (b) Average permeability of the sandstone matrix and individual deformation bands in host sandstone outside the fault zone, and sandstone matrix and deformation bands in the fault zone of fault 6 plotted for the three measured transects (A, B, and C). (c) Permeability of sandstone matrix plotted vs. that of nearby deformation bands in two transects (D and E) nearly perpendicular to fault 3. (d) Average permeability of sandstone matrix and individual deformation bands in host sandstone outside the fault zone, and sandstone matrix and deformation bands in the fault zone of fault 3 plotted for the two measured transects (D and E). Note that in (a) and (b), the 1:1 line represents the hypothetical case where both permeabilities are equal, but the data show an order of magnitude permeability reduction from sandstone matrix to individual deformation bands in the immediate vicinity of measured points. Also note that in (c) and (d), the sandstone matrix sample points in fault zones have lower permeability than the matrix points in host sandstone outside the fault zones.

Techniques for prediction of subseismic fault populations are important to characterize structural permeability in reservoirs. Currently, faults with displacements of less than 10 m (33 ft) on 3-D seismic images are below seismic resolution. It is plausible that a population of subseismic faults in a sandstone-mudstone sequence deformed by major normal faults originate from the reactivation of bed joints.

CONCLUSIONS

An outcrop study was carried out on a series of normal faults cutting a sandstone-mudstone sequence in Miri, Sarawak. Fault zones in these porous sandstones are characterized by a combination of shale smear and anastomosing deformation bands, both of which contribute to the sealing efficiency of the faults. A combination of shale smear and deformation band zone would produce lateral fault seals and, hence, compartmentalized reservoirs; this interpretation is consistent with the existence of subsurface fault seals in the Miri oil field and offshore Sarawak (Haile and Ho, 1991). Deformation bands increase in number and linkage from host sandstone toward the fault slip plane. The fault zone defined by anastomosing deformation bands has a lower permeability by at least one order of magnitude. Reactivation of bed joints by simple shear appears to be an important source of small faults in a sandstone-mudstone sequence.

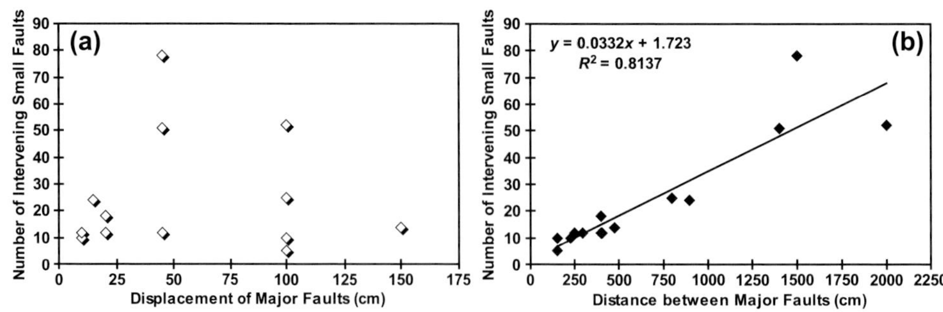

FIGURE 10. Number of small faults (with an offset of <5 cm [<2 in.]) between major faults counted along a scan line (80.9 m; 265.4 ft) in the Miri Airport Road outcrop. The data are plotted against the displacement of the bounding major faults (a) and against the distance between the major faults (b). Good correlation in (b) indicates that the frequency of the small faults is related to fracture spacing, and hence, these faults are reactivated joints.

ACKNOWLEDGMENTS

We are grateful to the Japan National Oil Corporation (presently Japan Oil, Gas and Metals National Corporation) for supporting this study, to Kiyofumi Suzuki, Andrey Rezanov (formerly of Japan National Oil Corporation), Abdullah Adli, Jalil Basiron of Petroliam Nasional Bhd. for a wonderful fieldwork in Sarawak, and to Mario Wannier of Shell and Charles Stuart (both of whom have first-hand knowledge of the region) for reviewing the chapter and for their comments to improve it.

REFERENCES CITED

Alexander, L. L., and J. W. Handschy, 1998, Fluid flow in a faulted reservoir system: Fault trap analysis for the Block 330 field in Eugene Island, south Addition, offshore Louisiana: AAPG Bulletin, v. 82, p. 387–411.

Antonellini, M., and A. Aydin, 1994, Effect of faulting on fluid flow in porous sandstones: Petrophysical properties: AAPG Bulletin, v. 78, p. 355–377.

Antonellini, M., and A. Aydin, 1995, Effect of faulting on fluid flow in porous sandstones: Geometry and spatial distribution: AAPG Bulletin, v. 79, p. 642–671.

Antonellini, M. A., A. Aydin, and D. D. Pollard, 1994, Microstructure of deformation bands in porous sandstones at Arches National Park, Utah: Journal of Structural Geology, v. 16, p. 941–959.

Aydin, A., 1978, Small faults formed as deformation bands in sandstone: Pure and Applied Geophysics, v. 116, p. 913–930.

Aydin, A., and Y. Eyal, 2002, Anatomy of a normal fault with shale smear: Implications for fault seal: AAPG Bulletin, v. 86, p. 1367–1381.

Bait, B., H. D. Johnson, J. Ranggon, and J. A. Lopez, 1991, Tatu–Bintulu–Miri area: A geological field guide: Geological Society of Malaysia Post-Conference Field Trip (May 6–8, 1991).

Becker, A., and M. R. Gross, 1996, Mechanism for joint saturation in mechanically layered rocks: An example from southern Israel: Tectonophysics, v. 257, p. 223–237.

Berg, R. R., and A. H. Avery, 1995, Sealing properties of Tertiary growth faults, Texas Gulf Coast: AAPG Bulletin, v. 79, p. 375–393.

Bruhn, R. L., W. A. Yonkee, and W. T. Pany, 1990, Structural and fluid-chemical properties of seismogenic normal faults: Tectonophysics, v. 175, p. 139–157.

Burhannudinnur, M., and C. K. Morley, 1997, Anatomy of growth fault zones in poorly lithified sandstones and shales: Implications for reservoir studies and seismic interpretation: Part 1. Outcrop study: Petroleum Geoscience, v. 3, p. 211–224.

Childs, C. J., J. J. Walsh, and J. Watterson, 1997, Complexity in fault zone structure and implications for fault seal prediction, *in* P. Møller-Pedersen and A. G. Koestler, eds., Hydrocarbon seals: Importance for exploration and production: Norwegian Petroleum Society Special Publication 7, p. 61–72.

Childs, C., T. Manzocchi, P. A. R. Nell, J. J. Walsh, J. A. Strand, A. E. Heath, and T. H. Lygren, 2002, Geological implications of a large pressure difference across a small fault in the Viking graben, *in* A. G. Koestler and R. Hunsdale, eds., Hydrocarbon seal quantification: Norwegian Petroleum Society Special Publication 11, p. 187–201.

Corbett, P. W. M., and J. L. Jensen, 1993, Application of probe permeametry to the prediction of two-phase flow performance in laminated sandstones (lower Brent Group, North Sea): Marine and Petroleum Geology, v. 10, p. 335–346.

Doughthy, P. T., 2003, Clay smear seals and fault sealing potential of an exhumed growth fault, Rio Grande rift, New Mexico: AAPG Bulletin, v. 87, p. 427–444.

Flodin, E., M. Prasad, and A. Aydin, 2003, Petrophysical constraints on deformation styles in Aztec Sandstone, southern Nevada, U.S.A.: Pure and Applied Geophysics, v. 160, p. 1589–1610.

Fossen, H., and J. Hesthammer, 1998, Deformation bands and their significance in porous sandstone reservoirs: First Break, v. 16, p. 21–25.

Fowles, J., and S. D. Burley, 1994, Textural and permeability characteristics of faulted, high porosity sandstones: Marine Petroleum and Geology, v. 11, p. 608–623.

Gabrielsen, R. H., and A. G. Koestler, 1987, Description and structural implications of fractures in Late Jurassic sandstones of the Troll field, northern North Sea: Norsk Geologisk Tidsskrift, v. 67, p. 371–381.

Garden, I. R., S. C. Guscott, S. D. Burley, K. A. Foxford, J. J. Walsh, and J. Marshall, 2001, An exhumed palaeohydrocarbon migration fairway in a faulted carrier system, Entrada Sandstone of SE Utah, U.S.A.: Geofluids, v. 1, p. 195–213.

Gibson, R. G., 1994, Fault-zone seals in siliclastic strata of the Columbus basin, offshore Trinidad: AAPG Bulletin, v. 78, p. 1372–1385.

Gibson, R. G., and P. A. Bentham, 2003, Use of fault-seal analysis in understanding petroleum migration in a complexly faulted anticlinal trap, Columbus basin, offshore Trinidad: AAPG Bulletin, v. 87, p. 465–478.

Goggin, D. J., 1993, Probe permeametry: Is it worth the effort?: Marine and Petroleum Geology, v. 10, p. 299–308.

Goggin, D. J., R. L. Thrasher, and L. W. Lake, 1988, A theoretical and experimental analysis of minipermeameter response including gas slippage and high velocity effects: In Situ, v. 12, p. 79–116.

Gross, M. R., 1993, The origin and spacing of cross joints: Examples from the Moterey formation, Santa Barbara, California: Journal of Structural Geology, v. 15, p. 737–751.

Gross, M. R., D. Bahat, and A. Becker, 1997, Relations between jointing and faulting based on fracture-spacing ratios and fault-slip profiles: A new method to estimate strain in layered rocks: Geology, v. 25, p. 887–890.

Haile, N. S., 1969, Geosynclinal theory and the organizational pattern of the north-west Borneo geosyncline: Quarterly Journal of Geological Society (London), v. 124, p. 171–194.

Haile, N. S., 1974, Borneo, *in* A. M. Spencer, ed., Mesozoic–Cainozoic orogenic belts: Geological Society (London) Special Publication 4, p. 333–347.

Haile, N. S., and C. K. Ho, 1991, Geological field guide: Sibu–Miri traverse, Sarawak (September 24–October 1, 1991): Kuala Lumpur, Petronas Research & Scientific Services, 62 p.

Hall, R., 1996, Reconstructing Cenozoic SE Asia, *in* R. Hall and D. Blundell, eds., Tectonic evolution of southeast Asia: Geological Society (London) Special Publication 106, p. 153–184.

Hammond, K. J., and J. P. Evans, 2003, Geochemistry, mineralization, structure, and permeability of a normal-fault zone, Casino mine, Alligator Ridge district, north central Nevada: Journal of Structural Geology, v. 25, p. 717–736.

Hasegawa, S., R. Sorkhabi, S. Iwanaga, N. Sakuyama, and O. A. Mahmud, 2005, Fault-seal analysis in the Temana field, offshore Sarawak, Malaysia, *in* R. Sorkhabi and Y. Tsuji, eds., Faults, fluid flow and petroleum traps: AAPG Memoir 85, p. 43–58.

Ho, K. F., 1978, Stratigraphic framework for oil exploration in Sarawak: Geological Society of Malaysia Bulletin, v. 10, p. 1–13.

Hussin, A. H., B. Bait, and J. J. Jau, 1992, Guide for field trip 1. Geology of Sarawak (December 3–5, 1992): Symposium on Tectonic Framework and Energy Resources of the Western Margin of the Pacific Basin: Geological Society of Malaysia, Kuala Lumpur, 55 p.

Hutchinson, C. S., 1989, Geological evolution of southeast Asia: Oxford Monograph on Geology and Geophysics, v. 13, 368 p.

Hutton, W., 1998, Insight pocket guides: Sarawak: Singapore, APA Publications, 93 p. plus map.

Kolddoye, B. A., A. Aydin, and E. May, 2003, A new process-based methodology for analysis of shale smear along normal faults in the Niger Delta: AAPG Bulletin, v. 87, p. 445–464.

Liechti, P., F. W. Roe, and N. S. Haile, 1960, The geology of Sarawak, Brunei, and the western part of north Borneo: Geological Survey Department, British Territories in Borneo, Bulletin, v. 3, 360 p.

Lindsay, N. G., F. C. Murphy, J. J. Walsh, and J. Watterson, 1993, Outcrop studies of shale smears on fault surfaces: International Association of Sedimentologists Special Publication 15, p. 113–123.

Naruk, S. J., et al., 2002, Common characteristics of proven and leaking faults: AAPG Hedberg Research Conference on Evaluating the Hydrocarbon Sealing Potential of Faults and Caprocks, December 1–5, 2002, Barossa Valley, Australia, p. 71–74.

Pittman, E. D., 1981, Effect of fault-related granulation on porosity and permeability of quartz sandstones, Simpson Group (Ordovician), Oklahoma: AAPG Bulletin, v. 65, p. 2381–2387.

Prosser, D. J., and R. R. Carter, 1997, Permeability heterogeneity within the Jerdung Formation: An outcrop analogue for subsurface Miocene reservoirs in Brunei, *in* A. J. Fraser, S. J. Matthews, and R. W. Murphy, eds., Petroleum geology of southeast Asia: Geological Society (London) Special Publication 126, p. 195–235.

Shipton, Z. K., J. P. Evans, K. R. Robeson, C. B. Forster, and S. Snelgrove, 2002, Structural heterogeneity and permeability in faulted eolian sandstone: Implications for subsurface modeling of faults: AAPG Bulletin, v. 86, p. 863–883.

Sibson, R. H., 1985, A note on fault reactivation: Journal of Structural Geology, v. 7, p. 751–754.

Van der Zee, W., 2001, Dynamics of fault gouge development in layered sand-clay sequences: Ph.D. thesis, Achen University, Germany, chapter 3, p. 29–66.

Wilkins, S. J., M. R. Gross, M. Wacker, Y. Eyal, and T. Engelder, 2000, Faulted joints: Kinematics, scaling relations, and criteria for their identification: Journal of Structural Geology, v. 23, p. 315–327.

Yielding, G., B. Freeman, and D. T. Needham, 1997, Quantitative fault seal prediction: AAPG Bulletin, v. 81, p. 897–917.

Davatzes, N. C., and A. Aydin, 2005, Distribution and nature of fault architecture in a layered sandstone and shale sequence: An example from the Moab fault, Utah, *in* R. Sorkhabi and Y. Tsuji, eds., Faults, fluid flow, and petroleum traps: AAPG Memoir 85, p. 153–180.

Distribution and Nature of Fault Architecture in a Layered Sandstone and Shale Sequence: An Example from the Moab Fault, Utah

N. C. Davatzes[1]
Department of Geological and Environmental Science, Stanford University, Stanford, California, U.S.A.

A. Aydin
Department of Geological and Environmental Science, Stanford University, Stanford, California, U.S.A.

ABSTRACT

We examined the distribution of fault rock and damage zone structures in sandstone and shale along the Moab fault, a basin-scale normal fault with nearly 1 km (0.62 mi) of throw, in southeast Utah. We find that fault rock and damage zone structures vary along strike and dip. Variations are related to changes in fault geometry, faulted slip, lithology, and the mechanism of faulting. In sandstone, we differentiated two structural assemblages: (1) deformation bands, zones of deformation bands, and polished slip surfaces and (2) joints, sheared joints, and breccia. These structural assemblages result from the deformation band-based mechanism and the joint-based mechanism, respectively. Along the Moab fault, where both types of structures are present, joint-based deformation is always younger. Where shale is juxtaposed against the fault, a third faulting mechanism, smearing of shale by ductile deformation and associated shale fault rocks, occurs. Based on the knowledge of these three mechanisms, we projected the distribution of their structural products in three dimensions along idealized fault surfaces and evaluated the potential effect on fluid and hydrocarbon flow. We contend that these mechanisms could be used to facilitate predictions of fault and damage zone structures and their permeability from limited data sets.

[1]*Present address:* Earthquake Hazards Team, U.S. Geological Survey, Menlo Park, California, U.S.A.

DOI:10.1306/1033722M853134

INTRODUCTION

Faults can act as highly permeable pathways that enhance fluid flow parallel to faults, low-permeability zones that inhibit fluid flow across faults, or complex conduit-barrier systems (Bredehoeft et al., 1982) that evolve temporally (Sibson, 1990; Caine et al., 1996) and spatially (Caine and Forster, 1999; Davatzes, 2003). This behavior is evident from field observations of localized diagenesis that indicates focused fluid flow along faults (Dholakia et al., 1998; Nelson et al., 1999; Sigda et al., 1999; Taylor et al., 1999; Eichhubl and Boles, 2000) or barrier behavior that inhibited cross-fault flow (Antonellini and Aydin, 1994, 1995; Antonellini et al., 1999; Aydin, 2000). The permeability of a fault is related to its architecture, i.e., the type, geometry, distribution, and density of structures composing a fault zone (Knipe, 1993; Caine et al., 1996; Heynekamp et al., 1999). Many authors (Matthai et al., 1998; Caine and Forster, 1999; Flodin et al., 2001; Jourde et al., 2002) have used the geometry and connectivity of fault zone structures with distinct hydrologic properties to determine the equivalent bulk (upscaled) permeability of a fault zone. Thus, predicting the distribution of different structures along a fault zone is a key step in predicting the overall permeability structure of a fault.

Along the Moab fault system, southeast Utah, we recognized three distinct sets of structures that might impact the function of the fault in fluid flow. Joints, networks of joints, and breccia may enhance permeability by introducing zones of increased secondary porosity that are well connected (Dholakia et al., 1998; Caine and Forster, 1999; Taylor et al., 1999; Flodin et al., 2001; Davatzes and Aydin, 2003). Conversely, in sandstone adjacent to the fault, cataclastic deformation bands are zones of well-connected porosity reduction that could reduce permeability (Foxford et al., 1998; Davatzes and Aydin, 2003). In general, joints always overprint deformation bands along the Moab fault system. The overprinting indicates a temporal change in the faulting mechanisms and resulting structures that comprise the fault and, consequently, a corresponding change in the fault zone permeability. Finally, clay or shale generally has low permeability (Skerlec, 1999) and is abundant along the Moab fault. Thus, incorporating these materials into or adjacent to a fault zone impacts the fault-parallel and fault-normal permeability (Weber et al., 1978; Lehner and Pilaar, 1991, 1997; Lindsay et al., 1993; Knipe, 1997; Caine and Forster, 1999; Sigda et al., 1999). This effect may be enhanced by foliation and gouge formation from shale-rich rocks and shale smear along the fault zone (Knipe 1993, 1997; Gibson, 1994; Faulkner and Rutter, 1998).

Traditional approaches to characterizing the cross-fault sealing potential of a fault zone have primarily focused on the function of low-permeability rocks juxtaposed against high-permeability rocks across a fault (Allan, 1989; Knipe, 1997) and the potential for shale-rich fault rocks to occur in the fault zone (Yielding et al., 1997). These methods include the following:

1) Juxtaposition analysis: This method assumes that the fault has no hydraulic properties, but sealing may result from the juxtaposition of low-permeability rocks (e.g., shale) against high-permeability rocks (Allan, 1989; Knipe, 1997).
2) Shale smear factor: This method compares the thickness of shale units in a faulted interval to the critical throw at which shale is no longer present in the fault zone (Lindsay et al., 1993; Gibson, 1994; Aydin and Eyal, 2002).
3) Shale gouge ratio: This ratio is calculated as the cumulative thickness of shale units in the faulted interval divided by the throw (Fristad et al., 1997; Yielding et al., 1997). Sealing is determined by calibrating this number to subsurface data on pore pressure differences across faults in the subsurface. Sealing is typically considered when the cumulative thickness of shale units in the faulted interval is about 18–20% of the throw.
4) Smear gouge ratio: This is the ratio between sand and shale in the faulted interval and is similar to the shale gouge ratio (Skerlec, 1999).
5) Statistical data sets based on subsurface estimates or outcrop analogs (e.g., Foxford et al., 1998).

However, none of these methods adequately consider and justify the physical processes responsible for fault development or consider the possibility of enhanced flow in the fault zone. In this contribution, we explore the benefits of using faulting processes to analyze, extrapolate, or predict fault characteristics in combination with other commonly cited geometric parameters.

We propose that three factors control the development of fault architecture along a fault in a layered sandstone and shale sequence: (1) the relative contributions of different faulting mechanisms to fault growth and slip; (2) the geometry of the fault; (3) the distribution of rock types. We quantified the relationships between these three factors along the Moab normal fault system. The architecture of this fault system in Jurassic sandstone units has been extensively characterized in previous work (Foxford et al., 1996, 1998; Davatzes, 2003; Davatzes and Aydin, 2003). However, the effect of shale on the architecture of the Moab fault has not been adequately investigated. In this study, we quantify the behavior of shale in the Moab fault zone and integrate these results with the previous results in sandstone to determine the three-dimensional architecture of the Moab fault. Our results demonstrate that despite the complexity of the Moab fault, fault zone structures are distributed in a consistent pattern determined by the

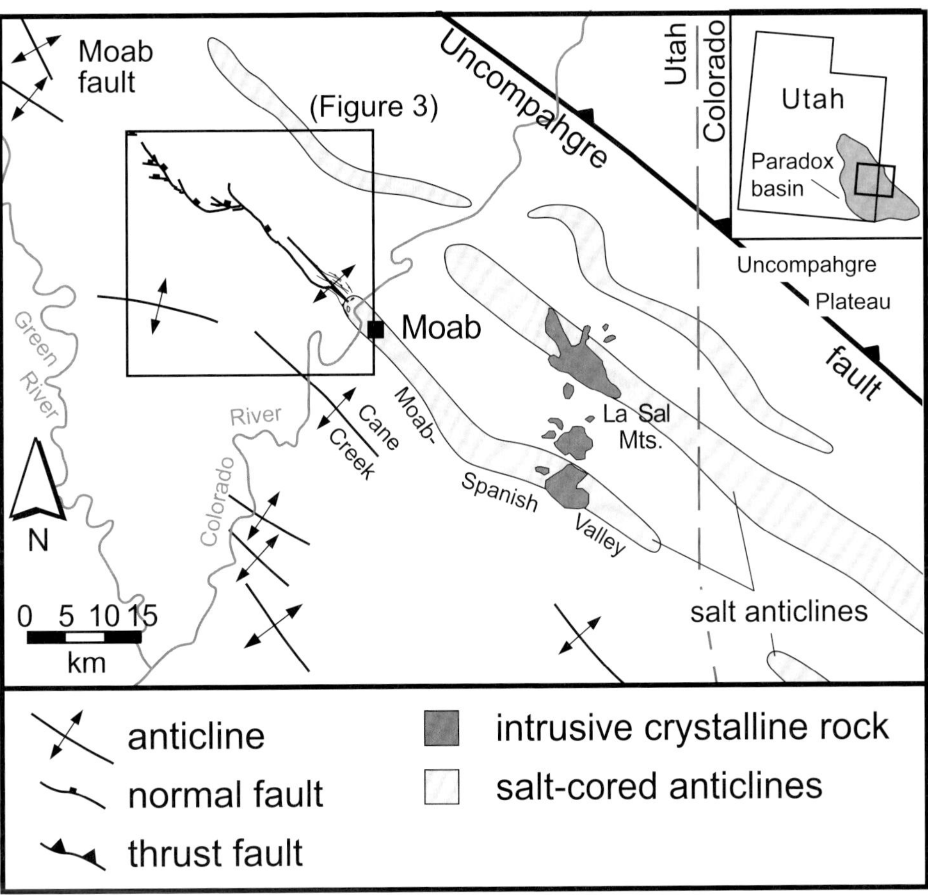

FIGURE 1. Tectonic map of part of the northeast Paradox basin in eastern Utah (modified from Doelling, 1988).

influence of the faulted stratigraphy, slip distribution, and fault geometry on the mechanisms of faulting.

GEOLOGIC SETTING

The Moab fault system is a basin-scale normal fault, approximately 45 km (28 mi) long, with nearly 1 km (0.62 mi) of throw located in the northeastern part of the Paradox basin in southeast Utah (Figure 1). The fault crops out in Pennsylvanian to Cretaceous stratigraphy overlying the Paradox Salt Formation (Figure 2). The southeastern part of the Moab fault (Figure 3) exposes stratigraphic units dominantly composed of interbedded sandstone and shale in varying ratios. These units include the Honaker Trail, Cutler, Moenkopi, and Chinle formations in the footwall and the Salt Wash Member of the Morrison Formation in the hanging wall (Figure 2).

Exposures along the northwestern extent of the Moab fault system are limited to Jurassic and younger units (Figure 3). The stratigraphy in this region is dominated by an abrupt change from a thick package of eolian sandstone units, including the Wingate, Kayenta, Navajo, and Entrada formations (Figure 2), to a shale-dominated sequence that includes the Late Jurassic Morrison Formation and the Cretaceous Cedar Mountain, Dakota Formation, and Mancos Shale. The sandstone units typically have high porosities from 15 to 25% (Dyer, 1983; Antonellini and Aydin, 1994; Foxford et al., 1996) and are exposed in the footwall, whereas the shale units are exposed primarily in the hanging wall, although the Tidwell and Salt Wash Members of the Morrison Formation intermittently crop out in the footwall along the fault as well. In these shale-dominated units, stratigraphic boundaries are demarcated by prominent, but relatively thin, sandstone layers.

Most of the data on the shale in the fault zone that are presented in this study are derived from observations in the Morrison Formation (Doelling, 1982, 1988), which is divided into three members. The Tidwell Member was deposited in quiet, shallow water on the margin of a flood plain and is composed of shale with some siltstone, limestone, and chert-bearing layers. The Salt Wash Member is a fluvial deposit dominated by overbank mud with channel sand deposits, forming sandstone sheets of highly variable thickness. Three to five sandstone layers occur in the shale and may range in thickness from 1 to 20 m (3 to 66 ft) but are typically 3–5 m (10–16 ft) thick. Finally, in the study area, the Brushy Basin Member is a distal flood-plain, shale-dominated deposit with isolated, coarse, and incised-channel sandstones. Ash layers make this unit rich in swelling clays such as smectite.

Faulting in the study area probably occurred between 60 and 43 Ma based on K-Ar dating of shale gouge in the Morrison Formation (Pevear et al., 1997; and other arguments summarized in Davatzes and Aydin, 2003). The faulting was associated with a period of salt movement (Doelling, 1988) during the Laramide orogeny and either occurred during maximum burial of the Entrada formation to a depth between 2000 and 2500 m (6600 and 8200 ft) (Pevear et al., 1997; Garden et al., 2001) or during a phase of subsidence immediately preceding maximum burial (Nuccio and Condon, 1996). Faulting possibly extended into the period of rapid uplift and exhumation following maximum burial.

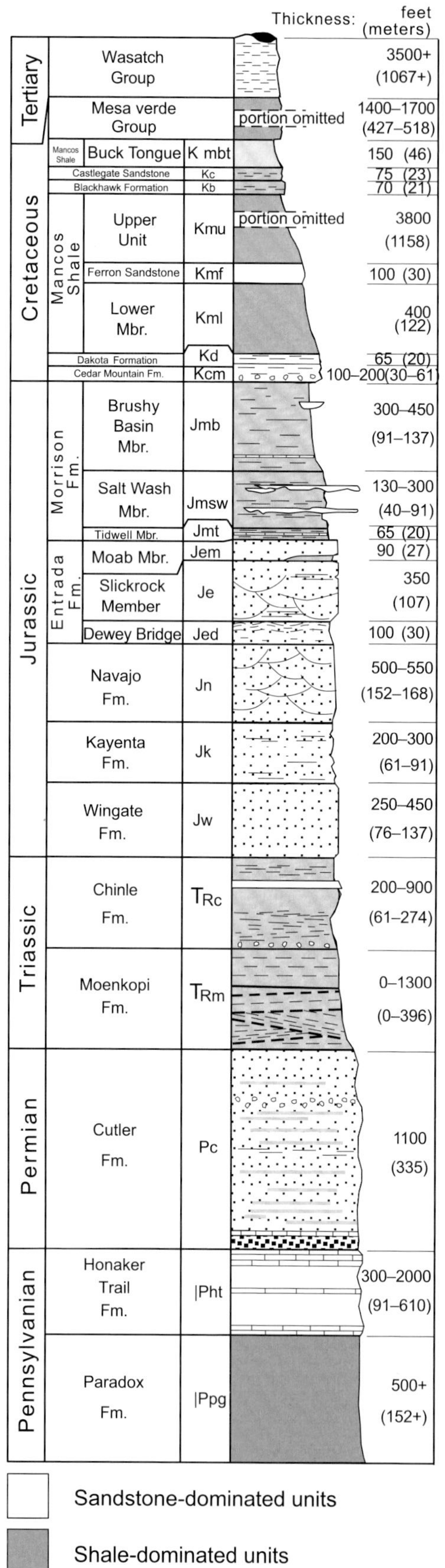

FIGURE 2. Stratigraphy of the study area (modified from Doelling, 1985).

FAULTING MECHANISMS, TERMINOLOGY, AND FIELD METHODS

Following previous workers, we divide the fault zone into a central core that accommodates the majority of offset and is bounded on either side by damage zones (Sibson, 1977; Chester and Logan, 1986; Caine et al., 1996). In this chapter, we use the term "fault rock" as a general term for all highly deformed rocks that comprise the fault core. Structures in the fault core and damage zone are associated with three distinct faulting mechanisms that evolve with increasing offset (Figure 4).

Faults that cut sandstone units along the Moab fault are comprised of two distinct sets of structures that result from different faulting mechanisms (see Davatzes and Aydin, 2003, for a detailed discussion). Cataclastic deformation bands, zones of deformation bands, and slip surfaces are products of the deformation band-based faulting mechanism (Figure 4a) (Aydin and Johnson, 1978; Antonellini and Aydin, 1994). This mechanism involves grain crushing and pore collapse in tabular zones a few millimeters thick, single deformation bands that each accommodate 1–10 mm (0.04–0.4 in.) of shear displacement.

In contrast, joints, sheared joints, splay fractures, zones of fragmented rock, and breccia are products of the joint-based mechanism (Figure 4b) (Myers, 1999; Flodin, 2003). This mechanism localizes shear across established discontinuities, such as joints, resulting in a stress perturbation at the tip of the sheared discontinuity (e.g., Anderson, 1995). Tensile quadrants of the stress perturbation promote the formation of new joints called splay fractures (e.g., Brace and Bombolakis, 1963). This mechanism proceeds by cyclic shearing of earlier joints and formation of new joints as splay fractures. Splay fractures are also known in the literature as secondary fractures, horsetail fractures, pinnate fractures, kink fractures, bridge fractures, and tail fractures (Segall and Pollard, 1980; Granier, 1985; Engelder, 1987; Martel, 1990; Cruikshank et al., 1991; Cruikshank and Aydin, 1994).

In layered sandstone and shale, a fault zone may be composed of smeared shale that remains continuous across the fault zone, sandwiched between brittle faults in surrounding sandstone units (Figure 4c) that form a relay. Smeared shale refers to the entire thickness of shale entrained in the fault core, including beds folded and attenuated parallel to fault slip surfaces as well as gouge (Weber et al., 1978; Lehner and Pilaar, 1997; Aydin and Eyal, 2002; Koledoye et al., 2003). Shale gouge refers to a fault rock consisting of the portion of entrained shale in which sedimentary layering is thoroughly disrupted and mixing may have occurred.

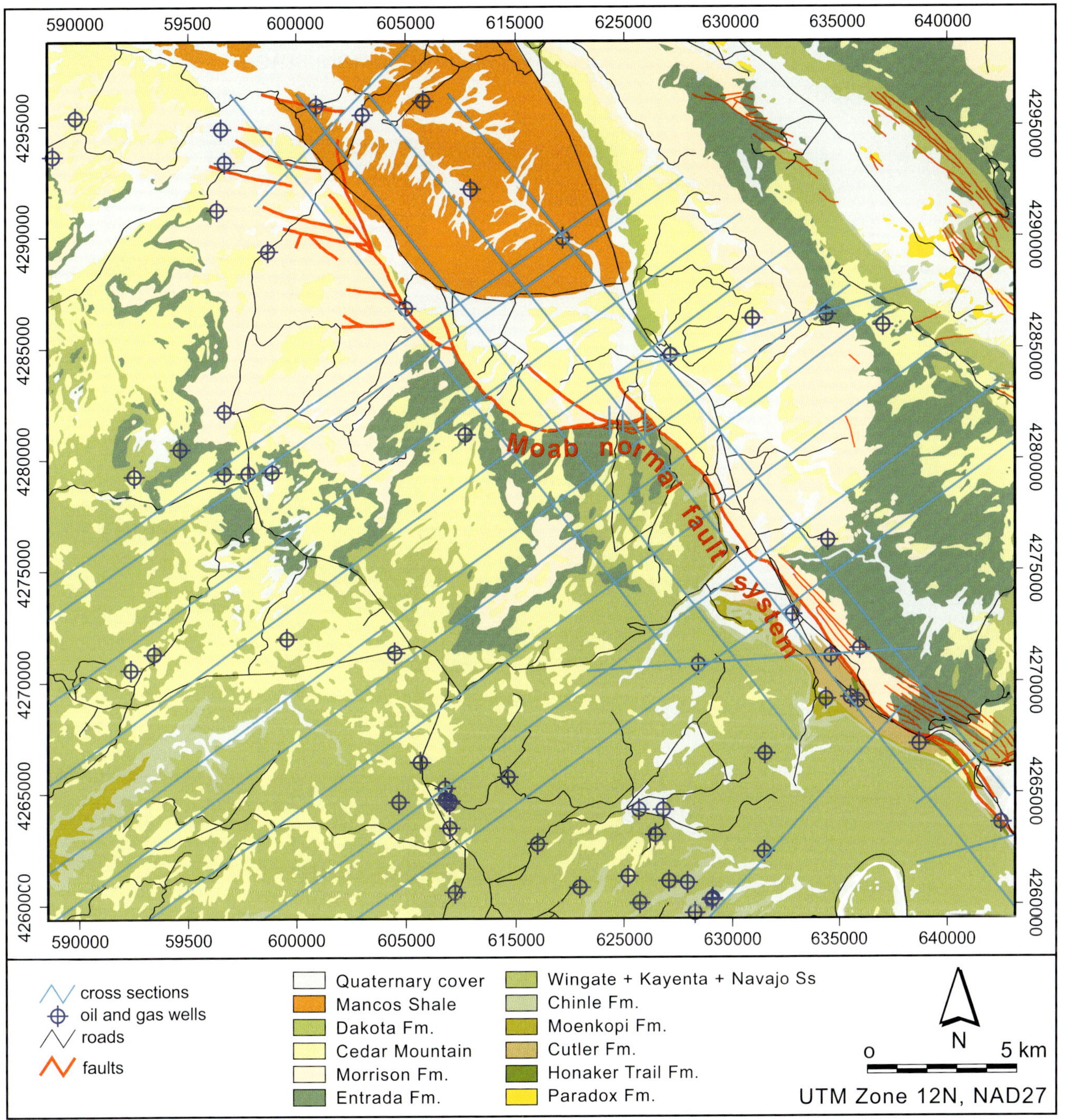

FIGURE 3. Geology of the Moab area including data sets used to constrain the three-dimensional fault geometry and stratigraphy. The basic geologic map was modified from Doelling (2002) by traditional mapping techniques and local GPS measurements. Map coordinates are Universal Transverse Mercator (UTM) zone 12, in the North American Datum established in 1927 (NAD27).

Data were collected in the field using several traditional methods. We conducted detailed outcrop mapping of fault zone structures, their distribution, relative geometry, and crosscutting relationship to infer the mechanisms of infaulting. Maps and cross sections were created using unrectified outcrop and rectified air photo base maps, global positioning surveying (GPS), and a 10-m (33-ft) resolution digital elevation model. We also quantified the thickness distribution of shale fault rocks in the fault zone using GPS data and a tape measure. We distinguished the source of shale fault rocks whenever possible. As a consequence of this analysis, we identified several parameters that appear to have controlled the distribution of fault zone structures.

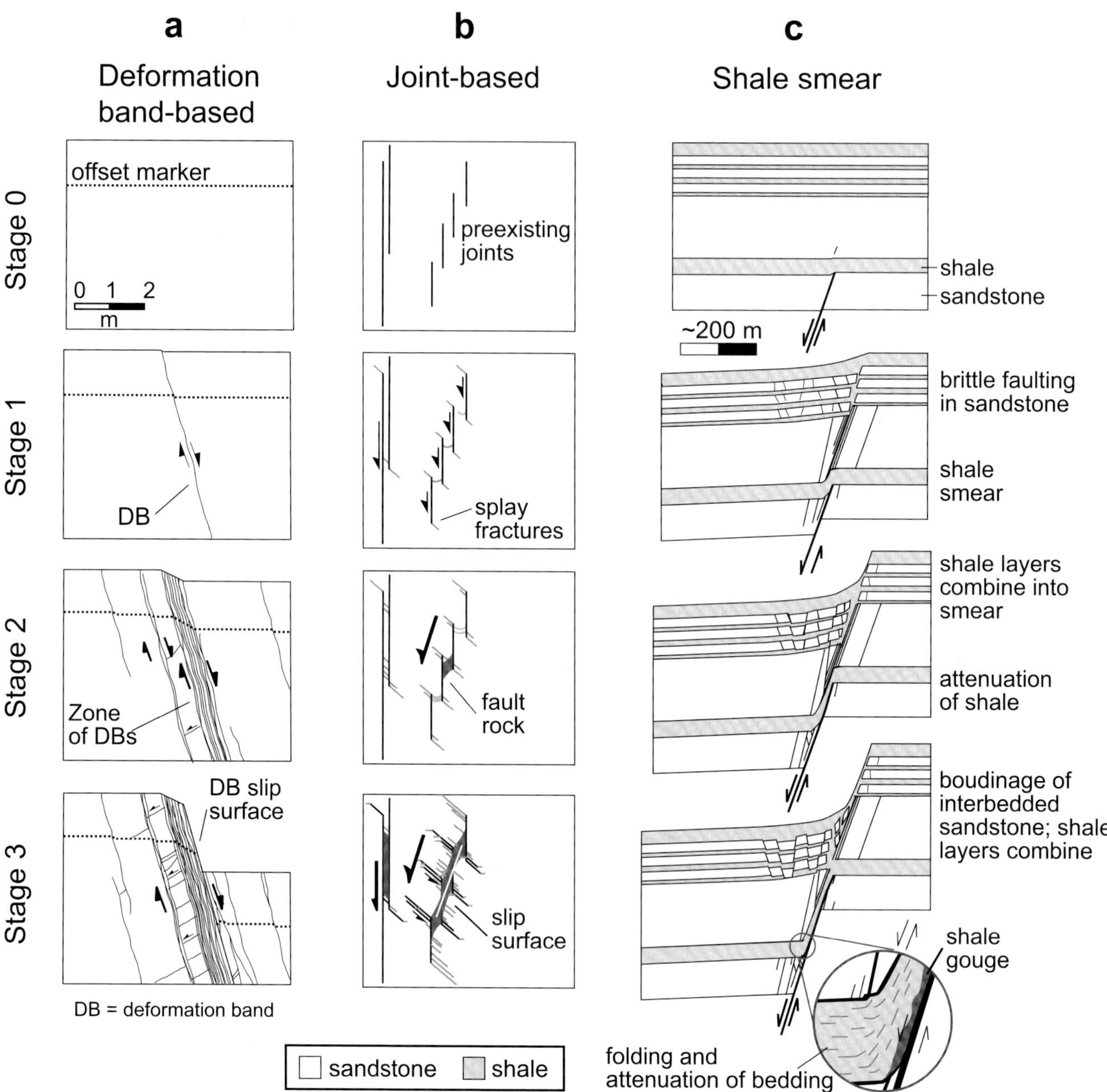

FIGURE 4. Evolution of fault architecture by brittle faulting mechanisms in sandstone including (a) deformation band-based (modified after Aydin and Johnson, 1978; Antonellini and Aydin, 1995) and (b) joint-based faulting (modified after Myers, 1999). (c) Evolution of fault architecture with shale smear includes ductile smearing of shale and brittle faulting in surrounding sandstone. The length scale in (b) is relative to the dimensions of the initial discontinuities. The length scale in (c) is relative to the shale bed thickness.

OBSERVATIONS AND RESULTS

Detailed Examples of Shale Deformation

Small faults in the Salt Wash Member demonstrate several key aspects of shale behavior in fault zones. First, despite brittle failure, indicated by deformation bands and some joints, in the surrounding sandstone layers (Figure 5), the shale units remain continuous from footwall to hanging wall, although they are drastically thinned. As a result, the fault surface from the upper sandstone unit to the lower sandstone unit is discontinuous and forms a vertical relay. As the shale layer approaches the relay, its internal bedding rotates to be parallel with the fault surfaces in the relay. A thin sandstone bed in the shale layer is also rotated with the shale and accommodates extension and thinning by brittle faulting (hanging wall of fault in Figure 5). In this case, the sandstone bed provides a marker that indicates that shale has accommodated extension without brittle

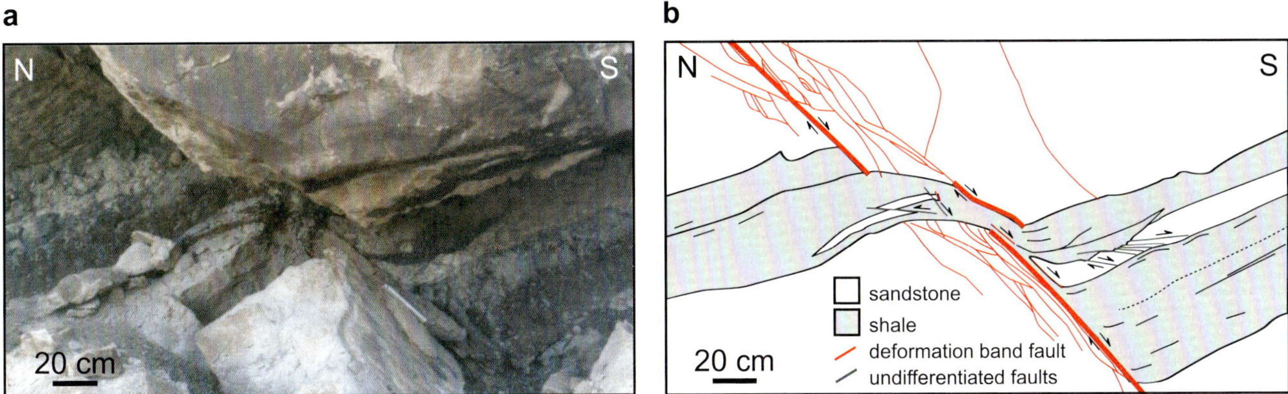

FIGURE 5. (a) Outcrop photo and (b) detailed cross section of small fault cutting coarse sandstone layers that sandwich a thinner shale layer, which is smeared through the relay in the Salt Wash Member. Outcrop located at 60°96′84″E, 42°85′17.6″N (Zone 12N, NAD27).

faulting, and the relatively thin sandstone layer has been moved with the shale. The minimum thickness of shale in the fault zone inside the relay is 10 cm (4 in.) and is thickest near the source of shale. This geometry suggests that the accumulation of throw and the distance from the source of shale might have both affected the thickness of shale in the core of this fault.

Second, parts of the Moab fault with larger offset contain a thin seam of shale gouge in the fault core adjacent to the fault surface, where shale is juxtaposed against the fault (e.g., Figure 6a, b). Multiple shale beds merge with this gouge layer (Figure 6b) and thus appear to constitute sources of shale. In addition, sandstone beds sandwiched between these shale beds are faulted and extended. The interbedded shale layers remain continuous between the sandstone blocks but commonly appear thinned. As a result, many fault-bounded sandstone blocks form boudins surrounded by shale. Both sandstone boudins and shale beds are nearly parallel to the fault slip surface. This geometry suggests that the entire package of thinly interbedded sandstone and shale has been extended and potentially attenuated in the direction of slip as part of the fault core.

In Figure 6c and d, two distinct fault cores consist of smeared shale and breccia bounded by slip surfaces in sandstone layers that are thick relative to shale layers. The fault core in the upper center of Figure 6d is thickest where a shale layer enters the fault core near the top of the exposure. The increased thickness of the core in Figure 6d and the entrainment of multiple shale layers in Figure 6b are consistent with continual incorporation of additional shale layers into the fault core. Conversely, the shale in the fault core, in the center of Figure 6c and d, is thinnest at the lower corner of a thick, hanging-wall sandstone unit. This geometry appears to be related to resistance by the relatively thick sandstone layer to folding and boudinage (e.g., thin sandstone layers in Figure 6b). Such resistance might limit rotation and cause the sandstone to impinge on the shale smear as the footwall and hanging wall slip past each other. Thus, the relative abundance and thickness of shale layers vs. sandstone layers might be a major factor in determining how shale is distributed in the fault core.

Geometry and Distribution of Shale in the Fault Zone

The northwestern part of the Moab fault system is characterized by folded shale units in the hanging wall juxtaposed against the thick sequence of Jurassic sandstone units in the footwall (Figure 7). Folds are asymmetric across the fault zone, with most of the change of bedding inclination in the hanging wall, whereas beds in the footwall exhibit the regional dip. In the hanging wall, the overall fold geometry is asymptotic; bed dip is greatest adjacent to the fault and approaches the regional dip over several hundred meters into the hanging wall (e.g., sections D, E, V, H, J3, L, and Q in Figure 7). This fold geometry broadens the area over which shale units in the hanging wall are juxtaposed against the fault core. In addition, sandstone marker beds in shale units converge as bed dip increases toward the fault over the length of these folds. This convergence indicates thinning of shale layers as they approach the fault core. Taken together, the rotation of bedding dip and thinning of shale into the fault core, which are absent in the footwall, might indicate that the shale in the fault core is preferentially derived from shale units in the hanging wall.

Section J3 in Figure 7 illustrates the geometry of folds when throw is small. In this case, the fold geometry has a smaller amplitude and wavelength and is monoclinal. The faults that cut sandstone do not appear to penetrate overlying shale units. Furthermore,

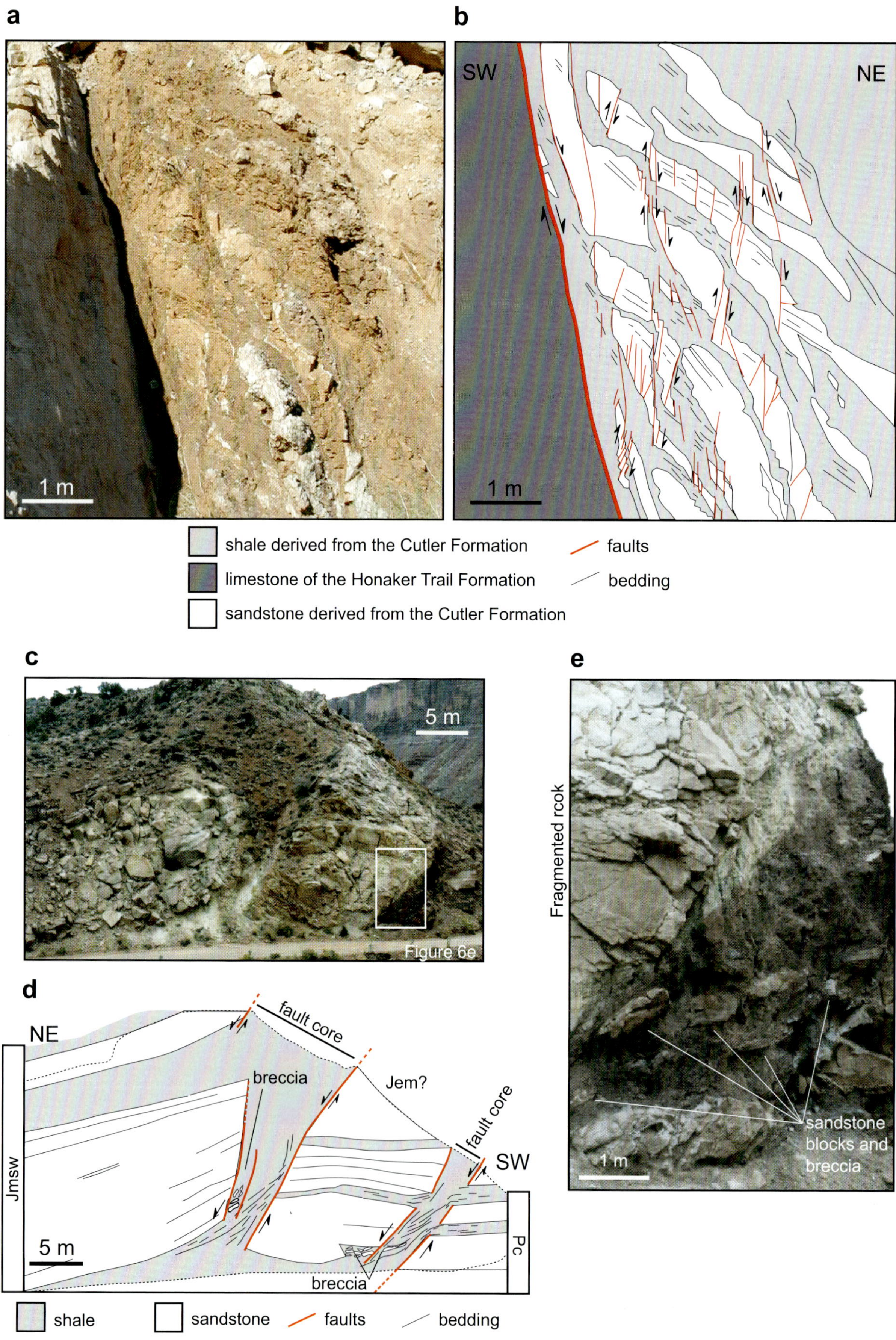
a
1 m
b
SW
NE
1 m
shale derived from the Cutler Formation
limestone of the Honaker Trail Formation
sandstone derived from the Cutler Formation
faults
bedding
c
5 m
Figure 6e
d
NE
fault core
breccia
Jem?
fault core
SW
Jmsw
5 m
Pc
breccia
shale
sandstone
faults
bedding
e
Fragmented rcok
sandstone blocks and breccia
1 m

the faults are located beneath the upper fold hinge on the footwall side of the monocline (e.g., section J2 in Figure 7), placing most of the fold on the hanging-wall side. This example illustrates the tendency for overlying shale units to fold instead of fault; thus, overlying shale units remain continuous across the fault.

Close to the fault, the units involved in folding typically kink sharply to a steeper dip angle, becoming subparallel to the fault slip surface (e.g., sections D, H, K, M, N, and S in Figure 7). This geometry is similar to the detailed cases (Figures 5, 6) where rotation, extension, and attenuation of beds are greatest in a narrow zone adjacent to the fault slip surface (which constitutes the fault core). As the distance from the kink point along the fault core increases, the thickness of the shale in the fault core slowly decreases. The result in these locations is a thin shale smear along the fault despite its proximity to a source of shale. In a small number of other locations, the broader scale folding locally and abruptly dies out over short distances (e.g., stations U and V in Figure 7). This geometric variation produces corresponding variations in the thickness of shale fault rocks along fault strike that are greatest close to the source of the shale in the hanging wall. However, in all of our examples, the kinked part of the fold is present and includes some shale separating sandstone in the footwall from sandstone in the hanging wall.

QUANTITATIVE MEASUREMENTS OF SHALE FAULT ROCK THICKNESS AND CONTROLLING PARAMETERS

We interpret the consistency of the shale geometry in the fault zone (Figures 5–7) to suggest that a systematic process, consistent with the conceptual model in Figure 4c, governs the thickness distribution and persistence of shale in the fault zone. Evidence from the detailed maps of the fault zone indicates that several parameters might impact the distribution of shale fault rocks. In Figures 5–7, multiple shale layers combine to form the shale smear in the fault core. This suggests that the thickness of the shale fault rock might be related to the thickness of shale available in the stratigraphic column. Similarly, the relative thickness of sandstone and shale units also appears to affect the shape of folding related to entrainment of the shale in the fault zone (e.g., Figure 6d). Furthermore, the thickness of shale fault rock consistently thins along the fault core away from the source of shale (Figures 5, 6d, 7). This geometry implies that both distance from the source of shale and, consequently, fault throw impact the distribution of shale fault rock. These two parameters are also practical in the field, because exposure is commonly limited, and therefore, it is not always possible to observe shale fault rocks at their thinnest point.

The stratigraphic distribution and relative thickness of sandstone and shale units are associated with two distinct types of fault architecture in the field examples. The first type occurs when shale layers are confined between thick, stiff sandstone layers that fold very little adjacent to the fault core (e.g., Figures 5, 6c) and is idealized in Figure 8a. The second type occurs when a thick sandstone layer is overlain by a shale-dominated sequence with a broad fold in the hanging wall (e.g., Figure 7) and is idealized in Figure 8b. In both cases, bedding in the shale is nearly parallel to the fault slip surfaces in the fault core. The first type is found where sandstone beds are typically as thick or thicker than shale beds (e.g., Curtis, Moenkopi, and Chinle formations and some small faults confined in the Salt Wash Member). The second type is more prevalent, where shale from the Morrison and Cedar Mountain formations and the Mancos Shale sits atop the thick, Jurassic sandstone sequence. We have distinguished these two types because it is likely that they may have different effects on the architecture of the fault zone and the incorporation of shale into the fault core.

To investigate the potential relationship between (1) the original thickness of shale incorporated in the fault rock, (2) the distance from the source of shale along the fault core, and (3) the fault throw with the thickness of shale in the fault core, we estimated the thickness of shale in each stratigraphic unit (Table 1) and made 162 measurements of the remaining parameters at 80 stations along the fault (Table 2). Most data were available from the northwestern portion of the fault (Figure 9a), where exposure is best, and the second type of fault zone architecture is typical (Figure 8b). Because the relative thickness of sandstone and shale units impacts the geometry of the shale smear (e.g.,

Figure 6. (a) Outcrop photo and (b) detailed cross section of deformed interbedded sandstone and shale derived from the Cutler Formation juxtaposed against limestone from the Honaker Trail Formation near the entrance to Arches National Park. Sandstone beds are faulted and extended, whereas shale beds have thinned but remain continuous. This outcrop is located at 62°00′86″E, 42°74′72.5″N (Zone 12N, NAD27). Throw on this segment of the Moab fault is approximately 250 m (820 ft). (c) Outcrop photo and (d) simplified cross section of two fault strands located at 61°76′67″E, 42°76′78.9″N (Zone 12N, NAD27). Combined throw across both segments is about 906 m (2970 ft). (e) Detailed photo of (c) and (d) depicting entrainment of sandstone blocks and breccia in shale.

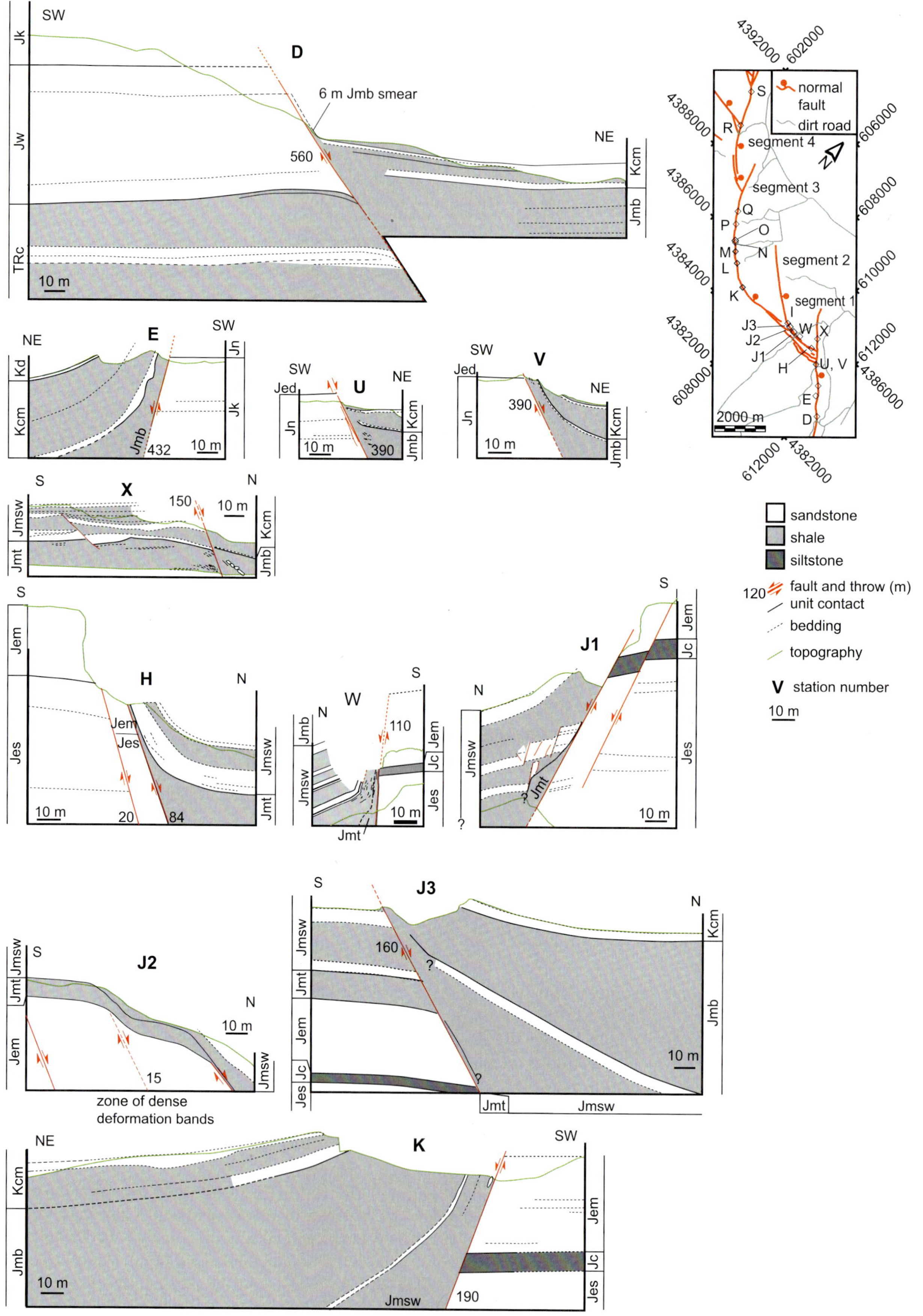

normal fault
dirt road
segment 4
segment 3
segment 2
segment 1
2000 m
sandstone
shale
siltstone
120 fault and throw (m)
unit contact
bedding
topography
V station number
10 m
6 m Jmb smear
zone of dense deformation bands

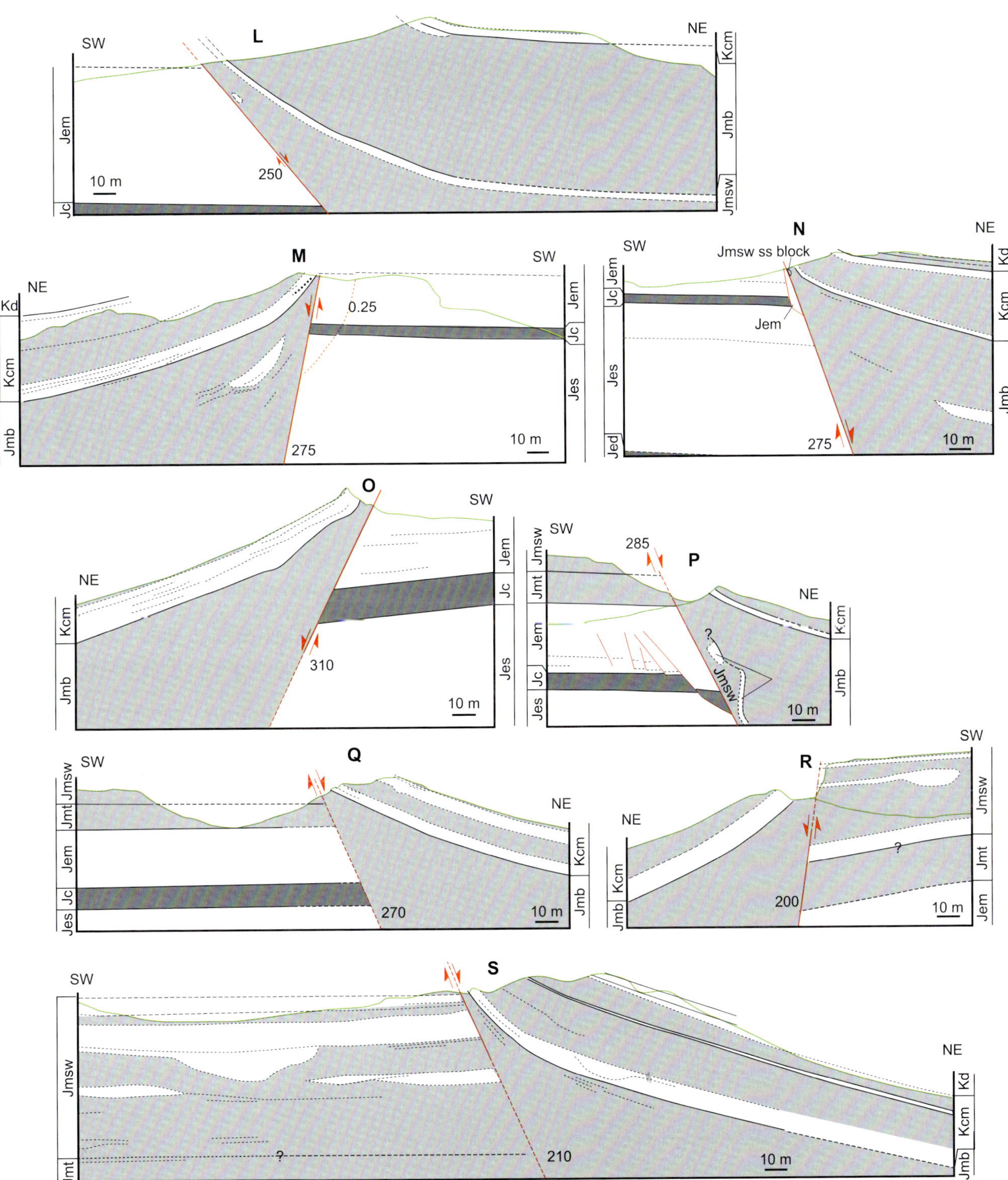

FIGURE 7. Cross sections of fault exposures along the northwestern portion of the Moab fault system. Section locations and letter designations are shown in an index map but, in general, are ordered from east to west. Downdip exposure at the selected locations allowed us to document the geometry of shale units as they approach the fault zone. Cross sections are limited to directly observable features, are not extrapolated into the subsurface, and do not show the detailed structure in damage zones. Note that abbreviations of stratigraphic names are defined in Figure 2.

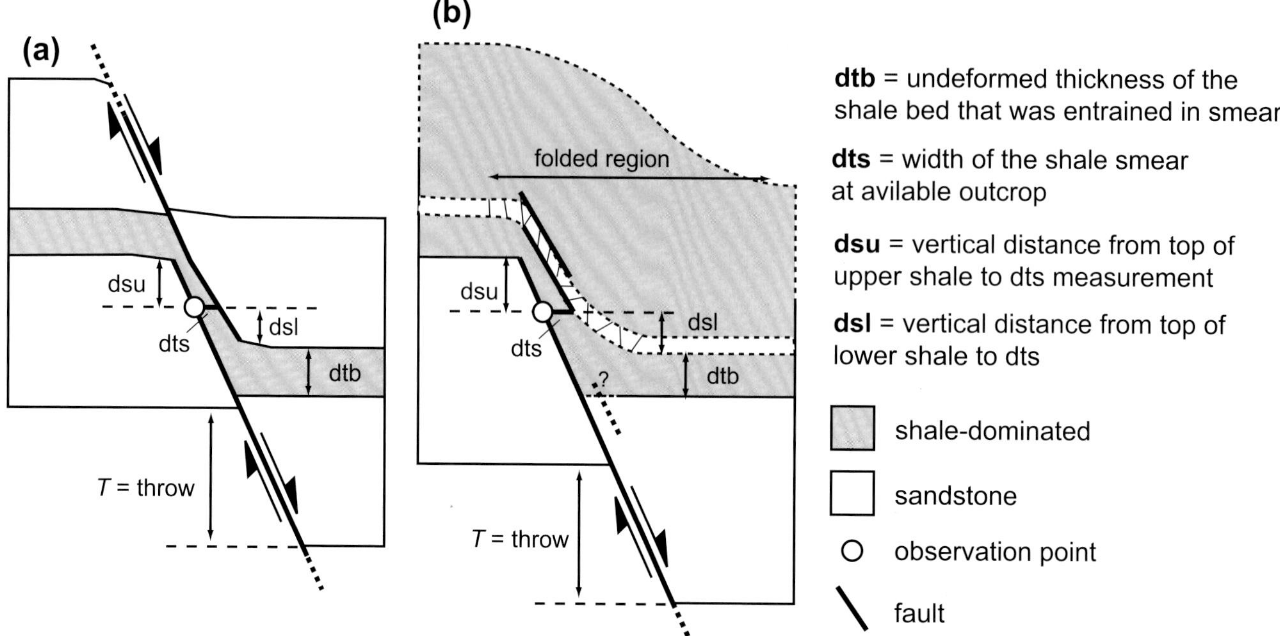

FIGURE 8. Schemes for quantifying the change in (or distribution of) shale smear width related to: throw, thickness of and distance from the source bed of shale, fault geometry, and geometry of stratigraphy. (a) Typical shale smear model (e.g., Aydin and Eyal, 2002) consisting of a compliant unit sandwiched between two stiffer units. (b) A two-layer system with thick sandstone below and dominantly shale above.

Figure 8), we have only plotted data from examples corresponding to case 2 in Figure 9.

For convenience in the field, we measured the width of shale fault rock along a horizontal line from the fault slip surface to the identifiable top of any shale-dominated stratigraphic unit. The location of this measurement was determined by outcrop availability. For each measurement of the width of the shale fault rock (dts in Figure 8), we also determined a distance from the nearest potential source of shale in the hanging wall. This distance (dsl in Figure 8) was measured by projecting the top of the stratigraphic unit from the nearest location where the unit lies at the regional strike and dip into the fault core. Finally, we estimated the total thickness of shale units that could have contributed to the shale fault rock (dtb in Figure 8) based on the measured stratigraphy, fault throw, and the units juxtaposed at the observation point.

The distribution of smeared shale width has a consistent shape (Figure 9b, c) for individual examples of shale smear. In each case, the width of the smeared shale decreases monotonically with increasing distance from the hanging-wall source of the shale (dsl) and defines a continuous shale smear. These curves have a generally asymptotic shape, indicating that thin shale smears (a few meters thick) persist for long distances in the fault core with only a slight tapering in thickness. The width of the smeared shale reaches a minimum as the distance from the hanging-wall source of shale approaches the throw (T). This might imply that shale in these smears is primarily derived from units in the hanging wall instead of the footwall.

Table 1. Potential shale sources estimated from field observation and data from Doelling (1982, 1988).

Stratigraphic Unit (See Figure 2)	*Thickness (dtb) (m)*	*Total shale (m)*	*Error (m; ±)*
Jmb	80	90	10
Jmsw	68	50	15
Jmt	20	20	5
Trc	110	70	15
TRm	125	70	35
Pc	320	120	50

However, a simple relationship that relates the change in width of the smear to throw or distance from sources of shale is not evident (Figure 9d). In each case, some shale fault rock is present in the fault zone, which is consistent with the nearly asymptotic curve defining the rate of change in the width of the shale smear observed in Figure 9b and c. However, when comparing examples from different parts of the fault, a great deal of heterogeneity exists. The major distinction appears to be the ratio of dts/T that marks the transition from rapid thinning in the smear to a slow rate of thinning with increasing dts/T, which is different for many of the examples (Figure 9b, c). In some cases, this heterogeneity is associated with complications in the fault geometry or the shape of the fold in the hanging wall. For instance, segmentation of the fault system along strike appears to be a factor in maintaining larger smear widths (e.g., Figure 6d and section J2 in Figure 7). In these cases, the segmentation distributes the throw among several faults with smaller throw. Consequently, each fault segment has a relatively wide smear. In other cases where a narrow fold hinge exists in the hanging wall, the value of dts may be very low for all ratios of dts/T. As a result, thin shale smears can occur at small throw.

We do find that shale from a given unit is absent from the fault zone where throw is beyond some maximum magnitude. The Tidwell Member (Jmt) vanishes when throw is about 150 m (490 ft) (T/dtb is larger than 7.5–8 in Figure 9), and the Salt Wash Member disappears when throw is about 270 m (886 ft) (T/dtb is larger than 3.7–4). The Brushy Basin Member is never visibly attenuated to zero width, which includes examples with throw as much as 560 m (1840 ft) (T/dtb is larger than 3.5). As a result, these measurements provide an empirical criterion for estimating the maximum distance from shale beds that shale fault rocks might be distributed in the fault core.

DISTRIBUTION OF FAULT ZONE STRUCTURES IN THREE DIMENSIONS

The fault zone along the Moab fault system in Permian to Cretaceous sandstone and shale consists of structures from three distinct faulting mechanisms (Figures 10, 11). In sandstone from all stratigraphic units, deformation band-based structures occur everywhere along the Moab fault (Davatzes, 2003) and form a core and damage zone of variable width in sandstone units. In folded hanging-wall sandstone layers, deformation bands are also distributed throughout folds. Joint-based structures overprint deformation bands at locations with complex fault geometry, such as intersections and relays along either strike or dip (Davatzes, 2003). In addition, joints occur in great density in all folded hanging-wall sandstone layers, again overprinting deformation bands.

Shale is incorporated into the fault zone over the area between a shale unit's position in the footwall and its corresponding position in the hanging wall (idealized in Figure 8). Individual shale units smeared in the fault core taper to zero at large throw. Broad folds in the hanging wall define wide damage zones in shale-dominated units that are absent in the footwall. Folds or faults in sandstone along the contact between the thick Jurassic sandstone package and overlying shale might be necessary to conserve the volume of smeared and folded shale in the hanging wall but are commonly poorly exposed. These kinds of folds and faults are evident in the relays in cross sections in Figure 11c and d and near the intersection in Figure 11b between segments 1 and 2.

We used data presented in this study, previously published data from the Moab fault (Foxford et al., 1998; Doelling, 2002; Davatzes, 2003; Davatzes and Aydin, 2003), and criteria for characterizing the distribution of fault zone structures (Davatzes; 2003) and smear thickness to map fault architecture onto juxtaposition diagrams (Allan, 1989; Knipe, 1997), thus representing the three-dimensional architecture of the Moab fault system. The fault plane geometry and juxtaposition diagrams were constructed using publicly available well data from the Utah Division of Oil, Gas, and Mining, published cross sections (Doelling, 1988; Doelling, 2002), detailed stratigraphic sections (Doelling, 1982), mapped formation contacts (Figure 3), a 10-m (33-ft) resolution digital elevation model, and measured strikes and dips of bedding. Additional cross sections both normal and parallel to the fault were constructed to constrain stratigraphic horizons produced in GoCAD®. The initial juxtaposition diagram was then constructed using Fault Analysis Projection System (FAPS®). GoCAD and FAPS are commercially available software used for three-dimensional reservoir model construction and fault plane analysis, respectively.

Traditional juxtaposition (also called Allan) diagrams plot the distribution of rock types that intersect and are juxtaposed across an idealized fault plane (Allan, 1989; Knipe, 1997). This analysis typically assumes that the fault zone itself has no impact on sealing and is a plane of zero thickness. We have used the distribution of lithologic units from the juxtaposition analysis as a first-order control on the distribution of fault zone structures. The resulting diagram is broken into three sets: (1) footwall damage zone immediately adjacent to the fault core (Figure 12a); (2) hanging-wall damage zone immediately adjacent to the fault core (Figure 12b); and (3) fault core (Figure 12c). The plane or map that represents each portion of the fault zone (damage zones

Table 2. Quantitative shale-smear data and potential controlling parameters. Easting and northing are expressed as UTM coordinates, Zone 12N, NAD27. Note that abbreviations of stratigraphic names are defined in Figure 2.

Station	*Segment*	*Easting (m)*	*Northing (m)*	*Elevation (Meters above Sea Level)*	*Throw (m)*	*Estimated Elevation of Shale Source Unit Top*	*dsl (m)*	*dsu (m)*	*Unit in Footwall*	*Unit in Hanging Wall*	*Potential Sources of Shale*	*dts (m)*	*dsl/T*	*dts/ dtb*
S1	4	601628	4290451	–	210	–	150	60	Jmsw bottom third	Jmb top	Jm	8.0	0.71	0.05
S3	4	601628	4290451	–	210	–	145	65	Jmsw bottom third	Jmb top	Jm	15.0	0.69	0.09
S4	4	601628	4290451	–	210	–	140	70	Jmsw bottom third	Jmb top	Jm	16.0	0.67	0.10
S5	4	601628	4290451	–	210	–	135	75	Jmsw bottom third	Jmb top	Jm	19.0	0.64	0.12
S6	4	601628	4290451	–	210	–	130	80	Jmsw bottom third	Jmb top	Jm	21.0	0.62	0.13
S7	4	601628	4290451	–	210	–	125	85	Jmsw bottom third	Jmb top	Jm	24.0	0.60	0.15
S8	4	601628	4290451	–	210	–	120	90	Jmsw bottom third	Jmb top	Jm	29.0	0.57	0.18
S9	4	601628	4290451	–	210	–	115	95	Jmsw bottom third	Jmb top	Jm	36.0	0.55	0.23
S10	4	601628	4290451	–	210	–	110	100	Jmsw bottom third	Jmb top	Jm	46.0	0.52	0.29
S11	4	601628	4290451	–	210	–	105	105	Jmsw bottom third	Jmb top	Jm	61.0	0.50	0.38
S12	4	601628	4290451	–	210	–	100	110	Jmsw bottom third	Jmb top	Jm	70.0	0.48	0.44
S13	4	601628	4290451	–	210	–	95	115	Jmsw bottom third	Jmb top	Jm	89.0	0.45	0.56
S14	4	601628	4290451	–	210	–	90	120	Jmsw bottom third	Jmb top	Jm	111.0	0.43	0.69
S15	4	601628	4290451	–	210	–	85	125	Jmsw bottom third	Jmb top	Jm	126.0	0.40	0.79
R1	4	602263	4288284	1463	200	1390	78	122	Jmsw bottom third	Kd bottom	Jm	3.0	0.39	0.03
R2	4	602263	4288284	1458	200	1390	73	127	Jmsw bottom third	Kd bottom	Jm	10.0	0.37	0.08
R3	4	602263	4288284	1453	200	1390	68	132	Jmsw bottom third	Kd bottom	Jm	14.0	0.34	0.12
R4	4	602263	4288284	1448	200	1390	63	137	Jmsw bottom third	Kd bottom	Jm	18.0	0.32	0.15
R5	4	602263	4288284	1443	200	1390	58	142	Jmsw bottom third	Kd bottom	Jm	25.0	0.29	0.21
R6	4	602263	4288284	1438	200	1390	53	147	Jmsw bottom third	Kd bottom	Jm	32.0	0.27	0.27
R7	4	602263	4288284	1433	200	1390	48	152	Jmsw bottom third	Kd bottom	Jm	39.0	0.24	0.33
R8	4	602263	4288284	1428	200	1390	43	157	Jmsw bottom third	Kd bottom	Jm	48.0	0.22	0.40
R9	4	602263	4288284	1423	200	1390	38	162	Jmsw bottom third	Kd bottom	Jm	57.0	0.19	0.48
R10	4	602263	4288284	1418	200	1390	33	167	Jmsw bottom third	Kd bottom	Jm	68.0	0.17	0.57
R11	4	602263	4288284	1413	200	1390	28	172	Jmsw bottom third	Kd bottom	Jm	72.0	0.14	0.60
Q1	3	604533	4286836	1472	270	1435	37	233	Jem top	Jmb top third	Jm	7.0	0.14	0.05
Q2	3	604533	4286836	1467	270	1435	32	238	Jem top	Jmb top third	Jm	17.0	0.12	0.12
Q3	3	604533	4286836	1462	270	1435	27	243	Jem top	Jmb top third	Jm	25.0	0.10	0.18
Q4	3	604533	4286836	1457	270	1435	22	248	Jem top	Jmb top third	Jm	35.0	0.08	0.25
Q1	3	604533	4286836	1452	270	1435	17	253	Jem top	Jmb top third	Jm	45.0	0.06	0.32
Q2	3	604533	4286836	1447	270	1435	12	258	Jem top	Jmb top third	Jm	60.0	0.04	0.43
Q3	3	604533	4286836	1442	270	1435	7	263	Jem top	Jmb top third	Jm	73.0	0.03	0.52
1	3	604533	4286836	1468	280	1426	42	238	Jem	Jmb	Jm	4.0	0.15	0.03
3	3	604830	4286433	1461	285	1426	35	250	Jes	Jmb top	Jm	5.6	0.12	0.04

Table 2. Quantitative shale-smear data and potential controlling parameters. Easting and northing are expressed as UTM coordinates, Zone 12N, NAD27. Note that abbreviations of stratigraphic names are defined in Figure 2 (cont.).

Station	*Segment*	*Easting (m)*	*Northing (m)*	*Elevation (Meters above Sea Level)*	*Throw (m)*	*Estimated Elevation of Shale Source Unit Top*	*dsl (m)*	*dsu (m)*	*Unit in Footwall*	*Unit in Hanging Wall*	*Potential Sources of Shale*	*dts (m)*	*dsl/T*	*dts/dtb*
P1	3	604830	4286433	1450	285	1430	25	260	Jes	Jmb top	Jm	16.0	0.09	0.10
P2	3	604830	4286433	1445	285	1430	20	265	Jes	Jmb top	Jm	25.0	0.07	0.16
P3	3	604830	4286433	1440	285	1430	15	270	Jes	Jmb top	Jm	35.0	0.05	0.22
P4	3	604830	4286433	1435	285	1430	10	275	Jes	Jmb top	Jm	47.0	0.04	0.29
P5	3	604830	4286433	1430	285	1430	5	280	Jes	Jmb top	Jm	60.0	0.02	0.38
4	3	605286	4285922	1433	285	1425	8	277	Jem top	Kcm	Jm	3.5	0.03	0.02
5	3	605318	4285916	1485	280	1425	60	220	Jem bottom	Jmb top	Jm	11.0	0.21	0.07
O1	3	605318	4285916	1468	310	1417	51	259	Jem mid	Jmsw top	Jmsw + Jmt	7.0	0.16	0.10
O2	3	605318	4285916	1463	310	1417	46	264	Jem mid	Jmsw top	Jmsw + Jmt	10.0	0.15	0.14
O3	3	605318	4285916	1458	310	1417	41	269	Jem mid	Jmsw top	Jmsw + Jmt	17.0	0.13	0.24
O4	3	605318	4285916	1453	310	1417	36	274	Jem mid	Jmsw top	Jmsw + Jmt	21.0	0.12	0.30
O5	3	605318	4285916	1448	310	1417	31	279	Jem mid	Jmsw top	Jmsw + Jmt	24.0	0.10	0.34
O6	3	605318	4285916	1443	310	1417	26	284	Jem mid	Jmsw top	Jmsw + Jmt	27.0	0.08	0.39
O7	3	605318	4285916	1438	310	1417	21	289	Jem mid	Jmsw top	Jmsw + Jmt	38.0	0.07	0.54
O8	3	605318	4285916	1433	310	1417	16	294	Jem mid	Jmsw top	Jmsw + Jmt	50.0	0.05	0.71
O9	3	605318	4285916	1428	310	1417	11	299	Jem mid	Jmsw top	Jmsw + Jmt	60.0	0.04	0.86
O10	3	605318	4285916	1423	310	1417	6	304	Jem mid	Jmsw top	Jmsw + Jmt	72.0	0.02	1.03
O11	3	605318	4285916	1418	310	1417	1	309	Jem mid	Jmsw top	Jmsw + Jmt	84.0	0.00	1.20
O12	3	605318	4285916	1413	310	1417	−4	314	Jem mid	Jmsw top	Jmsw + Jmt	96.0	−0.01	1.37
M1	3	605565	4285675	1486	275	1409	77	198	Jem top	Jmb top	Jm	2.0	0.28	0.01
M2	3	605565	4285675	1481	275	1409	72	203	Jem top	Jmb top	Jm	5.0	0.26	0.03
M3	3	605565	4285675	1476	275	1409	67	208	Jem top	Jmb top	Jm	7.0	0.24	0.04
M4	3	605565	4285675	1471	275	1409	62	213	Jem top	Jmb top	Jm	12.0	0.23	0.08
M5	3	605565	4285675	1466	275	1409	57	218	Jem top	Jmb top	Jm	17.0	0.21	0.11
M6	3	605565	4285675	1461	275	1409	52	223	Jem top	Jmb top	Jm	27.0	0.19	0.17
M7	3	605565	4285675	1456	275	1409	47	228	Jem top	Jmb top	Jm	40.0	0.17	0.25
M8	3	605565	4285675	1451	275	1409	42	233	Jem top	Jmb top	Jm	53.0	0.15	0.33
M9	3	605565	4285675	1446	275	1409	37	238	Jem top	Jmb top	Jm	68.0	0.13	0.43
M10	3	605565	4285675	1441	275	1409	32	243	Jem top	Jmb top	Jm	88.0	0.12	0.55
M11	3	605565	4285675	1436	275	1409	27	248	Jem top	Jmb top	Jm	95.0	0.10	0.59
M12	3	605565	4285675	1431	275	1409	22	253	Jem top	Jmb top	Jm	140.0	0.08	0.88
N1	3	605565	4285675	1476	275	1409	67	208	Jes top	Jmb top	Jm	3.0	0.24	0.02
N2	3	605565	4285675	1471	275	1409	62	213	Jes top	Jmb top	Jm	4.0	0.23	0.03

Table 2. Quantitative shale-smear data and potential controlling parameters. Easting and northing are expressed as UTM coordinates, Zone 12N, NAD27. Note that abbreviations of stratigraphic names are defined in Figure 2 (cont.).

Station	*Segment*	*Easting (m)*	*Northing (m)*	*Elevation (Meters above Sea Level)*	*Throw (m)*	*Estimated Elevation of Shale Source Unit Top*	*dsl (m)*	*dsu (m)*	*Unit in Footwall*	*Unit in Hanging Wall*	*Potential Sources of Shale*	*dts (m)*	*dsl/T*	*dts/dtb*
N3	3	605565	4285675	1466	275	1409	57	218	Jes top	Jmb top	Jm	9.0	0.21	0.06
N4	3	605565	4285675	1461	275	1409	52	223	Jes top	Jmb top	Jm	19.0	0.19	0.12
N5	3	605565	4285675	1456	275	1409	47	228	Jes top	Jmb top	Jm	33.0	0.17	0.21
N6	3	605565	4285675	1451	275	1409	42	233	Jes top	Jmb top	Jm	47.0	0.15	0.29
N7	3	605565	4285675	1446	275	1409	37	238	Jes top	Jmb top	Jm	63.0	0.13	0.39
N8	3	605565	4285675	1441	275	1409	32	243	Jes top	Jmb top	Jm	84.0	0.12	0.53
L1	3	605930	4285400	1476	250	1410	66	184	Jem top	Jmsw top	Jm	30.0	0.26	0.19
L2	3	605930	4285400	1476	250	1320	156	94	Jem top		Jmsw + Jmt	10.0	0.62	0.06
L3	3	605930	4285400	1471	250	1320	151	99	Jem top		Jmsw + Jmt	12.0	0.60	0.08
L4	3	605930	4285400	1466	250	1320	146	104	Jem top		Jmsw + Jmt	13.0	0.58	0.08
L5	3	605930	4285400	1461	250	1320	141	109	Jem top		Jmsw + Jmt	16.0	0.56	0.10
L6	3	605930	4285400	1456	250	1320	136	114	Jem top		Jmsw + Jmt	19.0	0.54	0.12
L7	3	605930	4285400	1451	250	1320	131	119	Jem top		Jmsw + Jmt	22.0	0.52	0.14
L8	3	605930	4285400	1446	250	1320	126	124	Jem top		Jmsw + Jmt	27.0	0.50	0.17
L9	3	605930	4285400	1441	250	1320	121	129	Jem top		Jmsw + Jmt	34.0	0.48	0.21
L10	3	605930	4285400	1436	250	1320	116	134	Jem top		Jmsw + Jmt	40.0	0.46	0.25
L11	3	605930	4285400	1431	250	1320	111	139	Jem top		Jmsw + Jmt	48.0	0.44	0.30
L12	3	605930	4285400	1426	250	1320	106	144	Jem top		Jmsw + Jmt	57.0	0.42	0.36
L13	3	605930	4285400	1421	250	1320	101	149	Jem top		Jmsw + Jmt	80.0	0.40	0.50
L14	3	605930	4285400	1416	250	1320	96	154	Jem top		Jmsw + Jmt	160.0	0.38	1.00
K1	3	606746	4284891	1486	190	1360	126	64	Jem top	Jmb top	Jmsw + Jmt	14.0	0.66	0.20
K2	3	606746	4284891	1481	190	1360	121	69	Jem top	Jmb top	Jmsw + Jmt	13.0	0.64	0.19
K3	3	606746	4284891	1476	190	1360	116	74	Jem top	Jmb top	Jmsw + Jmt	14.0	0.61	0.20
K4	3	606746	4284891	1471	190	1360	111	79	Jem top	Jmb top	Jmsw + Jmt	15.0	0.58	0.21
K5	3	606746	4284891	1466	190	1360	106	84	Jem top	Jmb top	Jmsw + Jmt	17.0	0.56	0.24
K6	3	606746	4284891	1461	190	1360	101	89	Jem top	Jmb top	Jmsw + Jmt	20.0	0.53	0.29
K7	3	606746	4284891	1456	190	1360	96	94	Jem top	Jmb top	Jmsw + Jmt	24.0	0.51	0.34
K8	3	606746	4284891	1451	190	1360	91	99	Jem top	Jmb top	Jmsw + Jmt	27.0	0.48	0.39
K9	3	606746	4284891	1446	190	1360	86	104	Jem top	Jmb top	Jmsw + Jmt	32.0	0.45	0.46
K10	3	606746	4284891	1441	190	1360	81	109	Jem top	Jmb top	Jmsw + Jmt	38.0	0.43	0.54
K2	3	606746	4284891	1486	190	1450	36	154	Jem top	Jmsw top	Jm	50.0	0.19	0.31
13	3	607565	4284847	1484	115	1411	73	42	Jem top	Jmsw top	Jmsw + Jmt	2.0	0.63	0.03
14	3	607633	4284869	1462	110	1416	46	64	Jem top	Jmsw top	Jmsw + Jmt	6.5	0.42	0.09
15	3	607670	4284885	1448	110	1416	32	78	Jem bottom	Jmsw top	Jmsw + Jmt	5.1	0.29	0.10

Table 2. Quantitative shale-smear data and potential controlling parameters. Easting and northing are expressed as UTM coordinates, Zone 12N, NAD27. Note that abbreviations of stratigraphic names are defined in Figure 2 (cont.).

Station	Segment	Easting (m)	Northing (m)	Elevation (Meters above Sea Level)	Throw (m)	Estimated Elevation of Shale Source Unit Top	dsl (m)	dsu (m)	Unit in Footwall	Unit in Hanging Wall	Potential Sources of Shale	dts (m)	dsl/T	dts/dtb
16	3	607968	4284929	1440	95	1416	24	71	Curtis	Jmsw bottom	Jmt	2.0	0.25	0.10
17	3	608122	4284922	1443	90	1416	27	63	Curtis	Jmsw bottom	Jmt	7.0	0.30	0.14
18	3	608483	4284938	1450	85	1420	30	55	Jem top	Jmsw bottom	Jmt	8.0	0.35	0.11
19	3	608732	4284945	1474	60	1440	34	26	Jem top	Jmsw bottom	Jmt	7.8	0.57	0.16
	2	608990	4285190	1464	155	1411	53	102	Jem top	Jmb/Kcm	Jm	76.0	0.34	0.48
20	2	609306	4285136	1415	115	1310	105	10	Jem top	Jmsw bottom	Jmsw + Jmt	4.1	0.91	0.21
21	2	609348	4285139	1410	110	1310	100	10	Jem bottom	Jmsw bottom	Jmt	3.0	0.91	0.15
23	2	609488	4285117	1390	110	1330	60	50	Jes top	Jmsw bottom	Jmsw/Jmt	2.7	0.55	0.05
24	2	609574	4285099	1395	140	1330	65	75	Jes top	Jem bottom	Jmsw/Jmt	0.3	0.46	0.01
25a	2	609607	4285091	1420	140	1319	101	39	Jes top	Jmsw bottom	Jmt	0.7	0.72	0.04
25b	2	609607	4285091	1420	140	1379	41	99	Jes top	Jmsw mid	Jmt + Jmsw part	7.0	0.29	0.18
T	2	609684	4285176	–	1	–	0.1	0.1	Jmsw	Jmsw	Jmsw internal	0.1	0.10	0.05
26	2	609696	4285108	1417	130	1379	38	92	Jes	Jmsw top	Jmt	10.0	0.29	0.15
28	2	609789	4285108	1415	125	1380	35	90	Jes	Jmsw bottom	Jmt	4.0	0.28	0.08
29	2	609906	4285066	1440	120	1380	60	60	Jes top	Jmsw bottom	Jmt + Jmsw part	5.0	0.50	0.10
30	2	610283	4285152	1415	95	1380	35	60	Jem top	Jmsw bottom	Jmt	1.8	0.37	0.09
31	2	610291	4285157	1408	95	1380	28	67	Jem top	Jmsw bottom	Jmt	1.5	0.29	0.08
32	2	610302	4285167	1402	90	1380	22	68	Jem top	Jmsw bottom	Jmt	0.8	0.24	0.04
H1	2	610324	4285185	1410	84	1380	30	54	Jes top	Jmt top	Jmt	0.8	0.36	0.04
H2	2	610324	4285185	1405	84	1380	25	59	Jes top	Jmt top	Jmt	1.0	0.30	0.05
H3	2	610324	4285185	1400	84	1380	20	64	Jes top	Jmt top	Jmt	1.2	0.24	0.06
H4	2	610324	4285185	1395	84	1380	15	69	Jes top	Jmt top	Jmt	1.0	0.18	0.05
H5	2	610324	4285185	1390	84	1380	10	74	Jes top	Jmt top	Jmt	2.0	0.12	0.10
H6	2	610324	4285185	1385	84	1380	5	79	Jes top	Jmt top	Jmt	4.0	0.06	0.20
H7	2	610324	4285185	1380	84	1380	0	84	Jes top	Jmt top	Jmt	7.0	0.00	0.35
H8	2	610324	4285185	1375	84	1380	–5	89	Jes top	Jmt top	Jmt	13.0	–0.06	0.65
H9	2	610324	4285185	1370	84	1380	–10	94	Jes top	Jmt top	Jmt	55.0	–0.12	2.75
G	1	610237	4285604	1381	150	1380	1	149	Jmt top	Kcm bottom	Jm	5.0	0.01	0.03
34	1	610435	4285392	1389	375	1379	10	90	Jem top	Jmb top	Jmt	3.0	0.03	0.15
36	1	610467	4285364	1397	85	1379	18	67	Jem top	Jmb top	Jmt	3.9	0.21	0.20
37	1	610501	4285341	1399	80	1379	20	60	Jem top	Jmb top	Jmt	1.0	0.25	0.05
39	1	610521	4285328	1398	290	1379	19	271	Jem top	Jmb top	Jmt	2.5	0.07	0.13
40	1	610569	4285287	1400	300	1379	21	279	Jem bottom	Jmb top	Jmt	1.6	0.07	0.08
41	1	610638	4285205	1407	335	1379	28	307	Jes bottom	Jmb top	Jmt	3.0	0.08	0.15

Table 2. Quantitative shale-smear data and potential controlling parameters. Easting and northing are expressed as UTM coordinates, Zone 12N, NAD27. Note that abbreviations of stratigraphic names are defined in Figure 2 (cont.).

Station	*Segment*	*Easting (m)*	*Northing (m)*	*Elevation (Meters above Sea Level)*	*Throw (m)*	*Estimated Elevation of Shale Source Unit Top*	*dsl (m)*	*dsu (m)*	*Unit in Footwall*	*Unit in Hanging Wall*	*Potential Sources of Shale*	*dts (m)*	*dsl/T*	*dts/ dtb*
42	1	610645	4285197	1402	340	1379	23	317	Jes top	Jmb top	Jmt	2.9	0.07	0.15
43	1	610880	4284905	1432	385	1396	36	349	Jn bottom	Kcm bottom	Jm	5.6	0.09	0.04
U	1	610895	4284883	1428	390	1394	34	356	Jn top	Kcm	Jm	5.0	0.09	0.03
V	1	610910	4284860	1429	390	1394	35	355	Jn top	Kcm	Jm	2.8	0.09	0.02
44	1	610973	4284802	1433	390	1396	37	353	Jn top	Kcm bottom	Jm	1.5	0.09	0.01
45	1	611000	4284771	1433	390	1396	37	353	Jn top	Kcm bottom	Jm	2.3	0.09	0.01
46	1	611067	4284712	1426	395	1396	30	365	Jn top	Kcm bottom	Jm	0.2	0.08	0.00
47	1	611108	4284701	1416	405	1398	18	387	Jn top	Kcm bottom	Jm	0.3	0.04	0.00
48	1	611199	4284628	1453	420	1398	55	365	Jn top	Kcm bottom	Jm	3.3	0.13	0.02
49	1	611293	4284560	1460	430	1400	60	370	Jn top	Kcm bottom	Jm	2.6	0.14	0.02
50	1	611376	4284492	1462	445	1400	62	383	Jn top	Kcm bottom	Jm	5.6	0.14	0.04
51	1	611481	4284432	1437	460	1400	37	423	Jk top	Kcm	Jm	2.0	0.08	0.01
E1	1	611550	4284350	1451	432	1400	51	381	Jk top	Jmb top	Jm	5.6	0.12	0.04
E2	1	611550	4284350	1446	432	1400	46	386	Jk top	Jmb top	Jm	5.0	0.11	0.03
E3	1	611550	4284350	1441	432	1400	41	391	Jk top	Jmb top	Jm	7.0	0.09	0.04
E4	1	611550	4284350	1436	432	1400	36	396	Jk top	Jmb top	Jm	7.0	0.08	0.04
E5	1	611550	4284350	1431	432	1400	31	401	Jk top	Jmb top	Jm	8.0	0.07	0.05
E6	1	611550	4284350	1426	432	1400	26	406	Jk top	Jmb top	Jm	12.0	0.06	0.08
E7	1	611550	4284350	1421	432	1400	21	411	Jk top	Jmb top	Jm	16.0	0.05	0.10
E8	1	611550	4284350	1416	432	1400	16	416	Jk top	Jmb top	Jm	23.0	0.04	0.14
	1	611760	4284010	1450	520	1414	36	484	Jk, 30 m below top	Kcm bot	Jm	8.0	0.07	0.05
D	1	612350	4283480	1472	560	1383	89	471	Jw bottom	Kd	Jm	6.0	0.16	0.04
C1	1	617667	4276789	1382	60	–	55	5	Jem	Jmsw bottom	Jmt	3.0	0.92	0.15
C2	1	617667	4276789	1382	880	–	?	?	Pc bottom	Jem	TRc + TRm + Jmt + Jmsw	3.0	?	0.00
C3	1	617637	4276798	1382	940		?	?	Pc bottom	Jmsw	TRc + TRm + Jmt + Jmsw	4.0	?	0.00
B	1	618661	4275392	1290	912		?	?	Pc	Jmsw bottom	Jmt	3.0	?	0.15
A	1	620086	4274725	1282	250		?	?	Pht top	Pc	Pc/Pht	4.0	?	0.02

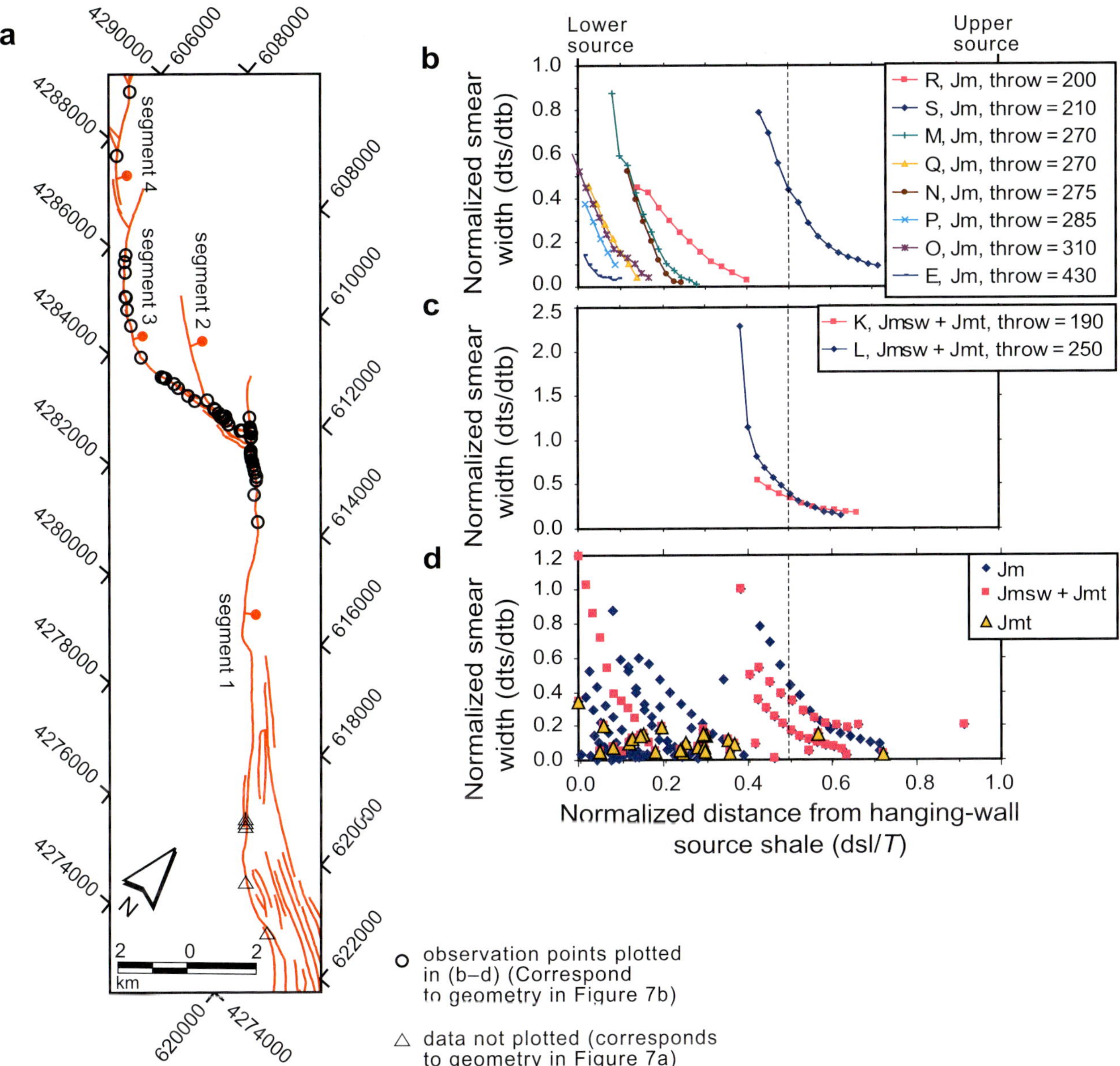

Figure 9. (a) Index map indicating locations of measurements. Variation of shale smear width vs. the distance from the source of shale in the hanging wall (dsl) normalized by the throw for shale derived from (b) the Brushy Basin Member, (c) the Salt Wash Member, and (d) for all units in the Morrison Formation. Smear width in (b–d) is normalized to the thickness of the inferred source of shale to facilitate comparisons between different observation points. Note that abbreviations of stratigraphic names are defined in Figure 2.

and core) maps the presence or absence of fault zone structures.

The architecture of the damage zones is largely dependent on the rock type in which they occur. The architecture of the fault core at any point is affected by any unit faulted past that position. Thus, the fault core at a point on the diagram might contain smeared shale even if shale is not currently juxtaposed against that portion of the fault because shale was faulted past that point. Consequently, distinguishing the fault cores from the damage zone provides a means of recognizing where these different parts of the fault zone might have different hydraulic properties. The fault core of most segments is dominated by shale-related fault rocks derived from either the Pennsylvanian to Triassic section or the Late Jurassic to Cretaceous section. Fault segments through the Jurassic sandstone section are composite zones of deformation bands, with locally developed breccia zones, and shale gouge resulting from offset of units rich in shale. In most locations, the fault is a composite of structures resulting from different faulting mechanisms.

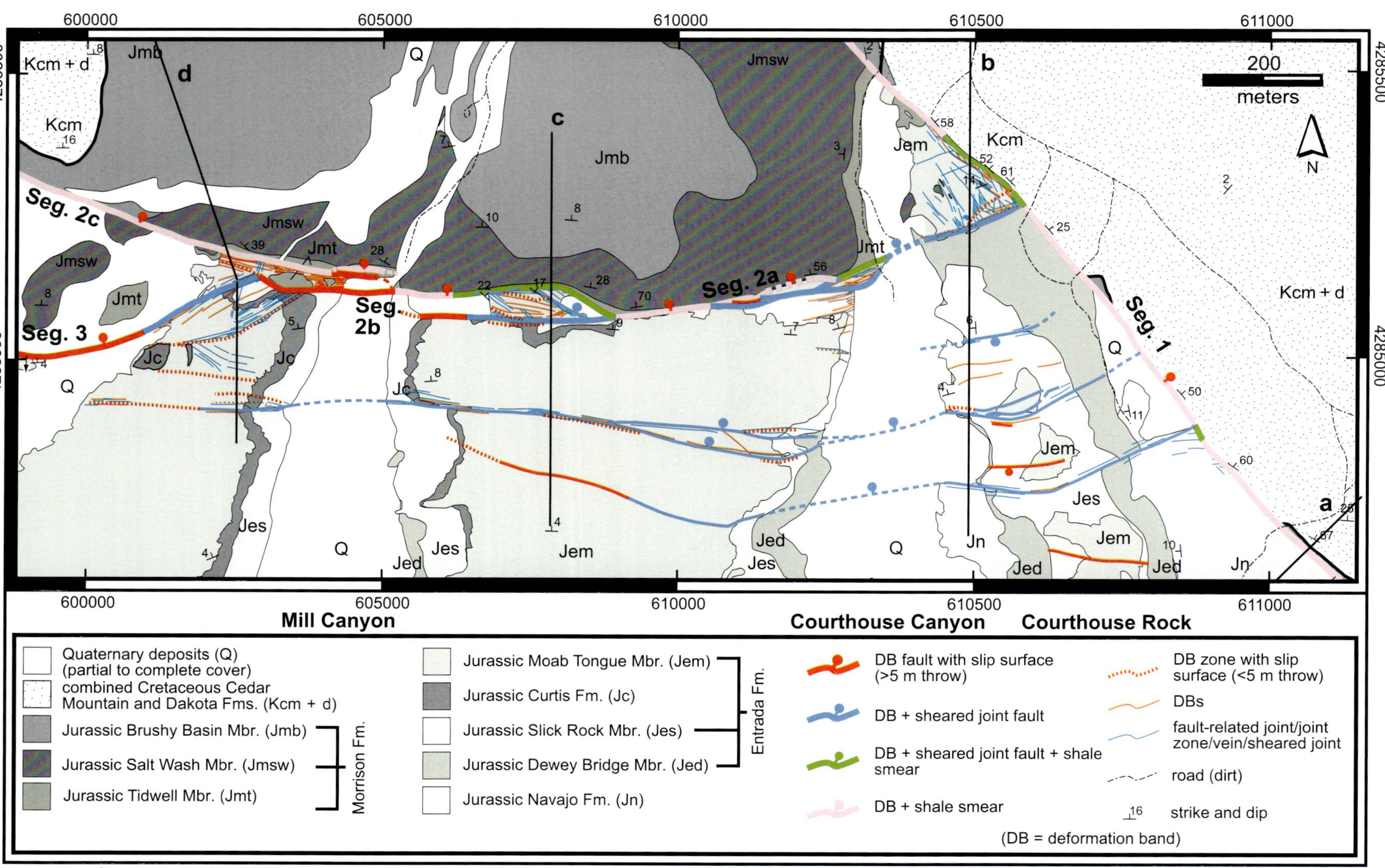

Figure 10. Map of the distribution of different types of fault zone structures resulting from distinct faulting mechanisms observed in sandstone and shale along a portion of the Moab fault. Locations of cross sections in Figure 11 are indicated.

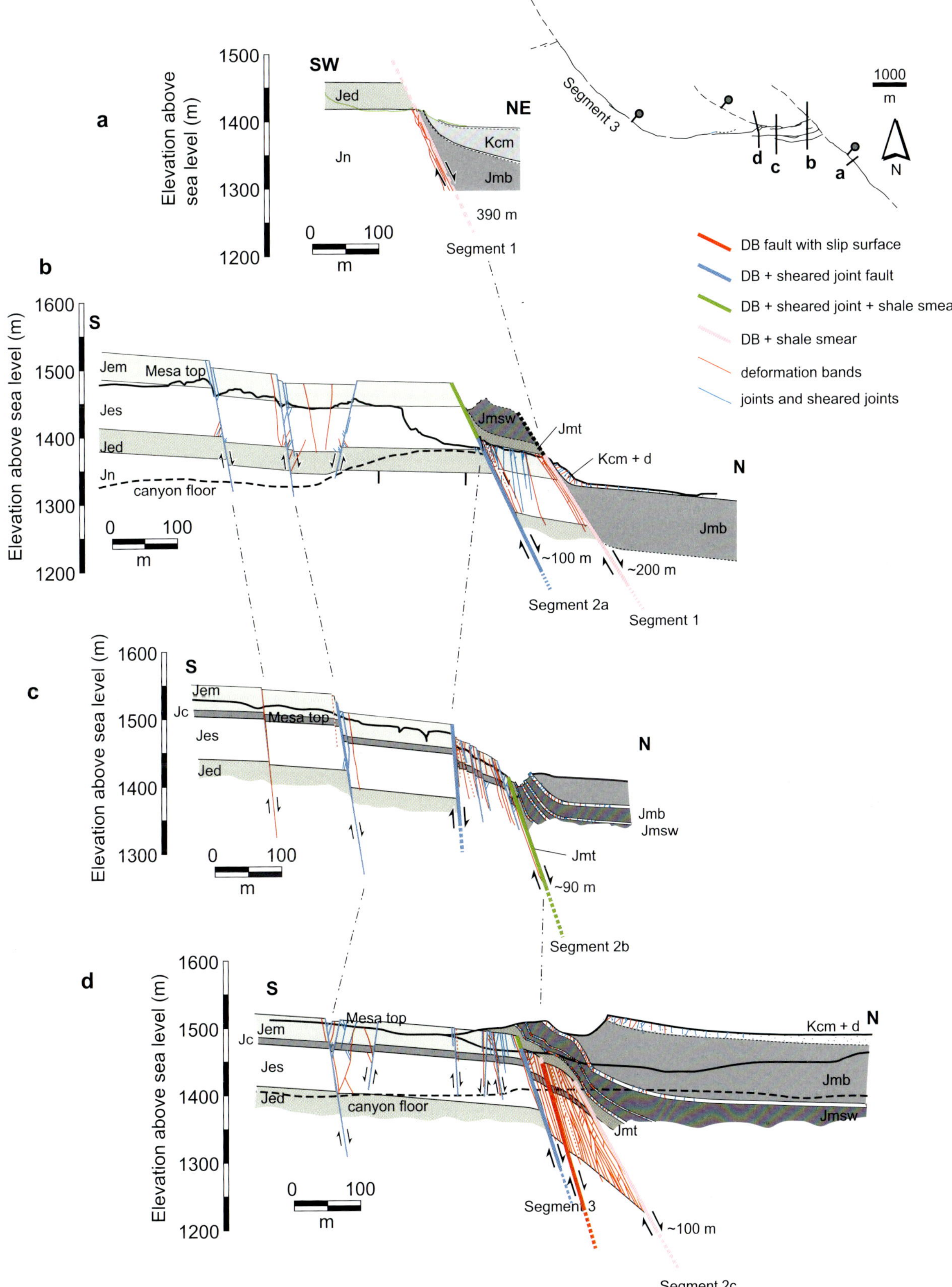

FIGURE 11. Cross sections along the Moab fault showing the dip-parallel distribution of different types of fault zone structures resulting from distinct faulting mechanisms as well as lateral variations in sandstone thickness associated with fault segmentation. Locations of sections (a–d) are indicated in Figure 10. Note that abbreviations of stratigraphic names are defined in Figure 2.

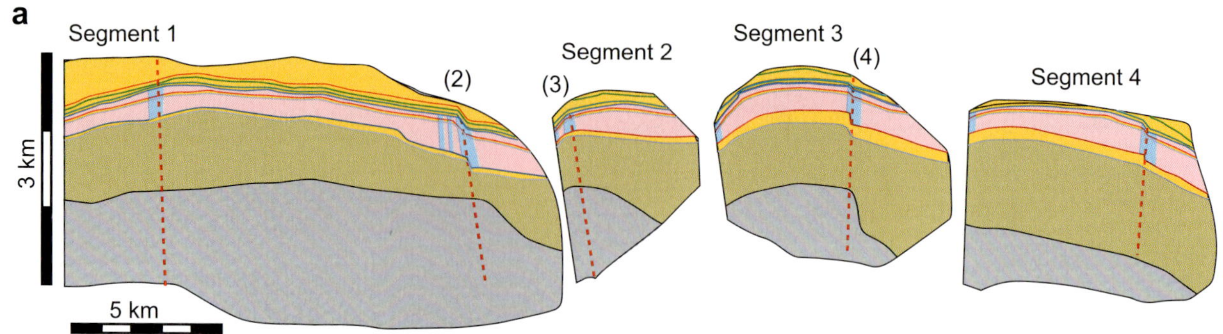
Footwall structures
a
Segment 1
(2)
(3)
Segment 2
Segment 3
(4)
Segment 4
3 km
5 km

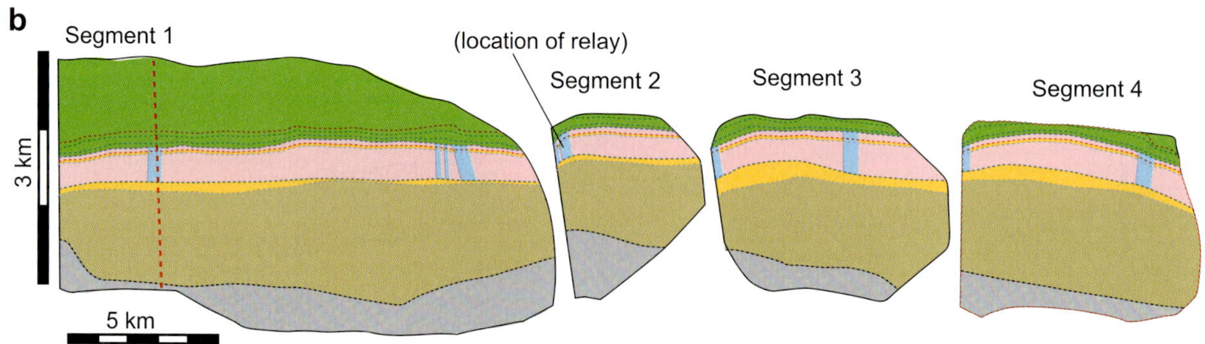
Hanging-wall structures
b
Segment 1
(location of relay)
Segment 2
Segment 3
Segment 4
3 km
5 km

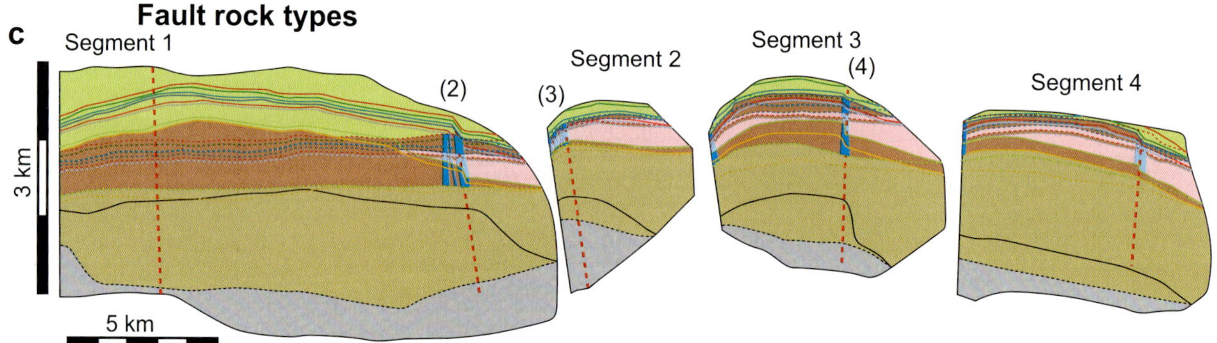
Fault rock types
c
Segment 1
(2)
(3)
Segment 2
Segment 3
(4)
Segment 4
3 km
5 km

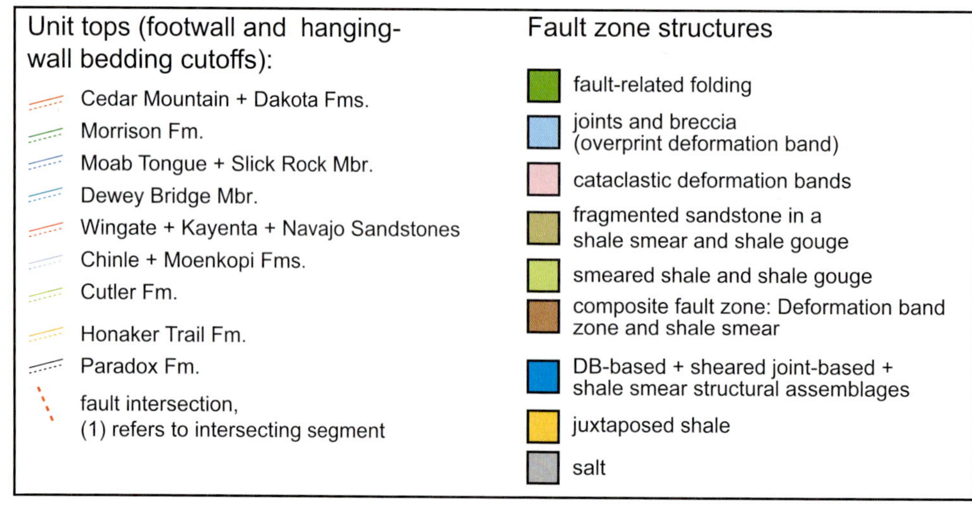
Unit tops (footwall and hanging-wall bedding cutoffs):
Cedar Mountain + Dakota Fms.
Morrison Fm.
Moab Tongue + Slick Rock Mbr.
Dewey Bridge Mbr.
Wingate + Kayenta + Navajo Sandstones
Chinle + Moenkopi Fms.
Cutler Fm.
Honaker Trail Fm.
Paradox Fm.
fault intersection,
(1) refers to intersecting segment
Fault zone structures
fault-related folding
joints and breccia
(overprint deformation band)
cataclastic deformation bands
fragmented sandstone in a
shale smear and shale gouge
smeared shale and shale gouge
composite fault zone: Deformation band
zone and shale smear
DB-based + sheared joint-based +
shale smear structural assemblages
juxtaposed shale
salt

DISCUSSION

Shale Deformation

Shale deformation associated with the Moab fault is dominated by folding and smearing similar to that described by many previous workers (Weber et al., 1978; Lehner and Pilaar, 1997; Aydin and Eyal, 2002; Koledoye et al., 2000). Relatively compliant shale units experience pervasive ductile extension and attenuation, which preserves bedding, as well as rotation of bedding to subparallelism with fault slip surfaces. Concurrently, gouge develops along the contact between the shale smear and fault slip surface. Gouge development suggests frictional wear along the contact of the fault surfaces with smeared shale; this process might contribute to thinning of the smeared shale as offset increases. Shale-related fault rocks are limited to the region between the footwall and hanging-wall sources of shale, which is consistent with the entrainment of shale into the fault core by either ductile attenuation or abrasion. Both processes probably contributed to the thinning of shale in the fault core. The measurements of smear width (Figure 9) and the field examples (e.g., Figures 5–7) indicate that individual shale units eventually attenuate to zero thickness as the distance from the source of shale becomes large at a large throw. Empirically, our estimates for the maximum distance from the source of shale and the throw that shale persists in the fault zone are consistent with the findings of other workers (Lindsay et al., 1993; Gibson, 1994; Aydin and Eyal, 2002). Because shale is primarily derived from shale units in the hanging wall, the gap in the shale smear will most likely first appear closer to the footwall cutoff.

Shale fault rocks persist in the fault core for relatively large distances from the source of shale and demonstrate a consistently low rate of thinning with distance from the source of shale (Figure 9). This might be explained if the fault core continues to incorporate shale from the source units into the smear as throw increases. For example, in Figure 4b, extension of the shale layer is observed well into the hanging wall. Similarly, shale layers in folds illustrated in Figure 7 thin toward the fault slip surface by extension parallel to bedding. This thinning and extension of the source layers adjacent to the fault might indicate that shale is extracted from the source layer into the fault core over a significant distance from the fault surface, providing a long-lasting supply of shale to the smear. Alternatively, broad fold hinges could increase the area that shale is juxtaposed against the fault and thus might contribute more shale to the fault rock, especially if the wear process is the primary mechanism of fault rock development. Both processes imply that the thickness of the sources of shale should be a significant controlling parameter. However, the measurements in Figure 9 and Table 2 do not clearly support this supposition.

A simpler possibility is that the smeared shale is uniformly thinned and spread out over a proportionately larger area as fault offset increases. This process requires no additional shale as the fault develops but does suggest that the initial input of shale into the fault zone is a primary control on the evolution of smear thickness. As a result, the shale smear would become discontinuous at an offset largely determined by the initial volume of shale in the fault zone. In turn, the initial volume of shale could be controlled by the geometry of the relay that initially forms around the smear (e.g., Figure 5). Thus, some of the variability in the measurement of smear width (Figure 9) could be attributed to the geometry of the fault system during an early stage of development.

In addition to geometric parameters, the competency contrast (Goodwin and Tikoff, 2002) and ductility contrast (Sperrevik et al., 2000; Clausen and Gabrielsen, 2002) between shale and sandstone units have been identified as important factors that determine what units can smear. The ductility of shale units might also impact the length and continuity of the smear from the source of shale. These properties will evolve because of compaction during burial, changes in the water content, and diagenesis of shale units and, thus, might be different at different burial depths or change during the life of the fault. Furthermore, Takahashi (2003) noted that the persistence of low cross-fault permeability in experimentally deformed samples containing a fine-grained layer depended on the effective normal stress. As offset increased, low permeability across the fault and the continuity of the fault seal persisted to greater throw when the effective normal stress was greater. These changes will ultimately determine the processes involved in faulting that can contribute to the sealing of faults.

Implications for Faults and Fluid Flow

The criteria used to estimate the sealing potential of a fault zone largely depend on the objective of the fault zone evaluation. For instance, if seal capacity for hydrocarbon trapping is estimated by the capillary entry pressure (Watts, 1987; Knipe et al., 1998) or failure by hydrofracture, then the simple presence or absence of shale and its petrophysical properties are the limiting

FIGURE 12. Fault plane diagrams depicting the distribution of fault zone components in the (a) footwall, (b) hanging wall, and (c) fault core. Fault planes are projected onto a vertical plane trending 315°. The vertical scale is exaggerated 2.5× the horizontal scale. Fault segments are numbered in the index map of Figures 7, 9, and 11.

factors. Conversely, if fluid flux is the limiting parameter, then the thickness of shale in the fault zone, its permeability, and the time interval are major limiting factors.

At the Moab fault, available outcrop evidence suggests that shale fault rock is continuous without physical gaps or windows. Consequently, sandstone beds internal to thick shale packages are highly unlikely to be directly juxtaposed against other sandstone units across the fault zone. Thus, communication of fluid from footwall sandstone units into sandstone units in the hanging wall would be inhibited by shale. This isolation would prevent charging of sandstone units in the hanging wall with hydrocarbons derived from outside the footwall side.

A similar case for fault sealing can be made for the distribution of deformation bands in the fault zone. Along the Moab fault, we argue that some minimum thickness of deformation bands is likely for a given throw (Davatzes, 2003), although the actual density and width of the zone are highly variable and may be much greater. Thus, the deformation bands present the potential to inhibit fluid flow across the fault and, in some circumstances, can be evaluated as a seal, or at least a baffle, to fluid flux.

In contrast, joint and breccia development near intersections in relays and in folded sandstone units of the Salt Wash Member along the fault system overprint deformation bands, suggesting the potential for increased permeability in these regions. This hypothesis is corroborated by oil staining, intense calcite and ankerite precipitation, and bleaching (Foxford et al., 1996) in the Entrada Formation at these locations, which we interpret to result from enhanced fluid flux. Thus, joint-based structures might have mitigated the impact of deformation bands in sandstone but might not breach shale seals unless the shale is very thin and the stress state is reasonable for hydrofracture. Local stress perturbations caused by slip on the fault segments that promote dilation at these locations are consistent with the distribution of joint-based structures (Davatzes, 2003). We were unable to directly observe evidence in the shale-rich units at these locations for joint-related structures. However, Foxford et al. (1998) does document the presence of veins at some locations within shale.

These effects are localized in the different parts of the fault zone, including the fault core and the damage zones (Figure 12), and thus, these zones may have disparate hydrologic properties. For instance, the fault core might be sealing, whereas the damage zone on either side of the fault may have enhanced permeability because of joint development (Figures 10, 11). Conversely, a permeable fault core may be bounded by shale or dense deformation band zones on one or both sides, resulting in an isolated fluid conduit such as at the relay indicated in Figure 12 and along portions of fault intersections.

Extrapolating Fault Architecture Using Faulting Mechanisms

Similar to most large, long-lived faults cutting a complicated stratigraphic package of sandstone and shale, the Moab fault system displays complex fault architecture with structures derived from many different faulting mechanisms. We have been able to identify specific parameters, including the stratigraphy (distribution and petrophysics of rock types), fault geometry, slip distribution, and deformation conditions, that controlled the involvement of these mechanisms during development of the Moab fault zone. These parameters provide criteria that can be used to predict or extrapolate the fault architecture from limited data sets common in petroleum exploration and production. As a result, we were able to map the distribution of different types of structures onto the fault plane diagram in Figure 12 and extrapolate them from surface observations along fault dip.

We suggest that this approach is an improvement over purely empirical or statistical approaches to predicting or extrapolating fault zone architecture because it can be checked for self-consistency, requires physically plausible structural assemblages, and is capable of predicting abrupt changes in fault characteristics. For instance, Foxford et al. (1998), who assessed the architecture of the Moab fault, tested the relationship between fault offset and the distribution of fault rocks. They found little consistency and, thus, predictability in the fault architecture. In analyzing the data, they did not consider the distance from potential sources of shale or the effects of fault geometry, such as intersections and relays. As a result, their data set makes it difficult to determine why so little correlation exists between parameters like throw and fault zone width or whether there should be. However, their results importantly emphasize that the spatial proximity of two data points does not insure similar properties. Furthermore, this problem will be compounded where data, samples, and observations are limited (e.g., boreholes, seismic, or other remote sensing data). Thus, changes in fault architecture that occur in small regions, such as joint development localized around intersections and some relays, will either be missed by this approach or remain unexplained.

In contrast, the criteria we have outlined have logical relationships to the faulting mechanisms. As a result, the potential for joint-based architecture could be tested using mechanical models (Davatzes, 2003) and borehole stress analyses. Furthermore, these predictions can be evaluated in the context of the deformation history to determine the potential temporal evolution of fault architecture, which is not possible using most statistical methods. For instance, at the Moab fault, joints postdate deformation bands, where both structures occur, and are associated with localized fluid flux. These circumstances indicate a change in faulting mechanism that produced a corresponding change in the hydraulic properties of the fault. Numerical simulations indicate that this might have occurred during a late stage of fault development, as the faults grow larger and interact, or if faulting continues into a period of exhumation (Davatzes, 2003). Thus, the potential for a faulted reservoir to charge with hydrocarbons could be linked not only to the timing of fault development but also to the evolution of fault properties. Finally, these logical relationships help clarify when uncertainty in controlling parameters might make an analysis of fault architecture difficult.

A variety of sources of error in this analysis must be mentioned because they impact the reliability of using faulting mechanisms to extrapolate fault zone properties. This analysis is sensitive to the geometry of the fault system and the faulted stratigraphy. In our study, the outcrop trace of the fault is very well constrained. However, salt redistribution during sedimentation leads to highly variable formation thickness. As a result, estimates of fault throw, unit juxtaposition across the fault, and the thickness of shale as a source for fault rock each has an associated uncertainty that increases with depth. In addition, folding is common in the shale-dominated units of the hanging wall of the Moab fault (Figure 7). Such folding has a profound impact on the construction of juxtaposition diagrams and the potential distribution of smeared shale. Because folding occurs within only a few hundred meters of the fault, it will be difficult to image in subsurface cases that rely on seismic data sets.

Similarly, downdip fault geometry is not well constrained. As a result, we assumed the simplest geometry while honoring surface data and available wells. Thus, the fault core is idealized as a single plane for the purpose of visualization and because of the resolution of downdip data. This assumption is of particular interest because the process of shale smear (Figure 4c) suggests vertical fault segmentation at a much smaller scale compared to the fault height and offset. Thus, we are aware of vertical segmentation, having observed it in some locations, but recognize that it is difficult to predict in the subsurface without very detailed, reliable stratigraphic data. In addition, the relative thickness of sandstone and shale seems to have affected the folding behavior of the stratigraphic package. Intervals of thicker sandstone and thinner shale are associated with small fold amplitudes and narrow hinges and a narrower fault zone. In some cases (e.g., Figure 7; station J1), sandstone layers abruptly truncate with the result that the fault zone is uncommonly wide in these regions for short distances along dip. Thus, sedimentological variation in the relative thickness of interbedded sandstone and shale in units such as the Salt Wash Member could control the local variation of shale smear thickness.

CONCLUSIONS

Three deformation mechanisms contributed to the development of the Moab fault, resulting in structural heterogeneity along the fault zone. Sandstone deformation is accommodated by the formation of deformation bands and related slip surfaces. At intersections and relays and in thin beds of folded sandstone layers, joints and sheared joints overprint deformation bands. In locations faulted past shale-rich stratigraphic horizons, shale smear and gouge, as well as fragmentation of sandstone layers, are prevalent fault zone components. Individual shale units persist as much as eight times their bed thickness along the fault zone, although shale smear thickness is highly variable along strike and dip. The development of these three types of structural assemblages suggests that the temporally and spatially variable architecture of the Moab fault should have behaved as a complex barrier and conduit system with a major impact on subsurface fluid flow. In addition, despite the complexity of the fault system, Figure 12 illustrates that the distribution of structures resulting from all three faulting mechanisms follows a consistent pattern related to the stratigraphy, juxtaposition, throw, and fault geometry. The present architecture of the fault demonstrates that faulting processes provide a self-consistent, physically realistic basis to interpolate the architecture of a fault zone in poorly constrained locations. Furthermore, this approach makes it possible to identify the sources of uncertainty that affect each faulting mechanism and thus assess the reliability of a sealing analysis.

ACKNOWLEDGMENTS

We acknowledge the financial support of the Rock Fracture Project in the Geological and Environmental Science Department at Stanford University. Useful feedback and discussions were provided by Peter Eichhubl. We also thank Bill Dunne, Jonathan Caine, Rasoul Sorkhabi, and Jennifer Wilson for careful reviews of the manuscript, which helped us improve the clarity of the presentation.

REFERENCES CITED

Allan, U. S., 1989, Model for hydrocarbon migration and entrapment within faulted structures: AAPG Bulletin, v. 73, p. 803–811.

Anderson, T. L., 1995, Fracture mechanics: Fundamentals and applications, 2d ed.: New York, CRC Press, 688 p.

Antonellini, M., and A. Aydin, 1994, Effect of faulting on fluid flow in porous sandstones; petrophysical properties: AAPG Bulletin, v. 78, no. 3, p. 355–377.

Antonellini, M., and A. Aydin, 1995, Effect of faulting on fluid flow in porous sandstones; geometry and spatial distribution: AAPG Bulletin, v. 79, no. 5, p. 642–671.

Antonellini, M., A. Aydin, and L. Orr, 1999, Outcrop-aided characterization of a faulted hydrocarbon reservoir: Arroyo Grande oil field, California, U.S.A., *in* W. C. Haneberg, P. S. Mozley, J. C. Moore, and L. B. Goodwin, eds., Faults and subsurface fluid flow in the shallow crust: American Geophysical Union Geophysical Monograph 113, p. 7–26.

Aydin, A., 2000, Fractures, faults, and hydrocarbon entrapment, migration and flow: Marine and Petroleum Geology, v. 17, p. 797–814.

Aydin, A., and Y. Eyal, 2002, Anatomy of a normal fault with shale smear: Implications for fault seal: AAPG Bulletin, v. 86, no. 8, p. 1367–1381.

Aydin, A., and A. M. Johnson, 1978, Development of faults as zones of deformation bands and as slip surfaces in sandstone: Pure and Applied Geophysics, v. 116, p. 931–942.

Bredehoeft, J. D., W. Back, and B. B. Hanshaw, 1982, Regional ground-water flow concepts in the United States; historical perspective, *in* T. N. Narasimhan, ed., Recent trends in hydrogeology: Geological Society of America Special Paper 189, p. 297–316.

Brace, W. F., and E. G. Bombolakis, 1963, A note on brittle crack growth in compression: Journal of Geophysical Research, v. 68, no. 12, p. 3709–3713.

Caine, J., J. Evans, and C. Forster, 1996, Fault zone architecture and permeability structure: Geology, v. 24, no. 11, p. 1025–1028.

Caine, S. C., and C. B. Forster, 1999, Fault zone architecture and fluid flow: insights from field data and numerical modeling, *in* W. C. Haneberg, P. S. Mozley, J. C. Moore, and L. B. Goodwin, eds., Faults and subsurface fluid flow in the shallow crust: American Geophysical Union Geophysical Monograph 113, p. 101–127.

Chester, F. M., and J. M. Logan, 1986, Implications for mechanical properties of brittle faults from observations of the Punchbowl fault zone, California: Pure and Applied Geophysics, v. 124, p. 77–106.

Clausen, J. A., and R. H. Gabrielsen, 2002, Parameters that control the development of clay smear at low stress states: An experimental study using ring-shear apparatus: Journal of Structural Geology, v. 24, p. 1569–1586.

Cruikshank, K. M., and A. Aydin, 1994, Role of fracture localization in arch formation, Arches National Park, Utah: Geological Society of America Bulletin, v. 106, no. 7, p. 879–891.

Cruikshank, K. M., G. Zhao, and A. M. Johnson, 1991, Analysis of minor fractures associated with joints and faulted joints: Journal of Structural Geology, v. 13, no. 8, p. 865–886.

Davatzes, N. C., 2003, Fault architecture as a function of deformation mechanism in clastic rocks with an emphasis on sandstone: Ph.D. thesis, Stanford University, California, 172 p.

Davatzes, N. C., and A. Aydin, 2003, Overprinting faulting mechanisms in high porosity sandstones of SE Utah: Journal of Structural Geology, v. 25, no. 11, p. 1795–1813.

Dholakia, S. K., A. Aydin, D. D. Pollard, and M. D. Zoback, 1998, Fault-controlled hydrocarbon pathways in the Monterey Formation, California: AAPG Bulletin, v. 82, no. 8, p. 1551–1574.

Doelling, H. H., 1982, Stratigraphic investigations of Paradox basin structures as a means of determining the rates and geologic age of salt-induced deformation: A preliminary study: Utah Geologic and Mineral Survey Open-file Report 29, 88 p., 9 plates.

Doelling, H. H., 1985, Geology of Arches National Park: Utah Geological and Mineral Survey, Map 74, 2 sheets, scale 1:50,000.

Doelling, H. H., 1988, Geology of Salt Valley anticline and Arches National Park, Grand County, Utah, *in* H. H. Doelling, C. G. Oviatt, and P. W. Huntoon, eds., Salt deformation in the Paradox region: Utah Geological and Mineral Survey, Salt Lake City, Utah, Bulletin, v. 122, p. 7–58.

Doelling, H. H., 2002, Geologic map of the Moab and eastern part of the San Rafael desert 30/X 60′ quadrangles, Grand and Emery Counties, Utah, and Mesa County Colorado: Utah Geological Survey Map 180, scale 1:100,000, 3 plates.

Dyer, J. R., 1983, Jointing in sandstones, Arches National Park, Utah: Ph.D. thesis, Stanford University, Stanford, California 275 p.

Eichhubl, P., and J. Boles, 2000, Focused fluid flow along faults in the Monterey Formation, coastal California: Geological Society of America Bulletin, v. 112, no. 11, p. 1667–1679.

Engelder, J. T., 1987, Joints and shear fractures in rock, *in* B. K. Atkinson, ed., Fracture mechanics of rock: London, Academic Press, p. 27–69.

Faulkner, D. R., and E. H. Rutter, 1998, The gas permeability of clay-bearing fault gouge under high pressure at 20°C: Journal of Geophysical Research, v. 105, no. 7, p. 16,415–16,426.

Flodin, E. A., 2003, Structural evolution, petrophysics, and large-scale permeability of faults in sandstone, Valley of Fire, Nevada: Ph.D. thesis, Stanford University, California, 180 p.

Flodin, E. A., A. Aydin, L. J. Durlofsky, and B. Yeten, 2001, Representation of fault zone permeability in reservoir flow models, *in* Society of Petroleum Engineers Annual Technical Conference and Exhibition, New Orleans, SPE Paper 71671, 10 p.

Foxford, K. A., I. R. Garden, S. C. Guscott, S. D. Burley, J. J. M. Lewis, J. J. Walsh, and J. Waterson, 1996, The field geology of the Moab fault, *in* A. C. Huffman, W. R. Lund, and L. H. Goodwin, eds., Geology and resources of the Paradox basin: Salt Lake City, Utah, Utah Geological Association Guidebook, v. 25, p. 265–283.

Foxford, K. A., J. J. Walsh, J. Watterson, I. R. Garden, S. C.

Guscott, S. D. Burley, Q. J. Fisher, and R. J. Knipe, 1998, Structure and content of the Moab fault zone, Utah, U.S.A., and its implications for fault seal prediction, *in* G. K. Jones, Q. J. Fisher, and R. J. Knipe, eds., Faulting, fault sealing and fluid flow in hydrocarbon reservoirs: Geological Society (London) Special Publication 147, p. 87–103.

Fristad, T., A. Groth, G. Yielding, and B. Freeman, 1997, Quantitative seal prediction: A case study from Oseberg Syd, *in* P. Meller-Pedersen and A. G. Koestler, eds., Hydrocarbon seals: Importance for exploration and production: Norwegian Petroleum Society Special Publication 7, p. 107–124.

Garden, I. R., S. C. Guscott, S. D. Burley, K. A. Foxford, J. J. Walsh, and J. Marshall, 2001, An exhumed palaeohydrocarbon migration fairway in the Entrada Sandstone of SE Utah, U.S.A.: Geofluids, v. 1, no. 3, p. 195–213.

Gibson, R. G., 1994, Fault-zone seals in siliciclastic strata of the Columbus basin, offshore Trinidad: AAPG Bulletin, v. 78, no. 9, p. 1372–1385.

Goodwin, L. B., and B. Tikoff, 2002, Competency contrast, kinematics, and the development of foliations and lineations in the crust: Journal of Structural Geology, v. 24, p. 427–444.

Granier, T., 1985, Origin, damping, and pattern of development of faults in granite: Tectonics, v. 4, no. 7, p. 721–737.

Heynekamp, M. R., L. B. Goodwin, P. S. Mozley, and W. C. Haneberg, 1999, Controls on fault-zone architecture in poorly lithified sediments, Rio Grande Rift, New Mexico; implications for fault-zone permeability and fluid flow, *in* W. C. Haneberg, P. S. Mozley, J. C. Moore, and L. B. Goodwin, eds., Faults and subsurface fluid flow in the shallow crust: American Geophysical Union Geophysical Monograph 113, p. 27–49.

Jourde, H., E. Flodin, A. Aydin, L. Durlofsky, and X. Wen, 2002, Computing permeability of fault zones in eolian sandstone from outcrop measurements: AAPG Bulletin, v. 86, no. 7, p. 1187–1200.

Knipe, R. J., 1993, The influence of fault zone processes and diagenesis on fluid flow, *in* A. D. Horbuyr and A. G. Robinson, eds., Diagenesis and basin development: AAPG Studies in Geology 36, p. 135–148.

Knipe, R. J., 1997, Juxtaposition and seal diagrams to help analyze fault seals in hydrocarbon reservoirs: AAPG Bulletin, v. 81, no. 2, p. 187–195.

Knipe, R. J., G. Jones, and Q. J. Fisher, 1998, Faulting, fault sealing, and fluid flow in hydrocarbon reservoirs; an introduction, *in* G. Jones, Q. L. Fisher, and R. J. Knipe, eds., Faulting, fault sealing and fluid flow in hydrocarbon reservoirs: Geological Society (London) Special Publication 147, p. vii–xxi.

Koledoye, A. B., A. Aydin, and E. May, 2000, Three-dimensional visualization of normal fault segmentation and its implication for fault growth: Leading Edge, v. 19, no. 7, p. 692, 696, 698, 700–701.

Koledoye, B. A., A. Aydin, and E. May, 2003, A new process-based methodology for analysis of shale smear along normal faults in the Niger Delta: AAPG Bulletin, v. 87, no. 3, p. 445–463.

Lehner, F. K., and W. F. Pilaar, 1991, On a mechanism of clay smear emplacement in synsedimentary normal faults: Inferences from field observations near Frechen, Germany, *in* P. Moller-Pedersen and A. G. Koestler, eds., Hydrocarbon seals: Importance for exploration and production: Norwegian Petroleum Society Special Publication 7, p. 39–50.

Lehner, F. K., and W. F. Pilaar, 1997, The emplacement of clay smears in synsedimentary normal faults: Inferences from field observations near Frenchen, Germany, *in* P. Moller-Pedersen and A. G. Koestler, eds., Hydrocarbon seals: Importance for exploration and production: Norwegian Petroleum Society Special Publication 15, p. 39–50.

Lindsay, N. G., F. C. Murphy, J. J. Walsh, J. Watterson, and I. D. Bryant, 1993, Outcrop studies of shale smears on fault surfaces, *in* S. S. Flint, ed., Geological modeling of hydrocarbon reservoirs and outcrop analogues: International Association of Sedimentologists Special Publication 15, p. 113–123.

Martel, S. J., 1990, Formation of compound strike-slip fault zones, Mount Abbot Quadrangle, California: Journal of Structural Geology, v. 12, no. 7, p. 869–882.

Matthai, S. K., A. Aydin, D. D. Pollard, S. G. Roberts, Q. J. Fisher, and R. J. Knipe, 1998, Numerical simulation of departures from radial drawdown in a faulted sandstone reservoir with joints and deformation bands, *in* G. Jones, Q. J. Fisher, and R. J. Knipe, eds., Faulting, fault sealing, and fluid flow in hydrocarbon reservoirs: Geological Society (London) Special Publication 147, p. 157–191.

Myers, R. D., 1999, Structure and hydraulic properties of brittle faults in sandstone: Ph.D. thesis, Stanford University, Stanford, California, 176 p.

Nelson, E. P., A. J. Kullman, M. H. Gardner, and M. Batzle, 1999, Fault-fracture networks and related fluid flow and sealing, Brushy Canyon Formation, west Texas, *in* W. C. Haneberg, P. S. Mozley, J. C. Moore, and L. B. Goodwin, eds., Faults and subsurface fluid flow in the shallow crust: American Geophysical Union Geophysical Monograph 113, p. 69–81.

Nuccio, V. F., and S. M. Condon, 1996, Burial and thermal history of the Paradox basin, Utah and Colorado, and petroleum potential of the middle Pennsylvanian Paradox basin: U.S. Geologic Survey Bulletin, Report B 2000-O, p. O1–O41.

Pevear, D. R., P. J. Vrolijk, and F. J. Lomgstaffe, 1997, Timing of Moab fault displacement and fluid movement integrated with burial history using radiogenic and stable isotopes: Geofluids II Extended Abstracts, p. 42–45.

Segall, P., and D. D. Pollard, 1980, Mechanics of discontinuous faults: Journal of Geophysical Research, v. 85, no. 8, p. 4337–4350.

Sibson, R. H., 1977, Fault rocks and fault mechanisms: Journal of the Geological Society (London), v. 133, p. 191–231.

Sibson, R. H., 1990, Conditions for fault-valve behavior, *in* R. J. Knipe and E. H. Rutter, eds., Deformation mechanisms, rheology and tectonics: Geological Society (London) Special Publication 54, p. 15–28.

Sigda, J. M., L. B. Goodwin, P. S. Mozley, and J. L. Wilson, 1999, Permeability alteration in small-displacement faults in poorly lithified sediments: Rio Grande rift, Central New Mexico, *in* W. C. Haneberg, P. S. Mozley, J. C. Moore, and L. B. Goodwin, Faults and subsurface fluid flow in the shallow crust: American Geophysical Union Geophysical Monograph 113, p. 51–68.

Skerlec, G. M., 1999, Evaluating top and fault seal, *in* E. A. Beaumont and N. H. Foster, eds., Exploring for oil and gas traps: AAPG Treatise of Petroleum Geology, Handbook of Petroleum Geology, p. 10-4–10-94.

Sperrevik, S. R. B., R. B. Faerseth, and R. H. Gabreilsen, 2000, Experiments on clay smear formation along faults: Petroleum Geoscience, v. 6, p. 113–123.

Takahashi, M., 2003, Permeability change during experimental fault smearing: Journal of Geophysical Research, v. 108, no. 5, p. 2235–2250.

Taylor, W. L., D. D. Pollard, and A. Aydin, 1999, Fluid flow in discrete joint sets; field observations and numerical simulations: Journal of Geophysical Research, v. 104, no. 12, p. 28,983–29,006.

Watts, N. L., 1987, Theoretical aspects of cap-rock and fault seals for single- and two-phase hydrocarbon columns: Marine and Petroleum Geology, v. 4, p. 274–307.

Weber, K. J., G. Mandl, F. Pilaar, F. K. Lehner, and R. G. Precious, 1978, The role of faults in hydrocarbon migration and trapping in Nigerian growth fault structures, *in* American Institute of Mining Metallurgical and Petroleum Engineers and Society of Petroleum Engineers Tenth Annual Offshore Technology Conference Proceedings, v. 4, p. 2643–2653.

Yielding, G., B. Freeman, and D. T. Needham, 1997, Quantitative fault seal prediction: AAPG Bulletin, v. 81, p. 897–917.

Shipton, Z. K., J. P. Evans, and L. B. Thompson, 2005, The geometry and thickness of deformation-band fault core and its influence on sealing characteristics of deformation-band fault zones, *in* R. Sorkhabi and Y. Tsuji, eds., Faults, fluid flow, and petroleum traps: AAPG Memoir 85, p. 181–195.

The Geometry and Thickness of Deformation-band Fault Core and its Influence on Sealing Characteristics of Deformation-band Fault Zones

Z. K. Shipton[1]
Department of Geology, Trinity College, Dublin 2, Ireland

J. P. Evans
Department of Geology and UF3-Innovation Campus, Utah State University, Logan, Utah, U.S.A.

L. B. Thompson
UF3, Innovation Campus, Utah State University, Logan, Utah, U.S.A.

ABSTRACT

Deformation-band faults in high-porosity reservoir sandstones commonly contain a fault core of intensely crushed rock surrounding the main slip surfaces. The fault core has a substantially reduced porosity and permeability with respect to both the host rock and individual deformation bands. Although fault core thickness is a large uncertainty in calculations of transmissibility multipliers used to represent faults in single-phase reservoir flow models, few data exist on fault core thickness in deformation-band fault zones. To provide accurate estimates of deformation-band fault petrophysical properties, we measured fault core thickness at six sites (each 4–15 m [13–49 ft] along strike) along the Big Hole fault in the Navajo Sandstone, central Utah. These data show that the thickness is highly variable and does not correlate with either the amount of slip or the number of slip surfaces. The thickness of the fault core is likely to be dependent on local growth processes, specifically the linkage of fault segments. This suggests that correlations of fault permeability with throw may not apply to deformation-band faults. Simple calculations of two-phase flow properties based on measured porosity and permeability values suggest that deformation-band faults

[1]*Present address:* Division of Earth Sciences, Centre for Geoscience, University of Glasgow, Glasgow, Scotland.

DOI:10.1306/1033723M853135

containing fault core are likely barriers to two-phase flow. More data on the variability of fault core thickness and its petrophysical properties need to be collected to characterize population statistics for models of deformation-band fault fluid-flow properties.

INTRODUCTION

Deformation bands are the dominant deformation elements in high-porosity sandstone (Aydin and Johnson, 1978). Deformation bands are zones of cataclasis commonly of the order of 1 mm (0.04 in.) thick, which accommodate offsets of less than a few millimeters. Increasing offset is accommodated by the progressive addition of deformation bands to form zones of bands, with offsets roughly proportional to the number of bands in the zone (Mair et al., 2000). Larger offsets are accommodated by throughgoing slip surfaces. In the model of Aydin and Johnson (1978), the evolution of deformation-band faults proceeds from the accumulation of single deformation bands to a cluster of deformation bands and to the presence of a slip surface. The Aydin and Johnston model does not address the process of slip-surface nucleation and propagation. Shipton and Cowie (2001) identified the fault core as a distinct deformation element in faults in the Navajo Sandstone, Utah, which continues to develop after the appearance of a throughgoing slip surface. The fault core consists of a narrow zone of intense cataclasis, accompanied by one or more slip surfaces. This fault core is surrounded by a damage zone consisting of deformation bands with occasional short segments of slip surfaces. This chapter presents preliminary data on the geometry and thickness of the fault core, their controlling growth processes, and their effect on the fluid-flow properties of deformation-band faults.

Deformation-band faults have been studied intensively in terms of their microstructures, fault geometries, petrophysical properties, influence on fluid flow, and their impact on the hydrocarbon industry (e.g., Pittman, 1981; Edwards et al., 1993; Antonellini and Aydin, 1994, 1995; Beach et al., 1997; Fisher and Knipe, 1998, 2001; Hesthammer et al., 2000; Jourde et al., 2002). They produce an order of magnitude or greater reduction of porosity and permeability with respect to the host rock. Shipton and Cowie (2001) showed that isolated deformation bands in the Navajo Sandstone (porosity 20–25%), Utah, have porosities of 3–13%, and the fault core has porosities as low as 1%. Antonellini and Aydin (1994) measured permeability values as low as 0.007 md adjacent to slip surfaces in the Entrada Sandstone, Utah. Deformation-band faults can therefore have a significant effect on fluid flow (Edwards et al., 1993; Knott, 1993; Gibson, 1994; Beach et al., 1997; Matthai et al., 1998; Antonellini et al., 1999; Fisher and Knipe, 2001; Ogilvie and Glover, 2001; Shipton et al., 2002).

Simple modeling of fluid flow in faults comprising clusters of deformation bands including a fault core allows the effect of varying parameters on the bulk permeability of the fault zone to be assessed (Shipton et al., 2002). Assuming that all fault elements are parallel, the bulk permeability is calculated in directions parallel and perpendicular to the fault using the weighted arithmetic mean and weighted harmonic mean, respectively. Sensitivity analyses of this fault model show that the thickness of the fault core is, by far, the most important variable for bulk fault zone permeability (Shipton et al., 2002). Therefore, the lack of data on fault core petrophysical properties and architecture makes it difficult to predict the permeability of deformation-band fault zones and to calculate representative transmissibility multipliers for use in reservoir models.

Most reservoir simulators represent faults as a transmissibility multiplier, T, assigned to the face of grid blocks that separate either side of the fault (Manzocchi et al., 1999). This assumes that the variability of fault zone properties can be upscaled to a homogenous effective value at the grid block scale. Determining appropriate values of T to assign across a fault surface is the focus of much research (Heinemann et al., 1998; Walsh et al., 1998; Manzocchi et al., 1999), but few geologic data exist that address how fault zone permeability or transmissibility varies along a fault plane. Most transmissibility values are assigned by a history-matching process (Heinemann et al., 1998). It would be preferable to make a physically based prediction of the transmissibility distribution on the fault surface prior to the start of production. Such analyses are critical, because in some cases, faults with transmissibility multipliers of less than 0.001 may retard fluid flow (Damsleth et al., 1998).

We examined the geometry, dimensions, and microstructures of fault core along the well-exposed Big Hole fault in the high-porosity Navajo Sandstone. The damage zone geometry of the Big Hole fault has been previously described in detail (Shipton and Cowie, 2001; Shipton et al., 2002), making it an ideal location for a study of fault core geometry and growth processes. Samples of fault core were collected for microstructural analysis at intervals along the fault zone with known displacement. We measured fault core thickness along the surface trace of the fault at several locations to characterize its spatial variability and its relationship to fault throw. We used these data to calculate likely permeability and capillary pressure values for the fault we measured. These data provide constraints on models of fluid flow in faults in high-porosity sandstones. We

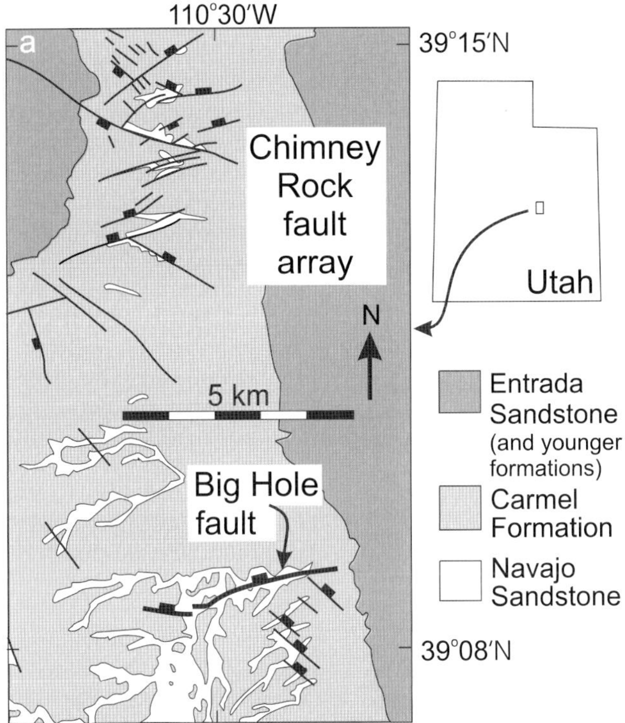

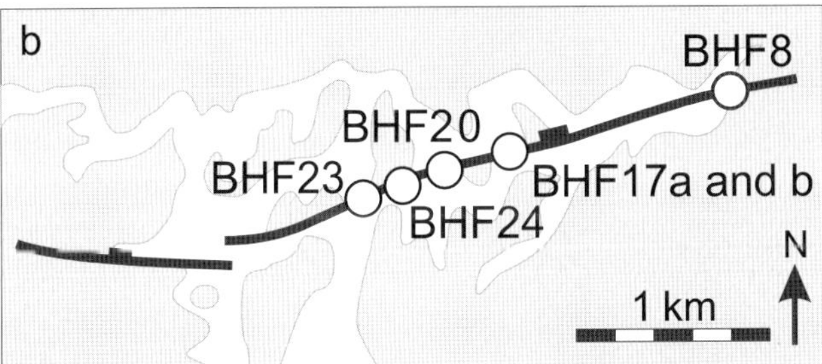

FIGURE 1. (a) Map of the Chimney Rock normal fault array showing the location of the Big Hole fault (BMF). Ticks mark the downthrown side of the faults. (b) Location of the survey sites along the Big Hole fault. Each site is referred to by the displacement on the main fault at that point, so BHF20 is the site at which the fault has 20 m (66 ft) displacement.

discuss flow parameters for faults in a specific setting and with a defined deformation mechanism (cataclasis in deformation bands). This approach has been successfully used elsewhere, for instance, in sheared-joint-based faults (Flodin et al., 2004) and clay-rich faults (Yielding et al., 1997).

GEOLOGICAL SETTING

The Big Hole fault is south of the Chimney Rock fault array in the northern San Rafael Swell, Utah (Figure 1). The faults are likely Late Cretaceous to early Tertiary in age, suggesting that faulting occurred under a lithostatic load of 40–80 MPa and temperatures of 75–90°C (Shipton and Cowie, 2001). Fault trace lengths range from 100 m (330 ft) to 4 km (2.5 mi), with displacements ranging up to 30 m (100 ft) (Krantz, 1988). Fault intersections in the Chimney Rock fault array create compartments of relatively undeformed host rock in the Navajo Sandstone (Figure 1), which are analogous to similar oil field- and aquifer-scale structures in porous sandstone (i.e., Gibson, 1994; Antonellini et al., 1999; Fossen and Hesthammer, 2000).

The faults cut the Jurassic Navajo Sandstone, a planar to cross-bedded, very fine- to fine-grained eolian arenite with about 90–95% quartz, 1–6% feldspar, and 1–3% clays. At this locality, the Navajo Sandstone is friable and weakly cemented by very low volumes of quartz overgrowths, iron oxide cements, and/or calcite cement. Primary porosity ranges from 13 to 25% (Shipton and Cowie, 2001). Permeabilities to water for the undeformed Navajo Sandstone, measured at room temperatures and pressures, are typically in the 100–1000-md range (Hood and Patterson, 1984) but can be greater than 4000 md (Shipton et al., 2002). Because of the relatively low volumes of cement in the host rock and the fault zone, we can investigate the mechanical aspects of fault sealing on fault-seal quality independently of the effects of diagenesis.

At the surface, the Big Hole fault strikes N70°E and dips 64°N with pure dip-slip slickenlines. Cowie and Shipton (1998) measured the fault throw from offset of the top Navajo Sandstone unconformity. A maximum of 29 m (95 ft) of throw at the fault center decreases approximately linearly toward each tip. The deeply incised, meandering Big Hole canyon provides sections through the fault zone structure. The fault typically consists of a fault core with one or more slip surfaces surrounded by clusters of deformation bands. The fault core sits in a damage zone consisting of deformation bands and occasional slip surfaces that are either synthetic or antithetic to the main fault.

FAULT CORE AND SLIP-SURFACE GEOMETRY

The fault core is a zone of fine-grained, pale-colored rock as much as 35 cm (14 in.) thick, bounded by or containing one or more slip surfaces (Figure 2). The fault core generally has a resistant, glassy, almost mylonitic appearance, although in thin section, it contains only cataclastic deformation (see next section). Occasionally, the fault core consists of a crumbly gouge (sample BHF24). Typically, several anastomosing slip-surface strands are present in the main fault zone. Slip surfaces are highly polished or mineralized and contain slickenlines. Both in outcrop and in drill core, these can be tightly mated surfaces (Figure 2a, b) or act as planes of parting (Figure 2c). Other slip surfaces that

Figure 2. (a) Example of the fault core from a drill hole through the Big Hole fault at BHF17a (from Shipton et al., 2002). The fault core is a white zone of densely packed deformation bands and is bounded on either side by slip surfaces (SS). (b) Fault core from a subhorizontal outcrop of the Big Hole fault at BHF23. Note the anastomosing deformation bands in the fault core (FC) and the slip surface at the margins. (c) Slip surfaces in a zone of deformation bands away from the main fault. Note that these are planes of parting. Thin knots of fault core occur along these slip surfaces and are associated with a deformation-band cluster on the right side of the photo.

surround these throughgoing surfaces tip out in zones of deformation bands. Only the surface or surfaces that cut through the entire outcrop can be accommodating a substantial majority of the offset.

In plan view, the throughgoing slip surface of the main fault plane is decorated with elongated pods of fault core (Figure 3a). These lozenge-shaped pods are separated by areas with almost zero thickness of fault core. In cross section view, the fault core material is localized at intersections between synthetic and antithetic deformation-band clusters that contain slip surfaces (Figure 3b). Thus, the fault core can be visuaized in three dimensions as being thickest at the intersection of synthetic and antithetic deformation-band clusters that are not totally planar, producing pods in plan view. In deformation-band clusters away from the fault, the fault core can be developed (as much as 10 cm [4 in.] thick) without slip surfaces (Figures 2c, 3b). Thus, fault core can develop without large amounts of slip, and large-offset slip surfaces can develop without a large thickness of fault core. Throughgoing slip surfaces are always polished and can even look glassy, showing that a thin veneer of very fine-grained fault core material surrounds all the slip surfaces.

FAULT CORE MICROSTRUCTURES

To investigate the evolution of fault core and slip-surface microstructures with increasing displacement, oriented samples were collected from surface outcrops with a rock drill and from drill core. Thin sections were examined under an optical microscope and with cathodoluminescence (CL) and backscatter (BSEM) modes on an electron microscope. Thin sections of fault rock were oriented perpendicular to the fault plane and parallel to the slip vector.

The fault core is highly comminuted (Figures 4, 5), and relatively intact grains lie in a tightly packed matrix of angular crushed grains, with a low resulting porosity (Figure 4a). Some of these uncrushed grains have been planed off at individual slip surfaces (Figure 4b),

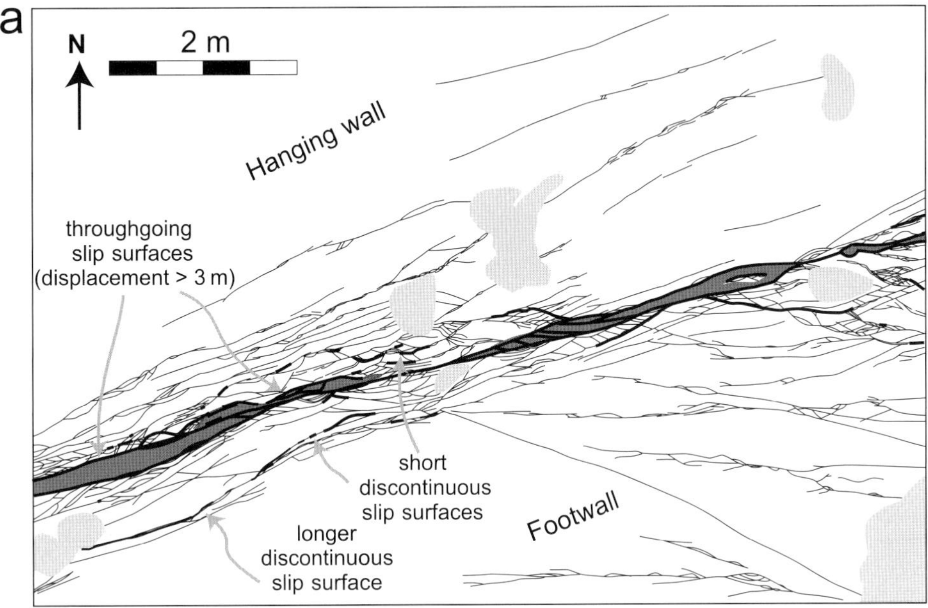

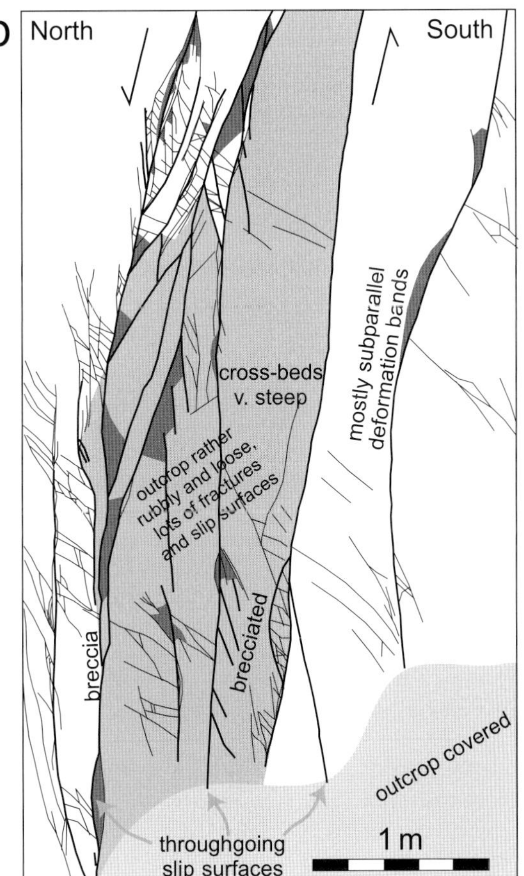

FIGURE 3. Fault core geometry. (a) Plan view of the outcrop at location BHF17. The fault core forms elongate lozenge-shaped pods (dark gray). Pale gray areas are sand on the outcrop. (b) Cross section view of a cliff outcrop at BHF23. Several slip surfaces cut the cliff face, and some can be seen to die out upward. Areas with very intense deformation, in which it is difficult to distinguish individual deformation bands, are shaded (medium gray). Some pods of breccia are marked. The fault core forms triangular areas at the intersection of synthetic and antithetic deformation band clusters (dark gray). A thin veneer of highly comminuted fault core material lines each of the slip surfaces but is too narrow to be represented on these figures.

indicating that transgranular fracturing is an important deformation mechanism along active slip surfaces. The fault core is therefore finer grained, with a larger variation in grain size than individual deformation bands. The thickness of the fault core can vary even on the scale of an individual thin section (Figure 4a). Aligned, unhealed transgranular microfractures are only seen in the zone of high grain crushing adjacent to the slip surfaces (Figure 4d). This is in contrast to microfractures in grains outside the fault core, which are rare and exclusively contained in individual grains.

Slip surfaces appear as thin (less than 1 mm [0.04 in.] thick), dark surfaces under plain polarized light. Under crossed polars, the areas of crushed grains, which surround all of the slip surfaces, become more apparent (Figure 4b). Slip surfaces can bifurcate and link at all angles, and wider zones of intense grain crushing are localized at the points where slip surfaces link (Figure 5a, b). Figure 5c shows two slip surfaces that curve toward one another. We interpret this as the interaction and propagation of two slip surfaces toward each other, and that therefore, the slip surfaces have propagated in the fault core.

Bands of highly comminuted grains occur in the fault core at an angle of 15° to the main fault (Figure 4a,b).

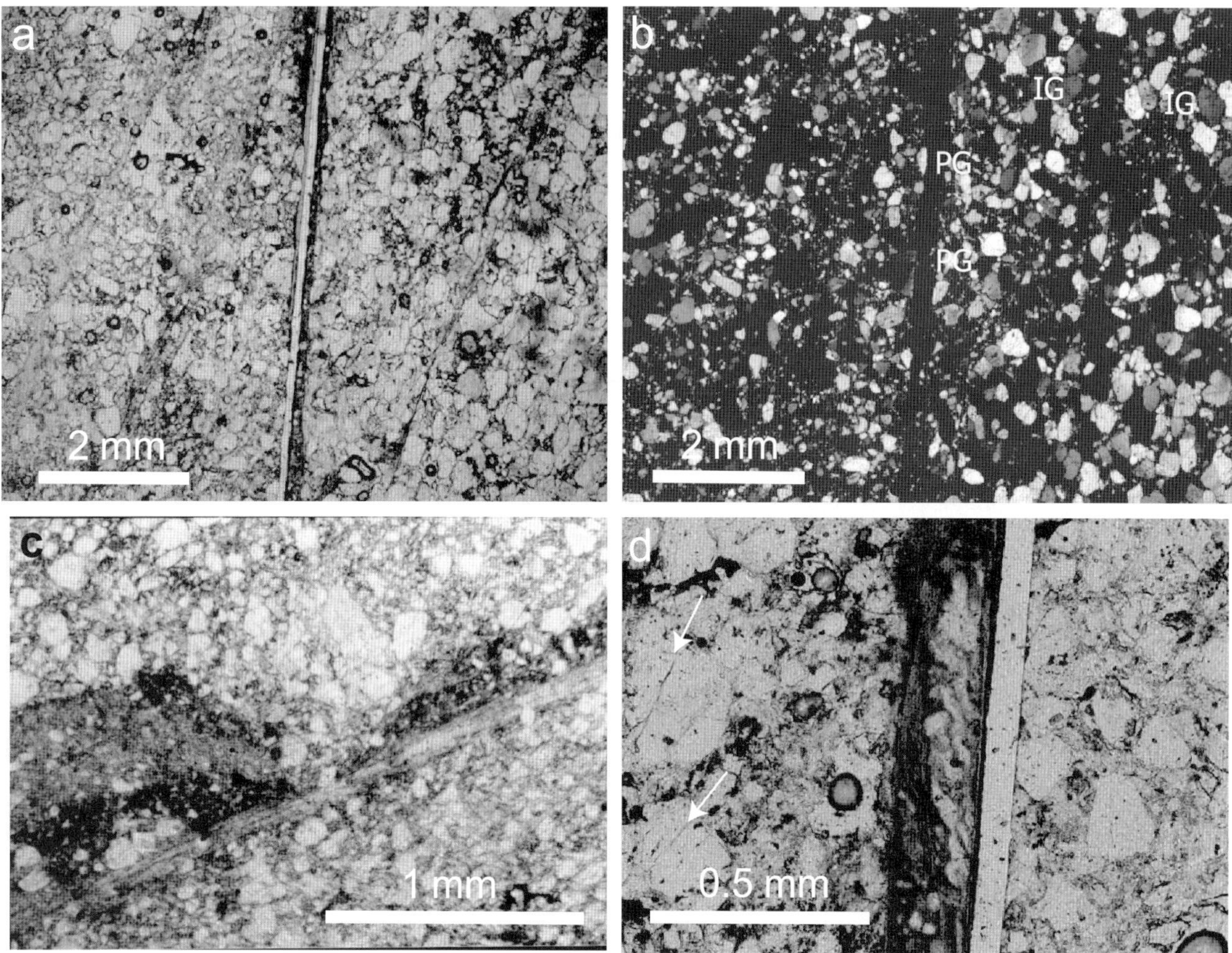

Figure 4. Photomicrographs of fault core microstructures. (a) Plane-polarized image and (b) crossed polars image of the main fault plane at BHF17. The highly crushed bands running at angles to the main fault are in the correct orientation to be Riedel shears formed in the progressively deforming crushed zone. IG marks indented grains formed by pressure solution. PG marks planed-off grains formed by transgranular fracture and movement along the slip surface. This slip surface parted during sample preparation. (c) Plane-polarized image of the fault from a drill core at displacement = 3 m (10 ft). Highly crushed material occurs all along the slip surface (pale, fine grained), but less-comminuted fault core material containing relatively intact grains surrounds this. The thickness of the fault core varies in thickness even on the scale of this thin section. Note that this material has been infiltrated by a dark cement adjacent to the slip surface. (d) Plane-polarized image of the main slip surface at BHF17 (just above a). The slip surface in this sample is not cemented, but note the dark, fine-grained cement in the fault core on the left side of the slip surface. In addition, note the microfractures in the grains on the left side of the image (arrowed).

These Riedel shearlike structures suggest that there has been some bulk movement in fault core material. The aligned microfractures and Riedel shears demonstrate a change in deformation mechanism from grain rolling and crushing in zones of deformation bands to brittle fracturing in the fault core. Grain crushing, microfracturing, and the formation of Riedel shears show that there has been bulk movement in the fault core after the nucleation of the slip surfaces and as further displacement is accumulated.

The traditional model for the growth of deformation-band faults suggests that a zone of deformation bands is a precursor to slip-surface development (Aydin and Johnson, 1978), but processes of slip-surface nucleation and growth are unclear from this model. Our observations suggest that the process of slip-surface growth is intimately linked with the growth and, therefore, geometry of the fault core. Understanding slip-surface growth is therefore crucial to predicting the likely variability of petrophysical properties across faults with sand-sand juxtapositions. Localized increase of granulation does occur in zones of deformation bands before slip-surface initiation, so grain crushing must be an early stage in the formation of slip surfaces. Shipton and Cowie (2001) noted that a high degree of grain crushing is seen ahead of fault tips where the slip surfaces

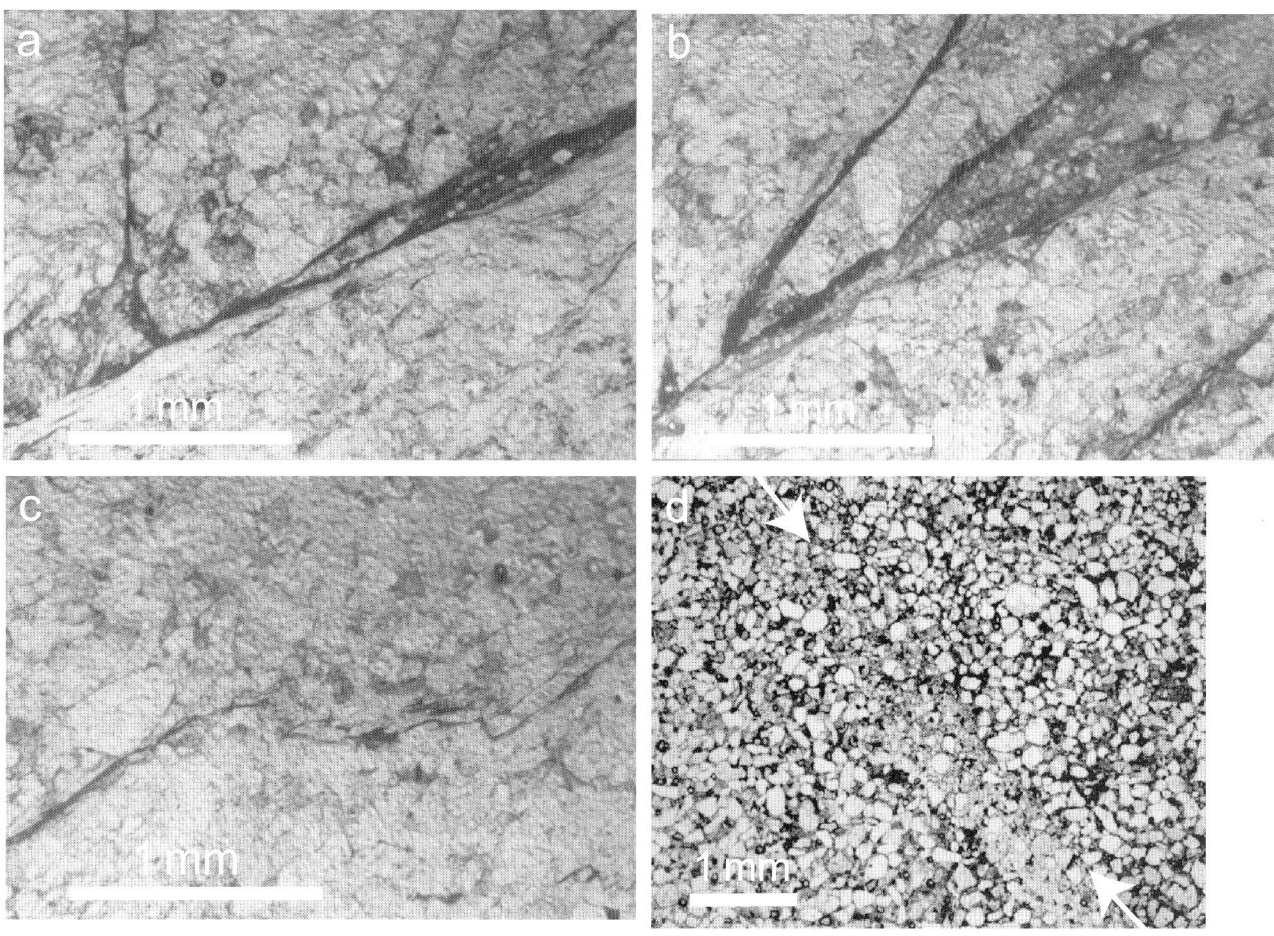

FIGURE 5. Plane polarized images of slip-surface geometries. (a) The main slip surface at BHF8. Surface bifurcates on the right side of the image. Note that the second slip surface joins at a high angle from the top. (b) Main fault plane slip surface at BHF8. Second slip surface joins at a lower angle from the top right, with increased grain crushing between these surfaces, forming a pod of fault core. (c) Linkage of two slip-surface segments away from the main fault slip surface at BHF8. Note how the slip surfaces are curving toward each other. Could further accumulation of displacement on this structure result in the formation of fault core material? (d) A surface in a deformation band cluster at BH20. Offset of eolian cross-beds in the Navajo Sandstone shows that this surface has accommodated 5 cm (2 in.) of slip. The grain crushing is more intense at the top of this deformation band (shown by the arrows). This region has the straight edges characteristic of slip surfaces. Is this a slip surface in the making?

are poorly developed (their figure 11). However, the development of fault core continues after the formation of a throughgoing surface.

EVIDENCE FOR FLUID FLOW IN THE FAULT CORE AND SLIP SURFACES

The fine-grained fault core is likely to be a barrier to fluid flow (Antonellini and Aydin, 1994; Shipton et al., 2002), but the permeability of slip surfaces is more complex. Slip surfaces can be either closely mated (Figure 2a) or planes of parting (Figure 2c), in both outcrop and drill core. The aperture of the slip surfaces is not simply related to surface weathering, and slip surfaces can have high along-fault permeability at depth. The reduction of porosity and permeability in deformation bands and fault core suggests that they would affect fluid flow through the highly porous Navajo Sandstone. However, cement (barite and oxides) is commonly localized along slip surfaces and in places has penetrated into the fault core (Figure 4c). These cements are sparse elsewhere in the deformation-band faults or host rock. Indented grain boundaries show that pressure solution has been active (Figure 4b). From CL observations, it can be seen that some fractured grains have been cemented with quartz after crushing, giving rise

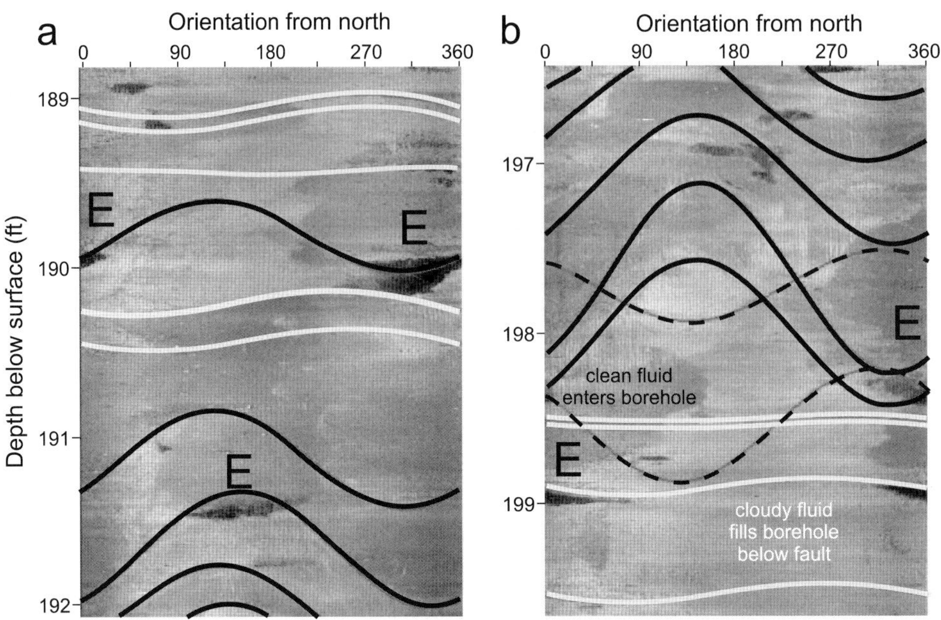

Figure 6. Downhole video image logs of a borehole drilled through the Big Hole fault where the displacement = 5 m (16 ft). Depths in feet on the *y*-axis, orientation from north on the *x*-axis. Black holes are borehole breakouts oriented north–south (E symbols). Note also that the vertical and horizontal scales on the images are not 1:1, elongating the breakouts. White lines are bedding; black lines are synthetic deformation bands or zones; and dashed lines are antithetic deformation bands. Cloudy fluid that has cuttings flour in it is pale. (a) At the top of the fault, this cloudy fluid is replaced by clear water coming into the borehole from the breakouts. More clean fluid is coming into the borehole along deformation bands further up the hole. Near the base of the fault (b), clean fluids are coming in from the breakouts along deformation bands at the bottom center and top left and right of the image.

to unusually shaped grains in thin section. These observations show that fluid has, at some time, flowed in the fault core despite its present-day low porosity and permeability.

Fluid flow associated with the Big Hole fault is observed on downhole video image logs that were run in boreholes drilled through the fault (Figure 6). The throw on the fault at this location is 5 m (16 ft). On downhole video logs, the visual difference between cloudy fluids with suspended solids and clear groundwater fluids can be viewed directly. As seen in Figure 6b, the image below 199 ft (61 m) is cloudy, with static borehole fluids filled with drill-bit "flour," but clean fluids move out of the formation at the base of the fault, and the image clears up noticeably above this point. It appears that the fluid is moving into the borehole along a bedding surface intersecting the well just above the fault core, but further fluid moves into the borehole from the base of the fault. Figure 6a, by comparison, shows a plume or column of cloudy material moving into the borehole from the top of the fault, at about 190 ft (58 m); other plumes also emanate from within the fault zone. Thus, the groundwater flow in the area is strongly impacted by both the highly permeable slip surfaces and the impermeable fault core.

FAULT CORE THICKNESS MEASUREMENTS

To characterize variations in fault core thickness along the strike of the fault, a tape was laid parallel to the fault surface, and the thickness of the fault core was measured at 30-cm (12-in.) intervals with 457 measurements in total. The number of slip surfaces in the fault core was also collected at each 30-cm (12-in.) step. These measurements were made at six outcrops along the fault strike at locations with increasing displacement (Table 1). This allows us to determine the relationship between fault core thickness variations and displacement. If more than one slip-surface strand occurs at an outcrop, the fault core thickness was measured along all strands.

Fault core thickness is highly variable (Figure 7). Along a single fault strand, the thickness can range from almost zero to 20 or 30 cm (8 or 12 in.) within a few meters (Figure 3a). The maximum thickness is typically 15–20 cm (6–8 in.) but can be as high as 36 cm (14 in.) (location BHF24). The average fault core thickness across all sites was 6.1 cm (2.4 in.), with a standard deviation of 5.5 cm (2.2 in.). The highest average fault core thickness at an individual site is seen at BHF24 (Figure 7). At

Table 1. Summary of the fault core thickness data, locations given in order of distance along the Big Hole fault from the eastern fault tip (Figure 1). The total length of transects at each location includes the length of multiple strands. The standard deviation of the average thickness values is given in parentheses.

	Main Fault Offset (m)	*Outcrop Style*	*Total Transect Length (m)/ Number of strands*	*Number of Thickness Measurements*	*Average Fault Core Thickness (cm)*	*Maximum Fault Core Thickness (cm)*
BHF8	8.4	cliff	12.9:2	46	4.75 (2.75)	14
BHF17a	16.5 (<3)	base of wash	24.9:1	81	6.15 (5.42)	23
BHF17b	16.5 (>14)	base of wash	60.9:8	220	4.45 (3.93)	21
BHF20	19.9	base of wash	14.1:1	45	6.34 (4.23)	15
BHF24	23.9	inclined 30–45°	17.4:2	55	13.59 (5.53)	36
BHF23	23.0	cliff 45–90°	1.0:1	10	2.89 (1.87)	5.5

this outcrop, the fault core is composed of a yellow or pink poorly indurated gouge. It is not clear if this change in fault core character is related to the nature of the exposure (inclined slope, prone to weathering) or to the nature of the host rock (i.e., different eolian subfacies or proximity to the subaerial unconformity at the top of the Navajo Sandstone). Alternatively, it could be caused by a change in fault zone processes at higher displacements of the fault. However, the BHF23 outcrop, which has an almost equivalent amount of slip and which is at the base of a cliff almost directly below the BHF24 outcrop, contains the regular resistant fault core material.

Neither the maximum nor the average fault core thickness at each location correlates with the displacement on the fault (Figure 7). The BHF17 outcrop has two strands, and the total offset is partitioned on the two strands in such a way that the northern strand (BHF17a) has more than 3 m (10 ft) of throw, and the southern strand (BHF17b) has less than 14 m (46 ft) of throw (Figure 7). If these two sites are plotted as if both strands had 16.45 m (53.96 ft) of throw (with no account for the offset partitioning), the correlation between throw and average fault core thickness would be even less significant.

The fault core thickness frequency plot for the data from all locations (Figure 8a) shows a left-skewed distribution with a peak at 5-cm (2-in.) thickness. With the exception of BHF24, plots for each location (Figure 8b) show a similar left-skewed distribution, even where many substrands were measured (location BHF17b). No consistent change is present in fault core thickness frequency distribution with displacement. The largest number of thicker fault cores is seen at localities with both the highest and lowest displacement. Fault core thickness also does not correlate with the number of slip surfaces seen in the fault core (Figure 9). Where several strands to the fault zone exist, subsidiary strands (i.e., not the main throughgoing strand) can have locations where fault core exists as much as 9.5 cm (3.7 in.) thick but with no slip surface.

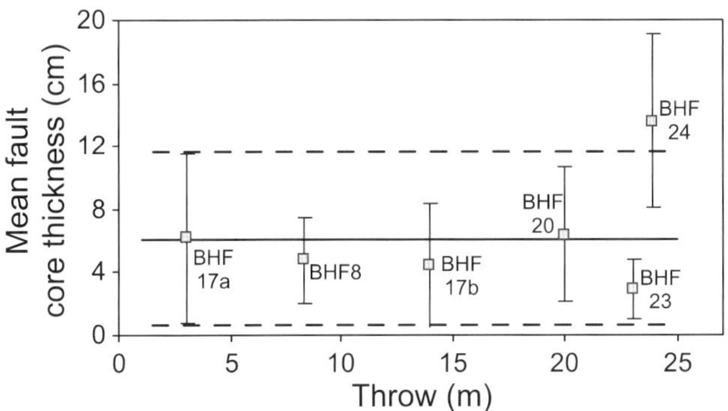

FIGURE 7. Mean fault core thickness vs. fault throw. The bars represent one standard deviation. The heavy line is the average for the entire data set, and the stippled lines are one standard deviation. No consistent change of mean fault core thickness occurred with increasing throw.

DISCUSSION

Evolution of Fault Core and Slip Surfaces: Preliminary Model

The model for the initiation of slip surfaces put forward by Aydin and Johnson (1978) did not distinguish whether the increased amount of crushing around slip surfaces was a precursor to slip-surface formation or occurred as displacement was accumulated (Figure 10a). The observations of fault core and slip-surface geometry presented here suggest that, for the Big Hole fault, a network of deformation bands does develop prior to the accumulation of slip on individual slip surfaces, and that localized intense grain crushing occurs before the nucleation of slip surfaces (Figure 10b). The

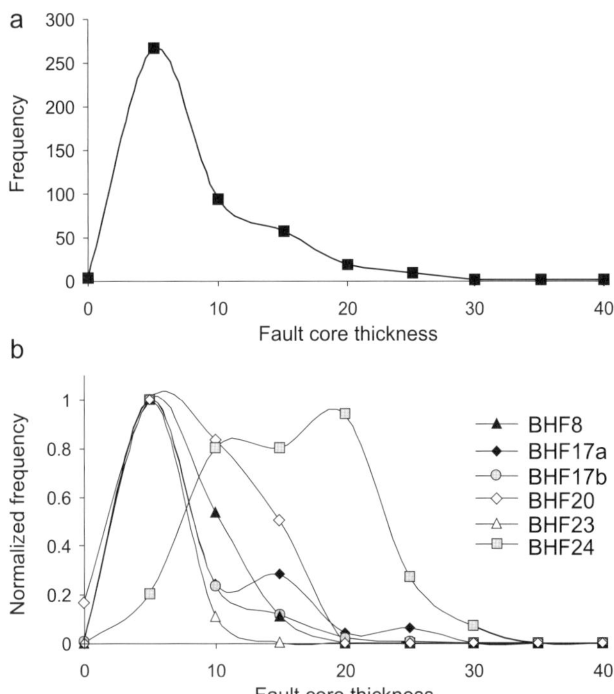

FIGURE 8. Fault core thickness frequency plots. (a) Global fault core thickness frequency plot showing a skewed distribution; (b) normalized fault core thickness frequency curves for each locality.

contrast in material properties between the highly comminuted fault core and host rock helps to localize slip surfaces at the boundaries of the low-porosity and, hence, denser fault core zone (Figure 10c). The main slip surface is always throughgoing, but on slip surfaces away from main faults, there can be gaps where no surfaces are present but where fault corelike material is developed (Figure 10d). However, the fault core continues to develop after localization of a throughgoing slip surface.

The critical parameters for the nucleation of slip surfaces have not been ascertained but could include critical strain, number of deformation bands, deformation-band zone thickness, and setting along the fault. The change of deformation mechanisms from grain rolling and crushing in zones of deformation bands to brittle fracture in the fault core is likely to be controlled by the drastic reduction in porosity in the fault core (Shipton and Cowie, 2001). Faults that form in low-porosity sandstones deform by brittle fracturing without grain rolling (Anders and Wiltschko, 1994; Vermilye and Scholz, 1998; Steen and Andresen, 1999), and this is also the case for continued deformation in the fault core material. This change in deformation mechanism results in a change of permeability, with the deformation of high-porosity host rock reducing permeabilities and the deformation of low-porosity rock (fault core) enhancing permeabilities (microfractures). Note that along the Big Hole fault, more than one of these mechanisms could be active at the same time.

Fault core is locally thicker where kinematic incompatibilities occur at intersections between slip surfaces (Figure 10b). In cross section, triangular areas of the fault core are localized at junctions of synthetic and antithetic slip surfaces or deformation-band clusters. In plan view, elongate lozenge-shaped pods of fault core occur along the main slip surfaces, suggesting that the synthetic and antithetic slip surfaces are not entirely strike parallel and have some curvature or corrugation. Antonellini and Aydin (1995) and Tindall and Davis (1999) show that deformation-band faults typically have thick zones of deformation bands at stepovers between fault segments. The thickness variations observed here may reflect some of these early-formed variations in fault structure (e.g., Figures 5c, 10b).

These observations emphasize the three-dimensional nature of the network of slip surfaces and show that predicting the thickness of fault core and, therefore, permeability at an individual location is complex. The thickness distribution of the fault core at the Big Hole fault is independent of the displacement on the main fault and the number of slip surfaces. The frequency of deformation bands and the number of slip surfaces are also independent of displacement. The width of the entire damage zone does seem to scale with displacement at the Big Hole fault (Shipton and Cowie, 2001) and at other deformation-band faults (Knott et al., 1996; Beach et al., 1997, 1999; Myers, 2004; Fossen and Hesthammer, 2000). Antonellini and Aydin (1995) demonstrated that a difference in microstructures exists between deformation bands in different host rocks; it is unclear to what extent this is true for slip surfaces and fault core. Local stress conditions, deformation history, and host rock variation should be considered when investigating the geometry and scaling of damage zones and fault core around deformation-band faults.

The Effect of the Fault Core on Single-phase Fluid Flow

The effect that a fault-related reduction of porosity and permeability has on fluid flow depends on the number of fluid phases in the rock. Reservoir simulators that incorporate faults either assign flow properties to a series of elements that represent the faults or, more commonly, have the fault modeled as a surface between two grid blocks and assign a single flow property value to the surface. By averaging material properties of the fault rock that affect the flow of either single- or two-phase flow, we can calculate an effective property that is constant on the scale of an individual grid block (Manzocchi et al., 2002).

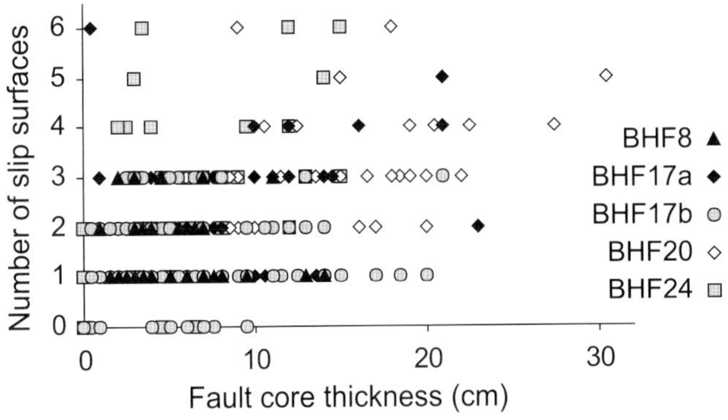

Figure 9. Fault core thickness vs. number of slip surfaces in or around the fault core. Each symbol represents data from a different outcrop. No correlation of fault core thickness with number of slip surfaces is present.

In single-phase flow (equivalent to flow through saturated rock [Manzocchi et al., 2002]), the controlling parameters are fault rock permeability, k_f, and fault thickness, t_f. Flow simulations commonly use a transmissibility multiplier, T, to represent the effect of fault rock on fluid flow between two grid blocks relative to host rock grid block permeabilities. Manzocchi et al. (1999) showed that for the case where grid blocks adjacent to the fault are the same size, L, and the matrix permeability of the grid blocks on either side of the fault, k_m, is the same, T is a function of fault thickness, t_f, and permeability, k_f, such that

$$\frac{1}{T} = 1 + \frac{t_f}{L}\left(\frac{k_m - k_f}{k_f}\right) \quad (1)$$

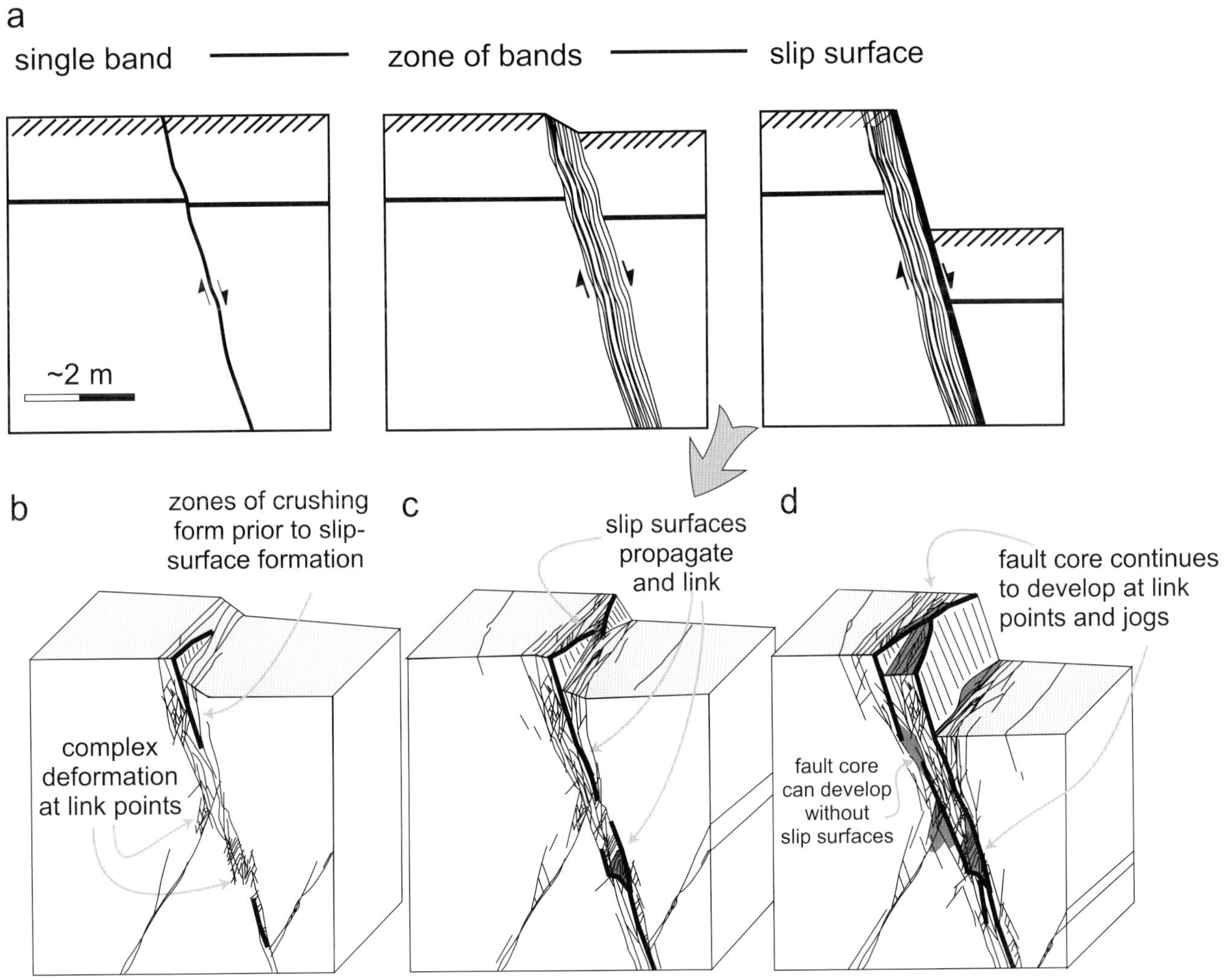

Figure 10. (a) Model of Aydin and Johnson (1978) modified to show development of slip surface and fault core. (b) Slip surfaces nucleate at local points of high-intensity grain crushing. (c) Slip surface segments propagate through developing zone of bands to link and eventually form throughgoing surface. (d) As displacement increases along a hroughgoing slip surface, further development is focused at kinematically incompatible linkage points. A thin veneer of highly comminuted fault core material lines each of the slip surfaces but is too narrow to be represented on these figures.

If we neglect the effect of faults in the damage zone, we can use this equation to examine the sensitivity of T to fault core thickness along the Big Hole fault. Damsleth et al. (1998), in a study of the effect of subseismic faults in the Njord field, Norway, calculated that faults with T values of less than 0.0005 have a sealing effect. If we use host rock permeability of 600 md (Shipton et al., 2002), fault core permeability of 0.007 md (Antonellini and Aydin, 1994), and an arbitrary grid block length of 15 m (49 ft) (see Shipton et al., 2002, for the effect of varying grid block size), then $T = 0.0005$ corresponds to fault core thicknesses above 35 cm (14 in.), close to the maximum thickness we measured. However, if the fault permeability is as low as 0.0001 md, or if host rock permeability was as high as 4000 md, the critical sealing thickness would be 5 cm (2 in.), close to the average measured fault core thickness.

This approach assumes a constant fault core thickness at the grid block scale; however, the along-strike variability of the fault core thickness is pronounced from meter to millimeter scales. The majority of reservoir simulators are not able to represent such fine-scale variability and rely on upscaling to simplify the transmissibility structure of the fault. Manzocchi et al. (1999) suggest that for situations where adjacent grid blocks have the same host rock and fault zone permeabilities, the area-weighted harmonic average of fault zone thickness (t_f) can be used in calculations of transmissibility multipliers, such that

$$t_f = A\left[\sum_{n=1}^{n=i} \frac{A_i}{t_{fi}}\right]^{-1} \quad (2)$$

where A is the total area of the fault surface, which is divided into sub-areas (A_i) with fault thickness (t_{fi}).

We calculate the area-weighted harmonic average of fault zone thickness for our data by assuming a unit depth of fault surface (the length-weighted harmonic average). The harmonic average is then used to calculate transmissibility for each location along the Big Hole fault (Table 2). This analysis shows that if the geometry of deformation bands around the main fault provides a pathway for fluid flow to the fault core, then a fault that juxtaposes sand against sand is unlikely to retard single-phase flow across the fault surface.

For deformation-band faults in high-permeability host rocks, the along-strike geometry and thickness variations of the fault core become important in determining if a fault acts as a seal. A low fault core thickness "hole" in an otherwise sealing fault core could provide a pathway for fluid flow through the fault. From our model of fault core development at the location of linkage between both coplanar and antithetic fault segments, we might expect the number of holes along a fault surface to decrease if the number of antithetic fault strands increases around higher displacement faults. It is of particular importance to quantify any scaling of fault permeability with offset at sand-sand juxtapositions, because this is commonly presumed by both geologists and reservoir engineers (Heinemann et al., 1998; Manzocchi et al., 1999).

The Effect of the Fault Core on Two-phase Fluid Flow

The main parameter controlling fault seal in two-phase flow systems is capillary entry pressure. Assuming that a fault is sealing by capillary pressure, the hydrocarbon column height that a fault can support, h, can be related to the radius of pore throats in the fault rocks, R, by

$$h = \frac{2\gamma}{Rg(\rho_w - \rho_c)} \quad (3)$$

where γ is the dynamic viscosity; ρ_w is the density of water; ρ_c is the density of oil; and g is gravitational acceleration. The empirical equations of Pittman (1992) can be used to estimate the pore-throat radius corresponding to the capillary entry pressures of fault rocks

Table 2. Calculated transmissibilities for each location along the Big Hole fault using the area-weighted harmonic average for the fault core thickness and assuming host rock grid blocks of 15-m (49-ft) length. The harmonic average is calculated for only the throughgoing strand at each location. Transmissibility is calculated for extreme values of host and fault permeability. The fault is assumed to leak if the transmissibility is more than 0.0005 (Damsleth et al., 1998).

	t (cm)	*T (k_m = 600; k_f = 0.007)*		*T (k_m = 600, k_f = 0.001)*		*T (k_m = 3000, k_f = 0.007)*	
BH8	1.91	0.0091	leak	0.0013	leak	0.0022	leak
BH17a	2.86	0.0061	leak	0.0009	leak	0.0015	leak
BH17b	3.16	0.0055	leak	0.0008	leak	0.0013	leak
BH20	7.00	0.0025	leak	0.0004	*seal*	0.0006	leak
BH24	9.71	0.0018	leak	0.0003	*seal*	0.0004	*seal*
BH23	1.23	0.0141	leak	0.0020	leak	0.0034	leak

based on their porosities and permeabilities. Using porosities of the fault core between 1 and 5% (Shipton and Cowie, 2001) and permeability of 0.007 md (Antonellini and Aydin, 1994), we calculate the pore aperture radius corresponding to the displacement pressure as 0.24–0.13 μm. These values correspond to hydrocarbon column heights of 103–191 m, respectively.

Data on the two-phase petrophysical properties of deformation-band faults are sparse. Gibson (1998) and Ogilvie and Glover (2001) measured values of *R* between 0.18 and 13.3 μm, corresponding to hydrocarbon column heights of 137 and 2 m, respectively. Knott (1993) found that 30% of sand-sand juxtapositions formed a seal in the North Sea. The hydrocarbon column heights measured by Gibson (1994) were approximately proportional to fault throw for sand-sand juxtapositions, with throws as much as 75 m (246 ft) (his figure 6), suggesting a relationship between fault offset and the sealing potential of the faults. These data suggest that deformation-band faults can be a barrier to two-phase fluid flow in hydrocarbon reservoirs, but that the petrophysical properties of the fault core are variable. Our data show that all slip surfaces are surrounded by a poorly sorted, fine-grained (and, therefore, small pore-throat size) fault core. Even if its thickness is very small, this should provide a capillary seal to two-phase flow. There does not seem to be an increase in grain crushing (producing a decrease in pore-throat size) with accommodation of slip, and there is some evidence for brittle fracture of the fault core occurring with increased slip, which would drastically decrease the sealing capacity of the fault zones. Further studies of the grain size distribution, pore-throat size, and capillary entry pressure of fault core material are necessary to fully quantify two-phase flow parameters for faults in sandstone.

The above analyses show that if a fluid pathway to the fault core exists through the damage zone, and the fault core contains open holes with little or no fault core material along them, then a fault that juxtaposes sand against sand is unlikely to significantly retard flow across the fault surface. Faults in the North Sea have a diagenetic component that helps them seal, but minimal diagenesis is present along the Big Hole fault. Much more petrophysical data need to be collected from deformation-band faults, and ideally, these data should be calibrated to cross-fault hydrocarbon column heights in reservoirs with known total hydrocarbon height (Fisher and Knipe, 2001; Fisher et al., 2001).

Flow along Slip Surfaces

The presence of minimal cements in the fault core and the observation of flow along slip surfaces in the image logs (Figure 6) suggest that slip surfaces have been and, in places, continue to be high-permeability conduits for fluid flow along the fault surface. Matthai et al. (1998) in Arches National Park, Utah, and Steen and Andresen (1999) in west Greenland have also described slip surfaces in deformation-band faults that contain neoformed carbonate cement. Matthai et al. (1998) use these observations to justify very high permeability values for slip surfaces in their flow models. The actual permeability values that would be appropriate to use for slip surfaces in flow models will vary depending on a large number of factors that are likely to be site specific and include the aperture and tortuosity of the fracture, geometry and linkage of fracture segments, and orientation with respect to applied stress.

High-permeability slip surfaces in this study are always surrounded by low-permeability and high-inferred capillary entry pressure fault core. Therefore, they are likely to represent barriers to cross-fault fluid flow, unless the fault core material is breached by fractures that develop during continued slip on the fault, at which point they will act as flow conduits. The majority of slip surfaces run parallel to the fault, so the predominant effect of open slip surfaces will be to conduct fluid parallel to the fault. Slip surfaces commonly cut through the fault core in the Big Hole fault (Figure 4) and, therefore, may act as pathways for across-fault flow (Jourde et al., 2002).

CONCLUSIONS

The Big Hole fault, Utah, is a deformation-band fault consisting of a fault core, damage zone, and surrounding high-porosity host rock (Navajo Sandstone). Similar to other faults in this lithology, the primary brittle deformation structures are deformation bands that dominate the damage zone. The fault core consists of deformation bands that are very densely spaced and are characterized by highly crushed grains and very little remaining pore space. This fault core has a reduced grain size and wider grain size distribution than the host rock and will therefore have reduced permeabilities and pore-throat sizes.

Fault core is wider in places where deformation bands or slip surfaces link, forming pods that are triangular-shaped cross sections and lozenge-shaped in plan view. Because slip surfaces are discrete planes along which significant slip can occur, they may represent pathways for fluid flow in a fault zone that otherwise has a very low permeability because of the presence of deformation bands and fault core. Slip surfaces nucleate in small patches at a relatively early stage in the development of a zone of deformation bands at points where grain crushing is locally increased. They then propagate and link up to form an anastomosing network. The development of fault core, the growth and linkage of slip surfaces in three dimensions, and the nature of these slip surfaces at depth (open or closed) have important implications for understanding the structural and permeability architecture of faults in this lithology. The Big Hole fault has relatively little diagenesis associated with it, and faults with only mechanical damage may be relatively poor seals compared with cemented cataclastic faults.

The thickness of the fault core is one of the primary controls on the bulk transverse permeability of these fault zones. Our data of fault core thickness along strike suggest that fault core thickness is highly variable, and that fault core thickness and thickness frequency distributions are not simply correlated to a single parameter, such as displacement or number of slip surfaces. Further work is necessary to understand the predictability of fault permeability with easily measurable parameters. To incorporate this fault zone element into accurate simulations of fault zone permeability, it is essential that a large database of fault core architecture and permeability properties be collected for faults in a given setting. Gathering data from several faults in different host rocks and fault displacements will maximize the usefulness of this database and allow any underlying relationships to be characterized fully.

ACKNOWLEDGMENTS

Comments from Jennifer Wilson, Rasoul Sorkhabi, and an anonymous reviewer substantially improved this chapter. Funding for this work was provided by the Office of Basic Energy Sciences–Department of Energy grants DE-FG03-00ER15042 and DE-FG03-95ER14526 and by Big Hole Drilling Project sponsors: ARCO and ARCO Alaska, Enterprise Oil, Exxon, Japan National Oil Corporation (presently Japan Oil, Gas and Metals National Corporation), Mobil, Schlumberger-Doll Research, Shell, and Statoil. Hoda Sondossi assisted in collecting the field data.

REFERENCES CITED

Anders, M. H., and D. V. Wiltschko, 1994, Microfracturing, palaeostress and the growth of faults: Journal of Structural Geology, v. 16, p. 795–815.

Antonellini, M., and A. Aydin, 1994, Effect of faulting on fluid flow in porous sandstones: Petrophysical properties: AAPG Bulletin, v. 78, p. 355–377.

Antonellini, M., and A. Aydin, 1995, Effect of faulting on fluid flow in porous sandstones: Geometry and spatial distribution: AAPG Bulletin, v. 79, p. 642–671.

Antonellini, M. A., A. Aydin, and L. Orr, 1999, Outcrop aided characterisation of a faulted hydrocarbon reservoir: Arroyo Grande oil field, California, U.S.A., *in* W. C. Haneberg, P. S. Mozley, C. J. Moore, and L. B. Goodwin, eds., Faults and subsurface fluid flow: American Geophysical Union Geophysical Monograph 113, p. 7–26.

Aydin, A., and A. M. Johnson, 1978, Development of faults as zones of deformation bands and as slip-surfaces in sandstones: Pure and Applied Geophysics, v. 116, p. 931–942.

Beach, A., J. L. Brown, A. L. Welbon, J. E. McCallum, P. Brockbank, and S. Knott, 1997, Characteristics of fault zones in sandstones from NW England: Application to fault transmissibility, *in* N. S. Meadows, S. P. Trueblood, M. Hardman, and G. Cowan, eds., Petroleum geology of the Irish Sea and adjacent areas: Geological Society (London) Special Publication 124, p. 315–324.

Beach, A., A. I. Welbon, P. J. Brockbank, and J. E. McCallum, 1999, Reservoir damage around faults: Outcrop examples from the Suez Rift: Petroleum Geoscience, v. 5, p. 109–116.

Damsleth, E., V. Sanglot, and G. Aamodt, 1998, Subseismic faults can seriously affect fluid flown in the Njord field off western Norway— A stochastic fault modelling case study: Presented at the Society of Petroleum Engineers Annual Technical Conference and Exhibition, New Orleans, September 27–30: SPE Paper 49024, 10 p.

Edwards, H. E., A. D. Becker, and J. A. Howell, 1993, Compartmentalization of an aeolian sandstone by structural heterogeneities: Permo-Triassic Hopeman Sandstone, Moray Firth, Scotland, *in* C. P. North and D. J. Prosser, eds., Characterization of fluvial and aeolian reservoirs: Geological Society (London) Special Publication 73, p. 339–365.

Fisher, Q. J., and R. J. Knipe, 1998, Fault sealing processes in siliciclastic sediments, *in* G. Jones, Q. J. Fisher, and R. J. Knipe, eds., Faulting, fault sealing, and fluid flow in hydrocarbon reservoirs: Geological Society (London) Special Publication 147, p. 117–133.

Fisher, Q. J., and R. J. Knipe, 2001, The permeability of faults in siliciclastic petroleum reservoirs of the North Sea and Norwegian continental shelf: Marine and Petroleum Geology, v. 18, p. 1063–1081.

Fisher, Q. J., S. D. Harris, E. McAllister, R. J. Knipe, and A. J. Bolton, 2001, Hydrocarbon flow across faults by capillary leakage revisited: Marine and Petroleum Geology, v. 18, p. 251–257.

Flodin, E. A., L. J. Durlofsky, and A. Aydin, 2004, Upscaled models of flow and transport in faulted sandstone: Boundary condition effects and explicit fracture modelling: Petroleum Geoscience, v. 10, p. 173–181.

Fossen, H., and J. Hesthammer, 2000, Possible absence of small faults in the Gullfaks field, northern North Sea: Implications for downscaling of faults in some porous sandstones: Journal of Structural Geology, v. 22, p. 851–863.

Gibson, R. G., 1994, Fault-zone seals in siliciclastic strata of the Columbus basin, offshore Trinidad: AAPG Bulletin, v. 78, p. 1372–1385.

Gibson, R. G., 1998, Physical character and fluid-flow

properties of sandstone-derived fault zones, *in* M. P. Coward, T. S. Daltaban, and H. Johnson, eds., Structural geology in reservoir characterization: Geological Society (London) Special Publication 127, p. 83–97.

Heinemann, Z. E., G. F. Heinemann, and B. M. Tranta, 1998, Modeling heavily faulted reservoirs: Presented at the Society of Petroleum Engineers annual technical conference and exhibition, New Orleans, September 27–30: SPE Paper 48998, 11 p.

Hesthammer, J., T. E. S. Johansen, and L. Watts, 2000, Spatial relations within fault damage zones in sandstone: Marine and Petroleum Geology, v. 17, p. 873–893.

Hood, J. W., and D. J. Patterson, 1984, Bedrock aquifers in the northern San Rafael Swell area, Utah, with special emphasis on the Navajo Sandstone: State of Utah Department of Natural Resources Technical Publication No. 78, 128 p.

Jourde, H., E. A. Flodin, A. Aydin, L. J. Durlofsky, and X. Wen, 2002, Computing permeabilities of fault zones in eolian sandstone from outcrop measurements: AAPG Bulletin, v. 86, p. 1187–1200.

Knott, S. D., 1993, Fault seal analysis in the North Sea: AAPG Bulletin, v. 77, p. 778–792.

Knott, S. D., A. Beach, P. J. Brockbank, J. L. Brown, J. E. McCallum, and A. I. Welbon, 1996, Spatial and mechanical controls on normal fault populations: Journal of Structural Geology, v. 18, p. 359–372.

Krantz, R. W., 1988, Multiple fault sets and three-dimensional strain: Theory and application: Journal of Structural Geology, v. 10, p. 225–237.

Mair, K., I. Main, and S. Elphick, 2000, Sequential growth of deformation bands in the laboratory: Journal of Structural Geology, v. 22, p. 25–42.

Manzocchi, T., J. J. Walsh, P. Nell, and G. Yielding, 1999, Fault transmissibility multipliers for flow simulation models: Petroleum Geoscience, v. 5, p. 53–63.

Manzocchi T., A. E. Heath, J. J. Walsh, and C. Childs, 2002, The representation of two phase fault-rock properties in flow simulation models: Petroleum Geoscience, v. 8, p. 119–132.

Matthai, S. K., A. Aydin, D. D. Pollard, and S. G. Roberts, 1998. Numerical simulation of departures from radial drawdown in a faulted sandstone reservoir with joints and deformation bands, *in* G. Jones, Q. J. Fisher, and R. J. Knipe, eds., Faulting, fault sealing and fluid flow in hydrocarbon reservoirs: Geological Society (London) Special Publication 147, p. 157–191.

Myers, R., and A. Aydin, 2004, The evolution of faults formed by shearing across joint zones in sandstone: Journal of Structural Geology, v. 26, p. 947–966.

Ogilvie, S. R., and P. W. J. Glover, 2001, The petrophysical properties of deformation bands in relation to their microstructure: Earth and Planetary Science Letters, v. 193, p. 129–142.

Pittman, E. D., 1981, Effect of fault-related granulation on porosity and permeability of quartz sandstones, Simpson Group (Ordovician) Oklahoma: AAPG Bulletin, v. 65, p. 2381–2387.

Pittman, E. D., 1992, Relationship of porosity and permeability to various parameters derived from mercury injection-capillary pressure curves for sandstone: AAPG Bulletin, v. 76, p. 191–198.

Shipton, Z. K., and P. A. Cowie, 2001, Damage zone and slip-surface evolution over μm to km scales in high-porosity Navajo Sandstone, Utah: Journal of Structural Geology, v. 23, p. 1825–1844.

Shipton, Z. K., J. P. Evans, K. Robeson, C. B. Forster, and S. Snelgrove, 2002, Structural heterogeneity and permeability in faulted eolian sandstone: Implications for subsurface modelling of faults: AAPG Bulletin, v. 86, p. 863–883.

Steen, Ø., and A. Andresen, 1999, Effects of lithology on geometry and scaling of small faults in Triassic sandstones, east Greenland: Journal of Structural Geology, v. 21, p. 1351–1368.

Tindall, S. E., and G. H. Davis, 1999, Monocline development by oblique-slip fault-propagation folding: The east Kaibab monocline, Colorado Plateau, Utah: Journal of Structural Geology, v. 21, p. 1303–1320.

Vermilye, J. M., and C. H. Scholz, 1998, The process zone: A microstructural view: Journal of Geophysical Research, v. 103, p. 12,223–12,237.

Walsh, J. J., J. Watterson, A. E. Heath, and C. Childs, 1998, Representation and scaling of faults in fluid flow models: Petroleum Geoscience, v. 4, p. 241–251.

Yielding, G., B. Freeman, and D. T. Needham, 1997, Quantitative fault seal prediction: AAPG Bulletin, v. 81, p. 897–917.

12

Flodin, E., M. Gerdes, A. Aydin, and W. D. Wiggins, 2005, Petrophysical properties and sealing capacity of fault rock, Aztec Sandstone, Nevada, *in* R. Sorkhabi and Y. Tsuji, eds., Faults, fluid flow, and petroleum traps: AAPG Memoir 85, p. 197–217.

Petrophysical Properties and Sealing Capacity of Fault Rock, Aztec Sandstone, Nevada

Eric Flodin[1]

Rock Fracture Project, Department of Geological and Environmental Sciences, Stanford University, Stanford, California, U.S.A.

Martha Gerdes

ChevronTexaco Exploration and Production Technology Company, San Ramon, California, U.S.A.

Atilla Aydin

Rock Fracture Project, Department of Geological and Environmental Sciences, Stanford University, Stanford, California, U.S.A.

William D. Wiggins

ChevronTexaco Exploration and Production Technology Company, San Ramon, California, U.S.A.

ABSTRACT

To refine flow models for sand-dominated fault rock, we present petrophysical data of host and fault rock samples from the eolian Aztec Sandstone, Valley of Fire State Park, Nevada, that has been deformed by strike-slip faults formed by progressive shearing along joint zones. The data include bulk mineralogy, porosity, permeability, grain-size distribution, and mercury-injection capillary pressure measurements of 40 host, fragmented, and fault rock samples. To investigate the impact of shear strain on fault zone properties, three sample localities with average shear strains of 28, 63, and 80 were investigated (25–160-m [82–525-ft] slip). No bulk mineralogical changes caused by fault zone cementation or mineral alteration were detected when comparing host and fault rock. Fault rock permeability is one to three orders of magnitude lower than median host rock permeability. Porosity reductions are less pronounced and show considerable overlap in values between the sample suites. Some

[1]*Present address:* Department of Geosciences, Indiana University Purdue University Fort Wayne (IPFW), Fort Wayne, Indiana, U.S.A.

DOI:10.1306/1033724M853136

fault rock samples appear to have dilated with respect to median host rock porosity. Median grain sizes for fault rock samples range from 3 to 51 μm, which is as much as two orders of magnitude reduction from host rock median grain sizes. There appears to be a lower limit of median grain size of 3 μm for fault rock samples irrespective of average fault shear strain. Fault rock capillary injection pressures range from one to almost two orders of magnitude higher than the host rock equivalent. For standard fluid properties, calculated maximum sealable hydrocarbon column heights range between 10 and 70 m (33 and 230 ft) of gas and 20–120 m (66–400 ft) of oil. These petrophysical data show that faults formed by shearing of joints in high-permeability, sand-prone systems will act as significant barriers to fluid flow during reservoir production and might be capable of sealing small to moderate hydrocarbon columns on an exploration timescale as well, assuming adequate continuity of the fault rock over large areas of the fault.

INTRODUCTION

The petrophysical characteristics of fault rocks are required input data for accurate modeling of fluid flow in faulted reservoirs. Specifically, fault rock petrophysical properties are used in calculations of fault transmissibility (Walsh et al., 1998; Manzocchi et al., 1999) and fault permeability upscaling models (Myers, 1999; Flodin et al., 2001; Jourde et al., 2002) to constrain reservoir performance predictions for transient flow problems. Fault rock capillary properties have recently been introduced to hydrocarbon migration and flow models (Childs et al., 2002; Manzocchi et al., 2002) and are more generally used in exploration to predict fault-seal capillary strength (e.g., Smith, 1966; Watts, 1987; Gibson, 1994; Knipe et al., 1997). Subsurface fault-seal applications benefit from calibration to fault rock petrophysical data to obtain realistic and predictive results, but such data are sparsely available from subsurface studies. In this regard, studies of well-exposed fault zones have improved our understanding of the variety of faulting styles and the details of fault damage zone architectures. Geometric and petrophysical characterizations from outcrop studies aid in understanding exploration risk and evaluating faulted reservoir performance. The primary focus has been given to shale-dominated fault zones, which are recognized to increase fault-sealing capacity on both exploration and production timescales. In contrast, the hydrodynamic behavior of sand-dominated fault zones is less understood and, therefore, not routinely considered as part of fault seal and flow prediction.

Two types of faulting mechanisms are recognized as operating in sandstone: deformation-band faulting (Aydin and Johnson, 1978; Jamison and Stearns, 1982) and sheared-joint-based faulting (Myers, 1999). These processes and the resultant fault architectures are distinct from each other. The petrophysical and flow properties of fault rock related to deformation-band-style faults have received much attention in the literature (e.g., Engelder, 1974; Pittman, 1981; Antonellini and Aydin, 1994; Fowles and Burley, 1994; Fisher and Knipe, 1998; Gibson, 1998; Main et al., 2000; Taylor and Pollard, 2000; Ogilvie and Glover, 2001; Shipton et al., 2002). In contrast, fault rocks associated with sheared-joint-style faults have been investigated only by Myers (1999). In this chapter, we focus on fault rock collected from faults formed by shearing across joint zones through splay jointing and subsequent shearing of splay joints.

Myers (1999), using image analysis techniques (Ehrlich et al., 1984), studied the porosity and grain-size characteristics of host and fault rock samples from the Aztec Sandstone at the Valley of Fire State Park, Nevada. His estimates of fault rock porosity from image analyses ranged between 0.5 and 7%, whereas mean fault rock grain-size estimates ranged between 20 and 40 μm. Based on these data, Myers estimated fault rock permeability using a modified Kozeny-Carmen relationship (Bear, 1972). His calculated permeability estimates ranged between 0.002 and 7 md. Myers (1999) also analyzed several samples using a conventional permeameter and found order-of-magnitude agreement between measured and estimated permeabilities.

In this chapter, we present new petrophysical data from 40 host and fault rock samples of the Aztec Sandstone collected from three outcrop localities that have different average shear strains, corresponding to displacements between 25 and 160 m (82 and 525 ft). We proceed by describing the structural elements that comprise sheared-joint-based faults, focusing on two elements, fragmented rock and fault rock. Maps of the three sample localities document the fault zone architecture and our sampling strategy. We then present mineralogical and petrophysical data, which include porosity and permeability, grain-size distribution, and capillary pressure curves from the fault zone and the host rock. We conclude with a discussion of our findings in relation to fault rock petrophysical evolution and the significance of sand-prone fault zones as barriers to hydrocarbon flow on production and exploration timescales.

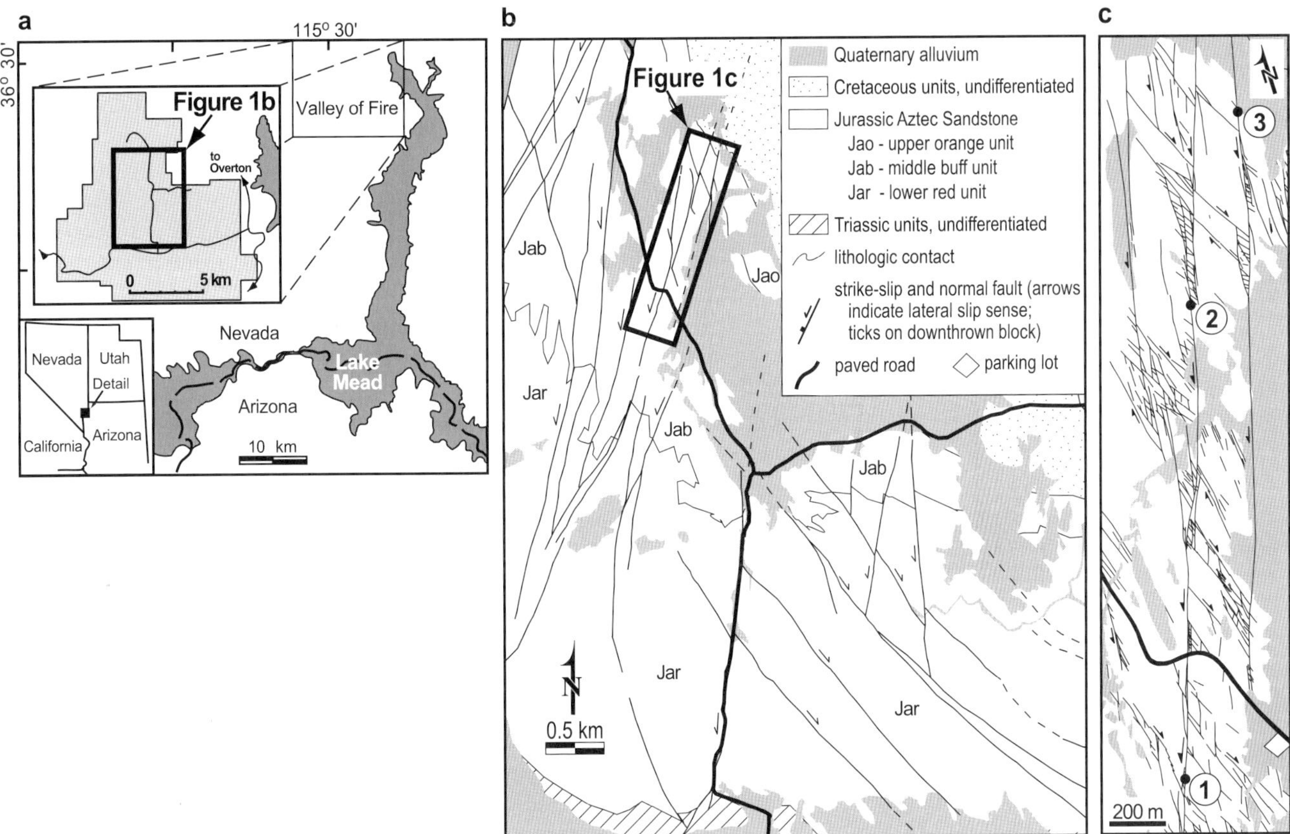

FIGURE 1. (a) Map showing the location of the study area in the Valley of Fire State Park, southern Nevada (modified after Myers, 1999). (b) Geologic map of the greater study area (modified after Myers, 1999; Taylor, 1999). (c) Detailed map showing sample locations 1, 2, and 3 (circled numbers). Northeast-oriented structures are left-lateral strike-slip faults, whereas northwest-oriented structures are right-lateral strike-slip faults. Some faults are labeled with arrows indicating slip direction (modified after Flodin, 2003).

Geologic Setting

This study focuses on predominantly strike-slip faults that occur in eolian Jurassic Aztec Sandstone exposed in the Valley of Fire State Park of southern Nevada (Figure 1). The Aztec Sandstone was deposited in the Early Jurassic in a back-arc basin and was part of a broad eolian erg system that included the Navajo Sandstone of the Colorado Plateau (Blakey, 1989). The Aztec Sandstone is a fine- to medium-grained subarkose to quartz arenite characterized by large-scale tabular-planar and wedge-planar cross-strata (Marzolf, 1983). The Aztec Sandstone porosity values range from 15 to 25%, and permeability values range from 100 to 5900 md (Myers, 1999; Flodin et al., 2003). Within the Valley of Fire, the Aztec Sandstone has a stratigraphic thickness of approximately 800 m (2600 ft) (Longwell, 1949).

Structures in the Aztec Sandstone record two phases of deformation: early thrust faulting related to the Cretaceous Sevier orogeny (Armstrong, 1968) and later strike-slip and minor normal faulting related to Miocene Basin and Range extensional tectonics (Bohannon, 1983). The strike-slip faults we studied formed during the later deformation event (Myers, 1999; Taylor, 1999) and were active between 12 and 4 Ma (Carpenter and Carpenter, 1994; Flodin, 2003). At the initiation of strike-slip faulting, the Aztec Sandstone in the vicinity of the Valley of Fire was buried beneath at least 1.6 km (1 mi) of sediments (based on stratigraphic thicknesses provided by Bohannon, 1983) and possibly by an additional 1–4 km (0.62–2.5 mi) of overlying Sevier-related thrust sheets (Brock and Engelder, 1977).

FAULT ZONE ELEMENTS

Strike-slip faults with a minor normal component in the Aztec Sandstone in the Valley of Fire were formed by a hierarchical process of shearing along preexisting joint zones. For sheared-joint-based faults, the initial shearing of planar joints is followed by the creation of zones of fragmented rock at joint stepovers and bends and newly formed splay joints near the ends of preexisting joints (Myers, 1999; Davatzes et al., 2003; Flodin, 2003). With increasing shear strain, this process is

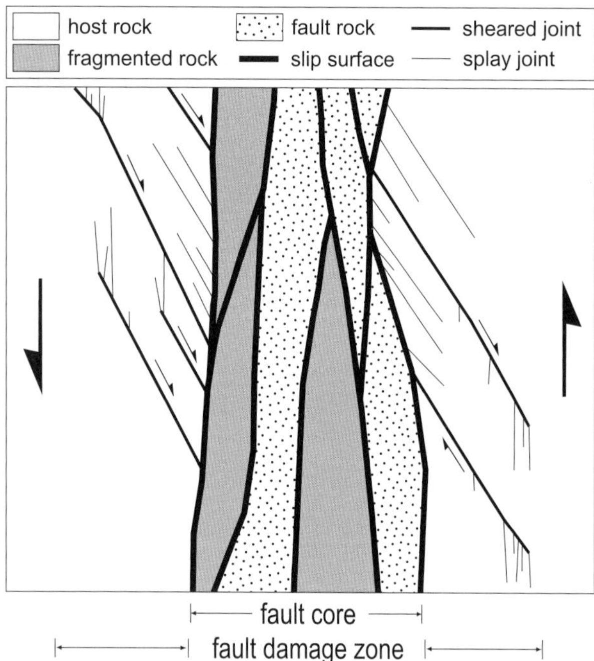

FIGURE 2. Schematic drawing (in map view) of a sheared-joint-based fault zone in sandstone. Arrows indicate slip sense along sheared structures.

progressively repeated as fragmented rock is further crushed to form isolated pockets of fault rock. Eventually, a throughgoing slip surface develops, and the discontinuous fault rock pockets coalesce to form a continuous seam along the fault.

The gross architecture of shear-joint faults can be divided into an inner fault core and an outer fault damage zone (Figure 2) using the terminology of Caine et al. (1996) and Aydin (2000). Structural elements in the fault core consist of sheared joints, deformation bands, fragmented rock, fault rock, and slip surfaces, whereas elements in the damage zone are in the form of splay joints and sheared splay joints (Myers, 1999). Because this study focuses on the flow properties at the scale of 2.5-cm (1-in.)-diameter core plugs, the individual hydrodynamic behavior of structural elements that are smaller than our measurement scale (i.e., joints, sheared joints, deformation bands, and slip surfaces) is not considered here. The characteristics of these distinct structural elements in relation to joint-based faults in sandstone are described by Myers (1999), Taylor (1999), and Aydin (2000). In this chapter, we focus on two elements that volumetrically comprise most of the fault core, fragmented rock, and fault rock, as well as the host rock from which these elements were derived.

In the field, the fault core is defined as a zone of fine-grained fault rock and intensely fragmented pockets of host rock that is divided up and bounded by major slip surfaces (Figure 3). In the study area, fault core widths range from centimeters to meters, whereas damage zone widths range from meters to tens of meters (Myers, 1999). At the scale of fault slip magnitudes at our sample localities (25–160 m; 82–525 ft), the damage zone is more or less symmetrically distributed with respect to the centrally located fault core. Fragmented rock is host rock divided by a network of fractures (joints and small faults) into blocks or clasts of various sizes. These pieces of host rock typically preserve sedimentary bedding. In some cases, fragmented rock is adjacent to the main body of host rock but is separated from it by sheared joints and slip surfaces. In other cases, the fragmented rock is completely isolated from the main host rock body by fault rock (Figure 3). Fragmented rock bodies have dimensions that range between centimeters and meters and are almost always internally deformed by joints, sheared joints, and deformation bands. Brecciated rock is a kind of fragmented rock with strongly angular clasts that are generally detached from the host rock. In contrast to fragmented rock, fault rock visually lacks recognizable sedimentary features. In outcrop, fault rock is very fine grained (as determined with a hand lens) and is commonly colored differently from the host rock (commonly lighter or white). Slip surfaces bound and sometimes cut across individual bodies of fault rock. The slip surfaces are planar in map view and are commonly identifiable by millimeter-wide seams of red alteration staining. Slip surface faces are commonly smooth to the touch and, in many instances, preserve kinematic indicators, such as slickensides and groove marks.

Differences between fragmented and fault rock samples representing various strain levels are apparent at the microscopic scale of observation (Figure 4). The sample shown in Figure 4a was collected from a fairly large and intact body of fragmented rock sandwiched between two bodies of fault rock. Many of the grains in this sample are fractured. However, they preserve their original shape, and nearly all of the primary porosity remains open. Elsewhere in this sample, we observed regions of nearly undeformed host rock as well as zones of localized deformation. Host rock samples collected in the vicinity of a slip surface (Figure 4b) typically show grain fracturing similar in intensity to the fragmented rock sample. However, in places, the primary porosity is filled with secondary pore-filling clay cements. An example of a higher level of strain in fragmented rock is shown in Figure 4c. Here, many of the grains are almost completely fragmented, and much of the primary porosity has collapsed or has been filled by smaller angular clasts; only a few grains remain intact.

Fault rock samples show a much larger reduction in grain size and primary porosity (Figure 4d). Thin sections of fault rocks exhibit areas with varying degrees of grain-size reduction and porosity loss, which could correspond to varying strains. Based on the greatest

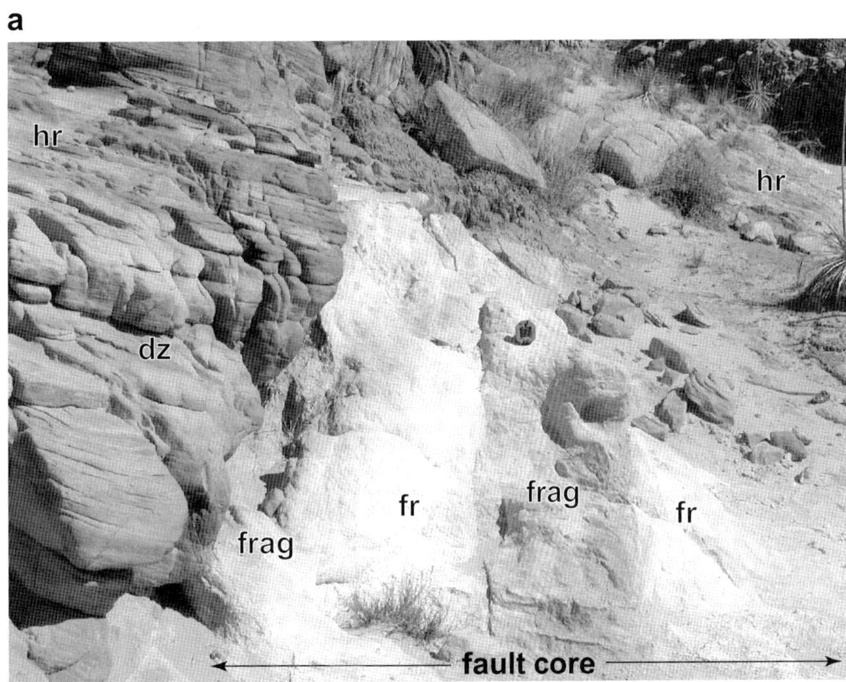

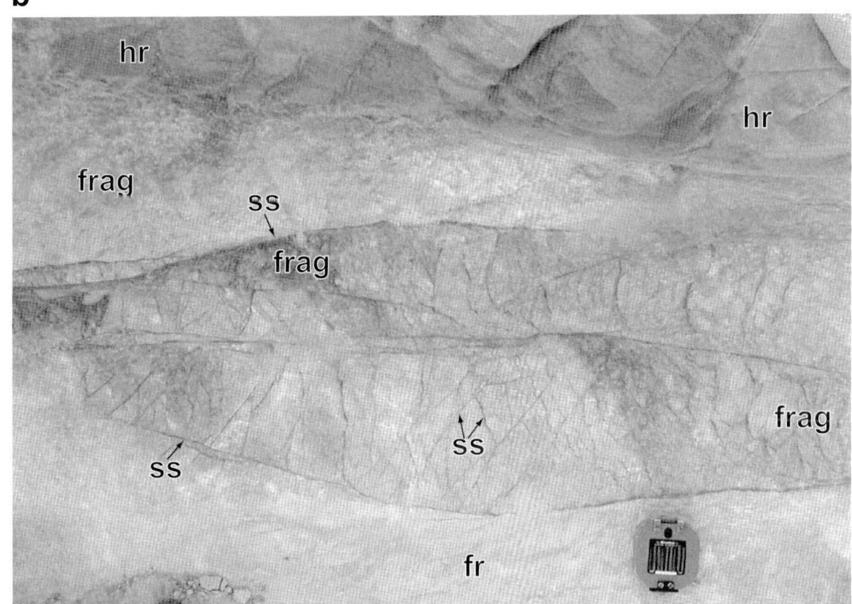

FIGURE 3. (a) Oblique cross-sectional view of a sheared-joint-based fault in Aztec sandstone. About 25 m (82 ft) of left-lateral slip has occurred across the width of the fault zone. View north. (b) Detailed map view of a fault core. dz = damage zone; fr = fault rock; frag = fragmented rock; hr = host rock; ss = slip surfaces. Brunton compass for scale in both photographs. Both photographs were taken in the vicinity of sample locality 2.

reduction in grain size, the region to the left of the slip surfaces (Figure 4d) appears to have accommodated more strain. Here, pore space (black) is nearly absent, and the broken grains are generally equant in shape and lack a preferred orientation. To the right of the slip surfaces, larger pores are evident, and individual grains are slightly more elongate in shape and show weak alignment. Most of the grains shown in Figure 4a–c do not appear to have undergone significant rotation and/or translation, because original grain roundness and outlines of relic grain-coating cements (observed elsewhere in these samples) are preserved. In contrast, the grains in the fault rock sample (Figure 4d) are angular and show evidence for translation in the form of dislocated cleavage in feldspars and spalling of edges on quartz grains. Grain rotation in the fault rocks must surely occur (e.g., Engelder, 1974), but evidence for it in thin section was not detected because of the fragmentation of the original grains and the lack of appropriate offset markers.

SAMPLE LOCALITIES WITH DIFFERENT AVERAGE SHEAR STRAIN

A total of 12 host rock, 8 fragmented rock, and 20 fault rock samples of the Aztec Sandstone in the Valley of Fire were collected in the vicinity of three fault localities (Figure 1c). Sample localities 1 and 2 occur along the same fault, whereas location 3 is located on the next major fault to the east (Figure 1). Total fault slip was estimated to be 25 m (82 ft) for both locations 1 and 2 and 160 m (525 ft) for location 3. Fault core thicknesses for locations 1, 2, and 3 range between 40 and 70, 160 and 175, and 210 and 240 cm (16 and 28, 63 and 69, and 83 and 94 in.), respectively (Figure 5). Total fault rock thicknesses [$\approx$ (fault core thickness) − (fragmented rock thickness)] across these same zones range between 30 and 50, 80 and 100, and 185 and 215 cm (12 and 20, 31 and 39, and 73 and 85 in.), respectively. We calculated the average shear strain (γ) accommodated across the fault core by dividing the

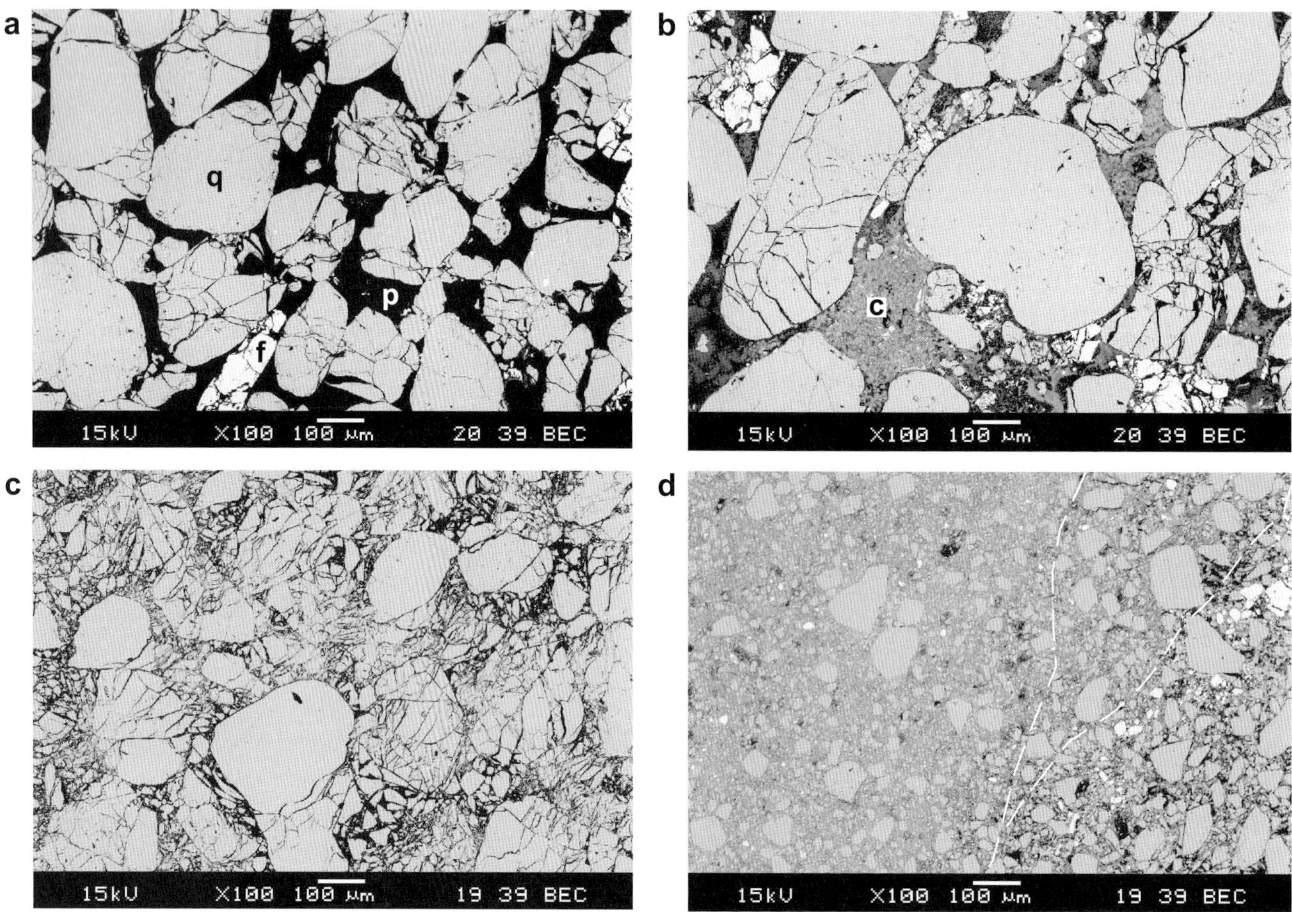

FIGURE 4. Compositional backscattered electron (BEC) images of variously deformed Aztec Sandstone. White minerals are potassium feldspar, light-gray minerals are quartz, dark-gray minerals are clays, and black is pore space. (a) Fragmented rock in fault core showing earliest signs of grain-scale fragmentation (sample 30). Nearly every grain shows varying degrees of fracturing. However, the primary pore space is preserved. f = potassium feldspar; p = pore space; q = quartz. (b) Host rock collected adjacent to the fault core (sample 36). Note that many of the grains are fractured, and that much of the pore space is occupied by clay minerals. c = clay. (c) Host rock collected adjacent to the fault core showing signs of severe grain-scale fragmentation, collapse of pores, and loss of primary porosity (sample 22). (d) Well-developed fault rock characterized by drastic grain-size reduction and a complete loss of primary porosity (sample 49). Dashed lines are approximate locations of slip surfaces. Note the variation of pore space (black) between the left and right sides of the image.

estimated fault slip magnitude by the average fault rock thickness. Using this relationship, we obtained average shear strains of 63, 28, and 80 for locations 1, 2, and 3, respectively.

Sample Collection Methods

Samples of seismic-scale faults are not commonly recovered from subsurface drill core. Therefore, exhumed large-offset faults provide an opportunity to study petrophysical and geometrical characteristics of fault zones. Our study area has an arid desert climate and a general lack of vegetation, both of which serve to minimize the effects of surficial weathering processes on outcrop samples. A series of samples was collected from the three localities along transects from undeformed host rock through the damage zone and fault core. Host rock samples were collected at different distances from the main fault in an attempt to detect fault-related damage of host rock away from the fault. Distances listed for the host rock in Table 1 (under column labeled Type) are made with reference to the nearest fault core and associated bounding slip surface. A few host rock samples were collected in the fault damage zone immediately adjacent to the fault core. All samples were carefully chosen to be the least affected by very near surface-weathering processes. Approximately 10 cm (4 in.) of surface material was removed from each sample area prior to collection (e.g., Dinwiddie et al., 1999). In most cases, hand samples were carefully chiseled from the outcrop. However, because of the difficulty of collecting small samples of

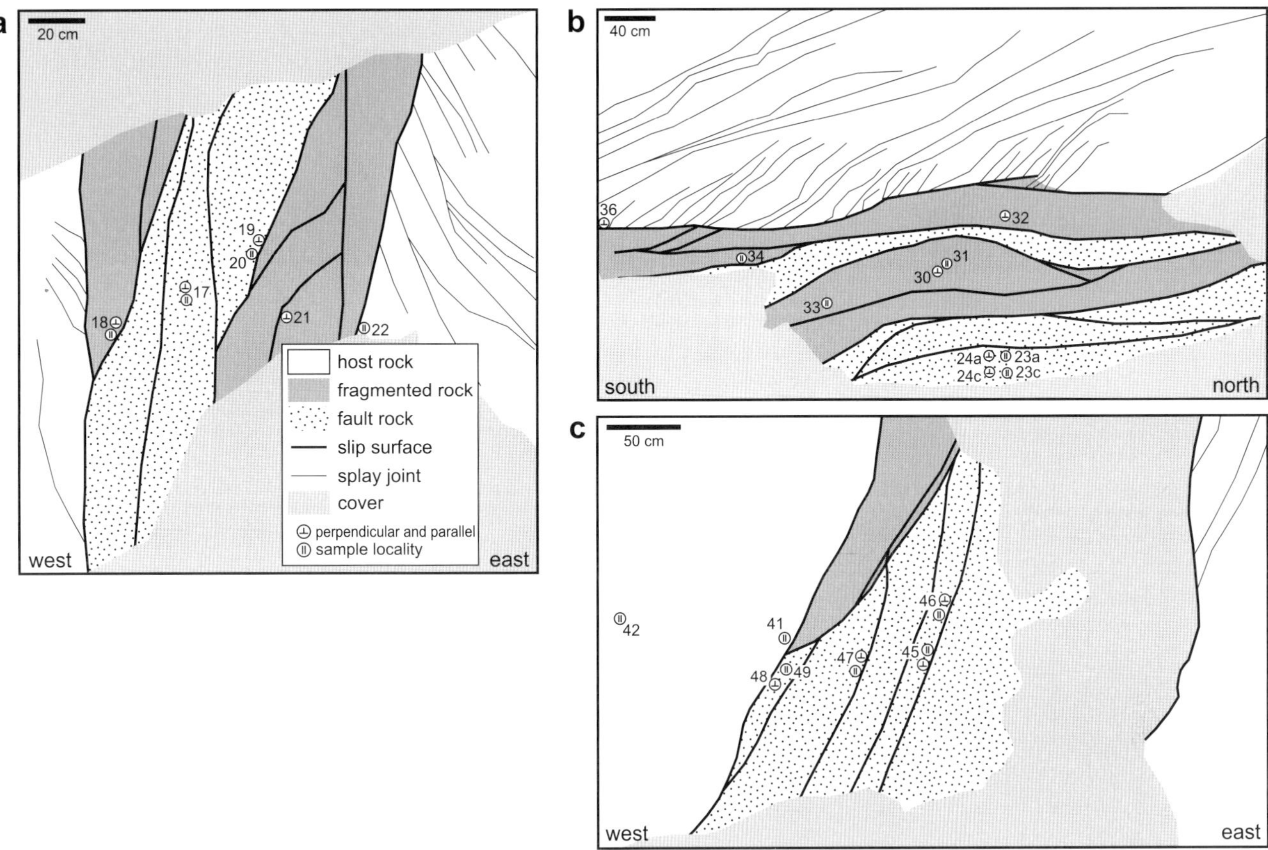

FIGURE 5. Fault zone maps for each sample locality. Note that some samples were collected outside of the areas shown in the maps. (a) Location 1 (cross section view north); (b) location 2 (oblique map view west); (c) location 3 (cross section view north).

the sometimes friable fault rock, some bulk samples as large as 6000 cm^3 (366 in.^{3}) were collected. Core plugs approximately 2.5 cm (1 in.) in diameter and 5 cm (2 in.) in length were then extracted from the central parts of the samples in a laboratory under dry conditions. For host and fragmented rock samples, plugs were cut both perpendicular and parallel to bedding. For fault rock samples, two orientations were considered. Eleven plugs were oriented perpendicular to the primary slip surface, and nine plugs were oriented parallel to both the fault slip vector and the primary slip surface.

MINERALOGY

Bulk mineralogy of host and fault rock samples was identified using an x-ray diffractometer (XRD) (Table 1; Figure 6). Thirteen samples were chosen from locations 1 and 3 to assess bulk mineralogical changes between the undeformed and deformed rocks. The bulk mineralogy of both host and fault rock samples consists of an average of 94% quartz, with a small range of distribution from 88 to 97%. The next most abundant mineral phase is potassium feldspar, which comprises about 3% of both host and fault rock. Authigenic clays also make up about 3% of both host and fault rock. Trace amounts (<1%) of dolomite, ferroan dolomite, or ankerite were detected in nearly all samples. Location 1 samples show slightly larger quartz abundance than those of location 3. These differences are negligible compared to overall host rock variability and likely indicate no bulk mineralogical change during the transformation of host rock to fault rock. However, some samples suggest a trend for slightly elevated clay abundance in the host rock with approaching distance to the fault core (discussed below).

Individual mineral grains were identified using a scanning electron microscope (SEM) equipped with an energy-dispersive spectrometer (EDS). Detrital grains in the host rock consisted almost entirely of quartz with minor amounts of potassium feldspar and trace occurrences of zircon, apatite, and monazite. In host and fault rock, nearly all potassium-feldspar grains showed signs of replacement by authigenic kaolinite and minor illite. Host rock cements, where present, included both iron oxide and clay minerals that generally occurred at grain-grain contacts and as grain coatings. These cements volumetrically comprised less than 1% of the total rock volume. Based on rock color and lack of magnetic properties in the hand sample,

Table 1. Summary of petrophysical data. hr = host rock (distance collected from fault core); hr-a = host rock adjacent to the fault core; frag = fragmented and damaged host rock; fr = fault rock; fr-s = fault rock with slip surface; para = parallel; perp = perpendicular.

Sample Identification	*Direction*	*Station*	*Type*	*Air Permeability (md)*	*Porosity (%)*	*Hg-air Threshold Pressure (psia)*		*Grain Size*		*XRD Mineralogy (%)*			
						7.5%	*10%*	*Median (μm)*	*Sorting (Folk)*	*Quartz*	*Dolomite*	*K-spar*	*Total Clay*
12	para	1	hr (1 m; 3.3 ft)	166	19.2								
13	para	1	hr (1 m; 3.3 ft)	2099	17.4			332	1.66	92.6	–	1.2	6.1
14	para	1	hr (10 m; 33 ft)	1173	16.7								
55	perp	1	hr (25 m; 82 ft)	123	22.7	11	11			91.9	0.5	4.5	3.0
56	perp	1	hr (50 m; 164 ft)	2620	22.0	7	7			96.5	0.4	2.7	–
62	perp	1	hr (50 m; 164 ft)	5991	23.5								
22	para	1	hr-a	82.1	24.4			139	1.85	96.8	0.3	1.4	1.5
18	para	1	frag	23.6	22.4					91.6	0.4	2.6	5.4
18	perp	1	frag	90.7	22.2	36	41	3.6	1.65				
19	perp	1	fr-s	0.353	13.3	496	544	3.4	1.77	95.8	–	1.2	3.0
20	para	1	fr-s	921	14.3								
78	perp	1	fr	3.07	16.4	77	89			96.1	–	1.4	2.5
21	perp	1	fr	4.17	15.7								
17	para	1	fr	36.1	19.4								
17	perp	1	fr	1.21	18.1			3.5	1.76	92.9	0.3	1.4	5.5
39	perp	2	hr (4 m; 13 ft)	166	19.7			170	1.61				
36	perp	2	hr-a	1.19	18.4			48	2.98				
30	perp	2	frag	1406	24.5			135	2.13				
31	para	2	frag	194	25.8								
33	para	2	frag	792	27.5								
32	perp	2	frag	133	23.3								
34	para	2	frag	93.7	23.1								
79	perp	2	frag	273	22.5								
80	perp	2	fr	21.5	21.0								
23a	para	2	fr	38.6	22.7								
23c	para	2	fr	12.0	22.7			4.0	1.73				
24a	perp	2	fr	2.88	21.4			50	2.81				
24c	perp	2	fr	21.1	21.7			51	2.95				
42	para	3	hr (1 m; 3.3 ft)	135	18.0			160	2.46	91.5	0.7	4.7	3.1
44	para	3	hr (10 m; 33 ft)	264	20.1			168	2.38				
41	para	3	hr-a	29.4	16.6	13	17	127	2.59	87.6	–	5.3	7.1
48	perp	3	fr-s	4.21	17.5								

Table 1. Summary of petrophysical data. hr = host rock (distance collected from fault core); hr-a = host rock adjacent to the fault core; frag = fragmented and damaged host rock; fr = fault rock; fr-s = fault rock with slip surface; para = parallel; perp = perpendicular (cont.).

Sample Identification	*Direction*	*Station*	*Type*	*Air Permeability (md)*	*Porosity (%)*	*Hg-air Threshold Pressure (psia)*		*Grain Size*		*XRD Mineralogy (%)*			
						7.5%	*10%*	*Median (µm)*	*Sorting (Folk)*	*Quartz*	*Dolomite*	*K-spar*	*Total Clay*
49	para	3	fr-s	2.31	17.5			3.6	1.74				
81	perp	3	fr	0.293	16.2	392	466			94.2	0.8	5.1	
45	para	3	fr	8.73	17.5								
45	perp	3	fr	5.28	19.7			6.0	1.81	92.6	0.7	5.0	1.7
46	para	3	fr	4.66	18.8								
46	perp	3	fr	3.32	19.4								
47	para	3	fr	0.284	15.7	97	113	3.1	1.78				
47	perp	3	fr	0.501	17.5	275	327	3.5	1.72	92.7	0.7	4.9	1.6

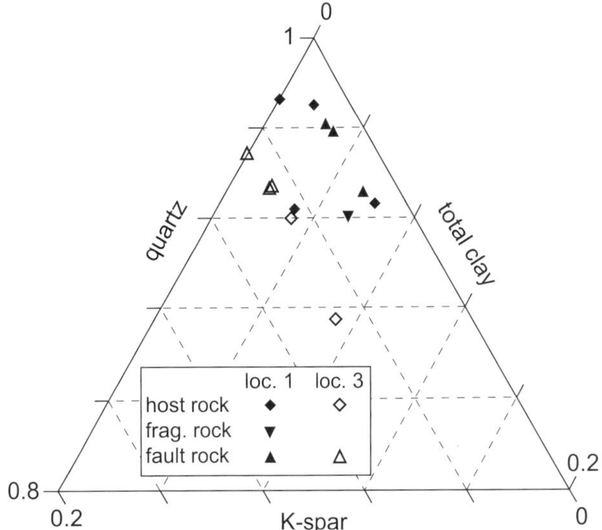

FIGURE 6. Quartz–total-clay–potassium-feldspar ternary plot of whole rock XRD data on selected samples of host rock, fragmented rock, and fault rock. Note that the ternary diagram displays a limited scale range of 20% for each phase.

the most abundant iron-oxide cement appears to be hematite with lesser amounts of goethite and limonite (Taylor, 1999). The majority of the clay cements appears to be kaolinite. Kaolinite was also sometimes found as a pore-filling phase not associated with potassium-feldspar alteration. For host rock away from fault zones, the occurrence of pore-filling kaolinite is generally limited to the stratigraphically lowest parts of the Aztec Sandstone (Flodin et al., 2003). However, elevated levels of pore-filling kaolinite occurred in a few host rock samples that were collected adjacent to the fault core and the bounding slip surface (samples 36, 41, and 44) (Figure 4b).

The majority of the fault rock samples were nearly uncemented. Exceptions to this observation included the occurrence of iron oxide and clay minerals that were localized in a reddish seam along individual slip surfaces and minor occurrences of silica-cemented zones in the fault core (Myers, 1999). We note that the clay and iron-oxide cements that were localized along slip surfaces were volumetrically minor and were difficult to detect with the SEM-EDS because of the overwhelming signal from detrital quartz.

PETROPHYSICAL CHARACTERISTICS

Porosity and Permeability

Porosity was measured with a helium porosimeter. Permeability was measured in a steady-state Hassler-sleeve air permeameter under a confining pressure of 2.8 MPa. Permeability data are presented uncorrected

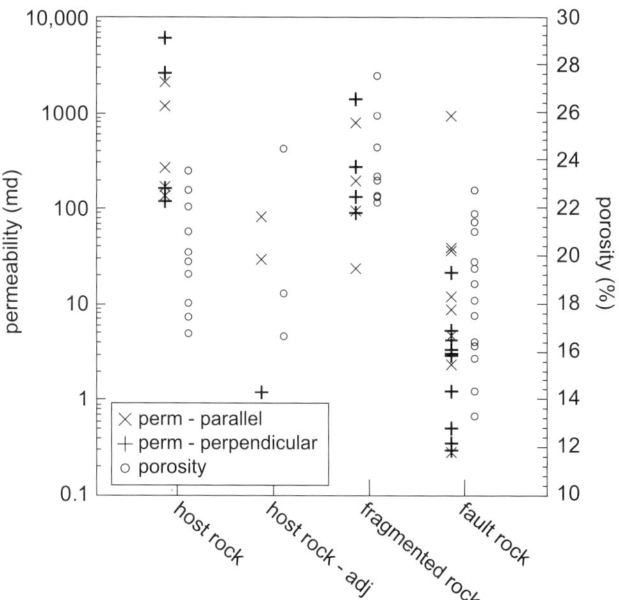

FIGURE 7. Summary plot by sample type for all porosity and air-permeability data; perm = permeability; adj = adjacent.

for Klinkenberg gas slippage effects, which become relevant at low flow velocities (low permeability) and tend to produce overestimated sample permeabilities (Goggin et al., 1988). The Klinkenberg effect is most pronounced for samples with less than 1 md permeability, although the overestimation is generally less than a factor of two.

Porosity and permeability data organized by sample genesis are presented in Figure 7 and Table 1. In each sample suite, excluding those collected immediately adjacent to the fault core, samples were oriented both perpendicular and parallel to either host rock bedding or the fault slip direction.

Measured host rock porosity values range between 16.6 and 24.4%. Three host rock samples (41, 42, and 44) collected along the same sedimentary bed at distances of 10 and 1 m (33 and 3.3 ft) and 1 cm (0.4 in.) from the fault core show a trend for decreasing porosity with proximity to the fault. Porosity values for the fragmented rock samples were the highest measured in this study, ranging from 22.2 to 27.5%. Fault rocks yielded the lowest porosity values, ranging between 13.3 and 22.7%. Although fault rock porosities include the lowest values measured, a considerable overlap is present in fault rock and host rock values (Figure 7).

Permeability data for pristine host rock samples range more than two orders of magnitude, from 123 to 5991 md. Given the relatively small number (11) of host rock samples, the true permeability range could even be greater. Host rock permeability variations ranging more than five orders of magnitude have been documented in other eolian sandstones (Chandler et al., 1989; Shipton et al., 2002). Reduced permeabilities ranging between 1.2 and 82 md were detected for the three host rock samples collected immediately adjacent to the fault core. Fragmented rock samples show permeabilities in the range of 93.7–1406 md. Permeability data for the fault rock samples range more than two orders of magnitude from 0.28 to 38.6 md. The single outlier fault rock sample that yielded a high permeability of 921 md is excluded from this range (see the discussion section). At the core plug scale of measurement, no significant permeability anisotropy was detected in any of the sample suites based on oriented sample plugs (Figure 7).

Porosity and Permeability vs. Confining Pressure

To investigate the impact of burial depth on porosity and permeability of host and fault rock, we selected a subset of six samples from location 3 for porosity and permeability analysis at stepwise increasing hydrostatic confining pressure (Figure 8). The data were collected from 6 MPa up to a maximum confining pressure of 60 MPa, which corresponds to approximately 3-km (2-mi) burial depth. Absolute porosity reductions between initial and final confining pressures range between 1.7 and 3.2%, with the majority of porosity reduction occurring over the first 20 MPa of pressure increase. Ratios between initial and final porosity values range from 0.83 to 0.9. Flodin et al. (2003) found similar porosity reductions across the same pressure range in a more regional study of the Aztec Sandstone host rock properties. In addition, that study found negligible hysteresis of porosity when the samples were returned to atmospheric confining conditions. This suggests that porosity reduction was an elastic process for these samples across the range of applied confining pressures. Similar to the porosity data, permeability values are most significantly reduced over the first 20 MPa of applied confining pressure. Ratios between initial and final permeability values range from 0.2 to 0.76 and have a median of 0.67.

Grain-size Analysis

Grain-size distributions were obtained by laser particle-size analysis (LPSA) for a subset of samples from each sampling locality. The analyses were made using a Coulter LS230 Series analyzer, which has lower and upper detection limits of 0.04 and 2000 μm, respectively. The reader is referred to Crawford (1998) for a thorough description of LPSA techniques. Sammis et al. (1987) warn of introducing a small grain-size fraction that is not present in the original sample because of grain comminution during the sometimes forceful disaggregation process. However, because the samples in

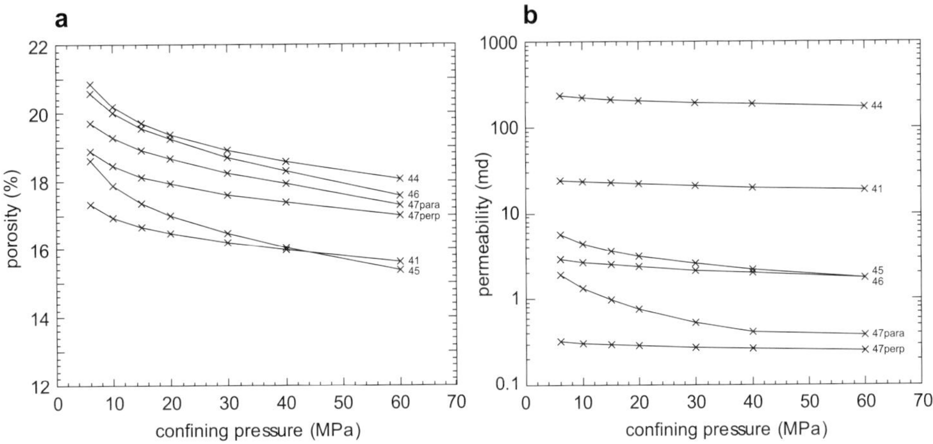

FIGURE 8. (a) Porosity and (b) air-permeability vs. hydrostatic confining pressure up to a maximum confining pressure of 60 MPa.

this study are poorly cemented to uncemented, we were able to disaggregate them with little effort.

Plots, which are organized by sample location number, of frequency vs. measured grain diameter are shown in Figure 9. Most host rock samples have bimodal distributions and are all strongly skewed toward the fine-grained fraction. Fault rock samples are weakly bimodal and are also fine skewed. Measured grain sizes range between 0.06 and 840 μm. At least one example from each of the sample suites possesses this full range of grain sizes (e.g., host, fragmented, and fault rock samples 36, 30, and 24c, respectively). However, in most cases, fault rock grain sizes span a much narrower range, between 0.06 and 20 μm, with respect to host rock samples. Calculated sorting statistics (Folk, 1968) for host rocks range between 1.61 and 2.98, which correspond to poorly sorted and very poorly sorted designations. For fault rocks, sorting ranges between 1.72 and 2.95, with a median value of 1.77. Median grain sizes for host rock samples range between 160 and 332 μm, whereas fault rock median values range between 3 and 51 μm (Table 1). Host rock samples adjacent to the faults possess lower median grain sizes (48–139 μm) than host rock samples collected away from the faults. Three host rock samples (41, 42, and 44) from location 3 show a trend for smaller median values and decreased sorting with proximity to the fault core (Table 1; Figure 9c). In terms of median grain size and sorting, one of the fragmented rock samples (18) has more affinity with fault rocks, whereas the other fragmented rock sample (30) is more similar to host rocks.

Capillary Pressure

Capillary pressure data were collected by conventional high-pressure mercury-injection porosimetry on plug samples using Micromeritics Auto Pore III 9420 instrumentation. Epoxy-sheathed, oriented mercury-injection experiments (Sneider et al., 1997) were not performed, because we had not detected any significant anisotropy in our oriented permeability data. The term "entry pressure" is used below to identify the pressure at which the mercury accesses the largest pore radius size (Pittman, 1992) and is signified by the first slope break in the mercury-air capillary pressure curve at low capillary pressure. The term "breakthrough pressure" is used to identify the lowest pressure at which a throughgoing fluid pathway forms in the sample. In this study, the breakthrough pressure is estimated at 7.5–10% of the cumulative percent of mercury intruded.

Figure 10 shows the mercury-air capillary pressure as a function of cumulative percent intruded mercury over a maximum applied pressure of 60,000 psi. The curves for the host rock were the most complex and were different from each other and from the curves representing the deformed samples. Fault rock sample curves possessed similar shapes and showed the highest initial entry pressures. These curves are characterized by high initial entry pressures followed by rapid increases in mercury saturation across relatively small pressure ranges. The fragmented rock sample showed a somewhat lower entry pressure followed by two slope changes around 10 and 30% injected mercury. Host rock samples displayed the lowest entry pressures. Following the initial slope break, fault-adjacent host rock sample 41 showed an additional slope break around 20% intruded mercury and a gradual steepening of the slope after that point. Curves for the two undamaged host rocks have shallow slopes for the first 60% injected mercury. Beyond this point, the curves record slower saturation changes with increasing mercury pressure and two slope breaks for each sample between 65 and 90% intruded mercury.

DISCUSSION

Porosity and Permeability

To assess relative changes in porosity and permeability between host and deformed rock, we normalized fault-adjacent host, fragmented, and fault rock porosity and permeability data using expanded median host rock data [φ_{median} = 22.4; k_{median} = 626 md

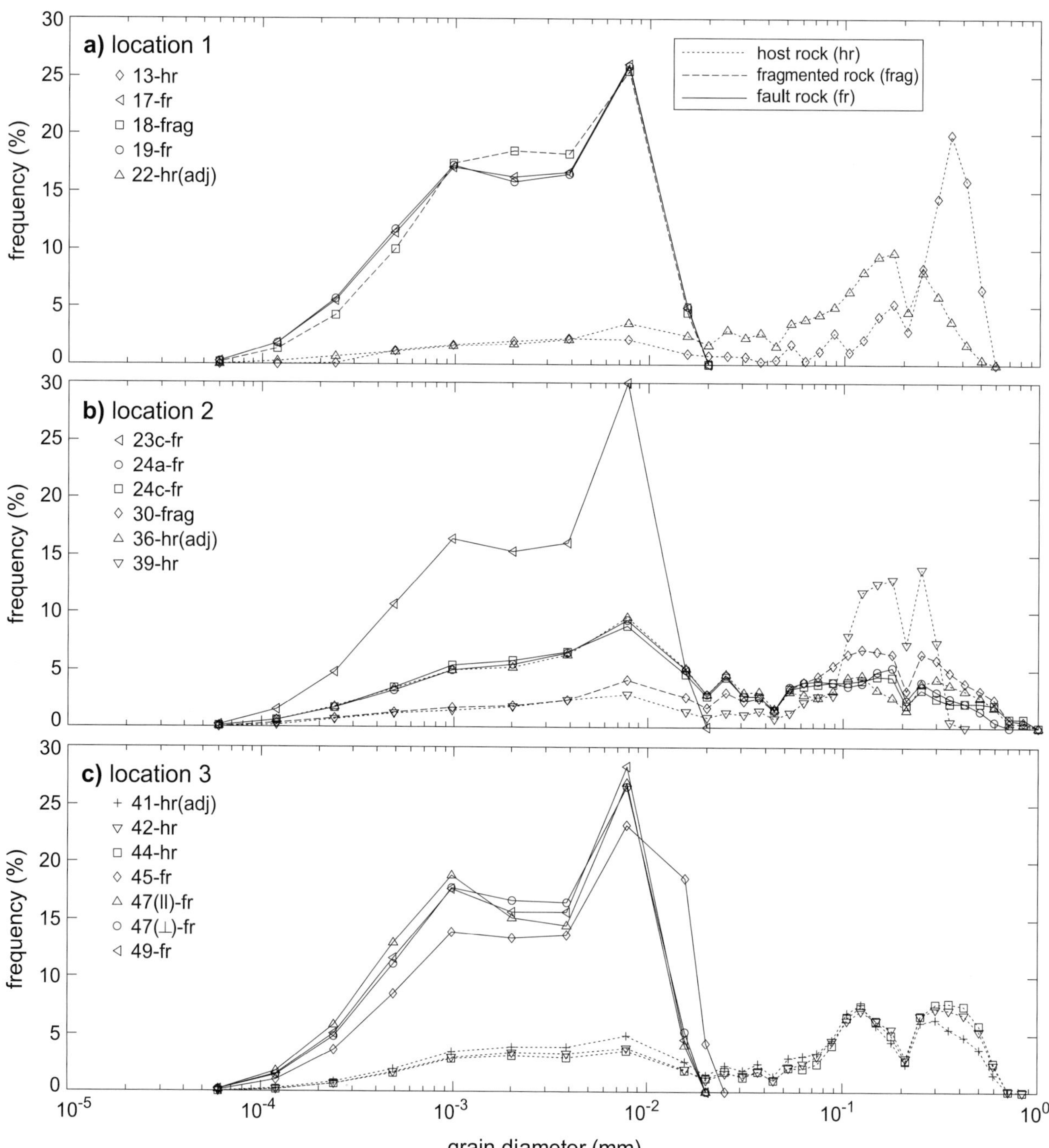

FIGURE 9. Grain-size distribution plots for select samples of host rock, fragmented rock, and fault rock, from location 1 (a), location 2 (b), and location 3 (c); adj = adjacent.

(calculated from an additional 20 host rock samples from Flodin et al., 2003)] (Figure 11). The single outlier fault rock sample (20), with a measured permeability of 921 md, is not included in Figure 11. This sample contained a slip surface that was throughgoing and parallel to the long-axis of the sample plug. The very high measured permeability might reflect flow through the fracturelike slip surface and, thus, might not represent the permeability of the fault rock itself. Slip surfaces were identified in three other petrographically similar samples (samples 19, 48, and 49). However, the slip surfaces in these samples were not throughgoing with respect to the long-axis of the permeability test direction and, thus, did not contribute significantly to flow through the plug.

Permeability reductions from host to fault rock ranged from one to more than three orders of magnitude. Given the great variability in host rock permeability, absolute permeability reductions between some host and fault rock could be as small as a factor of five and

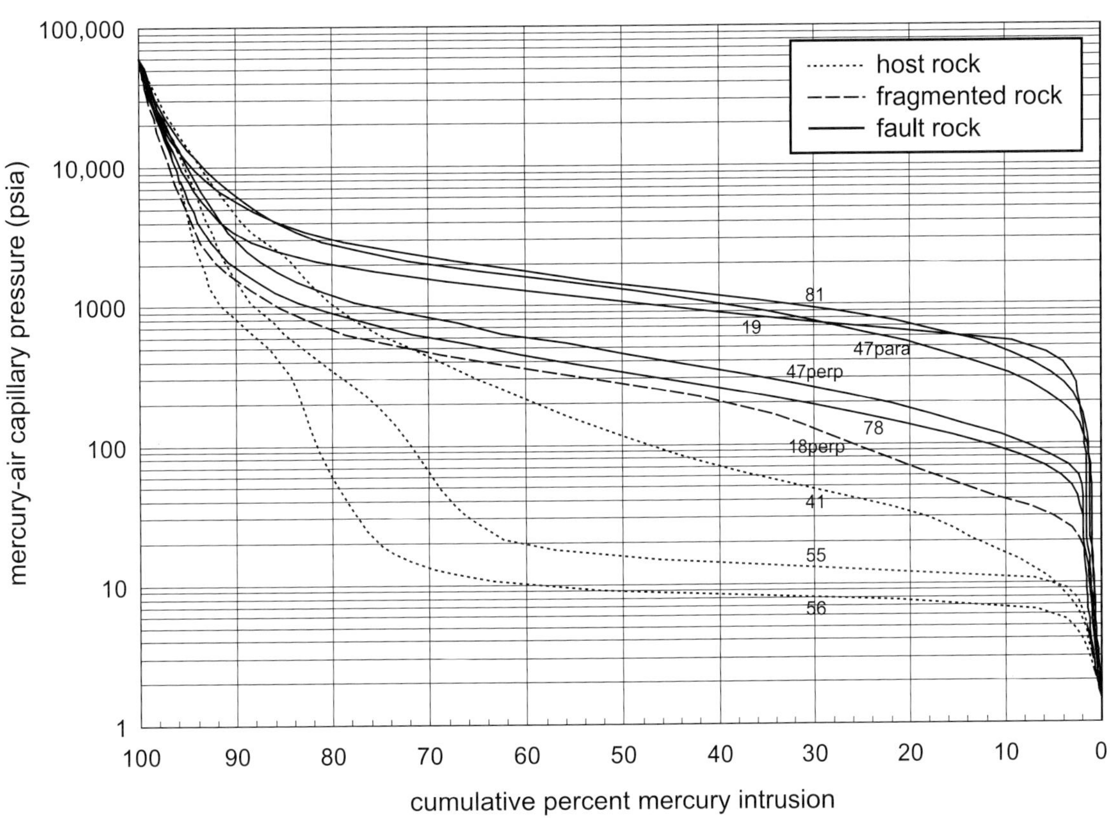

FIGURE 10. Capillary pressure vs. cumulative percent intrusion for select samples of host rock (dotted line), fragmented rock (dashed line), and fault rock (solid line). Sample numbers are also shown.

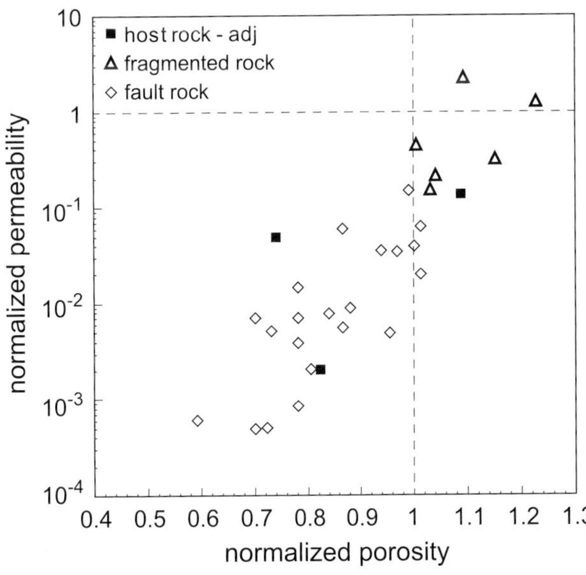

FIGURE 11. Plot showing relative change (increase or decrease) in permeability and porosity for fault rock and host rock collected adjacent to the fault core. See text for values used to normalize the permeability and porosity data; adj = adjacent.

as large as four orders of magnitude. These orders-of-magnitude permeability reductions are comparable to reductions found for cataclastic deformation bands formed in high-porosity, low clay-content sandstones. Using localized minipermeameter measurements, Antonellini and Aydin (1994), Ogilvie and Glover (2001), and Shipton et al. (2002) all found permeability reductions ranging from one to four orders of magnitude between deformation bands and adjacent host rock. Gibson (1998) reported similar ranges of permeability reduction for deformation bands involving only cataclasis from data collected on 1.9–2.5-cm (0.7–1-in.) core plugs. Antonellini and Aydin (1994) also found as much as seven orders of magnitude permeability reduction for wall rocks adjacent to slip planes. With the exception of sample 20, the slip surfaces contained in the fault rock samples of this study appeared not to have a significant effect on the plug permeability. Fault rock samples 19, 48, and 49 have slip surfaces that crossed the sample plug transverse to the permeability measurement direction. Measured permeabilities for these samples (0.35–4.2 md) fall within the range of permeabilities for fault rocks that do not have slip surfaces.

One surprising result of this study is the relative lack of porosity reduction found in the fault rock samples.

FIGURE 12. Secondary electron image (SEI) of freshly broken fault rock. Note the angular and somewhat equant nature of the fragmented quartz grains and the presence of microporosity.

The largest absolute porosity reduction between fault rock and median host rock is 9.1%, which equates to a 1.7 factor of reduction. However, most fault rock porosities display much less change. The lowest measured fault rock porosity is 13.3%. In contrast, Myers (1999) estimated fault rock porosities as low as 0.5% using image analysis techniques. Similarly, absolute porosity reductions obtained by image analysis for deformation-band style faults are reported to be as much as 20%, with relative changes approaching an order of magnitude (Antonellini and Aydin, 1994; Ogilvie and Glover, 2001; Shipton and Cowie, 2001). These discrepancies are likely caused by optical resolution limitations inherent with image analysis techniques. Myers (1999) reported a lower optical detection limit of 20 μm. However, our particle size data showed median fault rock grain sizes as small as 3 μm, indicating that the pore spaces between these grains are even smaller. Figure 12 is a high-magnification (1600×) secondary electron image of a freshly broken surface of fault rock. Grains in this image are very angular and range in size between submicrometer and tens of micrometers in size, whereas the pore spaces between the grains range from submicrometer to micrometers.

Despite the relative lack of porosity reduction measured in the fault rock samples, measured permeabilities are considerably lower than host rock values. This phenomenon is best understood by introducing the concept of microporosity. Based on a study of laboratory-deformed sandstones, Crawford (1998) concluded that the microporosity between the smallest grain sizes increased the tortuosity of the flow path and thus decreased the overall sample permeability. Myers (1999) similarly justified the exclusion of microporosity and instead measured only the macroporosity (or effective porosity), because it is the parameter that is thought to influence flow the most (Mavko and Nur, 1997). Indeed, Crawford et al. (2002) documented significant permeability reduction concomitant with less drastic porosity reduction in recent deformation experiments on fault rocks derived from end-member pure quartz sand.

Some samples show an increase in porosity with respect to median host rock (Figure 11). A porosity increase might indicate that the sample, in its original, undeformed state, had a somewhat higher porosity than median host rock, and with the reduction during deformation, the measured porosity is still greater than the calculated median value. Alternatively, the samples might have dilated relative to their original undeformed state. Thin-section analysis of these samples shows local regions containing a high density of intragranular microfractures. A high fracture density is particularly characteristic of fragmented rock samples 30 (Figure 4a) and 33, which show increased permeability with respect to median host rock. In most cases, however, these samples also contain localized damage zones in the form of sheared joints and deformation bands. The presence of localized damage in otherwise dilatant sample plugs appears to cause a net decrease in permeability (e.g., samples 22, 32, and 34).

A relationship between average shear strain and permeability was investigated by examining the distribution statistics of permeability data for each sample location (Figure 13). A relationship between shear strain and permeability is of interest because it could be used for predictive models of fault zone properties for subsurface applications, where only fault slip data are available. However, for this transform to be effective, the slip data would need to be used in conjunction with a fault rock thickness transform (Evans, 1990) to estimate the shear strain of subsurface faults. Samples from the three localities show an overall decrease in permeability with increasing shear strain. Median permeabilities are very similar between locations 1 and 3 (shear strains, $\gamma = 63$ and 80, respectively). However, location 3 samples are skewed toward lower values.

Grain-size Reduction

Grain-size distribution data show a consistent trend for grain-size reduction from host rock to fragmented rock to fault rock. There appears to be a lower limit of

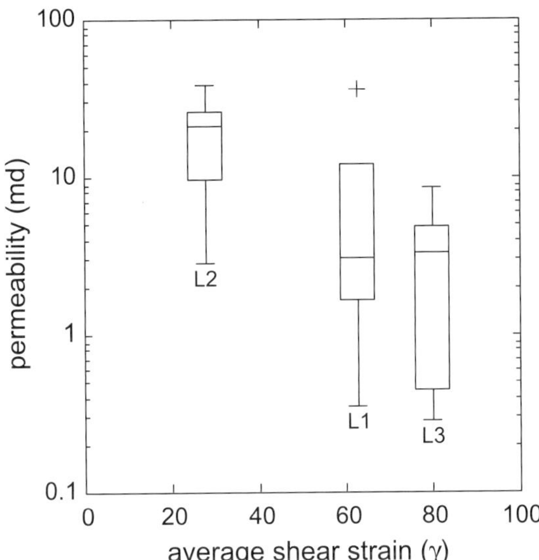

FIGURE 13. Box plot statistics of fault rock permeability vs. average shear strain. Each sample locality is indicated (e.g., L1 = location 1). The line at the center of the box is the sample median; box bottom and top are, respectively, the first and third quartiles; vertical lines span 1.5 times the interquartile range; a cross explicitly indicates outliers.

grain-size reduction for fault rock samples that is irrespective of average shear strain. Figure 14 shows a plot of cumulative weight percentage vs. grain diameter for all data collected in this study. Of the 10 analyzed fault rock samples, 7 occupy nearly the same area on the left side of this plot. At least one fault rock sample from each sample locality plots in this region. These data indicate that the fault rocks reach a level of grain-size maturity beyond which no new grains are fractured. Similar observations have been made of grain-size reductions for other faulting mechanisms. In a field study of deformation-band-style faults, Shipton and Cowie (2001) concluded that there is a maximum limit to grain crushing in the vicinity of slip surfaces. Mair et al. (2000) similarly found lower limits to grain-size reduction for deformation bands formed under controlled laboratory conditions. In a study of deformed crystalline rocks collected from faults related to the San Andreas system of southern California, Blenkinsop (1991) provided a conceptual model for grain-size reduction processes that predicts a lower limit of grain-size reduction.

The lower limit of grain-size reduction could be different for these faults had they formed under different confining conditions. Both Engelder (1974) and Marone and Scholz (1989) reported a decrease in median grain size with increasing confining stress in deformation experiments of sandstone and unlithified quartz sand, respectively. Similarly, Crawford (1998) suggested that the varying grain-size distributions between the deformed sandstones were caused by the different normal stresses under which the experiments were performed. However, the study concludes that the grain-size variability could also be caused by the fact that the samples had different total strains. Relationships between increasing confining pressure and decreasing lower grain-size limits have also been reported in studies of naturally deformed crystalline rock (Sammis et al., 1986; Blenkinsop, 1991).

Sorting

In general, the sample suites show a decrease in sorting from undamaged host rock to damaged host and fragmented rock and an increase in sorting from fragmented rock to fault rock (Table 1). Macroscopically, host rock samples 13 and 39 are among the best sorted rocks examined in this study, although none of the samples are as well sorted as is typical for eolian sandstones (Pettijohn et al., 1972). Based on petrographic analysis, these host rock samples showed weakly distributed intergranular microfractures. Disaggregation of the fractured grains in these two samples appears to have increased the fine-grained fraction, which led to the poorer sorting characteristics than might have been expected for undamaged host rock. Among the least sorted in this study are samples that show intense grain fragmentation and only moderate reduction in median grain size. Samples that fall into this category include host rock (samples 22, 36, 41, 42, and 44), fragmented rock (sample 30), and fault rock (samples 24a and 24c). In contrast, samples that show significant reduction in median grain size also possess better sorting characteristics relative to damaged host and fragmented rock. Most of the samples in this category

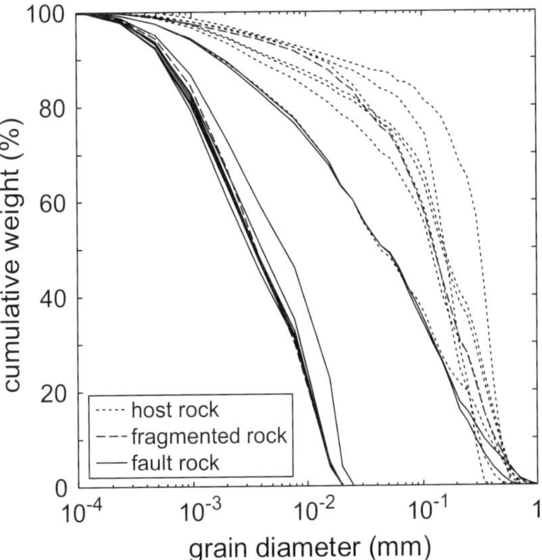

FIGURE 14. Cumulative weight percent of all grain-size distribution data shown in Figure 9.

are fault rocks that fall within a small range of sorting numbers (1.77 ± 0.03). One of the samples in this category (18perp) is classified as fragmented rock. In hand sample, this sample has preserved sedimentary features and, as such, is classified as fragmented rock. However, in thin section, damage in this sample most resembles that of the fault rocks.

Sorting characteristics can also be qualitatively inferred from the shape of the mercury-injection curves. Vavra et al. (1992) argue that better sorting leads to samples with similar pore-throat sizes. Uniform pore-throat sizes, in turn, create broad, flat plateaus on the mercury-injection curves that indicate a large volume of mercury intrusion occurs through a very narrow pressure window. With decreased sorting, the curve takes the shape of progressively increasing slope with no plateaus (see sample 41 in Figure 10). In all samples, the volume of mercury intruded decreases as smaller pore throats are invaded with increasing pressure, resulting in the highest slopes as 100% cumulative mercury intrusion volume is approached. If two distinct plateaus are present on the mercury-injection curve, a bimodal distribution is implied. Grain-size distributions inferred from the mercury-injection data (Figure 10) support our generalizations concerning grain sorting and sample genesis. Host rock samples 55 and 56 possess the shallowest initial curves, indicating a high degree of sorting as a large amount of the pore volume is invaded over a relatively narrow pressure range. Above 60% intruded mercury, saturations change more slowly with increasing mercury pressure, which we interpret to be caused by the invasion of the mercury into the small pore throats of the intergranular microfractures. Host rock sample 41, which was collected adjacent to the fault core, is characterized by a broad pressure range over which mercury invades the sample. Based on grain-size data (Table 1), this sample is the least sorted of all samples subjected to mercury-injection analysis. Note that, in addition to being fragmented, sample 41 has anomalously high clay content (~7%), which is present as a pore-filling cement. The fault rock samples are characterized by shallow but steady capillary pressure curves that are slightly steeper than the host rock slopes. Likewise, these samples are, in general, more poorly sorted than the host rock samples (Table 1).

Sealing Capacity

Mercury-injection data are used to estimate the resistance of a rock to invasion by a nonwetting fluid such as gas and most oils. The capillary strength of a rock is described in terms of its breakthrough pressure, which is the pressure at which the nonwetting fluid can form a throughgoing flow path in the rock (Sneider et al., 1997). This measure of fault seal has also been called displacement pressure (Schowalter, 1979) and threshold pressure (Katz and Thompson, 1987). Breakthrough pressure is estimated from mercury-injection curves using one of several methods. In this study, a range of breakthrough pressures for each sample was estimated by picking the mercury-air capillary pressure that corresponds to 7.5% (Sneider et al., 1997) and 10% (Schowalter, 1979) cumulative intruded mercury. Estimated breakthrough pressures range between 7 and 17 psi mercury-air for host rock samples, between 36 and 41 psi mercury-air for a fragmented rock sample, and between 77 and 544 psi mercury-air for fault rock samples (Table 1).

Based on the breakthrough pressures, we compute sealing capacity for five fault rocks, one fragmented rock, and three host rocks using representative fluid properties for both oil and gas and the following equations (Sneider et al., 1997):

$$P_{C,h/w} = \frac{(\sigma_{h/w} \cos \theta_{h/w})}{(\sigma_{Hg/air} \cos \theta_{Hg/air})} P_{C,Hg/air} \quad (1)$$

and

$$h = \frac{P_{C,h/w}}{\Delta\rho \cdot 0.433} \quad (2)$$

where P_C is the capillary breakthrough pressure (psi); h is the maximum hydrocarbon height (ft); σ is the interfacial tension between hydrocarbon and brine (dynes/cm); θ is the contact angle; and $\Delta\rho$ is the difference in density between the hydrocarbon and brine (g/cm^3). Input parameters for computing maximum sealable hydrocarbon column heights are given in Table 2, and the calculated column heights are presented in Table 3. Based on the input parameters, fault rock samples show considerable range for hydrocarbon seal potential from 10 to 69 m (33 to 226 ft) of gas or from 17 to 120 m (56 to 394 ft) of oil. Three out of five fault rock samples (19perp, 47perp, and 81) fall in the category of a class C seal (30–150 m [98–492 ft] oil), according to the Sneider et al.'s (1997) seal classification system, whereas the other two fault rock samples

Table 2. Input parameters used for calculation of hydrocarbon column heights (from Sneider et al., 1997).

Input Parameters	*Hg-air*	*Gas-brine*	*Oil-brine*
Wetting angle (θ)	140	0	0
Interfacial tension (σ) (dynes/cm)	480	70	30
Brine density (g/cm^3)		1.11	1.11
Hydrocarbon (g/cm^3)		0.05	0.8498

Table 3. Calculated maximum sealable hydrocarbon column heights using fluid properties given in Table 2. Low and high estimates are presented based on mercury-air capillary pressures at 7.5% and 10% mercury intrusion, respectively.

Sample	*Type*	*Maximum Sealable Column Height (m)*			
		Gas		*Oil*	
		7.5%	*10%*	*7.5%*	*10%*
56	hr	1	1	1	1
55	hr	1	1	2	2
41para	hr-adj	2	2	3	4
18perp	frag	5	5	8	9
78	fr	10	11	17	20
47para	fr	12	14	21	25
47perp	fr	35	41	60	72
81	fr	49	59	86	102
19perp	fr	63	69	109	120

(47para and 78) correspond to a class D seal (15–30 m [49–98 ft] oil). Not surprisingly, fragmented rock (18) and fault-adjacent host rock (41para) samples show no potential for sealing an economic accumulation of hydrocarbon. The fragmented rock sample is considered a class E seal (<15 m; <49 ft), whereas the fault-adjacent host rock sample is considered a class F rock (waste zone rock), because it is nonsealing and has a low-permeability nature. The two pristine host rock samples (55 and 56) are reservoir-quality rocks (based on porosity and permeability data) with no capacity for sealing (<2 m [<7 ft] oil column).

The samples we analyzed did not yield high sealing capacities comparable to class A or B of Sneider's seal rock types. Nevertheless, our data indicate that sand-on-sand sheared-joint faults in low clay-content sandstones are capable of sealing small to moderate hydrocarbon columns given that a seam of high capillary breakthrough pressure fault rock is laterally persistent. For sheared-joint faults in the Valley of Fire, continuous fault rock seams were found along faults with as little as 6 m (20 ft) of slip (Myers, 1999). Whether the samples with high capillary breakthrough pressures represent a laterally continuous zone of fault rock remains to be investigated by more detailed mapping and sampling. The column heights calculated in this study fall within the range of column heights calculated by Gibson (1998) for small-offset (centimeters to meters of slip), purely cataclastic deformation bands in clean sandstones. We note, however, that had conditions been favorable in the Aztec Sandstone for the postdeformational deposition of cements, the sealing potential of the fault rocks would be significantly higher (Fisher and Knipe, 1998; Gibson, 1998).

Equivalent Cross-fault Permeability

The fault core includes domains of deformed rocks with differing petrophysical properties. Even what is mapped or designated as fault rock is highly heterogeneous and displays a broad range of porosities, permeabilities, and capillary breakthrough pressures. Therefore, we adopted a numerical upscaling procedure to generalize our distinctly sampled permeability data. Following the methodology of Jourde et al. (2002), we use a finite-difference solution of the single-phase pressure equation subject to pressure and no-flow boundary conditions to calculate equivalent cross-fault permeability for the mapped areas shown in Figure 5. As an example, the model geometry used for the location 1 calculation is presented in Figure 15. Fault-parallel equivalent permeability is not considered, because the fault maps are of insufficient detail (cf. Myers, 1999) to provide accurate results for the fine-scale calculations. Because slip surfaces are oriented parallel to the flow direction (Figure 5), we do not include them in our upscaling models. High-permeability slip surfaces have been found to bear negligible influence on upscaled cross-fault permeability in cases where the slip surfaces do not cross the fault core (Jourde et al., 2002). In assigning fragmented and fault rock permeabilities to the flow model,

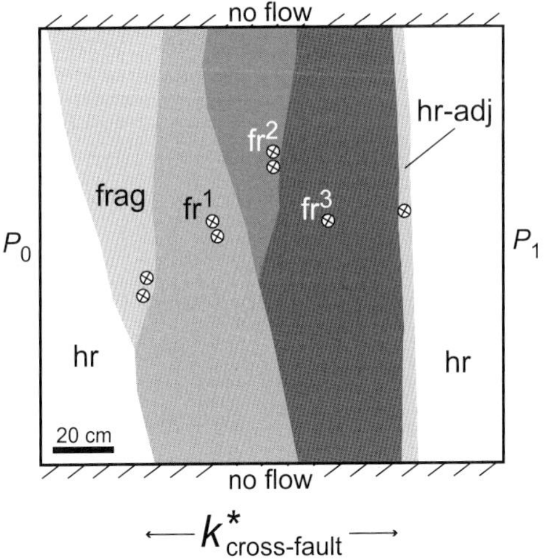

FIGURE 15. Input map used to calculate equivalent cross-fault permeability at location 1. No-flow boundary conditions are imposed along borders perpendicular to the fault zone, whereas fixed pressure boundary conditions are imposed across the fault zone. fr = fault rock; frag = fragmented rock; hr = host rock; hr-adj = fault-adjacent host rock. Crossed circles are sample localities; P_0 and P_1 = pressure boundary conditions.

we assume that the measured plug permeability values are laterally continuous with respect to the slip direction and thus assign those values over the length of the model domain. Only fault-perpendicular plug permeability data were used in this analysis (Table 1). A median value of 626 md is assigned to the host rock, except where measured host rock permeability values dictate otherwise, as is the case for the host rock samples collected adjacent to the fault core.

Calculated equivalent cross-fault permeabilities as a function of average shear strain for the three sample localities are plotted in Figure 16. As a reference point, median host rock permeability is plotted at $\gamma = 0$. The data show a trend for decreasing equivalent cross-fault permeability with increasing shear strain. For each upscaling model, the calculated permeability was most influenced by the least permeable and most continuous body of fault rock. Because the fault map areas are relatively simple in their geometry and structural stratification (Figure 5), the upscaled permeabilities closely approach the theoretical lower limit calculated using a harmonic average (Deutsch, 1989). However, for more complicated fault geometries, computed cross-fault permeabilities have been found to be greater than the harmonic average (Myers, 1999; Flodin et al., 2001). For reservoir confining conditions, upscaled cross-fault permeabilities could be expected to scale with the reductions indicated by the confining pressure analyses (Figure 8), that is, around a factor of three. We note that the calculated equivalent permeability values would be even lower if the relative permeability curves resulting from multiphase flow were considered (Manzocchi et al., 2002). Given the high breakthrough pressures for some of the fault rock and, therefore, small average pore-throat radii, the relative permeability of oil could be reduced significantly relative to the intrinsic permeability of the rock.

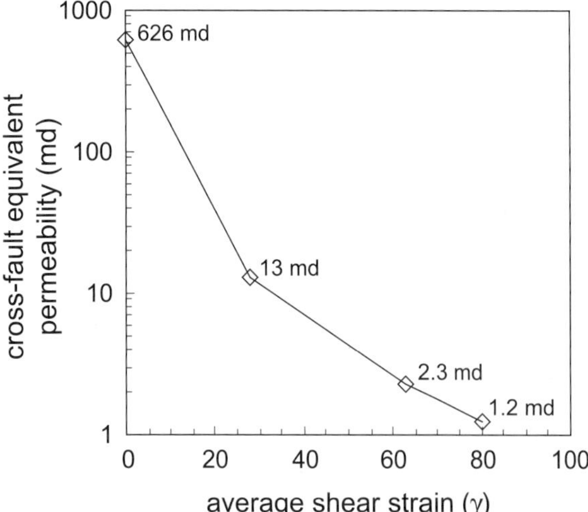

FIGURE 16. Calculated equivalent cross-fault permeability as a function of average shear strain. Median host rock is plotted as a reference starting point. Numbers next to symbols are the calculated upscaled permeability values.

The upscaled cross-fault permeability results show that sand-on-sand faults formed by shearing along joint zones in nearly pure quartz sandstones will act as baffles to subsurface fluid flow on production timescales (tens of years). These data would be useful for estimating the flow properties of sand-prone faults in reservoir models where the faults are explicitly included in the model. Typical fault transmissibility models that use only shale gouge ratio, in conjunction with permeability transforms, to predict fault baffles during production (e.g., Walsh et al., 1998; Manzocchi et al., 1999; Gerdes et al., 2002) will not predict the impact of purely cataclastic and sand-rich fault rocks.

CONCLUSIONS

The following conclusions can be drawn from this study of host, fragmented, and fault rock associated with variously strained sheared-joint-based faults in eolian Aztec Sandstone in Nevada:

1) No significant change in bulk mineralogy between host and fault rock was detected. This indicates that fault rock genesis in the study area was primarily a mechanical process and did not involve significant mass transfer or diagenesis. However, two regions of the fault zone do show anomalously higher pore-filling cement concentrations. Clay cements occur in greater abundance in host rock immediately adjacent to the fault core, whereas both clay and iron-oxide cements are localized around slip surfaces in the fault rock. Silica overgrowth cements occur in negligible quantities.
2) The greatest absolute porosity reduction of fault rock with respect to median host rock is 9.1% (down from 22% for median host), which equates to a factor 1.7 reduction. However, many fault rock samples show little to no change with respect to median host rock. All fragmented rock samples and two fault rock samples are dilated with respect to median host rock. Porosities range between 13.3 and 22.7% for fault rock, 22.5 and 27.5% for fragmented rock, and 16.7 and 23.5% for host rock.
3) Fault rock permeabilities are one to three orders of magnitude less than median host rock permeabilities. Permeabilities range from 0.28 to 39 md for fault rock, 23 to 1400 md for fragmented rock, and 123 to 6000 md for host rock.
4) Median grain size is reduced sequentially from host rock to fragmented rock to fault rock. Sorting

decreases from host to fragmented rock but increases from fragmented to fault rock.

5) High-pressure mercury-injection analyses on fault rocks that were derived from nearly pure quartz sandstone show that these rocks are capable of sealing small to moderate hydrocarbon columns if fault rock having high breakthrough pressure is present as a continuous interface. For typical fluid properties, calculated hydrocarbon column heights for fault rocks range between 10 and 69 m (33 and 226 ft) of gas and 17 and 120 m (56 and 394 ft) of oil.
6) Based on single-phase permeability upscaling calculations, faults formed by shearing along joint zones in nearly pure quartz sandstones will act as baffles to subsurface fluid flow. Upscaled cross-fault permeabilities range between 1.2 and 13 md, from low to high shear strain. These values equate to a one to two orders of magnitude reduction between fault core and median host rock permeability (626 md).
7) A relationship between average shear strain and changes in petrophysical properties show that higher shear strain leads to lower permeability, lower median grain size, and in general, higher capillary breakthrough pressure. Correlation of these petrophysical properties with fault slip alone is not a sufficient proxy. Data on fault slip and fault zone thickness together are required for accurate assessment of fault zone flow properties.
8) Finally, it should be noted that the fault rock, which is the most crucial element impacting fault perpendicular flow, appears to be continuous along all of the fault segments examined in this study. Therefore, accurate assessment of reservoir-scale fluid-flow behavior requires knowledge of fault rock distributions and flow properties, as well as knowledge of fault segment continuity and associated reservoir damage.

ACKNOWLEDGMENTS

Mercury-injection and x-ray diffraction analyses were performed by ChevronTexaco Geotechnical Center, San Ramon, California. Permeability and porosity data were collected by Core Laboratories, U.S.A. This work benefited from discussions with Peter Eichhubl and Rod Myers. Bob Jones assisted with scanning electron microscopy–energy-dispersive spectrometry data collection. Special thanks to John Popek, Bruce McCollom, and John Urbach at the Geotechnical Center for facilitating the collection of this data set, to the staff of the Valley of Fire State Park, and to Kurt Sternlof for transporting the samples from the field area. Zoe Shipton and Rasoul Sorkhabi are thanked for providing thoughtful reviews. Atilla Aydin and Eric Flodin were supported by the U.S. Department of Energy, Office of Basic Energy Sciences, Chemical Sciences, Geosciences, and Biosciences Division (award DE-FG03-94ER14462 to D. D. Pollard and A. Aydin).

REFERENCES CITED

Antonellini, M. A., and A. Aydin, 1994, Effect of faulting on fluid flow in porous sandstones: Petrophysical properties: AAPG Bulletin, v. 78, p. 355–377.

Armstrong, R. L., 1968, Sevier orogenic belt in Nevada and Utah: Geological Society of America Bulletin, v. 79, p. 429–458.

Aydin, A., 2000, Fractures, faults, and hydrocarbon entrapment, migration, and flow: Marine and Petroleum Geology, v. 17, p. 797–814.

Aydin, A., and A. M. Johnson, 1978, Development of faults as zones of deformation bands and as slip surfaces in sandstone: Pure and Applied Geophysics, v. 116, p. 931–942.

Bear, J., 1972, Dynamics of fluids in porous media: New York, Elsevier Press, 764 p.

Blakey, R. C., 1989, Triassic and Jurassic geology of the southern Colorado Plateau, *in* J. P. Jenny and S. J. Reynolds, eds., Geologic evolution of Arizona: Arizona Geological Society Digest, v. 17, p. 369–396.

Blenkinsop, T. G., 1991, Cataclasis and processes of particle size reduction: Pure and Applied Geophysics, v. 136, p. 59–86.

Bohannon, R. G., 1983, Mesozoic and Cenozoic tectonic development of the Muddy, North Muddy, and northern Black Mountains, Clark County, Nevada, *in* D. M. Miller, V. R. Todd, and K. A. Howard, eds., Tectonic and stratigraphic studies in the eastern Great Basin: Geological Society of America Memoir 157, p. 125–148.

Brock, W. G., and T. Engelder, 1977, Deformation associated with the movement of the Muddy Mountain overthrust in the Buffington window, southeastern Nevada: Geological Society of America Bulletin, v. 88, p. 1667–1677.

Caine, J. S., J. P. Evans, and C. B. Forster, 1996, Fault zone architecture and permeability structure: Geology, v. 24, p. 1025–1028.

Carpenter, D. G., and J. A. Carpenter, 1994, Fold-thrust structure, synorogenic rocks, and structural analysis of the North Muddy and Muddy Mountains, Clark County, Nevada, *in* S. W. Dobbs and W. J. Taylor, eds., Structural and stratigraphic investigations and petroleum potential of Nevada, with special emphasis south of the Railroad Valley producing trend: Nevada Petroleum Society Conference II, p. 65–94.

Chandler, M. A., G. Kocurek, D. J. Goggin, and L. W. Lake, 1989, Effects of stratigraphic heterogeneity in eolian sandstone sequence, Page Sandstone, northern Arizona: AAPG Bulletin, v. 73, p. 658–668.

Childs, C., Ø. Sylta, S. Moriya, J. J. Walsh, and T. Manzocchi, 2002, A method for including the capillary properties of faults in hydrocarbon migration models, *in* A. G. Koestler and R. Hunsdale, eds., Hydrocarbon seal quantification: Norwegian Petroleum Society Special Publication 11, p. 187–201.

Crawford, B. R., 1998, Experimental fault sealing: Shear band permeability dependency on cataclastic fault gouge characteristics, *in* M. P. Coward, T. S. Daltaban, and H. Johnson, eds., Structural geology in reservoir characterization: Geological Society (London) Special Publication 127, p. 27–47.

Crawford, B. R., R. D. Myers, A. Woronow, D. R. Faulkner, and E. H. Rutter, 2002, Porosity-permeability relationships in clay-bearing fault gouge: Presented at the Society of Petroleum Engineers/International Society of Rock Mechanics Rock Mechanics Conference, Irving, Texas, October 20–23, SPE Paper 78214, 13 p.

Davatzes, N. C., A. Aydin, and P. Eichhubl, 2003, Overprinting faulting mechanisms during the development of multiple fault sets in sandstone, Chimney Rock fault array, Utah, U.S.A.: Tectonophysics, v. 363, p. 1–18.

Deutsch, C., 1989, Calculating effective absolute permeability in sandstone/shale sequences: Society of Petroleum Engineers Formation Evaluation, v. 4, p. 343–348.

Dinwiddie, C. L., F. J. Molz, L. C. Murdoch, and J. W. Castle, 1999, A new mini-permeameter probe and associated analytical techniques for measuring the in-situ spatial distribution of permeability: EOS, Transactions, American Geophysical Union Fall Meeting, v. 80, p. F349.

Ehrlich, R., S. K. Kennedy, S. J. Crabtree, and R. L. Cannon, 1984, Petrographic image analysis: I. Analyses of reservoir pore complexes: Journal of Sedimentary Petrology, v. 54, p. 1365–1376.

Engelder, J. T., 1974, Cataclasis and the generation of fault gouge: Geological Society of America Bulletin, v. 85, p. 1515–1522.

Evans, J. P., 1990, Thickness-displacement relationships for fault zones: Journal of Structural Geology, v. 12, p. 1061–1065.

Fisher, Q. J., and R. J. Knipe, 1998, Fault sealing processes in siliciclastic sediments, *in* G. Jones, Q. J. Fisher, and R. J. Knipe, eds., Faulting, fault sealing, and fluid flow in hydrocarbon reservoirs: Geological Society (London) Special Publication 147, p. 117–134.

Flodin, E. A., 2003, Structural evolution, petrophysics, and large-scale permeability of faults in sandstone, Valley of Fire, Nevada: Ph.D. thesis, Stanford University, Stanford, California, 180 p.

Flodin, E. A., A. Aydin, L. J. Durlofsky, and B. Yeten, 2001, Representation of fault zone permeability in reservoir flow models: Presented at the Society of Petroleum Engineers Annual Technical Conference and Exhibition, New Orleans, Louisiana, September 30–October 3, SPE Paper 71671, 10 p.

Flodin, E. A., M. Prasad, and A. Aydin, 2003, Petrophysical constraints on deformation styles in Aztec Sandstone: Pure and Applied Geophysics, v. 160, p. 1589–1610.

Folk, R. L., 1968, Petrology of sedimentary rocks: Austin, University of Texas Publication, 170 p.

Fowles, J., and S. Burley, 1994, Textural and permeability characteristics of faulted, high porosity sandstones: Marine and Petroleum Geology, v. 11, p. 608–623.

Gerdes, M., K. Mabe, A. Bernath, W. Milliken, W. D. Wiggins, and W. Sumner, 2002, An integrated Earth modeling approach to simulating fault flow behavior, *in* AAPG Hedberg Research Conference, Evaluating the Hydrocarbon Sealing Potential of Faults and Caprocks, December 1–5, Barossa Valley, South Australia, p. 92–93.

Gibson, R. G., 1994, Fault-zone seals in siliciclastic strata of the Columbus basin, offshore Trinidad: AAPG Bulletin, v. 78, p. 1372–1385.

Gibson, R. G., 1998, Physical character and fluid-flow properties of sandstone-derived fault zones, *in* M. P. Coward, T. S. Daltaban, and H. Johnson, eds., Structural geology in reservoir characterization: Geological Society (London) Special Publication 127, p. 83–97.

Goggin, D. J., R. L. Thrasher, and L. W. Lake, 1988, A theoretical and experimental analysis of minipermeameter response including gas slippage and high velocity flow effects: In Situ, v. 12, p. 79–116.

Jamison, W. R., and D. W. Stearns, 1982, Tectonic deformation of Wingate Sandstone, Colorado National Monument: AAPG Bulletin, v. 66, p. 2584–2608.

Jourde, H., E. A. Flodin, A. Aydin, L. J. Durlofsky, and X.-H. Wen, 2002, Computing permeability of fault zones in aeolian sandstone from outcrop measurements: AAPG Bulletin, v. 86, p. 1187–1200.

Katz, A. J., and A. H. Thompson, 1987, Prediction of rock electrical conductivity from mercury injection measurements: Journal of Geophysical Research, v. 92, p. 599–607.

Knipe, R. J., Q. J. Fisher, G. Jones, M. B. Clennell, A. B. Farmer, B. Kidd, E. McAllister, J. R. Porter, and E. A. White, 1997, Fault seal prediction methodologies, applications and successes, *in* P. Møller-Pedersen and A. G. Koestler, eds., Hydrocarbon seals— Importance for exploration and production: Norwegian Petroleum Society Special Publication 7, p. 5–40.

Longwell, C. R., 1949, Structure of the northern Muddy Mountain area, Nevada: Geological Society of America Bulletin, v. 60, p. 923–967.

Main, I. G., O. K. Kwon, B. T. Ngwenya, and S. C. Elphick, 2000, Fault sealing during deformation band growth experiments in porous sandstone: Geology, v. 28, p. 1131–1134.

Mair, K., I. G. Main, and S. C. Elphick, 2000, Sequential growth of deformation bands in the laboratory: Journal of Structural Geology, v. 22, p. 183–186.

Manzocchi, T., J. J. Walsh, P. A. R. Nell, and G. Yielding, 1999, Fault transmissibility multipliers for flow simulation models: Petroleum Geoscience, v. 5, p. 53–63.

Manzocchi, T., A. E. Heath, J. J. Walsh, and C. Childs, 2002, The representation of two phase fault-rock properties in flow simulation models: Petroleum Geoscience, v. 8, p. 119–132.

Marone, C., and C. H. Scholz, 1989, Particle-size distribution and microstructures within simulated fault gouge: Journal of Structural Geology, v. 11, p. 799–814.

Marzolf, J., 1983, Changing wind and hydraulic regimes during deposition of the Navajo and Aztec Sandstones,

Jurassic(?) southwestern United States, *in* M. E. Brookfield and T.S. Ahlbrandt, eds., Eolian sediments and processes: Amsterdam, Netherlands, Elsevier, p. 635–660.

Mavko, G., and A. Nur, 1997, The effect of percolation threshold in the Kozeny-Carmen relation: Geophysics, v. 62, p. 1480–1482.

Myers, R. D., 1999, Structure and hydraulics of brittle faults in sandstone: Ph.D. Thesis, Stanford University, Stanford, California, 176 p.

Ogilvie, S. R., and P. W. J. Glover, 2001, The petrophysical properties of deformation bands in relation to their microstructure: Earth and Planetary Science Letters, v. 193, p. 129–142.

Pettijohn, F. J., P. E. Potter, and R. Siever, 1972, Sand and sandstone: Berlin, Springer-Verlag, 618 p.

Pittman, E. D., 1981, Effect of fault-related granulation on porosity and permeability of quartz sandstones, Simpson Group (Ordovician), Oklahoma: AAPG Bulletin, v. 65, p. 2381–2387.

Pittman, E. D., 1992, Relationship of porosity and permeability to various parameter derived from mercury injection-capillary pressure curves for sandstone: AAPG Bulletin, v. 76, p. 191–198.

Sammis, C., R. H. Osborne, J. L. Anderson, M. Badert, and P. White, 1986, Self-similar cataclasis and the formation of fault gouge: Pure and Applied Geophysics, v. 124, p. 53–78.

Sammis, C., G. King, and R. Biegel, 1987, The kinematics of gouge deformation: Pure and Applied Geophysics, v. 125, p. 777–812.

Schowalter, T. T., 1979, Mechanics of secondary hydrocarbon migration and entrapment: AAPG Bulletin, v. 63, p. 723–760.

Shipton, Z. K., and P. A. Cowie, 2001, Damage zone and slip-surface evolution over μm to km scales in high-porosity Navajo Sandstone, Utah: Journal of Structural Geology, v. 23, p. 1825–1844.

Shipton, Z. K., J. P. Evans, K. R. Robeson, C. B. Forster, and S. Snelgrove, 2002, Structural heterogeneity and permeability in faulted aeolian sandstone: Implications for subsurface modeling of faults: AAPG Bulletin, v. 86, p. 863–883.

Smith, D. A., 1966, Theoretical considerations of sealing and non-sealing faults: AAPG Bulletin, v. 50, p. 363–374.

Sneider, R. M., J. S. Sneider, G. W. Bolger, and J. W. Neasham, 1997, Comparison of seal capacity determinations: Conventional cores vs. cuttings, *in* R. C. Surdam, ed., Seals, traps, and the petroleum system: AAPG Memoir 67, p. 1–12.

Taylor, W. L., 1999, Fluid flow and chemical alteration in fractured sandstones: Ph.D. Thesis, Stanford University, Stanford, California, 411 p.

Taylor, W. L., and D. D. Pollard, 2000, Estimation of in situ permeability of deformation bands in porous sandstone, Valley of Fire, Nevada: Water Resources Research, v. 36, p. 2595–2606.

Vavra, C. L., J. G. Kaldi, and R. M. Sneider, 1992, Geological applications of capillary pressure: A review: AAPG Bulletin, v. 76, p. 840–850.

Walsh, J. J., J. Watterson, A. E. Heath, and C. Childs, 1998, Representation and scaling of faults in fluid flow models: Petroleum Geoscience, v. 4, p. 241–251.

Watts, N. L., 1987, Theoretical aspects of cap-rock and fault seals for single- and two-phase hydrocarbon columns: Marine and Petroleum Geology, v. 4, p. 274–307.

13

Kwon, O., B. T. Ngwenya, I. G. Main, and S. C. Elphick, 2005, Permeability evolution during deformation of siliciclastic sandstones from Moab, Utah, *in* R. Sorkhabi and Y. Tsuji, eds., Faults, fluid flow, and petroleum traps: AAPG Memoir 85, p. 219–236.

Permeability Evolution during Deformation of Siliciclastic Sandstones from Moab, Utah

O. Kwon[1]
Core Laboratories, Houston, Texas, U.S.A.

B. T. Ngwenya
School of Geosciences, Grant Institute, University of Edinburgh, Edinburgh, United Kingdom

I. G. Main
School of Geosciences, Grant Institute, University of Edinburgh, Edinburgh, United Kingdom

S. C. Elphick
School of Geosciences, Grant Institute, University of Edinburgh, Edinburgh, United Kingdom

ABSTRACT

We conducted triaxial deformation experiments on large (0.1-m; 0.33-ft)-diameter cores of four sandstones from the Moab area to investigate the effect of total axial strain and effective confining pressure on the evolution of bulk permeability of faulted samples. Sandstones with low bulk porosities (Dewey Bridge and Slickrock Subkha) exhibited an increase in permeability with increasing inelastic axial strain at low effective confining pressures, whereas those with high porosity (Navajo and Slickrock Aeolian) showed a decline in permeability. However, all samples showed permeability decline with increasing inelastic axial strain at high effective confining pressures. Meanwhile, microstructural observations revealed no systematic dependence of the width of the shear zone and the number of deformation bands on either strain or effective confining pressure, although grain-size reduction was more intense at high effective confining pressures. A new geometric model has

[1]*Present address:* OMNI Laboratories, Houston, Texas, U.S.A.

DOI:10.1306/1033725M853137

been developed based on these observations for a constant effective confining pressure and is shown to provide excellent agreement with the experimental data at all effective confining pressures. However, the parameters of the model depend only weakly on effective confining pressure for low-porosity sandstones, suggesting that cataclastic fault seals in low-porosity rocks have a low sensitivity to burial depth in the range studied here.

INTRODUCTION

Faults are an important factor in the exploration, production, and management of hydrocarbon prospects. On the one hand, they act as hydrocarbon migration pathways through which reservoir rocks are recharged from source rocks or through which hydrocarbons are lost from reservoirs. On the other, they provide seals for creating hydrocarbon traps. This contribution will focus on the evolution of permeability in cataclastic faults during active deformation of siliciclastic sandstones from Moab, Utah.

In a general case, seals can be classified as membrane or hydraulic seals based on their most likely failure mode (Watts, 1987). Membrane seals, also referred to as capillary seals, arise from high capillary entry pressures that prevent hydrocarbons moving through the interconnected pore throats (Ingram et al., 1997). Meanwhile, hydraulic seals form when the hydrocarbon entry pressure exceeds the mechanical strength of the seal (Yielding et al., 1997). Research so far has shown that faults forming hydrocarbon seals arise from several mechanisms operating singly or in combination (Smith, 1966; Berg, 1975; Watts, 1987; Knipe, 1992, 1997; Macaulay et al., 1997; Yielding et al., 1997; Ngwenya et al., 2000):

1) lithologic juxtaposition, in which the reservoir rock is in contact with a low-permeability rock with high capillary entry pressure
2) clay smearing, where clays and shales have been entrained into the fault plane and thus impart to the fault a high capillary entry pressure
3) cataclasis, in which grains are crushed to produce finer grained, poorly sorted, fault gouge with better packing and, hence, smaller pore throats
4) diagenesis, in which the fault is preferentially cemented, either during regional fluid flow or enhanced precipitation kinetics of pressure solution and Ostwald ripening resulting from the reduced grain size

Although these mechanisms are now well understood, the effective exploitation of this knowledge in hydrocarbon production and management is limited by our inability to predict sealing capacity of faults in the subsurface. Prediction of fault-seal intensity requires the development of mathematical algorithms that account for the effect of parent rock composition, stress, fault displacement, and fluid chemistry (Main et al., 1994; Knipe, 1997; Yielding et al., 1997).

In general, seismic data and production history give some information about which faults act as seals, but petrographic and petrophysical analysis at the core scale is commonly required to assess sealing mechanisms to aid development of predictive algorithms relating to host rock composition (Fisher and Knipe, 1998; Knipe et al., 1998). However, most of the petrographic and petrophysical analysis is carried out at ambient conditions. Consequently, it does not provide sufficient information to predict fault sealing in a production setting where changes in effective stress and fluid composition may lead to fault seals evolving because of fault reactivation (Holdsworth et al., 1997) and/or diagenetic processes (Ngwenya et al., 1995). Laboratory experiments can address some of these problems, because permeability of faulted and intact rocks can be measured directly in the laboratory, considering external variables. In addition, microstructural analysis of experimentally deformed samples provides a reasonable image of the geometry of faults, thus allowing sealing mechanisms to be correlated with in-situ conditions.

In this chapter, we present data from a suite of experiments on siliciclastic sandstone reservoir analogs sourced from the Moab area in Utah, United States. The experiments were designed to investigate (1) the evolution of permeability as a function of stress and fault displacement and the relationships to fault zone microstructures and (2) variations arising from subtle changes in host rock composition. The experimental data are analyzed in terms of a new geometric model of permeability evolution that depends on increasing slip displacement and is based on microstructural evolution of the faulted samples. The parameters of the model are sensitive to effective confining pressure and capture the style and geometry of deformation in rocks with different petrophysical properties.

Another key concern in the management of faulted hydrocarbon reservoirs, as well as in basin models, is the possible reactivation of previously sealed faults in response to changes in stress induced by changes in fluid pressure during production and tectonic loading. The resulting movement may breach a permeability barrier (seal) that is critical to the maintenance of hydrocarbon columns or indeed may be a desired effect that creates a more effective hydrocarbon pathway to production wells (e.g., Sibson, 1990). Thus, characterization of the evolution of permeability as a function of slip displacement should also provide understanding of the manner in which such reactivation episodes affect fault seals.

EXPERIMENTAL METHODS

Test Materials

Four different sandstones from the Moab area in Utah were tested for this study. The sandstones tested are Navajo Sandstone, Slickrock Aeolian, Slickrock Subkha, and Dewey Bridge sandstones. The latter three are members of the thicker and more extensive Entrada Sandstone Formation, which overlies the Navajo Sandstone formation (Foxford et al., 1998). Prior to testing, the sandstones were characterized for mineralogy by powder x-ray diffraction and in thin section using the petrographic microscope. Initial porosities were determined by the water saturation method.

All the sandstones are generally predominantly quartzo-feldspathic in mineralogy but differ in grain texture and/or proportions of diagenetic minerals, which affects their petrophysical properties. Navajo samples are fine to medium grained (with modal diameter of 160 μm) and consist of subangular to subrounded quartz (75%) and K-feldspar (25%), with trace amounts of intergranular calcite. No obvious clays are present other than perhaps those associated with heavily altered individual feldspar grains. Slickrock Aeolian samples are slightly larger grained (180 μm), with much more rounded grains ranging from 0.1 to 0.5 mm (0.004 to 0.02 in.) in diameter. K-feldspar content is around 10%, and it is heavily altered. Calcite content can be as much as 5%, whereas illite is a subordinate phase that can locally reach 1%. Meanwhile, Slickrock Subkha samples are finer grained (130 μm) relative to Slickrock Aeolian and contain about 84% quartz, 15% K-feldspar, and about 1% carbonate [ankerite(?)]. Occasional laths of mica and illite are present, and all the detrital grains, which are subangular to subrounded, are coated in a rim of hematite. This latter mineral imparts a red color to the samples. Lastly, Dewey Bridge samples contain 80% quartz and 20% K-feldspar with the same texture as Slickrock samples (bimodal at 105 and 262 μm), but the amount of hematite is higher. Most samples are also banded into light, coarser bands with more tightly packed grains and finer bands, with more angular grains and a higher content of intergranular hematite. Dewey Bridge samples did not contain much clay based on bulk rock x-ray diffraction analysis.

Initial connected porosity was determined by taking the difference in weight of whole samples or end slabs before and after saturation with distilled water under the vacuum for 60 min. The weight of the saturated sample was recorded, and then the sample was dried in an oven at 60°C until there was no change in weight. At this point, it was assumed that the sample was nominally free of pore fluid, and porosity was determined from the weight difference, assuming a value of 1000 kg/m^3 for the density of distilled water. The initial average porosities determined in this way are 21, 23, 17, and 16% for Navajo, Slickrock Aeolian, Slickrock Subkha, and Dewey Bridge sandstones, respectively.

Deformation Experiments

The role of effective stress and axial strain on permeability evolution was investigated using a large-capacity triaxial apparatus. This servo-controlled triaxial deformation apparatus was designed to apply axial load and radial confining pressure to right cylindrical samples of 100 mm (3.9 in.) diameter and 200–225 mm (7.9–8.9 in.) long. Such large sample sizes are required to simulate realistic analogs of natural deformation bands (Main et al., 2000; Mair et al., 2000; Ngwenya et al., 2003). The disadvantage, however, is to limit the number of tests that can be conducted for materials that are not locally available.

The triaxial deformation apparatus (Figure 1) consists of several components. These include (1) a linear actuator to provide axial load; (2) a large, cylinder-shaped pressure vessel in which the sample sits on a basal plinth;

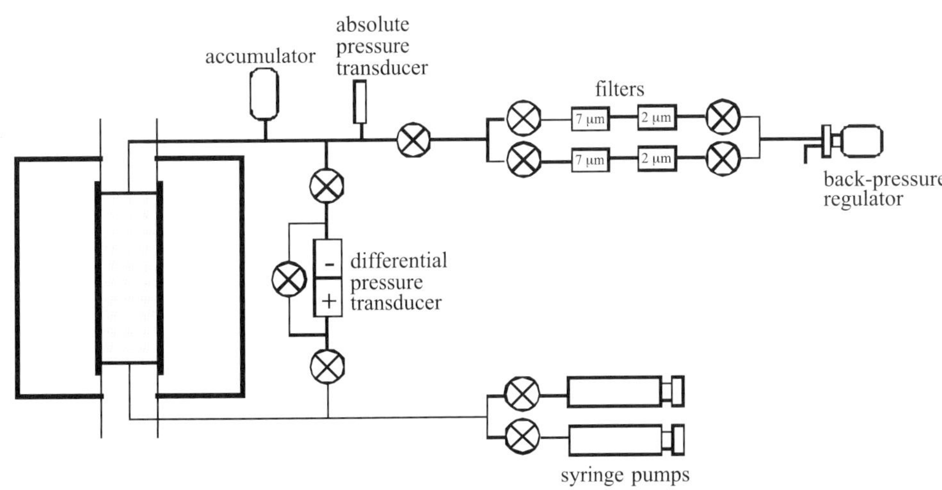

FIGURE 1. Schematic layout of the large-capacity deformation rig, with permeameter for steady-state permeability measurements.

(3) a confining pressure intensifier that supplies high-pressure confining fluid to the pressure vessel; (4) a stiff loading frame with a stiffness of 1.05×10^9 N m^{-1}, designed to minimize elastic stretching of the rig during deformation experiments; (5) pore-pressure pumps and control unit to apply pore pressure to the sample; (6) a computer data acquisition system; (7) an electronic control cabinet to control the apparatus; and (8) a large hydraulic pump unit to supply pressurized oil to the main hydraulic circuit and confining pressure intensifier. A maximum axial load of as much as 2 MN can be applied on the end surfaces of the sample by the actuator, and a maximum confining pressure of 70 MPa can be achieved in the pressure vessel. Maximum principal stress (σ_1) is applied axially to the end surfaces of the sample, and an equal amount of intermediate stress (σ_2) and minimum stress (σ_3) is applied radially as a confining pressure ($P_c = \sigma_2 = \sigma_3$). All tests were loaded initially to hydrostatic pressure ($\sigma_1 = \sigma_2 = \sigma_3$) conditions at room temperature, and then σ_1 was increased under constant strain rate loading, keeping σ_3 constant. The effective confining pressure (P_e) is defined as the difference between confining pressure (P_c) and pore pressure (P_p) (i.e., $P_e = P_c - P_p$). One suite of samples was deformed at a constant confining pressure to different ultimate axial strains of as much as 5.5%. A second suite of samples was deformed to about 5% axial strain as a function of confining pressure ranging from 20.7 to 55.5 MPa. In all cases, a perforated Melinex® sheet was placed on both ends of the sample to minimize frictional end effects.

Concurrent measurements of permeability starting at an initial hydrostatic permeability (k_0) were made during the deformation experiment with a specially designed flowthrough fluid-pressure system (Figure 1). Bulk permeability (k_{bulk}) was calculated using Darcy's law based on steady-state flow in the form

$$k_{bulk} = \frac{Q\mu L}{AdP} \quad (1)$$

where Q is the flow rate; μ is the dynamic viscosity of the permeant; L is the sample length; A is the cross-sectional area of the sample; and dP is the pressure difference across the sample. In this study, liquid permeabilities were measured under fully saturated conditions using multipar-H isoparaffinic hydrocarbon as permeant, which has a dynamic viscosity of 1.15 cp. Fluid pressure in the downstream reservoir was maintained constant at 6.9 MPa for all experiments using a backpressure regulator and a bladder-type accumulator.

The fluid delivered to the upstream reservoir was maintained at a constant flow rate using a single-stroke ISCO™ syringe pump, whereas the pressure difference to maintain this constant flow rate was measured continuously using a Validyne differential pressure transducer. Flow rates used varied depending on the initial permeability of each core sample, from 3.3 to 8.3×10^{-8} m^3 s^{-1}. Permeabilities were measured continuously over the entire deformation period, which was carried out at an axial strain rate of approximately 5×10^{-6} s^{-1}. It was assumed that this strain rate is low enough to maintain a steady-state condition during the experiment, so that the permeabilities can be measured continuously without stopping and holding the differential stress (e.g., Zhu and Wong, 1997). This was done to allow us to capture changes in permeability during the strain-hardening and strain-softening phases prior to dynamic failure of the sample. Although this assumption cannot be independently verified, the lack of differential pore-pressure transients during the quasi-static phases of loading suggests that it may have been satisfied. Under dynamic conditions, a strong suction pump may affect the upstream pore pressure, rendering Darcy's law invalid (Grueschow et al., 2003).

Volumetric strains of the sample were also determined by measuring fluid volume changes in the confining pressure system required to maintain the constant confining pressure during deformation. The displacement of the P_c intensifier piston that occurred to maintain constant P_c during deformation was converted to the volume assuming that changes in the confining pressure system volume are caused by changes in the volume of the sample. Volumetric strain was calculated from the change in sample volume divided by the initial sample volume before application of the initial hydrostatic pressure.

Microstructural Observations and Grain-size Analysis

After each test, deformed samples were dried and sliced in a direction perpendicular to the core axis while the sample was still encased in the rubber sleeves. The cut surfaces were photographed, and fracture patterns on the cut surfaces were examined. One-half of each cut sample was impregnated with polyester resin, and thin sections were prepared parallel to the sample axis and perpendicular to the fracture plane. Microscopic observations were carried out using polarized light microscopy to determine the mode of deformation and details of fracture geometry.

The fracture patterns on the lateral surfaces were also examined after the removal of the rubber sleeve from the second half. This second half sample was split along the fractures, and gouge materials were collected from the fracture surfaces. Grain-size analysis was carried out on particles collected from the fracture surfaces using a laser particle size analyzer with a range of 0.4–1000 μm. The grain-size distribution of the starting material was analyzed by carefully breaking off thin slabs of the core by hand. This provided a reference for the evolution of grain-size distribution in the fault gouge during deformation.

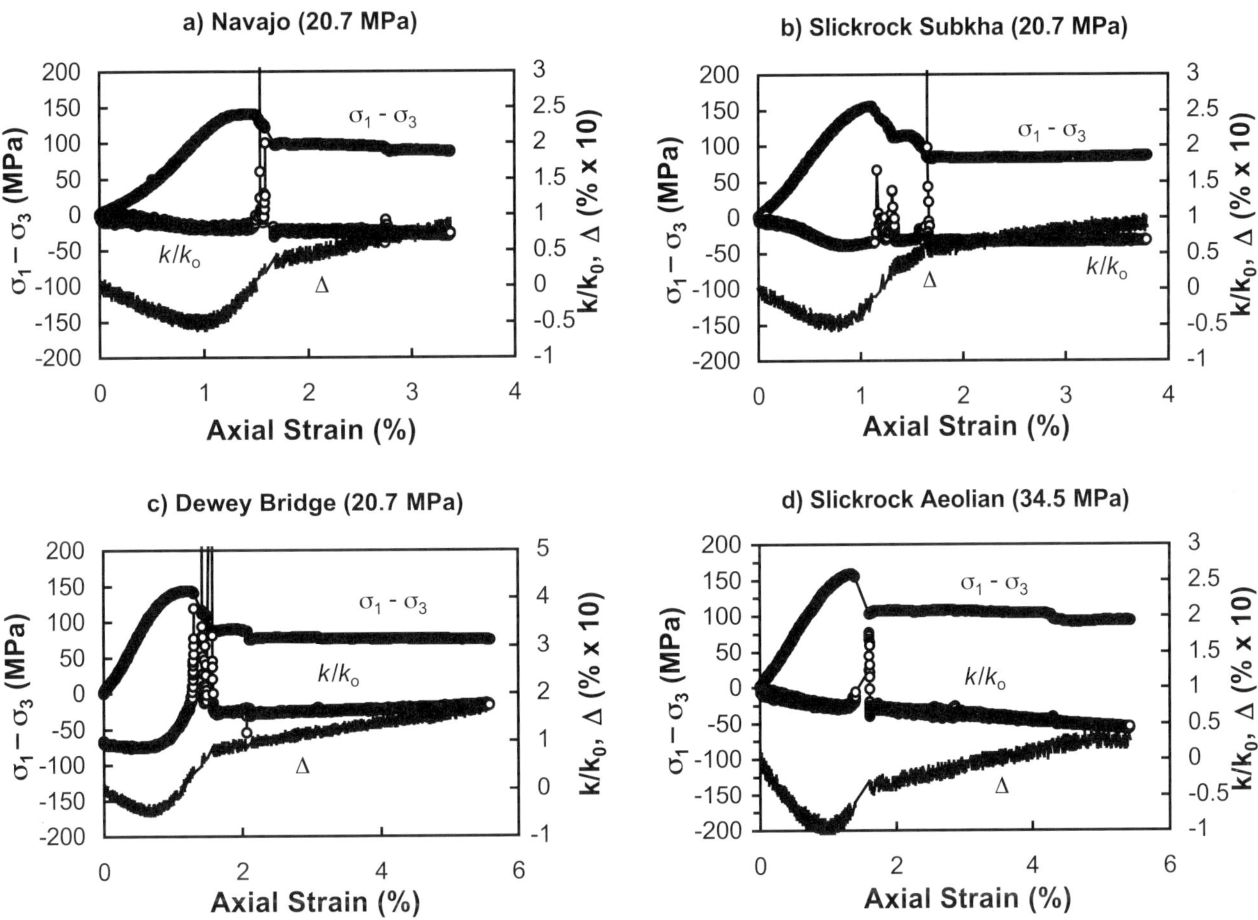

FIGURE 2. Graphs of differential stress, volumetric strain (**Δ**), and permeability, the latter plotted as k/k_0, where k_0 is permeability of intact sample at hydrostatic pressure. The graphs show data at 20.7 MPa effective confining pressure for (a) Navajo Sandstone, (b) Slickrock Subkha sandstone, and (c) Dewey Bridge sandstone ($k_0 = 2.1 \times 10^{-15}$ m^2), whereas the data for Slickrock Aeolian sandstone ($k_0 = 2.3 \times 10^{-13}$ m^2) are at 34.5 MPa. Initial permeabilities of Navajo and Slickrock Subkha sandstones are listed in Table 2. The four sandstones differ marginally in their porosity, but the permeability is significantly lower for the lower porosity sandstones.

RESULTS

Permeability Changes during Deformation

Figure 2 shows typical trends of the evolution of bulk sample permeability and volumetric strain during deformation for each Moab rock type for a relatively low and constant effective confining pressure of 20.7 MPa (34.5 MPa for Slickrock Aeolian). In general, the permeability evolution can be characterized by several stages: (1) a decrease in permeability at a nearly constant rate upon increase in differential stress in the linear elastic compaction regime, as also indicated by the linear decrease in volumetric strain; (2) permeability increase after yield stress and just prior to peak stress in association with sample dilatancy; (3) rapid increase in permeability during the strain-softening period, after peak stress, and before dynamic failure; (4) apparent transient peak in permeability coincident with dynamic stress drop at the time of failure, which can be attributed to nonsteady-state feedback between storativity and pore pressure during rapid dilatant slip (Grueschow et al., 2003); and (5) a postfailure permeability reduction with increasing axial strain. This trend is generally consistent with that reported for the Clashach sandstone (Ngwenya et al., 2003). Although the permeability at the time of dynamic failure shows dynamic transient response with a markedly increased apparent permeability, as predicted by the suction pump model, permeabilities right before and after the dynamic failure are real and reflect quasistatic processes occurring inside the sample.

All of the different formations tested here showed these five stages, but their amplitude and duration vary with respect to rock type. For example, stages 2 and 3 are more prominent in the lower permeability and lower porosity Dewey Bridge and Slickrock Subkha sandstones (Figure 2). In addition, the changes in permeability and volumetric strain in stage 5 do not always

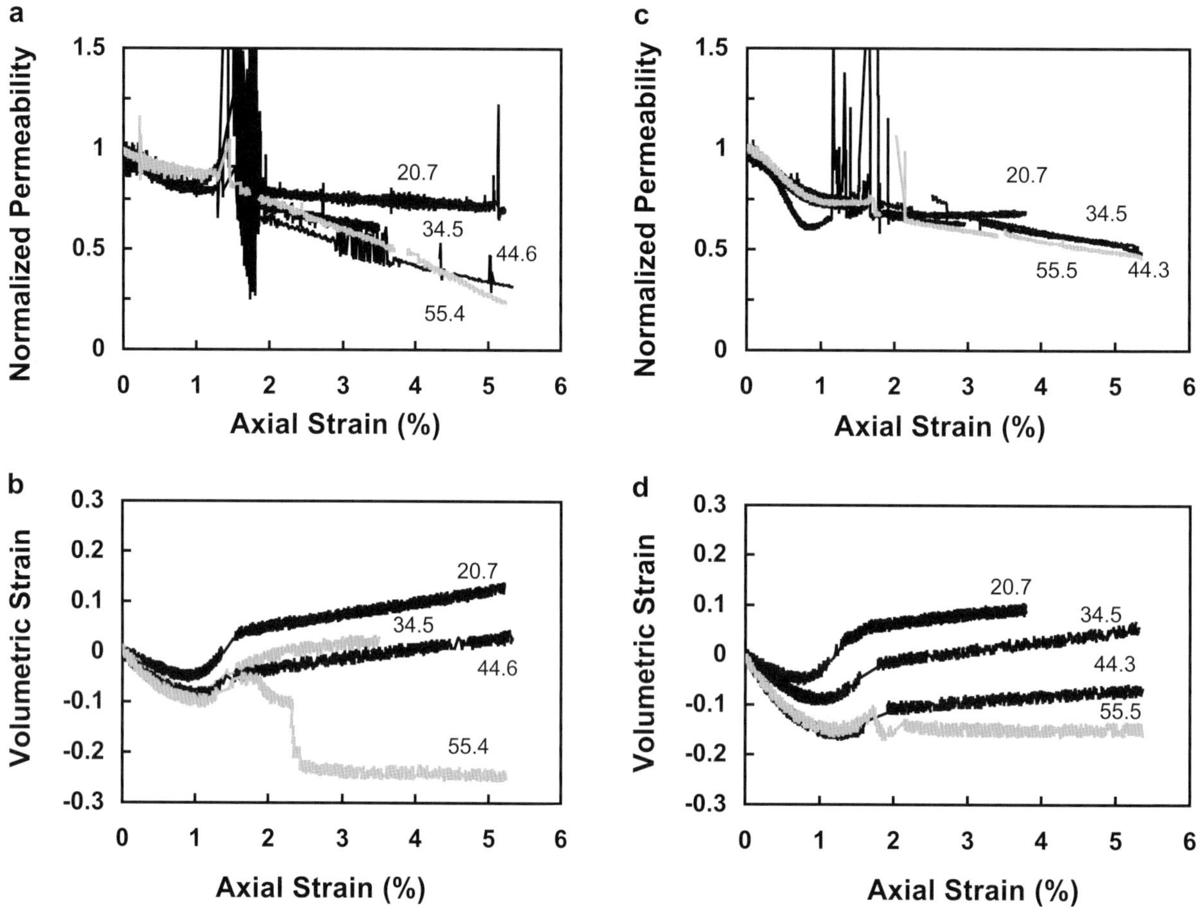

FIGURE 3. Graphs showing permeability evolution, normalized to the initial hydrostatic value and volumetric strain for Navajo (a and b) and Slickrock Subkha (c and d) samples at different effective confining pressures.

track one another. For example, the postfailure permeabilities for Navajo, Slickrock Aeolian, and Slickrock Subkha sandstones are nearly constant or decrease to only 50% of the initial hydrostatic values, while samples are dilating. This implies that fault sealing more or less balances or overcomes the increase in porosity caused by dilatancy. Meanwhile, the trend for the Dewey Bridge sample is of increasing permeability with increasing postfailure strain and positive dilatancy, implying only very weak fault sealing. This contrasts strongly with the permeabilities for Clashach sandstone, which were found to decrease by as much as three orders of magnitude with increasing axial strain at similar effective confining pressures (Ngwenya et al., 2003).

Changes in effective confining pressure also appear to affect each rock type differently. For example, the strain-dependent permeability of Navajo samples shows slightly more nonlinearity as effective confining pressure increases and as the volumetric response becomes compactive (Figure 3a, b). This is consistent with other studies (Teufel, 1987; Ngwenya et al., 2003) but contrasts with the trend for Slickrock Aeolian and Slickrock Subkha samples (Figure 3c, d), where the permeability change remains linear with increasing inelastic axial strain at different effective confining pressures and depends only weakly on confining pressure. For the Dewey Bridge and Slickrock Aeolian samples, it is not possible to establish any trend, because only one sample was deformed at a higher (34.5 MPa) effective confining pressure, at which the permeability also decreases linearly with inelastic axial strain. With the exception of the Navajo Sandstone, the permeability evolution for Moab sandstones has lower sensitivity to both inelastic axial strain and effective confining pressure relative to the Clashach sandstone (Ngwenya et al., 2003).

Fault Zone Microstructures and Gouge Grain Size Distribution

All sandstone samples deformed in this study failed by shear fracture (deformation band) with grain-scale mechanisms that included dilatant tensile cracking, crack extension by linkage of tensile cracks, cataclasis, and frictional sliding. Macroscopic shear fractures developed at inclinations of 15–35° to the maximum compression direction, and some samples exhibited incipient conjugate fractures. At low effective confining pressure,

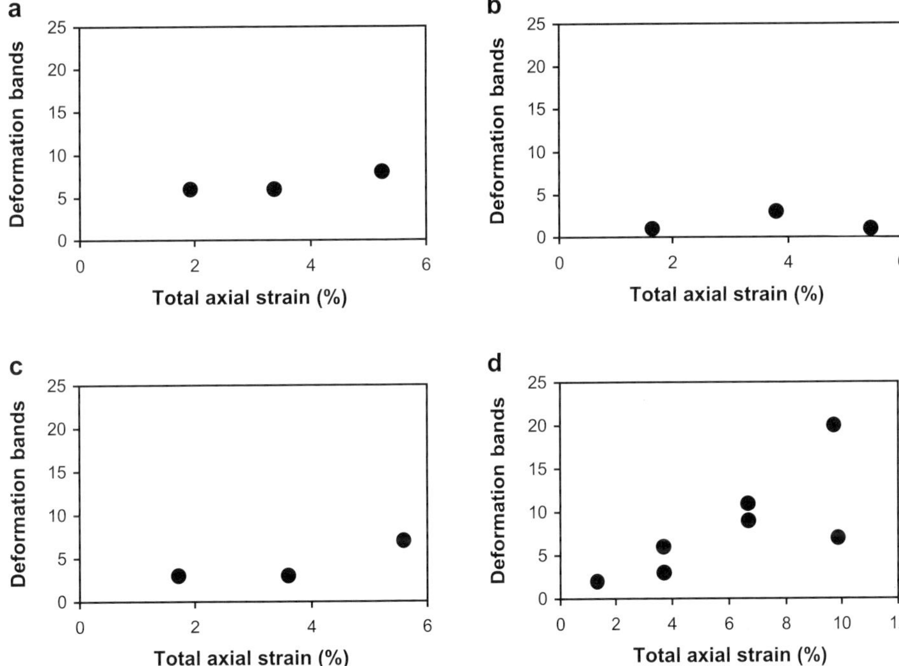

FIGURE 4. Graphs showing the relationship between the number of deformation bands and total axial strain for (a) Navajo, (b) Slickrock Subkha, and (c) Dewey Bridge sandstones. Unlike Clashach sandstone (d), no correlation exists between the number of deformation bands and the total axial strain for these sandstones.

the fault zone was inclined at a sufficiently steep angle to terminate against the stainless-steel platen on the top end of the sample.

Because of limitations in the core material, we have limited observations on the dependence of fault zone microstructures and grain-size distribution on total axial strain. These show that no systematic relationship exists between the number of deformation bands and the total axial strain for any of the three Moab sandstones (Figure 4a–c). In other rock types analyzed previously, the number of deformation bands increased linearly with increasing total axial strain, as shown in Figure 4d for the case of the Clashach sandstone (Ngwenya et al., 2003). In addition, shear fractures that developed in samples deformed to large total axial strain are only slightly wider than those at low total axial strains (Figure 5). However, there appears to be no systematic widening of the fault zone with increasing effective confining pressure (Figure 6). This contrasts with field observations of Shipton and Cowie (2001), who found that the width of damage zones consisting of deformation bands increased as offset on the fault was accumulated in Navajo Sandstone. These differences lend weight to the inference that microstructural characteristics are likely to depend on grain-size-to-sample-size ratio (Mair et al., 2000).

Observations on the radial surfaces of the cores commonly show that wider fracture zones consist of multiple strands (deformation bands) of highly comminuted cataclastic zones and dilated pods with a high density of extensional microcracks. However, when slabbed parallel to the axial direction and normal to the fault zone trace, it is commonly found that the distribution of strain is not homogeneous. For the most part, only two to three distinct strands are evident, but the fault zone can locally splay into several strands (Figure 7a). These observations were confirmed by microscopic examination of thin sections made perpendicular to the fault zone (Figure 7b). Thin-section observations also suggest that at high effective confining pressures, some of the strain may be accommodated by the formation of new strands within and at a low angle to existing strands with finer grain sizes. This may explain the apparent lack of positive correlation between the number of deformation bands with total axial strain and/or effective confining pressure. Additionally, deformation bands in samples containing hematite develop discontinuous stringers (Figure 7b) similar in structure to phyllosilicate domains described from clay-rich gouge (Fisher and Knipe, 1998).

The lack of variation in grain-size distribution with total axial strain (Figure 8a, b) is consistent with observations on other siliciclastic sandstones (Mair et al., 2000; Ngwenya et al., 2003). Similarly, analysis of the grain-size distribution for Slickrock Subkha samples deformed at varying effective confining pressure shows a shift toward increasing volume of finer grain size and also poorer sorting (Figure 8c). These observations are consistent with increasing grain comminution at high confining pressures in both small samples (e.g., Crawford, 1998) and large samples (Mair et al., 2002; Ngwenya et al., 2003).

INTERPRETATION AND DISCUSSION OF RESULTS

Our experiments were designed primarily to investigate the role of strain, effective confining pressure, and host rock composition on fault permeability evolution in different siliciclastic sandstones from the Moab area.

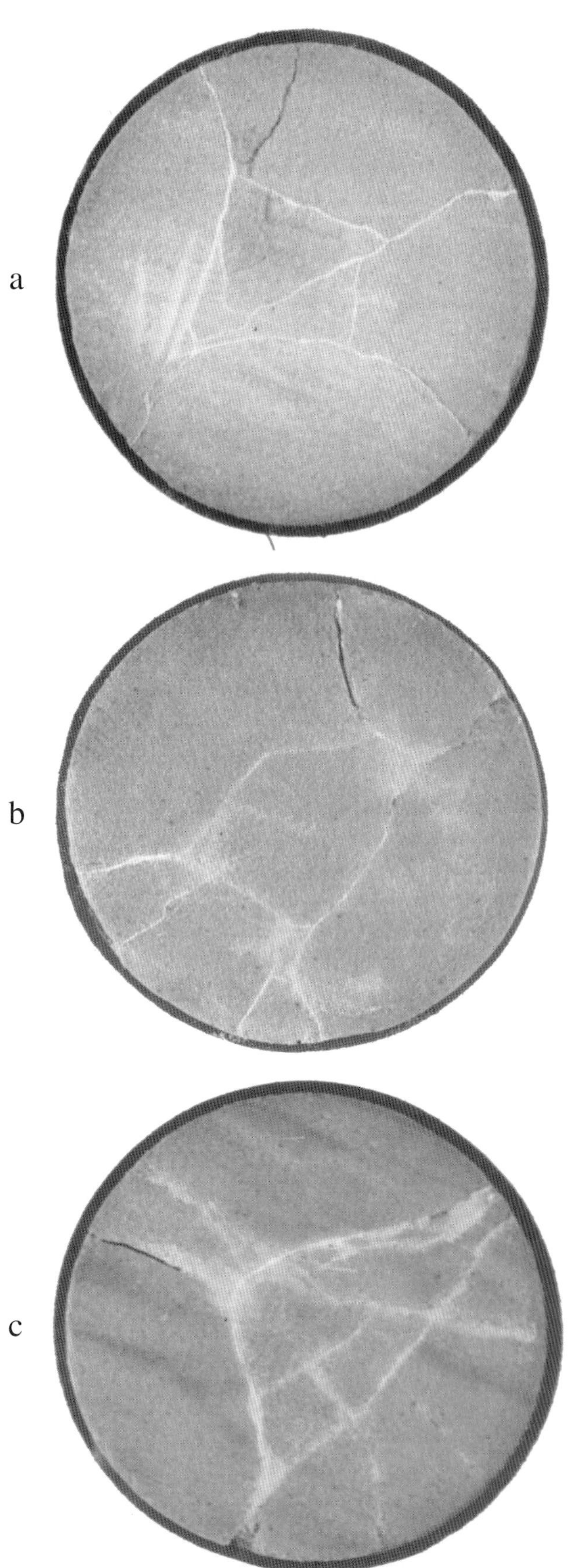

The results generally demonstrate a small variation in behavior for the different rock types analyzed. In this section, the implications of these observations are examined through the derivation of a geometric model of fault permeability evolution that may offer potential for predicting fault-seal intensity in siliciclastic hydrocarbon reservoirs in rocks with similar properties.

Geometric Model for Permeability Evolution

Main et al. (2000) developed a geometric model for evolution of permeability against axial strain in the postfailure phase. The model was based on the microstructural observations of Mair et al. (2000), notably the linearly increasing width of the fault zone caused by the formation of new deformation bands with increasing total axial strain. The model was later shown to be valid at different effective pressures (Ngwenya et al., 2003). Because our Moab data do not show systematic changes in the width of the shear band with postfailure strain, this model is not applicable in its present form. Nevertheless, the other assumptions on which the model was based are still valid. Thus, during deformation at constant effective confining pressure, (1) the bulk sample may either dilate or compact depending on the effective confining pressure. Consequently, the permeability of the matrix (k_m) can increase or decrease during dilation or compaction, respectively. Furthermore, (2) the porosity of each deformation band (db) is constant, so that k_{db} stays constant with increasing inelastic axial strain. Finally, (3) permeability varies exponentially with porosity over the small range of porosity changes measured. Based on the last assumption, the matrix permeability is defined by the equation

$$k_m = k_0 \exp[\eta(\phi - \phi_0)] \qquad (2)$$

where η is a constant whose significance is discussed later, whereas k_0 is the bulk permeability measured at hydrostatic conditions, and ϕ_0 is the initial porosity. For most of our data, the volumetric strain varies linearly with inelastic axial strain ($\delta = \varepsilon - \varepsilon_c$), where ε_c is the axial strain immediately after macroscopic failure (axial strain is defined positive for axial shortening). Thus, we define the porosity change in the postfailure regime by (Ngwenya et al., 2003)

$$\phi - \phi_0 = -\lambda\delta \qquad (3)$$

FIGURE 5. Radial view of Navajo samples deformed to different total axial strains, (a) 1.92%, (b) 3.37%, and (c) 5.22%, showing that the fault zone width does not scale linearly with total strain.

where λ is the rate of change of volumetric strain with inelastic axial strain, such that when λ is greater than 0 and δ is greater than 0, the sample compacts overall. Combining equations 2 and 3 yields

$$k_m = k_0 \exp(-\lambda\eta\delta) = k_0 \exp(-\gamma\delta) \quad (4)$$

where $\gamma = \lambda\eta$. This equation is identical to equation 4 of Main et al. (2000). Assuming that the bulk permeability (k_{bulk}) of the sample is a geometric mean of the permeability of the fault and matrix yields

$$\ln k_{bulk} = \frac{\sum_{i=1}^{n} (\ln k_i) l_i}{\sum_{i=1}^{n} l_i} = \frac{w \ln k_f + [l_0(1-\delta) - w] \ln k_m}{l_0(1-\delta)} \quad (5)$$

where k_f is the permeability of the shear band; l_i is the individual path length; and w is the width of the shear band, which, in this case, does not vary with postfailure axial strain. Substituting equation 4 into equation 5 and simplifying leads to

$$\ln k_{bulk} = \ln k_0 + \left[\frac{-\frac{w}{l_0}\ln\left(\frac{k_0}{k_f}\right) + \left(\frac{w}{l_0} - 1\right)\gamma\delta + \gamma\delta^2}{(1-\delta)}\right] \quad (6)$$

For $k_0 = k_c$, $\gamma = \gamma_I$, and $l_0 = l_c$, where l_c is the sample length at critical strain (ε_c), k_c is the bulk permeability at critical strain, and γ_I is the bulk inelastic strain sensitivity parameter for the sample (Main et al., 2000; Ngwenya et al., 2003), we have

$$\ln k_{bulk} = \ln k_c + \left[\frac{a_0 + a_1\delta + a_2\delta^2}{(1-\delta)}\right] \quad (7a)$$

where

$$a_0 = -\frac{w}{l_c}\ln\left(\frac{k_c}{k_f}\right) \quad (7b)$$

$$a_1 = \left(\frac{w}{l_c} - 1\right)\gamma_I \quad (7c)$$

and

$$a_2 = \gamma_I \quad (7d)$$

a

b

c

Figure 6. Radial view of Slickrock Subkha samples deformed to the same strain at different effective confining pressures: (a) 34.5 MPa, (b) 45.3 MPa, and (c) 55.5 MPa. Unlike Clashach (Ngwenya et al., 2003) and Locharbriggs sandstones (Mair et al., 2002), the fault zone width does not increase with increasing effective confining pressure.

a

b

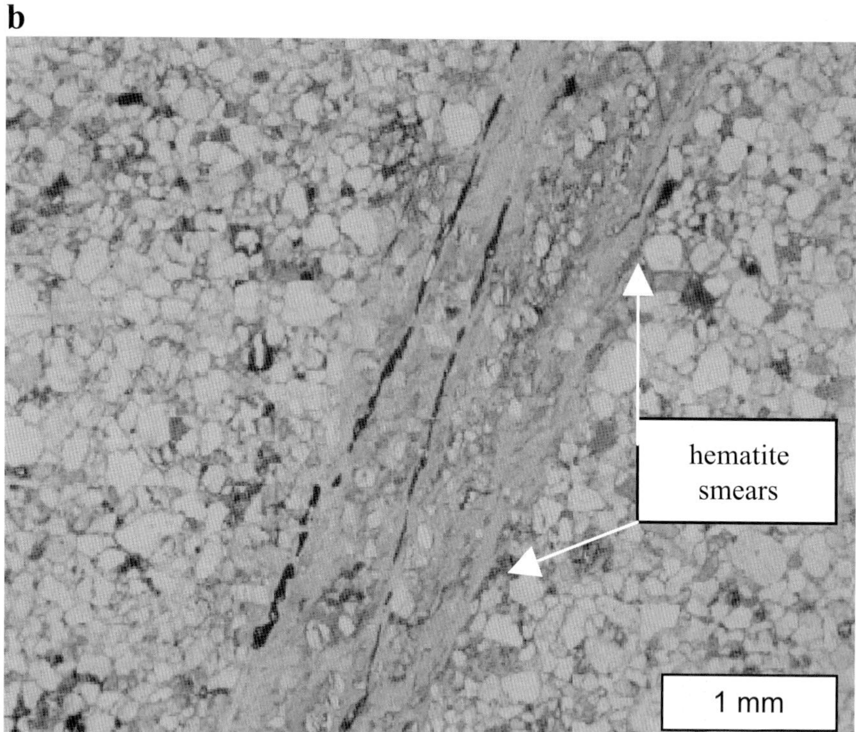

Figure 7. (a) Axial view of deformation bands in the Slickrock Subkha sample. The sample was fractured at 41.3 MPa and then slid at 55.5 MPa effective confining pressure. A single deformation band in the middle of the sample (bottom) splits into several strands against the piston (top end), and new strands conjugate to these strands start forming. (b) Thin-section photomicrograph showing that very few deformation bands develop in these samples despite high strains (5%). Note the hematite smears along the edges of the fault zone and the lack of dilatant microcracks around the fault zone.

Equation 7a describes the evolution of bulk permeability with increasing postfailure strain at a constant effective confining pressure and can be evaluated for samples deformed at different effective confining pressures to place constraints on the geometric significance of the different parameters.

The use of the geometric mean in equation 5 can be justified on physical grounds. By analogy with variations in hydraulic apertures, the harmonic mean corresponds to the case in which the permeability varies only along the flow direction, whereas the arithmetic mean describes variation transverse to the flow direction (Silliman, 1989; Manzocchi et al., 1999). The two means then form the lower and upper bounds, respectively, for the general case of three-dimensional heterogeneity (Zimmermann and Main, 2004). In this case, the bulk permeability is best represented by the intermediate geometric mean (Dagan, 1993) as used here.

Evaluating the Model

Equation 7a was fitted to the permeability data against inelastic axial strain for the Navajo and Slickrock Subkha sandstones for which we had at least four different effective confining pressures. The fitting was carried out using a nonlinear least-squares routine based on the Marquardt-Levenberg algorithm available in Numerical Recipes and elsewhere (Ngwenya et al., 2001) with k_c fixed. The correlation coefficients, together with the residual sum of squares, were used to assess goodness of fit to the data. Table 1 is a compilation of

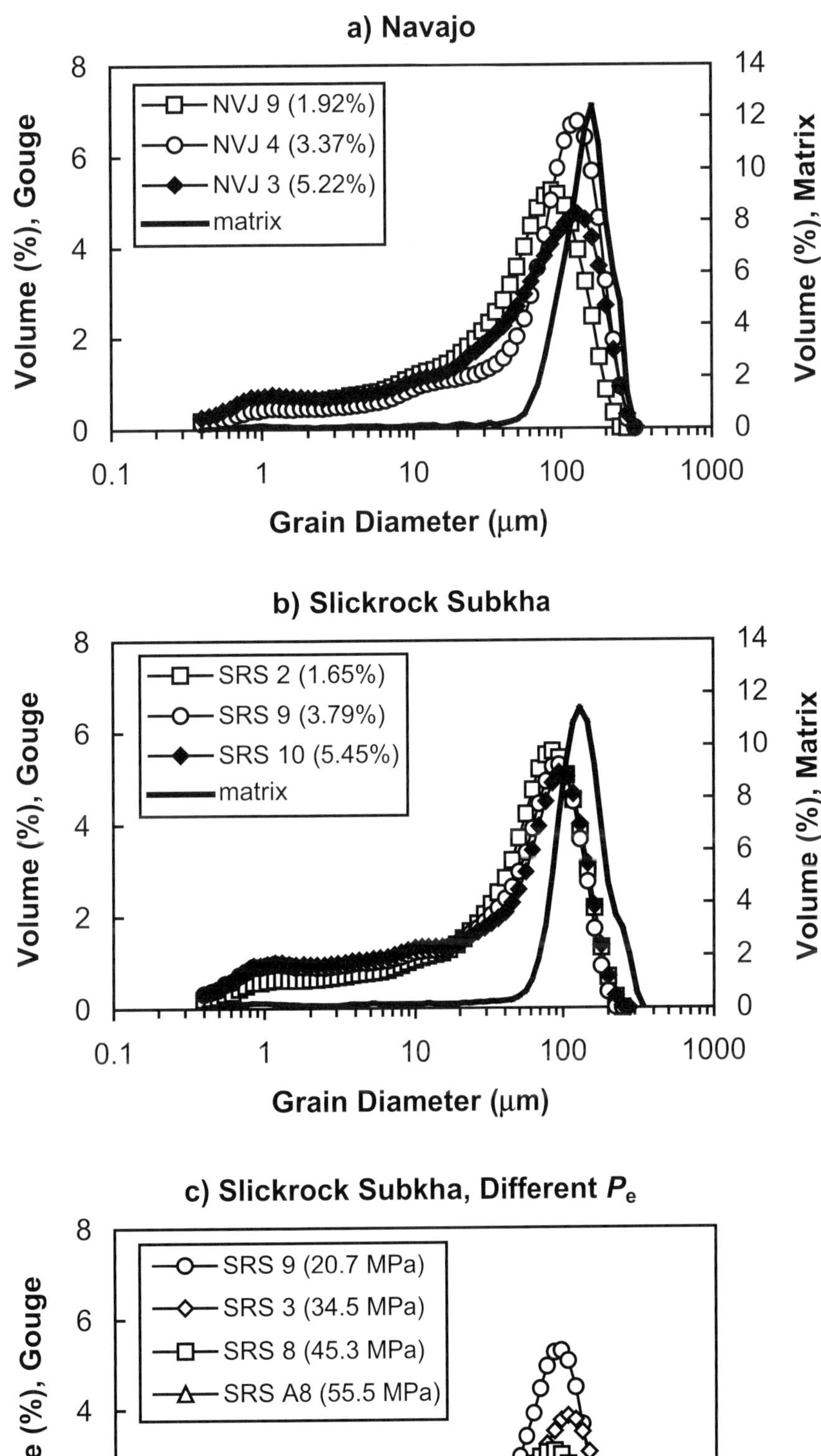

Figure 8. Graphs showing grain-size distribution at the same effective confining pressure but varying total axial strain for samples of (a) Navajo Sandstone and (b) Slickrock Subkha sandstone. Graph (c) shows grain-size distribution of the Slickrock Subkha sandstone deformed to similar total axial strain but different effective confining pressures. Note the increase in the volume fraction of finer grains at high effective confining pressures.

the constants a_0, a_1, and a_2 based on fitting equation 7a to the data by regression, along with regression coefficients in each case. Selected model curves are shown in Figures 9 and 10 for Navajo and Slickrock Subkha sandstones, respectively. The model provides a very accurate description of the experimental data, even in cases where there is no positive correlation caused by the low strain sensitivity.

Table 1 also shows that a_0 values are small. Maximum estimates of w/l_c are of the order of 9×10^{-3} and are based on shear band thickness and length at failure. With this estimate, the minimum a_0 value of -0.021 translates into a decrease in permeability of the fault by one order of magnitude relative to k_c. Generally, however, a_0 values are close to zero, suggesting that the permeability of the shear band does not change significantly with inelastic axial strain (i.e., $k_c \approx k_f$ in equation 7b). This is consistent with the similarity of grain-size distributions shown in Figure 8a and b and also validates assumption 2 of the model.

Because a_0 values are close to zero, we also tested a two-parameter model with $a_0 = 0$ and k_c again fixed, with the constants designated b_1 and b_2 instead of a_1 and a_2 being shown in Table 2. In all cases, the regression coefficient is greater for the three-parameter model. However, the regression coefficient alone is known to be biased in favor of models of greater complexity.

Table 1. Parameters, correlation coefficients, and Bayesian information criteria (BIC) obtained by fitting the experimental data to the three-parameter model (equation 7a). Note that a_o values are close to zero, suggesting that permeability of the fault zone (k_f) does not change significantly during deformation. Permeabilities are all in square meters.

Curve Fit Parameters for the Three-parameter Model of Equation 7a							
Sample	P_e *(MPa)*	*ln* k_c	a_0	a_1	a_2	r^2	*BIC*
NVJ 3	20.7	−28.8	0.004 ± 0.001	−3.48 ± 0.15	17.4 ± 3.8	0.735	334
NVJ 6	34.5	−29.4	0.011 ± 0.001	−14.3 ± 0.27	196.2 ± 14.5	0.934	390
NVJ 8	44.6	−28.9	0.031 ± 0.005	−25.0 ± 0.69	109.2 ± 17.8	0.987	119
NVJ 5	55.4	−29.1	−0.021 ± 0.002	−7.59 ± 0.34	−723.1 ± 9.9	0.995	212
SRS 9	20.7	−32.7	−0.012 ± 0.001	−1.23 ± 0.15	77.1 ± 6.9	0.125	1007
SRS 3	34.5	−30.7	0.012 ± 0.001	−7.24 ± 0.14	−72.1 ± 4.6	0.994	406
SRS 8	45.3	−31.8	−0.007 ± 0.001	−2.23 ± 0.14	−170.4 ± 4.0	0.988	632
SRS A8	55.5	−31.9	−0.003 ± 0.001	−8.24 ± 0.13	−45.9 ± 4.0	0.992	683

Hence, to compare the two fits, we used a modified version of Schwarz's information criterion, which uses Bayesian theory to penalize increased complexity in mathematical models (Main et al., 1999; Ngwenya et al., 2003) to demonstrate which of the two models is justified. To do this, it is necessary to compare Bayesian information criteria (BIC) values listed in the final column of Tables 1 and 2 for the same test (BIC fluctuations between tests contain no information; only comparisons between different models on the same data set have value). We conclude from this exercise that the three-parameter model was not justified by the data, because BIC_3 (Table 1) is less than BIC_2 (Table 2), where the subscript denotes the number of fitted parameters. This is illustrated visually in Figures 9 and 10, where the three-parameter model fails to improve the curve fitting to the data in a significant way. The superiority of the two-parameter model is further evidence that the permeability of the fault does not change significantly during postfailure deformation. In the discussion that follows, only values obtained from the two-parameter model of Table 2 are used.

The parameter γ_I lumps together all the effects that determine the evolving geometry of the flow path and, hence, encapsulates the whole geometric model for permeability evolution (cf. Ngwenya et al., 2003). By definition, $\gamma_I = \eta\lambda$ (equation 4), where η defines the contribution of the matrix to the overall rate of deformation. Comparison of equations 7c and d suggests that we should expect $b_2 \approx -b_1$ because w is much less than l_c. However, this is not borne out by values for b_1 and b_2 determined by the curved fits, except for two Navajo tests (NVJ3 and NVJ6). We are not able to resolve the reason for this discrepancy at present, suggesting that although equation 7a is a good statistical model for the data or for practical prediction of fault sealing after appropriate calibration, any physical interpretation of these two parameters should be treated with caution.

Table 2 also shows that γ_I is greater than 0 for all Navajo samples, except the one test at the highest effective confining pressure (NVJ5). Meanwhile, λ is less than 0 for all these tests (Figure 3b), although NVJ5 might be transitional. It follows that with the exception of NVJ5, η is less than 0 for all the Navajo samples, because $\gamma_I = \eta\lambda$. These values are, however, associated with decreasing permeability as a function of inelastic axial strain (Figure 3a). The apparent paradox of decreasing bulk permeability in the face of global dilation ($\lambda < 0$) has previously been interpreted to result from increasing tortuosity of the flow path (Zhu and Wong, 1996; Main et al., 2000; Ngwenya et al., 2003). Because the inferred permeability of the fault remains approximately constant during deformation, we infer that most of the dilation occurs in the matrix around the fault. This is consistent with microstructural observations of localization of aligned dilatant microcracks (e.g., Ngwenya et al., 2000).

Although empirical and limited in applicability to natural data set, the Kozeny-Carman relation in the form of equation 2 has been found to fit a large number of data from deformation experiments (Zhu and Wong, 1997; Ngwenya et al., 2003). Using this relation in the form given by Walsh and Brace (1984)

$$k_{\text{bulk}} = \frac{\phi^3}{b\tau^2 S^2} \tag{8}$$

where ϕ is the porosity; b is a shape factor (dependent on the geometry of the pores); τ is the tortuosity of the flow path; and S is the specific surface area per unit volume of the porous medium, we identify two other mechanisms to explain the apparent paradox of decreasing permeability in a dilating rock: (1) the geometry of the connected pores evolves from approximate circular tubes where $b = 2$ to planar cracks with $b = 3$; and (2) the development of microcracks increases the surface area-to-volume ratio in the rock (Ojala et al., 2003). In fact, mechanisms 1 and 2 are likely to be as important as changes in tortuosity, unless the axial microcracks that develop are highly inclined to the flow direction (see figure 6.3 in Middleton and Wilcock, 1994).

In contrast to the Navajo samples, Slickrock Subkha samples yield γ_l less than 0, except SRS9, which was deformed at low effective confining pressure, whereas λ is less than 0 for all these tests (Figure 3d). These trends lead to η greater than 0 for most Slickrock Subkha samples and η less than 0 for SRS9. As with Navajo samples, these values of η greater than 0 are associated with decreasing permeability as a function of inelastic axial strain, whereas the permeability of SRS9 increases slightly (Figure 3c). By contrast, with Navajo samples, for η greater than 0 to be coupled to reduction in permeability in the face of global dilation implies that the matrix must be compacting (equation 2), because k_f is approximately constant. It follows that η may be regarded as an inelastic strain sensitivity index for the matrix, such that when η is less than 0, the matrix dilates, whereas compaction is implied when η is greater than 0. Thus, Navajo samples dilate at most of the pressures used in this study, except at the highest effective confining pressure of 55.4 MPa. Meanwhile, only the sample at low effective confining pressure experiences dilation in the matrix in the Slickrock Subkha tests (Table 2).

The lack of multiple deformation bands in samples deformed at higher effective confining pressures in Figure 6 is further evidence that compaction dominates matrix deformation in Slickrock Subkha samples. However, these samples also dilate globally (Figure 3d); hence, the fault zone must be dilating overall to account for the observed global dilation. This occurs despite the compaction involved in producing a poorly sorted close-packed fault gouge and is not consistent with the model results, which suggest that the permeability of the fault zone does not change during deformation. Experimental studies have shown that dilation is commonly observed at high strains because of shear localization via formation of secondary Reidel shears in the fault zone (Zhang and Tullis, 1998). Although evidence for their presence is not convincing in the samples examined here, the enhanced grain comminution that accompanies this process can offset the expected increase in permeability. More studies, perhaps involving hypocentral mapping of acoustic emissions, are required to resolve the relative contribution of matrix and fault zone volume changes.

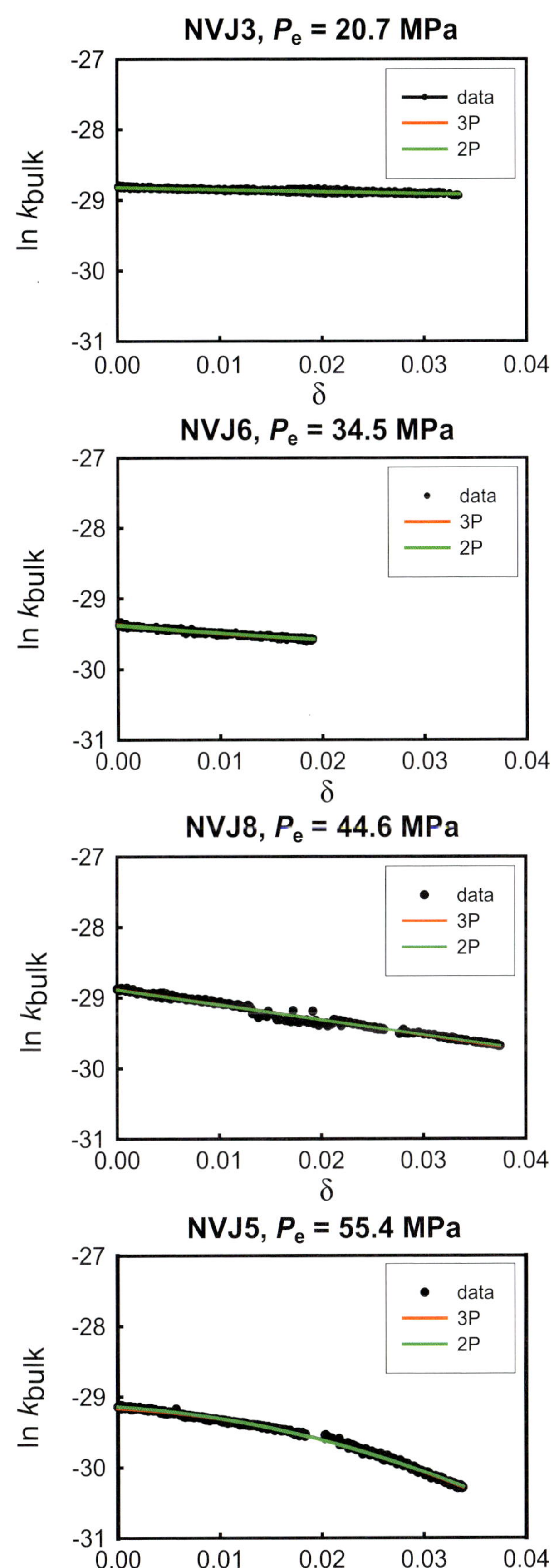

FIGURE 9. Graphs showing the comparison of experimental data in the postfailure phase with model predictions based on three-parameter and two-parameter evaluations of equation 7a for the Navajo Sandstone. Note that the three-parameter model does not improve data fitting, as shown by BIC data (Tables 1 and 2).

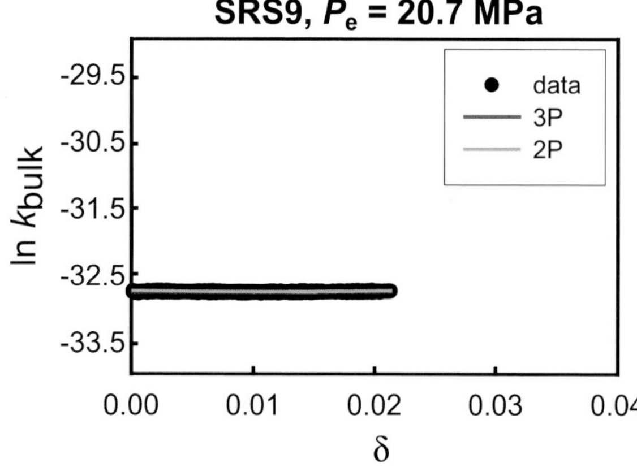

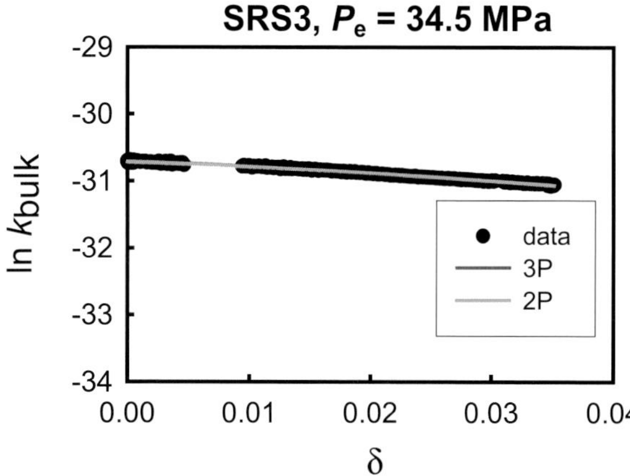

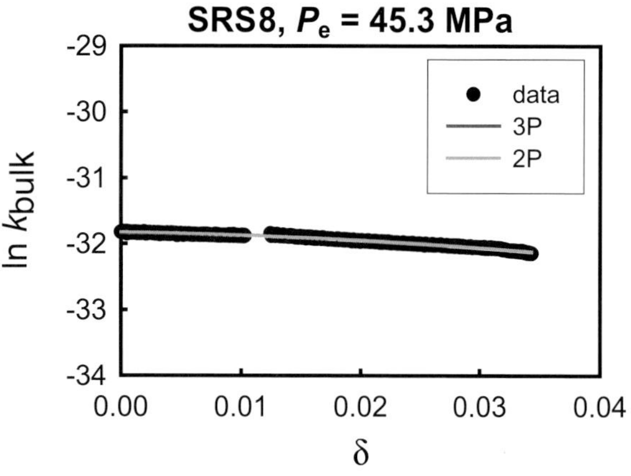

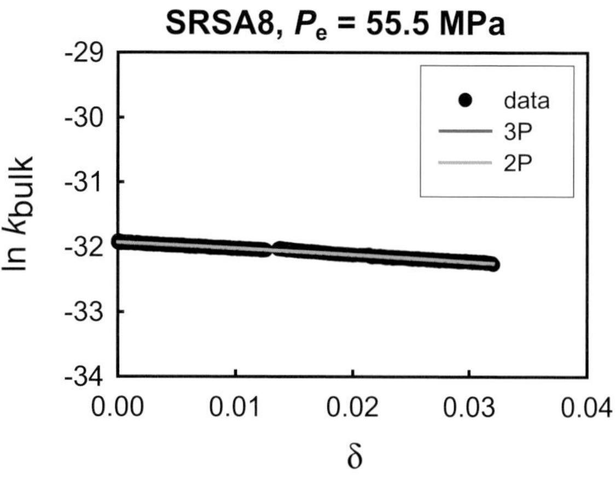

FIGURE 10. Graphs showing the comparison of experimental data in the postfailure phase with model predictions based on three-parameter and two-parameter evaluations of equation 7a for the Slickrock Subkha sandstone. Note here that the two-parameter model is also superior to the three-parameter model (Tables 1 and 2).

Implications for Fault-seal Evolution

Although we have not specifically quantified fault sealing in this study, the results have implications on predicting the evolution of fault seals in siliciclastic reservoirs of varying petrophysical properties. Previous petrographic analysis and petrophysical measurements of outcrop and core samples show that for a given stress state, the sealing properties of a fault are a function of lithologic juxtaposition, fault rock types (clay vs. cataclastic gouge), diagenetic cementation (Knipe, 1992), and host rock composition (e.g., Storti et al., 2003). Some of these measurements also reveal that clay gouge forms more effective seals relative to cataclastic faults, which support smaller columns (e.g., Smith, 1980; Knipe, 1992; etc.). Moreover, significant improvement in sealing capacity can be achieved with diagenetic overprinting of the cataclastic fault gouge (Fulljames et al., 1997; Ngwenya et al., 2000; Fisher and Knipe, 2001). Thus, lithology, and clay content in particular, exerts a primary control on fault-seal intensity (Fisher and Knipe, 1998). In addition, evidence from reservoir core studies indicates little variation in cataclastic sealing for siliciclastic rocks containing as much as 5% clay (Fisher and Knipe, 1998). We therefore expected little difference in behavior among Moab sandstones, because there was very little variation in clay content. The role of hematite cementation (Slickrock Subkha and Dewey Bridge) is unknown, but it may act as a lubricant for grain boundary sliding (Ngwenya et al., 2000). However, they may also form discontinuous stringers similar to phyllosilicate domains (Fisher and Knipe, 1998) in the deformation band, which are likely to accelerate permeability impairment. We have seen evidence of such stringers (smears) on the margins of deformation bands in thin section (see Figure 7).

We assume, therefore, that the increasing permeability with inelastic axial strain shown by Dewey Bridge samples at 20.7 MPa, which contrasts with other rock types, is caused primarily by petrophysical and/or textural properties of this rock. Previous studies have documented a log-linear relationship between permeability reduction (defined as k_f/k_m) and porosity of the matrix (Fulljames et al., 1997) for cataclastic deformation bands, where k_f is the permeability of the deformation band, and k_m is the permeability of the matrix. Available data show that for porosity below about 16%, the permeability of the deformation band is likely to be higher than that of the matrix ($k_f/k_m > 1$) for active

Table 2. Parameters, correlation coefficients, and Bayesian information criteria (BIC) obtained by fitting the experimental data to the two-parameter (without a_0) model of equation 7a (see text for details). Note the consistently higher BIC values compared with the three-parameter model, which shows that the two-parameter model is superior to the three-parameter one. Also included are the steady-state permeability values measured under hydrostatic conditions (k_o), which are used to normalize permeabilities in Figures 2 and 3.

Curve Fit Parameters for the Two-parameter Model of Equation 7a with $a_o = 0$

Sample	*P_e (MPa)*	*ln k_0*	*ln k_c*	*b_1*	*b_2*	*r^2*	*BIC*
NVJ 3	20.7	−28.6	−28.8	−2.94 ± 0.07	3.74 ± 2.96	0.733	500
NVJ 6	34.5	−29.1	−29.4	−11.8 ± 0.16	90.6 ± 11.1	0.929	535
NVJ 8	44.6	−28.5	−28.9	−21.6 ± 0.39	33.4 ± 12.9	0.985	168
NVJ 5	55.4	−28.8*	−29.1	−10.1 ± 0.20	−660.2 ± 7.2	0.995	286
SRS 9	20.7	−32.4	−32.7	−3.53 ± 0.09	167.5 ± 5.3	0.00	1364
SRS 3	34.5	−30.4	−30.7	−5.98 ± 0.15	−101.0 ± 5.4	0.990	524
SRS 8	45.3	−31.4	−31.8	−3.02 ± 0.08	−151.3 ± 2.83	0.987	933
SRS A8	55.5	−31.5*	−31.9	−8.64 ± 0.07	−35.3 ± 2.77	0.992	1025

**Measured at P_e = 41.7 MPa.*

faults. As porosity decreases further, deformation is dominated by sample dilation, particularly with increasing strain, which reduces the rate of permeability impairment (Zhu and Wong, 1996). It was not possible to investigate the dependence of permeability reduction on porosity in this study, mainly because of lack of sufficient samples and because only bulk permeability values were measured. Nevertheless, the small differences in the strain dependency of permeability between the Navajo Sandstone and other Moab rock types are likely to reflect porosity differences.

The field data of Fulljames et al. (1997) also show that for a given porosity, cataclastic sealing intensity increases with depth. This occurs as a result of increased grain comminution for active faults (Crawford, 1998) or because of diagenetic cementation of faults of shallow origin (Fisher and Knipe, 2001). Using the model of Main et al. (2000), Ngwenya et al. (2003) published an empirical equation for predicting permeability of cataclastic faults as a function of inelastic axial strain and effective confining pressure for Clashach sandstone. However, the parameters for both Navajo and Slickrock Subkha samples do not vary systematically with effective confining pressure. In general, both b_1 and b_2 decrease systematically with increasing effective confining pressure (Figure 11), but the correlation coefficients are very low ($r^2 < 0.514$). Nevertheless, the trends suggest that permeability will decrease with increasing effective confining pressure and inelastic axial strain for the rock types examined here. However, in comparison to high-porosity sandstones, cataclastic fault seals in the low-porosity sandstones examined here are less sensitive to burial depth. This can also be seen by comparing the two rock types in this study, where the sensitivity to effective stress is lower for the low-porosity Slickrock Subkha samples than for the Navajo samples (Figure 11). The low sensitivity to stress is most likely caused by the fact that deformation is dominated by shear-induced dilation of the shear band at high strain. This may indicate that a more complex physical model is needed, with matrix, fault gouge, and near-fault localized deformation all being separately included to interpret the data.

CONCLUSIONS

We have reported an experimental program designed to investigate the evolution of permeability during deformation of siliciclastic sandstones with varying petrophysical properties and/or mineralogy. The results show subtle differences in strain-dependent permeability evolution, with low-porosity sandstones exhibiting an increase in permeability after fault development at low effective confining pressure, whereas high-porosity sandstones show a permeability decline once the fault is formed. A geometric model based on the rate of growth of the shear band width as a function of inelastic axial strain provides a good statistical description of the data at different effective confining pressures. The observation of a slow increase in the width of the zone of deformation bands with increasing strain is borne out by the model parameters. Unlike previous results (Ngwenya et al., 2003), the parameters of the model do not vary systematically with effective confining pressure in the range studied here most likely because of sample dilation at high strains. Nevertheless, the model appears to

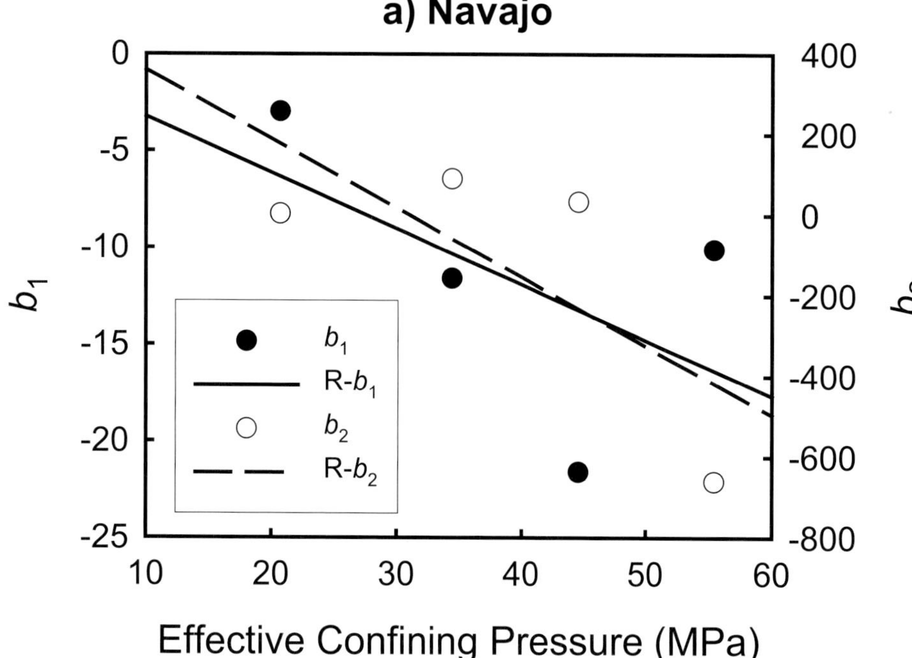

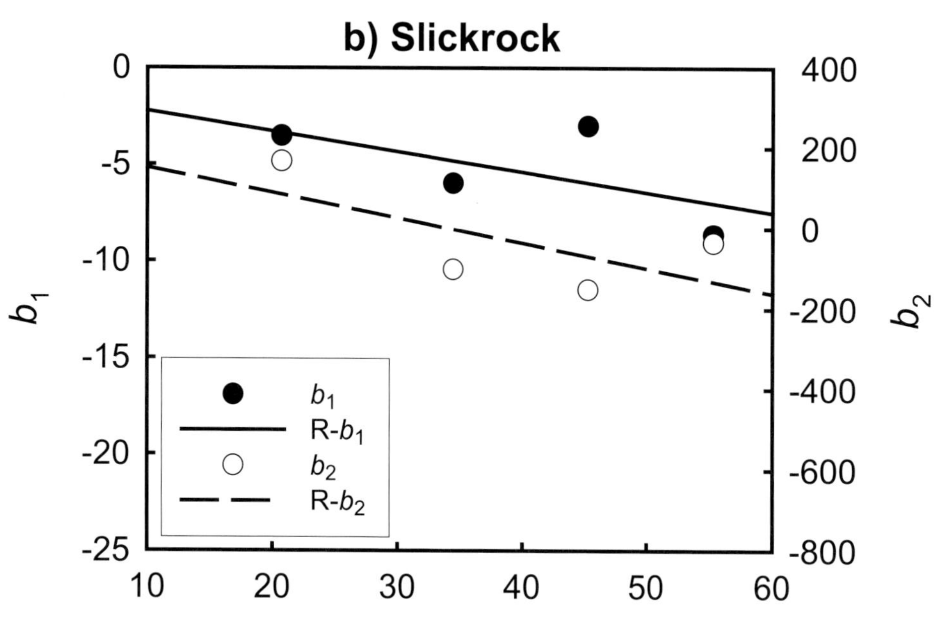

FIGURE 11. Plots of the parameters b_1 and b_2 obtained from the permeability evolution model (Table 2) against effective confining pressure for Navajo and Slickrock Subkha samples. Note that for these sandstones, no systematic variation of these parameters with effective confining pressure is present, although the Navajo Sandstone is more stress sensitive.

be robust and suggests that cataclastic faults in low-porosity sandstones have a lower sensitivity to burial depth than those with high porosity.

ACKNOWLEDGMENTS

Core material for this study was kindly provided by David Yale of ExxonMobil. The study was supported financially by BP-Amoco, ExxonMobil, Lasmo, Shell, Statoil, and the Japanese National Oil Corporation (presently Japan Oil, Gas and Metals National Corporation). Constructive discussions with Jan Konstanty, Jonathan Craig, Andrew McCann, Miki Takahashi, David Yale, Kes Heffer, Chris Townsend, Bill Shea, Brian Crawford, and Fredrick Dula were very helpful. Mike Hall and Alex Jackson provided valuable laboratory support, and Yvonne Fletcher helped with photography. Thoughtful reviews by Teng-Fong Wong and Yves Guegen helped us to clarify some of our arguments. Our final thanks go to Rasoul Sorkhabi for his support during the preparation of this contribution.

REFERENCES CITED

Berg, R. R., 1975, Capillary pressure in stratigraphic traps: AAPG Bulletin, v. 59, p. 939–956.

Crawford, B. R., 1998, Experimental fault sealing: Shear band permeability dependency on cataclastic fault gouge characteristics, *in* M. P. Coward, S. Daltaban, and H. Johnson, eds., Structural geology in reservoir characterization and field development: Geological Society (London) Special Publication 127, p. 27–47.

Dagan, G., 1993, High-order correction for effective permeability of heterogeneous isotropic formations of lognormal conductivity distribution: Transport in Porous Media, v. 12, p. 279–290.

Fisher, Q. J., and R. J. Knipe, 1998, Fault sealing processes in siliciclastic sediments, *in* G. Jones, Q. J. Fisher, and R. J. Knipe, eds., Faulting, fault sealing and fluid flow in hydrocarbon reservoirs: Geological Society (London) Special Publication 147, p. 117–134.

Fisher, Q. J., and R. J. Knipe, 2001, The permeability of faults within siliciclastic petroleum reservoirs of the North Sea and Norwegian continental shelf: Marine and Petroleum Geology, v. 18, p. 1063–1081.

Foxford, K. A., J. J. Walsh, J. Waterson, I. R. Garden, S. C. Guscott, and S. D. Burley, 1998, Structure and content of the Moab fault zone, Utah, U.S.A., and its implications for fault seal prediction, *in* G. Jones, Q. J. Fisher, and R. J. Knipe, eds., Faulting, fault sealing and fluid flow in hydrocarbon reservoirs: Geological Society (London) Special Publication 147, p. 87–103.

Fulljames, J. R., L. J. J. Zijerveld, and R. C. M. W. Franssen, 1997, Fault seal processes: Systematic analysis of fault seals over geological and production time scales, *in* P. Moller-Pedersen and A. G. Koestler, eds., Hydrocarbon seals: Importance for exploration and production: Norwegian Petroleum Society Special Publication 7, p. 51–59.

Grueschow, E., O. Kwon, I. G. Main, and J. W. Rudnicki, 2003, Observation and modeling of the suction pump effect during rapid dilatant slip: Geophysical Research Letters, v. 30, no. 5, 1226, doi: 10.1029/2002GL015905.

Holdsworth, R. E., W. R. Bailey, J. Imber, C. A. Butler, and G. E. Lloyd, 1997, The structural role played by fluids during reactivation of mid-crustal and upper mantle fault zones: Geofluids, v. 97, p. 103–106.

Ingram, G. M., J. L. Urai, and M. A. Naylor, 1997, Sealing processes and top seal assessment, *in* P. Moller Pedersen and A. G. Koestler, eds., Hydrocarbon seals: Importance for exploration and production: Norwegian Petroleum Society Special Publication 7, p. 165–174.

Knipe, R. J., 1992, Faulting processes and fault seal, *in* R. M. Larsen, H. Brekke, B. T. Larsen, and E. Talleraas, eds., Structural and tectonic modelling and its application to petroleum geology: Norwegian Petroleum Society Special Publication 1, p. 325–342.

Knipe, R. J., 1997, Juxtaposition and seal diagrams to analyse fault seals in hydrocarbon reservoirs: AAPG Bulletin, v. 81, p. 187–195.

Knipe, R. J., G. Jones, and Q. J. Fisher, 1998, Faulting, fault sealing and fluid flow in hydrocarbon reservoirs: An introduction, *in* G. Jones, Q. J. Fisher, and R. J. Knipe, eds., Faulting, fault sealing and fluid flow in hydrocarbon reservoirs: Geological Society (London) Special Publication 147, p. vii–xxi.

Macaulay, C., I. Boyce, A. J. Fallick, and R. S. Haszeldine, 1997, Quartz veins record vertical fluid flow at Graben Edge: Fulmar oil field, central North Sea: AAPG Bulletin, v. 81, p. 2024–2035.

Main, I. G., B. G. D. Smart, G. B. Shimmield, S. C. Elphick, B. R. Crawford, and B. T. Ngwenya, 1994, The effects of combined changes in pore fluid chemistry and stress state on permeability in reservoir rocks: Preliminary results from analogue materials, *in* J. O. Aasen, ed., North sea oil and gas reservoirs III: Dordrecht, Kluwer, p. 357–370.

Main, I. G., T. Leonard, O. Papastiliotis, C. G. Hatton, and P. G. Meredith, 1999, One slope or two? Detecting statistically significant breaks of slope in geophysical data, with application to fracture scaling relationships: Geophysical Research Letters, v. 26, p. 2801–2804.

Main, I. G., O. Kwon, B. T. Ngwenya, and S. C. Elphick, 2000, Fault sealing during deformation-band growth in porous sandstone: Geology, v. 28, p. 1131–1134.

Mair, K., I. G. Main, and S. C. Elphick, 2000, Sequential development of deformation bands in the laboratory: Journal of Structural Geology, v. 22, p. 25–42.

Mair, K. M., S. E. Elphick, and I. G. Main, 2002, Influence of confining pressure on the mechanical and structural evolution of laboratory deformation bands: Geophysical Research Letters, v. 29, no. 10, 1410, DOI: 10.1029/2001GL013964.

Manzocchi, T., J. J. Walsh, P. Nell, and G. Yielding, 1999, Fault transmissibility multipliers for flow simulation models: Petroleum Geoscience, v. 5, p. 53–63.

Middleton, G. V., and P. R. Wilcock, 1994, Mechanics in the earth and environmental sciences: Cambridge, Cambridge University Press, 459 p.

Ngwenya, B. T., S. C. Elphick, and G. B. Shimmield, 1995, Reservoir sensitivity to water flooding: An experimental study of seawater injection in North Sea reservoir analogs: AAPG Bulletin, v. 79, p. 285–304.

Ngwenya, B. T., S. C. Elphick, I. G. Main, and G. B. Shimmield, 2000, Experimental constraints on the diagenetic sealing faults in porous sandstones: Earth and Planetary Science Letters, v. 183, p. 187–199.

Ngwenya, B. T., I. G. Main, S. C. Elphick, B. R. Crawford, and B. G. D. Smart, 2001, A constitutive law for low-temperature creep of water-saturated sandstones: Journal of Geophysical Research, v. 106, p. 21,811–21,826.

Ngwenya, B. T., O. Kwon, S. C. Elphick, and I. G. Main, 2003, Permeability evolution during progressive development of deformation bands in porous sandstones: Journal of Geophysical Research, v. 108, no. B7, 2343, DOI: 10.1029/2002JB001854.

Ojala, I. O., B. T. Ngwenya, I. G. Main, and S. C. Elphick, 2003, Correlation of microseismic and chemical properties of brittle deformation in Locharbriggs sandstone: Journal of Geophysical Research, v. 108, no. B5, p. 2268, DOI: 10.1029/2002JB002277.

Shipton, Z. K., and P. A. Cowie, 2001, Damage zone and slip-surface evolution over μm to km scales in high-porosity Navajo Sandstone, Utah: Journal of Structural Geology, v. 23, p. 1825–1844.

Sibson, R., 1990, Conditions for fault valve behaviour, *in* R. J. Knipe and E. H. Rutter, eds., Deformation mechanisms, rheology and tectonics: Geological Society (London) Special Publication 54, p. 15–28.

Silliman, S. E., 1989, Interpretation of the difference between aperture estimates derived from hydraulic tracer tests in a single fracture: Water Resources Research, v. 25, p. 2275–2283.

Smith, D. A., 1966, Theoretical consideration of sealing and non-sealing faults: AAPG Bulletin, v. 50, p. 363–374.

Smith, D. A., 1980, Sealing and nonsealing faults in Louisiana Gulf Coast salt basin: AAPG Bulletin, v. 64, p. 145–172.

Storti, F., A. Billi, and F. Salvini, 2003, Particle size distribution in natural carbonate fault rocks: Insights for non-self similar cataclasis: Earth and Planetary Science Letters, v. 206, p. 173–186.

Teufel, L. W., 1987, Permeability changes during shearing deformation in fractured rock: Proceedings on 28th U.S. Rock Mechanics Symposium, Brookfield, A. A. Balkema, p. 473–480.

Walsh, J. B., and W. F. Brace, 1984, The effect of pressure on porosity and the transport properties of rock: Journal of Geophysical Research, v. 89, p. 9425–9431.

Watts, N. L., 1987, Theoretical aspects of cap-rock and fault seals for single and two-phase hydrocarbon columns: Marine and Petroleum Geology, v. 4, p. 274–307.

Yielding, G., B. Freeman, and D. T. Needham, 1997, Quantitative fault seal prediction: AAPG Bulletin, v. 81, p. 897–917.

Zhang, S., and T. E. Tullis, 1998, The effect of fault slip on permeability and permeability anisotropy in quartz gouge: Tectonophysics, v. 295, p. 41–52.

Zhu, W., and T.-F. Wong, 1996, Permeability reduction in a dilating rock: Network modelling of damage and tortuosity: Geophysical Research Letters, v. 23, p. 3099–3102.

Zhu, W., and T.-F. Wong, 1997, The transition from brittle faulting to cataclastic flow: Permeability evolution: Journal of Geophysical Research, v. 102, p. 3027–3041.

Zimmermann, R. W., and I. G Main, 2004, Hydromechanical behaviour of fractured rocks, *in* Y. Gueguen and M. Bouteca, eds., Mechanics of fluid-saturated rocks: London, Academic Press, p. 363–422.

14

Milliken, K. L., R. M. Reed, and S. E. Laubach, 2005, Quantifying compaction and cementation in deformation bands in porous sandstones, *in* R. Sorkhabi and Y. Tsuji, eds., Faults, fluid flow, and petroleum traps: AAPG Memoir 85, p. 237–249.

Quantifying Compaction and Cementation in Deformation Bands in Porous Sandstones

K. L. Milliken

Department of Geological Sciences, John A. and Katherine G. Jackson School of Geosciences, The University of Texas at Austin, Austin, Texas, U.S.A.

R. M. Reed

Bureau of Economic Geology, John A. and Katherine G. Jackson School of Geosciences, The University of Texas at Austin, Austin, Texas, U.S.A.

S. E. Laubach

Bureau of Economic Geology, John A. and Katherine G. Jackson School of Geosciences, The University of Texas at Austin, Austin, Texas, U.S.A.

ABSTRACT

Combined electron microbeam imaging techniques can be used to quantify cementation and compaction processes in deformation bands and their surrounding host rocks. Mixed secondary and backscattered electron signals can be used to definitively identify pore space, whereas scanned cathodoluminescence can be used to discriminate between detrital and authigenic quartz. Classic deformation bands from three porous sandstone units, the Cambrian Hickory Sandstone of central Texas (two bands and two host rocks), the Pennsylvanian Tensleep Sandstone of Wyoming (one band and one host rock), and the Pennsylvanian Weber Sandstone of northwestern Colorado (one band and one host rock), were examined using these imaging techniques.

Cathodoluminescence images demonstrate that bands develop through a combination of grain-scale brittle processes and cementation. Point counting of scanning electron microscopy image mosaics reveals that the intergranular volume in deformation bands is higher than is apparent from transmitted light microscopy. Cementation equals or exceeds compaction as a cause of porosity decline in both deformation bands and host rocks. No evidence exists for significant pressure solution during band development. The intergranular volumes of host rocks in the range of 31–38% suggest that all of these samples have experienced burial of 2 km (1.2 mi) or less.

DOI:10.1306/1033726M85252

Contrasts in the compactional and cementational states of bands and surrounding host rocks possibly reflect the differing availability of quartz nucleation surfaces in these different parts of the rock. Preferential emplacement of cement in the bands can lead to divergent paths of compactional behavior in bands relative to host rocks during the postkinematic phase of their burial history.

INTRODUCTION

Deformation bands form in porous sandstones and show a localized decrease in permeability and increased lithification relative to the surrounding sandstone (Aydin, 1978; Aydin and Johnson, 1978; Pittman, 1981; Hill, 1989; Edwards et al., 1993; Antonellini and Aydin, 1994, 1995; Antonellini et al., 1994a, b; Fowles and Burley, 1994; Davis, 1999; Cashman and Cashman, 2000; Taylor and Pollard, 2000). Hill (1989) and Mollema and Antonellini (1996) interpret some bands to be zones of localized compaction based on field relations that demonstrate that the tabular zones are not faults (have no offset) and petrographic evidence that porosity in the bands is reduced. Subsequently, triaxial deformation experiments (Olsson, 1999) and theoretical studies (Issen and Rudnicki, 2000) showed that localized compaction might readily occur in sandstone. Although natural bands form in rocks that are undergoing diagenetic alteration in reactive fluids (water), these experiments and models focus on the mechanics of band formation and have yet to consider coupling of mechanical and chemical processes in band development. Our observations of natural bands show that band formation likely proceeds through linked chemical and mechanical processes.

Petrophysical modifications in bands correlate with changes in rock texture that are readily observable in hand specimen and in light microscopy (Figure 1). These textural modifications are widely attributed to brittle particle-size reduction during band development. Based primarily on observations in transmitted light microscopy, porosity reduction is also postulated to be an important factor in permeability reduction in the bands (e.g., Antonellini and Pollard, 1995; Mollema and Antonellini, 1996; Cashman and Cashman, 2000; Issen and Rudnicki, 2000). Scanned cathodoluminescence imaging provides an exceptionally clear confirmation of the important function of brittle processes in deformation-band development and also reveals a significant component of quartz cementation in bands (Fowles and Burley, 1994; Milliken, 1996; Reed and Laubach, 1999; Milliken and Laubach, 2000; Besuelle, 2001; Milliken and Reed, 2002).

APPROACH

As a consequence of particle-size reduction in deformation bands, interparticle pore spaces are also reduced in size relative to those in the surrounding sandstone. Accurate assessment of pores having diameters less than the thickness of a thin section (30–35 μm) requires a methodology that yields porosity information through a thinner slice of the rock than does conventional transmitted light microscopy. Backscattered electron imaging represents such a technique and provides clear discrimination of porosity vs. mineral components at higher magnifications than does light microscopy (Krinsley et al., 1983; Henry and Toney, 1987; Primmer and Shaw, 1987; Nadeau and Hurst, 1991; Evans et al., 1994). Authigenic vs. detrital quartz cannot also be correctly assessed using transmitted light microscopy, because the cements form as overgrowths in optical continuity with the detrital grains. Relative differences in cathodoluminescence intensity between detrital and authigenic quartz overcome this problem, allowing accurate determination of both inter- and intragranular quartz cement in sandstones (e.g., Sipple, 1968; Zinkernagel, 1978; Houseknecht, 1991; Hogg et al., 1992; Evans et al., 1994; Milliken, 1994; Dickinson and Milliken, 1995; Laubach and Milliken, 1996; Milliken and Laubach, 2000). This study employs the above-referenced techniques, using a combined secondary and backscattered electron signal to assess porosity and a cathodoluminescence image to assess quartz cement volumes. Quantitative data on grain volumes, porosities, and cement volumes in bands and their surrounding host rocks are then used to compare the functions of compaction and cementation as causes of porosity loss in deformation bands and their enclosing host rocks (Tables 1, 2).

METHODOLOGY AND SAMPLING

Application of the above-described imaging techniques is demonstrated here using samples of classic deformation bands from the Hickory Sandstone (Riley

FIGURE 1. (A) Plane light image of a thin section showing a deformation band cutting a sandstone. Note the porosity reduction in the band relative to the host sandstone. (B) Plane polarized light photomicrograph of deformation band in a horizontal thin section orthogonal to the one shown in (A). Sample is BBD-4 from an outcrop of Tensleep Formation, western Wyoming, United States.

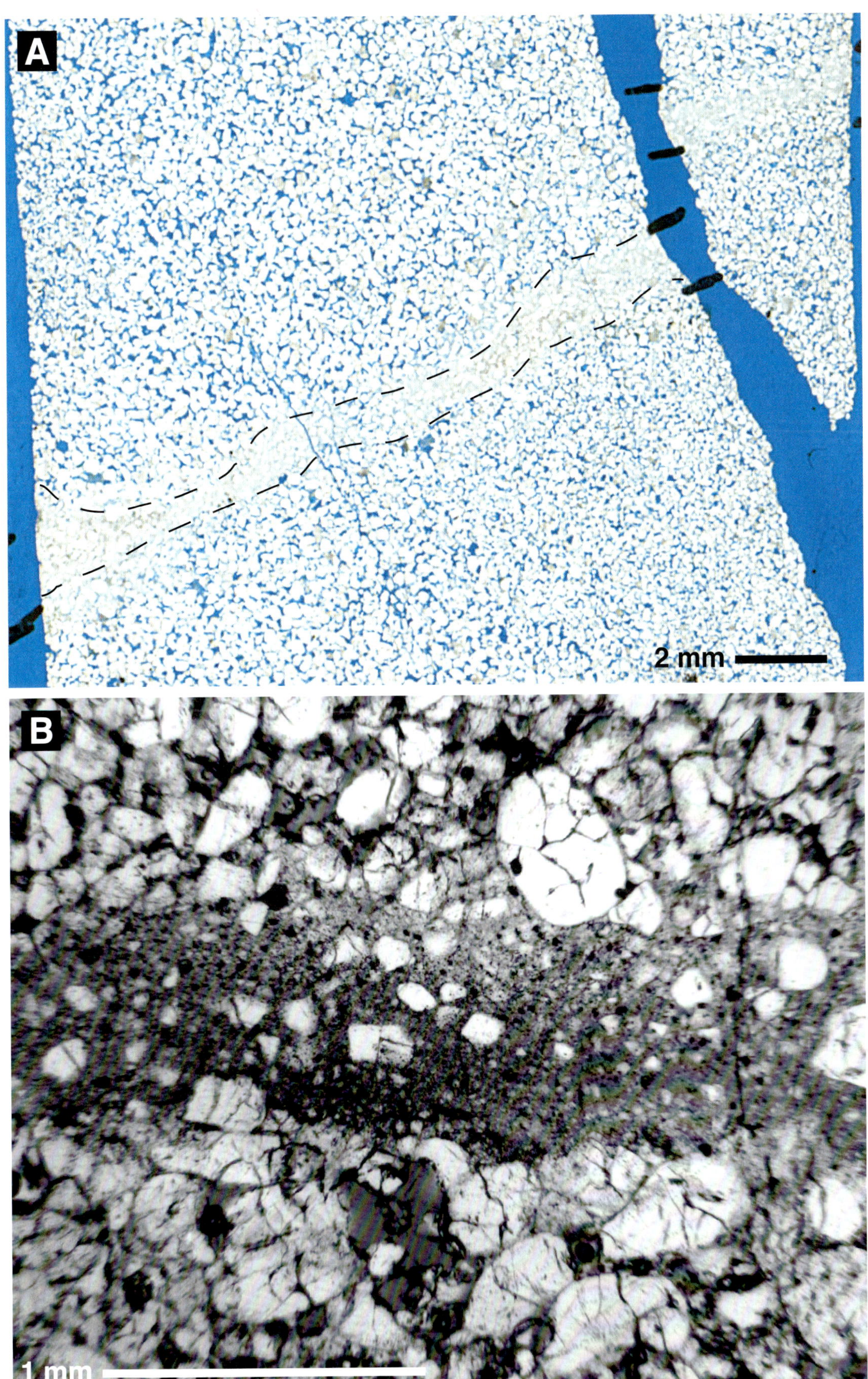
A
2 mm
B
1 mm

Table 1. Point-count data.

Formation	*Sample Number*	*Lithology*	*Points*	*Grain*	*Quartz Cement*	*Quartz Cement in Fracture*	*Other Cements*	*Primary Porosity*	*Secondary Porosity*		*Total Intergranular Cement*	*Grand Total Cement*
Hickory Sandstone	H-5A	subarkose and deformation band	1556	65.4	16.6	0	2.3	15.7	0	100	18.9	18.9
Hickory Sandstone		subarkose and host rock	1198	66.4	6.1	1.6	4.9	21.0	0	100	11	12.6
Hickory Sandstone	H-5B	subarkose and deformation band	2864	62.7	19.9	0.3	1.4	13.9	1.8	100	21.3	21.6
Hickory Sandstone		subarkose and host rock	1274	61.5	13.1	1.7	3.5	19.6	0.6	100	16.6	18.3
Tensleep	BBD-4c	band	1191	60.8	28.2	1.5	0.2	8.4	0.9	100	28.4	29.9
Tensleep		host rock	1234	68.7	8.3	5.0	1.0	16.9	0.1	100	9.3	14.3
Weber	W-3-98	band	1162	62.6	24.9	1.0	0.3	9.5	1.8	100	25.2	26.2
Weber		host rock	1853	61.8	16.3	3.2	0	14.7	3.9	100	16.3	19.5

Formation, Cambrian) of central Texas (McBride et al., 2002; Milliken and Reed, 2002), the Weber Formation of northwestern Colorado (Pennsylvanian–Permian), and the Tensleep Formation (Pennsylvanian–Permian) near Cody, Wyoming (Figure 2). Slabs for thin sectioning were cut approximately perpendicular to representative deformation bands. Samples were vacuum pressure impregnated with blue-dyed resin. Thin sections were ground to standard thickness (30 μm), polished, and carbon coated (25-nm thickness).

Imaging was performed using conventional petrographic light microscopy and scanning electron microscopy (SEM). An accelerating voltage of 10–15 kV with sample current set near 90% of the maximum for the SEM was employed for examining the luminescence variations in authigenic (relatively dark luminescent) and detrital quartz (relatively bright luminescent); secondary electron images were recorded under the same conditions. Color cathodoluminescence images were acquired using RGB (red, green, blue) color filters. In the SEM used, the secondary electron detector collects a sufficient component of backscattered electrons such that atomic weight contrasts are readily discernible in images of flat polished specimens (no topographic contrast).

CATHODOLUMINESCENCE POINT-COUNT TECHNIQUE

Mosaic images were collected of a large area using both cathodoluminescence and secondary electron modes. Care was taken to produce image mosaics without gaps and which covered a representative area of the sample. Separate mosaics were made of areas inside and outside the deformation bands. Different magnifications were used inside and outside the bands. Outside the bands, a lower magnification was used to cover a wide area without having to acquire an excessive number of images. Inside the bands, the fine particle size and the presence of microporosity required the use of higher magnification images.

Image manipulation and stacking procedures were done using Adobe Photoshop®. Individual cathodoluminescence and secondary electron images were placed as linked layers in a master mosaic file. Precise correspondence between cathodoluminescence and secondary electron images was maintained. The linked images were then properly positioned relative to adjacent mosaic segments.

Once the images were positioned and aligned, all of the cathodoluminescence images were combined into a single layer, and all of the secondary electron images were combined into another, resulting in separate layers containing the corresponding cathodoluminescence and secondary electron mosaics. When working with

Table 2. Intergranular Volume (IGV) Data. IGV = intergranular volume; COPL = compactional porosity loss; CEPL = cementational porosity loss; I_{comp} = index of compaction.

Sample	*Primary Porosity*	*Intergranular cement*	*IGV*	*COPL*	*CEPL*	I_{comp}
H5A band	15.7	18.9	34.6	8.3	17.3	0.32
H5A host rock	21.0	12.6	33.6	9.6	11.4	0.46
H5B band	13.9	21.6	35.5	7.0	20.1	0.26
H5B host rock	19.6	18.3	37.9	3.4	17.7	0.16
BBD-4C band	8.4	29.9	38.3	2.8	29.1	0.09
BBD-4C host	16.9	14.3	31.2	12.8	12.4	0.51
W-3-98 band	9.5	26.2	35.7	6.7	24.4	0.22
W-3-98 host	14.7	19.5	34.2	8.8	17.8	0.33

panchromatic gray-scale cathodoluminescence images, the secondary electron layer was tinted a contrasting color to allow clearer differentiation between cathodoluminescence and secondary electron features. The layer containing the secondary electron images was positioned above the layer containing the cathodoluminescence images. The secondary electron layer visibility was set to "screen" to create a combined image featuring both cathodoluminescence and secondary electron information.

A grid containing a suitable number of intersection points was created using Adobe Photoshop and placed over the mosaic. The intersection points of the grid lines were used as the points to be identified and counted.

By turning on and off the visibility of the layers in the layers palette, secondary electron–backscattered electron or cathodoluminescence images can be observed alone or in combination. This allowed a determination of the material present at each point to be made. In the samples examined for this study, it was not always possible to differentiate quartz and feldspar using combined secondary electron–backscattered electron and cathodoluminescence imaging, so in most samples, quartz and feldspar grains were not counted separately. Categories counted include quartz + feldspar grains, other grains, intergranular quartz cement, intragranular quartz cement (fracture fill), other cement (authigenic feldspar, carbonate, oxides, and clay), intergranular porosity, microporosity,

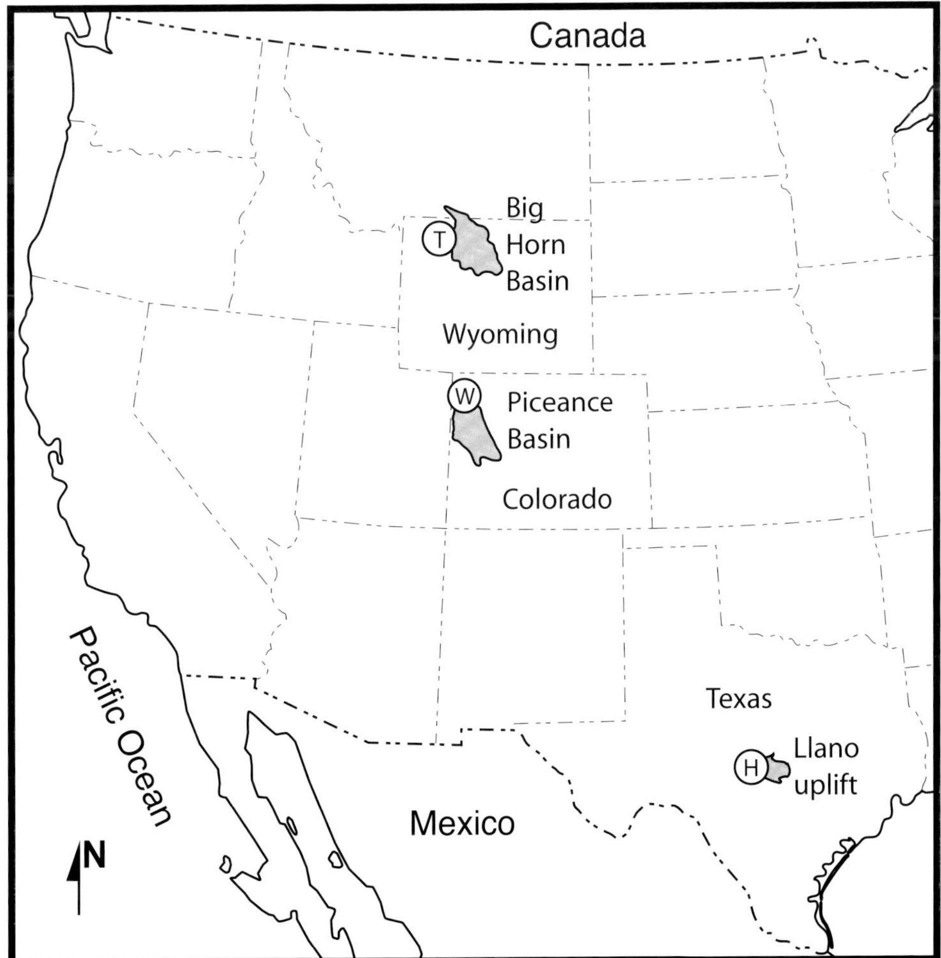

FIGURE 2. Geographic distribution of samples used in this study. H = Hickory Sandstone; W = Weber Formation; T = Tensleep Formation.

secondary porosity, and fracture porosity. In some samples, feldspar, siderite, or clay cements could be differentiated and were counted separately.

Although this method does not permit grain and cement identifications as specific as those in a light microscope observation, the combined secondary electron–backscattered electron and cathodoluminescence images provide sufficient discrimination between grains, cements, and porosity for an accurate assessment of intergranular volume (IGV). Importantly, quartz cement and microporosity are more readily and accurately identified by the techniques employed here. Rapid and accurate mineralogical identification of other cement types is also supported by the easy access to energy-dispersive system (EDS) x-ray analysis.

ASSESSMENT OF POROSITY AND VOLUME LOSS IN SANDSTONES

At deposition, sand has porosity in the range of 40–45%, a value that does not vary significantly with particle size or depositional environment in well-sorted sediments (Beard and Weyl, 1973; Pryor, 1973; Wilson and McBride, 1988; Atkins and McBride, 1992; Lundegard, 1992). A well-documented exception is the study by Dickinson and Ward (1994), who reported porosities of 34% in modern and Pleistocene eolian sandstones. Depositional porosity is affected by sorting (Beard and Weyl, 1973; Pryor, 1973; Wilson and McBride, 1988), although this effect is not apparent in the large data set of Lundegard (1992), in which sandstones range from very well to very poorly sorted (Lundegard, 1992, his figure 2, p. 254).

During diagenesis, compaction and cementation operate together to reduce porosity. The relative importance of compaction vs. cementation in causing porosity loss can be calculated from point-count data following established conventions (Lundegard, 1992; Ehrenberg, 1989):

$$\text{Compactional porosity loss (COPL)} = P_i - (((100 - P_i) \times \text{ingrV})/(100 - \text{ingrV}))$$

$$\text{Cementational porosity loss (CEPL)} = (P_i - \text{COPL}) \times (C/\text{ingrV})$$

This method of calculation accounts for the overall reduction in sediment volume by compaction. The index of compaction (I_{comp}) is simply the ratio of compactional porosity loss to total porosity loss: I_{comp} = COPL/(COPL + CEPL).

The parameter ingrV is reported conventionally as a percentage of the rock volume and is the sum of detrital matrix, primary porosity, and the volume of cement-filling primary pore space (Paxton et al., 2002). In the absence of detrital matrix (as in this study), ingrV is equivalent to precement porosity. In sandstones with substantial intragranular fracturing, intragranular cements may account for a significant part of the total cement volume (Reed and Laubach, 1999; Makowitz and Milliken, 2002, 2003). In such cases, correct assessment of compaction requires the addition of this intragranular material to the cement volume (as opposed to the grain volume), a determination that is made possible by the use of cathodoluminescence imaging (Makowitz and Milliken, 2002, 2003). In this study, COPL and CEPL are calculated from an assumed initial porosity (P_i = 40% and also P_i = 45%), the ingrV, and the total volume of cement, C (filling both primary and fracture porosity).

Compaction and cementation represent fundamentally different mechanisms of porosity decline in that compaction corresponds to an irreversible volume loss. Compactional volume loss includes both pore space reduction and, potentially, also losses from the solid volume, as in the case of secondary pore collapse and also in the case of pressure solution. In contrast, cementation solely involves occlusion of pores, either primary or secondary, in the case of replacement. Cementation is potentially reversible if a cement phase dissolves during later diagenesis.

In a kinematic sense, compactional volume loss is a component of strain that comprises changes in both bulk shape and volume. The kinematic history of a rock volume, just like the diagenetic history, begins at the sediment-water interface. In the early stages of burial, diagenesis, and kinematic history, compaction is a major component of strain. This early strain corresponds to a shortening perpendicular to bedding and parallel to the direction of the principal stress, which is overburden.

Compaction in sandstone is accommodated by grain rearrangement, pore collapse, grain deformation, and dissolution. Even under conditions of shallow burial, compaction in porous sandstone may involve brittle and ductile changes in grain shape. Initially, sandstone compaction by mechanical processes alone may equal as much as the total initial porosity, i.e., 40–45% if ingrV is decreased to 0, but under deeper burial conditions (higher temperature), dissolution (pressure solution) becomes increasingly important. However, at these depths, other diagenetic reactions, notably cementation, tend to strengthen sandstones and impede compaction, which may become negligible (except for thermal and elastic components) over considerable depth ranges.

Compaction bands, and deformation bands in general, are prevalent in porous sandstone buried in the depth ranges where diagenetic reactions such as quartz precipitation begin to affect rock properties.

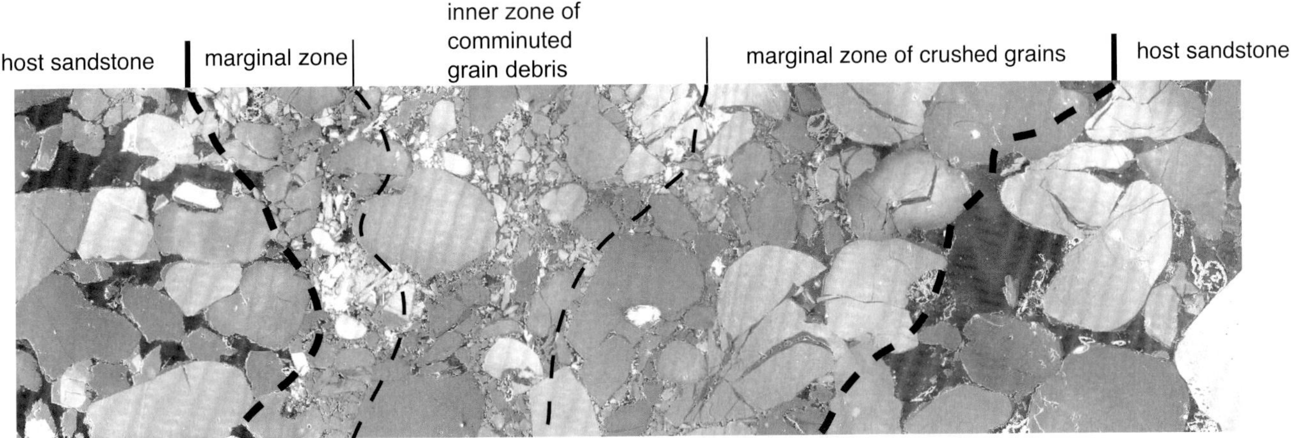

FIGURE 3. Panchromatic gray-scale cathodoluminescence image mosaic showing a traverse across a typical deformation band. In cathodoluminescence images, quartz grains typically are more luminescent than quartz cement. Sample is H5A from the Hickory Sandstone, central Texas, United States. Long dimension of image mosaic is 2 mm.

RESULTS

Imaging

Bands examined in this study show a two-layer structure as described previously by Milliken (1996) and Milliken and Reed (2002) (Figure 3). In SEM images, marginal zones have a clear boundary with adjacent less-deformed sandstone. In the marginal zone, grains are crushed, but grain fragments are separated by distinct opening-mode intragranular fractures, such that particles derived from a common grain can be readily identified. Within the interior zone of larger bands (bands > 5–10 grain widths), grain crushing has proceeded to an extreme degree, producing a bimodal particle-size distribution in which a few, large, unfractured survivor grains (fine to medium sand size) are surrounded by finely comminuted grain debris (coarse clay to silt size). Grain volume loss through pressure dissolution, clearly discernible in cathodoluminescence images as sutured boundaries between adjacent detrital grains, is very minor in both host rocks and deformation bands.

Many comminuted particles in the bands have highly angular, splintery shapes and range in sizes that are near the limit of resolution of this cathodoluminescence imaging technique (near 1 μm). The ingrV in bands includes substantial amounts of quartz cement and lesser volumes of microporosity (defined here as pores less than 30 μm in diameter) (Figure 4). In color cathodoluminescence images, the elongated distribution of related particle groups (particles derived from a common grain) shows that band development has involved elongation parallel to the long dimension of the bands (Figure 5). Whether this strain represents flattening or elongation in one direction has not been established. In Figure 5, the apparent elongation shown is parallel to bedding; in bands from another area, elongation perpendicular to bedding but parallel to the plane of the band has been observed. Initial strains of individual grains are probably strongly influenced by local pore geometries and, hence, may not truly reflect the overall strain of the band.

Point Counts

Scanning electron microscopy-based point counts of four deformation band and host rock pairs are given in Table 1. Table 2 gives calculated parameters related to porosity (and volume) loss through cementation and compaction. Cementation is greater in the bands in all of the sample pairs. Assuming an initial P_i of 40%, cementation equals or exceeds compaction as a cause of porosity loss in all of the samples (Figure 6A). Compaction is generally greater in the host rocks than in the bands (three of the four sample pairs). I_{comp} calculations are sensitive to the assumption of initial porosity, with higher initial values yielding larger values for compactional contribution to porosity loss (Figure 6B). Assuming $P_i = 45$, cementation still equals or exceeds compaction in all of the deformation band samples and one of the host rocks.

DISCUSSION

The small sizes of pores in deformation bands and the cryptic appearance of quartz cement in transmitted plane and polarized light lead to clear advantages for SEM-based imaging methods over transmitted light microscopy for assessment of compaction and cementation in deformation bands.

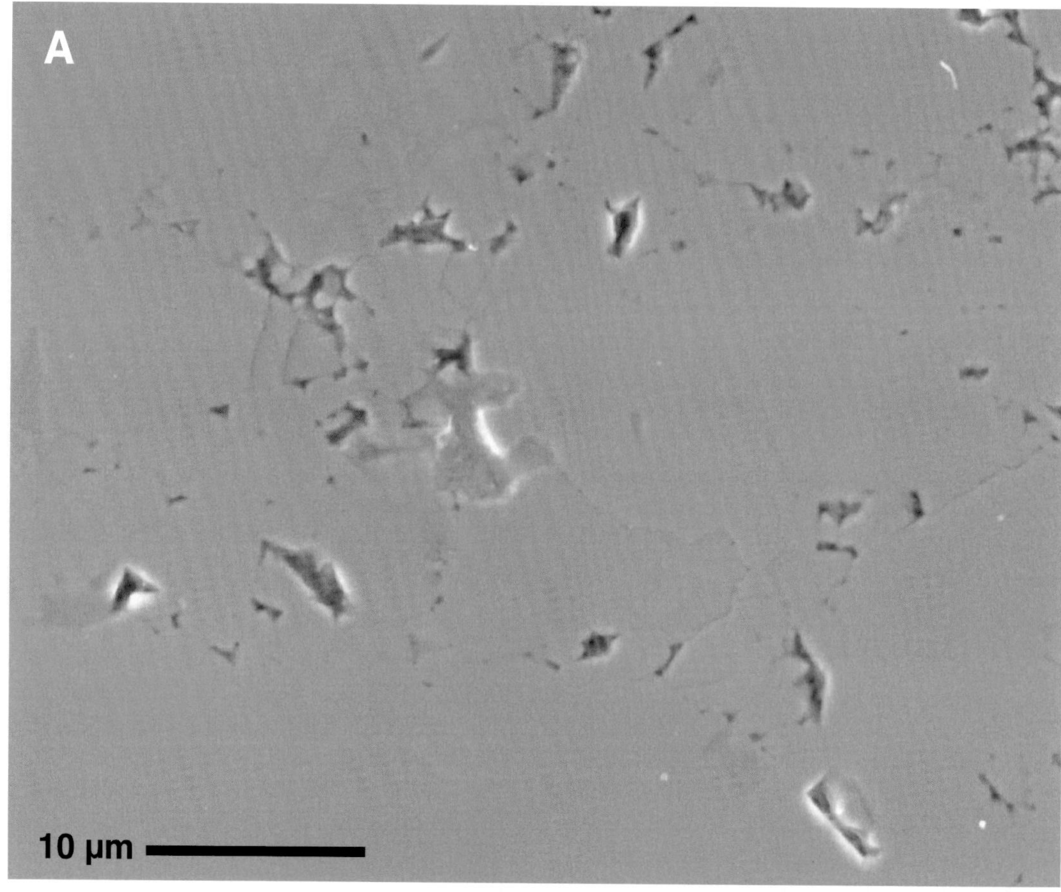
A
10 μm

B

In the samples examined in this study, cementation is a more important cause of porosity loss ($I_{comp} <$ 0.5) than compaction, confirming previous observations of substantial cementation in deformation bands (Milliken, 1996; Reed and Laubach, 1999; Milliken and Laubach, 2000). The importance of cementation in deformation bands contrasts with the average sandstone in which compaction exceeds cementation as a cause of porosity loss (Lundegard, 1992). However, given the enhanced visibility of quartz overgrowths and intragranular quartz cement, the relative importance of cementation vs. compaction in sandstone should not be evaluated without scanned cathodoluminescence imaging.

Intergranular volumes of the host sandstones suggest that these rocks have experienced only shallow burial. Intergranular volumes ranging from 34 to 38% suggest as much as 2 km (1.2 mi) of burial (Paxton et al., 2002) but very likely much less. The burial history of the Hickory Sandstone suggests that two of the samples used in this study underwent band formation at slightly less than 1 km (0.62 mi) in depth. Formation of deformation bands at relatively shallow burial depths is consistent with the idea that bands of this type form only in relatively porous sandstones (Smith, 1983; Olsson, 1999) and also with the observation that such bands are observed in very young, virtually unconsolidated sediments in tectonically active areas (Cashman and Cashman, 2000).

Given the prominent time-temperature controls on the progress of quartz cementation (Walderhaug, 1994; Lander and Walderhaug, 1999), the degree of quartz cementation observed in these samples suggests that they may have experienced elevated geothermal gradients at some point in their histories or, alternatively, contain quartz cementation of a type related to shallow groundwaters (Thiry and Marechal, 2001).

The contrasts in the degree of cementation between bands and host rocks highlight the important function of microscale textures in controlling the progress of quartz cementation. The tendency for greater cementation in the bands matches the pattern that is expected if quartz surface area is an important control on quartz cement localization (Lander and Walderhaug, 1999), and clean or fresh quartz surfaces created during fracturing represent favored sites for quartz nucleation (Reed and Laubach, 1996; Fisher et al., 2000; Chuhan et al., 2002).

In three of the four examples, compaction in the host rock is distinctly greater than that in the bands, suggesting that excess cementation in the bands put these two parts of the rock onto different tracks with respect to compaction. In three samples, postkinematic burial has led to further compaction in the host rocks that was not experienced by the more highly cemented deformation bands.

The degree of similarity in ingrV between bands and host rocks is somewhat surprising in view of the particle-size variation induced by grain comminution in the bands. However, as previously noted, COPL does not vary across a wide range of sorting in the large database reported by Lundegard (1992). This study possibly extends that conclusion into very poorly sorted materials of silt and sand size. Alternatively, the angular shapes of particles in the bands may have been a factor in limiting ingrV decline (Fraser, 1935).

There is no certain evidence that the brittle processes responsible for band formation were accompanied by volume loss in excess of what was experienced by the surrounding host rocks. Although some grains have been elongated by brittle fracture (Figure 4) possibly reflecting flattening or elongation in the plane of the deformation band, we have not conducted a complete strain or kinematic analysis of these samples. Possibly, at the time of band formation, the bands had, in fact, undergone more compaction than the host rock, and only during subsequent burial did the host rocks surpass the bands in degree of compaction. Scanning electron microscopy-based cathodoluminescence methods provide the resolution that would allow strain analysis of bands (using, for example, sets of mutually orthogonal thin sections) and even reconstruction of stain history (by sequential reconstruction of grain-scale deformation registered to the progress of quartz cementation).

Possibly, the formation of deformation bands depends on interacting mechanical and chemical processes. In the bands described here, fracture and cement precipitations are at least broadly contemporaneous. Our analysis of samples of compaction bands originally described by Hill (1989) and subsequently by Mollema and Antonellini (1996) shows that quartz cement is localized in grains and along grain contacts in these otherwise slightly cemented rocks (Reed and Laubach, 1999). Deformation bands are common in eolian sandstones, where grain coats and grain surface damage

Figure 4. (A) Secondary electron–backscattered electron image showing distribution of microporosity in the central region of a deformation band. (B) Combined secondary electron–backscattered electron–cathodoluminescence image that reveals the distribution of particles, cement, and porosity in the same field of view seen in (A). This field of view is made up of 50.5% grains, 43% quartz cement, 1.2% late clay cement, and 5.0 microporosity. At the lower edge of the image, a thick quartz overgrowth shows apparently fibrous cathodoluminescence zoning. Sample is BBD-4 from an outcrop of Tensleep Formation, western Wyoming, United States.

A
elongate particle assemblage
marginal intragranular fractures
B.
quartz cement
marginal intragranular fractures
200 μm

B
50 μm

Figure 6. Compactional porosity loss (COPL) vs. cementational porosity loss (CEPL) values for four deformation-host rock pairs, assuming initial intergranular porosities of 40% (A) and 45% (B). Samples below the 1:1 line have porosity loss dominated by cementation. Samples plotting above the line have porosity loss dominated by compaction.

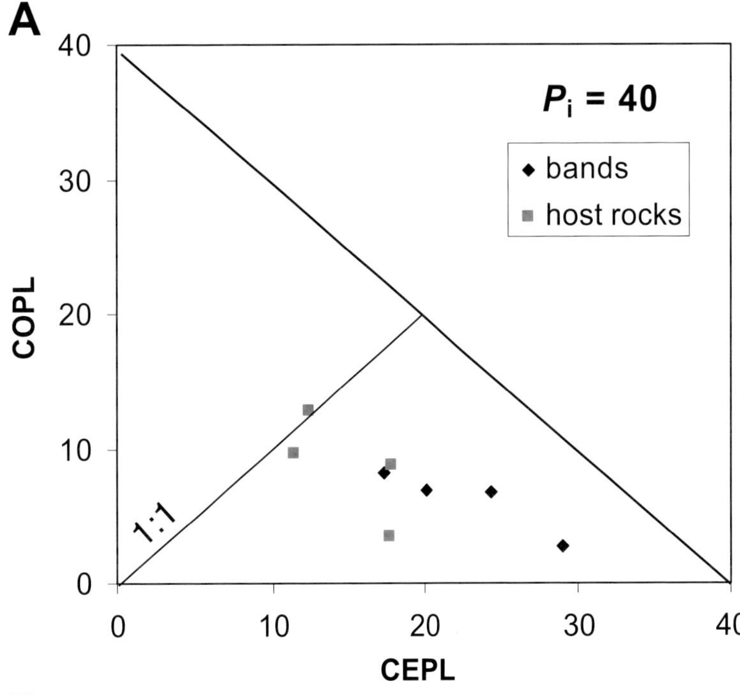

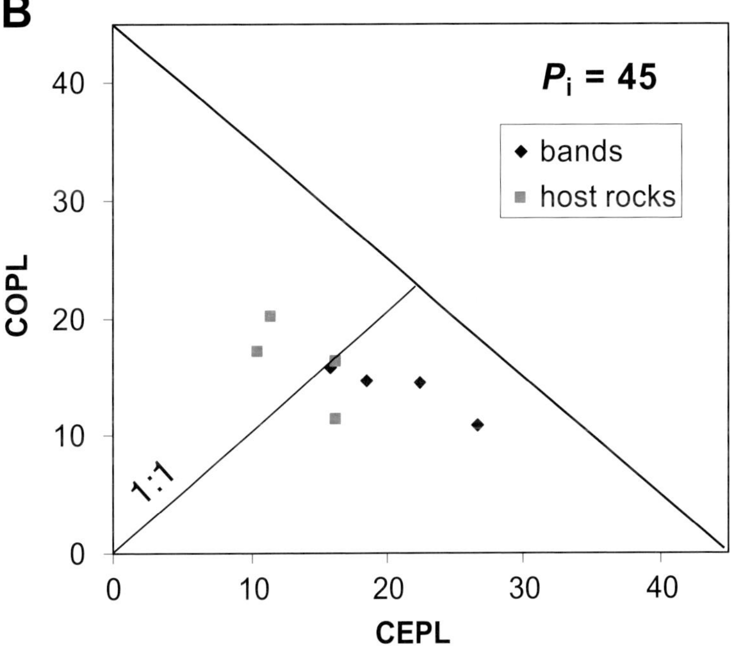

inhibit quartz cement precipitation, allowing anomalous porosity preservation. Breaks in grain coats caused by grain-grain impingement permit accumulation of cement on the fresh fracture surfaces and adjacent grain surfaces, which might be a strain-hardening process that could promote band formation. Mechanical interpretation of band development can benefit from knowledge of concurrent host rock compaction and cementation, information that is partly recorded in host rock ingrV.

CONCLUSIONS

Scanned cathodoluminescence imaging of quartz-cemented sandstones greatly contributes to determining accurately the amount of quartz cement present both inside and outside grains, in sandstone host rocks, and in deformation bands. These accurate measurements are necessary for determining ingrV and porosity loss, whether inside or outside of a deformation band.

Cementation exceeds compaction as a cause of porosity loss in the deformation bands and host rocks examined in this study.

Contrasts in ingrV between bands and host rocks are relatively small.

Preferential early cementation in the bands, as well as the highly angular shapes of the comminuted particles, may have inhibited later compaction relative to that in the surrounding host rocks.

Intergranular volumes in bands and host rocks constrain the burial depth of band formation to no more than 2 km (1.2 mi) and possibly much less.

ACKNOWLEDGMENTS

This study was supported by the industry sponsors of the Fracture Research and Application Consortium, The University of Texas at Austin. The senior author also

Figure 5. (A) Color cathodoluminescence image showing elongation of a group of comminuted grain fragments related to the destruction of a common grain (distinct blue cathodoluminescence). Arrow indicates the long axis of the band. Grains are luminescent in bright red or blue; cements are typically darker colors to virtually nonluminescent. Epoxy filling the porosity luminesces green. (B) Higher magnification color cathodoluminescence image of the area marked with a box in (A). Sample is BBD-4 from an outcrop of Tensleep Formation, western Wyoming, United States.

acknowledges support from E. F. McBride, Gregory Chair, Department of Geological Sciences, The University of Texas at Austin. We thank reviewers Stan Paxton and Sal Bloch for their insightful and constructive comments.

REFERENCES CITED

Antonellini, M., and A. Aydin, 1994, Effect of faulting on fluid flow in porous sandstones: Petrophysical properties: AAPG Bulletin, v. 78, p. 355–377.

Antonellini, M., and A. Aydin, 1995, Effect of faulting on fluid flow in porous sandstones: Geometry and spatial distribution: AAPG Bulletin, v. 79, p. 642–671.

Antonellini, M. A., and D. A. Pollard, 1995, Distinct element modeling of deformation bands in sandstone: Journal of Structural Geology, v. 17, p. 1165–1182.

Antonellini, A., A. Aydin, D. D. Pollard, and P. D'Onofro, 1994a, Petrophysical study of faults in sandstone using petrographic image analysis and x-ray computerized tomography: Pure and Applied Geophysics, v. 143, p. 181–201.

Antonellini, M. A., A. Aydin, and D. D. Pollard, 1994b, Microstructure of deformation bands in porous sandstones at Arches National Park, Utah: Journal of Structural Geology, v. 16, p. 941–959.

Atkins, J. E., and E. F. McBride, 1992, Porosity and packing of Holocene river, dune, and beach sands: AAPG Bulletin, v. 76, p. 339–355.

Aydin, A., 1978, Small faults formed as deformation bands in sandstones: Pure and Applied Geophysics, v. 116, p. 913–930.

Aydin, A., and A. M. Johnson, 1978, Development of faults as zones of deformation bands and as slip surfaces in sandstone: Pure and Applied Geophysics, v. 116, p. 931–942.

Beard, D. C., and P. K. Weyl, 1973, Influence of texture on porosity and permeability of unconsolidated sand: AAPG Bulletin, v. 57, p. 349–369.

Besuelle, P., 2001, Compacting and dilating shear bands in porous rocks: Theoretical and experimental conditions: Journal of Geophysical Research, B, Solid Earth and Planets, v. 106, p. 13435–13442.

Cashman, S., and K. Cashman, 2000, Cataclasis and deformation-band formation in unconsolidated marine terrace sand, Humboldt County California: Geology, v. 28, p. 111–114.

Chuhan, F. A., A. Kjeldstad, K. Bjørlykke, and K. Høeg, 2002, Porosity loss in sand by grain crushing— Experimental evidence and relevance to reservoir quality: Marine and Petroleum Geology, v. 19, p. 39–53.

Davis, G. H., 1999, Structural geology of the Colorado Plateau region of southern Utah with special emphasis on deformation bands: Geological Society of America Special Paper 342, 157 p.

Dickinson, W. W., and K. L. Milliken, 1995, The diagenetic role of brittle deformation in compaction and pressure solution, Etjo sandstone, Namibia: Journal of Geology, v. 103, p. 339–347.

Dickinson, W. W., and J. D. Ward, 1994, Low depositional porosity in eolian sands and sandstones, Namib desert: Journal of Sedimentary Research, v. A64, p. 226–232.

Edwards, H. E., A. D. Becker, and J. A. Howell, 1993, Compartmentalization of an aeolian sandstone by structural heterogeneities: Permo-Triassic Hopeman Sandstone, Moray Firth, Scotland, *in* C. P. North and D. J. Prosser, eds., Characterization of fluvial and aeolian reservoirs: Geological Society (London) Special Publication 73, p. 339–365.

Ehrenberg, S. N., 1989, Assessing the relative importance of compaction processes and cementation to reduction of porosity in sandstones: Discussion: AAPG Bulletin, v. 73, p. 1274–1276.

Evans, J., A. J. C. Hogg, M. S. Hopkins, and R. J. Howarth, 1994, Quantification of quartz cements using combined SEM, CL, and image analysis: Journal of Sedimentary Research, v. 64, p. 334–338.

Fisher, Q. J., R. J. Knipe, and R. H. Worden, 2000, Microstructures of deformed and non-deformed sandstones from the North Sea: Implications for the origins of quartz cement in sandstones, *in* R. H. Worden and S. Morad, eds., Quartz cementation in sandstones: International Association of Sedimentologists Special Publication 29, p. 129–146.

Fowles, J., and S. Burley, 1994, Textural and permeability characteristics of faulted, high porosity sandstones: Marine and Petroleum Geology, v. 11, p. 608–623.

Fraser, H. J., 1935, Experimental study of the porosity and permeability of clastic sediments: Journal of Geology, v. 43, p. 910–1010.

Henry, D. J., and J. B. Toney, 1987, Combined cathodoluminescence/backscatter electron imaging and trace element analysis with the electron microprobe: Application to geological materials, *in* R. H. Geiss, ed., Microbeam analysis: San Francisco, San Francisco Press, p. 339–360.

Hill, R. E., 1989, Analysis of deformation bands in the Valley of Fire State Park, Nevada: M.A. Thesis, University of Nevada, Las Vegas, Nevada, 118 p.

Hogg, A. J. C., E. Sellier, and A. J. Jordan, 1992, Cathodoluminescence of quartz cements in Brent Group sandstones, Alwyn South, UK North Sea, *in* A. C. Morton, R. S. Haszeldine, M. R. Giles, and S. Brown, eds., Geology of the Brent Group: Geological Society (London) Special Publication 61, p. 421–440.

Houseknecht, D. W., 1991, Use of cathodoluminescence petrography for understanding compaction, quartz cementation, and porosity in sandstones, *in* C. E. Barker and O. C. Kopp, eds., Luminescence microscopy and spectroscopy: Quantitative and qualitative applications: SEPM Short Course 25, p. 59–75.

Issen, K. A., and J. W. Rudnicki, 2000, Conditions for compaction bands in porous rock: Journal of Geophysical Research, B, Solid Earth and Planets, v. 105, p. 21,529–21,536.

Krinsley, D., B. Nagy, H. Dypvik, and M. Rigal, 1983, Microtextures in mudrocks as revealed by backscattered electron imaging: Precambrian Research, v. 61, p. 191–207.

Lander, R. H., and O. Walderhaug, 1999, Predicting porosity through simulating sandstone compaction and quartz cementation: AAPG Bulletin, v. 83, p. 433–449.

Laubach, S. E., and K. L. Milliken, 1996, Using CL observations to evaluate fractures in siliciclastic petroleum reservoirs: International Conference on Cathodoluminescence and Related Techniques in Geosciences and Geomaterials, Nancy, France, co-sponsored by the Society of Geology Applied to Mineral Deposits, Society for Luminescence Microscopy and Spectroscopy, and Société Française de Mineralogie et Cristallographie, p. 83–84.

Lundegard, P. D., 1992, Sandstone porosity loss— A 'big picture' view of the importance of compaction: Journal of Sedimentary Petrology, v. 62, p. 250–260.

Makowitz, A., and K. L. Milliken, 2002, Quantitative measurements of brittle deformation with burial compaction, Frio Formation, Gulf of Mexico basin: Transactions of the Gulf Coast Association of Geological Societies, v. 52, p. 695–706.

Makowitz, A., and K. L. Milliken, 2003, Quantification of brittle deformation in burial compaction, Frio and Mt. Simon sandstones: Journal of Sedimentary Petrology, v. 73, p. 1007–1021.

McBride, E. F., A. A. Abdel-Wahab, and K. L. Milliken, 2002, Petrography and diagenesis of half-billion-year-old cratonic sandstone (Hickory) Llano region, Texas: Bureau of Economic Geology Report of Investigations, v. 264, 74 p.

Milliken, K. L., 1994, The widespread occurrence of healed microfractures in siliciclastic rocks: Evidence from scanned cathodoluminescence imaging: Proceedings of the First North American Rock Mechanics Symposium, Austin, Texas, p. 825–832.

Milliken, K. L., 1996, Deformation bands formed by interrelated cementation and brittle fracture in porous sandstones, Hickory Formation, central Texas: Synkinematic diagenesis revealed by cathodoluminescence imaging: AAPG Annual Meeting, San Diego, California, v. 5, p. 825–832.

Milliken, K. L., and S. E. Laubach, 2000, Brittle deformation in sandstone diagenesis as revealed by scanned cathodoluminescence imaging with application to characterization of fractured reservoirs, *in* M. Pagel, V. Barbin, P. Blanc, and D. Ohnenstetter, eds., Cathodoluminescence in geosciences: Berlin, Springer, p. 225–243.

Milliken, K. L., and R. M. Reed, 2002, Internal structure of deformation bands as revealed by cathodoluminescence imaging, Hickory Sandstone (Cambrian), central Texas: Transactions of the Gulf Coast Association of Geological Societies, v. 52, p. 725–736.

Mollema, P.-N., and M. Antonellini, 1996, Compaction bands; a structural analog for anti-mode I cracks in aeolian sandstones: Tectonophysics, v. 267, p. 209–228.

Nadeau, P. H., and A. Hurst, 1991, Application of back-scattered electron microscopy to the quantification of clay mineral microporosity in sandstone: Journal of Sedimentary Petrology, v. 61, p. 921–925.

Olsson, W. A., 1999, Theoretical and experimental investigation of compaction bands: Journal of Geophysical Research, v. 104, p. 7219–7228.

Paxton, S. T., O. Szabo, J. M. Ajdukiewicz, and R. E. Klimentidis, 2002, Construction of an intergranular volume compaction curve for evaluating and predicting compaction and porosity loss in rigid-grain sandstone reservoirs: AAPG Bulletin, v. 86, p. 2047–2067.

Pittman, E. D., 1981, Effect of fault-related granulation on porosity and permeability of quartz sandstones, Simpson Group (Ordovician), Oklahoma: AAPG Bulletin, v. 65, p. 2381–2387.

Primmer, T. J., and H. F. Shaw, 1987, Diagenesis in shales: Evidence from back-scattered electron microscopy and electron microprobe analyses: Proceedings of the International Clay Conference, Denver, Colorado, v. 8, p. 135–143.

Pryor, W. A., 1973, Permeability-porosity patterns and variations in some Holocene sand bodies: AAPG Bulletin, v. 57, p. 162–189.

Reed, R. M., and S. E. Laubach, 1996, The role of mirofractures in the development of quartz overgrowth cements in sandstones: New evidence from cathodoluminescence studies: Geological Society of America Annual Meeting, p. A-280.

Reed, R. M., and S. E. Laubach, 1999, Fracturing and quartz cementation in sandstone compaction bands: Preliminary results from cathodoluminescence studies: Geological Society of America Annual Meeting, p. A87.

Sipple, R. F., 1968, Sandstone petrology, evidence from luminescence petrography: Journal of Sedimentary Petrology, v. 38, p. 530–554.

Smith, G. A., 1983, Porosity dependence of deformation bands in the Entrada Sandstones, La Plata County, Colorado: The Mountain Geologist, v. 20, p. 82–85.

Taylor, W. L., and D. D. Pollard, 2000, Estimation of in situ permeability of deformation bands in porous sandstone, Valley of Fire, Nevada: Water Resources Research, v. 36, p. 2595–2606.

Thiry, M., and B. Marechal, 2001, Development of tightly cemented lenses in uncemented sand; example of the Fontainebleau sand (Oligocene) in the Paris basin: Journal of Sedimentary Research, v. 71, p. 473–483.

Walderhaug, O., 1994, Temperatures of quartz cementation in Jurassic sandstones from the Norwegian continental shelf— Evidence from fluid inclusions: Journal of Sedimentary Petrology, v. 64, p. 311–323.

Wilson, J. C., and E. F. McBride, 1988, Compaction and porosity evolution of Pliocene sandstones, Ventura basin, California: AAPG Bulletin, v. 72, p. 664–681.

Zinkernagel, U., 1978, Cathodoluminescence of quartz and its application to sandstone petrology: Contributions to Sedimentology, v. 8: Stuttgart, E. Schweizerbart'sche Verlagsbuchandlung (Nägele und Obermiller), 69 p.

15

Sorkhabi, R., 2005, Geochemical signatures of fluid flow in thrust sheets: Fluid-inclusion and stable isotope studies of calcite veins in western Wyoming, *in* R. Sorkhabi and Y. Tsuji, eds., Faults, fluid flow, and petroleum traps: AAPG Memoir 85, p. 251–267.

Geochemical Signatures of Fluid Flow in Thrust Sheets: Fluid-inclusion and Stable Isotope Studies of Calcite Veins in Western Wyoming

Rasoul Sorkhabi[1]

Technology Research Center, Japan National Oil Corporation, Chiba, Japan

ABSTRACT

The fold and thrust belt of western Wyoming consists of a series of thrust sheets formed in the Sevier–Laramide orogenies of Late Cretaceous–early Eocene. Cambrian to Eocene rocks form thrust packages that have been transported several kilometers in a thin-skinned (basement-detached) tectonic regime. The major thrust faults in the region are east verging and become younger from the west toward the east with a decrease in displacement and an increase in thrust velocity. The thrust sheets contain calcite veins of various styles, including those in fractured limestones, conjugate veins, en echelon veins, and fracture fills of fault breccias. Calcite veins provide geochemical signatures of fluid flow during the development of the thrust faults. Fluid-inclusion data from the calcite veins demonstrate the flow of various generations of aqueous fluid and liquid hydrocarbons. Homogenization temperatures (minimum trapping temperatures) of aqueous inclusions define three populations: 110–125, 130–140, and 158–163°C. Waters with salinities ranging from about 1 to 21 wt. % NaCl equivalent were trapped. Petroleum-bearing inclusions record relatively lower homogenization temperatures (<100°C) compared to aqueous inclusions. Stable isotope analyses of limestone rocks indicate that they are typical marine limestones (with $\delta^{13}C$ values mainly in the range of +5 to −5‰ relative to Peedee belemnite) altered to various degrees by low $\delta^{18}O$ waters. Similarities in the $\delta^{13}C$ values of the host limestone rocks and calcite veins suggest that the carbon was derived from the same limestone formation. In contrast, the $\delta^{18}O$ values of calcite veins were found to be more negative than those of host limestones, indicating that vein calcite precipitated from water with either lower $\delta^{18}O$ composition or higher temperatures. Calculated for temperatures of 110–165°C (based on homogenization temperatures of

[1]*Present address:* Energy & Geoscience Institute, University of Utah, Salt Lake City, Utah, U.S.A.

DOI:10.1306/1033727M853128

aqueous inclusion obtained from fluid-inclusion microthermometry), the composition of vein water (with $\delta^{18}O_{water}$ values of 0 to +14‰ relative to the standard mean ocean water) appears to have been similar to diagenetic or formation water in the marine basin. This interpretation is also supported by the existence of highly saline aqueous inclusions in calcite veins. Micropermeability measurements of calcite veins in a fault breccia and in a calcite-cemented sandstone yielded a similar range of gas permeabilities between the host rock fragments and the calcite veins, indicating that mineralized veins do not seem to reduce single-phase permeability in carbonate reservoirs.

INTRODUCTION

Hickman et al. (1995), in a review of the history of studies on fluid flow and faulting, noted that Agricola in 1556 was probably the first naturalist to recognize veins as signatures of fluid flow through fractured rocks. Mineral veins in rocks exposed at the Earth's surface and in drill cores offer a unique opportunity to investigate the structural evolution and physical properties of paleofluid systems in rock fractures. Modern attention to fluid flow and rock fractures aims at quantifying these processes and constructing predictive models applicable to the petroleum industry and groundwater investigations and to evaluation of ore deposits, geothermal systems, and nuclear waste disposal sites. In recent decades, an abundant literature has emerged on the subject of fluid flow in faulted and fractured rocks; these studies report on field observations and the results of stress-strain analyses and numerical modeling (see, e.g., Parnell, 1994; Hickman et al., 1995; Jamtveit and Yardley, 1996; National Research Council, 1996; McCaffrey et al., 1999; Faybishenko et al., 2000). The application of geochemical results to the structural interpretation of paleofluid flow systems in rocks (e.g., Kerrich et al., 1984; Kerrich, 1986; Kerrich and Hyndman, 1986; Atnipp et al., 1987; Bruhn et al., 1990; Smith et al., 1991; Goddard and Evans, 1995; Hippler, 1997; Eichhubl and Boles, 2000; Bebout et al., 2001; Agosta and Kirschner, 2003) is of vital importance, especially in sedimentary basins where subsurface fluid movement has economic significance.

This chapter describes a study of calcite veins in rocks collected from various thrust sheets of the western Wyoming fold and thrust belt. Outcrop observations, fluid-inclusion and stable isotope analyses, and micropermeability measurements are presented. The significance of this study is that the data come from carbonate rocks in thrust sheets, which have been less studied in comparison to normal faults in clastic sequences.

WESTERN WYOMING THRUST TECTONICS

Western Wyoming exposes excellent sections of the Rocky Mountain fold and thrust belt and foreland basin (Figure 1). This region is a classic province not only for oil and gas reserves but also for coal, oil shale, and phosphate deposits. A century-old heritage of petroleum exploration and geological surveys in western Wyoming has provided a wealth of geophysical and geological data that has been used to unravel the subsurface structures and geological history of this overthrust belt and basin (e.g., Armstrong and Oriel, 1965; Royse et al., 1975; Dixon, 1982; Lamerson, 1982; Wiltschko and Dorr, 1983; Oriel, 1986; Coogan, 1992).

Figure 2 shows the stratigraphic column together with information on the thickness, depositional environment, and petroleum systems of western Wyoming (Dixon, 1982; Lamerson, 1982). The stratigraphic succession in the region ranges from Precambrian gneisses through Eocene shale and sandstones; intervening Paleozoic–Mesozoic rocks include both carbonate and clastic sediments (Figure 2). Cambrian to Middle Jurassic sediments were deposited on a passive margin along the western side of the North American craton. Middle Jurassic–Eocene sediments were deposited during the tectonic development of the Rocky Mountain belt and foreland basins (the Sevier orogeny).

The western Wyoming overthrust belt is characterized by thin-skinned tectonics. Geophysical data indicate that the basement dips gently (<4°) to the west below a regional sole decollement in the shale beds of the Cambrian Gros Ventre Formation. The major thrust faults rising from several lithostratigraphic boundaries (detachment horizons) have transported thrust packages (containing Paleozoic–Mesozoic rocks) up to several kilometers to the east. The major thrust faults from east to west are the Prospect, Darby, Hogsback, Absaroka, Mead–Crawford, and the Paris–Willard thrusts

FIGURE 1. (a) A simplified structural map of the western Wyoming fold and thrust belt (modified after Royse et al., 1975); (b) a geological cross section at the longitude of Afton [line AA′ in (a)] (modified after Dixon, 1982); (c) location of samples reported in this study [shaded area in (a)]. Note that numbers 1 to 100 in (c) refer to W1 to W100 sample codes in this study.

a

112°
44°
111°
110°
44°
Wyoming
N
0 10 20 km
0 10 20 30 mi
Teton Range
Teton F.
Jackson
Grand Valley F.
Hoback F.
Alpine
Absaroka T.
Hoback
Basin
43°
43°
Pinedale
Darby T.
Meade T.
Prospect T.
A
A'
Afton
AA' (Figure 1b)
Sheep Pass T.
Wyoming
Study Area (Figure 1c)
Precambrian basement
thrust fault
normal fault
fault inferred
Paris-Willard T.
Montpelier
Crawford T.
Tunp T.
La Barge
Green River Basin
Hogsback T
Idaho
Utah
42°
Moax Arch
Logan
Kemmerer
Tunp T.
Lazeart Syn.
western Wyoming overthrust belt
Evanston
Wyoming
Utah
41°
41°
112°
110°
Wasatch F.
Uintah thrust
Salt Lake City
Uinta Mtns.

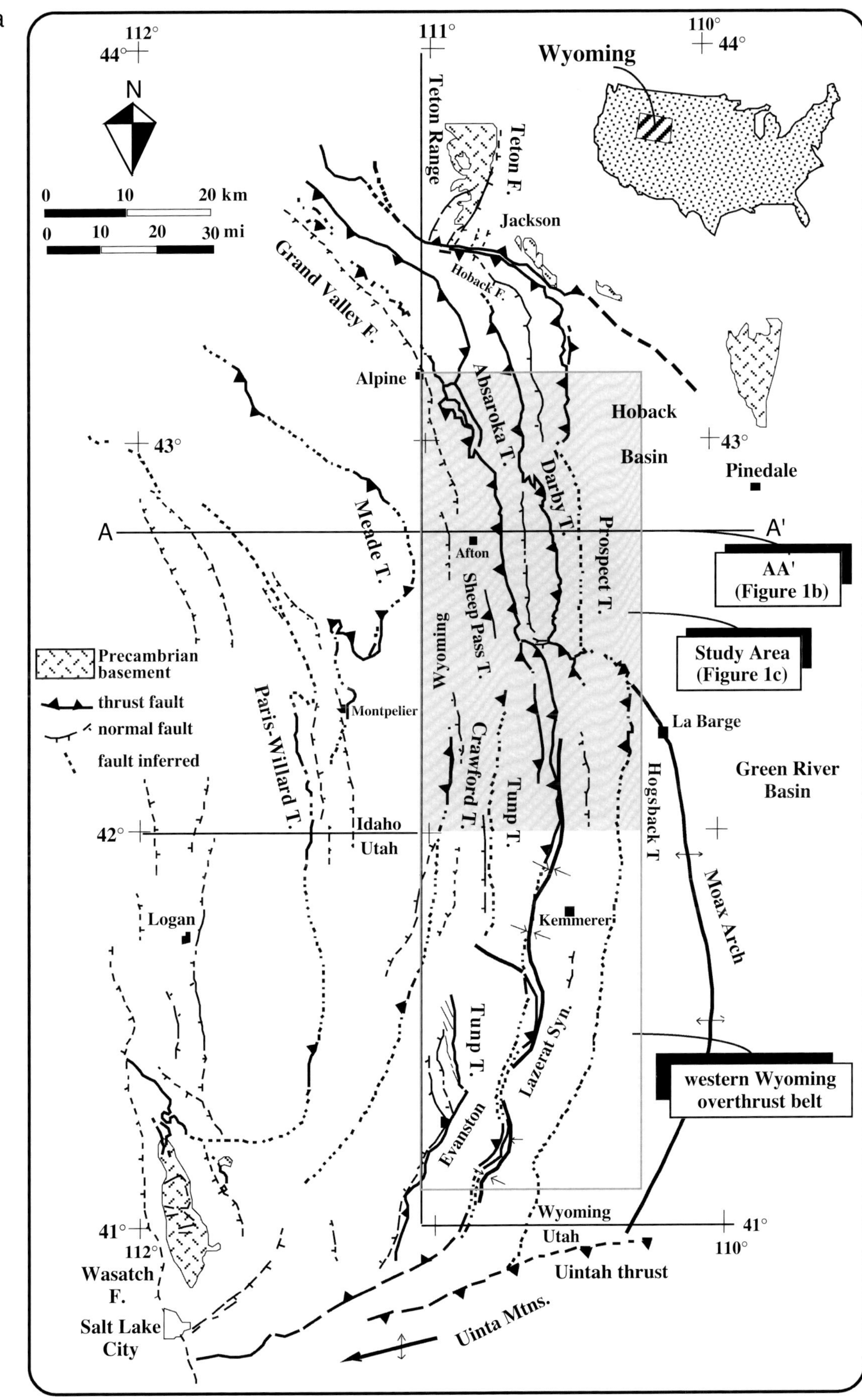

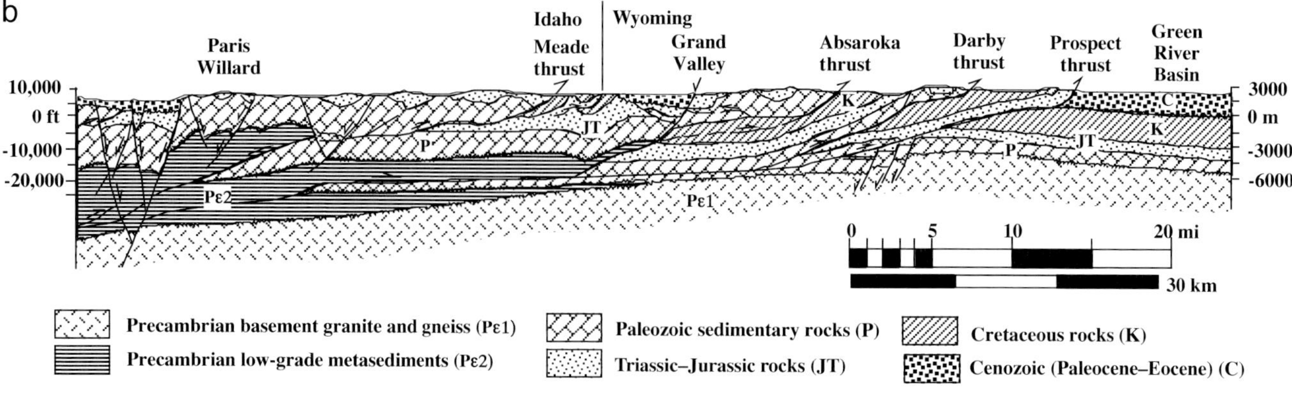

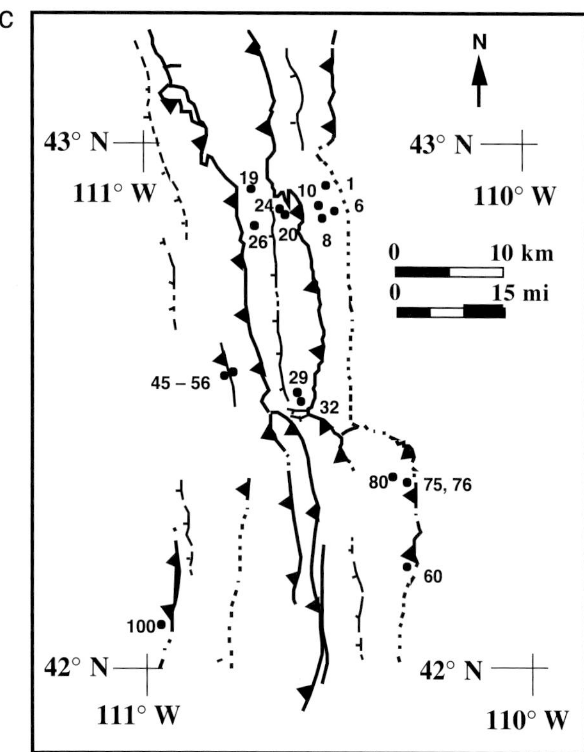

FIGURE 1. (cont.).

(Figure 1b). The thrust sheets are generally north–south trending and east verging. Dips are commonly gentle but become high angle (20–50°W) close to the exposed thrust contacts, commonly marking fault ramps.

Figure 3 shows the kinematics of the major thrust sheets based on the available tectonic and stratigraphic data (fault duration and age data are from Wiltschko and Dorr, 1983; fault displacement data are estimates by Craddock, 1992). Thrust faults become younger from the west to the east as deformation and rupture progressively shifted eastward from the older toward the younger thrust faults. Based on fault age-displacement data (Figure 3), it appears that the amount of displacement decreased, but displacement rates increased with time (from older to younger thrusts).

Western Wyoming thrust sheets and exhumed thrust zones are associated with rock fractures and mineral veins. Figure 4 shows various styles of calcite veins in the study region, including mineral fills in the fractured Twin Creek Limestone (Jurassic) exposed at the base of the hanging wall of the Prospect and Darby thrust faults, conjugate veins in calcareous sandstone of the Stump Formation (Jurassic) on the hanging wall of Darby thrust, en echelon veins (tension gashes) in the Sheep Pass thrust zone in Phosphoria limestone (Permian), and veins in fault breccia of the Hogsback thrust in Cambrian Gros Ventre limestone.

The calcite veins represent mineral-filled fractures with sharp fracture margins and a drusy fabric characterized by the growth of small crystals (mostly <100 μm in width) along the margin and large crystals (100–300 μm in width) in the interior of the vein. This fabric indicates crystal growth in a fluid-filled space generated by extension fracturing (Roberts, 1990). Fracturing, hydrothermal fluid flow, and vein mineralization were likely related to the thrust tectonics of the Sevier orogeny during the Late Cretaceous–early Eocene (Figure 3).

ANALYTICAL METHODS AND RESULTS

For this study, rock samples containing calcite veins were collected from various thrust sheets in western Wyoming. The veins represent open fractures filled with calcite. Therefore, geochemical analyses of the veins provide a record of the nature and properties of the subsurface fluids that have migrated through the rock at various times. Sample locations are given in Figure 1c; detailed descriptions are given in Table 1. Fluid-inclusion, stable (oxygen and carbon) isotope, and micropermeability studies were carried out on the samples.

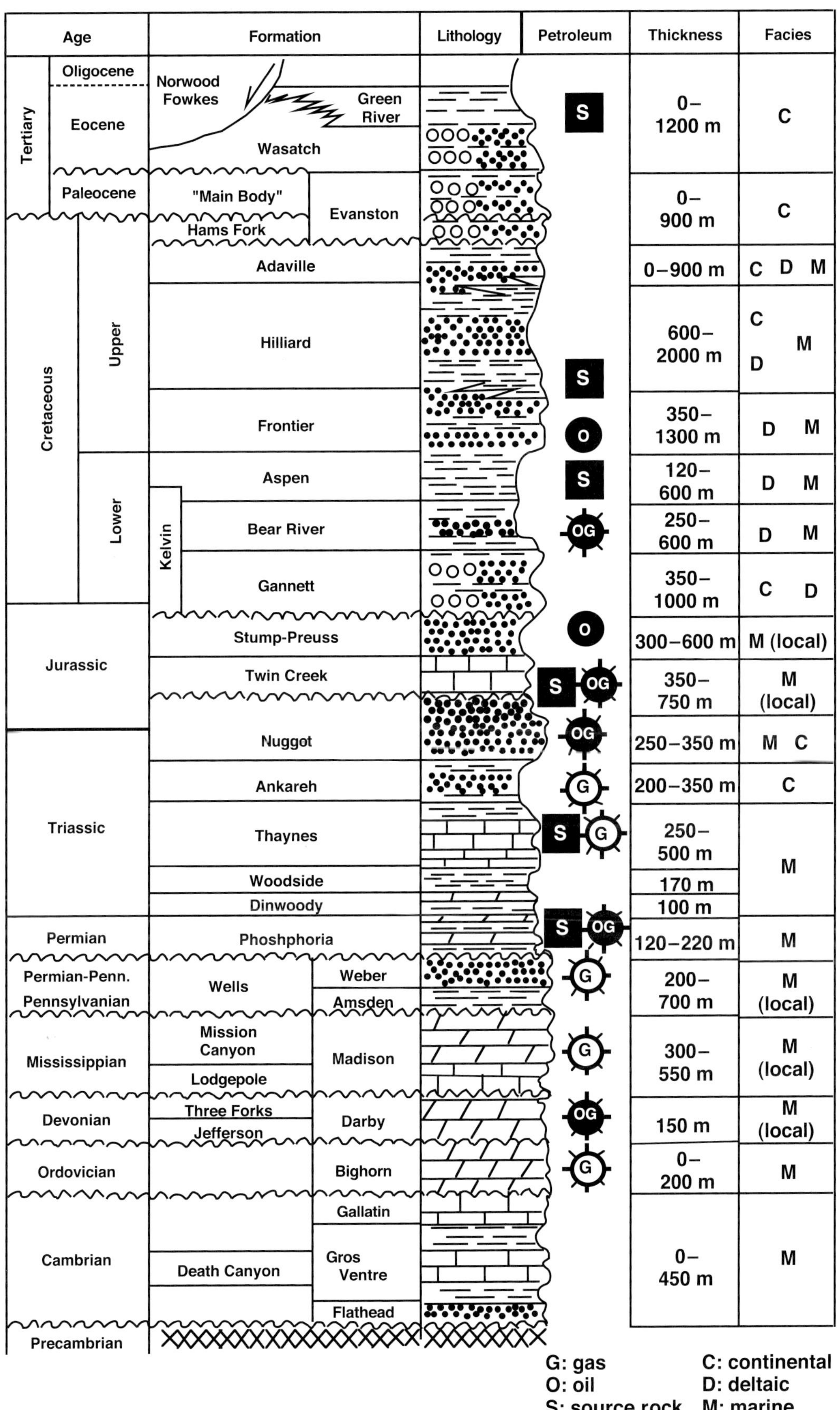

FIGURE 2. A composite stratigraphic column of the western Wyoming fold and thrust belt and information on the facies and petroleum systems (modified after Dixon, 1982; Lamerson, 1982).

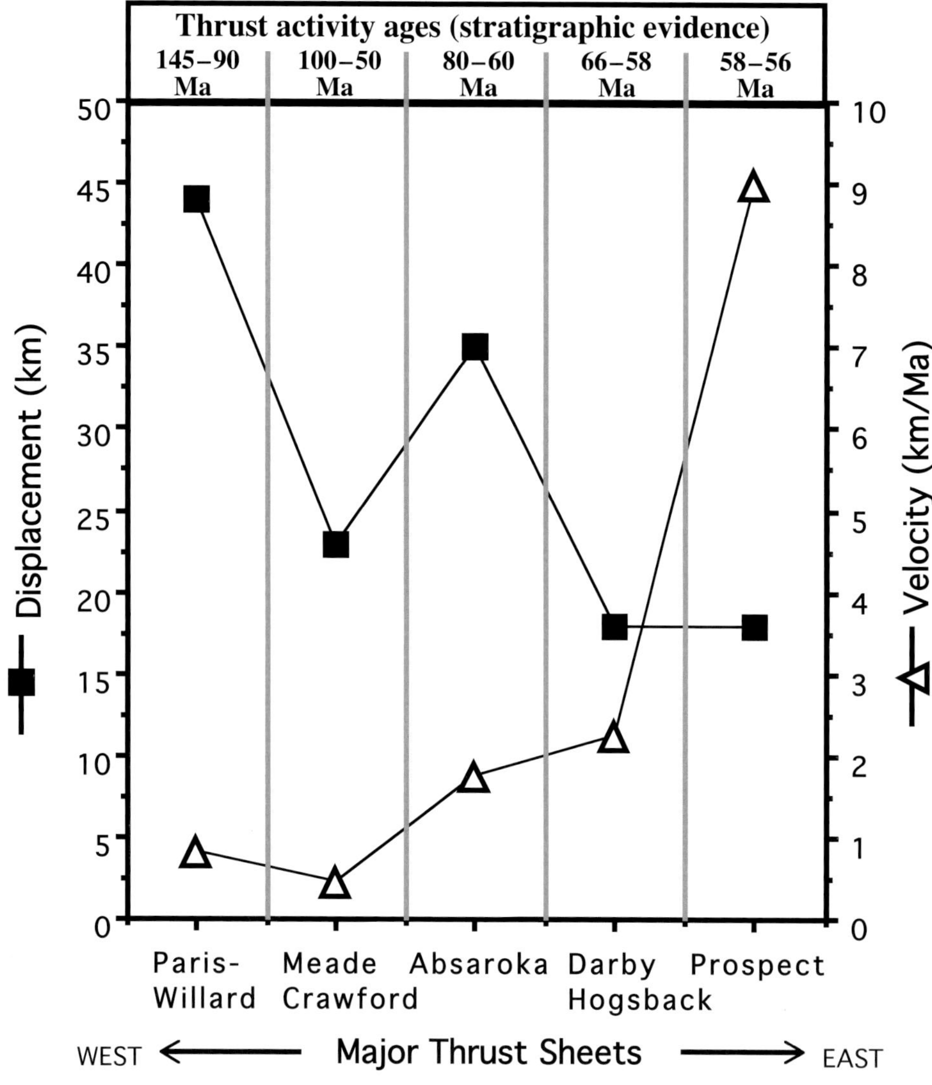

FIGURE 3. The age (initiation-termination period), displacement, and velocity of major thrust faults in western Wyoming based on tectonic and stratigraphic evidence (data sources: Wiltschko and Dorr, 1983; Craddock, 1992).

Fluid-inclusion Analyses of Calcite Veins

Fluid-inclusion analyses (Goldstein and Reynolds, 1994) were conducted on doubly polished sections using a U.S. Geological Survey–type gas flow heating-freezing stage attached to a petrographic microscope. Data accuracy, based on measurements of synthetic fluid inclusions, was ±0.1°C at temperatures below 0°C and ±3.0°C at temperatures above 195°C. The analyses included both aqueous inclusions (Table 2; Figure 5) and petroleum-bearing inclusions (Table 3; Figure 6). Ice-melting (freezing) and homogenization (heating) temperatures were measured on aqueous inclusions. Ice-melting temperatures were converted to weight percentage NaCl equivalent using the equation of Bodnar (1993). All of the fluid inclusions were secondary inclusions formed on fracture planes in calcite crystals. Nonetheless, the calcite veins and the fractures in them possibly belong to the same deformation phase, with fracture planes having developed in calcite veins during thrust tectonics.

For aqueous inclusions, the homogenization temperatures, which represent minimum estimates of the trapping temperatures, define three populations: 110–125, 130–140, and 158–163°C (Table 2). Salinity data show several populations: 0.18–0.35, 1.1–1.6, 1.7–2.8, 3.71–4.65, 11.7–13.9, and 17.8–21.0%. The low-salinity fluids (<3.5%) are interpreted as meteoric water trapped during the uplift to the surface (upthrusting), whereas very high-salinity fluids (>11%) are considered to indicate either migration of deep basin brines through the thrust zones or interaction of meteoric water with evaporate beds along their flow path. Because evaporate deposits constitute a minor portion of the Paleozoic–Mesozoic strata in the region, highly saline fluids were probably deep-basin brines.

Petroleum (liquid hydrocarbon) inclusions, which were observed in three vein samples, provide evidence of petroleum migration in the thrust sheets. Based on their fluorescence color, three distinct petroleum populations were distinguished (Table 3). Homogenization temperatures of the petroleum inclusions indicate various trapping temperatures that cluster into two groups: a low-temperature (63–70°C) group, possibly indicative of relatively shallow flow, and a higher temperature (87–105°C) group, suggesting deeper fluid migration. Sample 75B contains the three types of petroleum inclusions and the two regimes of entrapment temperatures (Table 3). Overall, the homogenization temperatures of petroleum inclusions are lower than those of the brine inclusions, although it should be noted that petroleum-bearing inclusions and aqueous inclusions were not observed in the same samples.

Calcite veins in fractured limestone of the Twin Creek Formation (Jurassic)

Sample W1 on hanging wall of Prospect thrust

Sample W19 on hanging wall of Darby thrust

Conjugate veins in calcareous sandstone of the Stump Formation (Jurassic)

(c) and (d) are on the hanging wall of Darby thrust; sample W29 was collected from this outcrop.

En echelon veins (tension gahes) in limestone of the Phosphoria Formation (Permian)

(e) Sample W 56 (f) and sample W 45 from the Sheep Pass thrust zone (a splay of Absaroka thrust)

Calcite veins in fault breccia from the Hogsback thrust (host rock: Cambrian limestone)

FIGURE 4. Photographs showing various styles of calcite veins observed in the thrust sheets of western Wyoming.

Table 1. Description of samples analyzed in this study.

Sample	*Latitude*	*Longitude*	*Host Rock Type*	*Formation*	*Age*	*Thrust Sheet and Comments*
W1	42°56.021′N	110°30.175′W	limestone	Twin Creek	Jurassic	fractured limestone on hanging wall of Prospect thrust; North Horse Creek
W6	42°48.107′N	110°29.357′W	limestone	Peterson (in Gannet)	Late Cretaceous	limestone on hanging wall of Prospect thrust; North Cottonwood Creek
W8B	42°48.382′N	110°30.548′W	limestone	Bear River	Late Cretaceous	Prospect (same comment as W6)
W10B	42°56.021′N	110°30.175′W	black shale	Aspen	Late Cretaceous	Prospect (same comment as W6)
W19	42°56.633′N	110°41.984′W	limestone	Twin Creek	Jurassic	fractured limestone on hanging wall of Darby thrust; Blind Bull Creek
W20H	42°50.512′N	110°36.379′W	limestone	Madison	Mississippian	thrust zone in the Darby sheet near McDougall Meadows
W24	42°50.948′N	110°37.747′W	quartzite	Weber (Wells)	Pennsylvanian	hanging wall of Darby thrust; Sheep Creek
W26	42°51.875′N	110°42.177′W	calcareous sandstone	Stump	Upper Jurassic	belemnite-rich rock on hanging wall of Darby thrust; Sheep Creek
W29	42°31.480′N	110°41.045′W	black shale	Aspen	Cretaceous	hanging wall of Darby; Greys River road
W32	42°31.756′N	110°41.365′W	limestone	Madison	Mississippian	hanging wall of Absaroka thrust near La Barge Meadows
W45A	42°35.934′N	110°45.404′W	limestone	Phosphoria	Permian	from hanging wall of Absaroka thrust
W46.5	42°35.907′N	110°45.484′W	sandstone	Amsden	Pennsylvanian	fractured sandstone from Sheep Pass thrust zone (splay of Absaroka thrust)
W50	42°36.215′N	110°45.541′W	black limestone	Madison	Mississippian	from hanging wall of Sheep Pass thrust
W54	42°36.200′N	110°45.500′W	fault breccia	Amsden Sandstone fragments	Pennsylvanian	from Sheep Pass thrust zone (splay of Absaroka thrust)
W56	42°36.200′N	110°45.436′W	limestone	Phosphoria	Permian	from hanging wall of Absaroka thrust
W60	42°13.156′N	110°18.600′W	fault breccia	Gros Ventre limestone fragments	Cambrian	from Hogsback thrust zone; La Barge Creek
W75A	42°18.853′N	110°18.784′W	dolomite	Bighorn	Ordovician	from Hogsback thrust zone; La Barge field
W76B	42°13.867′N	110°18.700′W	dolomite	Bighorn	Ordovician	from Hogsback thrust zone; La Barge field
W76C	42°13.867′N	110°18.700′W	limestone	Gros Ventre	Cambrian	from Hogsback thrust zone; La Barge field
W80	42°20.759′N	110°19.393′W	black limestone	Madison	Mississippian	from hanging wall of Hogsback thrust Hogsback Ridge
W100	42°05.847′N	110°56.289′W	phosphatic siltstone	Phosphoria	Permian	from Cokeville anticline on hanging wall of Crawford thrust; abandoned coke mine

Table 2. Fluid-inclusion data of aqueous inclusions in calcite veins.

Sample	*FIA*	*Number*	T_h (°C)	T_m (°C)	*Salinity (wt. %) NaCl Equivalent*
W10B	FIA-1	2	158–163	−0.1 to −0.2	0.18–0.35
	FIA-2	4	159–165	−2.2 to −2.8	3.71–4.65
W24	FIA-1	5	133–139	−8.0 to −10.0	11.7–13.9
W56	FIA-1	5	136–140	−15.0 to −18.0	18.6–21.0
	FIA-2	2	135–139	14.0 to −18.0	17.8–21.0
W60	FIA-1	4	110–118	−0.1 to −0.2	0.18–0.35
	FIA-2	6	116–125	−0.1 to −0.2	0.18–0.35
	FIA-3	3	111–118	−0.1 to −0.2	0.18–0.35
	FIA-4	6	116–125	−0.6 to −0.9	1.10–1.60
	FIA-5	7	115–125	−1.0 to −1.6	1.70–2.80
W76C	FIA-1	2	133–139	−14.0 to −17.0	17.8–20.2
	FIA-2	2	131–137	−15.0 to −18.0	18.6–21.0
	FIA-3	2	132–137	−15.0 to −18.0	18.6–21.0

FIA = fluid-inclusion assemblage; Number = number of inclusions in each FIA; T_h *= homogenization temperature (minimum fluid trapping temperature);* T_m *= final ice-melting temperature (~freezing temperature); salinities calculated from ice-melting temperatures (Bodnar, 1993).*

Stable Isotope Chemistry of Host Limestone Rocks and Calcite Veins

Twelve samples were analyzed for carbon and oxygen isotopes to compare the chemical signatures of the host rocks and calcite veins. Powder samples were recovered by an electric drill from selected points on rock matrix and calcite veins. Carbon dioxide gas was recovered from powder samples using standard extraction techniques and was analyzed with a stable isotope mass spectrometer (BMAT252 Thermo Electron Corporation). The measurement accuracy was ±0.1‰ for $\delta^{18}O$ and ±0.06‰ for $\delta^{13}C$. Isotopic measurements were calibrated relative to the Peedee belemnite (PDB). When $\delta^{18}O$ values are reported relative to standard mean ocean water (SMOW), they have been calculated from PDB values using the equation of $\delta^{18}O_{(SMOW)} = 1.03091\delta^{18}O_{(PDB)} + 30.91$ (Coplen et al., 1983).

The results of the isotopic analyses are shown in Table 4 and Figures 7–9. The majority of samples fall within the field of marine limestones (with $\delta^{13}C$ values of +5 to −5) (Figure 7) (Hudson, 1977; Hofes, 1997). The negative $\delta^{18}O$ values of the host rocks relative to PDB indicate either some alteration and diagenesis in the presence of low ^{18}O water or precipitation at elevated temperatures. The $\delta^{18}O$ values of calcite veins are more negative than those of host rocks, indicating that the calcite veins were more subjected to isotope exchange with low ^{18}O water or elevated temperatures.

Figure 8 depicts the carbon and oxygen values for both the host rocks and the calcite veins with reference to the stratigraphic age of the host rocks (from the Ordovician to the Cretaceous). The number of samples in Figure 8 is not sufficient to draw conclusions for such a long span of geologic time. Variations in $\delta^{13}C$ values of the host rocks probably reflect the real rock and water isotopic compositions in the marine basin because carbon values are independent of temperature (Hofes, 1997). Interpretation of oxygen isotope data is more complex. One trend observed in Figure 8 is that the $\delta^{18}O$ values of host rocks appear to decrease (become more negative) from older (deeper) to younger (shallower) rocks. This trend may be in contrast to a column of strata in which deeper rocks show more negative $\delta^{18}O$ values (Savin and Yeh, 1981), because they experienced higher burial temperatures. The trend in the $\delta^{18}O$ values of host rocks, therefore, seems to indicate that the isotope exchange of shallower strata with low ^{18}O water was greater than that of the deeper carbonate rocks. Note that in Figure 8, calcite veins are depicted in the same stratigraphic positions as the corresponding host rocks for the ease of comparison, but that the calcite veins are definitely younger (possibly formed during the Late Cretaceous–early Eocene upthrusting of the Rocky Mountains).

Figure 9 plots the carbon (Figure 9a) and oxygen isotope data (Figure 9b) for the rock matrix and vein samples. The $\delta^{13}C$ values for the majority of samples (10 out of 12) are very similar between the host limestones and the calcite veins with a good correlation of 0.9. This similarity is significant, considering the differences in the stratigraphic ages and thrust sheets of the samples (Table 1; Figures 8, 9a), as well as the fact that $\delta^{13}C$ values are not affected by temperature changes. It suggests that carbon in the fracture-filled calcite veins was derived from the same limestone formation. In other words, carbon extracted by hydrothermal fluids from

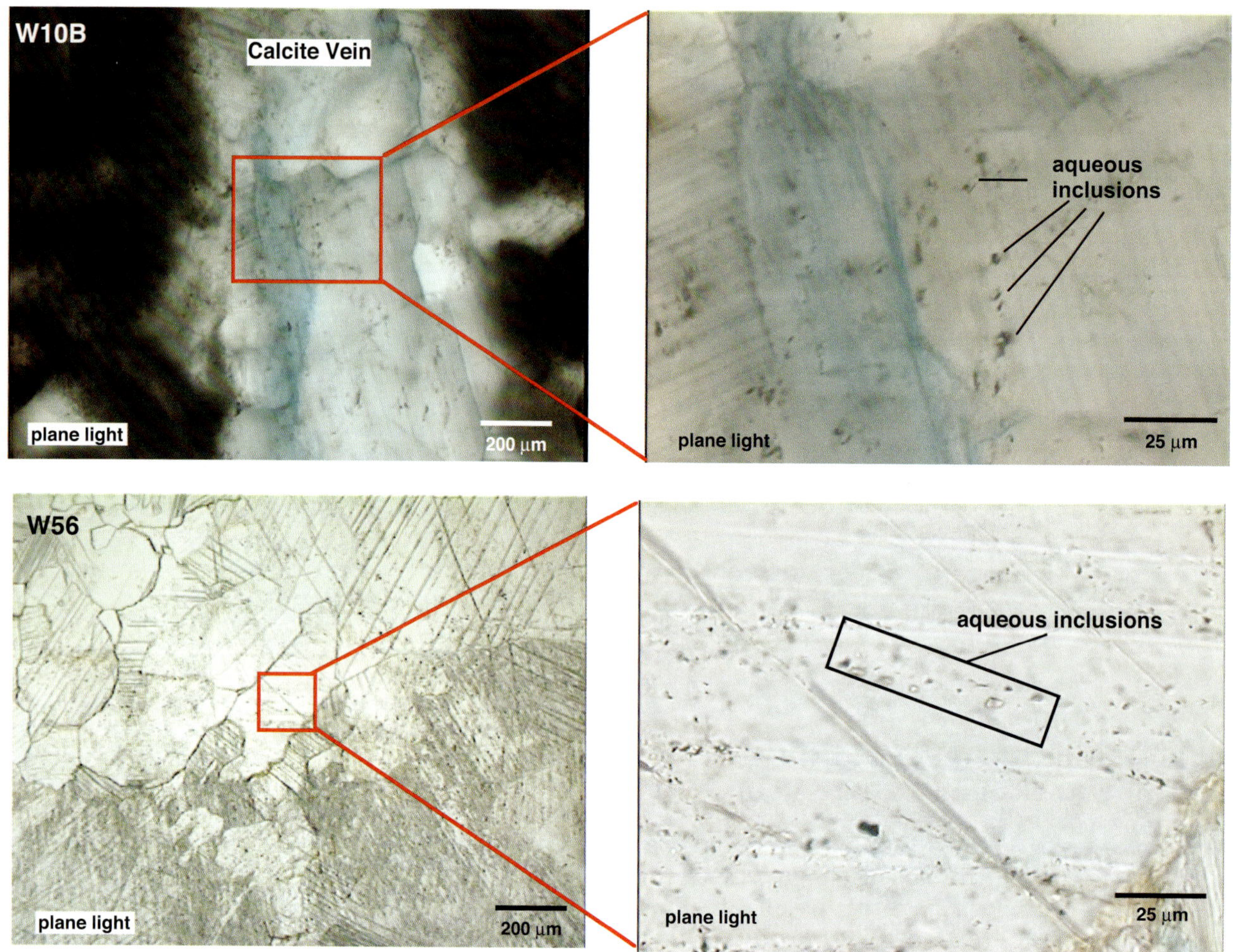

FIGURE 5. Microphotographs of fluid inclusions in calcite veins: aqueous inclusions in sample W10B (calcite vein in Cretaceous limestone) from the Prospect thrust sheet and sample W56 (calcite vein in the limestone of Phosphoria Formation, Permian age) from the Sheep Pass thrust zone (a splay of the Absaroka thrust sheet). See Table 2 for their fluid-inclusion data.

host limestone was not transported across formations but rather was precipitated in the vein in the same formation. In contrast, $\delta^{18}O$ values are markedly different between the host rocks and the calcite veins. The fact that most of the $\delta^{18}O$ values of calcite veins fall below the 1:1 line in Figure 9b and become more negative indicates that the water-precipitating vein calcite was either at higher temperature or more depleted in ^{18}O.

A full interpretation of $\delta^{18}O$ data requires knowledge of temperature because oxygen fractionation is temperature dependent. The relationship between temperature and the oxygen values of calcite and water is given by the following equation (Friedman and O'Neill, 1977):

$$10^3 \ln[(1030.91 + 1.03091\delta^{18}O_{calcite})/(10^3 + \delta^{18}O_{water})] = (2.78 \times 10^6/T^2) - 2.89$$

Table 3. Fluid-inclusion data from petroleum inclusions in calcite veins.

Sample	*FIA*	*Number*	T_h (°C)	*Fluorescence*	*°API Density*
W29	FIA-1	5	87–100	white	45–50
W54	FIA-1	1	63	white	45–50
W75B	FIA-1	6	97–105	blue-white	40–45
	FIA-2	4	66–70	yellow	25–35
	FIA-3	7	97–102	orange	15–25

FIA = fluid-inclusion assemblage; Number = number of inclusions in each FIA; T_h = homogenization temperature (minimum fluid-trapping temperature); API density is an approximate estimate from the color of fluorescence (Goldstein and Reynolds, 1994).

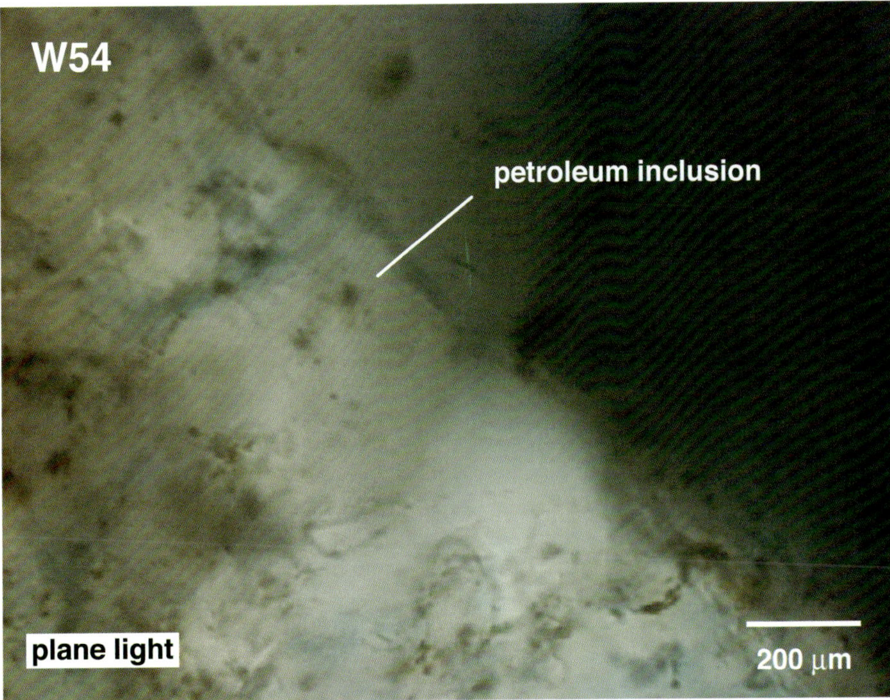

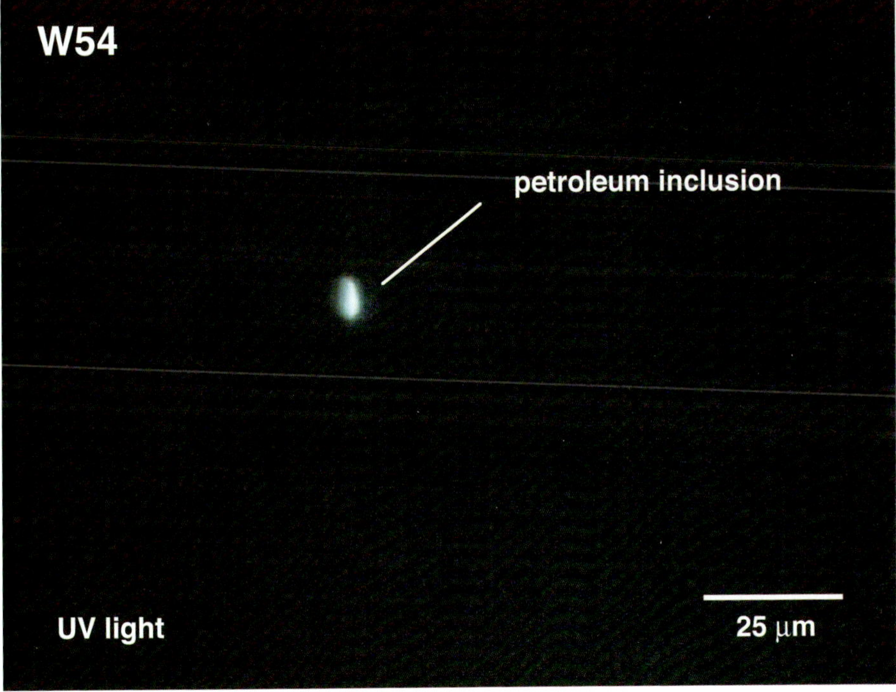

FIGURE 6. Microphotographs of petroleum fluid inclusion in a calcite vein: sample W54, Carboniferous limestone from the Sheep Pass thrust (a splay of the Absaroka thrust sheet). See Table 3 for the fluid-inclusion data of this sample.

where T is water temperature in Kelvin, $\delta^{18}O_{calcite}$ is given relative to PDB, and $\delta^{18}O_{water}$ is given relative to SMOW. The $\delta^{18}O_{calcite}$ values of calcite veins can be obtained from sample analyses (Table 4). Deciphering the origin of water from which calcite has precipitated in the vein is difficult. A significant complicating factor is temperature. If some constraints can be placed on the temperature of hydrothermal fluid migrating through the veins, we can obtain some idea about the composition of the water.

We investigated the impact of temperature on calcite vein water using the above equation. The results of these calculations are shown in Figure 10. Starting with the $\delta^{18}O$ values of calcite veins obtained from laboratory analyses (and converted to SMOW values in Figure 10), the first series of calculations was made using the ocean surface temperatures of 25°C (298.15°K). This yields $\delta^{18}O$ values of −3 to −13‰ (relative to SMOW), which represent meteoric water. Similar $\delta^{18}O$ values have been obtained from freshwater bivalves of the Cretaceous–Eocene sedimentary rocks of Wyoming (Dettman and Lohman, 2000). However, such a low temperature is not realistic for calcite veins through which focused hydrothermal solutions have migrated.

The second and third calculations were made using temperatures of 110 and 165°C (383.15 and 438.15°K). These are, respectively, the lowest and the highest homogenization temperature obtained from microthermometry of aqueous inclusions in the calcite veins (Table 2). These temperatures were used, although three uncertainties should be kept in mind. First, fluid-inclusion homogenization temperatures (in the absence of pressure data) are minimum fluid-trapping temperatures; second, the measured fluid inclusions are secondary (and hence, somewhat younger); and third, the fluid-inclusion and stable isotope data are not for the same samples (see Tables 2, 4). Nevertheless, because aqueous inclusion microthermometry indicates homogenization temperatures of more than 110°C for all the samples analyzed, it has been assumed that similar

Table 4. Stable isotope data on calcite veins and host rocks.

Sample	*Matrix δ¹³C (‰) PBD*	*Matrix δ¹⁸O (‰) PBD*	*Vein δ¹³C (‰) PBD*	*Vein δ¹⁸O (‰) PBD*
W1	+2.75	−6.96	+2.82	−9.81
W6	−6.84	−10.16	−6.87	−12.46
W8B	−6.67	−15.8	+1.32	−13.04
W19	+2.87	−9.32	+2.63	−9.46
W20H	+3.40	−3.57	+2.48	−5.14
W26	+1.75	−10.88	+0.86	−12.22
W32	+4.58	−5.23	+4.26	−6.93
W45A	−1.01	−8.55	−2.05	−9.64
W50	−0.29	−7.43	−2.63	−14.88
W75B	−0.72	−4.95	−2.91	−11.51
W80	+1.29	−3.99	+5.62	−10.43
W100	−11.48	−7.77	−8.83	−10.21

temperature ranges also prevailed for other samples (from which stable isotope data have been obtained). Calculating for temperatures of 110 and 165°C, we obtain $\delta^{18}O_{water}$ values ranging from 0 to +14‰ (relative to SMOW). These values indicate water $\delta^{18}O$ compositions above normal ocean water (with 0 SMOW). It is plausible that the aqueous solution flowing through the veins originated in the marine basin itself (such as diagenetic, pore-filled, or formation water). This interpretation is also consistent with high salinities (>11% NaCl equivalent) determined for the majority of the aqueous inclusions (Table 2).

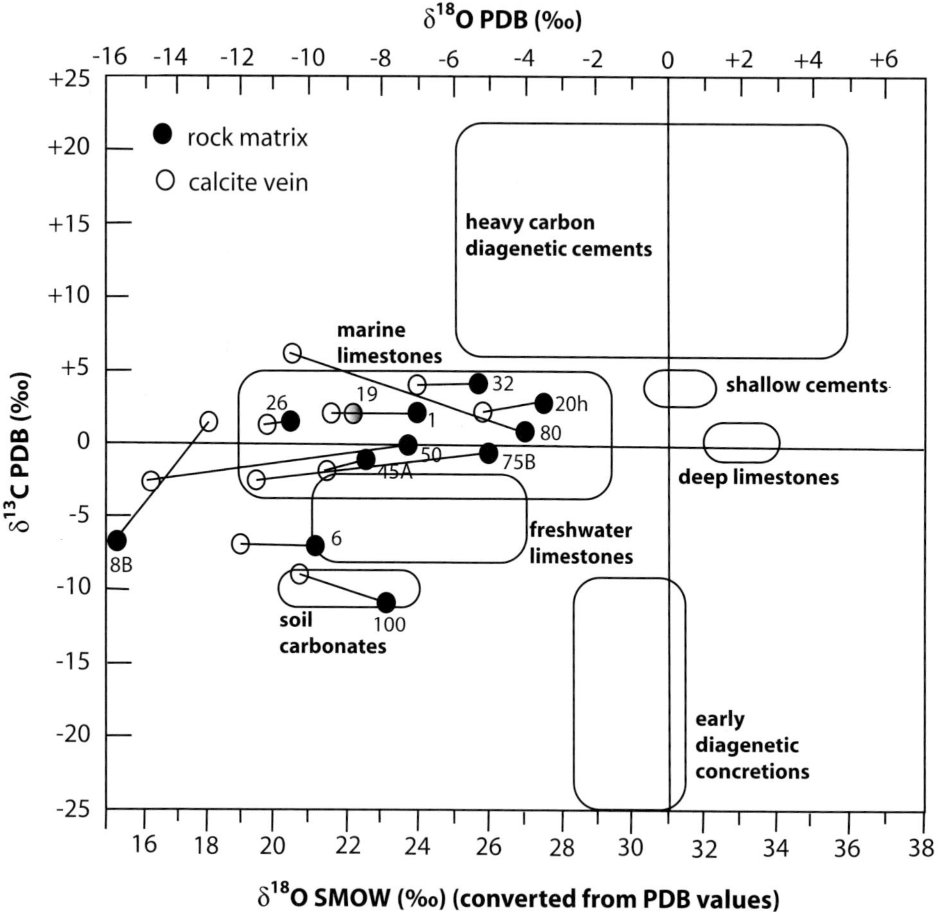

FIGURE 7. Plot of carbon and oxygen isotopic data obtained in this study. Square fields are for carbonate rocks from Hudson (1977). Data for host rock matrix are shown in solid circles, and those for calcite veins are shown in open circles. The gray-shaded circle indicates a sample with the same values for rock matrix and calcite veins. Lines show data for each sample pair (host rock and calcite vein). Note that oxygen values are plotted relative to the PDB (laboratory data) and relative to SMOW (converted from PDB values).

Thrust Faults and Fluid Flow

The motion of large thrust sheets through relatively weak rocks, especially anisotropic strata, has been an enigma in structural geology. Plate motion and convergence is now widely believed to be the primary stress engine for thrust

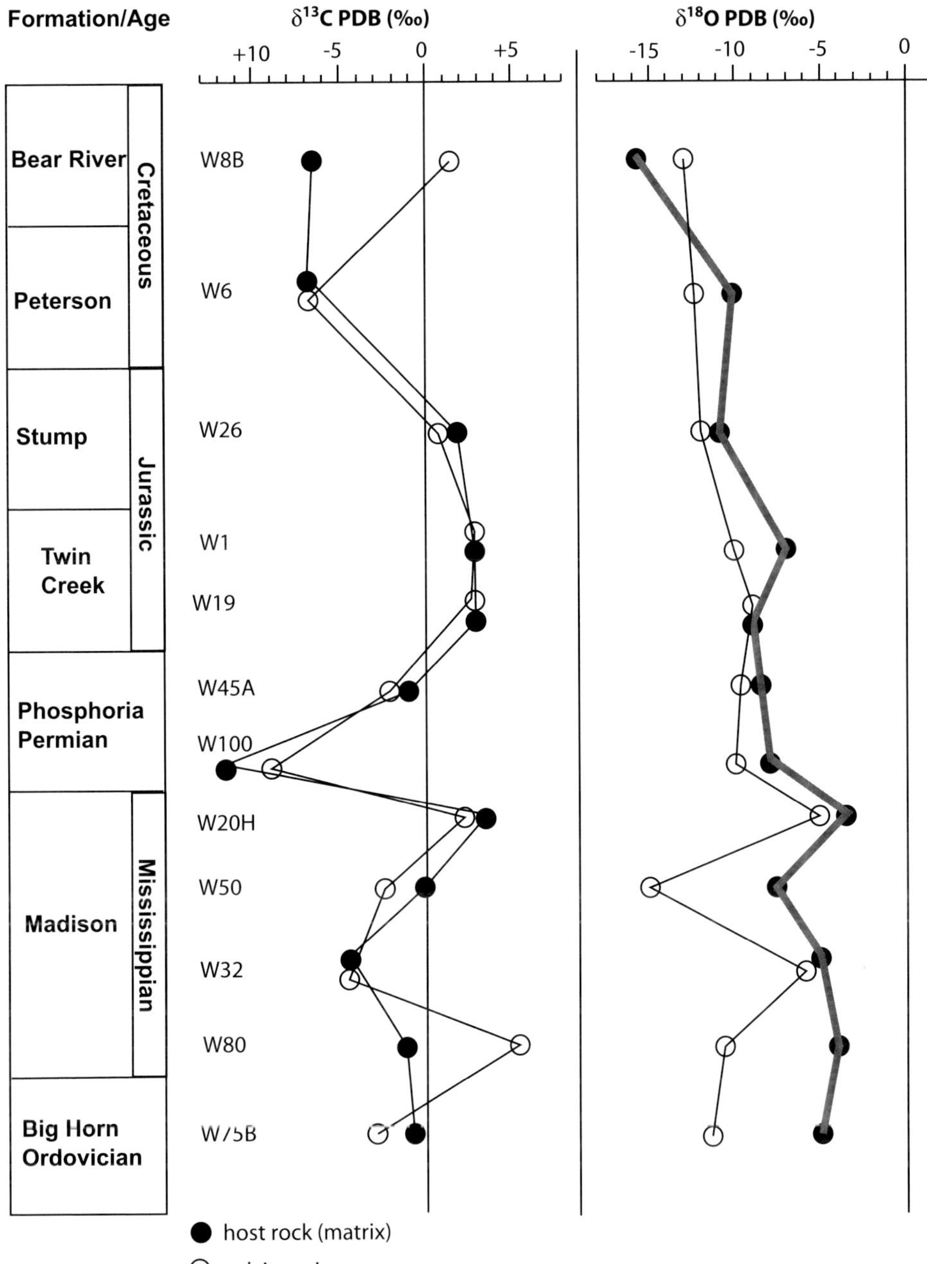

FIGURE 8. Plot of carbon and oxygen isotopic data for samples plotted against the stratigraphic (formation) age of the samples. Solid circles represent host rocks, and open circles show calcite veins. Calcite veins are depicted in the same stratigraphic formations as their host rocks for the sake of comparison, but their ages are younger than the stratigraphic ages. Gray thick line shows that host rocks show more negative $\delta^{18}O$ values as they become younger, indicating their enrichment in light oxygen probably because of exposure to meteoric waters.

tectonics. Hubbert and Rubey (1959) demonstrated theoretically that fluid flow along thrust faults aids the movement of thrust sheets by increasing pore-fluid pressure and decreasing the effective normal stress acting on faults. The fact that thrust faults in Wyoming have been associated with fluidized rocks, such as fault breccia and mineral veins, signifies the function of high-pressure fluid flow in addition to plate tectonic stress for the transportation of these thrust sheets.

In referring to Figure 3, it was pointed out that the velocity of the major thrust faults in western Wyoming appears to have increased through time (from the older to the younger thrusts). This can be explained as a combined effect of the following three causes:

1) Increase in the plate motion rates: A plate kinematic analysis of the Pacific Basin by Engebretson et al. (1985) has demonstrated that the velocity of the Farallon Plate in a direction normal to the trench axis increased from about 100 to nearly 150 km/m.y. from 75 to 45 Ma. This implies a dramatic increase in the convergence rate between the North American continental plate and the Farallon oceanic plate.
2) Decrease in the angle of the subduction slab: The distribution of subduction-related magmatic rocks in western America shows that magmatic activity was confined to a narrow belt in California–Oregon–Washington in the Late Jurassic–Early Cretaceous, but that it spread to Idaho–Nevada–Utah–Colorado during the Late Cretaceous. This eastward shift of the magmatic belt has been interpreted to be caused by flattening of the subducting oceanic slab beneath the western margin of North America (Atwater, 1989). This, in turn, would have led to a marked thermal contrast between the cold Precambrian continental crust of North America and the hot magmatic zone on its western rim and, hence, the subsidence (underthrusting) of the crystalline basement and the detachment and eastward thrusting of the

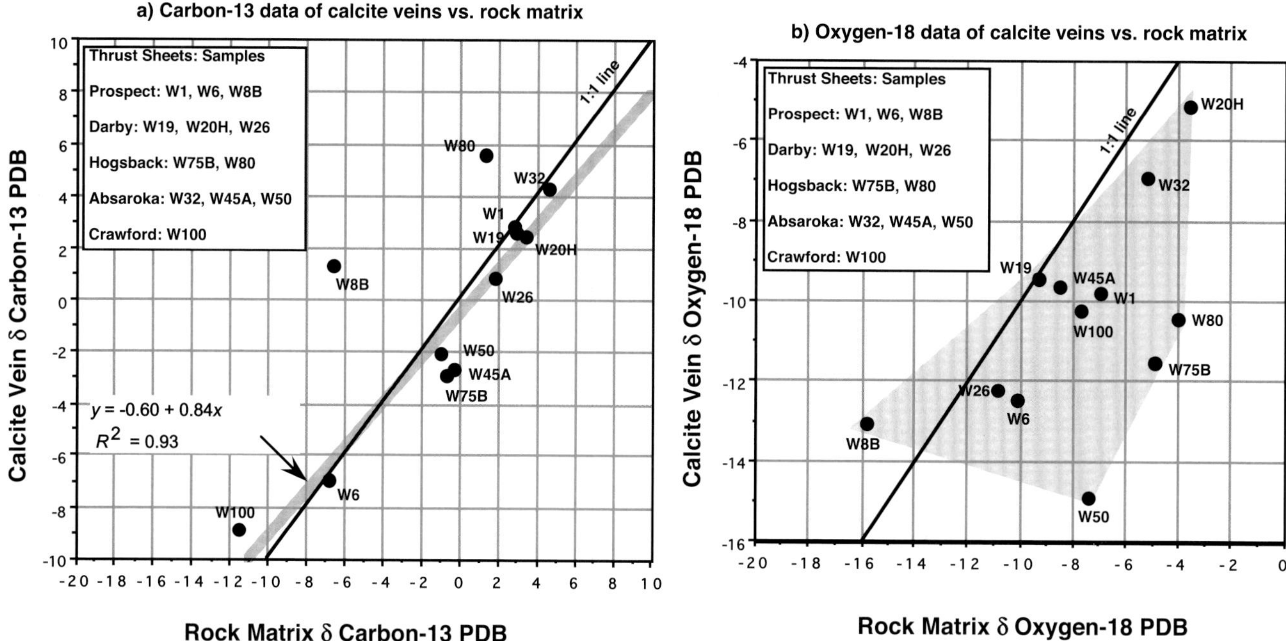

FIGURE 9. Plots of stable isotope data showing δ^{13}C % values for calcite veins and host rock matrix (a) and δ^{18}O % values for calcite veins and host rock matrix (b). Both carbon and oxygen data are relative to PDB. Overall, the carbon data are similar for both host rocks and calcite veins, but the oxygen data are different between host rocks and calcite veins, which show more negative δ^{18}O values. See Table 4 for data and text for interoperation.

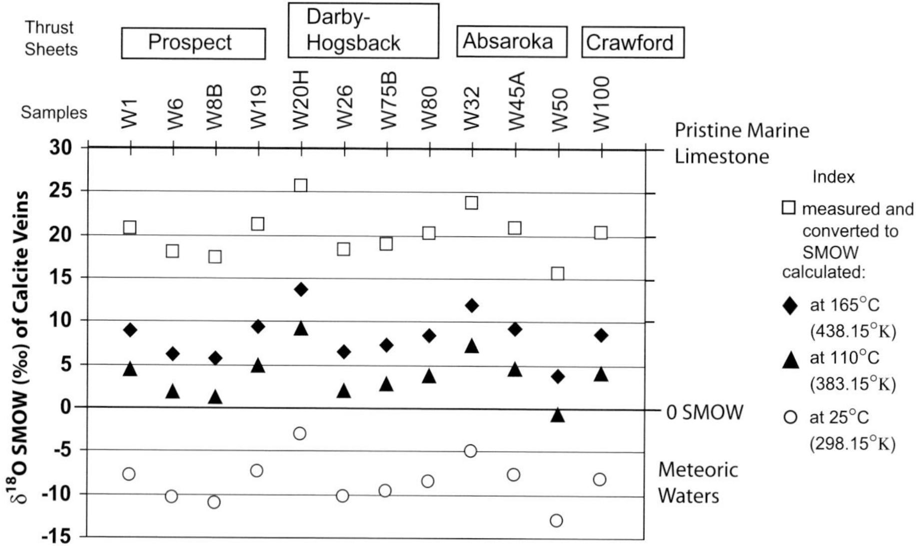

FIGURE 10. δ^{18}O % values for calcite vein samples as measured in the laboratory and converted to values relative to SMOW (open squares) and calculated for water temperatures of 25°C (open circles), 110°C (solid triangles), and 165°C (solid diamonds). The temperature 25°C is considered to be surface ocean temperature, whereas the temperatures 110 and 165°C are based on the homogenization temperatures obtained from aqueous fluid inclusions in the calcite veins. The higher temperatures are more realistic estimates and indicate that the vein water, from which calcite precipitated, was not meteoric water but a water with positive SMOW values between unmodified ocean water and diagenetic water in the rocks. The thrust sheets from which the samples have been collected are also defined in the figure, although the δ^{18}O values do not show any trend controlled by thrusts. It appears that all thrust sheets were subjected to the same hydrothermal solutions.

sedimentary package in Wyoming and other parts of the Sevier fold and thrust belt (Scholten, 1982).

3) Maastrichtian flooding: Concurrent with the increase in the motion of the Farallon Plate and the decrease in its subduction angle during the Late Cretaceous, an interior seaway developed to the east of the Sevier orogenic belt. This seaway (sometimes called the Mowry Sea), whose initiation is marked by the deposition of marine shale (e.g., the Mowry Shale in Colorado and Aspen Shale in Wyoming), extended from the Arctic as far south as the Gulf of Mexico and led to Maastrichtian flooding of basins lying in the eastern fronts of the Sevier orogen (Williams and Stelc, 1975).

It is reasonable to expect that combined increases in the tectonic, thermal, and hydraulic stresses contributed to the formation and accelerating transport of large thrust sheets along the western margin of North America, including the Wyoming fold and thrust belt.

Permeability Measurements on Host Rocks and Calcite Veins in Fault Zones

From a petroleum exploration point of view, it is significant to understand the impact of vein deposition in sedimentary rocks. Micropermeability measurements (using N_2 gas flow and a PDPK-400 Corelab instrument) were made on two samples: W60 (a fault breccia of Hogsback thrust in the Gros Ventre limestone of Cambrian age) and W46.5 (Amsden Sandstone of Pennsylvanian age collected from the Sheep Pass thrust zone) (Figure 11). The permeability of W60 limestone rock matrix is in the range of 0.001–0.02 md, and calcite veins in the same rock sample show similar permeability values. This implies that after vein mineralization, the permeability from the host rock to the mineral vein did not change. This result is consistent with the interpretation that the carbon material of the vein was derived from the host rock and was precipitated in the vein under similar confining pressures (same rock formations at similar depths).

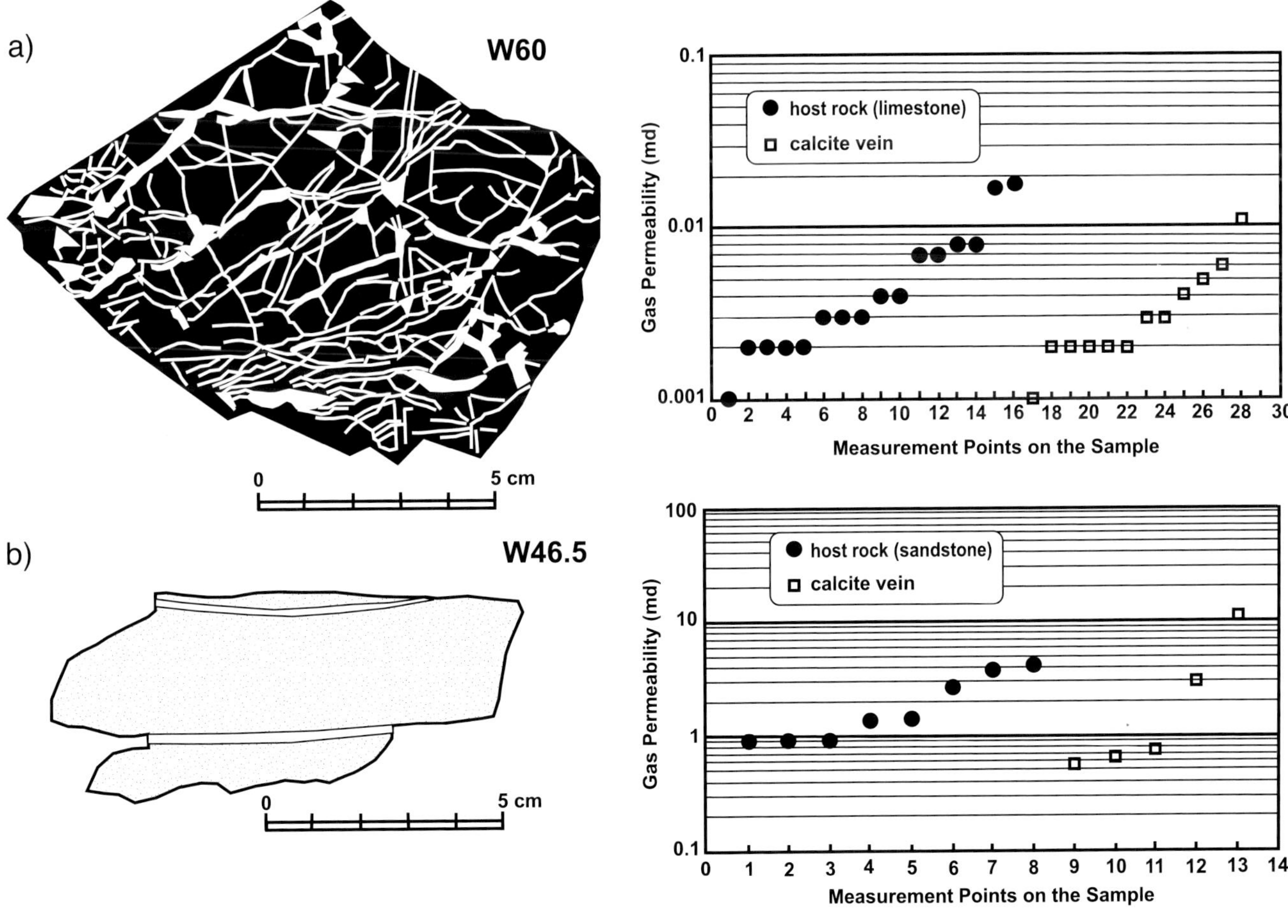

FIGURE 11. Gas permeability data (measured by micropermeameter on various points of host rocks and calcite veins): (a) sample W60 (fault breccia of the Hogsback thrust in the Cambrian Gors Ventre limestone); (b) sample W46.5 (calcite vein in Amsden Sandstone of Pennsylvanian age). The data have been arranged according to host rock and vein categories.

The permeability of W46.5 sandstone rock matrix is in the range of 0.9–4 md, and that of calcite veins in the same rock is similarly in the range of 0.5–10 md. W46.5 is a calcite-cemented sandstone, and the calcite vein material was probably also derived from the host rock. The permeability of W46.5 sandstone (host rock) is two orders of magnitude higher than that of W60 limestone (host rock), which is reasonable considering the differences in rock types (sandstone vs. limestone) and stratigraphic ages (Pennsylvanian vs. Cambrian).

Overall, permeability for a given host rock fragment or a calcite vein was found to be a range of values on the order of one magnitude instead of a single permeability value. These results have two important implications on permeability distribution in faulted carbonate reservoirs. First, it is important to consider permeability as a range of values in reservoir simulations. Second, it appears that in faulted carbonate reservoirs, the permeability of fault rocks with calcite veins is not different from that of the host rock because of the fact that the same host rock is reprecipitated as fracture-filling mineral veins or fault breccia. This is in contrast to porous sandstone reservoirs, in which fault-zone rocks show markedly lower permeability values compared to host rocks (e.g., Antonellini and Aydin, 1994; Fowles and Burley, 1994).

CONCLUDING REMARKS

Western Wyoming has long offered a natural laboratory for understanding the deformation style of thin-skinned fold and thrust belt. Thrust sheets in this region are associated with fractures and mineral veins that offer a unique opportunity to investigate the geochemical signatures of paleofluid systems associated with the development of thrust faults. Fluid-inclusion analyses of calcite veins indicated the flow of various generations of brine as well as the migration of liquid hydrocarbons through the fractures. Stable $\delta^{18}O$ isotope data calculated at temperatures of 110–165°C (based on the homogenization temperatures of aqueous inclusions obtained from fluid-inclusion thermometry) suggested that diagenetic or formation water (brines) circulated through the veins as indicated by positive $\delta^{18}O$ values relative to SMOW. $\delta^{18}C$ values were found to be remarkably similar between calcite veins and host limestone rocks, suggesting the derivation of carbon in the veins from the same formations. Gas permeability showed a similar range of values between calcite veins and host rock matrix, indicating little permeability change caused by vein mineralization in the rocks. Studies of the geochemical signature of fluids flushing through thrust zones in western Wyoming amplify the significance of interactions between thrust faulting and subsurface fluid flow.

ACKNOWLEDGMENTS

This study was supported by the Japan National Oil Corporation (presently Japan Oil, Gas and Metals National Corporation). I am grateful to Danniel Schelling, Kiyofumi Suzuki, and Andrey Rezanov for a wonderful field work in Wyoming and Emiko Shinbo for technical support of stable isotope measurements. The fluid-inclusion measurements were carried out at Fluid Inclusion Technologies Inc.; I thank Don Hall and Ashley Bigge very much for their technical support. This chapter has been reviewed by Ezat Heydari, Joe Moore, and David Dettman, whose comments and corrections greatly improved the interpretations and the text. I am also thankful to Yoshihiro Tusji and Doug Ekart for discussions and for reading an earlier version of the chapter. I alone, however, am responsible for any errors.

REFERENCES CITED

Agosta, F., and D. L. Kirschner, 2003, Fluid conduits in carbonate-hosted seismogenic normal faults of central Italy: Journal of Geophysical Research, v. 108B, p. 2221–2233.

Antonellini, M., and A. Aydin, 1994, Effect of faulting on fluid flow in porous sandstones: Petrophysical properties: AAPG Bulletin, v. 78, p. 355–377.

Armstrong, F. C., and S. S. Oriel, 1965, Tectonic development of the Idaho–Wyoming thrust belt: AAPG Bulletin, v. 49, p. 1847–1866.

Atnipp, V. A., J. M. Budai, and K. C. Lohmann, 1987, Evidence for multiple stages of fluid migration through the Wyoming overthrust belt (abs.): Geological Society of America Annual Meeting Abstracts with Program, v. 19, p. 576.

Atwater, T., 1989, Plate tectonic history of the northeast Pacific and western North America, *in* E. L. Winterer, D. M. Hussong, and R. W. Decker, eds., The eastern Pacific Ocean and Hawaii, Geological Society of America, The Geology of North America, v. N, p. 15–72.

Bebout, G. E., D. J. Anastasio, and J. E. Holl, 2001, Synorogenic crustal fluid infiltration in the Idaho–Montana thrust belt: Geophysical Research Letters, v. 28, p. 4295–4298.

Bodnar, R. J., 1993, Revised equation and table for determining the freezing point depression of H_2O-NaCl solution: Geochimica et Cosmochimica Acta, v. 57, p. 683–684.

Bruhn, R. L., W. A. Yonkee, and W. T. Parry, 1990, Structural and fluid-chemical properties of seismogenic normal faults: Tectonophysics, v. 175, p. 139–157.

Coogan, J. C., 1992, Structural evolution of piggyback basins in the Wyoming–Idaho–Utah thrust belt, *in* P. K. Link, M. A. Kuntz, and L. B. Platt, eds., Regional geology of eastern Idaho and western Wyoming: Geological Society of America Memoir 179, p. 55–81.

Coplen, T. B., C. Kendall, and J. Hopple, 1983, Comparison

of stable isotope reference samples: Nature, v. 302, p. 236–238.

Craddock, J. P., 1992, Transpression during tectonic evolution of the Idaho–Wyoming fold-and-thrust belt, *in* P. K. Link, M. A. Kuntz, and L. B. Platt, eds., Regional geology of eastern Idaho and western Wyoming: Geological Society of America Memoir 179, p. 125–139.

Dettman, D. L., and K. C. Lohman, 2000, Oxygen isotope evidence for high-altitude snow in the Laramide Rocky Mountains of North America during the Late Cretaceous and Paleogene: Geology, v. 28, p. 243–246.

Dixon, J., 1982, Regional structural synthesis, Wyoming salient of western overthrust belt: AAPG Bulletin, v. 66, p. 1560–1580.

Eichhubl, P., and J. R. Boles, 2000, Focused fluid flow along faults in the Monterey Formation, coastal California: Geological Society of America Bulletin, v. 112, p. 1667–1679.

Engebretson, D. C., A. Cox, and R. G. Gordon, 1985, Relative motions between oceanic and continental plates in the Pacific Basin: Geological Society of America Special Paper 206, 59 p.

Faybishenko, B., P. A. Witherspoon, and S. M. Benson, eds., 2000, Dynamics of fluids in fractured rocks: American Geophysical Union Geophysical Monograph 122, 400 p.

Fowles, J., and S. D. Burley, 1994, Textural and permeability characteristics of faulted, high porosity sandstones: Marine Petroleum and Geology, v. 11, p. 608–623.

Friedman, I., and J. R. O'Neill, 1977, Compilation of stable isotope fractionation factors of geochemical interest, *in* M. Fleischer, ed., Data of geochemistry: U.S. Geological Survey Professional Paper 440KK, p. 1–12.

Goddard, J. V., and J. P. Evans, 1995, Fluid-rock interactions in faults of crystalline thrust sheets, northwestern Wyoming, U.S.A.: Inferences from geochemistry of fault-related rocks: Journal of Structural Geology, v. 17, p. 533–549.

Goldstein, R. H., and T. J. Reynolds, 1994, Systematics of fluid inclusions in diagenetic minerals: SEPM Short Course 31, 199 p.

Hickman, S., R. Sibson, and R. Bruhn, eds., 1995, Mechanical involvement of fluids in faulting: Journal of Geophysical Research, v. 100B (special section), p. 12,831–13,132.

Hippler, S. J., 1997, Microstructures and diagenesis in North Sea fault zones: Implications for fault-seal potential and fault-migration rate, *in* R. C. Surdam, ed., Seals, traps, and the petroleum system: AAPG Memoir 67, p. 103–113.

Hofes, J., 1997, Stable isotope geochemistry, 4th ed.: Berlin, Springer-Verlag, 201 p.

Hubbert, M. K., and W. W., Rubey, 1959, Role of fluid pressure in mechanics of overthrust faulting: 1. Mechanics of fluid-filled porous solids and its application to overthrust faulting: Geological Society of America Bulletin, v. 70, p. 115–166.

Hudson, J. D., 1977, Stable isotopes and limestone lithifications: Journal of the Geological Society (London), v. 133, p. 637–660.

Jamtveit, B., and B. Yardley, eds., 1996, Fluid flow and transport in rocks: mechanisms and effects: London, Chapman & Hall, 408 p.

Kerrich, R., 1986, Fluid infiltration into fault zones: chemical, isotopic, and mechanical effects: Pure and Applied Geophysics, v. 124, p. 226–268.

Kerrich, R., and D. Hyndman, 1986, Thermal and fluid regimes in the Bitterrot lobe-Sapphire block detachment zone, Montana: Evidence from $^{18}O/^{16}O$ and geologic relations: Geological Society of America Bulletin, v. 97, p. 147–155.

Kerrich, R., T. E. La Tour, and L. Willmore, 1984, Fluid participation in deep fault zones: Evidence from geological, geochemical, and $^{18}O/^{16}O$ relations: Journal of Geophysical Research, v. 89B, p. 4331–4343.

Lamerson, P. R., 1982, The Fossil Basin and its relationship to the Absaroka thrust system, Wyoming and Utah, *in* R. D. Powers, ed., Geologic studies of the Cordilleran thrust belt: Denver, Rocky Mountain Association of Geologists, v. 1, p. 279–340.

McCaffrey, K., L. Lonergan, and J. Wilkinson, eds., 1999, Fractures, fluid flow, and mineralization: Geological Society (London) Special Publication 155, 328 p.

National Research Council, 1996, Rock fractures and fluid flow: Contemporary understanding and application: Washington, D.C., National Academy Press, 551 p.

Oriel, S. S., 1986, The Idaho–Wyoming salient of the North American Cordilleran foreland thrust belt: Bulletin Societe de Geologie France, v. 8II, no. 5, p. 755–765.

Parnell, J., ed., 1994, Geofluids: Origin, migration, and evolution of fluids in sedimentary basins: Geological Society (London) Special Publication 78, 372 p.

Roberts, G., 1990, Structural controls on fluid migration in foreland thrust belts, *in* J. Letouzey, ed., Petroleum and tectonics in mobile belts: Paris, Editions Technip, p. 193–210.

Royse, F. M., A. Warner, and D. L. Reese, 1975, Thrust belt structural geometry and related stratigraphic problems: Wyoming–Idaho–northern Utah, *in* D. W. Bolyard, ed., Deep drilling frontiers of the Central Rocky Mountains: Denver, Rocky Mountain Association of Geologists, p. 41–54.

Savin, S. M., and H.-W. Yeh, 1981, Stable isotopes in ocean sediments, in C. Emiliani, ed., The Sea: New York, John Wiley, v. 7, p. 1521–1554.

Scholten, R., 1982, Continental subduction in the northern U.S. Rockies— A model for back-arc thrusting in the Western Cordillera, *in* R. D. Powers, ed., Geologic studies of the Cordilleran thrust belt: Denver, Rocky Mountain Association of Geologists, v. 1, p. 123–136.

Smith, B. M., S. J. Reynolds, H. W. Day, and R. J. Bondar, 1991, Deep-seated fluid involvement and mineralization, South Mountains metamorphic core complex, Arizona: Geological Society of America Bulletin, v. 103, p. 559–569.

Williams, G. D., and C. R. Stelc, 1975, Speculations on the Cretaceous palaeogeography of North America: Geological Society of Canada Special Paper 13, Cretaceous System in the Western Interior of North America, p. 1–20.

Wiltschko, D. V., and J. A. Dorr, 1983, Timing of deformation in overthrust belt and foreland of Idaho, Wyoming and Utah: AAPG Bulletin, v. 67, p.1304–1322.

16

Tagami, T., and M. Murakami, 2005, Zircon fission-track thermochronology of the Nojima fault zone, Japan, *in* R. Sorkhabi and Y. Tsuji, Faults, fluid flow, and petroleum traps: AAPG Memoir 85, p. 269–285.

Zircon Fission-track Thermochronology of the Nojima Fault Zone, Japan

Takahiro Tagami
Division of Earth and Planetary Sciences, Graduate School of Science, Kyoto University, Kyoto, Japan

Masaki Murakami
Division of Earth and Planetary Sciences, Graduate School of Science, Kyoto University, Kyoto, Japan

ABSTRACT

Fission-track (FT) thermochronologic analysis was performed on zircon separates from rocks in and around the Nojima fault, which was activated during the 1995 Kobe earthquake. Samples were collected from the University Group 500-m (1600-ft) (UG 500) borehole, Geological Survey of Japan 750-m (2500-ft) (GSJ 750) borehole, the fault trench at Hirabayashi, and nearby outcrops. Zircon FT ages from the UG 500 borehole record about 2-Ma cooling age in the zircon partial annealing zone (ZPAZ) for samples within about 25-m (80-ft) distance from the fault plane, whereas those of the GSJ 750 borehole record about 30–40-Ma cooling ages in the same fault-zone width. On the basis of one-dimensional heat conduction modeling as well as the consistency between the degree of FT annealing and the degree of deformation and alteration of borehole rocks, these cooling ages in both boreholes are interpreted as consequences of ancient thermal overprints by heat transfer or dispersion via fluids in the fault zone. For the fault trench samples, zircon FTs of the 2–10-mm (0.08–0.4-in.)-thick pseudotachylyte layer were totally reset (or remained reset) and subsequently cooled at about 56 Ma, which is interpreted as the time of final cooling through ZPAZ immediately after the pseudotachylyte formation. It is therefore suggested that the present Nojima fault system was reactivated in the middle Quaternary from an ancient fault initiated at about 56 Ma at midcrustal depths.

INTRODUCTION

Quantitative assessment of heat generation and transfer along faults and associated with fault movement is of primary importance in understanding the dynamics and structural history of faulting (Scholz, 1996), as well as in constraining the heat budget and thermotectonic evolution of mobile belts (Fukahata and Matsu'ura, 2001). These thermal signatures also provide a tool for constraining the ages of faults. Among a range of methodologies available to investigate thermal regimes around faults, thermal history

DOI:10.1306/1033728M853138

analysis using radiometric dating methods is extremely useful because it reveals temporal changes of temperature in the fault-zone rocks. Fission-track (FT) analysis is especially valuable in this regard because of its following advantages:

1) Temperature is the only environmental factor to cause track fading. Recent laboratory hydrothermal annealing experiments of zircon FTs show that FT annealing is indistinguishable between atmospheric and hydrothermally pressured conditions (Brix et al., 2002; Yamada et al., 2003).
2) The FT method is useful for thermal history analysis of the upper crust and sedimentary basins because FTs in minerals such as apatite, zircon, and sphene are annealed at temperatures of about 100 to 300°C (Naeser, 1979; Wagner and Van den haute, 1992). Therefore, mineral FT ages record relatively low closure temperatures and are sensitive to low-temperature effects on rocks.
3) Fission-track-length distribution provides additional information for quantitative reconstruction of thermal history in partial annealing zones (Gleadow et al., 1986; Green et al., 1989; Hasebe et al., 1994).

The FT technique has been successfully applied by many researchers to thermal history analysis of faults in various regions (see Wagner and Van den haute, 1992; Gallagher et al., 1998, for references), e.g., the Median Tectonic Line, Japan (Tagami et al., 1988a), the Alpine fault, New Zealand (Kamp et al., 1989), the Pejo fault system, Italian eastern Alps (Viola et al., 2003), Castle Mountain fault, Alaska (Parry et al., 2001), and San Gabriel fault, California (d'Alessio et al., 2003).

This chapter examines zircon FT data of the Nojima fault rocks collected from the University Group 500-m (1600-ft) (UG 500; or Disaster Prevention Research Institute 500 m [1600 ft], after Murata et al., 2001; Ogura 500 m [1600 ft], Tagami et al., 2001) borehole, Geological Survey of Japan 750-m (2500-ft) (GSJ 750) borehole, the fault trench at Hirabayashi, and nearby outcrops on the Awaji Island, Japan (Figures 1, 2). The Nojima fault, which was activated during the 1995 Kobe earthquake (Hyogoken-Nanbu earthquake; magnitude [M] 7.2), has been investigated multifariously from the viewpoint of geology and geophysics (Ando, 2001). Together with such information, we discuss geologic processes producing the observed FT profiles around the fault.

Compared to apatite, the systematics of FT thermochronology in zircon is a relatively recent development. Moreover, case studies of thermal history in sedimentary basins using apatite far exceed those of zircon. Because our research group has largely contributed to the calibration and development of zircon FT thermochronology and because we have employed zircon in this case study, we first discuss the data constraining the FT annealing in zircon, which may be of further interest to the petroleum geologists studying thermal histories of sedimentary rocks and basins.

FISSION-TRACK THERMOCHRONOLOGY OF ZIRCON

A series of laboratory studies of temperature-dependent track retention in zircon has been conducted in the last decade, using confined spontaneous track lengths as a measure of annealing, as well as etched track widths for standardization of track revelation (Tagami et al., 1990, 1998; Yamada et al., 1993; 1995a, b; 1998; Hasebe et al., 1994). As a result of this research, kinetic parameters that describe FT annealing at laboratory timescales have been established for zircon (Yamada et al., 1995b; Galbraith and Laslett, 1997; Tagami et al., 1998). The fanning model formula for the mean length (μ) after annealing at temperature T (Kelvin) for time t (hours) is given by (Tagami et al., 1998)

$$\mu = 11.35[1 - \exp\{-6.502 + 0.1431(\ln t + 23.515)/(1000/T - 0.4459)\}]$$

Samples from deep boreholes have provided additional constraints on the FT annealing behavior in zircon on geological timescales (Vienna Basin, Austria, Tagami et al., 1996; Kola, Russia, Green et al., 1996; Kontinentale Tiefbohrprogramm der Bundesrepublik, Hauptbohrung, Germany, Coyle and Wagner, 1996; Ministry of International Trade and Industry-Nishikubiki, Japan, Hasebe et al., 2003). Figure 3 summarizes the results of the annealing experiments and the borehole sample studies. The validity of extrapolating the kinetic models to geological timescales is supported by the consistency between the model extrapolations and borehole constraints (Tagami et al., 1998) (Figure 3). In addition, the mineral parageneses and quartz microstructures of high-pressure–low-temperature rocks on Crete, Greece, indicate that the higher temperature limit of the zircon partial annealing zone (ZPAZ) is between 350 and 400°C for a heating duration of 4 ± 2 Ma (Brix et al., 2002). This constraint is consistent with the model extrapolation within errors. The zircon FT closure temperature directly calibrated from $^{40}Ar/^{39}Ar$ diffusion of K-feldspar and biotite is about 260–265°C and about 235°C for cooling rates of 30–50°C/m.y. and about 4°C/m.y., respectively (Foster et al., 1996). These estimated temperatures are both within the ZPAZ based

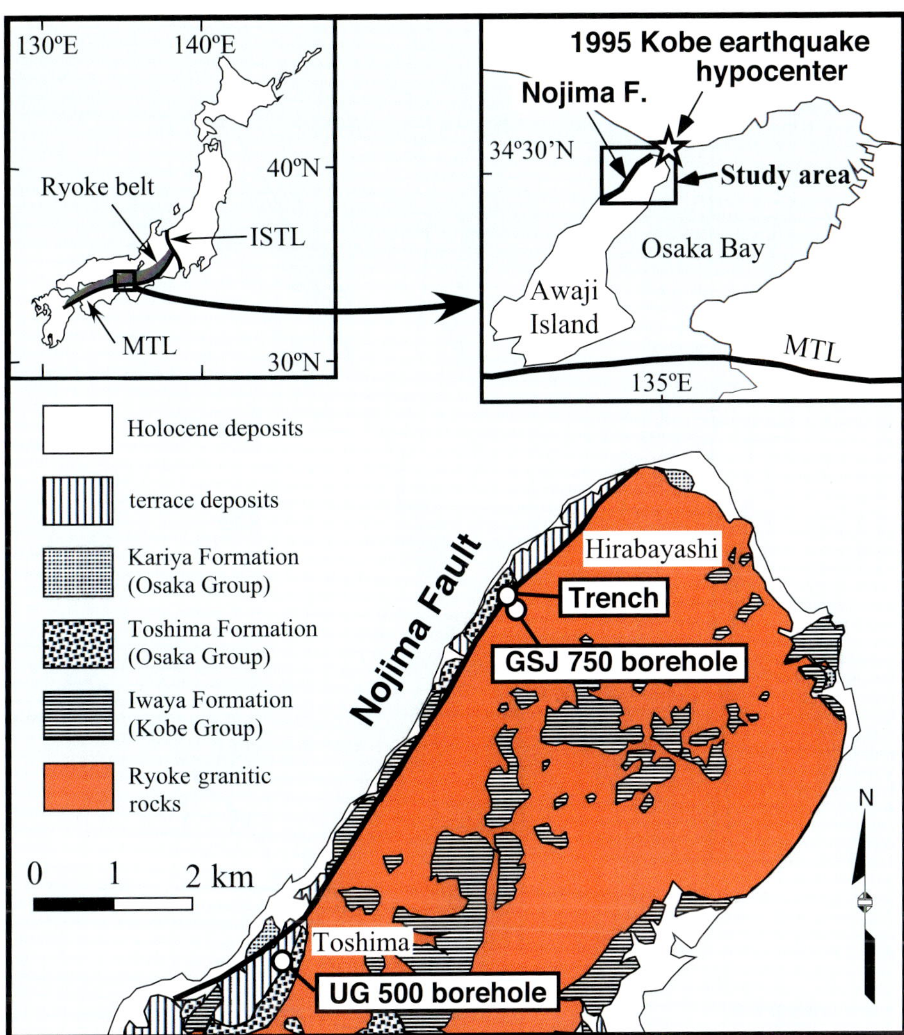

FIGURE 1. A geologic map showing sampling localities. MTL = Median Tectonic Line; ISTL = Itoigawa–Shizuoka Tectonic Line.

on the extrapolation (Figure 3) (see also Tagami et al., 1996; Brandon et al., 1998, for evaluation and compilation of other available data).

Track annealing properties of zircon have also been constrained through the FT analysis of thermal perturbation around a granitic pluton (Tagami and Shimada, 1996) (Figure 4). These data come from the contact metamorphic aureole in the Cretaceous Shimanto belt, southwest Japan, formed by the intrusion of the Takatsukiyama granite at 15 Ma, for which the heating duration is estimated as about 10^5–10^6 yr using one-dimensional heat conduction modeling (Tagami and Shimada, 1996).

Figure 4a presents the relationship between sample-mean zircon FT ages and their surface (two-dimensional) distances from the intrusion contact of the Takatsukiyama granite (Tagami and Shimada, 1996). The ages are generally similar to or older than Cretaceous depositional ages for samples more than 4 km (2.5 mi) away from the contact. Approaching the contact, however, the ages show a continuous reduction from about 100 to 15 Ma at about 3-km (1.8-mi) distance. They coincide with the zircon FT ages of the Takatsukiyama granite (15.0 ± 0.4 Ma) for localities less than 2.5 km (1.6 mi) from the contact. Figure 4b shows representative examples of single-grain age distributions using radial plots (Galbraith, 1990) and track-length histograms. Track-length distributions (Figure 4b) show variations consistent with increasing resetting caused by the heating at 15 Ma. In unreset samples, track lengths show a unimodal distribution and average about 10 μm. Approaching the contact, samples that experienced the thermal effects of the pluton intrusion were significantly annealed and shortened, thus yielding a bimodal distribution with a greater number of tracks less than 9 μm in length. Less than 2.5 km (1.6 mi) from the contact, thermal resetting is complete because of higher paleotemperatures, and the samples have a unimodal distribution of long tracks formed after the 15-Ma event. The mean lengths in unannealed or totally annealed (reset) samples are close to the reference value of 10.5 ± 0.1 μm determined for zircon age standards from rapidly cooled volcanic rocks (Hasebe et al. 1994).

Figure 5 illustrates a schematic relationship among (1) three temperature zones that approximately express the thermal stability of the zircon FT system, i.e., total stability zone, partial annealing zone, and total annealing zone; (2) three hypothetical thermal histories that are characterized by secondary heating to different paleomaximum temperatures; and (3) three sets of length and age distributions that resulted from the three different time-temperature pathways. Using this relationship, we can inversely reconstruct the thermal history of rocks from measured FT length and age data (for quantitative analysis, see Gallagher et al., 1998). Here, we note that the boundary temperatures (i.e., T_1 and T_2 in Figure 5) for the three temperature

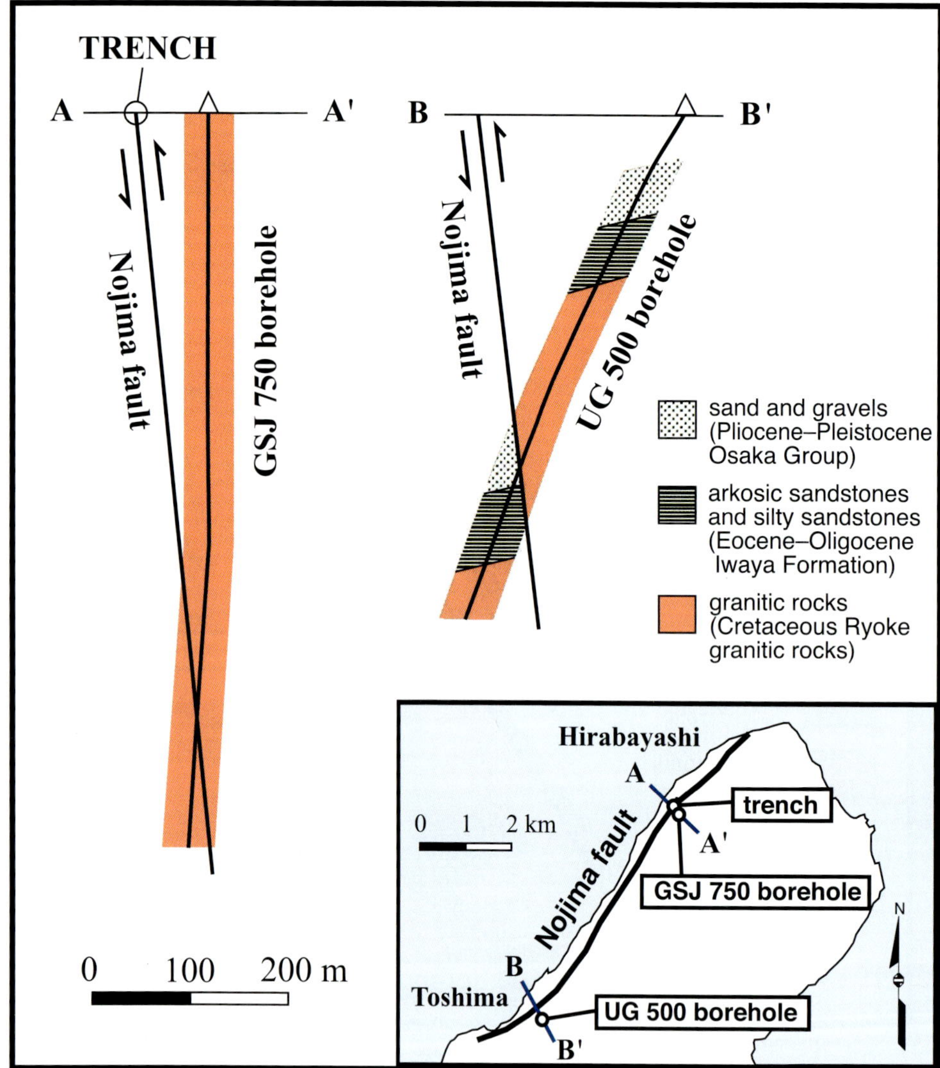

FIGURE 2. Schematic cross sections of the two boreholes analyzed. The section consists of granitic rocks throughout the GSJ 750 borehole, whereas sedimentary sequences covering the granitic basement are observed on both the hanging wall and footwall of the fault in the UG 500 borehole. The Nojima fault is a high-angle thrust with a dip of 83°SE.

zones substantially vary with heating time duration, as shown in Figure 3. This variation in partial annealing zone boundaries should be kept in mind when we analyze a heating phenomenon that has duration significantly shorter (<1 m.y.) than the ordinary geological timescales (1–1000 m.y.), the former of which may be the case of thermochronology of fault zones.

These lines of evidence suggest that the zircon FT system nicely quantifies the thermal history of rocks in nature. Fission-track investigations of rocks from fault zones require consideration of two factors that are worth special consideration. First, fault rocks may have been subjected to hydrothermally pressurized conditions at some stage during fault development. To evaluate this factor, laboratory heating experiments of zircon were carried out using a hydrothermal synthetic apparatus (Brix et al., 2002; Yamada et al., 2003). For reliable comparison, Yamada et al. (2003) annealed the same zircon samples using the same temperature monitor and experimental procedure as those employed in the previous experiment at atmospheric conditions (Yamada et al., 1995b). The observed FT annealing characteristics are indistinguishable between the heating carried out at atmospheric and hydrothermally pressurized conditions. This probably validates the application of annealing kinetics based on the experiments at atmospheric conditions to rocks subjected to hydrothermal conditions in nature, such as those in fault zone and plate subduction settings. Second, frictional heating along a fault is a short-term phenomenon with heating durations on the order of seconds, significantly shorter than the conventional laboratory heating having durations of about 0.1–10,000 hr. Thus, high-temperature and short-term annealing experiments were newly designed and conducted using a graphite furnace coupled with infrared radiation thermometry (M. Murakami, R. Yamada, T. Tagami, unpublished data). Their results show that the observed track-length reduction by 3.6–10 s heating at 599–912°C is similar to that predicted by the FT annealing kinetics based on the conventional laboratory heating for 4.5 min to 10,000 hr at 350–750°C (Yamada et al., 1995b; Tagami et al., 1998). In addition, it was found that spontaneous FTs in zircon are totally annealed at 850 ± 50°C for about 4 s. These physical conditions are similar to those of some pseudotachylyte formations in nature (e.g., Otsuki et al., 2003, for pseudotachylyte of the Nojima fault).

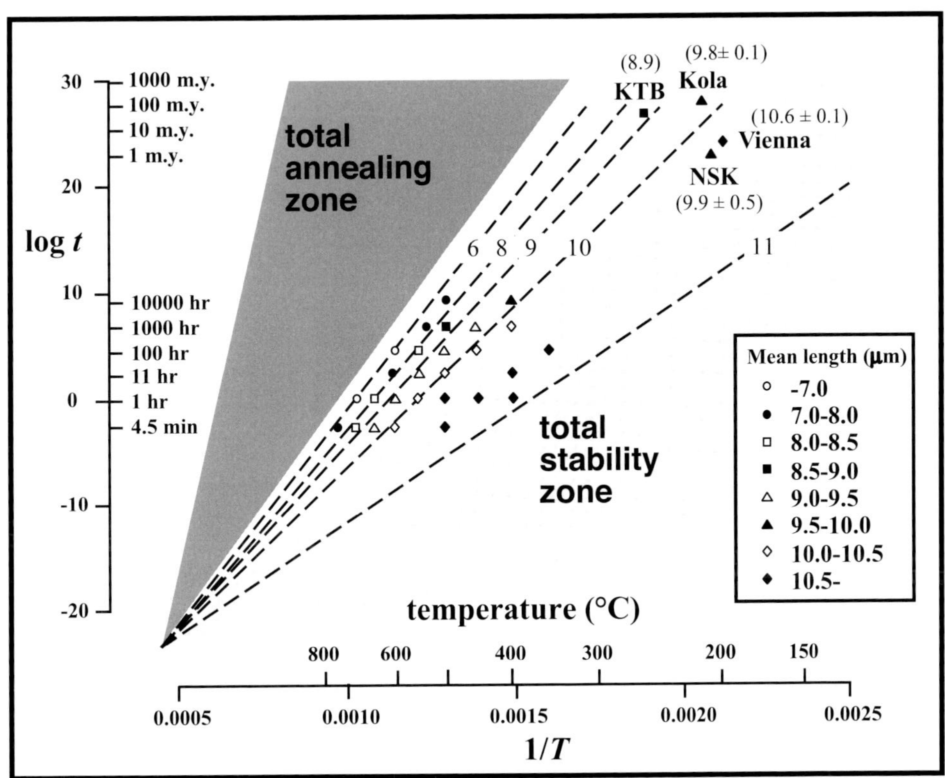

FIGURE 3. Arrhenius plot showing the design points of the laboratory annealing experiments and contour lines for the fitted fanning model extrapolated to geological time (Yamada et al., 1995b; Tagami et al., 1998). Also shown are the approximate locations of the four deep borehole samples subjected to long-term natural annealing, which were used to test the kinetic model (Tagami et al., 1998; Hasebe et al., 2003). Three temperature zones can be defined as a first-order approximation for annealing of the zircon FT system: the total stability zone, where tracks are thermally stable and, hence, accumulated as time elapsed; the partial annealing zone, where tracks are partially stable and slowly annealed and shortened; and the total annealing zone, where tracks are unstable and, thus, the crystallographic damages repair soon after their formation. Note that the partial annealing zone is defined here as a zone having mean lengths of about 4–10.5 µm (Yamada et al., 1995b) and shown as a white region intervened by the total stability zone and total annealing zone. KTB = Kontinentale Tiefbohrprogramm der Bundesrepublik; NSK = Nishikubiki.

GEOLOGIC OUTLINE OF THE NOJIMA FAULT

Awaji Island is located in the southern area of the inner zone of southwest Japan, with the southern tip of the island in contact with the Median Tectonic Line (Figure 1). The basement of the island is comprised by Cretaceous Ryoke granitic rocks consisting mainly of granodiorite (Mizuno et al., 1990), with hornblende and biotite K-Ar ages ranging from 88 ± 4 to 90 ± 5 and 70 ± 4 to 88 ± 4 Ma (1 standard error), respectively (Takahashi, 1992). No FT ages have been reported from the island before our studies. Zircon FT ages of the Ryoke granitic rocks in the Kii Peninsula, located about 80 km (50 mi) to the east of the Awaji Island, are 59 ± 3 to 78 ± 9 Ma (Tagami et al., 1988a; Hasebe and Tagami, 2001). In the northern part of the island, the Eocene–Oligocene Iwaya Formation partially overlies the basement (Mizuno et al., 1990; Yamamoto et al., 2000). The Pliocene–Pleistocene Osaka Group is distributed along the northwest coast of the island.

The 1995 Kobe earthquake caused great disaster in the southern parts of Kobe City and the northwest area of Awaji Island. The hypocenter was located in the Akashi strait, and the focal depth was reported to be 14 km (8.6 mi) (Figure 1). As a result, a greater than 10-km (6-mi)-long surface rupture was formed along the preexisting northeast–southwest, striking the Nojima fault, which is a high-angle reverse fault dipping 83°SE (Figure 1) (Awata et al., 1996; Murata et al., 2001). Maximum displacement of the surface rupture was observed as 180 cm (71 in.) right lateral and 130 cm (51 in.) reverse components at Hirabayashi, located at the northern part of the fault. The fault in the area forms a geological boundary between the Osaka Group and Iwaya Formation on the west

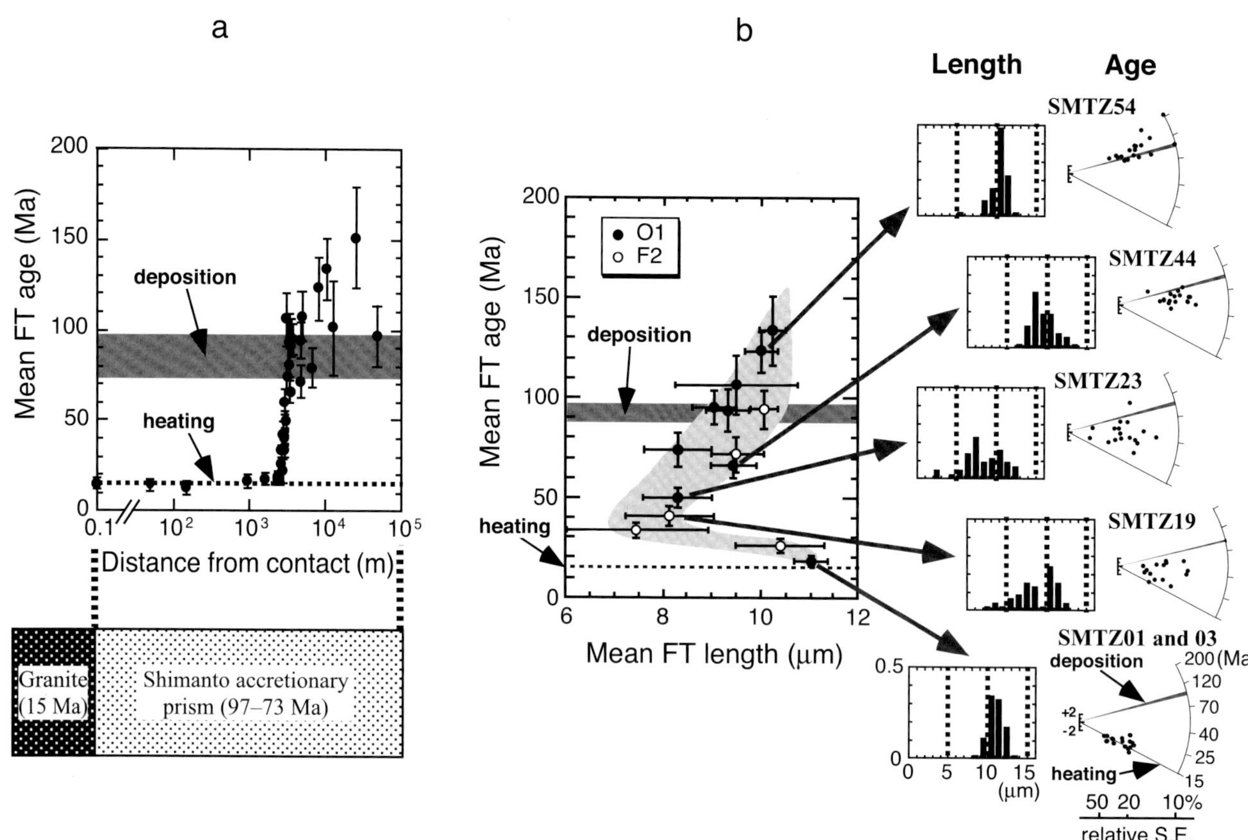

FIGURE 4. (a) Mean zircon FT ages vs. their distance from the contact of Takatsukiyama granite (Tagami and Shimada, 1996). The shaded zone and dashed line represent, respectively, 97–73-Ma deposition of the Shimanto accretionary complex and the 15-Ma rapid cooling of the granite subsequent to its intrusion and thermal overprinting. The transitional nature of track annealing is observed at about 3-km (2-mi) distance from the contact, as a result of 15-Ma heating of Mesozoic zircons in the Shimanto sandstones. (b) Mean zircon FT ages vs. mean track lengths for the samples from the transitional annealing zone in (a). The plotted data present a characteristic trend that forms a Boomerang-shape bend, with the upper end predating deposition and lower end coincident with the granite intrusion event. Also shown are distributions of track lengths and single-grain ages of five representative samples. Error bars are ±2 standard error O1 = O1 member of the Ogura formation; F2 = F2 member of the Furushiroyama formation.

and the Cretaceous Ryoke granitic rocks on the east (Figure 1). To the southwest around Toshima, however, the sedimentary sequences are observed on both sides of the fault, with no granitic basement outcrops nearby (Figure 1).

MATERIALS AND METHODS

Sample Description

After the 1995 earthquake, three boreholes were drilled, two of which hit the Nojima fault zone (Figure 2). One of these boreholes is the GSJ 750 borehole at Hirabayashi, drilled through the Cretaceous granodiorite with seven shear zones bearing fault gouges (Figure 6). Of the seven shear zones, the one named Main Shear Zone is regarded as the central (main) zone of the Nojima fault, with the active fault trace at 625.27 m (2051.4 ft) (Tanaka et al., 2000). The FT analysis was carried out on zircon separates from 10 samples of GSJ 750 borehole and an outcrop sample about 130 m (426 ft) distant from the borehole site to the east. Of the 10 borehole samples, 6 are from the fault-fracture zone and 4 are from fresh, undeformed granodiorite (Figure 6).

The second borehole used for FT analysis is the UG 500 borehole (Figure 7), with the following stratigraphic section (Murata et al., 2001): (1) sands and gravels (0–118.0 m [0–387.1 ft]; borehole apparent depth throughout), correlated to the Pliocene–Pleistocene Osaka Group; (2) arkosic sandstones and silty sandstones (118.0–191.6 m; 387.1–628.6 ft), correlated to the Eocene–Oligocene (after Yamamoto et al., 2000) Iwaya Formation of the Kobe Group; and (3) granitic rocks (191.6–389.4 m; 628.6–1277.6 ft), consisting mainly of granodiorite of the Cretaceous Ryoke granitic rocks. This sequence is observed again on the other side of the fault at 389.4-m (1277.6-ft) depth:

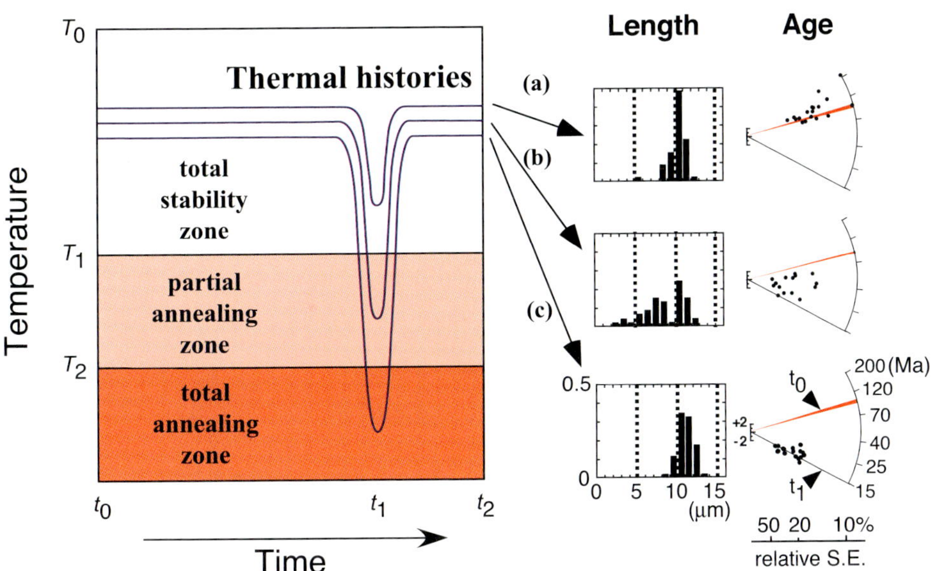

FIGURE 5. A schematic diagram showing a relationship between (1) three temperature zones that approximately express the thermal stability of the zircon FT system, i.e., total stability zone, partial annealing zone, and total annealing zone; (2) three hypothetical thermal histories characterized by secondary heating to different paleomaximum temperatures (T_{max}); and (3) three sets of zircon FT length and age distributions that resulted from the three different time-temperature pathways. Using this relationship, we can inversely reconstruct the thermal history of rocks from measured FT length and age data. T_0 = temperature at Earth's surface; T_1 = temperature at the boundary between total stability zone and partial annealing zone; T_2 = temperature at the boundary between partial annealing zone and total annealing zone; t_0 = time of deposition (or rock formation); t_1 = time of secondary heating; t_2 = present; (a) a thermal history with $T_{max} < T_1$; (b) a thermal history with $T_1 < T_{max} < T_2$; (c) a thermal history with $T_2 < T_{max}$. Note that T_1 and T_2 substantially vary with heating time duration (see Figure 3).

sands and gravels (389.4–410.7 m; 1277.6–1347.4 ft), arkose sandstones and silty sandstones (410.7–494.2 m; 1347.4–1621.4 ft), and granitic rocks (494.2–550.7 m; 1621.4–1806.7 ft). Boundaries between the Osaka Group and Kobe Group (118.0 and 410.7 m; 387.1 and 1347.4 ft) as well as those between the Kobe Group and the Ryoke granitic rocks (191.6 and 494.2 m; 628.6 and 1621.4 ft) are considered to be unconformable. All samples used for analysis were collected from granitic rocks (Figure 7). For comparison, samples were also taken from two surface exposures of granitic rocks: one adjacent (<30 cm; <12 in.) to a branch of the Nojima fault (Nojima branch fault; Murata et al., 2001) at a locality about 300 m (1000 ft) away from the borehole site to the southeast (NFL02) and the second surface sample 30 m (100 ft) southeast of the NFL02.

The Nojima fault was trenched at Hirabayashi (Figure 1). The exposed fault rocks consist of granitic cataclasite, a 2–10-mm (0.08–0.4-in.)-wide pseudotachylyte layer and siltstone of the Osaka Group, from the hanging wall (southeast) to the footwall (northwest). Otsuki et al. (2003) estimated the temperature of the pseudotachylyte formation as about 750–1280°C, based primarily on the observation of melting of K-feldspar and plagioclase. For FT analysis, a 50-cm (20-in.)-wide gray fault rock consisting of the following four layers was sampled from the footwall toward the hanging wall (Figure 7): (1) greenish-gray gouge (NT-LG); (2) pseudotachylyte (NT-PTa); (3) gray gouge (NT-UG); and (4) reddish granite (NF-HB1). For comparison, a sample was also analyzed from the same pseudotachylyte layer (NT-PTb) located about 7 m (23 ft) northeast of NT-PTa.

Experimental Method

Zircon crystals were separated by conventional heavy-liquid and magnetic separation techniques. Zircons were dated by the external detector method (Naeser, 1979), which allows determination of the FT ages of individual grains. Laboratory procedures are described by Tagami et al. (1988b). Neutron irradiation of samples was conducted at the thermal column pneumatic tube facility of Kyoto University Research Reactor, which has a cadmium ratio of 200 for Au and is thus well thermalized for FT use (Tagami and Nishimura, 1992). Ages were calibrated by the zeta method (Hurford and Green, 1983), with a zeta factor of 358.9 ± 7.0 (1 standard error) determined for a dosimeter glass National Bureau of Standard/National Institute of Standards and Technology-Standard Reference Materials (NBS/NIST-SRM) 962a (equivalent to 612) by multiple analyses of International Union of

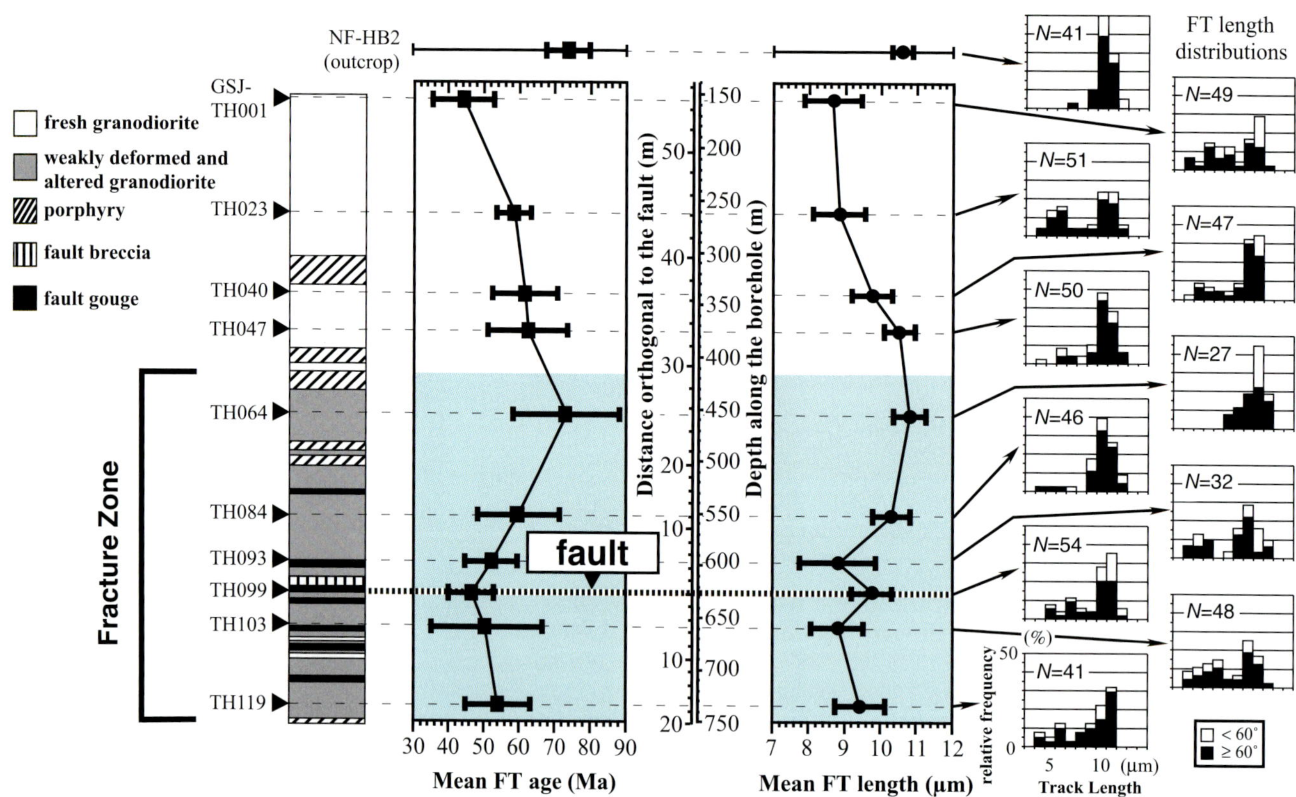

FIGURE 6. Zircon FT length and age vs. distance to the fault in the GSJ 750 borehole. Zircons from the outcrop are characterized by mean lengths of about 10–11 μm and unimodal length distributions, showing no signs of appreciable reduction of FT length. In contrast, those from adjacent to the fault at depth show significantly reduced means of about 6–8 μm and distributions having two components of long and short tracks. Error bars are ±2 standard error. Note that the track-length distribution is shown in a histogram with different patterns for tracks marking azimuth angles greater (solid column) and less (open column) than 60° to the crystallographic *c*-axis, because track lengths in zircon depend on etching and annealing properties that show an angular variation. Also shown are the distribution of fault rocks (Tanaka et al., 1999) and sample localities (Murakami et al., 2002).

Geological Science zircon standards, i.e., Buluk Member tuff, Fish Canyon tuff, and Mount Dromedary banatite zircons (Hurford, 1990). Errors were calculated using the conventional analysis given by Green (1981). The χ^2 analysis (Galbraith, 1981; Green, 1981) was employed to test the assumption that all analyzed grains are derived from a single population and have a common age with only Poissonian variation.

Track lengths were measured on horizontal confined tracks (HCTs; Laslett et al., 1982), revealed either as tracks in tracks by the conventional procedure (e.g., Tagami and Shimada, 1996; see also Yamada et al., 1995a) or as tracks in cleavages by the artificial fracturing method (Yamada et al., 1998). We adopted HCTs for measurement that have (1) orientations more than 60° to the crystallographic *c*-axis because of enhanced anisotropy in annealing and etching for tracks less than 60° (i.e., etched tracks subparallel to *c*-axis appear longer than those subperpendicular to *c*-axis) (Hasebe et al., 1994); and (2) widths of 1.0 ± 0.5 μm to minimize the over- or underetching effect on measured HCT length (Yamada et al., 1993, 1995a). The lengths of HCTs were measured by taking photographs of tracks using a digital camera installed on an optical microscope with a 100× dry objective, followed by counting the pixel length of the track using an image analysis software on a computer. Conversion from pixel length to true length was calibrated by measuring a Nikon microscale. Overall precision of measured length is about ± 0.1 μm.

DATA AND INTERPRETATION

GSJ 750 Borehole

Figure 6 presents zircon FT mean age, mean length, and length distributions plotted against distance to the fault (Murakami et al., 2002) (Table 1). Most tracks are not shortened in zircons from the outcrop sample (NF-HB2), with a mean length of 10.61 μm and unimodal length distribution with slightly negative skewness. This length is indistinguishable from the reference mean length of 10.5 ± 0.1 μm (1 standard

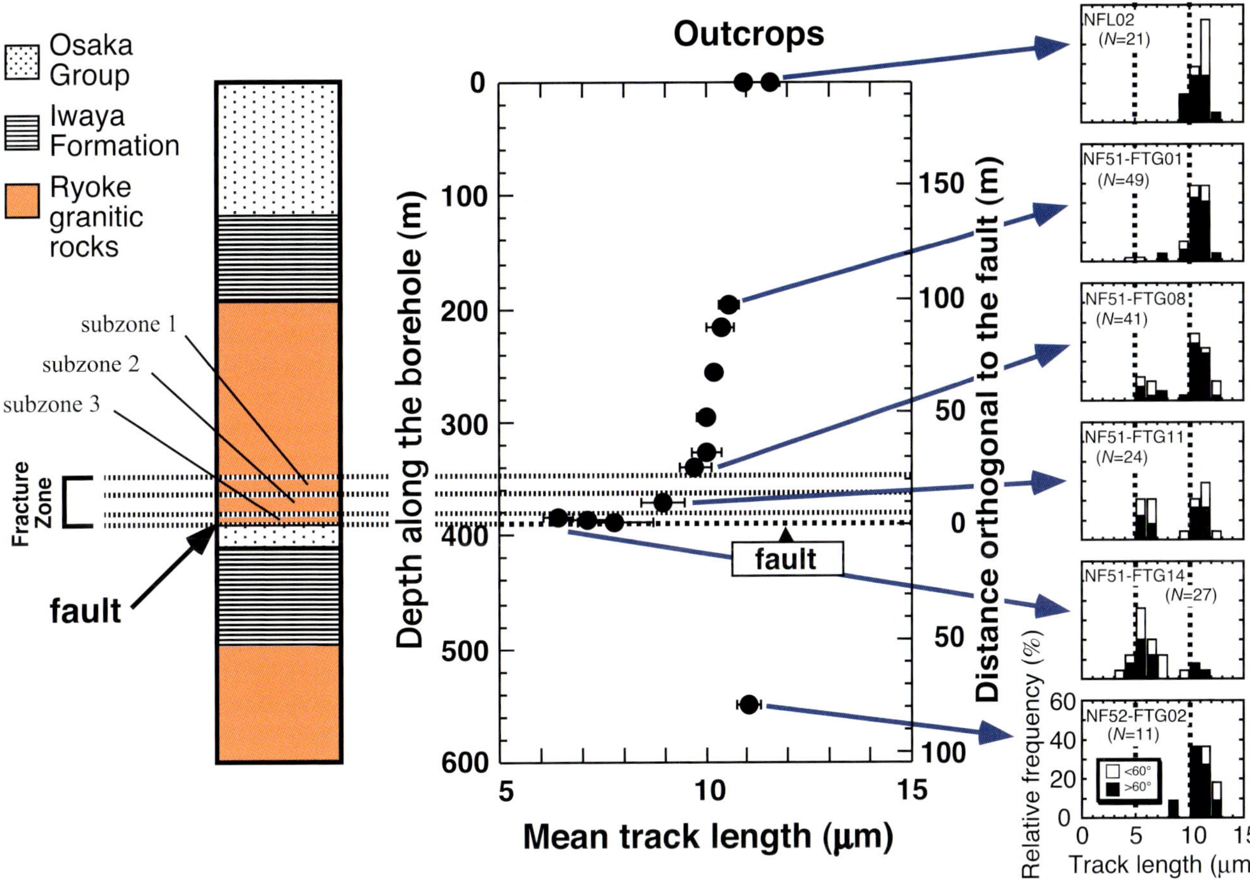

FIGURE 7. Zircon mean FT length vs. distance to the fault in the UG 500 borehole, with track length histograms of representative samples. Zircons from localities greater than 30 m (100 ft) away from the fault plane as well as those from outcrops are characterized by the mean lengths of about 10–11 μm and unimodal distributions with negative skewness, showing no signs of appreciable reduction of FT length. In contrast, those from adjacent to the fault at depth show significantly reduced means of about 6–8 μm and distributions having a dominant peak about 6–7 μm with rather positive skewness. Error bars are ±1 standard error. Also shown is the lithology of the borehole section (Murata et al., 2001), along with the three subzones in the fracture zone that represent dominant fault rock types (Tanaka et al., 2001); subzone 1 = non- to weakly deformed and altered rocks; subzone 2 = weakly deformed and altered rocks; subzone 3 = fault breccia and fault gouge.

error) (Hasebe et al., 1994). The mean age of this sample is 74 ± 3 Ma (1 standard error), consistent with zircon FT and biotite and hornblende K-Ar cooling ages of the Ryoke granitic rocks in the region (Tagami et al., 1988a; Takahashi, 1992; Hasebe and Tagami, 2001). These lines of evidence suggest that the zircon FT age is interpreted as the time of cooling of granodiorite from high temperatures. Among all borehole samples, GSJ-TH064 at 455 m (1493 ft) is the only sample in which the mean FT age and length are in agreement with those of outcrop samples. The mean FT ages and lengths of the other nine samples are younger and shorter than those of these two samples. The FT length distributions of some of nine samples (GSJ-TH001, GSJ-TH023, GSJ-TH093, and GSJ-TH103) are bimodal. This bimodal distribution is interpreted either by (1) residence in the ZPAZ (Figure 3) temperatures under the regional geothermal regime for an extended period of time, after the initial cooling of granodiorite to within the ZPAZ, or (2) secondary heating up to within the ZPAZ temperatures after the initial cooling below the ZPAZ. Model 2 is further divided into two cases: (2a) secondary heating as a result of tectonic downward motion of the regional geotherm (burial or underthrusting) and (2b) secondary heating by a thermal event that locally perturbed the geothermal structure. Model 2b is preferred to the others because the mean age and length data that indicate stronger annealing upward in the upper section (GSJ-TH001 to GSJ-TH064; Figure 6) are not consistent with the normal geothermal regime, and because the zone of stronger annealing downward (GSJ-TH064 to GSJ-TH099) is too narrow (about 25 m [82 ft] in distance orthogonal to the fault, which is equal to about 210 m [689 ft] in a vertical direction above the fault; Figure 2) to be formed by residence within or

Table 1. Fission-track analysis of zircon from GSJ borehole and outcrop at Hirabayashi.

Sample	D_e (m)	D_i (m)	$\rho_s(N_s)$ (10^6 cm^{-2})	$\rho_i(N_i)$ (10^6 cm^{-2})	$\rho_d(N_d)$ (10^6 cm^{-2})	*T* (Ma)	*n*	$P(\chi^2)$	*L*(*N*) (μm)	σ (μm)	$L_{all}(N_{all})$ (μm)	σ_{all} (μm)
NF-HB2			7.73 (915)	5.08 (601)	0.284 (6312)	77.2 ± 4.4	18	11	10.61 ± 0.16 (31)	0.89	10.74 ± 0.14 (41)	0.88
			5.51 (672)	5.21 (635)	0.380 (8443)	71.8 ± 4.3	27	100				
						74.4 ± 3.1*						
GSJ-TH001	154.7–155.0	55	2.99 (230)	2.23 (172)	0.186 (4138)	44.6 ± 4.6	10	80	8.69 ± 0.43 (32)	2.41	9.04 ± 0.33 (49)	2.34
GSJ-TH023	260.3–260.6	44	5.71 (902)	2.33 (368)	0.153 (4294)	66.8 ± 4.5	9	45	8.86 ± 0.39 (41)	2.50	9.15 ± 0.35 (51)	2.51
			4.43 (211)	2.54 (121)	0.148 (4174)	46.3 ± 5.4	6	50				
			6.29 (594)	5.40 (510)	0.284 (6312)	59.2 ± 3.8	17	36				
						58.7 ± 2.6*						
GSJ-TH040	339.5–339.8	36	5.61 (278)	3.25 (161)	0.186 (4138)	57.5 ± 5.9	7	92	9.80 ± 0.31 (36)	1.87	9.70 ± 0.29 (47)	2.02
			5.59 (236)	2.65 (112)	0.187 (4142)	70.2 ± 8.2	6	77				
						61.8 ± 4.8*						
GSJ-TH047	372.7–373.0	33	4.77 (521)	2.54 (277)	0.186 (4138)	62.6 ± 4.9	13	13	10.54 ± 0.24 (38)	1.49	10.18 ± 0.26 (50)	1.83
			4.81 (353)	2.55 (187)	0.187 (4142)	62.9 ± 5.9	10	40				
GSJ-TH064	455.5–455.8	25	8.63 (269)	3.17 (99)	0.153 (4294)	74.0 ± 8.9	6	77	10.85 ± 0.26 (20)	1.14	11.06 ± 0.20 (27)	1.05
			6.54 (84)	2.41 (31)	0.148 (4174)	71.8 ± 15.2	3	91				
						73.4 ± 7.7*						
GSJ-TH084	550.2–550.6	12	5.53 (208)	3.11 (117)	0.186 (4138)	59.2 ± 7.0	7	100	10.33 ± 0.29 (36)	1.73	10.36 ± 0.24 (46)	1.65
			8.52 (86)	4.56 (46)	0.187 (4142)	62.3 ± 11.5	3	96				
						60.0 ± 6.0*						
GSJ-TH093	595.1–595.5	4.7	6.69 (221)	3.76 (124)	0.186 (4138)	59.3 ± 6.8	6	18	8.85 ± 0.55 (21)	2.51	9.22 ± 0.45 (32)	2.53
			8.22 (98)	5.79 (69)	0.187 (4142)	47.4 ± 7.5	3	97				
			4.46 (139)	4.46 (139)	0.284 (6312)	50.8 ± 6.2	5	27				
						52.7 ± 3.9*						
GSJ-TH099	625.5–626.0	0	7.36 (189)	4.01 (103)	0.153 (4294)	50.1 ± 6.3	4	87	9.80 ± 0.32 (36)	1.95	9.99 ± 0.26 (54)	1.94
			2.51 (55)	1.27 (34)	0.148 (4174)	42.9 ± 9.4	4	34				
			3.87 (199)	4.28 (220)	0.284 (6312)	46.0 ± 4.6	11	68				
						46.8 ± 3.5*						
GSJ-TH103	658.3–658.8	5.2	8.18 (105)	5.37 (69)	0.187 (4142)	50.8 ± 8.0	3	94	8.84 ± 0.39 (34)	2.29	8.60 ± 0.34 (48)	2.36
GSJ-TH119	734.5–734.8	17	4.77 (210)	3.18 (140)	0.186 (4138)	50.0 ± 5.6	9	31	9.45 ± 0.38 (34)	2.23	9.28 ± 0.37 (41)	2.38
			7.03 (116)	5.33 (88)	0.284 (6312)	66.9 ± 9.6	4	83				
						54.3 ± 4.8*						

D_e = depth along the borehole; D_I = distance from the fault; ρ_s = spontaneous track density of a sample; N_s = number of tracks counted to determine ρ_s; ρ_i = induced track density of a sample measured in muscovite external detector; N_i = number of tracks counted to determine ρ_i; ρ_d = induced track density of glass dosimeter NBS-SRM962 measured in a muscovite external detector; N_d = number of tracks counted to determine ρ_d; T = FT age with its 1σ error; n = number of crystals counted; $P(\chi^2)$ = probability of obtaining the observed value of χ^2 parameter, for N degree of freedom, where N = (number of counted crystals) – 1 (Galbraith, 1981; Green, 1981); L = mean length of confined FTs greater than 60° to c-axis, with 1 standard error; N = number of measured tracks to determine L; σ = standard deviation for the length distribution of confined FTs greater than 60° to c-axis; L_{all} = mean length of confined FTs in all crystallographic orientations, with 1 standard error; N_{all} = number of measured tracks to determine L_{all}; σ_{all} = standard deviation for the length distribution of confined FTs in all crystallographic orientations.

*Weighted mean age.

Table 2. Fission-track analysis of zircon from UG borehole and outcrop at Toshima.

Sample	D_e *(m)*	D_i *(m)*	*L(N)* *(μm)*	*σ (μm)*	$L_{all}(N_{all})$ *(μm)*	σ_{all} *(μm)*
NFL02	0	0	10.64 ± 0.04 (14)	0.74	10.89 ± 0.16 (21)	0.75
NFL03	0	0	11.52 ± 0.30 (8)	0.85	11.59 ± 0.19 (15)	0.73
NF51-FTG01	195.3–195.6	100	10.77 ± 0.17 (38)	1.06	10.56 ± 0.21 (49)	1.47
NF51-FTG02	215.7–216.6	90	10.44 ± 0.28 (30)	1.55	10.36 ± 0.31 (37)	1.88
NF51-FTG04	234.6–235.0	70	9.97 ± 0.21 (69)	1.76	10.21 ± 0.18 (103)	1.82
NF51-FTG06	296.2–296.5	50	10.02 ± 0.23 (41)	1.83	10.37 ± 0.21 (71)	1.25
NF51-FTG19	326.3–326.7	30	10.43 ± 0.25 (19)	1.11	10.03 ± 0.35 (26)	1.77
NF51-FTG08	339.6–340.0	25	10.04 ± 0.35 (30)	1.93	9.76 ± 0.36 (41)	2.29
NF51-FTG11	370.2–370.5	10	9.09 ± 0.74 (13)	2.69	8.99 ± 0.53 (24)	2.61
NF51-FTG14	384.0–384.3	2.5	6.67 ± 0.58 (15)	2.25	6.42 ± 0.38 (27)	1.98
NF51-FTG16	387.9–388.1	1.2	7.13 ± 0.59 (9)	1.76	7.13 ± 0.59 (15)	1.76
NF51-FTG17	388.1–388.3	0.5	9.56 ± 0.89 (5)	2.01	7.81 ± 0.89 (9)	2.67
NF52-FTG02	550.0–550.3	80	10.81 ± 0.31 (9)	0.92	11.06 ± 0.30 (11)	1.00

D_e = depth along the borehole; D_l = distance from the faut; L = mean length of confined FTs greater than 60° to c-axis, with 1 standard error; N = number of measured tracks to determine L; σ = standard deviation for the length distribution of confined FTs greater than 60° to c-axis; L_{all} = mean length of confined FTs in all crystallographic orientations, with 1 standard error; N_{all} = number of measured tracks to determine L_{all}; σ_{all} = standard deviation for the length distribution of confined FTs in all crystallographic orientations. After Tagami et al., 2001.

reheating into the ZPAZ under the geothermal gradient in the study area (21–24°C/km, measured in the National Research Institute for Earth Science and Disaster Prevention 1800-m [5905-ft] borehole, 200 m [660 ft] southeast of the GSJ 750; Kitajima et al., 1998; note that no geological signs of magmatism-related thermal anomalies are present around the borehole site in the recent past).

GSJ-TH001 and GSJ-TH099 have youngest mean ages of about 45–47 Ma. Whereas the former has a mean track length shorter than the neighboring samples, the mean length of the latter is longer than neighboring samples. The lower mean track lengths may have accumulated before the secondary heating and were subjected to stronger annealing and, thus, almost removed from the sample (i.e., heating into the region lower than Figure 3b). Therefore, it is likely that the GSJ 750 borehole rocks record the heat signatures at two different locations, one less than 25 m (82 ft) perpendicular to the fault and the other at the shallower depth (>25 m; 82 ft).

The age of last cooling after the secondary heating is estimated for samples near the fault by the model age calculation (Murakami et al., 2002; Wallis et al., 2003). The observed track-length distribution was decomposed into two components so that the longer component has a mean length of 10.5 μm, i.e., the mean length of unannealed spontaneous tracks in zircon (Hasebe et al., 1994). The proportion of decomposed components, coupled with the observed spontaneous track density and theoretical density-length relationship, gives a density estimate of long tracks accumulated since the last cooling. The model age was thus calculated using the estimated density for three samples near the fault (GSJ-TH093, GSJ-TH099, and GSJ-TH103) as about 35, 39, and 30 Ma, respectively. They are in an approximate agreement with an independent age estimate of about 30–40 Ma, deduced from the shape of mean FT age vs. mean FT length relationship (Murakami et al., 2002). Further, track-length distributions and age data were modeled using the Monte Trax program of Gallagher (1995), which uses the maximum likelihood approach combined with a genetic algorithm for an initial stochastic search. The modeled ages of initiation of last cooling are 35 ± 1, 38 ± 2, and 31 ± 1 Ma (1 standard error) for GSJ-TH093, GSJ-TH099, and GSJ-TH103 samples, respectively (T. Tagami and M. Murakami, unpublished data).

UG 500 Borehole

Track lengths of 13 samples were analyzed from the outcrops and borehole section (Figure 7; Table 2) (Tagami et al., 2001). It was found that FT lengths in zircons from localities more than 30 m (100 ft) away from the fault plane as well as those from outcrops are characterized by concordant mean values of about 10–11 μm and unimodal distributions with negative skewness, which showed no signs of appreciable reduction in FT length. In contrast, those adjacent (<2.6 m; <8.5 ft) to the fault at depths (e.g., NF51-FTG14) showed significantly reduced mean track lengths of about 6–8 μm and distributions having a

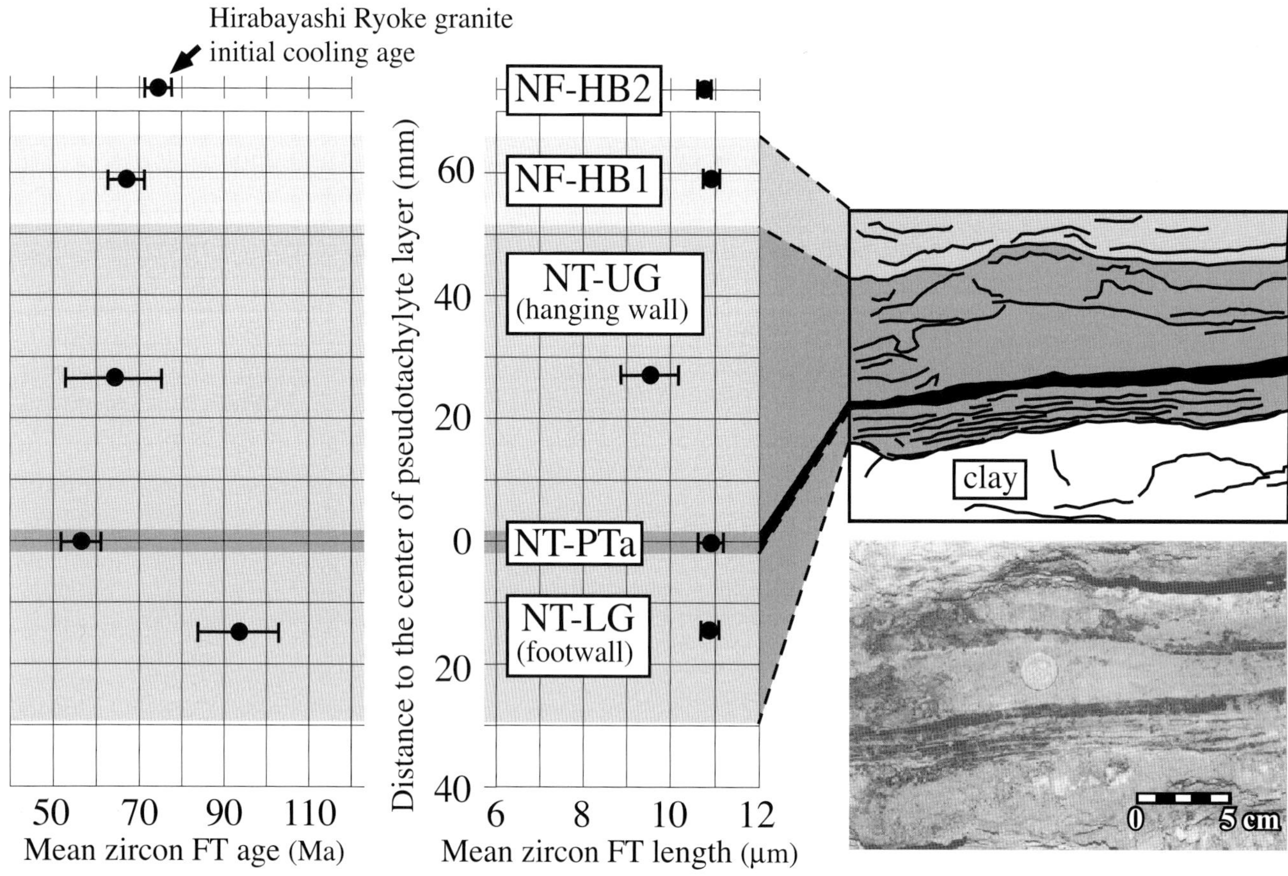

FIGURE 8. The photograph (below) and sketch (above) of sampled fault rock section from the Hirabayashi trench, with plots of mean zircon FT ages and lengths. The sample from the pseudotachylyte layer (NT-PTa) has an age significantly younger than that of initial cooling of the Ryoke host rock sample (NF-HB2). Error bars are ±1σ.

peak around 6–7 μm with rather positive skewness. Samples in between (i.e., NF51-FTG08 and NF51-FTG11) were likely to yield distributions of a transient pattern by mixing the two end-member components. The former pattern is similar in mean length to spontaneous tracks in unannealed zircons (Hasebe et al. 1994) and, thus, is interpreted to reflect cooling through the ZPAZ, without later, partial thermal overprints. The latter indicates substantial track shortening as a result of partial annealing and is interpreted either by the three models mentioned in the last section. The width of a high-paleotemperature region is as narrow as less than 25 m (82 ft) in distance orthogonal to the fault (Figure 7), which is equal to less than 210 m (690 ft) in a vertical direction above the fault (Figure 2). Hence, the heating cannot be explained by models 1 or 2a, both of which are accompanied by the ZPAZ under a subnormal geothermal regime. In addition, the region is about 200 m (660 ft) beneath the base of the Eocene–Oligocene Iwaya Formation, and this indicates that the post-Eocene uplift is of limited amount in the borehole site, probably less than 5 km (3.2 mi). These facts support model 2b, which calls on secondary heating by a thermal event. Together with a new zircon FT age of 84 ± 11 Ma (1σ) (T. Tagami and M. Murakami, unpublished data), the age of last cooling is estimated as about 2 Ma for NF51-FTG14 using the model age calculation approach of Murakami et al. (2002). This cooling age is consistent with the modeled age of 2.5 ± 0.4 Ma (1 standard error) obtained by the Monte Trax program (Gallagher, 1995).

Fault Trench at Hirabayashi

Figure 8 shows zircon FT data of the five samples from the Hirabayashi trench (Murakami and Tagami, 2003) (Table 3). The mean age of NF-HB1 is 67 ± 4 Ma, indistinguishable from the initial cooling age of the Ryoke basement of 74 ± 3 Ma (NF-HB2). NT-PTa, however, yielded an age of 56 ± 5 Ma, significantly younger than the initial cooling. The other pseudotachylyte sample (NT-PTb) also gave a younger age of 59 ± 8 Ma, approximately concordant with the former age. The age of NT-UG is relatively less precise and given as 64 ± 11 Ma, apparently between the initial cooling and the pseudotachylyte ages. The sample from the footwall, NT-LG, was dated as 94 ± 10 Ma, significantly older than the basement cooling age.

Table 3. Zircon FT analytical data of Nojima fault samples.

Sample	$\rho_s(N_s)$ (10^6 cm^{-2})	$\rho_i(N_i)$ (10^6 cm^{-2})	$\rho_d(N_d)$ (10^6 cm^{-2})	*T (Ma)*	*n*	$P(\chi^2)$	$L_{all}(N_{all})$ (μm)	σ_{all} (μm)
NF-HB1	6.97 (255)	6.67 (244)	0.364 (8086)	68.0 ± 6.3	13	95	10.96 ± 0.14 (24)	0.69
	5.86 (286)	5.27 (257)	0.332 (7368)	65.9 ± 5.9	14	66		
				66.9 ± 4.3				
NT-UG	4.33 (65)	4.40 (66)	0.364 (8086)	64.1 ± 11.3	5	15	9.55 ± 0.67 (3)	1.16
NT-PTa	5.11 (158)	2.97 (92)	0.193 (4294)	59.3 ± 7.9	5	58	10.94 ± 0.28 (10)	0.88
	7.46 (196)	8.87 (233)	0.364 (8086)	54.8 ± 5.5	8	13		
				56.3 ± 4.5				
NT-PTb	5.77 (146)	3.28 (83)	0.188 (4174)	59.1 ± 8.3	6	63		
NT-LG	8.32 (242)	5.78 (168)	0.364 (8086)	93.5 ± 9.6	8	60	10.93 ± 0.20 (24)	0.97

ρ_s = spontaneous track density of a sample; N_s = number of tracks counted to determine ρ_s; ρ_i = induced track density of a sample measured in muscovite external detector; N_i = number of tracks counted to determine ρ_i; ρ_d = induced track density of glass dosimeter NBS-SRM962 measured in a muscovite external detector; N_d = number of tracks counted to determine ρ_d; T = FT age with its 1σ error; n = number of crystals counted; $P(\chi^2)$ = probability of obtaining the observed value of χ^2 parameter, for N degree of freedom, where N = (number of counted crystals) – 1 (Galbraith, 1981; Green, 1981); L_{all} = mean length of confined FTs in all crytallographic orientations, with 1 standard error; N_{all} = number of measured tracks to determine L_{all}; σ_{all} = standard deviation for the length distribution of confined FTs in all crystallographic orientations.

Fission-track lengths were measured for four samples (Figure 7). Except NT-UG having only three measured tracks, the rest of the samples have mean lengths of 10.9–11.0 μm, approximately indistinguishable from the reference zircon mean length of 10.7 ± 0.1 μm (Hasebe et al., 1994; determined for whole crystallographic orientations). These results indicate that the ages of the three samples individually represent the time of cooling through the ZPAZ, with negligible influences of later, partial thermal perturbation. Coupled with the age data mentioned above, it is therefore suggested that the zircon FT system of the pseudotachylyte layer was totally reset (or kept reset) and subsequently cooled at about 56 Ma. It is also inferred that the greenish-gray gouge layer on the footwall side (NT-LG) was not heated into ZPAZ at that time. Its significantly older (about 94 Ma) age probably represents a dominant detrital age component of the host rock of the gouge, i.e., Osaka Group siltstone.

GEOLOGICAL IMPLICATIONS

Heat Source

Here, we examine two plausible models of the heat source responsible for the observed cooling ages: (1) frictional heating of fault motion and (2) heat transfer or dispersion via fluids in the fault zone. The range of detectable zircon FT annealing formed by model 1 is calculated as about 1 mm (0.04 in.) from the center of fault by one-dimensional heat conduction modeling (Murakami and Tagami, 2003), using the following conditions: (1) thermal diffusivity of 2 mm^2 s^{-1} (0.0031 $in.^2$ s^{-1}); (2) environmental temperature of 200°C at dry condition; (3) frictional heat of 1000°C (e.g., Otsuki et al., 2003) for 5 s (Kikuchi and Kanamori, 1996) at the fault plane, followed by heat dispersion caused by thermal conduction; (4) zircon FT annealing kinetics of Tagami et al. (1998), along with the definition of partial resetting (annealing) zone of Yamada et al. (1995b); and (5) frictional heating model of Cardwell et al. (1978). The predicted 1-mm (0.04-in.) range is approximately concordant with that of the about 56-Ma cooling age observed for the Hirabayashi trench. In the case of GSJ 750 and UG 500 boreholes, however, the spatial range of zircon FT annealing reaches at least about 5 and 3 m (16 and 10 ft), respectively, from the fault (Figures 6, 7). These observed ranges are too wide to be explained by the frictional heating even if earthquakes occurred repeatedly over an extended period of time.

With respect to model 2, four samples less than 12 m (39 ft) away from the fault in the GSJ 750 borehole (GSJ-TH084, GSJ-TH093, GSJ-TH099, and GSJ-TH103) are from deformed and altered rock. In particular, three of them (GSJ-TH093, GSJ-TH099, and GSJ-TH103) belong to shear zones deformed and altered prominently compared to other parts of the borehole section (Tanaka et al., 1999). Because these three samples show greater degrees of zircon FT annealing (Figure 6), it appears that the degree of annealing and deformation and alteration are positively correlated. A similar correlation was found for the UG 500 borehole: the region of greatest zircon FT annealing corresponds to the zone of cataclastic deformation of rocks and dissolution of heavy minerals (Figure 7) (Tagami et al., 2001). These lines of evidence favor model 2 as a heat source of the observed thermal anomalies. If we assume a constant fluid flow through the fault plane only (i.e., the width of fluid = 0 m) as a heat source, one-dimensional heat conduction calculation gives the duration and temperature of heating as 1–20 yr and 400–500°C, respectively, for the about 10-m (33-ft)-wide thermal anomaly zone. The estimated temperature becomes higher, and the duration becomes shorter, if the thermal anomaly zone is narrower, or the fluid flow pathway is wider. The paleodepth was probably shallower than about 7 km (4 mi) at the time of last cooling, because zircon FT system is closed at less than about 200°C for a geological timescale, which corresponds to less than about 7 km (4 mi) by assuming a geothermal gradient of 25°C/km.

Age Constraints on the Nojima Fault

The age data presented in the previous sections place important constraints on the history of Nojima fault. The age of the Nojima fault formation was previously estimated as about 1.2 Ma, on the basis of the spatial distribution of the Osaka Group in and around the Awaji Island (Murata et al., 2001). However, the approximately 56-Ma age of the pseudotachylyte suggests that the Nojima fault had already been initiated at about 56 Ma, much older than the previous estimate. Furthermore, microscopic observation using thin sections suggests that the pseudotachylyte layer was formed during at least two faulting stages (Otsuki et al., 2003). Thus, the approximately 56-Ma age is probably the time of final pseudotachylyte formation. It is likely, therefore, that the present Nojima fault system was formed by the middle Quaternary reactivation of an ancient fault, which was already initiated at about 56 Ma at the interior of crust. This new reconstruction is consistent with the approximately 30–40-Ma secondary heating found near the fault in the GSJ 750 borehole.

Another important aspect is the temporal and spatial variation in reconstructed thermal histories among the three sites:

1) UG 500 borehole site (Toshima): ancient heating into ZPAZ, followed by cooling at about 2 Ma within approximately 25 m (82 ft) from the fault
2) GSJ 750 borehole (Hirabayashi): ancient heating into ZPAZ, followed by cooling at about 30–40 Ma within approximately 25 m (82 ft) from the fault
3) trench (Hirabayashi): no thermal overprints for a rock from the footwall about 15 mm (0.6 in.) away from the fault as well as one from the hanging wall about 60 mm (2.4 in.) from the fault.

These lines of evidence demonstrate the heterogeneity of the Nojima fault in terms of ancient heat transfer or dispersion via fluids and their resultant thermal anomalies. In addition, the thermal anomaly observed in the shallower part of the GSJ 750 (i.e., GSJ-TH001, GSJ-TH023, GSJ-TH040, and GSJ-TH047) was not found for the UG 500. Because calcite layers and altered K-feldspars are widely observed in the four samples, they may represent a zone of ancient alteration possibly related to the past fault motion at Hirabayashi.

SUMMARY AND CONCLUSIONS

1) Zircon FT age and length data suggest that the GSJ 750 borehole rocks record the heat signatures at two different locations, one less than 25 m (82 ft) perpendicular to the fault and the other at the shallower depth (>25 m; >82 ft). The age of cooling after the heating event is estimated as about 30–40 Ma for samples within approximately 25 m (82 ft) from the fault using the model age calculation and mean FT age vs. mean FT length plot.
2) Zircon FT length data suggest that the UG 500 borehole rocks record a heat signature in the hanging wall less than 25 m (82 ft) from the fault. Together with new zircon age data, the age of cooling after the heating event is estimated as about 2 Ma using the model age calculation.
3) The plausible heat source of these thermal events is heat transfer or dispersion via fluids in the fault zone, on the basis of one-dimensional heat conduction modeling as well as the positive correlation between the degree of FT annealing and deformation and alteration of borehole rocks.
4) Zircon FT age and length data suggest that the zircon FT system of the pseudotachylyte layer at the Hirabayashi trench site was totally reset (or kept reset) and subsequently cooled at about 56 Ma. However, no signs of such cooling were found for a rock from the footwall about 15 mm (0.6 in.) away from the fault as well as one from the hanging wall about 60 mm (2.4 in.) from the fault. The spatial distribution of these data is concordant with that of detectable zircon FT annealing predicted by thermal modeling of the frictional heating of fault motion.
5) The present Nojima fault system was probably formed by the middle Quaternary reactivation of an ancient fault, which was already initiated at about 56 Ma at the interior of crust.
6) The Nojima fault shows a temporal and spatial variation in terms of the thermal anomalies recorded in the fault rocks, implying the heterogeneous heat transfer or dispersion via fluids in the fault zone.

ACKNOWLEDGMENTS

We thank John Garver, Mary Roden-Tice, and Rasoul Sorkhabi for reviewing this chapter and for valuable comments to improve it. This study was supported by a Grant-in-Aid (no. 12440137) from the Japanese Ministry of Education, Culture, Sports, Science, and Technology and by a visiting research program of the Kyoto University research reactor.

REFERENCES CITED

Ando, M., 2001, Geological and geophysical studies of the Nojima fault from drilling: An outline of the Nojima fault zone probe: Island Arc, v. 10, p. 206–214.

Awata, Y., K. Mizuno, Y. Sugiyama, R. Imura, R. Shimokawa, K. Okumura, and E. Tsukuda, 1996, Surface fault ruptures on the northwest coast of Awaji Island associated with the Hyogoken Nanbu earthquake of 1995, Japan (in Japanese with English abstract): Journal of Seismological Society of Japan, v. 49, p. 113–124.

Brandon, M. T., M. K. Roden-Tice, and J. I. Garver, 1998, Late Cenozoic exhumation of the Cascadia accretionary wedge in the Olympic Mountains, northwest Washington state: Bulletin of Geological Society of America, v. 110, p. 985–1009.

Brix, M. R., B. Stockhert, E. Seidel, T. Theye, S. N. Thomson, and M. Kuster, 2002, Thermobarometric data from a fossil zircon partial annealing zone in high pressure-low temperature rocks of eastern and central Crete, Greece: Tectonophysics, v. 349, p. 309–326.

Cardwell, R. K., D. S. Chinn, G. F. Moore, and D. L. Turcotte, 1978, Frictional heating on a fault zone with finite thickness: Geophysical Journal of Royal Astronomical Society, v. 52, p. 525–530.

Coyle, D. A., and G. A. Wagner, 1996, Fission-track dating of zircon and titanite from the 9101 m deep KTB: Observed fundamentals of track stability and thermal history reconstruction: Abstract of presentation at the International Workshop on Fission Track Dating, Gent, Belgium, August 26–30, 1996.

d'Alessio, M. A., A. E. Blythe, and R. Burgmann, 2003, No frictional heat along the San Gabriel fault, California; evidence from fission-track thermochronology: Geology, v. 31, p. 541–544.

Foster, D. A., B. P. Kohn, and A. J. W. Gleadow, 1996, Sphene and zircon fission track closure temperatures revisited; empirical calibrations from $^{40}Ar/^{39}Ar$ diffusion studies of K-feldspar and biotite, *in* International Workshop on Fission Track Dating, Gent 1996: Gent, University of Gent, August 26–30, 1996, p. 37.

Fukahata, Y., and M. Matsu'ura, 2001, Correlation between surface heat flow and elevation and its geophysical implication: Geophysical Research Letters, v. 28, p. 2703–2706.

Galbraith, R. F., 1981, On statistical models for fission track counts: Mathematical Geology, v. 13, p. 471–488.

Galbraith, R. F., 1990, The radial plot: Graphical display of spread in ages: Nuclear Tracks, v. 17, p. 207–214.

Galbraith, R. F., and G. M. Laslett, 1997, Statistical modeling of thermal annealing of fission tracks in zircon: Chemical Geology (Isotope Geoscience Section), v. 140, p. 123–135.

Gallagher, K., 1995, Evolving temperature histories from apatite fission-track data: Earth and Planetary Science Letters, v. 136, p. 421–435.

Gallagher, K., R. Brown, and C. Johnson, 1998, Fission track analysis and its applications to geological problems: Annual Review of Earth Planetary Sciences, v. 26, p. 519–572.

Gleadow, A. J. W., I. R. Duddy, P. F. Green, and J. F. Lovering, 1986, Confined fission track lengths in apatite: A diagnostic tool for thermal history analysis: Contribution to Mineralogy and Petrology, v. 94, p. 405–415.

Green, P. F., 1981, A new look at statistics in fission-track dating: Nuclear Tracks, v. 5, p. 77–86.

Green, P. F., I. R. Duddy, A. J. W. Gleadow, and J. F. Lovering, 1989, Apatite fission track analysis as a paleotemperature indicator for hydrocarbon exploration, *in* N. D. Naeser and T. H. McCulloh, eds., Thermal histories of sedimentary basins: Methods and case histories: New York, Springer, p. 181–195.

Green, P. F., K. A. Hegarty, I. R. Duddy, S. S. Foland, and V. Gorbachev, 1996, Geological constraints on fission track annealing in zircon: Abstract of presentation at International Workshop on Fission Track Dating, Gent, Belgium, August 26–30, 1996.

Hasebe, N., and T. Tagami, 2001, Exhumation of an accretionary prism; results from fission track thermochronology of the Shimanto belt, southwest Japan: Tectonophysics, v. 331, p. 247–267.

Hasebe, N., T. Tagami, and S. Nishimura, 1994, Towards zircon fission-track thermochronology: Reference framework for confined track length measurements: Chemical Geology (Isotope Geoscience Section), v. 112, p. 169–178.

Hasebe, N., N. Mori, T. Tagami, and R. Matsui, 2003, Geological partial annealing zone of zircon fission-track system: Additional constraints from the deep drilling MITI-Nishikubiki and MITI-Mishima: Chemical Geology, v. 199, p. 45–52.

Hurford, A. J., 1990, Standardization of fission track dating calibration: Recommendation by the Fission Track Working Group of the International Union of Geological Sciences Subcommission on Geochronology: Chemical Geology (Isotope Geoscience Section), v. 80, p. 171–178.

Hurford, A. J., and P. F. Green, 1983, The zeta age calibration of fission-track dating: Isotope Geoscience, v. 1, p. 285–317.

Kamp, P. J. J., P. F. Green, and S. H. White, 1989, Fission track analysis reveals character of collisional tectonics in New Zealand: Tectonics, v. 8, p. 169–195.

Kikuchi, M., and H. Kanamori, 1996, Rupture process of Kobe, Japan, earthquake of determined January 17, 1995, from teleseismic body waves: Journal Physics of Earth, v. 44, p. 429–436.

Kitajima, T., Y. Kobayashi, R. Ikeda, Y. Iio, and K. Omura, 1998, Heat flow measurement for the borehole at Hirabayashi, Awaji Island (in Japanese): Gekkan Chikyuu, v. 21, p. 108–113.

Laslett, G. M., W. S. Kendall, A. J. W. Gleadow, and I. R. Duddy, 1982, Bias in measurement of fission-track length distributions: Nuclear Tracks, v. 6, p. 79–85.

Mizuno, K., H. Hattori, A. Sangawa, and Y. Takahashi, 1990, Geology of the Akashi district, quadrangle-series (in Japanese with English abstract): Geological Survey of Japan, Tsukuba, Japan, scale 1:50,000, 1 sheet 90 p.

Murakami, M., and T. Tagami, 2004, Dating pseudotachylyte of the Nojima fault using the zircon fission-track method: Geophysical Research Letters, v. 31, DOI: 10.1029/2004GL020211, 4 p.

Murakami, M., T. Tagami, and N. Hasebe, 2002, Ancient thermal anomaly of an active fault system: Zircon fission-track evidence from Nojima GSJ 750 m borehole samples: Geophysical Research Letters, v. 29, DOI: 10.1029/2002GL015679, 4 p.

Murata, A., K. Takemura, T. Miyata, and A. Lin, 2001, Quaternary vertical offset and average slip rate of the Nojima fault on Awaji Island, Japan: The Island Arc, v. 10, p. 360–367.

Naeser, C. W., 1979, Fission-track dating and geologic annealing of fission tracks, *in* E. Jager and J. C. Hunziker, eds., Lectures in isotope geology: Berlin, Springer-Verlag, p. 154–169.

Otsuki, K., N. Monzawa, and T. Nagase, 2003, Fluidization and melting of fault gouge during seismic slip: Identification in the Nojima fault zone and implications for focal earthquake mechanisms: Journal of Geophysical Research, v. 108, no. B4, p. 2192, DOI: 10.1029/2001JB001711.

Parry, W. T., M. P. Bunds, R. L. Bruhn, C. M. Hall, and J. M. Murphy, 2001, Mineralogy, $^{40}Ar/^{39}Ar$ dating and apatite fission track dating of rocks along the Castle Mountain fault, Alaska: Tectonophysics, v. 337, p. 149–172.

Scholz, C. H., 1996, Faults without friction: Nature, v. 381, p. 556–557.

Tagami, T., and S. Nishimura, 1992, Neutron dosimetry and fission-track age calibration: Insights from intercalibration of uranium and thorium glass dosimeters: Chemical Geology (Isotope Geoscience Section), v. 102, p. 277–296.

Tagami, T., and C. Shimada, 1996, Natural long-term annealing of the zircon fission track system around a granitic pluton: Journal of Geophysical Research, v. 101, p. 8245–8255.

Tagami, T., N. Lal, R. B. Sorkhabi, and S. Nishimura, 1988a, Fission track thermochronologic analysis of the Ryoke belt and the Median Tectonic Line southwest Japan: Journal of Geophysical Research, v. 93, p. 13,705–13,715.

Tagami, T., N. Lal, R. B. Sorkhabi, H. Ito, and S. Nishimura, 1988b, Fission track dating using external detector method: A laboratory procedure: Memoirs Faculty of Science, Kyoto University, v. 53, p. 14–30.

Tagami, T., H. Ito, and S. Nishimura, 1990, Thermal annealing characteristics of spontaneous fission tracks in zircon: Chemical Geology (Isotope Geoscience Section), v. 80, p. 159–169.

Tagami, T., A. Carter, and A. J. Hurford, 1996, Natural long-term annealing of the zircon fission-track system in Vienna Basin deep borehole samples: Constraints upon the partial annealing zone and closure temperature: Chemical Geology, v. 130, p. 147–157.

Tagami, T., R. F. Galbraith, R. Yamada, and G. M. Laslett, 1998, Revised annealing kinetics of fission tracks in zircon and geological implications, *in* P. Van den haute and F. De Corte, eds., Advances in fission-track geochronology: Dordrecht, The Netherlands, Kluwer Academic Publishers, p. 99–112.

Tagami, T., N. Hasebe, H. Kamohara, and K. Takemura, 2001, Thermal anomaly around Nojima fault as detected by the fission-track analysis of Ogura 500 m borehole samples: Island Arc, v. 10, p. 457–464.

Takahashi, Y., 1992, K-Ar ages of the granitic rocks in Awaji Island with an emphasis on timing of mylonitization (in Japanese with English abstract): Gankou, v. 87, p. 291–299.

Tanaka, H., T. Higuchi, N. Tomida, K. Fujimoto, T. Ohtani, and H. Ito, 1999, Distribution, deformation and alteration of fault rocks along the GSJ core penetrating the Nojima fault, Awaji Island, southwest Japan (in Japanese with English abstract): Journal of Geological Society of Japan, v. 105, p. 72–85.

Tanaka, H., N. Tomida, N. Sekiya, Y. Tsukiyama, K. Fujimoto, T. Ohtani, and H. Ito, 2000, Distribution, deformation and alteration of fault rocks along the GSJ core penetrating the Nojima fault, Awaji Island, southwest Japan, *in* H. Ito, K. Fujimoto, H. Tanaka, and D. Lockner, eds., International Workshop of the Nojima Fault Core and Borehole Data Analysis: Tsukuba, Geological Survey of Japan, p. 81–101.

Tanaka, H., S. Hinoki, K. Kosaka, A. Lin, K. Takemura, A. Murata, and T. Miyata, 2001, Deformation mechanisms and fluid behavior in a shallow, brittle fault zone during coseismic and interseismic periods: Results from drill core penetrating the Nojima fault, Japan: Island Arc, v. 10, p. 381–391.

Viola, G., N. S. Mancktelow, D. Seward, A. Meier, and S. Martin, 2003, The Pejo fault system; an example of multiple tectonic activity in the Italian eastern Alps: Geological Society of America Bulletin, v. 115, p. 515–532.

Wagner, G. A., and P. Van den haute, 1992, Fission-track dating: Norwell, Massachusetts, Kluwer Academic Publishers, 285 p.

Wallis, S., Y. Moriyama, and T. Tagami, 2004, Chronology of the Sanbagawa metamorphism, SW Japan—New constraints from zircon fission track analysis: Journal of Metamorphic Geology, v. 22, p. 17–24.

Yamada, K., T. Tagami, and N. Shimobayashi, 2003, Experimental study on hydrothermal annealing of fission tracks in zircon: Chemical Geology, v. 201, p. 351–357.

Yamada, R., T. Tagami, and S. Nishimura, 1993, Assessment of overetching factor for confined fission track length measurement in zircon: Chemical Geology (Isotope Geoscience Section), v. 104, p. 251–259.

Yamada, R., T. Tagami, and S. Nishimura, 1995a, Confined fission-track length measurement of zircon: assessment of factors affecting the paleotemperature estimate: Chemical Geology (Isotope Geoscience Section), v. 119, p. 293–306.

Yamada, R., T. Tagami, S. Nishimura, and H. Ito, 1995b, Annealing kinetics of fission tracks in zircon: An experimental study: Chemical Geology (Isotope Geoscience Section), v. 122, p. 249–258.

Yamada, R., T. Yoshioka, K. Watanabe, T. Tagami, H. Nakamura, T. Hashimoto, and S. Nishimura, 1998, Comparison of experimental techniques to increase the number of measurable confined fission tracks in zircon: Chemical Geology (Isotope Geoscience Section), v. 149, p. 99–107.

Yamamoto, Y., H. Kurita, and T. Matsubara, 2000, Eocene calcareous nannofossils and dinoflagellate cysts from the Iwaya Formation in Awajishima Island, Hyogo Prefecture, southwest Japan, and their geologic implications (in Japanese with English abstract): Journal of Geological Society of Japan, v. 106, p. 379–382.

17

Takagi, H., A. Iwamura, D. Awaji, T. Itaya, and T. Okada, 2005, Dating of fault gouges from the major active faults in southwest Japan: Constraints from integrated K-Ar and XRD analyses, *in* R. Sorkhabi and Y. Tsuji, eds., Faults, fluid flow, and petroleum traps: AAPG Memoir 85, p. 287–301.

Dating of Fault Gouges from the Major Active Faults in Southwest Japan: Constraints from Integrated K-Ar and XRD Analyses

Hideo Takagi

Department of Earth Sciences, Faculty of Education and Integrated Arts and Sciences, Waseda University, Tokyo, Japan

Akira Iwamura[1]

Division of Earth, Environment and Resources Science and Engineering, Graduate School of Science and Engineering, Waseda University, Tokyo, Japan

Dohta Awaji

Division of Earth, Environment and Resources Science and Engineering, Graduate School of Science and Engineering, Waseda University, Tokyo, Japan

Tetsumaru Itaya

Research Institute of Natural Sciences, Okayama University of Science, Okayama, Japan

Toshinori Okada

Institute for Frontier Research on Earth Evolution, Japan Marine Science and Technology Center, Yokosuka, Japan; Research Institute of Natural Sciences, Okayama University of Science, Okayama, Japan

ABSTRACT

Fault gouges were mapped and collected along the Atotsugawa fault, one of the major active faults in Japan, and along the Mozumi–Sukenobu fault, branching off from the Atotsugawa fault. Most of the fault gouge samples contain mica clay minerals, chlorite, smectite, and quartz. To constrain the timing of faulting, K-Ar and x-ray diffraction analyses (XRD) were carried out on mica clay minerals separated from the gouge samples. Each gouge sample was divided into four grain-size fractions of 5–2, 2–1, 1–0.35, and 0.35–0.05 μm. Kübler illite crystallinity indices for the finer grain-size fractions (0.05–1 μm; illite crystallinity = 0.4–0.8) were found to be higher than those for the coarser fractions (1–5 μm; illite crystallinity = 0.3–0.6), indicating the relatively higher

[1]*Present address:* Weathernews, Inc., Chiba, Japan.

DOI:10.1306/1033729M853139

concentration of authigenic mica clay minerals. The genesis of the clay minerals was probably related to hydrothermal alteration events associated with fault activity in the finer fractions. K-Ar ages were younger for the finer fractions of all samples. One sample from the Atotsugawa fault yields the youngest age of 61 Ma for the finer two fractions, which probably dates the thermal activity associated with the fault because of a similar age for the finer two fractions. This also suggests that the contamination of protolith mica is negligible for these samples. The finest fraction of the Mozumi–Sukenobu fault gouge derived from the interbedded sandstone and mudstone of the Tetori Group (Upper Jurassic–Lower Cretaceous) gives a K-Ar age of 45 Ma. Although this gouge sample is derived from mudstone and sandstone and, thus, the contamination of protolith illite cannot be identified by illite crystallinity, the age probably approximates that of a hydrothermal alteration event associated with the fault activity, because the age is significantly younger than the sedimentary age of the protolith Tetori Group. These K-Ar ages from the Atotsugawa and Mozumi–Sukenobu faults are compared with previous K-Ar ages of fault gouges from major active faults in Japan, including the Median Tectonic Line in Shikoku–Kinki and the Atera fault. The K-Ar data from these faults indicate that the major active faults in the Inner Zone of Southwest Japan were initiated at 60–50 Ma. Heterogeneity in and around the Late Cretaceous granitic terrane, especially the boundary of rigid granitic body and soft accretionary complex, seems to be the preferred sites for fault initiation.

INTRODUCTION

Age determination of fault movements is essential to understanding structural evolution of both basement and basin areas. Moreover, knowledge of timing of fault movement is important for fault trap analysis. Stratigraphic, archeological, and isotopic methods have been applied to dating of fault activity events (see review by Murphy et al., 1979). K-Ar analysis of the fine-grained fraction (<2 μm) from fault gouge samples has proved to be a particularly helpful method for direct dating of fault activity (Lyons and Snellenburg, 1971; Kralik et al., 1987). Fine-grained fractions of fault gouge consist mainly of mica clay minerals that contain potassium such as illite and/or mixed-layer illite and smectite. We use the term "mica clay mineral" as an umbrella term instead of illite, which has been recently defined as a series name of dioctahedral interlayer-deficient micas (Rieder et al., 1998). These clay minerals in the core of a fault zone not only reduce the permeability of the fault rock but also provide a valuable material for age determination of the fault. The closure temperature of the K-Ar system in illite is estimated to be about 260°C (Hunziker et al., 1986). K-Ar dating of the illite that grew in the fault rock would indicate the time of fault activity associated with temperatures of about 260°C, and later fault movements with lower thermal effects would not modify the K-Ar system (Shibata and Takagi, 1988). The reliability of the K-Ar dating of fine-grained fractions in fault gouge has also been supported by K-Ar analyses of adularia cements in the fractures of cataclasite collected from the fault exposures, including dated fault gouges in Europe (Bonhomme et al., 1983) and in Japan (Figure 1) (Shibata and Takagi, 1988).

The illite K-Ar dating method can be problematic if mica minerals from the protolith contaminate authigenic mica clay minerals in the fault gouge. Illite crystallinity defined by Kübler (1984) (which is obtained as the half-height width of 10-Å diffraction peak) and illite polytype compositions help to sort out this protolith contamination. The illite crystallinity method has been widely used to determine the thermal maturity in sedimentary-metasedimentary rocks especially in accretionary complexes (Underwood et al., 1993; Awan and Kimura, 1996; Tanabe and Kano, 1996). The Kübler's illite crystallinity index has been divided into three: the diagenetic zone (>0.42), the anchi (-metamorphic) zone (0.42 > illite crystallinity >0.25), and the epi (-metamorphic) zone (<0.25) in ascending order of metamorphic grade (Kübler, 1968; Blenkinsop, 1988; Kisch, 1990). These zones are correlative to the metamorphic grade of zeolite facies, prehnite-pumpellyite facies, and pumpellyite-actinolite facies or higher grade, respectively (Kisch, 1987). The illite crystallinity is also a good indicator of authigenic mica clay minerals formed by fault-related hydrothermal alteration (Tanaka et al., 1995). Analyses of grain-size fractionated fault gouge samples show that finer fractions of fault gouge have generally larger illite crystallinity index values and younger K-Ar ages, suggesting a higher concentration of authigenic mica clay minerals in the finer fractions (Vrolijk and van der Pluijm, 1999; Cho et al., 2001). K-Ar ages of fault gouge

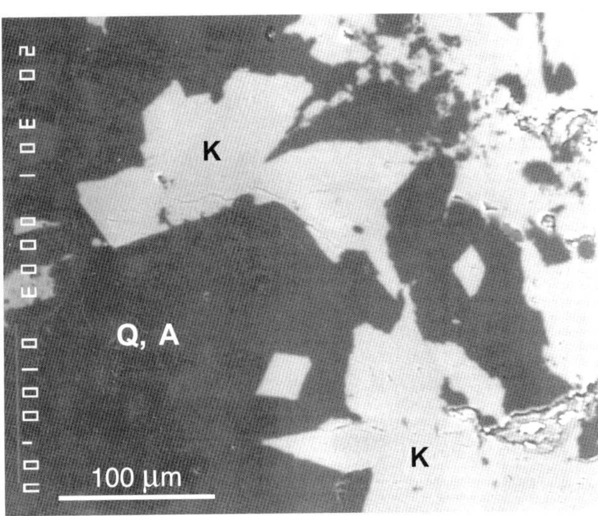

FIGURE 1. Backscattered electron image of an adularia (K)-albite (A)-quartz (Q) vein cemented on a fracture of pelitic schist in the Sambagawa belt close to the MTL (after Takagi and Shibata, 1992). The K-Ar age of this vein is 20.7 Ma, similar to the K-Ar age of 18.5 Ma of the fine fraction (<2 μm) in the fault gouge from the same MTL outcrop (Kitagawa, Oshika village, Chubu region).

materials with the illite crystallinity index >0.42 are indicative of diagenetic thermal conditions. It is known that mica polytypism, that is, the difference in the manner of stacking of layered atomic structural units of mica clay minerals, reflects the physicochemical conditions of their formation. Velde (1965) showed that the sequential formation of mica polytypes, $1Md \rightarrow 1M \rightarrow 2M_1$, is expected with an increase of time, temperature, and/or pressure. In a study of the Lewis thrust in the Canadian Rockies, Van der Pluijm et al. (2001) estimated percentages of detrital illite in different grain-size fractions based on polytype analyses ($2M_1$ vs. 1M or 1Md) and found a linear relationship between illite Ar-Ar age and detrital illite percentage. By extrapolating the Ar-Ar age line to 0% detrital illite, they obtained a reliable age for the fault activity.

Unlike dating of metamorphic-igneous rocks, K-Ar age from fault gouge in sedimentary rocks probably gives a maximum age for the fault, because the illite crystallinity or polytype analyses are not able to differentiate mica clay minerals formed in the fault core from the diagenetic illites formed in the sedimentary basin.

In Japan, K-Ar dating of fault rocks has been applied to the Median Tectonic Line (MTL), which divides southwest Japan into the Inner and Outer Zones (Figure 2) (Shibata and Takagi, 1988; Shibata et al., 1988, 1989; Takagi et al., 1989, 1992), the Atera fault (Yamada et al., 1992), and the Akaishi Tectonic Line (ATL) that merges into the MTL (Tanaka et al., 1995). Kanaori et al. (1988) also reported K-Ar whole rock ages of less than 2-μm fractions of fault gouge samples in the Atotsugawa fault.

This chapter presents K-Ar and illite crystallinity data for fault gouge samples collected from the Atotsugawa fault and from the Mozumi–Sukenobu fault, which is a branch of the Atotsugawa fault (Figure 3), in the Inner Zone of southwest Japan. The data were determined for four fractions (5–2, 2–1, 1–0.35, and 0.35–0.05 μm in grain size) from each sample. The Atotsugawa fault is one of the major inland active faults in Japan and has been investigated by several researchers from the viewpoint of its seismicity; however, no systematic attempt has been made to constrain its formational age. Based on our new results and the previous K-Ar age data on fault rocks, the formation ages of major active faults in Japan, including the MTL in Shikoku and Kinki regions and the Atera fault in Chubu region (Figure 2), are discussed.

TECTONIC SETTING OF THE INNER ZONE OF SOUTHWEST JAPAN

The Median Tectonic Line

Excepting older terranes such as Hida and Hida marginal terranes, the pre-Miocene basement rocks of southwest Japan are composed mainly of upper Paleozoic to Cenozoic accretionary complexes. The accretionary terranes young toward the Pacific Ocean. The MTL is the longest trench-linked strike-slip fault (~1000 km; ~620 mi) in Japan, running from Kanto to Chubu, Kinki, Shikoku, and to western Kyushu. Along much of its length, the MTL marks the southern margin of the Ryoke belt composed of Cretaceous high-temperature metamorphic and associated granitic rocks and the northern margin of the Sambagawa belt that is made up of Cretaceous high-pressure metamorphics. The MTL also divides southwest Japan into the Inner Zone (Japan Sea side) and the Outer Zone (Pacific side). The Inner Zone of southwest Japan is characterized by a wide distribution of Cretaceous–Paleogene granitic rocks and a local distribution of Triassic–Jurassic granitic rocks only in the Hida belt (see Figure 2). Cretaceous volcanic rocks are also widely distributed only in the Inner Zone of southwest Japan.

The timing of the initiation of the MTL can be dated as middle Cretaceous (ca. 100 Ma) on the basis of Chemical Th-U-total Pb isochron method (CHIME) monazite age data (Suzuki and Adachi, 1998) from sheet bodies of foliated and deformed older Ryoke granite along the MTL (Takagi, 1996). Structural analyses along the MTL (Miyata et al., 1980; Takagi, 1986) indicate a sinistral shear sense during the Late Cretaceous to Paleogene, probably related to oblique subduction of Izanagi–Pacific Plate. However, a dextral sense of shear along the active fault segments of the MTL (shown as solid lines in Figure 2) in

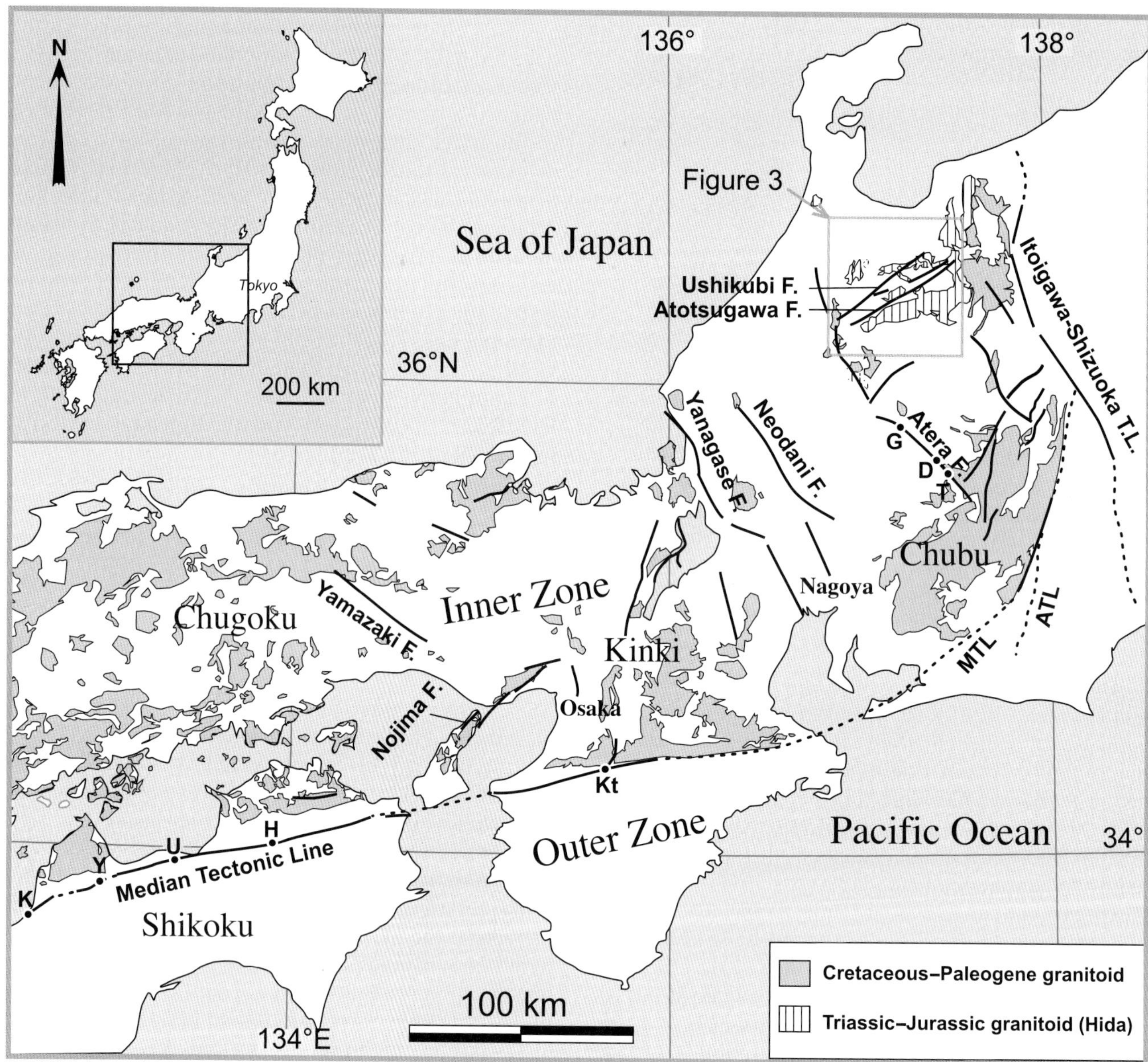

FIGURE 2. Distribution of major active faults and Cretaceous–Paleogene granitic rocks in the Chubu, Kinki, and Shikoku regions, southwest Japan, simplified from active faults and seismicity in and around the Japanese islands (scale 1:3,000,000; Research Group for Active Faults of Japan, 1991) and the geological map of Japan (scale 1:1000,000; Geological Survey of Japan). Fault gouge locations along the MTL and the Atera fault: K = Konogawa; Y = Yuyaguchi; U = Urayamagawa; H = Hiruma; Kt = Kitayama; T = Taze; D = Daimon; G = Gero.

Shikoku, western Kinki, and parts of Chubu is probably caused by Quaternary oblique subduction of the Philippine Sea Plate (Maruyama and Seno, 1986). The active fault segments have an average displacement rate of about 2–7 m/1000 yr (6.6–23 ft/1000 yr) (Research Group for Active Faults of Japan, 1991).

The Atera Fault

Besides the MTL, inland active faults are concentrated almost only in the Inner Zone of southwest Japan, especially in Chubu and Kinki regions (Figure 2). Dextral active faults, such as the Atotsugawa, Ushikubi, and Nojima faults striking east-northeast–west-southwest to northeast–southwest in the Inner Zone of southwest Japan are considered to form a conjugate system with the northwest–southeast-striking sinistral active faults, such as the Atera, Neodani, Yanagase, and Yamazaki faults (Figure 2) under the Quaternary east–west regional compressive stress field in southwest Japan (Sugimura and Matsuda, 1965). These conjugate fault systems have been active since the late Neogene on the basis of topographic features of active faults and stratigraphic data (Matsuda, 1976).

The Atera fault, which is one of the major active faults in central Japan, extends for about 55 km (34 mi)

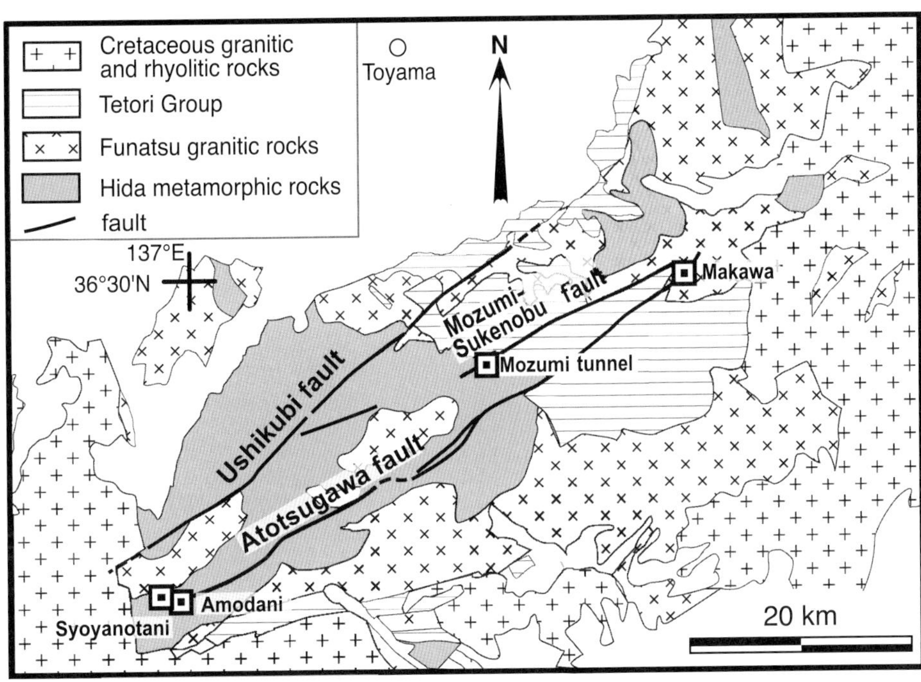

FIGURE 3. Geological sketch map of the Atotsugawa fault area simplified from the 1:500,000 scale of the geological map (Yamada et al., 1974) and neotectonic map (Kato and Sugiyama, 1985) of Kanazawa (Geological Survey of Japan, 1974, 1985).

from Nakatsugawa town, Nagano Prefecture, to Gero town, Gifu Prefecture, in a northwest–southeast trend. It is a sinistral fault having an average displacement rate of about 1–5 m/1000 yr (3–16 ft/1000 yr) (Research Group for Active Faults of Japan, 1991). The wall rocks around the Atera fault are mainly composed of Late Cretaceous Nohi rhyolitic rocks and Ryoke granitic rocks, which intruded the rhyolitic rocks.

CHARACTERISTICS OF THE ATOTSUGAWA AND MOZUMI–SUKENOBU FAULTS

The Atotsugawa Fault

The Atotsugawa fault is a major active fault in central Japan extending for about 60 km (37 mi) from the foot of Mt. Tateyama at Toyama Prefecture to Amo Pass at Gifu Prefecture with an east-northeast–west-southwest trend (Figure 3). The fault shows dextral displacement with an average displacement rate of about 2–4 m/1000 yr (7–13 ft/1000 yr) (Research Group for Active Faults of Japan, 1991). The wall rocks along the Atotsugawa fault are largely composed of late Paleozoic Hida metamorphic rocks, Triassic–Jurassic Funatsu granitic rocks (part of the Hida belt), and clastic sedimentary rocks of the Middle Jurassic to Lower Cretaceous Tetori Group.

The characteristics of intrafault materials along the Atotsugawa fault were first described by Kanaori et al. (1982a, b), who also determined K-Ar ages of 53.7 and 87.7 Ma on less-than-2-μm fractions separated from fault gouges (Kanaori et al., 1990). Mapping of fault gouge and collection of samples for this study were conducted at three good exposures: Amo-dani (Figure 4a), Shoyanotani, and the Makawa River (Figure 4b). Foliation, lineation, and asymmetric composite planar fabrics (P-Y-R_1 fabrics; Rutter et al., 1986) observed at these outcrops, as well as in polished slabs of fault gouges from east-northeast–west-southwest-striking fault zones, clearly indicate a dextral shear sense.

The best exposure is located at Amo-dani, the western termination of the Atotsugawa fault. The thickness of the fault zone is about 30 m (100 ft), and fault rocks are exposed along the Amo-dani River (Figure 5). Several gouge zones and minor faults are identified in the fault zone. The protoliths of the brittle fault rocks are calcareous gneiss, biotite gneiss, and Funatsu granitic rocks. The foliations and lineations measured from oriented fault gouge samples are N79°E/72°N and N71°E/20° (location A) and N59°E/68°N and S54°W/12° (location G), longitude and latitude, respectively, and the composite planar fabric (P-R_1) of the fault zone shows dextral shear (Figure 6a). North–south-striking minor fault zones (locations B–D) oblique to the general trend of the Atotsugawa fault (Figure 5) give a complex fault attitude. Five samples of oriented fault gouges (A1–A5; Figure 5) were collected, two of which (A1 and A2) were dated in this study.

The Mozumi–Sukenobu Fault

Two major active faults running parallel to the Atotsugawa fault are the Mozumi–Sukenobu fault and the Ushikubi fault (Figure 3). Earthquake evidence for recent activity on the Mozumi–Sukenobu fault has been obtained from the eastern domain at the Mozumi Pass (Wada et al., 1996). The 481-m (1578-ft)-long Mozumi research tunnel was constructed in the Kamioka mine for examination of an active fault in 1997 (Figure 7) (Ito et al, 1998). Three typical fracture zones can be observed at distances of 80, 190, and 260 m (262, 623, and

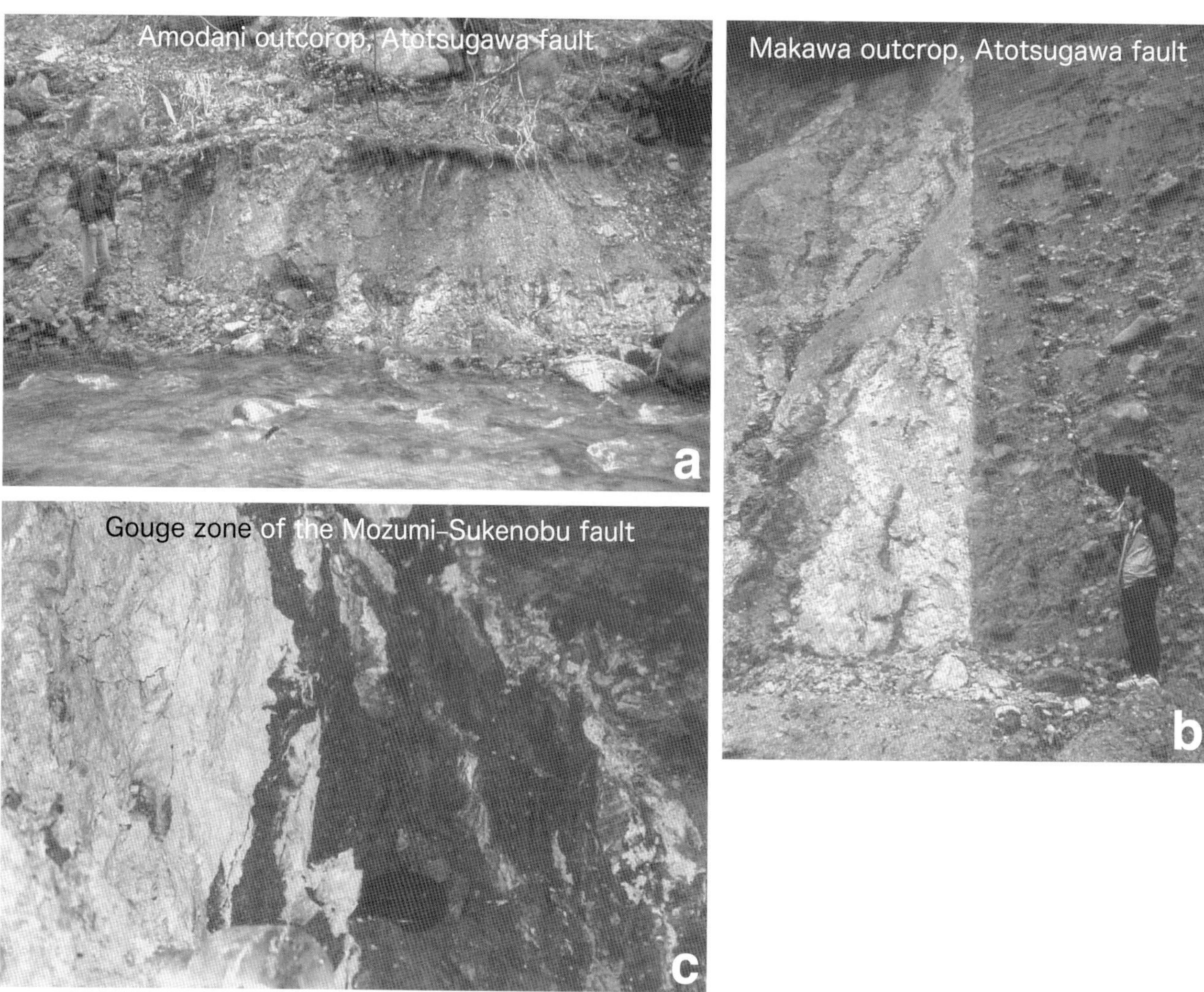

Figure 4. Photographs of the outcrops of the Atotsugawa fault (a and b) and the Mozumi–Sukenobu fault (c). (a) Amodani outcrop of the Atotsugawa fault (points E–G in Figure 5). (b) Well-known Makawa outcrop of the Atotsugawa fault sharply bounding Early Jurassic Funatsu granitic rocks (left) and Quaternary terrace gravel (right). (c) Photograph of the fracture zone A at a vertical wall in the Mozumi tunnel. Camera lens cap for scale at the bottom of photograph.

853 ft) (zones A, B, and C, respectively) from a crossroad, and thus, the total thickness of the Mozumi–Sukenobu fault is about 200 m (660 ft). The fracture zone A (Figure 4c) has a thickness of about 22 m (72 ft) (Ito et al., 1998). Fault gouges and fault breccias have been derived from interbedded sandstone and mudstone of the Inotani formation (Lower Cretaceous), Tetori Group. The oriented sample of fault gouge from zone A has the foliation of N59°E/75°S, with a stretching lineation of N65°E/12°. The fracture zone B is partly composed of soft gouge, yellowish white in color. The fracture zone C has a 3-m (10-ft)-thick bluish gray fault gouge. The attitude of the main slip surface is N54°E/75°S, with slickenlines of N60°E/20° (Figure 6b). Groove structures on the side of fragments on the slickenside show a dextral sense of movement. Three samples (M1–M3) of fault gouge from each zone (Figure 7) were collected, and one sample (M1) was dated for this study.

X-RAY DIFFRACTION ANALYSES AND POTASSIUM-ARGON DATING

Sample Preparation

Fault gouge samples were washed with deionized water to remove organic substances. Four grain-size fractions, 5–2, 2–1, 1–0.35, and 0.35–0.05 μm, were separated from samples using hydraulic elutriation, centrifuge, and supercentrifuge separators.

X-ray Powder Diffraction Analyses

X-ray diffraction (XRD) analysis of each fraction was carried out using a Rigaku Geiger Flex RAD-1B at Waseda University, under conditions of 40 kV and 20 mA

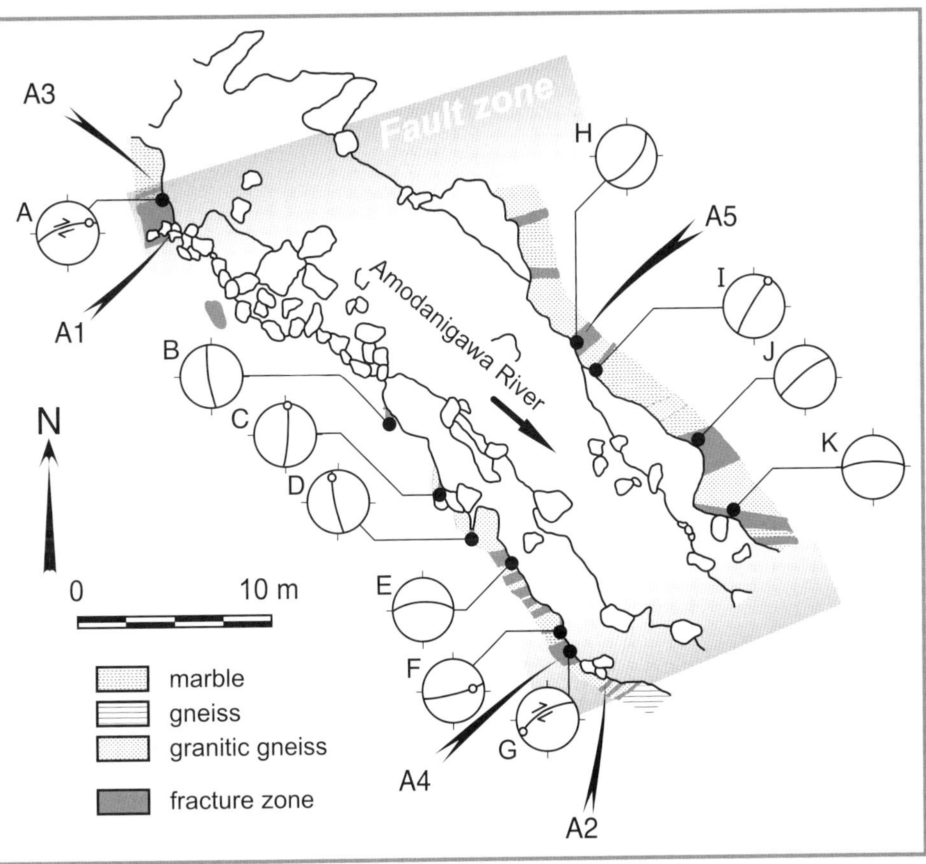

Figure 5. Schematic showing of the Amodani outcrop of the Atotsugawa fault. Inset stereograms project orientations of a fault surface (great circle) and slickenlines or cataclastic lineations (dot). Arrows in stereogram A and G indicate the slip sense.

and a step-scanning mode as previously described in Matsuda et al. (1998). Measurements were made on the random powder and oriented aggregate samples. For measurements of oriented aggregates, the amount of specimen was prepared in a range of 5–10 mg to avoid data fluctuation caused by variations in clay thickness of the specimens (Moore and Reynolds, 1989; Kisch, 1991).

The illite crystallinity index (Kübler, 1968, 1984) was determined as a half-height width ($\Delta 2\theta$) of the basal reflection of mica clay minerals at 10 Å on the oriented specimens after the ethylene glycol (EG) treatment. Step mode is 1 s for 0.01° (2θ) step spacing (Hara and Kimura, 2000) using 1/2°–1/2°–0.3 mm width slit system. Illite crystallinity values are calculated with the attached software of Rad-1B system. Amounts of expandable layers in mica clay minerals were determined from a shift of basal 10-Å reflection upon glycolation (Shimoda, 1977). Polytype analysis of mica clay minerals followed the method described in Togashi (1979) that

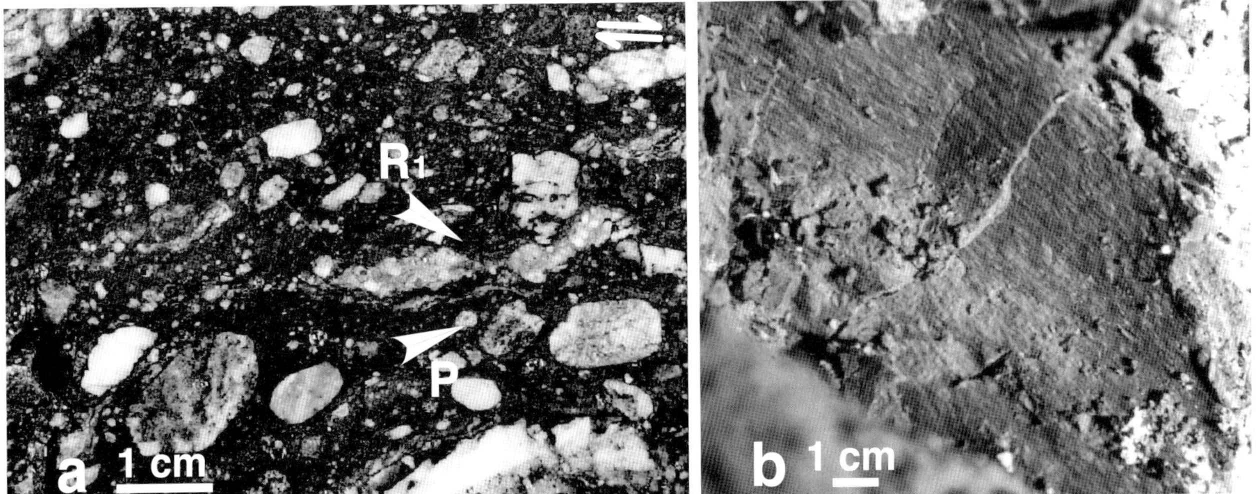

Figure 6. Fault gouge structures. (a) *XZ* polished slice of foliated fault gouge (location G) from Amodani. Dextral sense of shear can be identified from P-R_1 fabric in the fault gouge. (b) Slickenline on the surface (slickenside) of fault gouge (zone A) from the Mozumi tunnel.

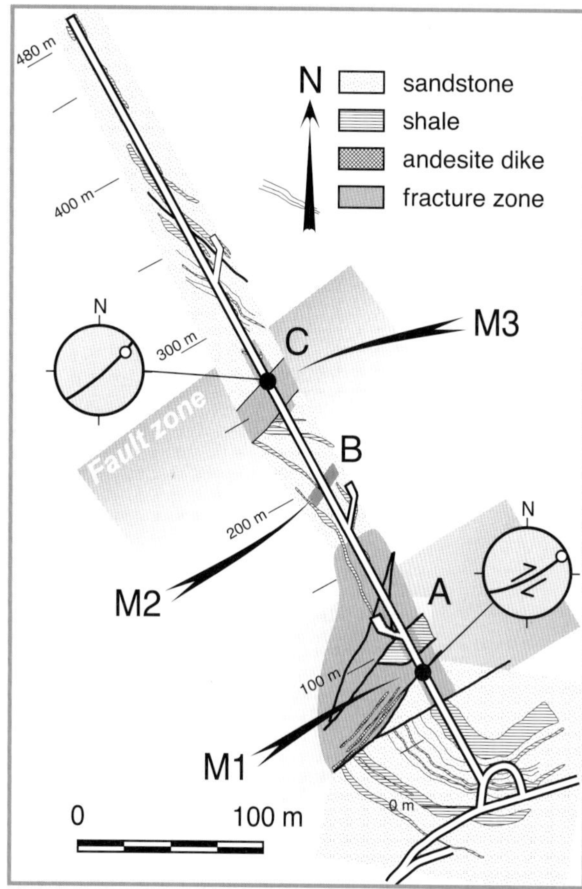

FIGURE 7. Schematic map of the Mozumi tunnel (partly modified after Ito et al. 1998). Two stereograms indicate orientations of a fault surface (great circle) and slickenlines (dots). The arrows in stereogram A indicate the slip sense.

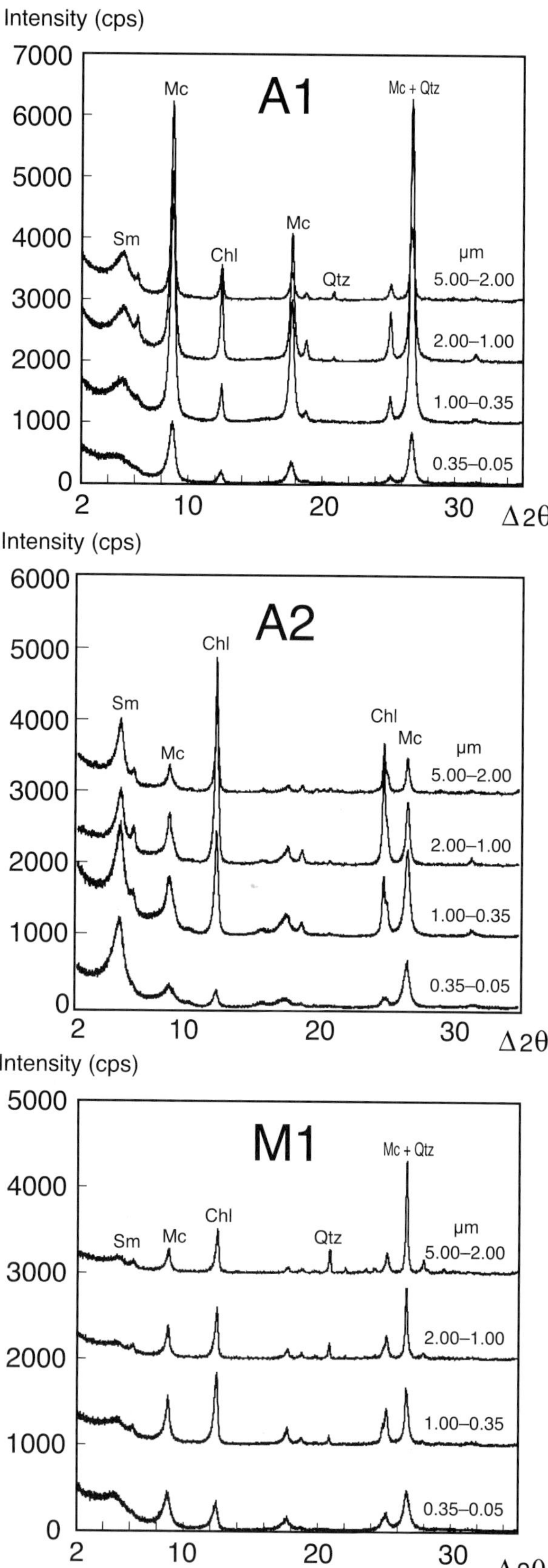

FIGURE 8. XRD charts of grain-size fractions of fault gouge samples. Sm = smectite; Mc = mica clay mineral; Chl = chlorite; Qtz = quartz; Mc + Qtz = mica clay mineral and quartz.

used a comparison of the ratio of integrated intensity of (112) reflection (3.08 Å) for 1M and 1Md types and (025) reflection (3.00 Å) for $2M_1$ type.

X-ray charts of oriented samples after EG treatment are shown in Figure 8. The mineral composition and illite crystallinity index for each fraction from three dated samples are given in Table 1. The fine fractions are mainly composed of mica clay minerals, smectite (montmorillonite), and chlorite. Minor amounts of quartz and feldspar occur in the coarser fractions.

The illite crystallinity index generally increases with decreasing grain size (Figure 9), except for a few fractions in some samples (A3–A5). The exceptionally high illite crystallinity index in the finest fraction for A3 may be caused by a high content of smectite. In this case, it is difficult to determine the real half wavelength of illite (001) because of the interference of smectite (001) peak even after the EG treatment.

The polytypes of mica clay minerals (illite) range from 1Md through 1M to $2M_1$ with an increase of either temperature or time. The (112) and (025) peaks of mica clay minerals for all samples are not distinct, and thus,

Table 1. K-Ar ages of each grain-size fraction of fault gouges from the Atotsugawa fault and the Mozumi–Sukenobu fault.

Number	*Protolith*	*Color*	*Fraction (μm)*	*Mineral Composition*	*Illite Crystallinity*	*d (Å)*	*Polytype*	*Potassium (%)*	*Rad.⁴⁰Ar (10^{-8} cm^3 STP/g)*	*K-Ar Age (Ma)*	*Non Rad. Ar (%)*
					Atotsugawa Fault						
A1	granitic gneiss	bluish gray	5.00–2.00	MC, Chl, sm; q	0.37	10.03	–	4.000 ± 0.080	1334 ± 14	84.0 ± 1.8	4.1
			2.00–1.00	MC, Chl, sm; q	0.45	10.05	$2M_1$	6.025 ± 0.121	1886 ± 21	78.9 ± 1.8	3.7
			1.00–0.35	MC, chl, sm; q	0.49	10.04	$2M_1$	6.946 ± 0.139	1973 ± 21	71.8 ± 1.6	4.0
			0.35–0.05	MC, sm, chl	0.55	10.06	$2M_1$, 1M	6.159 ± 0.123	1672 ± 18	68.6 ± 1.5	9.0
A2	granitic gneiss	green	5.00–2.00	Chl, Sm, MC; q	0.41	10.02	–	2.760 ± 0.055	809.6 ± 8.5	74.0 ± 1.6	3.5
			2.00–1.00	Chl, Sm, MC; q	0.45	10.04	–	3.239 ± 0.065	840.4 ± 9.0	65.7 ± 1.5	5.9
			1.00–0.35	Sm, Chl, MC	0.64	10.04	1Md	4.188 ± 0.084	1006 ± 11	60.8 ± 1.4	10.8
			0.35–0.05	Sm, Chl, MC	0.78	10.01	1Md	4.057 ± 0.081	979 ± 11	61.2 ± 1.4	10.3
					Mozumi–Sukenobu Fault						
M1	sandstone and shale	bluish gray	5.00–2.00	Chl, MC, sm; q, f	0.34	10.05	–	1.721 ± 0.034	467.2 ± 5.0	68.6 ± 1.5	8.4
			2.00–1.00	Chl, MC, sm; q, f	0.31	10.07	–	3.257 ± 0.065	778.1 ± 9.0	60.5 ± 1.4	14.4
			1.00–0.35	Chl, MC, sm; q, f	0.36	10.06	–	4.073 ± 0.081	905 ± 11	56.4 ± 1.3	16.2
			0.35–0.05	MC, Chl, sm	0.59	10.09	1M	4.603 ± 0.092	817 ± 12	45.1 ± 1.1	27.0

Rad. = radiogenic; Non Rad. = nonradiogenic; MC = mica clay mineral; Sm = smectite; Chl = chlorite; q = quartz; f = feldspar. Traceable minerals are indicated in lower case.

we could not estimate the ratio of polytypes using the method by Togashi (1979).

The amount of expandable layers can be estimated from Δd (Å) (d_{001}, natural – d_{001}, glycolated). The Δd was almost negligible, indicating that the amount of expandable layers is less than 5% for all specimens.

The sample Am collected from the well-known Makawa outcrop (Figure 4b) at the northeastern end of the Atotsugawa fault has the illite crystallinity index of 0.24–0.28 for the coarser fractions (Figure 9). It contains dominantly kaolinite with minor mica clay minerals. This indicates that the clay minerals have formed at shallow crustal levels, and considerable contamination of mica clay minerals with pulverized mica derived from the protolith is inferred. Most of the fault gouges from other locations are composed of smectite, mica clay minerals, and chlorite but are devoid of kaolinite. The illite crystallinity index of mica clay minerals shows anchi zone or diagenetic zone. The polytypes determined in finer fractions are dominantly 1M or 1Md in the finest fraction, whereas only specimen A1 contains $2M_1$ polytype even in the finest fraction. The content of the mixed layer is rather low because the ratio of expandable layers is less than 5%.

K-Ar Dating

Analyses of potassium and argon and calculations of ages and errors follow the procedures given in Nagao et al. (1984) and Itaya et al. (1991). Potassium was analyzed by flame photometry using a 2000-ppm Cs buffer and has an analytical error within 2% at 95% (2σ) confidence level. Argon was analyzed on a 15-cm (6-in.) radius sector-type mass spectrometer with a single collector system using an isotopic dilution method and argon-38 spike (Itaya et al., 1991). Multiple runs of a standard (JG-1 biotite of 91 Ma) indicate that the error of argon analysis is about 1% at the 95% confidence level. The physical constants used in the age calculation are from Steiger and Jäger (1977); the decay constants ^{40}K to ^{40}Ar, ^{40}K to ^{40}Ca, and ^{40}K content in potassium are 0.581×10^{-10}/yr, 4.962×10^{-10}/yr, and 0.0001167, respectively.

The K-Ar ages of the three gouge samples are given in Table 1. Finer fractions give younger ages for all samples as also previously noted by studies of Takagi and Shibata (1992), Matsuda et al. (1998), Vrolijk and van der Pluijm (1999), and Cho et al. (2001). Sample A2 gives similar K-Ar age of ca. 61 Ma for two finer fractions, indicating that it is the likely age of the fault gouge (Figure 10a). Sample A1 does not have the asymptotic line parallel with the horizontal axis; however, the inclination of the line decreases into the finest grain-size fraction (Figure 10a). Accordingly, the K-Ar age of 68.6 Ma for the finest fraction is probably related to

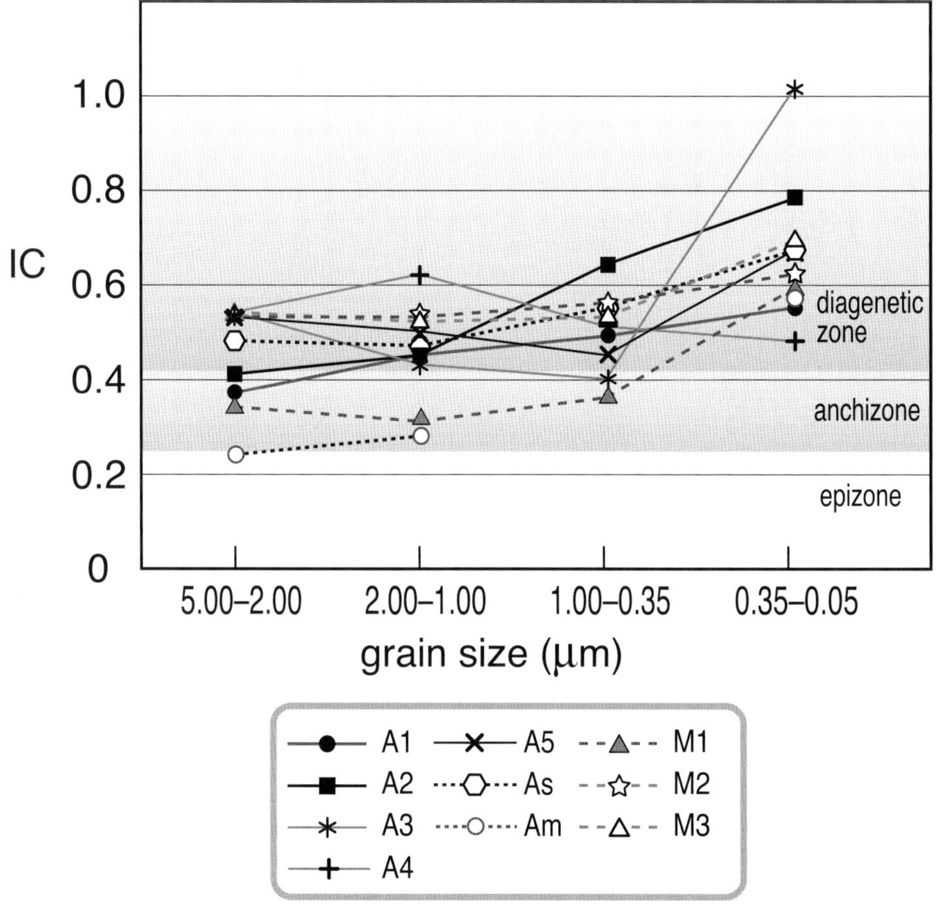

FIGURE 9. Illite crystallinity vs. grain-size diagrams of all fault gouge samples from the Atotsugawa fault and Mozumi–Sukenobu fault.

Late Cretaceous granitic rocks widely distributed in the Inner Zone of Southwest Japan.

In the Hida terrane, basement rocks are mainly composed of the Hida metamorphic rocks and the Triassic–Jurassic (ca. 180 Ma) Funatsu granitic rocks. The isotopic ages of the Hida metamorphic rocks vary about 500, 420, 340, 220–240, and 180 Ma, but the K-Ar biotite ages of the rocks concentrate at 180 Ma because of rejuvenation by intrusion of Funatsu granitic rocks (Ota and Itaya, 1989). Late Cretaceous granitic and volcanic rocks are widely distributed on both sides of the terrane (Figure 3) (Kano, 1990).

the age of a hydrothermal alteration event. Sample M1 does not have the asymptotic line parallel with the horizontal axis, and the inclination of the line increases toward the finest grain-size fraction (Figure 10a). Thus, it seems difficult to interpret the age data for this sample. The relationship between K-Ar age vs. illite crystallinity values is shown in Figure 10b.

DISCUSSION

Significance of the K-Ar Ages of Fault Gouges from the Atotsugawa and Mozumi–Sukenobu Faults

The K-Ar age of 61 Ma obtained from sample A2 suggests that the thermal activity associated with fault movement and responsible for clay mineral growth in the fault rock occurred in the early Paleocene. As mentioned above, the dextral strike-slip movement along the Atotsugawa fault was reactivated in the Neogene and the Quaternary, although the fault reactivation has not reset the K-Ar system in the mica clay minerals. The age of 61 Ma corresponds to the biotite K-Ar ages of 60–70 Ma (e.g., Shibata and Takagi, 1988; Suzuki and Adachi, 1998) reported from the

It is difficult to provide a reason why the finest fraction for M1 has an abrupt drop on the age–grain-size diagram (Figure 10a). The protolith of the fault gouge is the interbedded mudstone and sandstone of the Tetori Group in which authigenic illite and/or detrital muscovite are probably included. For this gouge sample, we cannot sort out the contribution of preexisting illite using the illite crystallinity. However, the obtained K-Ar age of 45 Ma for the finest fraction is much younger than the sedimentary (diagenesis) age determined by stratigraphic correlation (Lower Cretaceous Nagatogawa Formation, Itoshiro Subgroup; Yamashita et al., 1988) of the Tetori Group in this area. Therefore, the K-Ar age of 45 Ma probably reflects the major hydrothermal alteration event and authigenic clay mineral growth associated with the activity of the Mozumi–Sukenobu fault.

Correlation to the K-Ar Age of Fault Gouges from the Major Active Faults in Southwest Japan

K-Ar dating of fine fractions (<2 μm) of fault gouges along the MTL from the Chubu, Kinki, and Shikoku regions has previously been reported (Table 1) (Shibata and

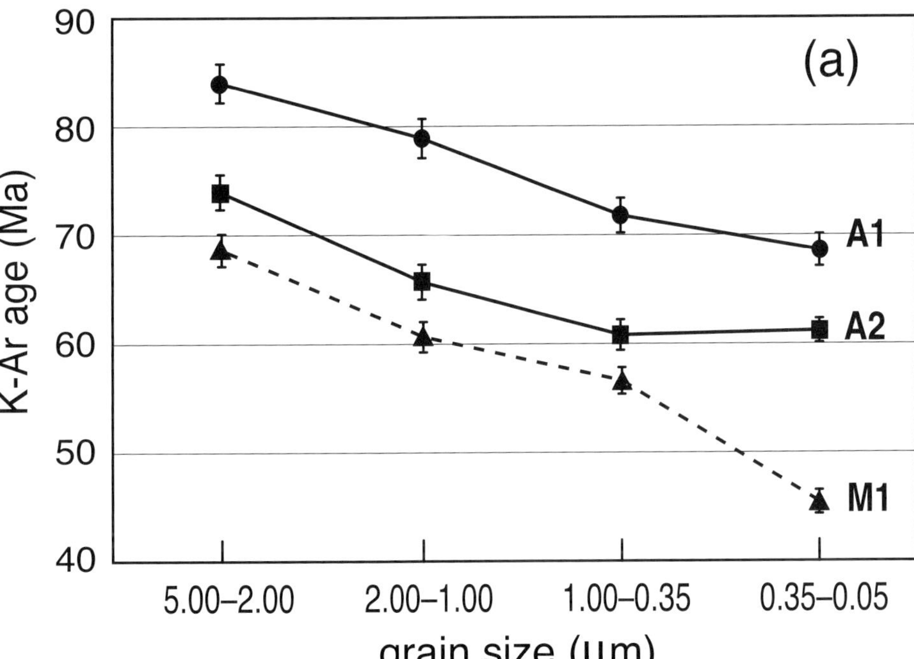

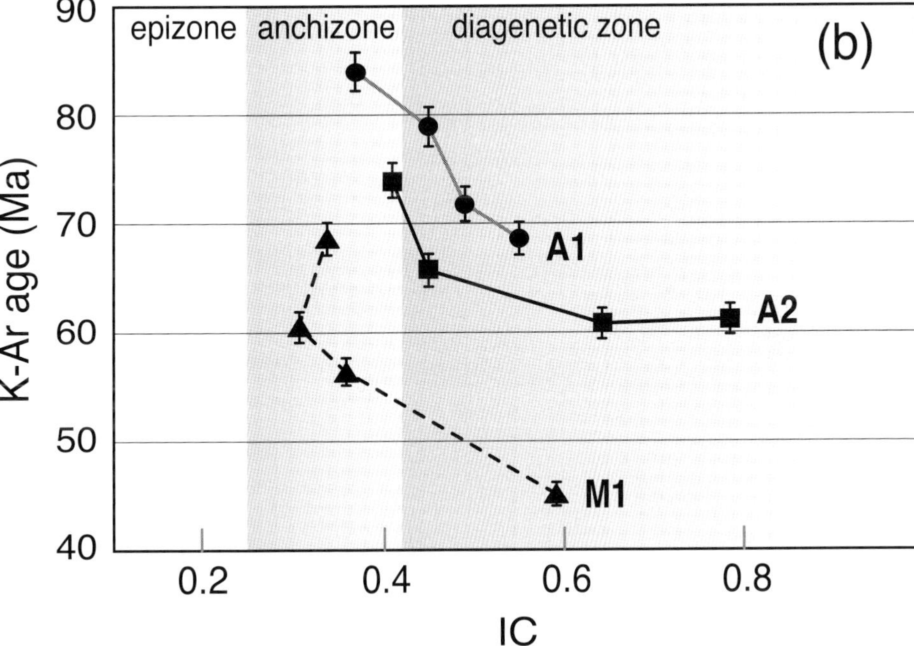

FIGURE 10. Variation of K-Ar ages with respect to the grain size (a) and illite crystallinity (b) of fault gouge samples from the Atotsugawa fault and Mozumi–Sukenobu fault.

Takagi, 1988; Shibata et al., 1988, 1989; Takagi et al., 1989, 1992). Most of the gouge samples are predominantly derived from the Sambagawa pelitic schist or uppermost Cretaceous Izumi Group mudstones, but some are derived from Ryoke granitic rocks. Compiling these data, where illite crystallinity data indicate the anchi or diagenetic zone (Takagi and Shibata, 1992), together with the data (15 Ma; Tanaka et al., 1995) from the Akaishi Tectonic Line (ATL in Figure 2) branching off from the MTL in the Chubu region, the K-Ar ages cluster in four groups: 58–63 Ma (in Shikoku), 34–45 Ma (in Chubu, Kinki, and Shikoku), 15–27 Ma (only in Chubu), and 10–14 Ma (Chubu and Shikoku) (Table 2; see Figure 2 for location). The oldest age (ca. 60 Ma) of fault gouges taken from only active fault domains of the MTL is correlative to K-Ar ages of fault gouges from the Atotsugawa fault. It suggests that the oldest hydrothermal alteration associates with brittle fault movement along the MTL. This age is correlative to the initial deformation phase of the uppermost Cretaceous Izumi Group along the MTL called "Ichinokawa phase" (Kobayashi, 1941). The second older age group, 34–45 Ma, is related to the hydrothermal alteration associated with fault reactivation after the deposition of Eocene Kuma Group along the MTL in Shikoku and is correlative to "Tobe phase" (Kobayashi, 1941). The third age group of 15–27 Ma comes only from the Chubu region and presumably reflects the fault activity related to the collision of the Izu-Bonin arc with the Honshu arc in Miocene time. The K-Ar age of 20.7 Ma from the adularia-bearing vein (Figure 1) in fractured pelitic schist from the Chubu region is also included in this age group. This collision-related fault reactivation at Chubu is called the "Akaishi phase" (Takagi and Shibata, 1992). The youngest age group of 10–14 Ma probably reflects the intrusions of acidic-intermediate dykes along the MTL in an extensional tectonic regime. This phase is related to the volcanic activities of Mt. Ishizuchi in Shikoku and is thus correlative with "Ishizuchi phase" (Suyari and Akojima, 1973).

Table 2. K-Ar ages of fine fractions (<2 μm) of fault gouges from the active fault segments of the Median Tectonic Line (Shibata et al., 1989) and from the Atera fault (Yamada et al., 1992).

Sample Number	*Location*	*Protolith*	*Color*	*Mineral composition*	*Illite Crystallinity*	*d (Å)*	*2M (%)*	*K_2O (%)*	*Age (Ma)*	*Reference*
				MTL-Kinki						
Ng1B	Kitayama	granite	light yellow	MC, Sm, k; q	0.42	10.06	36	5.74	52.2 ± 1.6	1
				MTL-Shikoku						
87S4	Hiruma	pelitic schist	brownish gray	MC, k; q	1.12	10.41	10	5.22	60.1 ± 1.4	2
87S9.5A*	Urayamakawa	Izumi Group	gray	MC, k; q	0.82	10.26	22	6.00	38.1 ± 1.3	2
87S9.5B	Urayamakawa	Izumi Group	brownish gray	MC, k; q	0.75	10.25	30	6.04	45.2 ± 1.4	2
87S25	Yuyaguchi	Izumi Group	brownish gray	MC, k	0.64	10.15	10	6.50	59.3 ± 1.9	2
87S30A	Konokawa	Izumi Group	brownish gray	MC, k; q	0.88	10.28	12	5.22	62.4 ± 2.1	2
87S30B	Konokawa	gravel	brownish gray	MC, k, Sm; q	0.80	10.22	8	5.12	57.8 ± 1.9	2
				Atera Fault						
63	Taze	granite	light yellow	Sm, MC, k; q, f	0.71	10.17	38	5.30	52.8 ± 1.7	3
64	Daimon	granite	white	MC, M, k; q	0.31	10.02	31	4.47	54.6 ± 1.8	3
66	Gero	welded tuff	light yellow	ML, Sm, k; q	0.43	10.08	13	6.01	51.7 ± 1.6	3

*MC = mica clay mineral; ML = mixed layer; Sm = smectite; k = kaolinite; q = quartz; f = feldspar; *<0.5 μm. References: (1) Shibata and Uchiumi, 1992; (2) Shibata et al. 1989; (3) Yamada et al. 1992.*

The Atera fault occurs in the Late Cretaceous plutonovolcanic terranes, such as the Ryoke granite and the Nohi rhyolite. The K-Ar ages of the fine fraction (<2 μm) of fault gouge samples from this fault belong to the anchi or diagenetic zone and are 52.8 and 54.6 Ma (Cretaceous granitic protolith) and 51.7 Ma (Cretaceous rhyolitic welded tuff protolith) (Yamada et al. 1992) (Table 2). These ages are younger than those of the gouge samples from the Atotsugawa fault.

We also attempted K-Ar age dating of fault gouge that was derived from Ryoke granite along the Nojima fault (Figure 2) at Hirabayashi on Awaji Island along which the Hyogoken–Nanbu earthquake (Kobe earthquake: $M = 7.2$) occurred in January 1995. However, no mica clay minerals for K-Ar analyses could be separated from the sample. Murakami et al. (2002) obtained fission-track ages of 52–55 Ma for zircon crystals from pseudotachylyte veins along the Nojima fault. This indicates that the formation of the Nojima fault dates back to early Eocene.

Our new data for the Atotsugawa fault and the Mozumi–Sukenobu fault, as well as the previous data from the Atera fault, indicate that the formation of these faults in southwest Japan can be placed at 60–50 Ma (early Paleogene), and that despite their Neogene–Quaternary reactivation, the thermal effects of these faults have not been sufficiently high to reset the K-Ar system in mica clay minerals. Permeability of clay-bearing fault gouge along a fault core is quite low on the order of 10^{-17}–10^{-22} m^2 compared to the damage zone or fractured zone of a fault (10^{-14}–10^{-16} m^2) under confining pressures of as much as 200 MPa (Evans et al., 1997; Faulkner and Rutter, 2000; Wibberley and Shimamoto, 2003). Accordingly, once clay gouge has formed, hydrothermal fluid finds it difficult to permeate into the fault core to form new clay minerals during fault reactivation, and this is a reason why old (initial) ages from fault gouge samples could be preserved.

Relationship between the Basement Geology and the Sites of Major Active Faults in Southwest Japan

Based on the previous K-Ar age data of fault gouges along the Atotsugawa and Atera faults, Kanaori et al. (1990, 1992) suggested the possibility that faults would provide a favorable tectonic setting for Late Cretaceous granite intrusion in central Japan. However, CHIME monazite ages marking the emplacement of Ryoke granitic rocks in the Chubu region range between 95 and 67 Ma (Suzuki and Adachi, 1998), whereas most K-Ar biotite ages recording the cooling of the granitic rock are 60–70 Ma. The K-Ar ages of fault gouges along major active faults in central Japan are 60–50 Ma, suggesting that major hydrothermal alteration events that are associated with fracturing along the faults are closely related to the postintrusive tectonic activities.

In the Inner Zone of southwest Japan, major active faults are distributed mainly in or around the margin of the Cretaceous granitic terrane (Ryoke and San-yo belts) and also occur in Jurassic accretionary complexes (Tanba–Mino–Ashio belts). Some domains of the superficial Jurassic accretionary complexes in the inner zone have suffered thermal metamorphism by Late Cretaceous granitic intrusions (Hayasaka, 1999). However, active faults are scarce in the Jurassic–Paleogene accretionary terranes in the Outer Zone (Figure 2), where only small bodies of Miocene granite are distributed. Therefore, the crustal distribution of Late Cretaceous granitic rocks (including acidic volcanic terrane) seems to be a key factor in localizing major active faults (Hayasaka, 1999). Especially, the boundary of rigid granitic body and soft accretionary complex seems to provide preferred sites for fault initiation on the basis of basement geological map made by Hayasaka (1999).

ACKNOWLEDGMENTS

We thank Akira Takeuchi of Toyama University, who kindly arranged a field sampling from the Mozumi tunnel, and Ken Shibata of Nagoya Bunri University, who organized the joint research on the K-Ar dating of fault gouge during 1987–1992. Thanks are also due to R. Sorkhabi, R. Maddock, and an anonymous referee for reviewing this chapter and correcting the English. This research was supported partly by Waseda University Grant for Special Research Projects (no. 1999B-006).

REFERENCES CITED

Awan, M. A., and K. Kimura, 1996, Thermal structure and uplift of the Cretaceous Shimanto belt, Kii Peninsula, southwest Japan, and illite crystallinity and illite b_0 lattice spacing study: The Island Arc, v. 5, p. 69–88.

Blenkinsop, T. G., 1988, Definition of low-grade metamorphic zones using illite crystallinity: Journal of Metamorphic Geology, v. 6, p. 623–636.

Bonhomme, M. G., D. Buhmann, and Y. Besnus, 1983, Reliability of K-Ar dating of clays and silicifications associated with vein mineralizations in western Europe: Geologische Rundschau, v. 72, p. 105–117.

Cho, K.-H., H. Takagi, A. Iwamura, D. Awaji, T.-W. Chang, S.-W. Shon, T. Itaya, and T. Okada, 2001, Timing of the hydrothermal alteration associated with the fault activities along the Ulsan fault zone, southwest Korea: Journal of the Korea Society of Economic and Environmental Geology, v. 34, p. 583–593 (in Korean with English abstract).

Evans, J. P., C. B. Forster, and J. V. Goddard, 1997, Permeability of fault-related rocks, and implications for hydraulic structure of fault zones: Journal of Structural Geology, v. 19, p. 1393–1404.

Faulkner, D. R., and E. H. Rutter, 2000, Comparisons of water and argon permeability in natural clay-bearing fault gouge under high pressure at 20°C: Journal of Geophysical Research, v. 105, p. 16,415–16,426.

Hara, H., and K. Kimura, 2000, Estimation of errors in measurement of illite crystallinity: The limits and problem of application to accretionary complexes: Journal of the Geological Society of Japan, v. 106, p. 264–279.

Hayasaka, Y., 1999, Geological characteristics of seismic basement of the southwest Japan (in Japanese): Abstract of the Japan Earth and Planetary Science Joint Meeting, La-001, Tokyo, (CD-ROM).

Hunziker, J. C., M. Frey, N. Clauer, R. D. Dallmeyer, H. Friedrichsen, W. Flehmig, K. Hochstrasser, P. Roggwiler, and H. Schwander, 1986, The evolution of illite to muscovite: Mineralogical and isotopic data from Glarus Alps, Switzerland: Contributions to Mineralogy and Petrology, v. 92, p. 157–180.

Itaya, T., K. Nagao, K. Inoue, Y. Honjou, T. Okada, and A. Ogata, 1991, Argon isotopic analysis by a newly developed mass spectrometric system for K-Ar dating: Mineralogical Journal, v. 15, p. 203–221.

Ito, T., A. Takeuchi, H. Tanaka, Y. Nishikawa, K. Sakogaichi, and Y. Okada, 1998, Underground observatory of an active fault: Preliminary geological reports of a tunnel across the Mozumi–Sukenobu fault (in Japanese): Earth Monthly, v. 20, p. 182–187.

Kanaori, Y., Y. Inohara, K. Miyakoshi, and Y. Satake, 1982a, Characteristics of intrafault materials within the Atotsugawa fault of central Japan: Part I (in Japanese with English abstract): Journal of the Japan Society of Engineering Geology, v. 23, p. 137–155.

Kanaori, Y., K. Miyakoshi, Y. Inohara, and Y. Satake, 1982b, Characteristics of intra-fault materials within the Atotsugawa fault of central Japan: Part II (in Japanese with English abstract): Journal of the Japan Society of Engineering Geology, v. 23, p. 201–213.

Kanaori, Y., K. Yairi, S. Kawakami, and K. Miyakoshi, 1988, Deformation microstructures and their genetic processes of granitic rocks in the northeastern region of the Atotsugawa fault, central Japan (in Japanese with English abstract): Journal of the Geological Society of Japan, v. 94, p. 887–901.

Kanaori, Y., K. Yairi, S. Kawakami, and T. Takeshita, 1990, Granite intrusion tectonics induced by fault motion in central Japan (in Japanese with English abstracts): Zisin (Journal of the Seismological Society of Japan), v. 43, p. 77–90.

Kanaori, Y., S. Kawakami, and K. Yairi, 1992, The block structures and Quaternary strike-slip block rotation of central Japan: Tectonics, v. 11, p. 47–56.

Kano, T., 1990, Granitic rocks in the Hida complex, central Japan (in Japanese with English abstract): Mining Geology, v. 40, p. 397–413.

Kato, H., and Y. Sugiyama, 1985, The neotectonic map of Kanazawa: Geological Survey of Japan, scale 1:500,000, 1 sheet.

Kisch, H. J., 1987, Correlation between indicators of very

low-grade metamorphism, *in* M. Frey, ed., Low temperature metamorphism: Glasgow, Blackie & Son, p. 227–300.

Kisch, H. J., 1990, Calibration of the anchizone: A critical comparison of illite "crystallinity" scales used for definition: Journal of Metamorphic Geology, v. 8, p. 31–46.

Kisch, H. J., 1991, Illite crystallinity: Recommendations on sample preparation, x-ray diffraction setting, and interlaboratory samples: Journal of Metamorphic Geology, v. 9, p. 665–670.

Kobayashi, T., 1941, The Sakawa orogenic cycle and its bearing on the origin of the Japanese Islands: Journal of Faculty of Science, Imperial University of Tokyo, sec. II, v. 5, no. 7, p. 1–578.

Kralik, M., K. Klima, and G. Riedmüller, 1987, Dating fault gouges: Nature, v. 327, p. 315–317.

Kübler, B., 1968, Evaluation quantitative du métamorphism par la crystallinité de l'llite: Bulletin du Centre de Recherches de pau-SNPA, v. 2, p. 385–397.

Kübler, B., 1984, Les indicateurs des transformations physiques et chimiques dans la diagenèse, tempèrature et calorimètrie, *in* M. Lagache, ed., Thermomètrie et baromètrie gèologiques: Société Francaise de Minéralogie Crystallographie, Paris, v. 2, p. 489–596.

Lyons, J. B., and J. Snellenburg, 1971, Dating faults: Geological Society of America Bulletin, v. 82, p. 1749–1752.

Maruyama, S., and T. Seno, 1986, Orogeny and relative plate motions: Example of the Japanese islands: Tectonophysics, v. 127, p. 305–329.

Matsuda, T., 1976, Empirical rules on sense and rate of recent crustal movements: Journal of the Geodetic Society of Japan, v. 22, p. 252–263.

Matsuda, T., M. Koizumi, Y. Sugiyama, and Y. Saka, 1998, Influence of detrital mica grains on measurement of illite crystallinity (in Japanese with English abstract): Journal of the Geological Society of Japan, v. 104, p. 718–721.

Miyata, T., H. Ui, and K. Ichikawa, 1980, Paleogene left-lateral wrenching on the Median Tectonic Line in southwest Japan: Memoirs of the Geological Society of Japan, v. 18, p. 51–68.

Moore, D. M., and R. C. Reynolds, Jr., 1989, X-ray diffraction and the identification and analysis of clay minerals: New York, Oxford University Publishing, 332 p.

Murakami, M., R. Yamada, and T. Tagami, 2002, Detection of frictional heating of fault motion by the fission track analysis: Abstracts of Japan Earth and Planetary Science Joint Meeting S039-P005, Tokyo, (CD-ROM).

Murphy, P. J., J. Briedis, and J. H. Peck, 1979, Dating techniques in fault investigations: Reviews in Engineering Geology, Geological Society of America, v. IV, p. 153–168.

Nagao, K., H. Nishido, T. Itaya, and K. Ogata, 1984, K-Ar age determination method (in Japanese with English abstract): Bulletin of the Hiruzen Research Institute, v. 9, p. 19–38.

Ota, K., and T. Itaya, 1989, Radiometric ages of granitic and metamorphic rocks in the Hida metamorphic belt, central Japan (in Japanese with English abstract): Bulletin of the Hiruzen Research Institute, Okayama University of Science, v. 15, p. 1–25.

Research Group for Active Faults of Japan, 1991, Active faults in Japan: Sheet maps and inventories (revised edition) (in Japanese with English abstract): Tokyo, University of Tokyo Press, 448 p.

Rieder, M., et al., 1998, Nomenclature of the micas: The Canadian Mineralogist, v. 36, p. 41–48.

Rutter, E. H., R. H. Maddock., S. H. Hall, and S. H. White, 1986, Comparative microstructures of natural and experimentally produced clay-bearing fault gouges: Pure and Applied Geophysics, v. 124, p. 3–30.

Shibata, K., and H. Takagi, 1988, Isotopic ages of rocks and intrafault materials along the Median Tectonic Line— An example in the Bungui-toge area, Nagano Prefecture (in Japanese with English abstract): Journal of the Geological Society of Japan, v. 94, p. 35–50.

Shibata, K., and S. Uchiumi, 1992, K-Ar age results: 4. New data from the Geological Survey of Japan (in Japanese): Bulletin of the Geological Survey of Japan, v. 43, p. 359–367.

Shibata, K., Y. Sugiyama, H. Takagi, and S. Uchiumi, 1988, Isotopic ages of rocks along the Median Tectonic Line in the Yoshino area, Nara Prefecture (in Japanese with English abstract): Bulletin of the Geological Survey of Japan, v. 39, p. 759–781.

Shibata, K., T. Nakajima, A. Sangawa, S. Uchiumi, and H. Aoyama, 1989, K-Ar ages of fault gouges from the Median Tectonic Line in Shikoku (in Japanese with English abstract): Bulletin of the Geological Survey of Japan, v. 40, p. 661–671.

Shimoda, S., 1977, Chemical composition and crystal structures of mica clay minerals (in Japanese with English abstract): Journal of the Mineralogical Society of Japan, v. 13 (special issue), p. 27–37.

Steiger, R. H., and E. Jäger, 1977, Subcommission on geochronology: Convention on the use of decay constants in geo- and cosmochronology: Earth and Planetary Science Letters, v. 36, p. 359–362.

Sugimura, A., and T. Matsuda, 1965, Atera fault and its displacement vectors: Geological Society of America Bulletin, v. 76, p. 509–522.

Suyari, K., and I. Akojima, 1973, Fault movements of the region adjacent to the median line and their significance with the topographic evolution of Shikoku, Japan, *in* R. Sugiyama, ed., Median Tectonic Line (in Japanese with English abstract): Tokyo, Tokai University Press, p. 179–189.

Suzuki, K., and M. Adachi, 1998, Denudation history of the high T/P Ryoke metamorphic belt, southwest Japan: Constraints from CHIME monazite ages of gneisses and granitoids: Journal of Metamorphic Geology, v. 16, p. 23–37.

Takagi, H., 1986, Implications of mylonitic microstructures for the geotectonic evolution of the Median Tectonic Line, central Japan: Journal of Structural Geology, v. 8, p. 3–14.

Takagi, H., 1996, Timing of mylonitization in the Ryoke belt, Chubu region (in Japanese): Earth Monthly, v. 19, p. 111–116.

Takagi, H., and K. Shibata, 1992, K-Ar dating of fault gouge— Examples along the Median Tectonic Line (in Japanese with English abstract): Memoirs of the Geological Society of Japan, v. 40, p. 31–38.

Takagi, H., K. Shibata, Y. Sugiyama, S. Uchiumi, and A. Matsumoto, 1989, Isotopic ages of rocks along the Median Tectonic Line in the Kayumi area, Mie Prefecture (in Japanese with English abstract): Journal of the Association of Petrology, Mineralogy and Economic Geology, v. 84, p. 75–88.

Takagi, H., T. Takeshita, K. Shibata, S. Uchiumi, and M. Inoue, 1992, Middle Miocene normal faulting along the Tobe thrust in western Shikoku (in Japanese with English abstract): Journal of the Geological Society of Japan, v. 98, p. 1069–1072.

Tanabe, H., and K. Kano, 1996, Illite crystallinity study of the Cretaceous Shimanto belt in the Akaishi Mountains, eastern southwest Japan: The Island Arc, v. 5, p. 56–68.

Tanaka, H., N. Uehara, and T. Itaya, 1995, Timing of the cataclastic deformation along the Akaishi Tectonic Line, central Japan: Contributions to Mineralogy and Petrology, v. 120, p. 150–158.

Togashi, Y., 1979, Polytypes and expandability of sericite from the Itaya kaolin clay deposit, northeast Japan: Journal of Japanese Associations of Mineralogy, Petrology and Mining Geology, v. 74, p. 100–113.

Underwood, M. B., M. M. Laughland, and S. M. Kang, 1993, A comparison among organic and inorganic indicators of diagenesis and low-temperature metamorphism, Tertiary Shimanto belt, Shikoku, Japan: Geological Society of America Special Paper 273, p. 45–61.

van der Pluijm, B. A., C. M. Hall, P. J. Vrolijk, D. R. Pevear, and M. C. Covey, 2001, The dating of shallow faults in the Earth's crust: Nature, v. 412, p. 172–175.

Velde, B., 1965, Experimental determination of muscovite polymorph stabilities: American Mineralogist, v. 50, p. 436–449.

Vrolijk, P., and B. A. van der Pluijm, 1999, Clay gouge: Journal of Structural Geology, v. 21, p. 1039–1048.

Wada, H., K. Ito, M. Ando, and K. Wada, 1996, Observation of earthquakes in the Mozumi tunnel at Kamioka mine, central Honshu, Japan (in Japanese): Annuals of Disasters Prevention Research Institute, v. 39, B-1, p. 241–250.

Wibberley, C. A. J., and T. Shimamoto, 2003, Internal structure and permeability of major strike-slip fault zones: The Median Tectonic Line in Mie Prefecture, southwest Japan: Journal of Structural Geology, v. 25, p. 59–78.

Yamada, N., T. Sakamoto, T. Nozawa, and T. Enda, 1974, The geological map of Kanazawa: Geological Survey of Japan, scale 1:500,000, 1 sheet.

Yamada, N., K. Shibata, E. Tsukuda, S. Uchiumi, A. Matsumoto, H. Takagi, and H. Akahane, 1992, Radiometric ages of igneous rocks around the Atera fault, central Japan, with special reference to the age of activity of the Atera fault (in Japanese with English abstract): Bulletin of the Geological Survey of Japan, v. 43, p. 759–779.

Yamashita, N., Y. Kaseno, and J. Itoigawa, eds., 1988, Regional geology of Japan: Part 5. Chubu II (in Japanese): Tokyo, Kyoritsu Shuppan, 310 p.

Sorkhabi, R., (Compiler), 2005, A bibliographic guide to fault traps, fault seals, and fault-related fluid flow in sedimentary basins, *in* R. Sorkhabi and Y. Tsuji, eds., Faults, fluid flow, and petroleum traps: AAPG Memoir 85, p. 303–342.

Appendix

A Bibliographic Guide to Fault Traps, Fault Seals, and Fault-related Fluid Flow in Sedimentary Basins

Compiled by Rasoul Sorkhabi[1]

Technology Research Center, Japan National Oil Corporation, Chiba, Japan

The following bibliography is a list by subject of selected journal articles and books on the function of faults in fluid-flow migration and entrapment in sedimentary basins. It is a bibliographic guide for further reading and research; it is not an exhaustive bibliography. In selecting references, the following criteria were considered:

1) references on fault traps, fault sealing, and fault-related fluid flow in sedimentary basins (not igneous or metamorphic terrains)
2) full articles (not abstracts) published in English-language mainstream journals, so that the references have an international accessibility
3) under each category where a book or a special journal issue has been referenced, individual papers from that volume have not been listed for the sake of space
4) for some categories specified with an asterisk (*), only book volumes and journal special issues are given because of the broad scope of these topics
5) the bibliography is up to March 2003 (including AAPG Bulletin's thematic issue on fault seals).

The bibliographic guide has been arranged as follows:

1. Fault trap structures
   - 1.1 Faults and the classification of traps
   - 1.2 Structural styles and closure of fault traps
   - 1.3 Normal faults, extensional basins, and passive margins
   - 1.4 Reverse faults, compressional basins, and active margins
   - 1.5 Strike-slip faults and wrench basins
   - 1.6 Faults of salt tectonics
   - 1.7 Faults of inversion tectonics and inverted basins
   - 1.8 Subsurface mapping of faults
   - 1.9 Faults on seismic images
   - 1.10 Construction of fault surface geometry and fault juxtaposition
2. Fault sealing processes
   - 2.1 Sealing mechanisms: General aspects
   - 2.2 Fault rocks and fault sealing processes: General aspects
   - 2.3 Cataclastic rocks
   - 2.4 Shale smear in clastic sequences
   - 2.5 Deformation bands in porous sandstone
   - 2.6 Pressure solution seals
   - 2.7 Mineral veins and crack seals associated with faulting
   - 2.8 Fault-related diagenesis and fluid-rock interactions in fault zones
3. Fault-related fluid flow
   - 3.1 Fluid flow in rocks
   - 3.2 Fault-related fluid flow: Field studies
   - 3.3 Fluid flow and natural seismicity

[1] *Present address:* Energy & Geosciences Institute, University of Utah, Salt Lake City, Utah, U.S.A.

DOI:10.1306/1033730M853128

3.4 Fluid flow and induced seismicity in petroleum basins
3.5 Structural controls on petroleum seepage

4. Petrophysical properties of faults
   4.1 Empirical measurements
   4.2 Experimental measurements
   4.3 Numerical simulation

5. Analysis of fault stress, stability, and failure
   5.1 Rock mechanics, rock stress, and fault mechanics
   5.2 Borehole technologies for fault stress analysis
   5.3 Field studies

1. FAULT TRAP STRUCTURES

1.1. Faults and the Classification of Traps

Biddle, K. T., and C. C. Wielchowsky, 1994, Hydrocarbon traps, *in* L. B. Magoon and W. G. Dow, eds., The petroleum system—From source to trap: AAPG Memoir 60, p. 219–235.

Brod, I. O., 1945, Geological terminology in classification of oil and gas accumulations: AAPG Bulletin, v. 29, p. 1738–1755.

Clapp, F. G., 1910, A proposed classification of petroleum and natural gas fields based on structure: Economic Geology, v. 5, p. 503–521.

Clapp, F. G., 1917, Revision of the structural classification of petroleum and natural gas fields: Geological Society of America Bulletin, v. 28, p. 553–602.

Clapp, F. G., 1929, The role of geologic structure in the accumulation of petroleum, *in* D. Powers, ed., Structure of typical American oil fields: Tulsa, AAPG, v. II, p. 667–716.

Harding, T. P., and J. D. Lowell, 1979, Structural styles, their plate-tectonic habitats, and hydrocarbon traps in petroleum provinces: AAPG Bulletin, v. 63, p. 1016–1058.

Heald, K. C., 1940, Essentials for oil pools, *in* E. DeGolyer, ed., Elements of the petroleum industry: American Institute of Mining and Metallurgical Engineers, p. 26–62.

Leverson, A. I., 1967, Geology of petroleum, 2d ed.: San Francisco, W. H. Freeman, chapters 6–7, p. 232–384.

Milton, N. J., and G. T. Betram, 1992, Trap styles—A new classification based on sealing surfaces: AAPG Bulletin, v. 76, p. 983–999.

North, F. K., 1985, Trapping mechanism for oil and gas, *in* Petroleum geology: Boston, Allen & Unwin, chapter 16, p. 253–341.

Prison, S. J., 1945, Genetic and morphologic classification of reservoirs: Oil Weekly, v. CXVIII (June 18), p. 54–59.

Sales, J. K., 1993, Closure vs. sealing strength—A fundamental control on the distribution of oil and gas, *in* A. G. Dore, J. H. Auguston, C. Hermanrud, O. Sylta, and D. J. Stewart, eds., Basin modelling: Advances and applications: Norwegian Petroleum Society Special Publication 3, p. 399–414.

Sales, J. K., 1997, Seal strength vs. trap closure—A fundamental control on the distribution of oil and gas, *in* R. C. Surdam, ed., Seals, traps, and the petroleum system: AAPG Memoir 67, p. 57–83.

Ver Wiebe, W. A., 1929, Tectonic classification of oil fields in the United States: AAPG Bulletin, v. 13, p. 409–440.

Vincelette, R. R., E. A. Beaumont, and N. H. Foster, 1999, Classification of exploration traps, *in* E. A. Beaumont and N. H. Foster, eds., Exploring for oil and gas traps: AAPG Treatise of Petroleum Geology, chapter 2, p. 1–42.

Wilhelm, O., 1945, Classification of petroleum reservoirs: AAPG Bulletin, v. 29, p. 1537–1579.

Willis, D. G., 1961, Entrapment of petroleum, *in* G. B. Moody, ed., Petroleum exploration handbook: New York, McGraw-Hill, p. 61–68.

Wilson, W. B., 1934, Proposed classification of oil and gas reservoirs, *in* W. E. Wrather and F. H. Lahee, eds., Problems of petroleum geology: Tulsa, AAPG, p. 433–445.

Wilson, W. B., 1942, Classification of oil reservoirs: AAPG Bulletin, v. 26, p. 1291–1292.

1.2. Structural Styles and Closure of Fault Traps

Anastasio, D. J., E. A. Erslev, and D. M. Fisher, eds., 1997, Fault-related folding: Journal of Structural Geology, v. 19, nos. 3–4, special issue, p. 243–602.

Bailey, R. J., and R. Stonneley, 1981, Petroleum: Entrapment and conclusions, *in* D. H. Tarling, ed., Economic geology and geotectonics: Oxford, Blackwell, p. 73–97.

Beaumont, E. A., and N. H. Foster, compilers, 1990–1993, Structural traps, v. I–VIII: AAPG Treatise of Petroleum Geology, Atlas of Oil and Gas Fields.

Burk, C. A., and C. C. Drake, eds., 1974, The geology of continental margins: Berlin, Springer-Verlag, 1009 p.

Busby, C. J., and R. V. Ingersoll, eds., 1995, Tectonics of sedimentary basins: Oxford, Blackwell Science, 579 p.

Casey, M., D. Dietrich, M. Ford, and A. J. Watkinson, eds., 1991, The geometry of naturally deformed rocks: Journal of Structural Geology, v. 15, nos. 3–5, special issue, p. 243–671.

Cloetingh, S., M. Fernandez, J. M. Munoz, W. Sassi, and F. Horvath, eds., 1997, Structural controls on sedimentary basin formation: Tectonophysics, v. 282, special issue, p. 1–442.

Cloetingh, S., B. D'Argenio, R. Catalano, F. Horvath, and W. Sassi, eds., 1999, Basin dynamics and basin fill: Models and constraints: Part I: Tectonophysics, v. 315, special issue, p. 1–384.

Cloetingh, S., B. D'Argenio, R. Catalano, F. Horvath, and W. Sassi, eds., 2000, Basin dynamics and basin fill: models and constraints: Part II. Tectonophysics, v. 324, no. 4, special issue, p. 203–336.

Cloetingh, S., M. Marzo, J. A. Marzo, and J. Verges, eds., 2002, Tectonics of sedimentary basins: From crustal structure to basin fill: Tectonophysics, v. 346, nos. 1–2, special issue, p. 1–135.

Cloetingh, S., L. O. Boldreel, B. T. Larsen, M. Heinsen, and C. Mortensen, eds., 2003, Tectonics of sedimentary basin formation, models and constraints: The Ziegler volume: Tectonophysics, v. 300, special issue, p. 1–387.

Cosgrove, J. W., and M. S. Ameen, eds., 2000, Forced folds and fractures: Geological Society (London) Special Publication 169, 225 p.

Downey, M. W., 1990, Faulting and hydrocarbon entrapment: Leading Edge, v. 9 (January), p. 20–22.

Eremenko, N. A., and I. M. Michailov, 1974, Hydrocarbon pools at faults: Bulletin of Canadian Petroleum Geology, v. 22, p. 106–118.

Foster, N., and E. A. Beaumont, compilers, 1988a, Traps and seals: I. Structural/fault-seal and hydrodynamic traps: AAPG Treatise of Petroleum Geology Reprint Series 6, 554 p.

Foster, N., and E. A. Beaumont, compilers, 1988b, Traps and seals: II. Stratigraphic/capillary traps: AAPG Treatise of Petroleum Geology Reprint Series 7, 410 p.

Foster, N., and E. A. Beaumont, compilers, 1988c, Structural concepts and techniques: I. Basic concepts, folding and structural techniques: AAPG Treatise of Petroleum Geology Reprint Series 9, 723 p.

Foster, N., and E. A. Beaumont, compilers, 1989a, Structural concepts and techniques: II. Basement-involved deformation: AAPG Treatise of Petroleum Geology Reprint Series 10, 479 p.

Foster, N., and E. A. Beaumont, compilers, 1989b, Structural concepts and techniques: III. Detached deformation: AAPG Treatise of Petroleum Geology Reprint Series 11, 651 p.

Frostick, L. E., and R. J. Steel, eds., 1993, Tectonic controls and signatures in sedimentary succession: Oxford, Blackwell Science Publications, 520 p.

Harding, T. P., 1974, Petroleum traps associated with wrench faults: AAPG Bulletin, v. 58, p. 1290–1304.

Harding, T. P., and J. D. Lowell, 1979, Structural styles, their plate-tectonic habitats, and hydrocarbon traps in petroleum provinces: AAPG Bulletin, v. 63, p. 1016–1058.

Harding, T. P., and A. C. Tuminas, 1988, Interpretation of footwall (lowside) fault traps sealed by reverse faults and convergent wrench faults: AAPG Bulletin, v. 72, p. 738–757.

Harding, T. P., and A. C. Tuminas, 1989, Structural interpretation of hydrocarbon traps sealed by basement normal block faults at stable flanks of foredeep basins and at rift basins: AAPG Bulletin, v. 73, p. 812–840.

Howell, J. V., ed., 1948, A symposium on the relation of oil accumulation to structure, v. 3: Tulsa, AAPG, 516 p.

Jaroszweski, W., 1984, Fault and fold tectonics: Chichester, Ellis Homwood, 565 p.

Jenyon, M. K., 1990, Oil and gas traps: Chichester, John Wiley and Sons, 398 p.

Larsen, R. M., H. Brekke, B. T. Larsen, and E. Talleraas, 1992, Structural and tectonic modelling and its application to petroleum geology: Norwegian Petroleum Society Special Publication 1, 549 p.

Lowell, J. D., 2002, Structural styles in petroleum geology, 4th ed.: Tulsa, Oil and Gas Consultants Inc., 311 p.

Magara, K., 1986, Geological models of petroleum entrapment: London, Elsevier Applied Science Publishers, 328 p.

McCann, T., and A. Saintot, eds., 2003, Tracing tectonic deformation using the sedimentary record: Geological Society (London) Special Publication 208, 368 p.

Nelson, R. A., T. L. Patton, and S. Serra, 1999, Exploring for structural traps, *in* E. A. Beaumont and N. H. Foster, eds., Exploring for oil and gas traps: AAPG Treatise of Petroleum Geology, chapter 20, p. 1–70.

Nieuwland, D. A., ed., 2003, New insights into structural interpretation and modelling: Geological Society (London) Special Publication 212, 320 p.

Perez-Estaun, A., and M. P. Coward, eds., 1991, Deformation and plate tectonics: Tectonophysics, v. 191, nos. 3–4, special issue, p. 185–449.

Perkins, H., 1961, Fault-closure type fields, southeast Louisiana: Transactions, Gulf Coast Association of Geological Societies, v. 11, p. 177–196.

Porter, S., ed., 1929, Structure of typical American oil fields, 2 volumes: Tulsa, AAPG, 510 p. (v. 1); 780 p. (v. 2).

Skerlec, G. M., 1999, Evaluating top and fault seal, *in* E. A. Beaumont and N. H. Foster, eds., Exploring for oil and gas traps: AAPG Treatise of Petroleum Geology, chapter 10, p. 1–94.

Stewart, I., and C. Vita Finzi, eds., 1999, Coastal tectonics: Geological Society (London) Special Publication 146, 352 p.

Watkins, J. S., and C. L. Drake, eds., 1982, Studies in continental margin geology: AAPG Memoir 34, 801 p.

Watkins, J. S., L. Montadert, and P. W. Dickerson, eds., 1979, Geological and geophysical investigations of continental margins: AAPG Memoir 29, 472 p.

Watkins, J. S., F. Zhiqiang, and K. J. McMillan, eds., 1992, Geology and geophysics of continental margins: AAPG Memoir 53, 419 p.

Watterson, J., O. R. Clausen, J. A. Peterson, T. McCann, B. M. O' Reilly, P. M. Shannon, C. B. Howard, P. J. Mason, and J. J. Walsh, 1994, Systematics of faults and fault arrays, *in* K. Helbig, ed., Modeling the Earth for oil exploration: Oxford, Pergamon Press, p. 205–316.

Weber, K. J., G. Mandl, W. F. Pilaar, F. Lehner, and R. G. Precious, 1978, The role of faults in hydrocarbon migration and trapping in Nigerian growth fault structures: Society of Petroleum Engineers 10th Annual Offshore Technology Conference Proceedings, v. 4, p. 2643–2653.

Weeks, L. G., ed., 1958, Habitat of oil: Tulsa, AAPG, 1384 p.

Wilkerson, M. S., M. P. Fischer, and T. Apotria, eds., 2002, Fault-related folds: The transition from 2-D to 3-D: Journal of Structural Geology, v. 24, special issue, p. 591–904.

1.3. Normal Faults, Extensional Basins, and Passive Margins*

Bally, A. W., P. L., Bender, T. R. McGetchin, and R. I. Walcott, eds., 1980, Dynamics of plate interiors: American Geophysical Union Geodynamic Series 1, 162 p.

Bott, M. H. P., ed., 1976, Sedimentary basins of continental margins and cratons: Tectonophysics, v. 36, special issue, p. 1–314.

Cloetingh, S., B. D'Argenio, R. Catalano, F. Horvath, and W. Sassi, 1995, Interplay of extension and compression in basin formation: Tectonophysics, v. 252, special issue, p. 1–484.

Cloetingh, S., Z. Ben-Avraham, W. Sassi, and F. Horvath, eds., 1996a, Dynamics of extensional basin formation and inversion: Tectonophysics, v. 240, special issue, p. 1–323.

Cloetingh, S., O. Eldholm, B. T. Larsen, R. H. Gabrielsen, and W. Sassi, 1996b, Dynamics of extensional basin formation and inversion tectonics: Tectonophysics, v. 266, special issue, p. 1–523.

Coward, M. P., J. F. Dewey, and P. L. Hancock, eds., 1987, Continental extensional tectonics: Geological Society (London) Special Publication 28, 637 p.

Edwards, J. D., and P. A. Santogrossi, eds., 1990, Divergent/passive margin basins: AAPG Memoir 48, 252 p.

Gangi, A. F., ed., 1991, World rift systems: Tectonophysics, v. 197, nos. 2–4, special issue, p. 99–392.

Holdsworth, R. E., and J. P. Turner, eds., 2002a, Extensional tectonics: Part I. Regional scale processes: Geological Society (London) Key Issues in Earth Sciences, 352 p.

Holdsworth, R. E., and J. P. Turner, eds., 2002b, Extensional tectonics: Part II. Faulting and related processes: Geological Society (London) Key Issues in Earth Sciences, 328 p.

Illies, E., 1970, Graben problems: Berlin, Springer Verlag, 405 p.

Illies, J. H., ed., 1981, Mechanism of graben formation: Tectonophysics, v. 73, nos. 1–3, special issue, p. 1–266.

Irvine, T. N., ed., 1966, The world rift system: Report of a symposium: Geological Survey of Canada Paper 66-14, 456 p.

Lambiase, J. J., ed., 1995, Hydrocarbon habitat in rift basins: Geological Society (London) Special Publication 80, 381 p.

Landon, S. M., ed., 1994, Interior rift basins: AAPG Memoir 59, 276 p.

Leighton, M. W., D. R. Kolata, D. F. Oltz, and J. J. Eidel, eds., 1990, Interior cratonic basins: AAPG Memoir 51, 819 p.

Marshak, S., B. A. Van der Pluim, and M. Hamburger, eds., 1999, Tectonics of continental interiors: Tectonophysics, v. 305, nos. 1–3, special issue, p. 1–417.

Morgan, P., and B. H. Baker, eds., 1983, Processes of continental rifting: Tectonophysics, v. 94, special issue, p. 1–678.

Olsen, K. H., ed., 1995, Continental rifts: Evolution, structure, tectonics: Amsterdam, Elsevier, 466 p.

Palmason, G., ed., 1982, Continental and oceanic rifts: American Geophysical Union Geodynamic Series 8, 309 p.

Quennell, A. M., ed., 1985, Continental rifts: Pennsylvania, Hutchinson Ross, 346 p.

Ramberg, I. B., and E. E. Neumann, eds., 1978, Tectonics and geophysics of continental rifts: Drodrecht, D. Reidel, 459 p.

Roberts, A. M., G. Yielding, and B. Freeman, eds., 1991, The geometry of normal faults: Geological Society (London) Special Publication 56, 264 p.

Scruton, R. A., ed., 1982, Dynamics of passive margins: American Geophysical Union Geodynamics Series 6, 200 p.

Taylor, B., ed., 1995, Backarc basins: Tectonics and magmatism: New York, Plenum Press, 524 p.

Vine, F. J., and A. G. Smith, eds., 1981, Extensional tectonics associated with convergent plate boundaries: Philosophical Transactions of Royal Society, London, v. A300, special issue, 224 p.

Wilson, R. C. L., R. B. Whitmarsh, B. Taylor, and N.

Froitzheim, eds., 2001, Non-volcanic rifting of continental margins: A comparison of evidence from land and sea: Geological Society (London) Special Publication 187, 586 p.

Ziegler, P. A., ed., 1992, Geodynamics of rifting, 3 volumes: Tectonophysics, v. 208, nos. 1–3, special issue, p. 1–363; Tectonophysics, v. 213, nos. 1–2, special issue, p. 1–284; Tectonophysics, v. 215, nos. 1–2, special issue, p. 1–253.

1.4. Reverse Faults, Compressional Basins, and Active Margins*

Allen, P. A., and P. Homewood, eds., 1986, Foreland basins: International Association of Sedimentologists Special Publication 8, 453 p.

Biddle, K. T., ed., 1991, Active margin basins: AAPG Memoir 52, 324 p.

Cloetingh, S., B. D'Argenio, R. Catalano, F. Horvath, and W. Sassi, 1995, Interplay of extension and compression in basin formation: Tectonophysics, v. 252, special issue, p. 1–484.

Coward, M. P., and A. C. Ries, eds., 1986, Collisional tectonics: Geological Society (London) Special Publication 19, 415 p.

Dorobek, S. L., and G. M. Ross, eds., 1995, Stratigraphic evolution of foreland basins: SEPM Special Publication 52, 303 p.

Legget, J. K., ed., 1982, Trench-forearc geology: Sedimentation and tectonics on modern and ancient active plate margins: Geological Society (London) Special Publication 10, 576 p.

Macqueen, R. W., and D. A. Leckie, eds., 1992, Foreland basins and fold belts: AAPG Memoir 55, 460 p.

McClay, K. R., ed., 1992, Thrust tectonics: London, Chapman & Hall, 447 p.

McClay, K. R., and N. J., Price, eds., 1981, Thrust and nappe tectonics: Geological Society (London) Special Publication 9, 539 p.

Merie, O., 1998, Emplacement mechanics of nappes and thrust sheets: Dordrecht, Kluwer Academic Publishers, 159 p.

Mitra, G., and St. Wojtal, eds., 1988, Geometries and mechanisms of thrusting, with special reference to the Appalachians: Geological Society of America Special Paper 222, 236 p.

Mitra, S., and G. W. Fisher, eds., 1992, Structural geology of fold and thrust belts: Baltimore, John Hopkins University Press, 254 p.

Perry, W. J., D. H. Roeder, and D. R. Lageson, eds., 1984, North American thrust-faulted terranes: AAPG Reprints Series 27, 466 p.

Platt, J. P., M. P. Coward, J. Deramond, and J. Hossack, eds., 1984, Thrusting and deformation: Journal of Structural Geology, v. 8, nos. 3–4, special issue, p. 215–492.

Voight, R., ed., 1976, Mechanics of thrust faults and decollement: Benchmark Papers in Geology 32, Stroudsburg, Pennsylvania: Dowden, Hutchinson & Ross, 473 p.

1.5. Strike-Slip Faults and Wrench Basins*

Ballance, P. F., and H. G. Reading, eds., 1980, Sedimentation in oblique-slip mobile zones: International Association of Sedimentologists Special Publication 4, 265 p.

Biddle, K. T., and N. Christie-Blick, eds., 1985, Strike-slip deformation, basin formation, and sedimentation: SEPM Special Publication 37, 386 p.

Holdsworth, R. E., R. A. Strachan, and J. F. Dewey, eds., 1998, Continental transpressional and transtensional tectonics: Geological Society (London) Special Publication 135, 360 p.

Storti, F., R. E. Holdsworth, and F. Salvini, eds., 2003, Intraplate strike-slip deformation: Geological Society (London) Special Publication 210, 242 p.

Sylvester, A. G., compiler, 1984, Wrench fault tectonics: AAPG Reprint Series 28, 374 p.

Zolnai, G., 1991, Continental wrench-tectonics and hydrocarbon habitat (revised): AAPG Continuing Education Course Note Series 30, variably paginated.

1.6. Faults of Salt Tectonics*

Alsop, G. I., D. J. Blundell, and I. Davison, eds., 1996, Salt tectonics: Geological Society (London) Special Publication 100, 310 p.

Braunstein, J., and G. D. O'Brien, eds., 1968, Diapirism and diapers: AAPG Memoir 8, 444 p.

Cobbold, P. R., ed., 1993, New insights into salt tectonics: Tectonophysics, v. 228, nos. 3–4, special issue, p. 141–448.

Halbouty, M. T., 1979, Salt domes, Gulf region, United States and Mexico, 2d ed.: Houston, Gulf Publishing Co., 561 p.

Jackson, M. P. A., 1997, Conceptual breakthroughs in salt tectonics: A historical review, 1856–1993: Bureau of Economic Geology, University of Texas at Austin, Report of Investigations 226, 51 p.

Jackson, M. P. A., and S. J. Seni, eds., 1984, Atlas of salt domes in the east Texas basin: Bureau of Economic Geology, University of Texas at Austin, Report of Investigations 140, 102 p.

Jackson, M. P. A., and C. J. Talbot, 1991, A glossary of salt tectonics: Bureau of Economic Geology, University of Texas at Austin, Geological Circular 9104, 44 p.

Jackson, M. P. A., and B. C. Vendeville, 1994, Initiation of salt diapirism by regional extension: Global setting, structural style, and mechanical models:

Bureau of Economic Geology, University of Texas at Austin, Report of Investigations 215, 39 p.

Jackson, M. P. A., C. J. Talbot, and R. R. Cornelius, 1988, Centrifuge modeling of the effects of aggradation and progradation on syndepositional salt structures: Bureau of Economic Geology, University of Texas at Austin, Report of Investigations 224, 67 p.

Jackson, M. P. A., R. R. Cornelius, C. H. Craig, A. Gansser, J. Stöcklin, and C. J. Talbot, 1990, Salt diapirs of the Great Kavir, central Iran: Geological Society of America Memoir 177, 139 p.

Jackson, M. P. A., D. G. Roberts, and S. Snelson, eds., 1996, Salt tectonics: A global perspective: AAPG Memoir 65, 454 p.

Jenyon, M. K., 1986, Salt tectonics: London, Elsevier Applied Science Publishers, p. 191.

Lerche, I., and K. Petersen, 1995, Salt and sediment dynamics: Boca Raton, Florida, Chemical Rubber Company Press, 336 p.

Moore, R. C., ed., 1926, Geology of salt dome oil fields: A symposium: Tulsa, AAPG, 874 p.

Schultz-Ela, D. D., M. P. A. Jackson, and B. C. Vendeville, 1994, Mechanics of active salt diapirism: Bureau of Economic Geology, University of Texas at Austin, Report of Investigations 224, 56 p.

Vendeville, B. C., and M. P. A. Jackson, 1992, The rise and fall of diapers during thin-skinned extension: Bureau of Economic Geology, University of Texas at Austin, Report of Investigations 209, 51 p.

1.7. Faults of Inversion Tectonics and Inverted Basins*

Buchanan, J. G., and P. G. Buchanan, eds., 1995, Basin inversion: Geological Society (London) Special Publication 88, 596 p.

Cloetingh, S., Z. Ben-Avraham, W. Sassi, and F. Horvath, eds., 1996a, Dynamics of extensional basin formation and inversion: Tectonophysics, v. 240, special issue, p. 1–323.

Cloetingh, S., O. Eldholm, B. T. Larsen, R. H. Gabrielsen, and W. Sassi, 1996b, Dynamics of extensional basin formation and inversion tectonics: Tectonophysics, v. 266, special issue, p. 1–523.

Cooper, M. A., and G. D. Williams, eds., 1989, Inversion tectonics: Geological Society (London) Special Publication 44, Classics, 375 p.

Zeigler, P. A., ed., 1987, Compressional intra-plate deformations in the Alpine foreland (selected papers from the Symposium on Inverted Mesozoic Sedimentary Basins in the Alpine Foreland, held at European Union of Geoscience Meeting 3, Strasbourg, April 1–4, 1985; International Lithosphere Program Publication No. 0134): Tectonophysics, v. 137, p. 1–420.

1.8. Subsurface Mapping of Faults*

Badgley, P. C., 1959, Structural methods for the exploration geologist: New York, Harper & Row, 280 p.

Bishop, M. S., 1960, Subsurface mapping: New York, John Wiley, 198 p.

Haun, J. D., and L. W. LeRoy, eds., 1958, Subsurface geology in petroleum exploration: Denver, Colorado School of Mines, 887 p.

LeRoy, L. W., and J. W. Low, 1954, Graphical problems in petroleum geology: New York, Harper & Row, 238 p.

LeRoy, L. W., D. O. Leroy, J. W. Raese, and S. D. Schwochow, eds., 1987, Subsurface geologic methods, 5th ed.: Golden, Colorado School of Mines, 1002 p.

Moore, C. A., 1963, Handbook of subsurface geology: New York, Harper & Row, 235 p.

Russell, W. L., 1955, Structural geology for petroleum geologists: New York, McGraw-Hill, 427 p.

Tearpock, D. J., and R. E. Bischke, 2002, Applied subsurface geologic mapping with structural methods, 2d ed.: Englewood Cliffs, Prentice-Hall, 854 p.

1.9. Faults on Seismic Images*

Bally, A. W., ed., 1983a, Seismic expressions of structural styles—A picture and work atlas, v. 1. The layered Earth: AAPG Studies in Geology 15, 195 p.

Bally, A. W., ed., 1983b, Seismic expressions of structural styles—A picture and work atlas, v. 2. Tectonics of extensional provinces: AAPG Studies in Geology 15, 336 p.

Bally, A. W., ed., 1983c, Seismic expressions of structural styles—A picture and work atlas, v. 3. Tectonics of compressional provinces/strike slip tectonics: AAPG Studies in Geology 15, 411 p.

Barazangi, M., and L. Brown, eds., 1986a, Reflection seismology: A global perspective: American Geophysical Union Geodynamic Series 13, 311 p.

Barazangi, M., and L. Brown, eds., 1986b, Reflection seismology: The continental crust: American Geophysical Union Geodynamic Series 14, 339 p.

Carbonell, R., J. Gallart, and M. Torne, eds., 2000, Deep seismic profiling of the continents and their margins: Tectonophysics, v. 329, special issue, p. 1–359.

Clowes, R. M., and A. G. Green, eds., 1994, Seismic reflection probing of the continents and their margins: Tectonophysics, v. 232, special issue, p. 1–450.

Klemperer, S. L., and W. D. Moores, eds., 1998a, Deep seismic profiling of the continents: General results and new methods: Tectonophysics, v. 286, special issue, p. 1–298.

Klemperer, S. L., and W. D. Moores, eds., 1998b, Deep

seismic profiling of the continents: A global survey: Tectonophysics, v. 287, special issue, p. 1–321.

Leven, J. H., D. M. Finlayson, C. Wright, J. C. Dooley, and B. L. N. Kennett, eds., 1990, Seismic probing of continents and their margins: Tectonophysics, v. 173, special issue, p. 1–641.

Matthews, D. H., and C. Smith, eds., 1987, Deep seismic reflection profiling of the continental lithosphere: Geophysical Journal of Royal Astronomical Society, v. 89, special issue, p. 1–447.

Meissner, R., L. Brown, H.-J. Dürbaum, W. Franke, K. Fuchs, and F. Siefert, editors, 1991, Continental lithosphere: Deep seismic reflections: American Geophysical Union Geodynamics Series 22, p. 1–450.

Thybo, H., 2002, Deep seismic probing of the continents and their margins: Tectonophysics, v. 355, special issue, p. 1–263.

Weimer, P., and T. L. Davis, eds., 1996, Applications of 3-D seismic data to exploration and production: AAPG Studies in Geology 42, p. 1–270.

White, D. J., T. J. Bodoky, and Z. Hajnal, eds., 1996, Seismic reflection probing of the continents and their margins: Tectonophysics, v. 264, special issue, p. 1–392.

1.10. Construction of Fault Surface Geometry and Fault Juxtaposition

Allan, U. S., 1989, Model for hydrocarbon migration and entrapment within faulted structures: AAPG Bulletin, v. 73, p. 803–811.

Bailey, R. J., and R. Stonneley, 1981, Petroleum: Entrapment and conclusions, *in* D. H. Tarling, ed., Economic geology and geotectonics: Oxford, Blackwell, p. 73–97.

Barnett, J. A. M., J. Mortimer, J. Rippon, J. J. Walsh, and J. Watterson, 1987, Displacement geometry in the volume containing a single normal fault: AAPG Bulletin, v. 71, p. 925–937.

Brenneke, J. C., 1995, Analysis of fault traps: World Oil, v. 216, no. 12, p. 63–71.

Brown, A. R., G. S. Edwards, and R. E. Howard, 1987, Fault slicing—A new approach to the interpretation of fault detail: Geophysics, v. 52, p. 1319–1327.

Buchanan, P. G., and D. A. Nieuwland, 1996, Modern developments in structural interpretation, validation and modelling: Geological Society (London) Special Publication 99, 369 p.

Childs, C. J., J. J. Walsh, and J. Watterson, 1997, Complexity in fault zone structure and implications for fault seal prediction, *in* P. Møller-Pedersen and A. G. Koestler, eds., Hydrocarbon seals: Importance for exploration and production: Norwegian Petroleum Society Special Publication 7, p. 61–72.

Cooper, M. A., and P. M. Trayner, 1986, Thrust-surface geometry: Implications for thrust belt evolution and section balancing techniques: Journal of Structural Geology, v. 8, p. 305–312.

Dalley, R. M., E. C. A. Gevers, G. M. Stampfli, D. J. Davies, C. N. Gastaldi, P. A. Ruijtenberg, and G. J. O. Vermeer, 1989, Dip and azimuth displays for 3D seismic interpretation: First Break, v. 7, p. 86–95.

Freeman, B., G. Yielding, and M. Badley, 1990, Fault correlation during seismic interpretation: First Break, v. 8, p. 87–95.

Hesthammer, J., and H. Fossen, 1997a, Seismic attribute analysis in structural interpretation of the Gullfaks field, northern North Sea: Petroleum Geoscience, v. 3, p. 13–26.

Hesthammer, J., and H. Fossen, 1997b, The influence of seismic noise on structural interpretation of seismic attribute maps: First Break, v. 4, p. 209–219.

Hoffman, K. S., D. R. Taylor, and R. T. Schnell, 1996, 3-D improves/speeds up fault plane analysis: Leading Edge, v. 15 (February), p. 117–122.

Jackson, P., 1987, The corrugation and bifurcation of fault surfaces by cross-slip: Journal of Structural Geology, v. 9, p. 247–250.

Jones, G., and R. J. Knipe, 1996, Seismic attribute maps; application to structural interpretation and fault seal analysis in the North Sea Basin: First Break, v. 14, p. 449–461.

Knipe, R. J., 1997, Juxtaposition and seal diagrams to help analyze fault seals in hydrocarbon reservoirs: AAPG Bulletin, v. 81, p. 187–195.

Larsen, R. M., H. Brekke, B. T. Larsen, and E. Talleraas, 1992, Structural and tectonic modelling and its application to petroleum geology: Norwegian Petroleum Society Special Publication 1, 549 p.

Lee, J. J., and R. L. Bruhn, 1996, Structural anisotropy of normal-fault surfaces: Journal of Structural Geology, v. 18, p. 1043–1059.

Marchal, D., M. Guiraud, and T. Rives, 2003, Geometric and morphologic evolution of normal fault planes and traces from 2D and 4D data: Journal of Structural Geology, v. 25, p. 135–158.

Nicol, A., J. J. Walsh, J. Watterson, and P. Bretan, 1995, Three dimensional geometry and growth of conjugate normal faults: Journal of Structural Geology, v. 17, p. 847–862.

Nicol, A., J. Watterson, J. J. Walsh, and C. Childs, 1996, The shapes, major axis orientations and displacement patterns of fault surfaces: Journal of Structural Geology, v. 18, p. 235–248.

Rippon, J. H., 1985, Contoured patterns of the throw and hade of normal faults in the Coal Measures (Westphalian) of north-east Derbyshire: Proceeding of the Yorkshire Geological Society, v. 45, p. 147–161.

Roberts, A. M., G. Yielding, and B. Freeman, eds., 1991, The geometry of normal faults: Geological Society (London) Special Publication 56, 264 p.

Roubey, D., H. Xiao, and J. Suppe, 2000, 3-D restoration of complexly folded and faulted surfaces using multiple unfolding mechanisms: AAPG Bulletin, v. 84, p. 805–829.

Walsh, J. J., and J. Watterson, 1987, Distributions of cumulative displacement and seismic slip on a single normal fault surface: Journal of Structural Geology, v. 9, p. 1039–1046.

Walsh, J. J., and J. Watterson, 1990, New methods of fault projection for coal mine planning: Proceedings of the Yorkshire Geological Society, v. 48, p. 209–219.

Watterson, J., 1986, Fault dimensions, displacements and growth: Pure and Applied Geophysics, v. 124, p. 365–373.

Watterson, J., O. R. Clausen, J. A. Peterson, T. McCann, B. M. O'Reilly, P. M. Shannon, C. B. Howard, P. J. Mason, and J. J. Walsh, 1994, Systematics of faults and fault arrays, *in* K. Helbig, ed., Modeling the Earth for oil exploration: Oxford, Pergamon Press, p. 205–316.

White, N. J., J. A. Jackson, and D. P. McKenzie, 1986, The relationship between the geometry of normal faults and that of the sedimentary layers in their hanging walls: Journal of Structural Geology, v. 8, p. 897–909.

Williams, G., and I. Vann, 1987, The geometry of listric normal faults and deformation in their hanging walls: Journal of Structural Geology, v. 9, p. 789–795.

2. FAULT SEALING PROCESSES

2.1. Sealing Mechanisms: General Aspects

Berg, R. R., 1975, Capillary pressure in stratigraphic traps: AAPG Bulletin, v. 59, p. 939–956.

Bjørkum, P. A., O. Walderhaug, and P. H. Nadeau, 1998, Physical constraints on hydrocarbon leakage and trapping revisited: Petroleum Geoscience, v. 43, p. 239–273.

Boult, P. J., 1993, Membrane seal and tertiary migration pathways in the Bodalla South oilfield, Eromanga Basin, Australia: Marine and Petroleum Geology, v. 10, p. 3–13.

Downey, M. W., 1984, Evaluating seals for hydrocarbon accumulations: AAPG Bulletin, v. 68, p. 1752–1763.

Downey, M. W., 1994, Hydrocarbon seal rocks, *in* L. B. Magoon and W. G. Dow, eds., The petroleum system—From source to trap: AAPG Memoir 60, p. 159–164.

England, W. A., 1994, Secondary migration and accumulation of hydrocarbons, *in* L. B. Magoon and W. G. Dow, eds., The petroleum system—From source to trap: AAPG Memoir 60, p. 211–217.

England, W. A., A. S. Mackenzie, D. M., Mann, and T. M. Quigley, 1987, The movement and entrapment of petroleum fluids in the subsurface: Journal of the Geological Society (London), v. 144, p. 327–347.

Grunau, H. R., 1987, A worldwide look at the cap-rock problem: Journal of Petroleum Geology, v. 10, p. 245–266.

Gussow, W. C., 1954, Differential entrapment of oil and gas: A fundamental principle: AAPG Bulletin, v. 38, p. 816–853.

Heum, O. R., 1996, A fluid dynamic classification of hydrocarbon entrapment: Petroleum Geoscience, v. 2, p. 145–158.

Hubbert, M. K., 1953, Entrapment of petroleum under hydrodynamic conditions: AAPG Bulletin, v. 37, no. 8, p. 1954–2026.

Ingram, G. M., J. L. Urai, and M. A. Naylor, 1997, Sealing processes and top seal assessment, *in* P. Møller-Pedersen and A. G. Koestler, eds., 1997, Hydrocarbon seals: Importance for exploration and production: Norwegian Petroleum Society Special Publication 7, p. 165–174.

Jennings, J. B., 1987, Capillary pressure techniques: Application to exploration and development geology: AAPG Bulletin, v. 71, p. 1106–1209.

Myers, J., 1968, Differential pressures—A trapping mechanism in Gulf Coast oil and gas fields: Transactions of the Gulf Coast Association of Geological Studies, v. 18, p. 56–80.

Sales, J. K., 1993, Closure vs. sealing strength—A fundamental control on the distribution of oil and gas, *in* A. G. Dore, J. H. Auguston, C. Hermanrud, O. Sylta, and D. J. Stewart, eds., Basin modelling: Advances and applications: Norwegian Petroleum Society Special Publication 3, p. 399–414.

Sales, J. K., 1997, Seal strength vs. trap closure—A fundamental control on the distribution of oil and gas, *in* R. C. Surdam, ed., Seals, traps, and the petroleum system: AAPG Memoir 67, p. 57–83.

Schowalter, T. T., 1979, Mechanisms of secondary hydrocarbon migration and entrapment: AAPG Bulletin, v. 63, p. 723–760.

Vavra, C. L., J. G. Kaldi, and R. M. Sneider, 1992, Geological application of capillary pressure: A review: AAPG Bulletin, v. 76, p. 840–850.

Willis, D. G., 1961, Entrapment of petroleum, *in* G. B. Moody, ed., Petroleum exploration handbook: New York, McGraw-Hill, p. 61–68.

Ziegler, D. L., 1992, Hydrocarbon columns, buoyancy

pressures, and seal efficiency: Comparisons of oil and gas accumulations in California and the Rocky mountain area: AAPG Bulletin, v. 76, p. 501–508.

2.2. Fault Rock and Fault Sealing Processes: General Aspects

Aki, K., 1979, Characteristics of barriers on earthquake faults: Journal of Geophysical Research, v. 84, p. 6140–6148.

Aydin, A., 2000, Fractures, faults, and hydrocarbon migration and flow: Marine and Petroleum Geology, v. 17, p. 797–814.

Beach, A., J. L. Brown, P. J. Brockbank, S. D. Knott, J. E. McCallum, and A. O. Welbon, 1997, Fault seal analysis of SE Asian basins with examples from west Java, *in* A. J. Fraser, S. J. Matthews, and R. W. Murphy, eds., Petroleum geology of southeast Asia: Geological Society (London) Special Publication 126, p. 185–194.

Blecker, W., C. Elliot, S. Lin, and C. R. Van Staal, eds., 1998, Evolution of structures in deforming rocks: Journal of Structural Geology, v. 23, nos. 6–7, special issue, p. 843–1178.

Boland, J., and A. Ord, eds., 2001, Deformation processes in the Earth's crust: Tectonophysics, v. 335, nos. 1–2, special issue, p. 1–228.

Brandon, M. T., J. R. Henderson, W. D. Means, and S. Paterson, eds., 1994, Applications of strain: From microstructures to orogenic belts: Journal of Structural Geology, v. 16, no. 4, special issue, p. 437–612.

Broussard, M. J., and B. E. Lock, 1995, Modern analytical techniques for fault surface seal analysis: A Gulf Coast case history: Gulf Coast Association of Geological Societies Transactions, v. XLV, p. 87–93.

Brown, A., 2003, Capillary effects on fault-fill sealing: AAPG Bulletin, v. 87, p. 381–396.

Burg, J.-P., D. Mainprice, and J. P. Petit, eds., 1992, Mechanical instabilities in rocks and tectonics: Journal of Structural Geology, v. 14, nos. 8–9, special issue, p. 893–1109.

Carreras, J., P. R. Cobbold, J. G. Ramsay, and S. H. White, eds., 1980, Shear zone in rocks: Journal of Structural Geology, v. 2, nos. 1–2, special issue, p. 1–287.

Chester, F. M., T. Engelder, and T. Shimamoto, eds., 1998, Rock deformation: The Logan volume: Tectonophysics, v. 295, nos. 1–2, special issue, p. 1–257.

Cobbold, P. R., and W. M. Schwerdtner, 1983, Strain patterns in rocks: Journal of Structural Geology, v. 5, nos. 3–4, special issue, p. 255–470.

Cobbold, P. R., D. Gapais, W. D. Means, and S. H. Treagus, eds., 1987, Shear strain in rocks: Journal of Structural Geology, v. 9, nos. 5–6, special issue, p. 521–778.

Coward, M. P., T. S. Daltaban, and H. Johnson, eds., 1998, Structural geology in reservoir characterization: Geological Society (London) Special Publication 127, 266 p.

Dainelli, J., and A. Vignolo, 1993, Regional assessment of sealing faults through pore-pressure analysis in the northern Adriatic: First Break, v. 11, p. 287–294.

De Meer, S., M. R. Drury, J. H. P. DePresser, and G. M. Pennock, eds., 2002, Deformation mechanisms, rheology and tectonics: Current status and future perspectives: Geological Society (London) Special Publication 200, 424 p.

Dincau, A. R., 1998, Prediction and timing of production induced fault seal breakdown in the south Marsh island 66 gas field: Gulf Coast Association of Geological Societies Transactions, v. 48, p. 21–32.

Duba, A. G., W. B. Durham, J. W. Handin, and H. F. Wu, eds., 1990, The brittle-ductile transition in rocks ("Heard volume"): American Geophysical Union Geophysical Monograph 56, 243 p.

Edwards, H. E., A. D. Becker, and J. A. Howell, 1993, Compartmentalisation of an eolian sandstone by structural heterogeneity: Permo-Triassic Hopeman Sandstone, Moray Firth, Scotland, *in* C. P. North and D. J. Prosser, eds., Characterisation of fluvial and aeolian reservoirs: Geological Society (London) Special Publication 73, p. 339–365.

Fisher, Q. J., S. D. Harris, E. MacAllister, R. J. Knipe, and A. J. Bolton, 2001, Hydrocarbon flow across faults by capillary leakage revisited: Marine and Petroleum Geology, v. 18, p. 251–257.

Galehouse, J. S., 1990, Effect of the Loma Prieta earthquake on surface slip along the Calaveras fault in the Hollister area: Geophysical Research Letters, v. 17, p. 1219–1222.

Griggs, D. T., and J. Handin, eds., 1960, Rock deformation: Geological Society of America Memoir 79, 382 p.

Groshong, R. H., 1988, Low temperature deformation mechanisms and their interpretation: Geological Society of America Bulletin, v. 100, p. 1329–1360.

Hancock, P. C., and C. McA. Powell, eds., 1984, Multiple deformations in ductile and brittle rocks: Journal of Structural Geology, v. 7, nos. 3–4, special issue, p. 269–501.

Hancock, P. L., E. M. Kalper, N. S. Macktelow, and J. G. Ramsay, eds., 1984, Planar and linear fabrics of deformed rocks: Journal of Structural Geology, v. 6, nos. 1–2, special issue, p. 1–287.

Haneberg, W. C., P. S. Mozley, J. C. Moore, and L. B. Goodwin, eds., 1999, Faults and subsurface fluid flow in the shallow crust: American Geophysical Union Geophysical Monograph 113, 222 p.

Heard, H. C., I. Y. Borg, N. L. Carter, C. B. Raleigh, eds., 1972, Flow and fracture of rocks ("Griggs volume"): American Geophysical Union Geophysical Monograph 16, 352 p.

Hesthammer, J., and H. Fossen, 2000, Uncertainties associated with fault sealing analysis: Petroleum Geoscience, v. 6, p. 37–45.

Hesthammer, J., P. A. Bjørkum, and L. Watts, 2002, The effect of temperature on sealing capacity of faults in sandstone reservoirs: Examples from the Gullfaks and Gullfaks Sor fields, North Sea: AAPG Bulletin, v. 86, p. 1733–1751.

Hobbs, B. E., and H. C. Heard, eds., 1986, Mineral and rock deformation: Laboratory studies: American Geophysical Union Geophysical Monograph 36, 324 p.

Holdsworth, R. E., R. A. Strachan, J. F. Magloughlin, and R. J. Knipe, eds., 2001, The nature and tectonic significance of fault zone weakening: Geological Society (London) Special Publication 186, 342 p.

Iliffe, J. E., I. Lerche, and K. Nakayama, 1992, Physical changes in fault blocks caused by rotation and their implications for hydrocarbon accumulations, *in* R. M. Larsen, H. Brekke, B. T. Larsen, and E. Talleraas, eds., Structural and tectonic modeling and its application to petroleum geology: Norwegian Petroleum Society Special Publication 1, p. 253–267.

Johnson, S. E., and M. L. Williams, eds., 2002, Microstructural processes: Journal of Structural Geology, v. 24, nos. 6–7, special issue, p. 997–1214.

Jones, M. E., and M. F. Preston, eds., 1987, Deformation of sediments and sedimentary rocks: Geological Society (London) Special Publication 29, 350 p.

Jones, R. M., and R. R. Hillis, 2003, An integrated, quantitative approach to assessing fault-seal risk: AAPG Bulletin, v. 87, p. 507–524.

Jones, G., Q. J. Fisher, and R. J. Knipe, eds., 1998, Faulting, fault sealing and fluid flow in hydrocarbon reservoirs: Geological Society (London) Special Publication 147, 320 p.

Jones, R. M., P. Boult, R. R. Hillis, S. D. Mildren, and J. Kaldi, 2000, Integrated hydrocarbon seal evaluation in the Penola trough, Otway basin: Australian Petroleum Production and Exploration Association Journal, v. 20, p. 194–212.

Kirby, S. H., A. G. Sylvester, J. Tullis, and H. R. Wenk, eds., 1991, Microstructures and rheology of rocks and rock-forming minerals: Journal of Structural Geology, v. 15, nos. 9–10, special issue, p. 1061–1271.

Knipe, R. J., 1986, Deformation mechanism path diagrams for sediments undergoing lithification, *in* J. C. Moore, ed., Structural fabrics in Deep Sea Drilling Project cores from forearcs: Geological Society of America Memoir 166, p. 151–160.

Knipe, R. J., 1989, Deformation mechanisms—Recognition from natural tectonites: Journal of Structural Geology, v. 11, p. 127–146.

Knipe, R. J., 1992, Faulting processes and fault seal, *in* R. M. Larsen, H. Brekke, B. T. Larsen, and E. Talleraas, eds., Structural and tectonic modelling and its application to petroleum geology: Norwegian Petroleum Society Special Publication 1, p. 325–342.

Knipe, R. J., 1993, The influence of fault zone processes and diagenesis on fluid flow, *in* A. D. Horbury and A. G. Robinson, eds., Diagenesis and basin development: AAPG Studies in Geology 36, p. 135–154.

Knipe, R. J., and A. M. McCaig, 1994, Microstructural and microchemical consequences of fluid flow in deforming rocks, *in* J. Parnell, ed., Geofluids: Origin, migration and evolution of fluids in sedimentary basins: Geological Society (London) Special Publication 78, p. 99–111.

Knipe, R. J., and E. H. Rutter, eds., 1990, Deformation mechanics, rheology and tectonics: Geological Society (London) Special Publication 54, 535 p.

Knipe, R. J., S. M. Agar, and D. J. Prior, 1991, The microstructural evolution of fluid flow paths in semi-lithified sediments from subduction complexes: Philosophical Transactions of the Royal Society of London, v. 355A, p. 261–273.

Knipe, R. J., Q. J. Fisher, G. Jones, M. R. Clennell, A. B. Farmer, A. Harrison, B. Kidd, E. McAllister, J. R. Porter, and E. A. White, 1997, Fault seal analysis: Successful methodologies, application and future directions, *in* P. Møller-Pedersen and A. G. Koestler, eds., Hydrocarbon seals: Importance for exploration and production: Norwegian Petroleum Society Special Publication 7, p. 15–40.

Knott, S. D., 1993, Fault seal analysis in the North Sea: AAPG Bulletin, v. 77, p. 778–792.

Koestler, A. G., and R. Hunsdale, eds., 2002, Hydrocarbon seal quantification: Norwegian Petroleum Society Special Publication 11, 263 p.

Leiss, B., K. Ullemeyer, and K. Weber, eds., 2000, Textural and physical properties of rocks: Journal of Structural Geology, v. 22, nos. 11–12, special issue, p. 1527–1873.

Lister, G. S., P. F. Williams, H. J. Zwart, and R. J. Lisle, eds., 1977, Fabrics, microstructures and microtectonics: Tectonophysics, v. 39, special issue, p. 1–487.

Lister, G. S., H.-J. Behr, K. Weber, and H. J. Zwart, eds., 1981, The effect of deformation on rocks: Tectonophysics, v. 78, special issue, p. 1–698.

Magnavita, L. P., 2000, Deformation mechanisms in porous sandstones: Implications for development of fault seal and migration paths in the Reconcave basin, Brazil, *in* M. R. Mello and B. J. Katz, eds., Petroleum systems of south Atlantic margins: AAPG Memoir 73, p. 195–212.

Maltman, A., ed., 1994, The geological deformation of sediments: London, Chapman & Hall, 362 p.

McClay, R. W., and T. A. Small, 1983, Hydrostratigraphic subdivisions and fault barriers of the Edwards aquifer, south-central Texas, U.S.A.: Journal of Hydrology, v. 61, p. 127–146.

Mildren, S. D., R. R. Hillis, and J. Kaldi, 2002, Calibrating predictions of fault seal reactivation in the Timor Sea: Australian Petroleum Production and Exploration Association Journal, v. 42, p. 187–202.

Mitra, S., 1988, Effects of deformation mechanisms on reservoir potential in central Appalachian overthrust belt: AAPG Bulletin, v. 72, p. 536–554.

Møller-Pedersen, P., and A. G. Koestler, eds., 1997, Hydrocarbon seals: Importance for exploration and production: Norwegian Petroleum Society Special Publication 7, 250 p.

Nybakken, S., 1991, Sealing fault traps—An exploration concept in a mature petroleum province: Tampen Spur, northern North Sea: First Break, v. 9, p. 209–222.

Omre, H., K. Sølna, N. Dahl, and B. Tørudbakken, 1994, Impact of fault heterogeneity in fault zones on fluid flow: North Sea oil and gas reservoirs III, Norwegian Institute of Technology: Dordrecht, Kluwer Academic Publishers, p. 185–200.

Ord, A., ed., 1989, Deformation of crustal rocks: Tectonophysics, v. 158, special issue, p. 1–354.

Ortoleva, P. J., ed., 1994, Basin compartments and seals: AAPG Memoir 61, 477 p.

Passchier, C. W., and R. A. J. Trouw, 1996, Microtectonics: Berlin, Springer-Verlag, 289 p.

Paterson, M. S., 1978, Experimental rock deformation: The brittle field: Berlin, Springer-Verlag, 254 p.

Richardson, S. M., N. Vivian, R. J. Cook, M. Wilkes, and H. Hussein, 1998, Application of fault seal analysis techniques in the western desert, Egypt, *in* D. S. MacGregor, R. T. J. Moody, and D. D. Clark-Lowes, eds., Petroleum geology of north Africa: Geological Society (London) Special Publication 132, p. 297–315.

Rudnicki, J. W., and T. C. Hzu, 1988, Pore pressure changes induced by slip on permeable and impermeable faults: Journal of Geophysical Research, v. 93B, p. 3275–3285.

Russell, K. D., and J. W. Handschy, eds., 2003, Fault seals: AAPG Bulletin, v. 87, no. 3, theme issue, p. 377–527.

Rutter, E. H., 1986, On the nomenclature of mode of failure transitions in rocks: Tectonophysics, v. 122, p. 381–387.

Rutter, E. H., A. Boriani, K. H. Brodie, and L. Burlini, eds., 1998, Structures and properties of high strain zones in rocks: Journal of Structural Geology, v. 20, nos. 2–3, special issue, p. 111–320.

Schmid, S. M., and M. R. Handy, 1991, Towards a genetic classification of fault rocks: Geological usage and tectonophysical implications, *in* D. W. Müller, J. McKenzie, and H. Weissert, eds., Controversies in modern geology: New York, Academic Press, p. 339–361.

Schmid, S. M., R. Heilbronner, and H. Stünitz, eds., 1999, Deformation mechanisms in nature and experiment: Tectonophysics, v. 303, special issue, p. 1–319.

Sibson, R. H., 1977, Fault rocks and fault mechanisms: Journal of the Geological Society, v. 133, p. 199–213.

Sibson, R. H., 1986, Earthquakes and rock deformation in crustal fault zones: Annual Reviews of Earth and Planetary Sciences, v. 14, p. 149–175.

Skerlec, G. M., 1999, Evaluating top and fault seal, *in* E. A. Beaumont and N. H. Foster, eds., Exploring for oil and gas traps: AAPG Treatise of Petroleum Geology, chapter 10, p. 1–94.

Sleep, N. H., and M. L. Blanpied, 1994, Ductile creep and compaction: A mechanism for transiently increasing fluid pressure in mostly sealed fault zones: Pure and Applied Geophysics, v. 143, p. 9–40.

Smith, D. A., 1966, Theoretical considerations of sealing and non-sealing faults: AAPG Bulletin, v. 50, p. 363–374.

Smith, D. A., 1980, Sealing and non-sealing faults in Louisiana Gulf Coast salt basin: AAPG Bulletin, v. 64, p. 145–172.

Snoke, A. W., J. Tullis, and V. R. Todd, 1998, Fault-related rocks: A photographic atlas: Princeton, Princeton University Press, 618 p.

Spencer, A. M., and V. B. Larsen, 1990, Fault traps in the northern North Sea, *in* R. F. P. Hardman and J. Brooks, eds., Tectonic events responsible for Britain's oil and gas reserves: Geological Society (London) Special Publication 55, p. 281–298.

Spiers, C. J., and T. Takeshita, eds., 1995, Influence of fluids on deformation processes in rocks: Tectonophysics, v. 254, nos. 3–4, special issue, p. 121–297.

Spray, J. G., and P. J. Hudleston, eds., 1989, Friction phenomena in rock: Journal of Structural Geology, v. 11, no. 7, special issue, p. 783–931.

Surdam, R. C., ed., 1997, Seals, traps, and the petroleum system: AAPG Memoir 67, 317 p.

Treadwell, D. D., J. D. Rietman, K. D. Weaver, and G. J. Murphy, 1980, Offshore faulting in the Gulf of the Farallones: Offshore Technology Conference Proceedings 12, v. 4, p. 611–616.

Trevena, A. S., 1989, Small faults as barriers to permeability in sandstone gas reservoirs, Pattani Basin, Thailand: AAPG Bulletin, v. 73, p. 553.

Wallbercher, E., W. Unzong, and M. Brandmayr, eds., 1994, Structures and tectonics at different lithospheric levels: Journal of Structural Geology, v. 16, p. 1495–1575.

Watts, N. L., 1987, Theoretical aspects of cap-rock and fault seals for single- and two-phase hydrocarbon columns: Marine and Petroleum Geology, v. 4, p. 274–307.

Weber, K. J., 1986, How heterogeneity affects oil recovery, *in* L. W. Lake and H. B. Carroll, eds., Reservoir characterisation: Orlando, Academic Press, p. 487–544.

Weber, K. J., 1997, A historical overview of the efforts to predict and quantify hydrocarbon trapping features in the exploration phase and in field development planning, *in* P. Møller-Pedersen and A. G. Koestler, eds., Hydrocarbon seals: Importance for exploration and production: Norwegian Petroleum Society Special Publication 7, p. 1–13.

Wehr, F. L., L. H. Fairchild, M. R. Hudec, R. K. Shafto, W. T. Shea, and J. P. White, 2000, Fault seal: Contrasts between the exploration and production problem, *in* M. R. Mello and B. J. Katz, eds., Petroleum systems of south Atlantic margins: AAPG Memoir 73, p. 121–132.

Weiss, L. E., 1972, Minor structures of deformed rocks: A photographic atlas: Berlin, Springer-Verlag, 431 p.

Wise, D. U., D. E. Dunn, J. T. Engelder, P. A. Geiser, R. D. Hatcher, S. A. Kish, A. L. Odom, and S. Schamel, 1984, Fault-related rocks: Suggestions for terminology: Geology, v. 12, p. 391–394.

Ziegler, D. L., 1992, Hydrocarbon columns, buoyancy pressures, and seal efficiency: Comparisons of oil and gas accumulations in California and the Rocky Mountain area: AAPG Bulletin, v. 76, p. 501–508.

Zwart, H. J., M. Martens, I. Van Der Molen, C. W. Passchier, C. J. Spiers, and R. L. M. Vissers, eds., 1987, Tectonic and structural processes on a macro-, meso- and micro-scale: Tectonophysics, v. 135, special issue, p. 1–251.

2.3. Cataclastic Rocks

Babaie, H. A., J. Hadizadeh, and A. Babaei, 1991, Timing and temperature of cataclastic deformation along segments of the Towaliga fault zone, western Georgia: Journal of Structural Geology, v. 13, p. 579–586.

Babaie, H. A., J. Hadizadeh, and A. Babaei, 1995, Self-similar cataclasis in the Saltville thrust zone, Knoxville, Tennessee: Journal of Geophysical Research, v. 100B, p. 18,075–18,084.

Blenkinsop, T. G., 1991, Cataclasis and processes of particle size reduction: Pure and Applied Geophysics, v. 136, p. 60–86.

Blenkinsop, T. G., and E. H. Rutter, 1986, Cataclastic deformation of quartzite in the Moine thrust zone: Journal of Structural Geology, v. 8, p. 669–682.

Chester, F. J., and J. M. Logan, 1986, Implications for mechanical properties of brittle faults from observations of the Punchbowl fault zone, California: Pure and Applied Geophysics, v. 124, p. 80–106.

Chester, F. M., M. Friedman, and J. Logan, 1985, Foliated cataclasites: Tectonophysics, v. 111, p. 139–146.

Chester, F. M., J. P. Evans, and R. L. Biegel, 1993, Internal structure and weakening mechanisms of the San Andreas fault: Journal of Geophysical Research, v. 98B, p. 771–786.

Dewhurst, D. N., M. R. Jones, R. R. Hillis, and S. D. Mildren, 2002, Microstructural and geomechanical characterisation of fault rocks from the Carnrvon and Otway basins: Australian Petroleum Production and Exploration Association Journal, v. 42, p. 167–186.

Draper, G., 1976, Grain size as an indication of shear strain in brittle fault zones: Tectonophysics, v. 35, p. T7–T13.

Dresen, G., and M. Handy, eds., 2001, Deformation mechanisms, rheology and microstructures: International Journal of Earth Sciences (Geologische Rundschau), v. 90, p. 1–210.

Dunn, D. E., L. J. LaFountain, and R. E. Jackson, 1983, Porosity dependence and mechanism of brittle fracture in sandstones: Journal of Geophysical Research, v. 78B, p. 2403–2417.

Engelder, J. T., 1974, Cataclasis and the generation of fault gouge: Geological Society of America Bulletin, v. 85, p. 1515–1522.

Fischer, U., B. Kulli, and H. Fluhler, 1998, Constitutive relationships and pore structure of undisturbed fracture zone samples with cohesionless fault gouge layers: Water Resources Research, v. 34, p. 1695–1701.

Hadizadeh, J., 1994, Interaction of cataclasis and pressure solution in a low-temperature carbonate shear zone: Pure and Applied Geophysics, v. 143, p. 255–280.

Hadizadeh, J., and E. H. Rutter, 1982, Experimental study of cataclastic deformation in a quartzite: Proceedings of 3rd U.S. Symposium on Rock Mechanics, Berkeley, Society of Mining Engineers, p. 372–379.

Hadizadeh, J., and E. H. Rutter, 1983, The low temperature brittle ductile transition in a quartzite and the occurrence of cataclastic flow in nature: Geologische Rundschau, v. 72, p. 493–509.

Hadizadeh, J., H. A. Babaie, and A. A. Babaei, 1991, Development of interlaced mylonites, cataclasites and breccias: Examples from the Towaliga fault zone, south-central Appalachians: Journal of Structural Geology, v. 13, p. 63–70.

Higgins, M. W., 1971, Cataclastic rocks: U.S. Geological Survey Professional Paper 687, 97 p.

House, W. M., and D. R. Gray, 1982, Cataclasites along the Saltville thrust, U.S.A., and their implications for thrust-sheet emplacement: Journal of Structural Geology, v. 4, p. 257–269.

Lade, P. V., J. A. Yamamuro, and P. A. Bopp, 1996, Significance of particle crushing in granular materials: Journal of Geotechnical Engineering, v. 122, p. 309–316.

Lloyd, G. E., and R. J. Knipe, 1992, Deformation mechanisms accommodating faulting of quartzite under upper crustal conditions: Journal of Structural Geology, v. 14, p. 127–144.

Logan, J. M., M. Friedman, N. G. Higgs, C. Dengo, and T. Shimamoto, 1979, Experimental studies of simulated gouge and their application to studies of natural fault zones: Proceedings of Conference VIII: Analysis of actual fault zones in bedrock: U.S. Geological Survey Open-file Report 70-1239, p. 305–343.

Logan, J. M., C. A. Dengo, N. G. Higgs, and Z. Z. Wang, 1992, Fabrics of experimental fault zones: Their development and relationship to mechanical behavior, *in* B. Evans and T.-F. Wong, eds., Fault mechanics and transport properties of rocks: New York, Academic Press, p. 33–68.

Magloughlin, J. F., 1992, Microstructural and chemical changes associated with cataclasis and frictional melting at shallow crustal levels: The cataclasite-pseudotachylyte connection: Tectonophysics, v. 204, p. 243–260.

Mandl, G., L. N. J. de Jong, and A. Maltha, 1977, Shear zones in granular material (an experimental study of their structure and mechanical genesis): Rock Mechanics, v. 9, p. 95–114.

Marone, C., and C. H. Shultz, 1989, Particle-size distribution and microstructures within simulated fault gouge: Journal of Structural Geology, v. 11, p. 799–814.

McCaig, A. M., and R. J. Knipe, 1990, Mass-transport mechanisms in deforming rocks: Recognition using microstructural and microchemical criteria: Geology, v. 18, p. 824–827.

Meglis, I., and T. Engelder, 1994, The mechanical properties of rock through an ancient transition zone in the Appalachian basin, *in* P. J. Ortoleva, ed., Basin compartments and seals: AAPG Memoir 61, p. 459–469.

Menéndez, B., W. Zhu, and T.-F. Wong, 1996, Micromechanics of brittle faulting and cataclastic flow in Berea Sandstone: Journal of Structural Geology, v. 18, p. 1–16.

Mitra, G., J. M. Hull, W. A. Yonkee, and G. M. Protzman, 1988, Comparison of mesoscopic and microscopic deformational styles in the Idaho–Wyoming thrust belt and the Rocky Mountain foreland, *in* C. J. Schmidt and W. J. Perry Jr., eds., Interaction of the Rocky Mountain foreland and Cordilleran thrust belt: Geological Society of America Memoir 171, p. 119–141.

Rutter, E. H., 1974, The influence of temperature, strain rate and interstitial water in the experimental deformation of calcite rocks: Tectonophysics, v. 22, p. 311–334.

Rutter, E. H., and J. Hadizadeh, 1991, On the influence of porosity on the low temperature brittle-ductile transition in siliciclastic rocks: Journal of Structural Geology, v. 13, p. 609–614.

Rutter, E. H., and S. H. White, 1979, The microstructures and rheology of fault gouges produced experimentally under wet and dry conditions at temperatures up to 400 degrees centigrade: Bulletin de Mineralogie, v. 102, p. 102–109.

Rutter, E. H., R. H. Maddock, S. H. Hall, and S. H. White, 1986, Comparative microstructures of natural and experimentally produced clay-bearing fault gouges: Pure and Applied Geophysics, v. 124, p. 3–30.

Sammis, C. G., R. H. Osborne, J. L. Anderson, M. Banerdt, and P. White, 1986, Self-similar cataclasis in the formation of fault gouge: Pure and Applied Geophysics, v. 124, p. 191–213.

Sammis, C., G. King, and R. Biegel, 1987, The kinematics of gouge deformation: Pure and Applied Geophysics, v. 125, p. 777–812.

Scholz, C. H., 1987, Wear and gouge formation in brittle fracturing: Geology, v. 15, p. 493–495.

Scott, T. E., and K. C. Nielson, 1991, The effects of porosity on the brittle-ductile transition in sandstones: Journal of Geophysical Research, v. 96B, p. 405–414.

Shimamoto, T., and J. M. Logan, 1981, Effects of simulated fault gouge on the sliding behavior of Tennessee sandstone, nonclay gouges: Journal of Geophysical Research, v. 86B, p. 2902–2914.

Sibson, R. H., 1986, Brecciation processes in fault zones—Inferences from earthquake rupturing: Pure and Applied Geophysics, v. 124, p. 159–175.

Simpson, C., 1986, Fabric development in brittle-to-ductile shear zones: Pure and Applied Geophysics, v. 124, p. 269–288.

Stel, H., 1985, Crystal growth in cataclasites: Diagnostic microstructures and implications: Tectonophysics, v. 78, p. 585–600.

Tanaka, H., 1992, Cataclastic lineations: Journal of Structural Geology, v. 14, p. 1239–1252.

Vrolijk, P., and B. A. van der Pluijm, 1999, Clay gouge: Journal of Structural Geology, v. 21, p. 1039–1048.

Wallace, R. E., and H. T. Morris, 1986, Characteristics of faults and shear zones in deep mines: Pure and Applied Geophysics, v. 124, p. 107–126.

White, S. H., 1976, The effects of strain on the microstructures, fabrics and deformation mechanisms in quartzites: Philosophical Transactions of the Royal Society of London, v. 283A, p. 69–86.

White, J. C., and S. H. White, 1983, Semi-brittle deformation within the Alpine fault zone, New Zealand: Journal of Structural Geology, v. 5, p. 579–589.

Wong, T.-F., C. David, and W. Zhu, 1997, The transition from brittle faulting to cataclastic flow in porous sandstones: Mechanical deformation: Journal of Geophysical Research, v. 102B, p. 3009–3025.

Wotjal, S., and G. Mitra, 1988, Nature of deformation in some fault rocks from Appalachian thrusting, *in* G. Mitra and S. Wotjal, eds., Geometries and mechanisms of thrusting with special reference to the Appalachians: Geological Society of America Special Paper 222, p. 17–33.

Zhang, J., T.-F. Wong, and D. M. Davis, 1990, Micromechanics of pressure-induced grain crushing in porous rocks: Journal of Geophysical Research, v. 95B, p. 341–352.

2.4. Shale Smear in Clastic Sequences

Alexander, L. L., and J. W. Handschy, 1998, Fluid flow in a faulted reservoir system: Fault trap analysis for the Block 330 field in Eugene island, south Addition, offshore Louisiana: AAPG Bulletin, v. 82, p. 387–411.

Aydin, A., and Y. Eyal, 2002, Anatomy of a normal fault with shale smear: Implications for fault seal: AAPG Bulletin, v. 86, p. 1367–1381.

Bentley, M. R., and J. J. Barry, 1991, Representation of fault sealing in a reservoir simulation: Cormorant block IV UK North Sea: 66th Annual Technical Conference & Exhibition of the Society of Petroleum Engineers, Dallas, Texas, p. 119–126.

Bouvier, J. D., C. H. Kaars-Sijpersteijn, D. F. Kluesner, C. C. Onyejekwe, and R. C. van der Pal, 1989, Three-dimensional seismic interpretation and fault sealing investigations, Nun River field, Nigeria: AAPG Bulletin, v. 73, p. 1397–1414.

Bretan, P., Yielding, G., and Jones, H., 2003, Using calibrated shale gouge ratio to estimate hydrocarbon column heights: AAPG Bulletin, v. 87, p. 396–414.

Clausen, J. A., and R. H. Gabrielsen, 2002, Parameters that control the development of clay smear at low stress states: An experimental study using ring-shear apparatus: Journal of Structural Geology, v. 24, p. 1569–1586.

Davies, R. K., L. An, P. Jones, A. Mathis, and C. Cornette, 2003, Fault-seal analysis South Marsh island 36 field, Gulf of Mexico: AAPG Bulletin, v. 87, p. 479–492.

Doughthy, P. T., 2003, Clay smear seals and fault sealing potential of an exhumed growth fault, Rio Grande rift, New Mexico: AAPG Bulletin, v. 87, p. 427–444.

Freeman, B., G. Yielding, D. T. Needham, and M. E. Badley, 1998, Fault seal prediction: The gouge ratio method, *in* M. P. Coward, T. S. Daltaban, and H. Johnson, eds., Structural geology in reservoir characterization: Geological Society (London) Special Publication 127, p. 19–25.

Fristad, T., A. Groth, G. Yielding, and B. Freeman, 1997, Quantitative fault seal prediction: A case study from Oseberg Syd., *in* P. Møller-Pedersen and A. G. Koestler, eds., Hydrocarbon seals: Importance for exploration and production: Norwegian Petroleum Society Special Publication 7, p. 107–124.

Fulljames, J. R., L. J. J. Zijerveld, R. C. M. W. Franssen, G. M. Ingram, and P. D. Richard, 1997, Fault seal processes, *in* P. Møller-Pedersen and A. G. Koestler, eds., Hydrocarbon seals: Importance for exploration and production: Norwegian Petroleum Society Special Publication 7, p. 51–59.

Gibson, R. G., 1994, Fault-zone seals in siliclastic strata of the Columbus basin, offshore Trinidad: AAPG Bulletin, v. 78, p. 1372–1385.

Gibson, R. G., 1998, Physical character and fluid-flow properties of sandstone-derived fault zones, *in* M. P. Coward, T. S. Daltaban, and H. Johnson, eds., Structural geology in reservoir characterization: Geological Society (London) Special Publication, v. 127, p. 83–97.

Gibson, R. G., and P. A. Bentham, 2003, Use of fault-seal analysis in understanding petroleum migration in a complexly faulted anticlinal trap, Columbus basin, offshore Trinidad: AAPG Bulletin, v. 87, p. 465–478.

Harris, D., G. Yielding, P. Levine, G. Maxwell, P. T. Rose, and P. Nell, 2002, Using shale gouge ratio (SGR) to model faults as transmissibility barriers in reservoirs: An example from the Strathspey field, North Sea: Petroleum Geoscience, v. 8, p. 167–176.

Jev, B. I. C., C. H. Kaars-Sijpesteijn, M. P. A. M. Peters, N. L. Watts, and J. T. Wilkie, 1993, Akaso field, Nigeria: Use of integrated 3-D seismic, fault slicing, clay smearing, and RFT pressure data on fault trapping and dynamic leakage: AAPG Bulletin, v. 77, p. 1389–1404.

Karakouzian, M., and N. Hudyma, 2002, A new apparatus for analog modeling of clay smears: Journal of Structural Geology, v. 24, p. 905–912.

Koledoye, B., A. Aydin, and E. May, 2000, 3-D visualization of fault segmentation and shale smearing in the Niger Delta: Leading Edge, v. 19, p. 692–701.

Koledoye, B. A., A. Aydin, and E. May, 2003, A new process-based methodology for analysis of shale smear along normal faults in the Niger Delta: AAPG Bulletin, v. 87, p. 445–464.

Lehner, F. K., and W. F. Pilaar, 1997, The emplacement of clay smears in synsedimentary normal faults: Inferences from field observations near Frechen, Germany, *in* P. Møller-Pedersen and A. G. Koestler, eds., Hydrocarbon seals: Importance for exploration and production: Norwegian Petroleum Society Special Publication 7, p. 39–50.

Lindsay, N. G., F. C. Murphy, J. J. Walsh, and J. Watterson, 1993, Outcrop studies of shale smears on fault surfaces: International Association of Sedimentologists Special Publication 15, p. 113–123.

Perkins, H., 1961, Fault-closure type fields, southeast Louisiana: Transactions, Gulf Coast Association of Geological Societies, v. 11, p. 177–196.

Rutter, D. J., R. H. Maddock, and S. H. Hall, 1986, Comparative microstructures of natural and experimentally produced clay-bearing fault gouges: Pure and Applied Geophysics, v. 124, p. 3–30.

Sorkhabi, R. B., S. Hasegawa, S. Iwanaga, and M. Fujimoto, 2002, Sealing assessment of normal faults in clastic reservoirs: The role of fault geometry and shale smear parameters: Journal of Japanese Association of Petroleum Technology, v. 67, p. 576–589.

Sperrevick, S., R. B. Farseth, and R. H. Cabrielsen, 2000, Experiments on clay smear formation along faults: Petroleum Geoscience, v. 6, p. 113–123.

Sperrevick, S., P. A. Gillespie, Q. J. Fisher, T. Halvorsen, and R. J. Knipe, 2002, Empirical estimation of fault rock properties, *in* A. G. Koeslter and R. Hunsdale, eds., Hydrocarbon seal quantification: Norwegian Petroleum Society Special Publication 11, p. 109–125.

Vrolijk, P., and B. A. van der Pluijm, 1999, Clay gouge: Journal of Structural Geology, v. 21, p. 1039–1048.

Wang, C. Y., N. H. Mao, and F. T. Wu, 1980, Mechanical properties of clays at high pressure: Journal of Geophysical Research, v. 85, p. 1462–1468.

Weber, K. J., G. Mandl, W. F. Pilaar, F. Lehner, and R. G. Precious, 1978, The role of faults in hydrocarbon migration and trapping in Nigerian growth fault structures: Society of Petroleum Engineers 10th Annual Offshore Technology Conference Proceedings, v. 4, p. 2643–2653.

Yielding, G., 2002, Shale gouge ratio—Calibration by geohistory, *in* A. G. Koeslter and R. Hunsdale, eds., Hydrocarbon seal quantification: Norwegian Petroleum Society Special Publication 11, p. 1–15.

Yielding, G., B. Freeman, and D. T. Needham, 1997, Quantitative fault seal prediction: AAPG Bulletin, v. 81, p. 897–917.

Yielding, G., J. A. Overland, and G. Byberg, 1999, Characterization of fault zones for reservoir modeling: An example from the Gullfaks fields, northern North Sea: AAPG Bulletin, v. 83, p. 925–951.

2.5. Deformation Bands in Porous Sandstone

Antonellini, M., and A. Aydin, 1994, Effect of faulting on fluid flow in porous sandstones: Petrophysical properties: AAPG Bulletin, v. 78, p. 355–377.

Antonellini, M., and A. Aydin, 1995, Effect of faulting on fluid flow in porous sandstones: Geometry and spatial distribution: AAPG Bulletin, v. 79, p. 642–671.

Antonellini, M. A., A. Aydin, and D. D. Pollard, 1994, Microstructure of deformation bands in porous sandstones at Arches National Park, Utah: Journal of Structural Geology, v. 16, p. 941–959.

Antonellini, M., A. Aydin, and L. Orr, 1999, Outcrop-aided characterization of a faulted hydrocarbon reservoir: Arroyo Grande oil field, California, U.S.A., *in* W. C. Haneberg, P. S. Mozley, J. C. Moore, and L. B. Goodwin, eds., Fault and subsurface fluid flow in the shallow crust: American Geophysical Union Geophysical Monograph 113, p. 7–26.

Aydin, A., 1978, Small faults formed as deformation bands in sandstone: Pure and Applied Geophysics, v. 116, p. 913–930.

Aydin, A., and A. M. Johnson, 1978, Development of faults as zones of deformation bands and as slip surfaces in sandstone: Pure and Applied Geophysics, v. 116, p. 931–942.

Aydin, A., and A. M. Johnson, 1983, Analysis of faulting in porous sandstones: Journal of Structural Geology, v. 5, p. 19–31.

Fossen, H., and J. Hesthammer, 1997, Geometric analysis and scaling relations of deformation bands in porous sandstone: Journal of Structural Geology, v. 19, p. 1479–1493.

Fossen, H., and J. Hesthammer, 1998, Deformation bands and their significance in porous sandstone reservoirs: First Break, v. 16, p. 21–25.

Fowles, J., and S. D. Burley, 1994, Textural and permeability characteristics of faulted, high porosity sandstones: Marine Petroleum and Geology, v. 11, p. 608–623.

Gibson, R. G., 1998, Physical character and fluid-flow properties of sandstone-derived fault zones, *in* M. P. Coward, T. S. Daltaban, and H. Johnson, eds., Structural geology in reservoir characterization: Geological Society (London) Special Publication 127, p. 83–97.

Main, I., O. Kwon, B. Ngwenya, and S. C. Elphick, 2000, Fault sealing during deformation band growth in porous sandstone: Geology, v. 28, p. 1131–1134.

Main, I., K. Mair, O. Kown, S. Elphick, and B. Ngwenya, 2001, Experimental constraints on the mechanical and hydraulic properties of deformation bands in porous sandstones: A review, *in* R. E. Holdsworth, R. A. Strachan, J. F. Magloughlin, and

R. J. Kinpe, eds., The nature and tectonic significance of fault zone weakening: Geological Society (London) Special Publication 86, p. 43–63.

Mair, K., I. G. Main, and S. C. Elphick, 2000, Sequential growth of deformation bands in the laboratory: Journal of Structural Geology, v. 22, p. 183–186.

Matthäi, S. K., A. Aydin, D. D. Pollard, and S. G. Roberts, 1998, Numerical simulation of departures from radial drawdown in a faulted sandstone reservoir with joints and deformation bands, *in* G. Jones, Q. J. Fisher, and R. J. Knipe, eds., Faulting, fault sealing and fluid flow in hydrocarbon reservoirs: Geological Society (London) Special Publication 147, p. 157–191.

Ogilvie, S. R., and P. W. J. Glover, 2001, The petrophysical properties of deformation bands in relation to their microstructure: Earth and Planetary Science Letters, v. 193, p. 129–142.

Pitman, E. D., 1981, Effect of fault-related granulation on porosity and permeability of quartz sandstones, Simpson Group (Ordovician), Oklahoma: AAPG Bulletin, v. 65, p. 2381–2387.

Smith, G. A., 1983, Porosity dependence of deformation bands in the Entrada Sandstone, La Plata County, Colorado: The Mountain Geologist, v. 20, p. 82–85.

Shipton, Z. K., and P. A. Cowie, 2001, Damage zone and slip-surface evolution over μm to km scales in high-porosity Navajo Sandstone, Utah: Journal of Structural Geology, v. 23, p. 1825–1844.

Shipton, Z. K., J. P. Evans, K. R. Robeson, C. B. Forster, and S. Snelgrove, 2002, Structural heterogeneity and permeability in faulted eolian sandstone: Implications for subsurface modeling of faults: AAPG Bulletin, v. 86, p. 863–883.

Swierczewska, A., and A. K. Tokarski, 1998, Deformation bands and the history of folding in the Magura nappe, western outer Carpathians (Poland): Tectonophysics, v. 297, p. 73–90.

Taylor, W. L., and D. D. Pollard, 2000, Estimation of in-situ permeability of deformation bands in porous sandstone, Valley of Fire, Nevada: Water Resources Research, v. 36, p. 2595–2606.

Underhill, J. R., and N. H. Woodcock, 1987, Faulting mechanisms in high-porosity sandstones, New Red sandstone, Arran, Scotland, *in* M. E. Jones and R. M. F. Preston, eds., Deformation of sediments and sedimentary rocks: Geological Society (London) Special Publication 29, p. 91–105.

2.6. Pressure Solution Seals

Angevine, C. L., and D. L. Turcotte, 1983, Porosity reduction by pressure solution: A theoretical model for quartz arenites: Geological Society of America Bulletin, v. 94, p. 1129–1134.

Angevine, C. L., D. L. Turcotte, and M. D. Furnish, 1982, Pressure solution lithification as a mechanism for the stick-slip behavior of faults: Tectonics, v. 1, p. 151–160.

Beach, A., 1979, Pressure solution as a metamorphic process in deformed terrigenous sedimentary rocks: Lithos, v. 12, p. 51–59.

Bernabé, Y., W. F. Brace, and B. Evans, 1982, Permeability, porosity and pore geometry of hot pressed calcite: Mechanical Mathematics, v. 1, p. 173–183.

Carrio-Schaffhauser, E., S. Raynaud, H. J. Latière, and F. Mazerolle, 1990, Propagation localization of stylolites in limestones, *in* R. J. Knipe and E. H. Rutter, eds., Deformation mechanisms, rheology and tectonics: Geological Society (London) Special Publication 54, p. 193–200.

De Boer, R. D., P. J. C. Nagtegaal, and E. M. Duyvus, 1977a, Pressure solution experiments on quartz sand: Geochimica et Cosmochimica Acta, v. 41, p. 249–256.

De Boer, R. B., P. J. C. Nagtegaal, and E. M. Duyvis, 1977b, Pressure solution experiments on quartz sand: Geochimica et Cosmochimica Acta, v. 41, p. 257–264.

Dewers, T., and P. J. Ortoleva, 1990, Interaction of reaction, mass transport, and rock deformation during diagenesis: Mathematical modeling of integranular pressure solution, stylolites, and differential compaction/cementation, *in* I. D. Meshri and P. J. Ortoleva, eds., Prediction of reservoir quality through chemical modeling: AAPG Memoir 49, p. 147–160.

Dewers, T., and P. J. Ortoleva, 1991, Influences of clay minerals on sandstone cementation and pressure solution: Geology, v. 19, p. 1045–1048.

Elias, B. P., and A. Hajash Jr., 1992, Changes in quartz solubility and porosity due to effective stress: An experimental investigation of pressure solution: Geology, v. 20, p. 451–454.

Fletcher, R. C., and D. D. Pollard, 1981, Anticrack model for pressure solution surfaces: Geology, v. 9, p. 419–425.

Gratier, J. P., 1987, Pressure solution-deposition creep and associated tectonic differentiation in sedimentary rocks, *in* M. E. Jones and R. M. F. Preston, eds., Deformation of sediments and sedimentary rocks: Geological Society (London) Special Publication 29, p. 25–38.

Gratier, J. P., and R. Guiguet, 1986, Experimental pressure solution-deposition on quartz grains: The crucial effect of the nature of the fluid: Journal of Structural Geology, v. 8, p. 845–856.

Gratier, J. P., T. Chen, and R. Hellmann, 1994, Pressure solution as a mechanism for crack sealing around faults, *in* S. Hickman, R. Sibson, and R. Bruhn, eds., Proceedings of the U.S. Geological Survey

Red Book Conference on the Mechanical Involvement of Fluids in Faulting: U.S. Geological Survey Open-file Report 94-228, p. 279–300.

Hadizadeh, J., 1994, Interaction of cataclasis and pressure solution in a low-temperature carbonate shear zone: Pure and Applied Geophysics, v. 143, p. 255–280.

Heald, M. T., 1955, Stylolites in sandstones: Journal of Geology, v. 63, p. 101–114.

Heald, M. T., 1959, Significance of stylolites in permeable sandstones: Journal of Sedimentary Petrology, v. 29, p. 251–253.

Hickman, S. H., and B. Evans, 1991, Experimental pressure solution in halite: The effect of grain/interphase boundary structure: Journal of Geological Society (London), v. 148, p. 549–560.

Hickman, S. H., and B. Evans, 1995, The kinetics of pressure solution at halite-silica interfaces and intergranular clay films: Journal of Geophysical Research, v. 100B, p. 13,113–13,132.

Houseknecht, D. W., 1988, Intergranular pressure solution in four quartzose sandstones: Journal of Sedimentary Petrology, v. 58, p. 228–246.

Mullis, A. M., 1991, The role of silica precipitation kinetics in determining the rate quartz pressure solution: Journal of Geophysical Research, v. 96B, p. 10,007–10,013.

Mullis, A. M., 1993, Determination of the rate limiting mechanism for quartz pressure solution: Geochimica et Cosmochimica Acta, v. 57, p. 1499–1503.

Peacock, D. C. P., Q. J. Fisher, E. J. M. Willemse, and A. Ayden, 1998, The relationship between faults and pressure solution seams in carbonate rocks and the implications for fluid flow, *in* G. Jones, Q. J. Fisher, and R. J. Knipe, eds., Faulting fault sealing and fluid flow in hydrocarbon reservoirs: Geological Society (London) Special Publication 147, p. 105–115.

Renard, F., J. P. Gratier, and B. Jamtveit, 2000, Kinetics of crack-sealing, intergranular pressure solution, and compaction around active faults: Journal of Structural Geology, v. 22, p. 1395–1407.

Robin, P.-Y., 1978, Pressure solution at grain-to-grain contacts: Geochimica et Cosmochimica Acta, v. 42, p. 1383–1389.

Rutter, E. H., 1976, The kinetics of rock deformation by pressure solution: Philosophical Transactions of the Royal Society of London, v. 283A, p. 203–220.

Rutter, E. H., 1983, Pressure solution in nature, theory and experiment: Journal of Geological Society (London), v. 144, p. 725–740.

Spiers, C. J., P. M. T. M. Schutjens, R. H. Brzesowsky, C. J. Peach, J. L. Liezenberg, and H. J. Zwart, 1990, Experimental determination of constitutive parameters governing creep of rocksalt by pressure solution, *in* R. J. Knipe and E. H. Rutter, eds., Deformation mechanisms, rheology and tectonics: Geological Society (London) Special Publication 54, p. 215–228.

Sprunt, E. S., and A. Nur, 1977, Destruction of porosity through pressure solution: Geophysics, v. 42, p. 726–741.

Stockdale, P. B., 1922, Stylolites: Their nature and origin: Bloomington, Indiana University Studies, v. 9, 97 p.

Sverdrup, E., and E. Prestholm, 1990, Synsedimentary deformation structures and their implications for stylolitization during deeper burial: Sedimentary Geology, v. 68, p. 201–210.

Tada, R., and R. Siever, 1989, Pressure solution during diagenesis: Annual Reviews of Earth Planetary Sciences, v. 17, p. 89–118.

Tada, R., R. Maliva, and R. Siever, 1987, A new mechanism for pressure solution in porous quartzose sandstone: Geochimica et Cosmochimica Acta, v. 51, p. 2295–2301.

Thompson, A., 1959, Pressure solution and porosity, *in* H. A. Ireland, ed., Silica in sediments: SEPM Special Publication 7, p. 92–110.

Wright, T. O., and L. B. Platt, 1982, Pressure dissolution and cleavage in the Matinsburg shale: American Journal of Science, v. 282, p. 122–135.

2.7. Mineral Veins and Crack Seals Associated with Faulting

Boullier, A. M., and F. Robert, 1992, Paleoseismic events recorded in Archaean gold-quartz vein networks, Val d'Or, Albitibi, Quebec, Canada: Journal of Structural Geology, v. 14, p. 161–179.

Brantley, S. L., B. Evans, S. Hickman, and D. Crerar, 1990, Healing of microcracks in quartz: Implications for fluid flow: Geology, v. 18, p. 136–139.

Budai, J. M., A. M. Martini, L. M. Walter, and T. C. W. Ku, 2002, Fracture-fill calcite as a record of microbial methoangenesis and fluid migration: A case study from the Devonian Antrim Shale, Michigan basin: Geofluids, v. 2, p. 163–183.

Cox, S. F., 1987, Antitaxial crack-seal vein microstructures and their relationship to displacement paths: Journal of Structural Geology, v. 9, p. 779–787.

Cox, S. F., and M. A. Etheridge, 1983, Crack-seal fiber growth mechanisms and their significance in the development of oriented layer silicate microstructures: Tectonophysics, v. 92, p. 147–170.

Garven, G., S. W. Bull, and R. R. Large, 2001, Hydrothermal fluid flow models of stratiform ore genesis in the McArthur basin, Northern Territory, Australia: Geofluids, v. 1, p. 289–311.

Haynes, F. M., and S. R. Titley, 1980, The evolution of fracture-related permeability within the Ruby

Star granodiorite, Sierrita porphyry copper deposit, Pima County, Arizona: Economic Geology, v. 75, p. 673–683.

Hulin, C. D., 1925, Structural control of ore deposition: Economic Geology, v. 24, p. 15–49.

Labaume, P., D. Craw, M. Lespinasse, and P. Muchez, eds., 2002, Tectonic processes and the flow of mineralizing fluids: Tectonophysics, v. 348, nos. 1–3, special issue, p. 1–185.

Macaulay, C. I., A. J. Boyce, A. E. Fallick, and R. S. Haszeldine, 1997, Quartz veins record vertical flow at a graben edge: Fulmar oil field, central North Sea: AAPG Bulletin, v. 81, p. 2024–2035.

Miller, L. D., R. J. Goldfarb, G. E. Gehrels, and L. W. Snee, 1994, Genetic links among fluid cycling, vein formation, regional deformation, and plutonism in the Juneau gold belt, southeastern Alaska: Geology, v. 22, p. 203–206.

Ramsay, J. G., 1980, The crack-seal mechanism of rock deformation: Nature, v. 284, p. 135–139.

Robert, F., A. C. Brown, and A. J. Audet, 1983, Structural control of gold mineralization at the Sigma mine, Val d'Or, Quebec: Canadian Institute of Mineralogical and Metallurgical Bulletin, v. 76, p. 72–80.

Robert, F., A.-M. Boullier, and K. Firdaous, 1995, Gold-quartz veins in metamorphic terranes and their bearing on the role of fluids in faulting: Journal of Geophysical Research, v. 100B, p. 12,861–12,879.

Sibson, R. H., 1987, Earthquake rupturing as a hydrothermal mineralizing agent: Geology, v. 15, p. 701–704.

Sibson, R. H., F. Robert, and K. H. Poulsen, 1988, High angle reverse faults, fluid pressure cycling and mesothermal gold-quartz deposits: Geology, v. 16, p. 551–555.

Smith, B. M., S. J. Reynolds, H. W. Day, and R. J. Bodnar, 1991, Deep-seated fluid involvement in ductile-brittle deformation and mineralization, South Mountains metamorphic core complex, Arizona: Geological Society of America Bulletin, v. 103, p. 559–569.

Valenta, R. K., 1989, Vein geometry in the Hilton area, Mount Sia, Queensland; implications for fluid behavior during deformation: Tectonophysics, v. 158, p. 191–207.

Valenta, R., I. Cartwright, and N. H. S. Oliver, 1994, Structurally controlled fluid flow associated with breccia vein formation: Journal of Metamorphic Geology, v. 12, p. 197–206.

2.8. Fault-related Diagenesis and Fluid-rock Interactions in Fault Zones

Note: The Water-Rock Interaction Group of the International Association of Geochemistry and Cosmochemistry organizes international symposia on water-rock interactions (WRI) every 3 yr (WRI-1, Prague, Czechoslovakia, 1974; WRI-2, Strasbourg, France, 1977; WRI-3, Edmonton, Canada, 1980; WRI-4, Misasa, Japan, 1983; WRI-5, Reykjavik, Iceland, 1986; WRI-6, Malvern, England, 1989; WRI-7, Park City, Utah, 1992; WRI-8, Vladiwostok, Russia, 1995; WRI-9, Taupo, New Zealand, 1998; WRI-10, Villasimius, Italy, 2001). The proceedings volumes of these symposia (published by A. A. Balkema Publishers, Rotterdam; not listed below) are also informative sources on fluid-rock interaction in deformed rocks.

Aharonov, E., E. Tenthorey, and C. H. Scholz, 1998, Precipitation sealing and diagenesis: 2. Theoretical analysis: Journal of Geophysical Research, v. 103B, p. 23,969–23,981.

Bjørkum, P. A., 1996, How important is pressure in causing dissolution of quartz in sandstone?: Journal of Sedimentary Research, v. 66, p. 147–154.

Bruhn, R. L., W. A. Yonkee, and W. T. Parry, 1990, Structural and fluid-chemical properties of seismogenic normal faults: Tectonophysics, v. 175, p. 139–157.

Bruhn, R. L., W. T. Parry, W. A. Yonkee, and T. Thompson, 1994, Fracturing and hydrothermal alteration in normal fault zones: Pure and Applied Geophysics, v. 142, p. 609–643.

Burley, S., and J. Rowe, 1994, The influence of faults on fluid migration and the resultant diagenesis in Triassic sandstones at Alderley Edge, north-eastern Cheshire: A field excursion guide for Amerada Hess Ltd., Diagenesis Research Group, Manchester.

Carter, K. E., and S. I. Dworkin, 1990, Channelized fluid flow through shear zones during fluid-assisted dynamic recrystallization, northern Apennines, Italy: Geology, v. 18, p. 720–723.

Chester, F. M., 1989, Dynamic recrystallization in semi-brittle faults: Journal of Structural Geology, v. 11, p. 847–858.

Eichhubl, P., ed., 1998, Diagenesis, deformation, and fluid flow in the Miocene Monterey Formation of coastal California: Society for Sedimentary Geology (SEPM) Pacific Section Special Publication, v. 83, 98 p.

Evans, J. P., and F. M. Chester, 1995, Fluid-rock interaction and weakening of faults of the San Andreas system: Inferences from San Gabriel fault rock geochemistry and microstructures: Journal of Geophysical Research, v. 100B, p. 13,007–13,020.

Feucht, L. J., and J. M. Logan, 1987, Effects of chemically active solutions on shearing behavior of a sandstone: Tectonophysics, v. 175, p. 159–176.

Garden, I. R., S. C. Guscott, S. D. Burley, K. A. Foxford, J. J. Walsh, and J. Marshall, 2001, An exhumed palaeo-hydrocarbon migration fairway in a faulted carrier system, Entrada Sandstone of SE Utah, U.S.A.: Geofluids, v. 1, p. 195–213.

Gaupp, R., A. Matter, J. Platt, K. Ramseyer, and J. Walzebuck, 1993, Diagenesis and fluid evolution of deeply buried Permian (Rotliegende) gas reservoirs, northwest Germany: AAPG Bulletin, v. 77, p. 1111–1128.

Goddard, J. V., and J. P. Evans, 1995, Fluid-rock interactions in faults of crystalline thrust sheets, northwestern Wyoming, U.S.A.: Inferences from geochemistry of fault-related rocks: Journal of Structural Geology, v. 17, p. 533–549.

Hadizadeh, J., and F. F. Foit, 2000, Feasibility of estimating cementation rate in a brittle fault zone using some precepts of sedimentary diagenesis: Journal of Structural Geology, v. 22, p. 401–409.

Hadizadeh, J., and R. D. Law, 1991, Water weakening of sandstone and quartzite deformed at various stress and strain rates: International Journal of Rock Mechanics and Mining Sciences, v. 28, p. 431–440.

Hammond, K. J., and J. P. Evans, 2003, Geochemistry, mineralization, structure, and permeability of a normal-fault zone, Casino mine, Alligator ridge district, north central Nevada: Journal of Structural Geology, v. 25, p. 717–736.

Hickman, S., R. Sibson, and R. Brhun, eds., 1994, Proceedings of the U.S. Geological Survey Red Book Conference on the Mechanical Involvement of Fluids in Faulting: U.S. Geological Survey Open-file Report 94-228, 615 p.

Hickman, S., R. Sibson, and R. Bruhn, eds., 1995, Mechanical involvement of fluids in faulting: Journal of Geophysical Research, v. 100B, special section, p. 12,831–13,132.

Hippler, S. J., 1993, Deformation microstructures and diagenesis in sandstone adjacent to an extensional fault: Implication for the flow and entrapment of hydrocarbons: AAPG Bulletin, v. 77, p. 625–637.

Hippler, S. J., 1997, Microstructures and diagenesis in North Sea fault zones: Implications for fault-seal potential and fault-migration rate, *in* R. C. Surdam, ed., Seals, traps, and the petroleum system: AAPG Memoir 67, p. 103–113.

Jansma, P. E., and R. C. Speed, 1993, Deformation, dewatering, and decollement development in the Antler foreland basin during the Antler orogeny: Geology, v. 21, p. 1035–1038.

Kamineni, D. C., R. Kerrich, and A. Brown, 1993, Effects of differential reactivity of minerals on the development of brittle to semi-brittle structures in granitic rocks; textural and oxygen isotope evidence: Chemical Geology, v. 105, p. 215–232.

Kerrich, R., 1986, Fluid infiltration into fault zones: Chemical, isotopic, and mechanical effects: Pure and Applied Geophysics, v. 124, p. 226–268.

Kerrich, R., and I. Allison, 1978, Flow mechanisms in rocks: Microscopic and mesoscopic structures, and their relation to physical conditions of deformation in the crust: Geoscience Canada, v. 5, p. 109–118.

Kerrich, R., I. Allison, R. L. Barnett, S. Moss, and J. Starkey, 1980, Microstructural and chemical transformations accompanying deformation of granite in a shear zone at Mieville, Switzerland; with implications for stress corrosion cracking and superplastic flow: Contributions to Mineralogy and Petrology, v. 73, p. 221–242.

Kerrich, R., T. E. La Tour, and L. Willmore, 1984, Fluid participation in deep fault zones: Evidence from geological, geochemical, and $^{18}O/^{16}O$ relations: Journal of Geophysical Research, v. 89B, p. 4331–4343.

Knipe, R. J., 1993, The influence of fault zone processes and diagenesis on fluid flow, *in* A. D. Horbury and A. G. Robinson, eds., Diagenesis and basin development: AAPG Studies in Geology 36, p. 135–154.

Knipe, R. J., and A. M. McCaig, 1994, Microstructural and microchemical consequences of fluid flow in deforming rocks, *in* J. Parnell, ed., Geofluids: Origin, migration and evolution of fluids in sedimentary basins: Geological Society (London) Special Publication 78, p. 99–111.

Knipe, R. J., S. M. Agar, and D. J. Prior, 1991, The microstructural evolution of fluid flow paths in semi-lithified sediments from subduction complexes: Philosophical Transactions of the Royal Society of London, v. 355A, p. 261–273.

Kraishan, G. M., and N. M. Lemon, 2000, Fault-related calcite cementation: Implications for timing of hydrocarbon generation and migration and secondary porosity development, Barrow subbasin, North West Shelf: Australian Petroleum Production and Exploration Association Journal, v. 40, p. 215–229.

Labaume, P., and I. Moretti, 2001, Diagenesis-dependence of cataclastic thrust fault zone sealing in sandstones: Example from the Bolivian sub-Andean zone: Journal of Structural Geology, v. 23, p. 1659–1675.

Lachenbruch, A. H., 1980, Frictional heating, fluid pressure, and the resistance to fault motion: Journal of Geophysical Research, v. 85B, p. 6097–6112.

Langseth, M., and J. C. Moore, 1990, Introduction to special section on the role of fluids in sediment accretion, deformation, diagenesis and metamorphism in subduction zones: Journal of Geophysical Research, v. 95B, p. 8737–8742.

Losh, S., 1989, Fluid-rock interaction in evolving ductile shear zone and across the brittle-ductile transition, central Pyrenees, France: American Journal of Science, v. 289, p. 600–648.

Mase, C. W., and L. Smith, 1985, Pore fluid pressures

and frictional heating on a fault surface: Pure and Applied Geophysics, v. 122, p. 583–607.

Mase, C. W., and L. Smith, 1987, Effects of frictional heating on the thermal, hydrologic and mechanical response of a fault: Journal of Geophysical Research, v. 92B, p. 6249–6272.

McCaig, A. M., D. M. Wayne, J. D. Marshall, D. Banks, and I. Hendersen, 1995, Isotopic and fluid inclusion studies of fluid movement along the Gavarnie thrust, central Pyrenees: Reaction fronts in carbonate mylonites: American Journal of Science, v. 295, p. 309–343.

McCants, C. Y., and S. D. Burley, 1996, Reservoir archtecture and diagenesis in downthrown fault block plays; the Lowlander prospect of block 14/20b, Witch Ground graben, Outer Moray Firth, UK North Sea, *in* A. Hurst, H. D. Johnson, S. D. Burley, A. C. Canham, and D. S. Mackertich, eds., Geology of the Humber Group, Central Graben and Moray Firth, UKCS: Geological Society (London) Special Publication 114, p. 251–285.

Mozley, P. S., and L. B. Goodwin, 1995, Patterns of cementation along a Cenozoic normal fault: A record of paleoflow orientations: Geology, v. 23, p. 539–542.

Ngwenya, B. T., S. C. Elphick, I. G. Main, and G. B. Shimmield, 2000, Experimental constraints on the diagenetic self-sealing capacity of faults in high porosity rocks: Earth and Planetary Science Letters, v. 183, p. 187–199.

O'Brien, G. W., and E. P. Woods, 1995, Hydrocarbon-related diagenetic zones (HRDZs) in the Vulcan subbasin, Timor Sea: Recognition and exploration implications: Australian Petroleum Exploration Association Journal, v. 35, p. 220–253.

O'Brien, G. W., M. Liks, I. R. Duddy, P. J. Eadington, S. Cadman, and M. Fellows, 1996, Late Tertiary fluid migration in the Timor sea: A key control on thermal and diagenetic histories? Australian Petroleum Exploration Association Journal, v. 36, p. 399–427.

Porter, K. W., and R. J. Weimer, 1982, Diagenetic sequence related to structural history and petroleum accumulation: Spindle field, Colorado: AAPG Bulletin, v. 66, p. 2543–2560.

Sample, J. C., M. R. Reid, H. J. Tobin, and C. J. Moore, 1993, Carbonate cements indicate channeled fluid flow along a zone of vertical faults at the deformation front of the Cascadia accretionary wedge (northwest U.S. Coast): Geology, v. 21, p. 507–510.

Sibson, R. H., 1987, Earthquake rupturing as a mineralizing agent in hydrothermal systems: Geology, v. 15, p. 701–704.

Smith, L., C. B. Forster, and J. P. Evans, 1990, Interaction of fault zones, fluid flow, and heat transfer at the basin scale, *in* S. P. Newman and I. Neretniek, eds., Hydrogeology of low permeability environments: International Association of Hydrogeologists, v. 2, p. 41–67.

Smith, B. M., S. J. Reynolds, H. W. Day, and R. J. Bodnar, 1991, Deep-seated fluid involvement in ductile-brittle deformation and mineralization, South Mountains metamorphic core complex, Arizona: Geological Society of America Bulletin, v. 103, p. 559–569.

Stel, H., 1985, Crystal growth in cataclasites: Diagnostic microstructures and implications: Tectonophysics, v. 78, p. 585–600.

Stierman, D. J., 1984, Geophysical and geological evidence for fracturing, water circulation, and chemical alteration in granitic rocks adjacent to major strike-slip faults: Journal of Geophysical Research, v. 89B, p. 5849–5857.

Stober, I., and K. Bucher, eds., 2002, Water-rock interaction: Dordrecht, Kluwer, 244 p.

Sverdrup, E., and K. Bjørlykke, 1992, Small fault in sandstones from Spitsbergen and Haltenbanken: A study of diagenetic and deformational structures and their relation to fluid flow, *in* R. M. Larsen, H. Brekke, B. T. Larsen, and E. Talleraas, eds., Structural and tectonic modelling and its application to petroleum geology: Norwegian Petroleum Society Special Publication 1, p. 507–517.

Sverdrup, E., and K. Bjørlykke, 1997, Fault properties and the development of cemented fault zones in sedimentary basins: Field examples and predictive models, *in* P. Møller-Pedersen and A. G. Koestler, eds., Hydrocarbon seals: Importance for exploration and production: Norwegian Petroleum Society Special Publication 7, p. 91–106.

Tenthorey, E., E. Aharonov, and C. H. Scholz, 1998, Precipitation sealing and diagenesis: 1. Experimental results: Journal of Geophysical Research, v. 103B, p. 23,951–23,967.

Whitworth, T. M., W. C. Haneberg, P. S. Mozley, and L. B. Goodwin, 1999, Solute-sieving-induced calcite precipitation on pulverized quartz sand: Experimental results and implications for the membrane behavior of fault gouge, *in* W. C. Haneberg, P. S. Mozley, J. C. Moore, and L. B. Goodwin, eds., Fault and subsurface fluid flow in the shallow crust: American Geophysical Union Geophysical Monograph 113, p. 149–158.

Wintsch, R. P., R. Christofferson, and A. K. Kronenberg, 1995, Fluid-rock interaction weakening of fault zones: Journal of Geophysical Research, v. 100B, p. 13,021–13,032.

Zhu, W., C. David, and T. F. Wong, 1995, Network modeling of permeability evolution during cementation and hot isostatic pressing: Journal of Geophysical Research, v. 100B, p. 15,451–15,464.

3. FAULT-RELATED FLUID FLOW

3.1. Fluid Flow in Rocks*

Bear, J., 1972, Dynamics of fluids in porous media: Amsterdam, Elsevier, 764 p.

Chen, Z., and R. E. Ewing, eds., 2002, Fluid flow and transport in porous media, mathematical and numerical treatment: Proceedings of an American Mathematical Society–Institute of Mathematical Statistics–Society for Industrial and Applied Mathematics joint summer research conference, June 17–21, 2001, American Mathematical Society, 524 p.

Corey, T. A., 1994, Mechanics of immiscible fluids in porous media, 3d ed.: Highlands Ranch, Colorado, Water Resources Publications, 252 p.

Doligez, B., ed., 1987, Migration of hydrocarbon in sedimentary basins: Paris, Technip, 681 p.

England, W. A., and A. J. Fleet, eds., 1991, Petroleum migration: Geological Society (London) Special Publication 59, p. 280.

Entov, V. M., V. M. Ryzhik, and G. I. Barenblatt, 1990, Theory of fluid flows through natural rocks: Dordrecht, Kluwer Academic Publishers, 395 p.

Evans, B., and T.-F. Wong, eds., 1992, Fault mechanics and transport properties of rocks: New York, Academic Press, 542 p.

Evans, D. D., T. J. Nicholson, and T. C. Rasmussen, eds., 2001, Flow and transport through unsaturated fractured rock, 2d ed.: American Geophysical Union Geophysical Monograph 42, 196 p.

Faybishenko, B., P. A. Witherspoon, and S. M. Benson, eds., 2000, Dynamics of fluids in fractured rocks: American Geophysical Union Geophysical Monograph 122, 400 p.

Fyfe, W. S. N., and A. B. Thompson, 1978, Fluids in the Earth's crust: Their significance in metamorphic, tectonic, and chemical transport processes: Amsterdam, Elsevier, 383 p.

Goff, J. C., and B. P. J. Williams, eds., 1987, Fluid flow in sedimentary basins and aquifers: Geological Society (London) Special Publication 34, 230 p.

Hickman, S., R. Sibson, and R. Bruhn, eds., 1994, Proceedings of the U.S. Geological Survey Red Book Conference on the Mechanical Involvement of Fluids in Faulting: U.S. Geological Survey Open-file Report 94-228, 615 p.

Hickman, S., R. Sibson, and R. Bruhn, eds., 1995, Mechanical involvement of fluids in faulting: Journal of Geophysical Research, v. 100B, special section, p. 12,831–13,132.

Indararatna, B., and R. Ranjith, eds., 2002, Hydromechanical aspects and unsaturated flow in jointed rock: Rotterdam, A. A. Balkema, 200 p.

Jamtveit, B., and B. Yardley, eds., 1996, Fluid flow and transport in rocks: mechanisms and effects: London, Chapman & Hall, 408 p.

Kostura, J. H., and J. H. Ravenscroft, compilers, 1977, Fracture-controlled production: AAPG Reprint Series 21, 221 p.

Kyser, K., ed., 2001, Fluids and basin evolution: Mineralogical Association of Canada Short Course 28, 262 p.

Lawrence, S. R., and C. Cornford, eds., 1995, Basin geofluids: Basin Research, v. 7, thematic issue, p. 1–108.

Lee, C.-H., and I. Farmer, 1993, Fluid flow in discontinuous rocks: London, Chapman & Hall, 169 p.

McCaffrey, K., L. Lonergan, and J. Wilkinson, eds., 1999, Fractures, fluid flow and mineralization: Geological Society (London) Special Publication 155, 328 p.

National Research Council, 1996, Rock fractures and fluid flow: Contemporary understanding and application: Washington, D.C., National Academy Press, 551 p.

Nuclear Energy Agency/European Community (NEA-EC), eds., 1998, Fluid flow through faults and fractures in argillaceous formations: Proceedings of a joint NEA-EC workshop held in Berne, Switzerland, June 10–12, 1996: Organization for Economic Cooperation and Development, 404 p.

Parnell, J., ed., 1994, Geofluids: Origin, migration and evolution of fluids in sedimentary basins: Geological Society (London) Special Publication 78, 372 p.

Parnell, J., ed., 1998, Dating and duration of fluid flow and fluid-rock interaction: Geological Society (London) Special Publication 144, 284 p.

Phillips, O. M., 1991, Flow and reaction in permeable rocks: Cambridge, Cambridge University Press, 277 p.

Rossmanith, H. P., ed., 1990, Mechanics of jointed and faulted rock: Proceedings of the International Conference, Vienna, April 18–20, 1990: Rotterdam, A. A. Balkema, 1008 p.

Rossmanith, H. P., ed., 1995, Mechanics of jointed and faulted rock: Proceedings of the 2nd International Conference, Vienna, April 10–14, 1995: Rotterdam, A. A. Balkema, 1049 p.

Rossmanith, H. P., ed., 1998, Mechanics of jointed and faulted rock: Proceedings of the 3rd International Conference, Vienna, April 6–9, 1998: Rotterdam, A. A. Balkema, 658 p.

Sahimi, M., 1995, Flow and transport in porous media and fractured rock: New York, Weinheim, 482 p.

Scheidegger, A. E., 1960, The physics of flow through porous media: Toronto, University of Toronto Press, 313 p.

Shapiro, S., ed., 2002, Seismic signatures of fluid transport: Geophysics, v. 67, special section, p. 197–323.

Spiers, C. J., and T. Takeshita, eds., 1995, Influence of

fluids on deformation processes in rocks: Tectonophysics, v. 254, nos. 3–4, special issue, p. 121–297.

Verweij, J. M., 1993, Hydrocarbon migration systems analysis: Amsterdam, Elsevier, 276 p.

Vigneresse, J. L., ed., 2001, Fluids and fractures in the lithosphere: Tectonophysics, v. 336, special issue, p. 1–244.

3.2. Fault-related Fluid Flow: Field Studies

Anderson, R., P. Flemings, S. Losh, J. Austin, and R. Woodhams, 1994, Gulf of Mexico growth fault drilled, seen as oil, gas migration pathway: Oil & Gas Journal, v. 92 (June 6, 1994), no. 23, p. 97–104.

Badertscher, N. P., G. Beaudoin, R. Therrien, and M. Burkhard, 2002, Glarus overthrust: A major pathway for the escape of fluids out of the Alpine orogen: Geology, v. 30, p. 875–878.

Billeaud, L. B., R. N. Anderson, P. B. Flemings, and J. Austin, 1994, Active gas and oil migration sought in a growth fault zone: Petroleum Engineer International, v. 67 (December 4), p. 17–22.

Bodvarsson, G. S., S. M. Benson, and P. A. Witherspoon, 1982, Theory of the development of geothermal systems charged by vertical faults: Journal of Geophysical Research, v. 87B, p. 9317–9328.

Bredehoeft, J. D., J. B. Wesley, and T. D. Fouch, 1994, Simulations of the origin of fluid pressure, fracture generation, and the movement of fluids in the Uinta basin, Utah: AAPG Bulletin, v. 78, p. 1729–1747.

Brown, K. M., 1995, The variation of the hydraulic conductivity structure of an overpressured thrust zone with effective stress, *in* B. Carson, G. K. Westbrook, R. J. Musgrave, and E. Suess, eds., Proceedings of the Ocean Drilling Program, Scientific Results, v. 146: Part 1. Cascadia margin: College Station, Ocean Drilling Program, Texas A&M University, chapter 17, p. 281–289.

Byrne, T., and D. Fisher, 1990, Evidence for a weak and overpressured decollement beneath sediment-dominated accretionary prisms: Journal of Geophysical Research, v. 98B, p. 9081–9097.

Cox, S. F., 1995, Faulting processes at high fluid pressures: An example of fault valve behavior from the Wattle Gully fault, Victoria, Australia: Journal of Geophysical Research, v. 100B, p. 12,841–12,859.

Cox, S. F., 1999, Deformational controls on the dynamics of fluid flow in mesothermal gold systems, *in* K. J. W. McCaffrey, L. Lonergan, and J. J. Wilkinson, eds., Fractures, fluid flow and mineralization: Geological Society (London) Special Publication 155, p. 123–140.

Dholakia, S. N., A. Aydin, D. D. Pollard, and M. D. Zoback, 1998, Fault-controlled hydrocarbon pathways in the Monterey Formation, California: AAPG Bulletin, v. 82, p. 1551–1574.

Douglas, T. A., C. P. Chamberlain, M. A. Poage, M. Abruzzese, S. Schultz, J. Hennerberry, and P. Layer, 2003, Fluid flow and the Heart Mountain fault: A table isotopic, fluid inclusion, and geochronologic study: Geofluids, v. 3, p. 13–32.

Dugan, B., and P. B. Flemings, 2000, Overpressure and fluid flow in the New Jersey continental slope: Implications for slope failure and cold seeps: Science, v. 289, p. 288–291.

Eichhubl, P., and J. R. Boles, 2000a, Rates of fluid flow in fault systems—Evidence from episodic fluid flow in the Miocene Monterey Formation, coastal California: American Journal of Science, v. 300, p. 571–600.

Eichhubl, P., and J. R. Boles, 2000b, Focused fluid flow along faults in the Monterey Formation, coastal California: Geological Society of America Bulletin, v. 112, p. 1667–1679.

Eichhubl, P., H. G. Greene, T. Naehr, and N. Maher, 2000, Structural control of fluid flow: Offshore seepage in the Santa Barbara basin, California: Journal of Geochemical Exploration, v. 69–70, p. 545–549.

Elmore, R. D., J. Parnell, M. H. Engel, M. Baron, S. Woods, M. Abraham, and M. Davidson, 2002, Palaeomagnetic dating of fluid-flow events in dolomotized rocks along the highland boundary fault, central Scotland: Geofluids, v. 2, p. 299–314.

Garden, R., S. C. Guscott, S. D. Burley, A. Foxford, J. Marshall, J. J. Walsh, and J. Watterson, 2001, The geometry and secondary hydrocarbon migration pathways in faulted carrier systems: An exhumed palaeo-fairway in the Entrada Sandstone, Utah: Geofluids, v. 1, p. 195–213.

Giggenbach, W. F., Y. Sano, and H. Wakita, 1993, Isotopic composition of helium and CO_2, and CH_4 contents in gases produced along the New Zealand part of a convergent plate boundary: Geochimica et Cosmochimica Acta, v. 57, p. 3427–3455.

Gordon, D. S., and P. B. Flemings, 1998, Generation of overpressure and compaction-driven fluid flow in a Plio-Pleistocene growth-faulted basin, Eugene Island 330, offshore Louisiana: Basin Research, v. 10, p. 177–196.

Goyal, K. P., and D. R. Kassoy, 1980, Fault zone controlled charging of a liquid-dominated geothermal reservoir: Journal of Geophysical Research, v. 85B, p. 1867–1875.

Grauls, D., and J. M. Baleix, 1994, Role of overpressures and in-situ stresses in fault-controlled hydrocarbon migration: A case study: Marine and Petroleum Geology, v. 11, p. 734–742.

Grindley, G. W., and P. R. L. Browne, 1976, Structural and hydrological factors controlling the permeability of some hot-water geothermal fields: U.S. Government Printing Office, Proceedings of the 2nd United Nations Symposium on Development and Use of Geothermal Resources, Washington, D.C., v. 1, p. 377–386.

Harris, S. D., L. Elliot, and R. J. Knipe, 1999, The pulsed migration of hydrocarbons across inactive faults: Hydrology and Earth System Sciences, v. 3, p. 151–175.

Hindle, A. D., 1997, Petroleum migration pathways and charge concentration: A three-dimensional model: AAPG Bulletin, v. 81, p. 1451–1481.

Hooper, E. C. D., 1991, Fluid migration along growth faults in compacting sediments: Journal of Petroleum Geology, v. 14, p. 161–180.

Huntoon, P. A., and D. A. Lundy, 1979, Fracture-controlled ground-water circulation and well siting in the vicinity of Laramie, Wyoming: Ground Water, v. 17, p. 463–469.

Kastning, E. H., 1977, Faults as positive and negative influences on groundwater flow and conduit enlargement, *in* R. R. Dilamarter and S. C. Csallany, eds., Hydrologic problems in karst regions: Bowling Green, Kentucky, West Kentucky University, p. 193–201.

Kirschner, L. A., H. Masson, and Z. D. Sharp, 1999, Fluid migration through thrust faults in the Helvetic nappes (western Swiss Alps): Contributions to Mineralogy and Petrology, v. 136, p. 169–183.

Knipe, R. J., 1993, Micromechanisms of deformation and fluid behaviour during faulting, *in* S. Hickman, R. Sibson, and R. Bruhn, eds., Proceedings of the U.S. Geological Survey Red Book Conference on the Mechanical Involvement of Fluids in Faulting: U.S. Geological Survey Open-file Report 94-228, p. 301–310.

Knipe, R. J., and A. M. McCaig, 1994, Microstructural and microchemical consequences of fluid flow in deforming rocks, *in* J. Parnell, ed., Geofluids: Origin, migration and evolution of fluids in sedimentary basins: Geological Society (London) Special Publication 78, p. 99–111.

Knipe, R. J., S. M. Agar, and D. J. Prior, 1991, The microstructural evolution of fluid flow paths in semi-lithified sediments from subduction complexes: Philosophical Transactions of the Royal Society of London, v. 355A, p. 261–273.

Knipe, R. J., G. Jones, and Q. J. Fisher, 1998, Faulting, fault sealing and fluid flow in hydrocarbon reservoirs: An introduction, *in* G. Jones, Q. J. Fisher, and R. J. Knipe, eds., Faulting, fault sealing and fluid flow in hydrocarbon reservoirs: Geological Society (London) Special Publication 147, p. vii–xxi.

Kolm, K. E., and J. S. Downey, 1994, Diverse flow patterns in the aquifers of the Amargosa Desert and vicinity, southern Nevada and California: Bulletin of the Association of Engineering Geologists, v. 31, p. 33–47.

Levens, R. L., R. E. Williams, and D. R. Ralston, 1994, Hydrogeologic role of geologic structures: Part I. The paradigm: Journal of Hydrology, v. 156, p. 227–243.

Lisk, M., M. M. Faiz, E. B. Bekele, and T. E. Ruble, 2000, Transient fluid flow in the Timor Sea, Australia: Implications for prediction of fault seal integrity: Journal of Geochemical Exploration, v. 69–70, p. 607–613.

Losh, S., 1998, Oil migration in a major growth fault: Structural analysis of the pathfinder core, South Eugene Island Block 330, offshore Louisiana: AAPG Bulletin, v. 82, p. 1694–1710.

Losh, S., L. Eglinton, M. Schoell, and J. Wood, 1999, Vertical and lateral fluid flow related to a large growth fault, South Eugene Island Block 330 Field, offshore Louisiana: AAPG Bulletin, v. 83, p. 244–276.

Magara, K., 1983, Physical constraints of faulting in a geopressured sedimentary basin: Journal of Petroleum Geology, v. 5, p. 417–426.

Manzocchi, T., P. S. Ringrose, and J. R. Underhill, 1998, Flow through fault systems in high porosity sandstones, *in* M. P. Coward, T. S. Daltaban, and H. Johnson, eds., Structural geology in reservoir characterisation: Geological Society (London) Special Publication 127, p. 65–82.

McCaig, A. M., 1989, Fluid flow through fault zones: Nature, v. 340, p. 600.

McCaig, A. M., 1998, Deep fluid circulation in fault zones: Geology, v. 16, p. 867–870.

Moore, J. C., and P. Vorlijk, 1992, Fluids in accretionary prisms: Reviews of Geophysics, v. 30, p. 113–135.

Nun, J. A., and P. Meulbroek, 2002, Kilometer-scale upward migration of hydrocarbon in geopressured sediments by buoyancy-driven propagation of methane-filled fractures: AAPG Bulletin, v. 86, p. 907–918.

O'Brien, G. W., P. Quife, R. Cowley, M. Morse, D. Wilson, M. Fellows, and M. Lisk, 1998, Evaluating trap integrity in the Vulcan subbasin, Timor Sea, Australia, using integrated remote sensing geochemical technologies, *in* P. G. Purcell and R. R. Purcell, eds., Petroleum Exploration Society of Australia Western Australian Basins Symposium, v. 2, p. 237–254.

O'Brien, G. W., M. Lisk, I. R. Duddy, J. Hamilton, P. Woods, and R. Cowley, 1999, Plate convergence, foreland development and fault reactivation: Primary controls on brine migration, thermal histories and trap breach in the Timor Sea, Australia: Marine and Petroleum Geology, v. 16, p. 533–560.

O'Brien, G. W., R. Cowley, P. Quaife, and M. Morse, 2002, Characterizing hydrocarbon migration and fault-seal integrity in Australia's Timor Sea via multiple, integrated remote-sensing technologies, *in* D. Schumacher and L. A. LeSchack, eds., Surface exploration case histories: Applications of geochemistry, magnetics, and remote sensing: AAPG Studies in Geology 48, p. 393–413.

Oliver, J., 1986, Fluids expelled tectonically from orogenic belts: Their role in hydrocarbon migration and other geologic phenomena: Geology, v. 14, p. 99–102.

Oliver, N. H. S., and P. D. Bons, 2001, Mechanisms of fluid flow and fluid-rock interaction in fossil metamorphic-hydrothermal systems inferred from vein-wallrock patterns, geometry, and microstructure: Geofluids, v. 1, p. 137–163.

Omre, H., K. Sølna, N. Dahl, and B. Tørudbakken, 1994, Impact of fault heterogeneity in fault zones on fluid flow: North Sea oil and gas reservoirs III, Norwegian Institute of Technology: Dordrecht, Kluwer Academic Publishers, p. 185–200.

Parry, W. T., 1994, Fault fluid composition from fluid inclusion observations, *in* S. Hickman, R. Sibson, and R. Bruhn, eds., Proceedings of the U.S. Geological Survey Red Book Conference on the Mechanical Involvement of Fluids in Faulting: U.S. Geological Survey Open-file Report 94-228, p. 334–348.

Pederson, T., and K. Bjørlykke, 1994, Fluid flow in sedimentary basins: Model of pore water flow in a vertical fracture: Basin Research, v. 6, p. 1–16.

Roberts, S. J., and J. A. Nunn, 1995, Episodic fluid expulsion from geopressured sediments: Marine and Petroleum Geology, v. 12, p. 195–204.

Roberts, S. J., J. A. Nunn, L. Cathles, and F.-D. Cipriani, 1996, Expulsion of abnormally pressured fluids along faults: Journal of Geophysical Research, v. 101B, p. 28,231–28,252.

Rudnicki, J. W., and T. C. Hzu, 1988, Pore pressure changes induced by slip on permeable and impermeable faults: Journal of Geophysical Research, v. 93B, p. 3275–3285.

Upton, P., D. Craw, T. G. Caldwell, P. O. Koons, Z. James, P. E. Wannamaker, G. J. Jiracek, and C. P. Chamberlain, 2003, Upper crustal fluid flow in the outboard region of the southern Alps, New Zealand: Geofluids, v. 3, p. 1–12.

3.3. Fluid Flow and Natural Seismicity

Aki, K., 1979, Characteristics of barriers on earthquake faults: Journal of Geophysical Research, v. 84B, p. 6140–6148.

Aki, K., 1995, Interrelation between fault zone structures and earthquake process: Pure and Applied Geophysics, v. 145, p. 647–676.

Blanpied, M. L., D. A. Lockner, and J. D. Byerlee, 1992, An earthquake mechanism based on rapid sealing of faults: Nature, v. 358, p. 574–576.

Boullier, A. M., and F. Robert, 1992, Paleoseismic events recorded in Archaean gold-quartz vein networks, Val d'Or, Albitibi, Quebec, Canada: Journal of Structural Geology, v. 14, p. 161–179.

Byerlee, J. D., 1993, A model for episodic flow of high-pressure water in fault zones before earthquakes: Geology, v. 21, p. 303–306.

Cox, S. F., M. A. Etheridge, and V. J. Wall, 1986, The role of fluids in syntectonic mass transport, and the localization of metamorphic vein-type ore deposits: Ore Geology Review, v. 2, p. 65–86.

Evans, D. M., 1966, Man-made earthquakes in Denver: Geotimes, v. 10, no. 9, p. 11–18.

Henderson, J. R., and B. Maillot, 1997, The influence of fluid flow in fault zones on patterns of seismicity: A numerical investigation: Journal of Geophysical Research, v. 102B, p. 2915–2924.

Hillis, R., 2000, Pore pressure/stress coupling and its implications for seismicity: Exploration Geophysics, v. 31, p. 448–454.

Irwing, W. P., and I. Barnes, 1980, Tectonic relations of carbon dioxide discharge and earthquakes: Journal of Geophysical Research, v. 85B, p. 3115–3121.

Johnson, P. A., and T. V. McEvilly, 1995, Parkfield seismicity: Fluid-driven?: Journal of Geophysical Research, v. 100B, no. B7, p. 12,937–12,950.

King, C.-Y., S. Azuma, G. Igarashi, M. Ohno, H. Saito, and H. Wakita, 1999, Earthquake-related water level changes at 16 closely clustered wells in Tono, central Japan: Journal of Geophysical Research, v. 104B, p. 13,073–13,082.

Lockner, D. A., and J. D. Byerlee, 1995, An earthquake instability model based on faults containing high fluid-pressure compartments: Pure and Applied Geophysics, v. 145, p. 717–745.

Muir Wood, R., 1994, Earthquakes, strain-cycling and the mobilization of fluids, *in* J. Parnell, ed., Geofluids: Origin, migration and evolution of fluids in sedimentary basins: Geological Society (London) Special Publication 78, p. 85–98.

Muir Wood, R., and G. C. P. King, 1993, Hydrological signatures associated with earthquake strain: Journal of Geophysical Research, v. 98B, p. 22,035–22,068.

Nicholson, C., and R. L. Wesson, 1990, Earthquake hazard associated with deep well injection: U.S. Geological Survey Bulletin, v. 1951, 74 p.

Nur, A., and J. R. Booker, 1972, Aftershocks caused by pore fluid flow?: Science, v. 175, p. 885.

Rojstaczer, S., and S. Wolf, 1992, Permeability changes associated with large earthquakes: An example from Loma Prieta, California: Geology, v. 20, p. 211–214.

Sato, T., R. Sakai, K. Furuya, and T. Kodama, 2000,

Coseismic spring flow changes associated with the 1995 Kobe earthquake: Geophysical Research Letters, v. 27, p. 1219–1222.

Shipley, T. H., G. F. Moore, N. L. Bangs, J. C. Moore, and P. L. Stoffa, 1994, Seismically inferred dilatancy distribution, northern Barbados ridge decollement: Implications for fluid migration and fault strength: Geology, v. 22, p. 411–414.

Sibson, R. H., 1981, Fluid flow accompanying faulting: Field evidence and models, *in* D. W. Simpson and P. G. Richards, eds., Earthquake prediction—An international review: American Geophysical Union Monograph 4, p. 592–603.

Sibson, R. H., 1986, Earthquakes and rock deformation in crustal fault zones: Annual Reviews of Earth and Planetary Sciences, v. 14, p. 149–175.

Sibson, R. H., 1992, Implications of fault valve behaviour for rupture nucleation and recurrence: Tectonophysics, v. 211, p. 283–293.

Sibson, R. H., J. M. Moore, and A. H. Rankin, 1975, Seismic pumping—A hydrothermal fluid transport mechanism: Journal of the Geological Society (London), v. 131, p. 653–659.

Stark, C. P., and J. A. Stark, 1991, Seismic fluids and percolation theory: Journal of Geophysical Research, v. 96B, p. 8417–8426.

Streit, J. E., and S. F. Cox, 2001, Fluid pressures at hypocenters of moderate to large earthquakes: Journal of Geophysical Research, v. 106B, p. 2235–2243.

3.4. Fluid Flow and Induced Seismicity in Petroleum Basins

Cornet, F. H., and P. Julien, 1989, Stress determination from hydraulic test data and focal mechanisms of induced seismicity: International Journal of Rock Mechanics, Mining Science and Geomechanics, v. 26, p. 235–248.

Cornet, F. H., and O. Scotti, 1993, Analysis of induced seismicity for fault zone identification: International Journal of Rock Mechanics, Mining Science and Geomechanics, v. 30, p. 789–795.

Cornet, F. H., and J. Yin, 1995, Analysis of induced seismicity for stress field determination and pore pressure mapping: Pure and Applied Geophysics, v. 145, p. 677–700.

Davis, S. D., 1989, Induced seismic deformation in the Cogdell oil field of west Texas: Bulletin of the Seismological Society of America, v. 79, p. 1477–1495.

Doser, D. I., M. R. Baker, and D. B. Mason, 1991, Seismicity in the War-Wink gas field, west Texas, and its relationship to petroleum production: Bulletin of Seismological Society of America, v. 81, p. 971–986.

El-Hussain, I. W., and P. J. Carpenter, 1990, Reservoir induced seismicity near Heron and El Vado reservoirs, northern New Mexico: Association of Engineering Geologists Bulletin, v. 27, p. 51–59.

Fabre, D., J.-R. Grasso, and Y. Orengo, 1992, Mechanical behavior of deep rock core sample from a seismically active gas field: Pure and Applied Geophysics, v. 137, p. 201–219.

Feigner, B., and J.-R. Grasso, 1990, Seismicity induced by gas production: I. Correlation of focal mechanisms and dome structure: Pure and Applied Geophysics, v. 134, p. 405–426.

Gibowicz, S. J., 1995, Scaling relations for seismic events induced by mining: Pure and Applied Geophysics, v. 145, p. 717–745.

Gibowicz, S. J., and A. Kijko, 1994, An introduction to mining seismology: New York, Academic Press, 399 p.

Grasso, J.-P., and B. Feigner, 1990, Seismicity induced by gas production: II. Lithology correlated events, induced stresses and deformation: Pure and Applied Geophysics, v. 134, p. 427–450.

Grasso, J.-P., and G. Wittlinger, 1990, 10 years of seismic monitoring over a gas field: Bulletin of Seismological Society of America, v. 80, p. 450–473.

Gupta, H. K., 1992, Reservoir induced earthquakes: Amsterdam, Elsevier, 364 p.

Gupta, H. K., 2000, Induced earthquakes: Dordrecht, Kluwer, 324 p.

Gupta, H. K., and R. K. Chandra, eds., 1995, Induced seismicity: Pure and Applied Geophysics, v. 145, no. 1, special issue, p. 1–217.

Gupta, H. K., B. K. Rastogi, and H. Narain, 1972, Common features of reservoir-associated seismicity: Bulletin of Seismological Society of America, v. 62, p. 481–492.

Guyoton, F., J.-R. Grasso, and P. Volant, 1992, Interrelation between induced seismic instabilities and complex geological structure: Geophysical Research Letters, v. 19, p. 705.

Horner, R. B., J. E. Barclay, and J. M. MacRae, 1994, Earthquakes and hydrocarbon production in the Fort St. John area of northeastern British Columbia: Canadian Journal of Exploration Geophysics, v. 30, p. 39–50.

Knoll, P., ed., 1992, Induced seismicity: Vermont, A. A. Balkema, 469 p.

Maury, V. M. R., J.-R. Grasso, and G. Wittlinger, 1992, Monitoring of subsidence and induced seismicity in the Lacq gas field (France): The consequences on gas production and field operation: Engineering Geology, v. 32, p. 123–135.

McGarr, A., 1991, On a possible connection between three major earthquakes in Californian and oil production: Bulletin of Seismological Society of America, v. 81, p. 948–970.

McGarr, A., ed., 1992, Induced seismicity: Pure and

Applied Geophysics, v. 139, nos. 3–4, special issue, p. 347–800.

Milne, W. G., ed., 1976, Induced seismicity: Engineering Geology, v. 10, nos. 2–4, special issue, p. 83–385.

Pennington, W. D., S. D. Davis, S. M. Carlson, J. Dupree, and T. E. Ewing, 1986, The evolution of seismic barriers and asperities caused by the depressuring of fault planes in oil and gas fields of south Texas: Bulletin of Seismological Society of America, v. 76, p. 939–948.

Peppin, W. A., and C. G. Bufe, 1980, Induced versus natural earthquakes: Search for a seismic discriminant: Bulletin of Seismological Society of America, v. 70, p. 269–281.

Roeloffs, E. A., 1988, Fault stability changes induced beneath a reservoir with cyclic variations in water level: Journal of Geophysical Research, v. 93B, p. 2107–2124.

Scotti, O., and F. H. Cornet, 1994, In situ evidence for fluid induced aseismic slip events along fault zones: International Journal of Rock Mechanics, Mining Science and Geomechanics, v. 316, p. 347–358.

Segall, P., 1985, Stress and subsidence resulting from fluid withdrawal in the epicentral region of the 1983 Coalinga earthquake: Journal of Geophysical Research, v. 90B, p. 6801–6816.

Segall, P., 1989, Earthquakes triggered by fluid extraction: Geology, v. 17, p. 942–946.

Segall, P., J.-R. Grasso, and A. Mossop, 1994, Poroelastic stressing and induced seismicity near the Lacq gas field, southwestern France: Journal of Geophysical Research, v. 99(B), p. 15,423–15,438.

Simpson, D. W., 1976, Seismicity changes associated with reservoir loading: Engineering Geology, v. 10, p. 123–150.

Simpson, D. W., 1986, Triggered earthquakes: Annual Reviews of Earth and Planetary Sciences, v. 14, p. 21–24.

Simpson, D. W., and S. K. Negmattullaev, 1981, Induced seismicity at Nurek reservoir: Bulletin of Seismological Society of America, v. 71, p. 1561–1586.

Simpson, D. W., W. S. Leith, and C. H. Scholz, 1988, Two types of reservoir-induced seismicity: Bulletin of Seismological Society of America, v. 78, p. 2025–2040.

Shtengelov, E. S., 1980, Effect of well exploitation of the upper Jurassic aquifer on seismicity of the Crimea: Water Resources, v. 7, p. 132–139.

Talwani, P., and S. Acree, 1985, Pore pressure diffusion and the mechanism of reservoir-induced seismicity: Pure and Applied Geophysics, v. 122, p. 947–965.

Talebi, S., ed., 1997, Seismicity associated with mines, reservoirs and fluid injections: Pure and Applied Geophysics, v. 150, nos. 3–4, special issue, p. 379–720.

Trifu, C. I., ed., 2002, The mechanism of induced seismicity: Pure and Applied Geophysics, v. 159, nos. 1–3, special issue, p. 1–617.

Volant, P., J.-R. Grasso, J.-L. Chatelain, and M. Frogneux, 1992, b-value, aseismic deformation and brittle failure within an isolated geological object; evidence from a dome structure loaded by fluid extraction: Geophysical Research Letters, v. 19, p. 1149–1152.

Wetmiller, R. J., 1986, Earthquakes near Rocky Mountain house, Alberta, and their relationship to gas production facilities: Canadian Journal of Earth Sciences, v. 23, p. 172–181.

Zoback, M. D., and S. Hickman, 1982, In situ study of the physical mechanisms controlling the induced seismicity at Monticello reservoir, south Carolina: Journal of Geophysical Research, v. 87B, p. 6959–6974.

3.5. Structural Controls on Petroleum Seepage

Barenblatt, G. I., I. P. Zheltov, and I. N. Kocina, 1960, Basic concepts in the theory of seepage of homogeneous liquids in fissured rocks (strata): Journal of Applied Mathematics and Mechanics, v. 24, p. 1286–1303.

Brooks, J. M., et al., 1990, Salt, seeps and symbiosis in the Gulf of Mexico: EOS, Transactions of the American Geophysical Union, v. 71, p. 1772–1773.

Clarke, R. H., and R. W. Cleverly, 1991, Petroleum seepage and post-accumulation migration, *in* W. A. England and A. J. Fleet, eds., Petroleum migration: Geological Society (London) Special Publication 59, p. 265–271.

De Gregorio, S., I. S. Diliberto, S. Giammanco, S. Gurrieri, and M. Valenza, 2002, Tectonic control over large-scale diffuse degassing in eastern Sicily (Italy): Geofluids, v. 2, p. 273–284.

Donovan, T. J., 1974, Petroleum microseepage at Cement, Oklahoma—Evidence and mechanisms: AAPG Bulletin, v. 58, p. 429–446.

Dugan, B., and P. B. Flemings, 2000, Overpressure and fluid flow in the New Jersey continental slope: Implications for slope failure and cold seeps: Science, v. 289, p. 288–291.

Eichhubl, P., H. G. Greene, T. Naehr, and N. Maher, 2000, Structural control of fluid flow: Offshore seepage in the Santa Barbara basin, California: Journal of Geochemical Exploration, v. 69–70, p. 545–549.

Link, W. K., 1952, Significance of oil and gas seeps in world oil exploration: AAPG Bulletin, v. 36, p. 1505–1539.

McGregger, D. S., 1993, Relationships between seepage, tectonism, and subsurface petroleum reserves: Marine and Petroleum Geology, v. 10, p. 606–619.

Parnell, J., ed., 2002, Fluid seeps at continental margins: Towards an integrated plumbing system: Geofluids, v. 2, thematic issue, p. 57–161.

Sassen, R., P. Grayson, G. Cole, H. H. Roberts, and P. Aharon, 1991, Hydrocarbon seepage and salt dome related carbonate reservoir rocks of the U.S. Gulf Coast: AAPG Bulletin, v. 75, p. 9, 1537–1538.

Saunders, D. F., K. R. Burson, and C. K. Thompson, 1999, Model for microseepage and its related near-surface alterations: AAPG Bulletin, v. 83, p. 170–185.

Schumacher, D., and M. A. Abrams, eds., 1996, Hydrocarbon migration and its near-surface expression: AAPG Memoir 66, 446 p.

Schumacher, D., and L. A. LeSchack, eds., 2002, Surface exploration case histories: Applications of geochemistry, magnetics, and remote sensing: AAPG Studies in Geology 48, 486 p.

Struckmeyer, H. I. M., H. A. Williams, R. Cowley, J. Totterdell, G. Lawrence, and G. O'Brien, 2002, Evaluation of hydrocarbon seepage in the Great Australian Bight: Australian Petroleum Production and Exploration Association Journal, v. 42, p. 371–386.

Wilson, R. D., P. H. Monaghan, A. Osanik, L. C. Price, and M. A. Rogers, 1974, Natural marine oil seepage: Science, v. 184, p. 857–865.

4. PETROPHYSICAL PROPERTIES OF FAULTS

4.1. Empirical Measurements

Antonellini, M., and A. Aydin, 1994, Effect of faulting on fluid flow in porous sandstones: Petrophysical properties: AAPG Bulletin, v. 78, p. 355–377.

Berg, R. R., and A. H. Avery, 1995, Sealing properties of Tertiary growth faults, Texas Gulf Coast: AAPG Bulletin, v. 79, p. 375–393.

Bernard, D., M. Danis, and M. Quintard, 1989, Effects of permeability anisotropy and throw on the transmissivity in the vicinity of a fault, *in* A. E. Beck, G. Garven, and L. Stegena, eds., Hydrogeological regimes and their subsurface thermal effects: American Geophysical Union Geophysical Monograph 47, p. 119–128.

Bredehoeft, J. D., 1997, Fault permeability near Yucca Mountain: Water Resources Research, v. 33, p. 2459–2463.

Bredehoeft, J. D., and D. L. Norton, 1990, The role of fluids in crustal processes: Washington, D.C, National Academy Press, 170 p.

Bredehoeft, J. D., K. Belitz, and S. Sharp-Hansen, 1992, The hydrodynamics of the Big Horn basin: A study of the role of faults: AAPG Bulletin, v. 76, p. 530–546.

Brown, K. M., 1995, The variation of the hydraulic conductivity structure of an overpressured thrust zone with effective stress, *in* B. Carson, G. K. Westbrook, R. J. Musgrave, and E. Suess, eds., Proceedings of the Ocean Drilling Program, Scientific Results, v. 146, part 1. Cascadia margin: College Station, Texas A&M University, Ocean Drilling Program, chapter 17, p. 281–289.

Bruhn, R. L., and W. A. Yonkee, 1988, Fracture networks: Implications for fault zone permeability and mechanics: EOS, Transactions of American Geophysical Union, v. 69, p. 484.

Caine, J. S., and C. B. Forster, 1999, Fault zone architecture and fluid flow: Insights from field data and numerical modeling, *in* W. C. Haneberg, P. S. Mozley, J. C. Moore, and L. B. Goodwin, eds., Fault and subsurface fluid flow in the shallow crust: American Geophysical Union Geophysical Monograph 113, p. 101–127.

Caine, J. S., J. P. Evans, and C. B. Forster, 1996, Fault zone architecture and permeability structure: Geology, v. 24, p. 1025–1028.

Childs, C., T. Manzocchi, P. A. R. Nell, J. J. Walsh, J. A. Strand, A. E. Heath, and T. H. Lygren, 2002, Geological implications of a large pressure difference across a small fault in the Viking graben, *in* A. G. Koestler and R. Hunsdale, eds., Hydrocarbon seal quantification: Norwegian Petroleum Society Special Publication 11, p. 187–201.

Evans, J. P., C. B. Forster, and J. V. Goddard, 1997, Permeability of fault-related rocks, implications for hydraulic structure of fault zones: Journal of Structural Geology, v. 19, p. 1393–1404.

Fisher, Q. J., and R. J. Knipe, 1998, Fault sealing processes in siliciclastic sediments, *in* R. J. Knipe, G. Jones, and Q. J. Fisher., eds., Faulting, fault sealing and fluid flow in hydrocarbon reservoirs: Geological Society (London) Special Publication 147, p. 117–134.

Flodin, E., M. Prasad, and A. Aydin, 2003, Petrophysical constraints on deformation styles in Aztec Sandstone, southern Nevada, U.S.A.: Pure and Applied Geophysics, v. 160, p. 1589–1610.

Forster, C. B., and J. P. Evans, 1994, Permeability structure of a thrust fault, *in* S. Hickman, R. Sibson, and R. Brhun, eds., Proceedings of the U.S. Geological Survey Red Book Conference on the Mechanical Involvement of Fluids in Faulting: U.S. Geological Survey Open-file Report 94-228, p. 216–223.

Fowles, J., and S. D. Burley, 1994, Textural and permeability characteristics of faulted, high porosity sandstones: Marine Petroleum Geology, v. 11, p. 608–623.

Fulljames, J. R., L. J. J. Zijerveld, R. C. M. W. Franssen, G. M. Ingram, and P. D. Richard, 1997, Fault seal processes, *in* P. Møller-Pedersen and A. G. Koestler, eds., Hydrocarbon seals: Importance for exploration and production: Norwegian Petroleum Society Special Publication 7, p. 51–59.

Fyfe, W. S. N., and A. B. Thompson, 1978, Fluids in the Earth's crust: Amsterdam, Elsevier, 383 p.

Ganser, D. R., 1988, Hydrogeological characteristics of the Ramapo fault, northern New Jersey: Ground Water, v. 25, p. 664–671.

Gibson, R. G., 1998, Physical character and fluid-flow properties of sandstone-derived fault zones, *in* M. P. Coward, T. S. Daltaban, and H. Johnson, eds., Structural geology in reservoir characterization: Geological Society (London) Special Publication 127, p. 83–97.

Harper, T. R., and E. R. Lundin, 1997, Fault seal analysis: Reducing our dependence on empiricism, *in* P. Møller-Pedersen and A. G. Koestler, eds., Hydrocarbon seals: Importance for exploration and production: Norwegian Petroleum Society Special Publication 7, p. 149–165.

Heynekamp, M. R., L. B. Goodwin, P. S. Mozley, and W. C. Haneberg, 1999, Controls on fault-zone architecture in poorly-lithified sediments, Rio Grande rift, New Mexico: Implications for fault-zone permeability and fluid flow, *in* W. C. Haneberg, P. S. Mozley, J. C. Moore, and L. B. Goodwin, eds., Fault and subsurface fluid flow in the shallow crust: American Geophysical Union Geophysical Monograph 113, p. 27–49.

Jourde, H., E. A. Flodin, A. Aydin, L. J. Durlofsky, and X. H. Wen, 2002, Computing permeability of fault zones in aeolian sandstone from outcrop measurements: AAPG Bulletin, v. 86, p. 1187–1200.

Kim, J.-W., R. R. Berg, J. S. Watkins, and T. T. Tieh, 2003, Trapping capacity of faults in the Eocene Yegua Formation, East Sour Lake field, southeast Texas: AAPG Bulletin, v. 87, p. 415–426.

Knai, T. A., and R. J. Knipe, 1998, The impact of faults on fluid flow in the Heidrun field, *in* R. J. Knipe, G. Jones, and Q. J. Fisher, eds., Faulting, fault sealing and fluid flow in hydrocarbon reservoirs: Geological Society (London) Special Publication 147, p. 269–282.

Knipe, R. J., Q. J. Fisher, G. Jones, M. R. Clennell, A. B. Farmer, A. Harrison, B. Kidd, E. McAllister, J. R. Porter, and E. A. White, 1997, Fault seal analysis: Successful methodologies, application and future directions, *in* P. Møller-Pedersen and A. G. Koestler, eds., Hydrocarbon seals: Importance for exploration and production: Norwegian Petroleum Society Special Publication 7, p. 15–40.

Kopf, A., 2001, Permeability variation across an active low-angle detachment, western Woodlark Basin (ODP Leg 180) and its implication for fault activation, *in* R. E. Holdsworth, R. A. Strachan, J. F. Magloughlin, and R. J. Kinpe, eds., The nature and tectonic significance of fault zone weakening: Geological Society (London) Special Publication 86, p. 23–41.

Leveille, G. P., R. J. Knipe, C. More, D. Ellis, G. Dudley, G. Jones, and Q. J. Fisher, 1997, Compartmentalization of Rotliegendes gas reservoirs by sealing faults, Jupiter area, southern North Sea, *in* K. Ziegler, P. Turner, and S. R. Daines, eds., Petroleum geology of the southern North Sea: Future potential: Geological Society (London) Special Publication 123, p. 87–104.

Manzocchi, T., J. J. Walsh, P. Nell, and G. Yielding, 1999, Fault transmissibility multipliers for flow simulation models: Petroleum Geoscience, v. 5, p. 53–63.

Manzocchi, T., A. E. Heath, J. J. Walsh, and C. Childs, 2002, The representation of two phase fault-rock properties in flow simulation models: Petroleum Geoscience, v. 8, p. 119–132.

Ogilvie, S. R., and P. W. J. Glover, 2001, The petrophysical properties of deformation bands in relation to their microstructure: Earth and Planetary Science Letters, v. 193, p. 129–142.

Ottesen Ellevset, S., R. J. Knipe, T. S. Olsen, Q. J. Fisher, and G. Jones, 1998, Fault controlled communication in the Sleipner Vest field, Norwegian continental shelf; detailed, quantitative input for reservoir and well planning, *in* R. J. Knipe, G. Jones, and Q. J. Fisher, eds., Faulting, fault sealing and fluid flow in hydrocarbon reservoirs: Geological Society (London) Special Publication 147, p. 283–297.

Pittman, E. D., 1981, Effect of fault-related granulation on porosity and permeability of quartz sandstones, Simpson Group (Ordovician), Oklahoma: AAPG Bulletin, v. 65, p. 2381–2387.

Pittman, E. D., 1992, Relationship of porosity and permeability to various parameters derived from mercury injection-capillary pressure curves for sandstone: AAPG Bulletin, v. 76, no. 2, p. 191–198.

Rawling, G. C., L. B. Goodwin, and J. L. Wilson, 2001, Internal architecture, permeability structure, and hydrologic significance of contrasting fault zone types: Geology, v. 29, p. 43–46.

Shipton, Z. K., J. P. Evans, K. R. Robeson, C. B. Forster, and S. Snelgrove, 2002, Structural heterogeneity and permeability in faulted eolian sandstone: Implications for subsurface modeling of faults: AAPG Bulletin, v. 86, p. 863–883.

Sigda, J. M., L. B. Goodwin, P. S. Mozley, and J. L. Wilson, 1999, Permeability alteration in small-displacement faults in poorly lithified sediments: Rio Grande rift, central New Mexico, *in* W. C. Haneberg, P. S. Mozley, J. C. Moore, and L. B. Goodwin, eds., Fault and subsurface fluid flow in the shallow crust: American Geophysical Union Geophysical Monograph 113, p. 51–68.

Sorkhabi, R. B., S. Iwanaga, M. Fujimoto, and S. Hasegawa, 2003, Sealing assessment of normal faults in clastic reservoirs: Modeling the petrophysical and stress attributes of faults: Journal of Japanese Association of Petroleum Technology, v. 68, p. 291–304.

Sperrevik, S., P. A. Gillespie, Q. J. Fisher, T. Halvorsen, and R. J. Knipe, 2002, Empirical estimation of fault rock properties, *in* A. G. Koeslter and R. Hunsdale, eds., Hydrocarbon seal quantification: Norwegian Petroleum Society Special Publication 11, p. 109–125.

Taylor, W. L., and D. D. Pollard, 2000, Estimation of in-situ permeability of deformation bands in porous sandstone, Valley of Fire, Nevada: Water Resources Research, v. 36, p. 2595–2606.

4.2. Experimental Measurements

Arch, J., and A. J. Maltman, 1990, Anisotropic permeability and tortuosity in deformed wet sediments: Journal of Geophysical Research, v. 95B, p. 9035–9046.

Axen, G. J., 1992, Pore pressure, stress increase, and fault weakening in low-angle normal faulting: Journal of Geophysical Research, v. 97B, p. 8979–8991.

Bolton, A. J., A. Maltman, and M. B. Clennell, 1998, The importance of overpressure timing and permeability evolution in fine-grained sediments undergoing shear: Journal of Structural Geology, v. 20, p. 1013–1022.

Brown, S. R., and R. L. Bruhn, 1998, Fluid permeability of deformable fracture networks: Journal of Geophysical Research, v. 103, p. 2489–2500.

Bruno, M. S., 1994, Micromechanics of stress-induced permeability anisotropy and damage in sedimentary rock: Mechanics of Materials, v. 18, p. 31–48.

Chu, C. L., C. Y. Wang, and W. Lin, 1981, Permeability and frictional properties of San Andreas fault gouges: Geophysical Research Letters, v. 8, p. 565–568.

Connolly, P., and J. Cosgrove, 1999, Prediction of fracture-induced permeability and fluid flow in the crust using experimental stress data: AAPG Bulletin, v. 83, p. 757–777.

Crawford, B. R., 1998, Experimental fault sealing: Shear band permeability dependency on cataclastic fault gouge characteristics, *in* M. P. Coward, T. S. Daltaban, and H. Johnson, eds., Structural geology in reservoir characterization: Geological Society (London) Special Publication 127, p. 27–47.

Dewhurst, D. N., and R. M. Jones, 2002, Geomechanical, microstructural, and petrophysical evolution in experimentally reactivated cataclasites: Applications to fault seal prediction: AAPG Bulletin, v. 86, p. 1383–1405.

Dewhurst, D. N., K. M. Brown, M. B. Clennell, and G. K. Westbrook, 1996a, A comparison of the fabric and permeability anisotropy of consolidated and sheared silty clay: Engineering Geology, v. 42, p. 253–267.

Dewhurst, D. N., M. B. Clennell, K. M. Brown, and G. K. Westbrook, 1996b, Fabric and hydraulic conductivity of sheared clays: Geotechnique, v. 46, p. 761–768.

Faulkner, D. R., and E. H. Rutter, 1998, The gas permeability of clay-bearing fault gouge at 20°C, *in* G. Jones, Q. J. Fisher, and R. J. Knipe, eds., Faulting, fault sealing and fluid flow in hydrocarbon reservoirs: Geological Society (London) Special Publication 147, p. 147–156.

Faulkner, D. R., and E. H. Rutter, 2000, Comparisons of water and argon permeability in natural clay-bearing fault gouge under high pressure at 20°C: Journal of Geophysical Research, v. 105B, p. 16,415–16,426.

Fischer, G. J., and M. S. Paterson, 1992, The determination of permeability and storage capacity: Pore pressure oscillation method, *in* B. Evans and T.-F. Wong, eds., Fault mechanics and transport properties of rocks: New York, Academic Press, p. 187–211.

Main, I., O. Kwon, B. Ngwenya, and S. C. Elphick, 2000, Fault sealing during deformation band growth in porous sandstone: Geology, v. 28, p. 1131–1134.

Main, I., K. Mair, O. Kown, S. Elphick, and B. Ngwenya, 2001, Experimental constraints on the mechanical and hydraulic properties of deformation bands in porous sandstones: A review, *in* R. E. Holdsworth, R. A. Strachan, J. F. Magloughlin, and R. J. Kinpe, eds., The nature and tectonic significance of fault zone weakening: Geological Society (London) Special Publication 86, p. 43–63.

Mordecai, M., and L. H. Morris, 1971, An investigation into the changes of permeability occurring in a sandstone when failed under triaxial conditions: Proceedings of the 12th U.S. Rock Mechanics Symposium: Sacramento, California, Association of Engineering Geologists, p. 221–239.

Morrow, C., L. Q. Shi, and J. Byerlee, 1981, Permeability and strength of San Andreas fault gouge under high pressure: Geophysical Research Letters, v. 8, p. 325–328.

Morrow, C. A., L. Q. Shi, and J. D. Byerlee, 1984, Permeability of fault gouge under confining pressure and shear stress: Journal of Geophysical Research, v. 89B, p. 3193–3200.

Peach, C. J., and C. J. Spiers, 1996, Influence of crystal plastic deformation on dilatancy and permeability development in synthetic salt rock: Tectonophysics, v. 256, p. 101–128.

Scholz, C. H., and M. H. Anders, 1994, The permeability of faults, *in* S. Hickman, R. Sibson, and R. Bruhn, R., eds., Proceedings of the U.S. Geological Survey Red Book Conference on the Mechanical Involvement of Fluids in Faulting: U.S. Geological Survey Open-file Report 94-228, p. 247–253.

Senseny, P. E., P. J. Cain, and G. D. Callahan, 1983, Influence of deformation history on permeability and specific storage of Mesaverde sandstone: Proceedings of the 24th U.S. Rock Mechanics Symposium: Sacramento, California, Association of Engineering Geologists, p. 525–531.

Seront, B., T.-F. Wong, J. S. Caine, C. B. Forster, R. L. Bruhn, and J. T. Fredrich, 1998, Laboratory characterization of hydrodynamical properties of a seismogenic normal fault system: Journal of Structural Geology, v. 20, p. 865–881.

Stormont, J. C., and J. J. K. Daemen, 1992, Laboratory study of gas permeability changes in rock salt during deformation: International Journal of Rock Mechanics and Mining Sciences, v. 29, p. 323–342.

Takahashi, M., 2003, Permeability change during experimental fault smearing: Journal of Geophysical Research, v. 108B, p. 2235–2250.

Teufel, L. W., 1987, Permeability changes during shearing deformation of fractured rock, *in* I. W. Farman, J. Daeman, C. S. Desai, C. E. Glass, and S. P. Neuman, eds., Proceedings of the 28th U.S. Rock Mechanics Symposium, Brookfield: Rotterdam, A. A. Balkema, p. 473–480.

Warpinski, N. R., and L. W. Teufel, 1992, Determination of the effective stress law for permeability and deformation in low-permeability rocks: Society of Petroleum Engineers Formation Evaluation, v. 7, p. 123–131.

Wibberley, C. A. J., and T. Shimamoto, 2003, Internal structure and permeability of major strike-slip fault zones: The Median Tectonic Line in the Mie prefecture, southwest Japan: Journal of Structural Geology, v. 25, p. 59–78.

Wong, T.-F., and W. Zhu, 1999, Brittle faulting and permeability evolution: Hydromechanical measurement, microstructural observation, and network modeling, *in* W. C. Haneberg, P. S. Mozley, J. C. Moore, and L. B. Goodwin, eds., Fault and subsurface fluid flow in the shallow crust: American Geophysical Union Geophysical Monograph 113, p. 83–99.

Zhang, S., and T. E. Tullis, 1998, The effect of fault slip on permeability and permeability anisotropy in quartz gouge: Tectonophysics, v. 295, p. 41–52.

Zhang, S., and S. F. Cox, 2000, Enhancement of fluid permeability during shear deformation of a synthetic mud: Journal of Structural Geology, v. 22, p. 1385–1393.

Zhang, S., S. F. Cox, and M. S. Paterson, 1994a, Porosity and permeability evolution during hot isostatic pressing of calcite aggregates: Journal of Geophysical Research, v. 99B, p. 15,741–15,760.

Zhang, S., S. F. Cox, and M. S. Paterson, 1994b, The influence of room temperature deformation on porosity and permeability in calcite aggregates: Journal of Geophysical Research, v. 99B, p. 15,761–15,775.

Zhang, S., T. E. Tullis, and V. J. Scruggs, 1999, Permeability anisotropy and pressure dependency of permeability in experimentally sheared gouge materials: Journal of Structural Geology, v. 21, p. 795–806.

Zhang, S., T. E. Tullis, and V. J. Scruggs, 2001, Implications of permeability and its anisotropy in mica gauge for pore pressures in fault zones: Tectonophysics, v. 335, p. 37–50.

Zhu, W., and T.-F. Wong, 1996, Permeability reduction in a dilatant rock: Network modeling of damage and tortuosity: Geophysical Research Letters, v. 23, p. 3099–3102.

Zhu, W., and T.-F. Wong, 1997, The transition from brittle faulting to cataclastic flow: Permeability evolution: Journal of Geophysical Research, v. 102B, p. 3027–3041.

Zhu, W., and T.-F. Wong, 1999, Network modeling of the evolution of permeability and dilatancy in compact rock: Journal of Geophysical Research, v. 104B, p. 2963–2971.

Zhu, W., C. David, and T. F. Wong, 1995, Network modeling of permeability evolution during cementation and hot isostatic pressing: Journal of Geophysical Research, v. 100B, p. 15,451–15,464.

Zhu, W., L. G. J. Montési, and T.-F. Wong, 2002, Effects of stress on the anisotropic development of permeability during mechanical compaction of porous sandstones, *in* S. De Meer, M. R. Drury, J. H. P. De Brusser, and G. M. Pennock, eds., Deformation mechanisms, rheology and tectonics: Current status and future perspectives: Geological Society (London) Special Publication 200, p. 119–136.

Zoback, M. D., and J. D. Byerlee, 1975a, The effect of microcrack dilatancy on the permeability of westerly granite: Journal of Geophysical Research, v. 80B, p. 752–755.

Zoback, M. D., and J. D. Byerlee, 1975b, Permeability and effective stress: AAPG Bulletin, v. 59, p. 154–158.

Zoback, M. D., and J. D. Byerlee, 1976a, Effect of high-pressure deformation on permeability of Ottawa sand: AAPG Bulletin, v. 60, p. 1531–1542.

Zoback, M. D., and J. D. Byerlee, 1976b, A note on the deformational behaviour and permeability of crushed granite: International Journal of Rock Mechanics Mining Science and Geomechanics, v. 13, p. 291–294.

4.3. Numerical Simulation

Abu-Elbasher, O. B., T. S. Daltaban, C. G. Wall, and J. S. Archer, 1991, Effect of stochastic faults on reservoir permeabilities in presence of stochastic shales, *in* Proceedings of the 7th Middle East Oil Show, Bahrain: Society of Petroleum Engineers Paper 21395, p. 497–504.

Caine, J. S., and C. B. Forster, 1999, Fault zone architecture and fluid flow: Insights from field data and numerical modeling, *in* W. C. Haneberg, P. S. Mozley, J. C. Moore, and L. B. Goodwin, eds., Fault and subsurface fluid flow in the shallow crust: American Geophysical Union Geophysical Monograph 113, p. 101–127.

Chang, F. T., G. P. Lennon, S. Pamukcu, and B. Carson, 1993, Modelling of fluid expulsion and deformation behavior of dewatering sediments: International Journal for Numerical and Analytical Methods in Geomechanics, v. 17, p. 531–551.

Chen, Z., and R. E. Ewing, eds., 2002, Fluid flow and transport in porous media, mathematical and numerical treatment: Proceedings of an American Mathematical Society–Institute of Mathematical Statistics–Society for Industrial and Applied Mathematics joint summer research conference, June 17–21, 2001, American Mathematical Society, 524 p.

Childs, C., O. Sylta, S. Moriya, J. J. Walsh, and T. Manzocchi, 2002, A method for including the capillary properties of faults in hydrocarbon migration models, *in* A. G. Koestler, and R. Hunsdale, eds., Hydrocarbon seal quantification: Norwegian Petroleum Society Special Publication 11, p. 127–139.

Foley, L., T. S. Daltaban, and J. T. Wang, 1998, Numerical simulation of fluid flow in complex faulted regions, *in* M. P. Coward, T. S. Daltaban, and H. Johnson, eds., Structural geology in reservoir characterization: Geological Society (London) Special Publication 127, p. 121–132.

Forster, C. B., and J. P. Evans, 1988, Modeling microscale fluid flux within fault zones: EOS, Transactions of the American Geophysical Union, v. 69, p. 485.

Forster, C. B., and L. Smith, 1988, Groundwater flow systems in mountainous terrain: 2. Controlling factors: Water Resources Research, v. 24, p. 1011–1023.

Forster, C. B., and J. B. Evans, 1991, Fluid flow in thrust faults and crystalline thrust sheets: Results of combined field and modeling studies: Geophysical Research Letters, v. 18, p. 979–982.

Forster, C. B., and J. P. Evans, 1994, Permeability structure of a thrust fault, *in* S. Hickman, R. Sibson, and R. Bruhn, eds., Proceedings of the U.S. Geological Survey red book conference on the mechanical involvement of fluids in faulting: U.S. Geological Survey Open-file Report 94-228, p. 216–223.

Ge, S., and G. Garven, 1994, A theoretical model for thrust-induced deep groundwater expulsion with application to the Canadian Rocky Mountains: Journal of Geophysical Research, v. 99B, p. 13,851–13,868.

Haneberg, W. C., 1995, Steady-state groundwater flow across idealized faults: Water Resources Research, v. 31, p. 1815–1820.

Heath, A. E., J. J. Walsh, and J. Watterson, 1994, Estimation of the effects of subseismic sealing faults on effective permeabilities in sandstone reservoirs: North Sea oil and gas reservoirs III, Norwegian Institute of Technology: Dordrecht, Kluwer Academic Publishers, p. 173–183.

Khalili, N. N., and S. Valliappan, 1991, Flow through fissured porous media with deformable matrix: Water Resources Research, v. 27, p. 1703–1709.

Lopez, D. L., and L. Smith, 1995, Fluid flow in fault zones: Analysis of the interplay of convective circulation and topographically-driven groundwater flow: Water Resources Research, v. 31, p. 1489–1503.

Lopez, D. L., and L. Smith, 1996, Fluid flow in fault zones: Influence of hydraulic anisotropy and heterogeneity on the fluid flow and heat transfer regime: Water Resources Research, v. 32, p. 3227–3235.

Matthai, S. K., and G. Fischer, 1996, Quantitative modeling of fault-fluid discharge and fault-dilation-induced fluid pressure variations in the seismogenic zone: Geology, v. 24, p. 183–186.

Matthai, S. K., and S. G. Roberts, 1996, The influence of fault permeability on single-phase fluid flow near fault-sand intersections: Results from steady-state high-resolution models of pressure-driven fluid flow: AAPG Bulletin, v. 80, p. 1763–1779.

Matthai, S. K., A. Aydin, D. D. Pollard, and S. G. Roberts, 1998, Simulation of transient well-test signatures for geologically realistic faults in sandstone reservoirs: Society of Petroleum Engineers Journal, SPE Paper 38442, p. 62–76.

Morland, L. W., 1992, Flow of viscous fluid through a porous deformable matrix: Surveys in Geophysics, v. 13, p. 209–268.

Rivenæs, J. C., and C. Dart, 2002, Reservoir compartmentalisation by water-saturated faults—Is evaluation possible with today's tools?, *in* A. G. Koeslter and R. Hunsdale, eds., Hydrocarbon seal quantification: Norwegian Petroleum Society Special Publication 11, p. 173–186.

Shan, C., I. Javandel, and P. A. Witherspoon, 1995, Characterization of leaky faults: Study of water flow in aquifer-fault-aquifer systems: Water Resources Research, v. 31, p. 2897–2904.

Shan, C., I. Javandel, and P. A. Witherspoon, 1999, Characterization of leaky faults: Study of air flow in faulted vadose: Water Resources Research, v. 35, p. 2007–2013.

Verkhovskiy, M. S., 1992, Simulation of stationary filtering near fault zones by finite elements method: Physics of the Solid Earth, v. 27, p. 542–546.

Walsh, J. J., J. Watterson, A. E. Heath, and C. Childs, 1998a, Representation and scaling of faults in fluid flow models: Petroleum Geoscience, v. 4, p. 241–251.

Walsh, J. J., J. Watterson, A. Heath, P. A. Gillespie, and C. Childs, 1998b, Assessment of the effects of subseismic faults on bulk permeabilities of reservoir sequences, *in* M. P. Coward, H. Johnson, and T. S. Daltaban, eds., Structural geology in reservoir characterization: Geological Society (London) Special Publication 127, p. 99–114.

Wangen, M., 2001, Communication between overpressured compartments: Geofluids, v. 1, p. 273–287.

Wieck, J., M. Person, and L. Strayer, 1995, A finite element method for simulating fault block motion and hydrothermal fluid flow within rifting basins: Water Resources Research, v. 31, p. 3241–3258.

Zhang, X., and D. J. Sanderson, 2002, Numerical modelling and analysis of fluid flow and deformation of fractured rock masses: Amsterdam, Elsevier, 300 p.

5. ANALYSIS OF FAULT STRESS, STABILITY, AND FAILURE

Note: The proceedings volumes of the following meetings (not listed below) also provide a wealth of information on rock mechanics, including fault mechanics. The International Society for Soil and Rock Mechanics organizes congresses every four years (Lisbon, 1966; Belgrade, 1970; Denver, 1974; Montreux, 1979; Melbourne, 1983; Montreal, 1987; Aachen, 1991; Tokyo, 1995; Paris, 1996; South Africa, 2003). The United States Symposia on Rock Mechanics (sponsored by the U.S. National Commission on Rock Mechanics) are held annually (39th symposium in Cambridge, Massachusetts, 2003).

5.1. Rock Mechanics, Rock Stress, and Fault Mechanics*

Amadei, B., and O. Stephansson, 1997, Rock stress and its measurement: London, Chapman & Hall, 512 p.

Anderson, E. M., 1951, Dynamics of faulting and dyke formation: Edinburgh, Oliver and Boyd, 191 p.

Atkinson, B. K., ed., 1987, Fracture mechanics of rock: London, Academic Press, 534 p.

Aubertin, M., F. Hassani, and H. Mitri, eds., 1996, Rock mechanics: Tools and techniques, 2 volumes: Proceedings of Second North American Rock Mechanics Symposium, Montreal, Canada: Rotterdam, A. A. Balkema, 2050 p.

Barla, G., ed., 1996, Predictions and performance in rock mechanics and rock engineering: Proceedings of International Society for Soil and Rock Mechanics/Society of Petroleum Engineers Regional Symposium/EUROCK 96, Turin, Italy, 3 volumes): Rotterdam, A. A. Balkema, 2100 p.

Barton, N., and O. Stephansson, eds., 1990, Rock joints: Proceedings of International Symposium on Rock Joints, Leon, Norway: Rotterdam, A. A. Balkema, 814 p.

Bayly, B., 1992, Mechanics in structural geology: Springer-Verlag, New York, 253 p.

Carter, N. L., M. Friedman, J. M. Logan, D. W. Stearns, eds., 1981, Mechanical behavior of crustal rocks ("Handin volume"): American Geophysical Union Geophysical Monograph 24, 236 p.

Charlez, P. A., Rock mechanics: Volume 1. Theoretical fundamentals, 333 p.; and Volume 2. Petroleum applications, 704 p.: Paris, Editions Technip.

Cohen, S. C., and P. Vanicek, eds., 1987, Slow deformation and transmission of stress in the Earth: International Union of Geodesy and Geophysics Geophysical Monograph 49, 138 p.

Cristescu, N., 1989, Rock rheology: Dordrecht, Kluwer, 336 p.

Engelder, T., 1993, Stress regime in the lithosphere: New Jersey, Princeton University Press, 457 p.

EUROCK 94 Staff, ed., 1994, Rock mechanics in petroleum engineering: Proceedings of International Society for Soil and Rock Mechanics/Society of Petroleum Engineers Regional Symposium/EUROCK 94, Delft: Rotterdam, A. A. Balkema, 992 p.

EUROCK 98 Staff, ed., 1998, Proceedings of Society of Petroleum Engineers/International Society for Soil and Rock Mechanics International Symposium/EUROCK 98, Trondheim, Norway, July 8–10, 1998, 2 volumes: Richardson: Society of Petroleum Engineers, 952 p.

Fjaer, E., R. M. Holt, A. M. Horsud, and A. M. Raacn, 1992, Petroleum related rock mechanics: Amsterdam, Elsevier, 339 p.

Girard, J., M. Liebman, C. Breeds, and T. Doe, eds., 2000, Pacific rims, rocks around the Pacific: Proceedings of 4th North American Rock Mechanics Symposium, Seattle, Washington: Rotterdam, A. A. Balkema, 1379 p.

Goodman, R. E., 1989, Introduction to rock mechanics, 2d ed.: New York, John Wiley & Sons, 562 p.

Haimson, B. C., and K. Kim, eds., 1993, A short course in modern in-situ stress measurements: The 35th U.S. Symposium on Rock Mechanics (June 27–30, 1993): Madison, University of Wisconsin, 284 p.

Herget, G., 1987, Stress in rock: Rotterdam, A. A. Balkema, 200 p.

Hudson, J. A., ed., 1992, Rock characterization: Proceedings of International Society for Soil and Rock Mechanics Symposium/EUROCK 92, Chester, United Kingdom: London, British Geotechnical Society: Telford, Thames Ltd., 487 p.

Jaeger, J. C., 1969, Elasticity, fracture and flow, 3d ed.: London, Meuthen, 268 p.

Jaeger, J. C., and N. G. W. Cook, 1979, Fundamentals of rock mechanics, 3d ed.: London, Chapman & Hall, 593 p.

Jaroszeweski, W., 1984, Fault and fold tectonics: Chichester, Ellis Honwood, 565 p.

Jumikis, A. R., 1983, Rock mechanics, 2d ed.: Houston, Gulf Publishing Co., 613 p.

Lehner, F. K., and J. L. Urai, eds., 2000, Aspects of tectonic faulting (in honor of Georg Mandl): Berlin, Springer-Verlag, 226 p.

Mandl, G., 1988, Mechanics of tectonic faulting: Amsterdam, Elsevier, 407 p.

Mandl, G., 2000, Faulting in brittle rocks: An introduction to the mechanics of tectonic faults: Berlin, Springer-Verlag, 434 p.

Maury, V., and D. Fourmaintaux, eds., 1990, Rock at great depth: Proceedings of International Society for Soil and Rock Mechanics/Society of Petroleum Engineers International Symposium, Pau, France, 3 volumes: Rotterdam, A. A. Balkema, 1620 p.

Means, W. D., 1976, Stress and strain: Basic concepts of continuum mechanics for geologists: New York, Springer-Verlag, 339 p.

Myer, L. R., C.-F. Tsang, N. G. W. Cook, and R. E. Goodman, eds., 1995, Fractured and jointed rock masses: Proceedings of International Conference on Fractured and Jointed Rock Masses, Lake Tahoe, California: Rotterdam, A. A. Balkema, 772 p.

Nelson, R. P., and S. E. Laubach, eds., 1994, Rock mechanics: Models and measurements: Proceedings of First North American Rock Mechanics Symposium, Austin, Texas: Rotterdam, A. A. Balkema, 1155 p.

Nicolas, A., 1987, Principles of rock deformation: Dordrecht, D. Reidel Publishing, 208 p.

Price, N. J., 1966, Fault and joint development in brittle and semi-brittle rock: Oxford, Pergamon Press, 176 p.

Ranalli, G., 1987, Rheology of the Earth: Deformation and flow processes in geophysics and geology: Boston, Allen & Unwin, 366 p.

Rossmanith, H. P., ed., 1983, Rock fracture mechanics: Wien, Springer-Verlag, 484 p.

Rossmanith, H. P., ed., 1990, Mechanics of jointed and faulted rock: Proceedings of International Conference on Mechanics of Jointed and Faulted Rock, Vienna: Rotterdam, A. A. Balkema, 994 p.

Rossmanith, H. P., ed., 1995, Mechanics of jointed and faulted rock: Proceedings of 2nd International Conference, Vienna: Rotterdam, A. A. Balkema, 1068 p.

Rossmanith, H. P., ed., 1998, Mechanics of jointed and faulted rock: Proceedings of 3rd International Conference, Vienna: Rotterdam, A. A. Balkema, 658 p.

Särkkä, P., and E. Eloranta, eds., 2001, Rock mechanics a challenge for society: Proceedings of International Society for Soil and Rock Mechanics/EUROCK 2001 Symposium, Espoo, Finland: Rotterdam, A. A. Balkema, 881 p.

Scholz, C. H., 1990, Mechanics of earthquakes and faulting: Cambridge, Cambridge University Press, 439 p.

Shimada, M., 2000, Mechanical behaviour of rocks under high pressure conditions: Rotterdam, A. A. Balkema, 186 p.

Stephansson, O., ed., 1988, Rock stress and rock stress measurements: Proceedings of International Symposium on Rock Stress and Rock Stress Measurements, Stockholm: Rotterdam, A. A. Balkema, 704 p.

Sugawara, K., and Y. Obara, eds., 1997, Rock stress: Proceedings of the 2nd International Symposium on Rock Stress, Kumamoto, Japan: Rotterdam, A. A. Balkema, 552 p.

Weijermars, R., 1997, Principles of rock mechanics: Amsterdam, Alboran Science Publishers, 360 p.

Wittke, W., 1990, Rock mechanics: Theory and applications with case histories Berlin, Springer-Verlag, 1075 p.

Xie, H., and M. Kwasniewski, eds., 1003, Fractals in rock mechanics: Rotterdam, A. A Balkema, 464 p.

5.2. Borehole Technologies for Fault Stress Analysis*

Ameen, M., ed., 2003, Fracture and in-situ stress characterization in hydrocarbon reservoirs: Geological Society (London) Special Publication 209, 224 p.

Asquith, G. B., and C. R. Gibson, 1982, Basic well log analysis for geologists: AAPG Methods in Exploration Series 3, 216 p.

Culig, M. J., and J. T. Nelson, 1997, Borehole imaging technology: Golden, Colorado, Colog.

Earlougher Jr., R. C., 1977, Advances in well test analysis: Dallas, Texas, Society of Petroleum Engineers, 264 p.

Harvey, P. K., and Lovell, M. A., eds., 1998, Core-log integration: Geological Society (London) Special Publication 136, 422 p.

Hearst, J. R., P. H. Nelson, and F. L. Paillett, 2000, Well logging for physical properties, 2d ed.: Chichester, John Wiley, 483 p.

Hurst, A., M. A. Lovell, and A. C. Morton, eds., 1990, Geological applications of wireline logs: Geological Society (London) Special Publication 48, 357 p.

Hurst, A., C. M. Griffiths, and P. F. Worthington, eds., 1992, Geological applications of wireline logs II:

Geological Society (London) Special Publication 65, 406 p.

Lovell, M., G. Williamson, and P. Harvey, eds., 1999, Borehole imaging: Applications and case histories: Geological Society (London) Special Publication 159, 270 p.

Rider, M., 1996, The geological interpretation of well logs, 2d ed.: Houston, Gulf Publishing Company, 280 p.

Valko, P., and M. J. Economides, 1995, Hydraulic fracture mechanics: Chichester, John Wiley, 298 p.

Zoback, M. D., and B. C. Haimson, eds., 1983, Hydraulic fracturing stress measurements: Washington, D.C., National Academy Press, 270 p.

5.3. Field Studies

Aadnoy, B. S., 1990, In-situ stress directions from borehole fracture traces: Journal of Petroleum Science and Engineering, v. 4, p. 143–153.

Aadnoy, B. S., R. K. Bratli, and C. D. Llndholm, 1994, In-situ stress modelling of the Snorre field: EUROCK '94 Proceedings of Rock Mechanics in Petroleum Engineering: Rotterdam, A. A. Balkema, p. 871–878.

Addis, M. A., N. C. Last, and N. A. Yassir, 1994, The estimation of horizontal stresses at depth in faulted regions and their relationship to pore pressure variation: EUROCK '94 Proceedings of Rock Mechanics in Petroleum Engineering: Rotterdam, A. A. Balkema, p. 887–895.

Ahmed, U., M. E. Markley, S. F. Crary, and O. Y. Liu, 1989, Enhances in-situ stress profiling using microfrac, core and sonic-logging data: Society of Petroleum Engineers Joint Rocky Mountain Regional/Low Permeability Reservoirs Symposium Proceedings, SPE Paper 19004, p. 707–719.

Amato, A., and P. Montone, 1995, State of stress in southern Italy from borehole breakout and focal mechanism data: Geophysical Research Letters, v. 22, p. 3119–3122.

Barton, C. A., and M. D. Zoback, 1988, Determination of in situ stress orientation from borehole guided waves: Journal of Geophysical Research, v. 93(B), p. 7834–7844.

Barton, C. A., and M. D. Zoback, 1994, Stress perturbations associated with active faults penetrated by boreholes; possible evidence for near-complete stress drop and a new technique for stress magnitude measurement: Journal of Geophysical Research, v. 99(B), p. 9373–9390.

Barton, C. A., and M. D. Zoback, 1998, Earth stress, rock fracture and wellbore failure—Wellbore imaging technologies applied to reservoir geomechanics and environmental engineering: Society of Exploration Geophysics of Japan International Symposium, December, Tokyo, Japan, p. 49–56.

Barton, C. A., M. D. Zoback, and K. L. Burns, 1988, In-situ stress orientation and magnitude at the Fenton geothermal site, New Mexico, determination from wellbore breakouts: Geophysical Research Letters, v. 15, p. 467–470.

Barton, C. A., M. D. Zoback, and D. Moos, 1995a, Fluid flow along potentially active faults in crystalline rock: Geology, v. 23, p. 683–686.

Barton, C. A., M. D. Zoback, and D. Moos, 1995b, In-situ stress and permeability in fractured and faulted crystalline rock, *in* H. P. Rossmanith, ed., Proceedings of the 2nd International Conference on the Mechanics of Jointed and Faulted Rock, Vienna, Austria: Rotterdam, A. A. Balkema, p. 381–386.

Barton, C. A., M. D. Zoback, D. Moos, and J. H. Sass, 1995c, In-situ stress and fracture permeability in the Long Valley Caldera, *in* J. K. Daemen and R. A. Schultz, eds., Rock mechanics: Proceedings of the 35th U.S. Symposium: Rotterdam, A. A. Balkema, p. 225–229.

Barton, C. A., M. D. Zoback, D. Moos, and J. Sass, 1995d, In-situ stress and fracture permeability in the Long Valley Caldera, *in* J. K. Daemen and R. A. Schultz, eds., Proceedings of the 35th Symposium on Rock Mechanics, Reno, Nevada: Rotterdam, A. A. Balkema, p. 225–230.

Barton, C. A., S. Hickman, M. D. Zoback, R. Morin, T. Finkeiner, J. Sass, and D. Benoit, 1996, Fracture permeability and in situ stress in the Dixie Valley, Nevada, geothermal reservoir: Proceedings of the 8th International Symposium on the Continental Crust through Drilling, Tsukiba, Japan, February 1996, p. 210–215.

Barton, C. A., D. Moos, and M. D. Zoback, 1997a, In situ stress measurements can help define local variations of fracture hydraulic conductivity at shallow depth: Leading Edge, v. 16 (November), p. 1653–1656.

Barton, C. A., D. Moos, P. Peska, and M. D. Zoback. 1997b, Utilizing wellbore image data to determine the complete stress tensor: Application to in permeability anisotropy and wellbore stability: Log Analyst, v. 37 (November–December), p. 21–33.

Barton, C. A., S. Hickman, R. Morin, M. D. Zoback, T. Finkeiner, J. Sass, and D. Benoit, 1997c, In situ stress and fracture permeability along the Stillwater fault zone, Dixie Valley, Nevada: Proceedings of the 22nd Workshop on Geothermal Reservoir Engineering, SGP-TR-156, Stanford University, Stanford, California, January 27–29, 1997, p. 147–152.

Barton, C., D. A. Castillo, D. Moos, P. Peska, and M. D. Zoback, 1998, Characterizing the full stress tensor based on observations of drilling-induced wellbore failures in vertical and inclined boreholes leading to improved wellbore stability and permeability prediction: Australian Petroleum Production and Exploration Association Journal, v. 38, p. 466–488.

Barton, C., S. Hickman, R. Morin, M. D. Zoback, and D. Benoit, 1998, Reservoir scale fracture permeability in the Dixie Valley, Nevada geothermal field: Proceedings of Society of Petroleum Engineers/ International Society for Soil and Rock Mechanics Rock mechanics in petroleum engineering, Trondheim, 2: Society of Petroleum Engineering, Richardson, Texas, SPE Paper 47371, p. 315–322.

Bell, J. S., and E. A. Babcock, 1986, The stress regime of the western Canadian basin and implications for hydrocarbon production: Bulletin of Canadian Petroleum Geology, v. 34, p. 364–378.

Bell, J. S., and M. B. Dusseault, 1990, Scale effects and the use of borehole breakouts as stress indicators, *in* A. P. Da Cunha, ed., Scale effects in rock masses: Rotterdam, A. A. Balkema, p. 327–337.

Bell, J. S., and D. I. Gough, 1979, Northeast-southwest compressive stress in Alberta—Evidence from oil wells: Earth and Planetary Science Letters, v. 45, p. 475–482.

Bell, J. S., and D. I. Gough, 1981, Intraplate stress orientations from Alberta oil wells, *in* R. J. O'Connell and W. S. Fyfe, eds., Evolution of the Earth: American Geophysical Union Geodynamics Series 5, p. 96–104.

Bell, J. S., and D. I. Gough, 1982, The use of borehole breakouts in the study of crustal stress, *in* M. D. Zoback and B. C. Haimson, eds., Proceedings of Workshop on Hydraulic Fracturing Stress Measurements: U.S. Geological Survey Open-file Report 82-1075, p. 539–557.

Bell, J. S., G. Caillet, and G. Le Marrec, 1992, The present-day stress regime of the southwestern part of the Aquitaine basin, France, as indicated by oil well data: Journal of Structural Geology, v. 14, p. 1019–1032.

Blumley, J., R. Kuhlman, H. Abass, C. Christiansen, and L. Nydahl, 1994, In-situ stress field determination and formation characterization—Offshore Qater case history: EUROCK '94 Proceedings of Rock Mechanics in Petroleum Engineering: Rotterdam, A. A. Balkema, p. 905–919.

Blumling, P., K. Fuchs, and T. Schneider, 1983, Orientation of the stress field from breakouts in a crystalline well in a seismic active area: Physics of the Earth and Planetary Interiors, v. 33, p. 250–254.

Bradshaw, G. A., and M. D. Zoback, 1988, Listric normal faulting, stress refraction and the state of stress in the Gulf basin: Geology, v. 16, p. 271–274.

Brereton, R., and B. Mueller, 1991, European stress—Contributions from borehole breakouts: Philosophical Transactions of the Royal Society of London, v. 337A, p. 165–179.

Brudy, M., and M. D. Zoback, 1999, Drilling-induced tensile wall fractures: Implications for determination of in-situ stress orientation and magnitude: International Journal of Rock Mechanics and Mining Science, v. 36, p. 191–215.

Brudy, M., M. D. Zoback, K. Fuchs, F. Rummel, and J. Baumgartner, 1997, Estimation of the complete stress tensor to 8 km depth in the KTB scientific drill holes: Implications for crustal strength: Journal of Geophysical Research, v. 102(B), p. 18,453–18,475.

Burns, K. L., 1988, Televiewer measurement of the orientation of in situ stress at the Fenton Hill hot dry rock site, New Mexico: Geothermal Resources Council Transactions, v. 12, p. 229–235.

Castillo, D. A., and M. A. Zoback, 1994, Systematic variations in stress state in the southern San Joaquin Valley—Inferences based on well-bore data and contemporary seismicity: AAPG Bulletin, v. 78, p. 1257–1275.

Castillo, D. A., D. J. Bishop, I. Donaldson, D. Kuek, M. de Ruig, M. Trupp, and M. W. Shuster, 2000, Trap integrity in the Laminaria High–Nancar trough region, Timor Sea: Prediction of fault seal failure using well-constrained stress tensors and fault surfaces interpreted from 3D seismic: Australian Petroleum Production Exploration Association Journal, v. 40, no. 1, p. 151–173.

Chery, J., M. D. Zoback, and R. Hassani, 2001, An integrated mechanical model of the San Andreas fault in central and northern California: Journal of Geophysical Research, v. 106B, p. 22,051–22,061.

Dart, R. L., and H. S. Swolfs, 1992, Subparallel faults and horizontal-stress orientations—An evaluation of in-situ stresses inferred from elliptical wellbore enlargements, *in* R. M. Larsen, H. Brekke, B. T. Larsen, and E. Talleraas, eds., Structural and tectonic modelling and its application to petroleum geology: Norwegian Petroleum Society Special Publication 1, p. 519–529.

Dart, R. L., and M. L. Zoback, 1989, Wellbore breakout stress analysis within the central and eastern continental United States: Log Analyst, v. 30 (January–February), p. 12–24.

Dunne, W. M., and P. L. Hancock, 1994, Palaeostress analysis of small-scale brittle structures, *in* P. L. Hancock, ed., Continental deformation: Oxford, Pergamon Press, p. 101–120.

Ferrill, D. A., J. Winterle, G. Witmeyer, D. Sims, S. Colton, A. Armstrong, and A. P. Morris, 1999, Stressed rock strains groundwater at Yucca Mountain, Nevada: Geological Society of America Today, v. 9, no. 5, p. 1–8.

Finkbeiner, T., C. A. Barton, and M. D. Zoback, 1997, Relationship between in-situ stress, fractures and faults, and fluid flow in the Monterey Formation,

Santa Maria basin, California: AAPG Bulletin, v. 81, p. 1975–1999.

Finkbeiner, T., M. D. Zoback, B. Stump, and P. Flemings, 2001, Stress, pore pressure and dynamically-constrained hydrocarbon column heights in the South Eugene island 330 field, Gulf of Mexico: AAPG Bulletin, v. 85, p. 1007–1031.

Gough, D. I., and J. S. Bell, 1981, Stress, orientations from oil-well fractures in Alberta and Texas: Canadian Journal of Earth Sciences, v. 18, p. 638–645.

Gough, D. I., and J. S. Bell, 1982, Stress orientations from borehole fractures with examples from Colorado, east Texas, and northern Canada: Canadian Journal of Earth Sciences, v. 19, p. 1358–1370.

Grauls, D., and C. Cassignol, 1993, Identification of a zone of fluid pressure-induced fractures from log and seismic data—A case history: First Break, v. 11, no. 2, p. 59–68.

Grauls, D. J., and J. M. Baleix, 1994, Role of overpressures and in situ stresses in fault-controlled hydrocarbon migration: A case study: Marine and Petroleum Geology, v. 11, p. 734–742.

Grollimund, B., and M. D. Zoback, 2003, Impact of glacially induced stress changes on fault-seal integrity: AAPG Bulletin, v. 87, p. 493–506.

Grollimund, B., M. D. Zoback, D. J. Wiprut, and L. Arnesen, 2001, Stress orientation, pore pressure and least principal stress data in the Norwegian sector of the North Sea: Petroleum Geoscience, v. 7, p. 173–180.

Guenot, A., 1989, Borehole breakouts and stress fields: International Journal of Rock Mechanics and Mining Sciences, and Geomechanics, v. 26, p. 185–195.

Haimson, B. C., 1993, The hydraulic fracturing method of stress measurement—Theory and practice, *in* J. A. Hudson, ed., Rock testing and site characterization: Compressive Rock Engineering, v. 3, p. 395–412.

Haimson, B. C., 1995, Hydraulic fracturing stress measurements in an inclined hole in South Carolina and their application to underground cavern design, *in* J. J. K. Daemen and R. A. Schultz, eds., Rock mechanics: Proceedings of the 35th U.S. Symposium: Rotterdam, A. A. Balkema, p. 231–236.

Haimson, B. C., and C. G. Herrick, 1985, In situ stress evaluation from borehole breakouts, experimental studies, *in* E. Ashworth, ed., Research and engineering applications in rock masses: Proceedings of the 26th U.S. Symposium on Rock Mechanics: Boston, A. A. Balkema, v. 2, p. 1207–1218.

Haimson, B. C., and C. G. Herrick, 1986, Borehole breakouts—A new tool for estimating in situ stress, *in* I. Stephansson, ed., Rock stress and rock stress measurements: Proceedings of International Symposium on Rock Stress and Rock Stress Measurements: Stockholm, Centek Publishers, p. 271–280.

Haimson, B. C., and C. G. Herrick, 1989, Borehole breakouts and in situ stress, *in* J. C. Rowley, ed., Drilling symposium 1989: Proceedings of 12th Annual Energy Sources Technology Conference and Exhibition, Houston: New York, American Society of Mechanical Engineers, v. 22, p. 17–22.

Haimson, B. C., and M. Y. Lee, 1987, The state of stress and natural fractures in a jointed Precambrian rhyolite in south-central Wisconsin, *in* I. W. Farmer, J. J. K. Daemen, C. S. Desai, C. E. Glass, and S. P. Neuman, eds., Rock mechanics: Proceedings of the 28th U. S. Symposium: Boston, A. A. Balkema, p. 231–240.

Hickman, S. H., J. H. Healy, and M. D. Zoback, 1985, In situ stress, natural fracture distribution, and borehole elongation in the Auburn geothermal well, Auburn, New York: Journal of Geophysical Research, v. 90B, p. 5497–5512.

Hickman, S. H., M. D. Zoback, and J. H. Healy 1988, Continuation of a deep borehole stress measurement profile near the San Andreas fault: 1. Hydraulic fracturing stress measurements at Hi Vista, Mojave desert, California: Journal of Geophysical Research, v. 93B, p. 15,183–15,195.

Hickman, S., M. D. Zoback, L. Younker, and W. Ellsworth, 1994, Deep scientific drilling in the San Andreas fault zone: EOS, Transactions of American Geophysical Union, v. 75, p. 137, 140, 142.

Hickman, S. H., L. W. Younker, M. D. Zoback, and G. A. Cooper, 1995, The San Andreas fault zone drilling project: Scientific objectives and technological challenges: Journal of Energy Resources and Technology, v. 117, p. 263–270.

Hickman, S., C. A. Barton, M. D. Zoback, R. Morin, J. Sass, and R. Benoit, 1997, In-situ stress and fracture permeability along the Stillwater fault zone, Dixie Valley, Nevada: International Journal of Rock Mechanics and Mining Sciences, v. 34, p. 3–4.

Hickman, S., M. D. Zoback, and R. Benoit, 1998, Tectonic controls on fault-zone permeability in a geothermal reservoir at Dixie Valley, Nevada: Society of Petroleum Engineers/International Society for Soil and Rock Mechanics Rock Mechanics in Petroleum Engineering: Richardson, Texas, Society of Petroleum Engineering, p. 79–86.

Hillis, R. R., 1997, Does the in-situ stress field control the orientation of open natural fractures in sub surface reservoirs?: Exploration Geophysics, v. 28, p. 80–87.

Hillis, R. R., 1998, Mechanisms of dynamic seal failure in the Timor Sea and central North Sea, *in* P. G. Purcell and R. R. Purcell, eds., The sedimentary basins of western Australia 2; Proceedings of

Petroleum Exploration Society of Australia Symposium, Perth, Western Australia: Bath, Geological Society, p. 313–324.

Hillis, R. R., and A. F. Williams, 1993, The contemporary stress field of the Barrow-Dampier sub-basin and its implications for horizontal drilling: Exploration Geophysics, v. 24, p. 567–576.

Hillis, R. R., D. D. Mildren, and J. J. Meyer, 2000, Borehole geomechanics in petroleum exploration and development: From controlling wellbore stability to predicting fault seal integrity: Preview, v. 89, p. 31–34.

Hubbert, M. K., and D. G. Willis, 1957, Mechanics of hydraulic fracturing: American Institute of Mining and Metallurgical Engineers Transactions, v. 210, p. 153–168.

Hubbert, M. K., and W. W. Rubey, 1959, Role of fluid pressure in mechanics of overthrust faulting: Parts I and II. Geological Society of America Bulletin, v. 70, p. 115–205.

Ito, T., and M. D. Zoback, 2000, Fracture permeability and in situ stress to 7 km depth in the KTB scientific drillhole: Geophysical Research Letters, v. 27, p. 1045–1048.

Ito, T., K. Hayashi, and M. D. Zoback, 1992, Analysis of stress-induced wellbore breakouts in an arbitrarily oriented tectonic stress field: Japan Society of Mechanical Engineers International Journal, v. 35, p. 265–270.

Lau, J. S. O., L. F. Auger, and J. G. Bisson, 1987, Subsurface fracture surveys using a borehole television camera and acoustic televiewer: Canadian Geotechnical Journal, v. 24, p. 499–508.

Laubach, S. E., R. W. Baumgardner Jr., E. R. Monson, E. Hunt, and K. J. Meador, 1988, Fracture detection in low-permeability reservoir sandstone—A comparison of BHTV and FMS logs to core: Society of Petroleum Engineers Annual Technical Conference and Exhibition Proceedings: Formation Evaluation and Reservoir Geology: SPE Paper 18119, p. 129–139.

Li, Y., and D. R. Schmitt, 1997, Well-bore bottom stress concentration and induced core fractures: AAPG Bulletin, v. 81, p. 1909–1925.

Li, Y., and D. R. Schmitt, 1998, Drilling-induced fractures and in-situ stress: Journal of Geophysical Research, v. 103B, p. 5225–5239.

Lorenz, J. C., S. J. Finley, and N. R. Warpinski, 1990, Significance of coring induced fractures in Mesaverde core, northwestern Colorado: AAPG Bulletin, v. 74, p. 1017–1029.

Luthi, S. M., and P. Souhaite, 1990, Fracture apertures from electrical borehole scans: Geophysics, v. 55, p. 821–833.

Magee, M., and M. L. Zoback, 1993, Evidence for a weak interplate thrust fault along the northern Japan subduction zone and implications for the mechanics of thrust faulting and fluid expulsion: Geology, v. 21, p. 809–812.

Mariucci, M. T., A. Amato, R. Gambini, M. Gigorgioni, and P. Montone, 2002, Along-depth stress rotations and active faults: An example in a 5-km deep well of southern Italy: Tectonics, v. 21, no. 4, DOI:10.1029/2001TC001338.

Marsch, F., G. Wessely, and W. Sackmaier, 1990, Borehole-breakouts as geological indications of crustal tensions in the Vienna basin, *in* H. P. Rossmanith, ed., Mechanics of jointed and faulted rock: Rotterdam, A. A. Balkema, p. 113–120.

Mastin, L., 1988, Effect of borehole deviation on breakout orientations: Journal of Geophysical Research, v. 93(B), p. 9187–9195.

Mastin, L. G., B. Heinemann, A. Krammer, K. Fuchs, and M. D. Zoback, 1991, Stress orientation in the KTB pilot hole determined from wellbore breakouts: Scientific Drilling, v. 2, p. 1–12.

Mildren, S. D., and R. R. Hillis, 2000, In situ stress in the southern Bonaparte basin, Australia: Implications for first- and second-order controls on stress orientation: Geophysical Research Letters, v. 27, p. 3413–3416.

Mildren, S. D., R. R. Hillis, and J. Kaldi, 2002, Calibrating predictions of fault seal reactivation in the Timor Sea: Australian Petroleum Production and Exploration Association Journal, v. 42, p. 187–202.

Montone, P., A. Amato, and S. Pondrelli, 1999, Active stress map of Italy: Journal of Geophysical Research, v. 104B, p. 25,595–25,610.

Moos, D., and M. D. Zoback, 1990, Utilization of observations related to wellbore failure to constrain the orientation and magnitude of crustal stresses: Application to continental, DSDP and ODP boreholes: Journal of Geophysical Research, v. 95, p. 9305–9325.

Moos, D., and R. H. Morin, 1991, Observations of wellbore failure in the Toa Baja well; implications for the state of stress in the north coast Tertiary basin, Puerto Rico: Geophysical Research Letters, v. 18, p. 505–508.

Morin, R. H., 1990, Information on stress conditions in the oceanic crust from oval fractures in a deep borehole: Geophysical Research Letters, v. 17, p. 1311–1314.

Paillet, F. L., and D. Goldberg, 1991, Acoustic televiewer log images of natural fractures and bedding planes in the Toa Baja borehole, Puerto Rico: Geophysical Research Letters, v. 18, p. 501–504.

Paillet, F. L., and K. Kim, 1987, The character and distribution of borehole breakouts their relationship to in situ stresses in deep Columbia River basalts: Journal of Geophysical Research, v. 92, p. 6223–6234.

Peska, P., and M. D. Zoback, 1995, Compressive and tensile failure of inclined wellbores and determination of in situ stress and rock strength: Journal of Geophysical Research, v. 100B, p. 12,791–12,811.

Phillips, R. J., ed., 1988, Scientific drilling near the San Andreas fault: Geophysical Research Letters, v. 15, no. 9, special issue, p. 933–1076.

Plumb, R. A., 1989, Fracture patterns associated with incipient wellbore breakouts, *in* V. Maury and D. Fourmaintraux, eds., Rock at great depth: Rotterdam, A. A. Balkema, v. 2, p. 761–768.

Plumb, R. A., and J. W. Cox, 1987, Stress directions in eastern North America determined to 4.5 km from borehole elongation measurements: Journal of Geophysical Research, v. 92B, p. 4805–4816.

Plumb, R. A., and S. H. Hickman, 1985, Stress-induced borehole elongation—A comparison between the four-arm dipmeter and the borehole televiewer in the Auburn geothermal well: Journal of Geophysical Research, v. 90B, p. 5513–5521.

Prensky, S. E., 1992, Borehole breakouts and in-situ stress—A review: Log Analyst, v. 33, no. 3, p. 304–312.

Qian, W., and L. B. Pedersen, 1991, Inversion of borehole breakout orientation data: Journal of Geophysical Research, v. 96B, p. 20,091–20,107.

Reynolds, S. D., and R. R. Hillis, 2000, The in situ stress field of the Perth basin, Australia: Geophysical Research Letters, v. 27, p. 3421–3424.

Rice, J. R., 1992, Fault stress states, pore pressure distributions, and the weakness of the San Andreas fault, *in* B. Evans and T. F. Wong, eds., Fault mechanics and transport properties of rocks: San Diego, Academic Press, p. 475–503.

Schmitt, D. R., 1993, Fracture statistics derived from digital from ultrasonic televiewer logging: Journal of Canadian Petroleum Technology, v. 32, p. 34–43.

Shamir, G., and M. D. Zoback, 1989a, Detailed analysis of wellbore breakouts in the Cajon Pass scientific drillhole *in* J. C. Rowley, ed., Drilling Symposium 1989: Proceedings of the 12th Annual Energy Sources Technology Conference and Exhibition (Houston, January 22–25): New York, American Society of Mechanical Engineers, v. 22, p. 11–15.

Shamir, G., and M. D. Zoback, 1989b, The stress orientation profile in the Cajon Pass, California, scientific drillhole, based on detailed analysis of stress induced borehole breakouts, *in* V. Maury and D. Fourmaintraux, eds., Rock at great depth: Rotterdam, A. A. Balkema, v. 2, p. 1041–1048.

Shamir, G., and M. D. Zoback, 1992, Stress orientation profile to 3.5 km depth near the San Andreas fault at Cajon Pass, California: Journal of Geophysical Research, v. 97B, p. 5059–5080.

Shamir, G., M. D. Zoback, and C. A. Barton, 1988, In situ stress orientation near the San Andreas fault: Preliminary results to 2.1 km depth from the Cajon Pass scientific drillhole: Geophysical Research Letters, v. 15, p. 989–992.

Sibson, R. H., 1990, Conditions for fault-valve behavior, *in* R. J. Knipe and E. H. Rutter, eds., Deformation mechanisms, rheology and tectonics: Geological Society (London) Special Publication 54, p. 15–28.

Sibson, R. H., 1994, Crustal stress, faulting and fluid flow, *in* J. Parnell, ed., Geofluids: Origin, migration and evolution of fluids in sedimentary basins: Geological Society (London) Special Publication 78, p. 69–84.

Sibson, R. H., 1996, Structural permeability of fluid-drive fault-fracture meshes: Journal of Structural Geology, v. 18, p. 1031–1042.

Springer, J. E., 1987, Stress orientations from wellbore breakouts in the Coalinga region: Tectonics, v. 6, p. 667–676.

Standen, E., 1991, Tips for analyzing fractures on electrical wellbore images: World Oil, v. 212 (April), p. 99–117.

Stock, J. M., J. H. Healy, S. H. Hickman, and M. D. Zoback, 1985, Hydraulic fracturing stress measurements at Yucca Mountain, Nevada and relationship to the regional stress field: Journal of Geophysical Research, v. 90, p. 8691–8706.

Straub, A., U. Kruckel, and Y. Gros, 1991, Borehole electrical imaging and structural analysis in a granitic environment: Geophysical Journal International, v. 106, p. 635–646.

Teufel, L. W., 1985, Insights into the relationship between wellbore breakouts, natural fractures, and in situ stress, *in* E. Ashworth, ed., Research and engineering applications in rock masses: Proceedings of 26th U.S. Symposium on Rock Mechanics, South Dakota School of Mines and Technology, Rapid City: Boston, A. A. Balkema, v. 2, p. 1199–1206.

Wiprut, D., and M. D. Zoback, 2000a, Fault reactivation and fluid flow along a previously dormant normal fault in the Norwegian North Sea: Geology, v. 28, p. 595–598.

Wiprut, D., and M. D. Zoback, 2000b, Constraining the full stress tensor in the Visnud field, Norwegian North Sea: Application to wellbore stability and sand production: International Journal of Rock Mechanics and Mining Science, v. 37, p. 317–336.

Wiprut, D., and M. D. Zoback, 2002, Fault reactivation, leakage potential, and hydrocarbon column heights in the northern North Sea, *in* A. G. Koestler and R. Hunsdale, eds., Hydrocarbon seal quantification: Norwegian Petroleum Society Special Publication 11, p. 203–219.

Yale, D. P., J. M. Rodriguez, T. B. Mercer, and D. W. Blaisdell, 1994, In-situ stress orientation and the effects of local structure—Scott field, North Sea, *in* EUROCK '94 Proceedings of Rock Mechanics in Petroleum Engineering: Rotterdam, A. A. Balkema, p. 945–953.

Yassir, N. A., and M. B. Dusseault, 1991, Stress trajectories in southwestern Ontario using wellbore breakout orientations, *in* J.-C. Roegiers, ed., Rock mechanics as a multidisciplinary science: Proceedings of the 32nd U.S. Symposium on Rock Mechanics: Rotterdam, A. A. Balkema, p. 83–92.

Zheng, Z., L. Meyer, and N. G. W. Cook, 1988, Borehole, breakout and stress measurements, *in* P. Cundall, R. Sterling, and A. Starfield, eds., Proceedings of the 29th U. S. Rock Mechanics Symposium: Rotterdam, A. A. Balkema, p. 471–478.

Zheng, Z., J. Kemeny, and N. G. W. Cook, 1989, Analysis of borehole breakouts: Journal of Geophysical Research, v. 94(B), p. 7171–7182.

Zoback, M. D., 1979, Recurrent faulting in the vicinity of Reelfoot Lake, northwestern Tennessee: Geological Society of America Bulletin, v. 90, p. 1019–1024.

Zoback, M. D., 1981, In-situ study of the mechanism of reservoir triggered earthquakes in the southeastern United States: Proceedings of the International Conference on Intra-continental Earthquakes, September 17–21, 1981, Ohrid, Yugoslavia, p. 273–289.

Zoback, M. L., 1992, First- and second-order patterns of stress in the lithosphere: The world stress map project: Journal of Geophysical Research, v. 97B, p. 11,703–11,728.

Zoback, M. D., and G. Beroza, 1993, Evidence for near frictionless faulting in the 1989 (M = 6.9) Loma Prieta, California earthquake and its aftershocks: Geology, v. 21, p. 181–185.

Zoback, M. D., and J. H. Healy, 1984, Friction, faulting and in-situ stress: Annales Geophysicae, v. 2, p. 689–698.

Zoback, M. D., and J. H. Healy, 1992, In situ stress measurements to 3.5 km depth in the Cajon Pass scientific research boreholes—Implications for the mechanics of crustal faulting: Journal of Geophysical Research, v. 97B, p. 5039–5057.

Zoback, M. D., and S. Hickman, 1982, In-situ study of the physical mechanisms controlling induced seismicity at Monticello reservoir, South Carolina: Journal of Geophysical Research, v. 87B, p. 6959–6974.

Zoback, M. D., and P. Peska, 1995, In situ rock strength in the GBRN/DOE "Pathfinder" well, South Eugene Island, Gulf of Mexico: Journal of Petroleum Technology, v. 47, p. 582–585.

Zoback, M. D., and J. C. Roller, 1979, Magnitude of shear stress on the San Andreas fault: Implications from a stress measurement profile at shallow depth: Science, v. 206, p. 445–447.

Zoback, M. D., and J. Townend, 2001, Implications of hydrostatic pore pressures and high crustal strength for the deformation of intraplate lithosphere: Tectonophysics, v. 336, p. 19–30.

Zoback, M. D., and J. Zinke, 2001, Production-induced normal faulting in the Valhall and Ekofisk oil fields: Pure and Applied Geophysics, v. 159, p. 403–420.

Zoback, M. L., and M. D. Zoback, 1980a, Faulting patterns in north-central Nevada and strength of the crust: Journal of Geophysical Research, v. 85B, p. 275–284.

Zoback, M. L., and M. D. Zoback, 1980b, State of stress in conterminous United States: Journal of Geophysical Research, v. 85B, p. 6113–6156.

Zoback, M. D., and M. L. Zoback, 1991, Tectonic stress field of North America and relative plate motions, *in* D. B. Slemmons, E. R. Engdahl, M. D. Zoback, and D. D. Blackwell, eds., Neotectonics of North America: Geological Society of America Decade Map, v. 1, p. 339–366.

Zoback, M. D., J. H. Healy, J. C. Roller, G. S. Gohn, and B. B. Higgins, 1978, Normal faulting and in-situ stress in the South Carolina coastal plain near Charleston: Geology, v. 6, p. 147–152.

Zoback, M. D., H. Tsukahara, and S. Hickman, 1980a, Stress measurements at depth in the vicinity of the San Andreas fault: Implications for the magnitude of shear stress at depth: Journal of Geophysical Research, v. 85B, p. 6157–6173.

Zoback, M. D., R. M. Hamilton, A. J. Crone, D. P. Russ, F. A. McKeown, and S. R. Brockman, 1980b, Recurrent intraplate tectonism in the New Madrid Seismic Zone: Science, v. 209, p. 971–976.

Zoback, M. D., D. Moos, L. Mastin, and R. N. Anderson, 1985, Wellbore breakouts and in-situ stress: Journal of Geophysical Research, v. 90B, p. 5523–5530.

Zoback, M. D., L. Mastin, and C. Barton, 1986, In-situ stress measurements in deep borehole using hydraulic fracturing, wellbore breakouts, and Stoneley-wave polarization, *in* O. Stephansson, ed., Rock stress and rock stress measurements: Proceedings of International Symposium on Rock Stress and Rock Stress Measurements, Stockholm: Stockholm, Cenek Publishers, p. 289–299.

Zoback, M. L., et al., 1989, Global patterns of tectonic stress: Nature, v. 341, p. 291–298.

Zoback, M. D., et al., 1993, Upper crustal strength inferred from stress measurements to 6 km depth in the KTB borehole: Nature, v. 365, p. 633–635.

Zoback, M. D., C. Barton, C. Chang, D. Moos, P. Peska, and L. Vernik, 1995, A review of some new methods for determining the in situ stress state from observations of borehole failure with applications to borehole stability and enhanced production in the North Sea, *in* M. Fejerskov and A. Myrvang, eds., Proceedings of Workshop on Rock Stresses in the North Sea, Trondheim, Norway: Trondheim, Norway, Senter for Industriforsking (SINTEF), p. 6–21.

Zoback, M. D., A. W. Chan, and J. C. Zinke, 2001, Production-induced normal faulting: Proceedings of the 38th U.S. Symposium on Rock Mechanics, Washington, D.C.: Lisse, Netherlands, A. A. Balkema, p. 157–163.

Zoback, M. D., J. Townend, and B. Grollimund, 2002, Steady-state failure equilibrium and deformation of intraplate lithosphere: International Geological Review, v. 44, p. 383–401.